Advanced Vibrations

Reza N. Jazar

Advanced Vibrations

Theory and Application

Second Edition

 Springer

Reza N. Jazar 🆔
School of Engineering
RMIT University
Melbourne, VIC, Australia

School of Civil Engineering
and Architecture
Xiamen University of Technology
Xiamen, Fuzhou, China

ISBN 978-3-031-16358-6   ISBN 978-3-031-16356-2   (eBook)
https://doi.org/10.1007/978-3-031-16356-2

This Springer imprint is published by the registered company Springer Nature Switzerland AG
The registered company address is: Gewerbestrasse 11, 6330 Cham, Switzerland

بتی سازم
از برگ گل، نوای رود، بوی شراب
از خواب خوش، نم نم باران، از کتاب
شویم از دانش و از دفتر،
خرافه و پندار، از دوزخ، از سراب
از دین و دیوار، مسجد و فریب، از سجده از طناب
زنده کنم فریاد امید،
ایرج و کاوه، فریدون و فرهاد، شاپور، شهاب
بی زاهد و دروغ،
سازم ایرانی از لبخند، از شادی، از آفتاب

I make an idol
from the leaves of flowers, from the sound of
rivers, from the smell of wine,
from a sweet nap, from the rain, from benign.
I wash lies from wisdom, heavenly books,
superstition, fantasy, hell, divine,
religion and wall, mosque and deception,
prostration, rope, and confine.
I revive the scream of hope, Iraj and Kaveh,
Fereydoun and Farhad, Shapoor and Shahab.
Without ascetic lies
I make a land full of smile, happiness, and
sunshine.

Rumi

082904;0131404;475513;1065502;470122;0285007;343131;
0691042;334135;0730350;373336;1265924;513026;000073
9;090400;0072900;402500;0034300;
083100;1791200;475513;1065502;093033;0134244;015638;
0300334;143545;1205838;452529;0754142;174945;003130
8;090148;0384424;255800;0323500;315659;0355558;23455
0;0902320;

Do not trust any science that justifies but fails to predict.

to:
Kavosh
Vazan
Mojgan

Preface

For several years, my students complained that the available books in the market on the topic of vibrations are all treating this science inconsistent from different viewpoints. They were asking me to write a book on vibrations that covers all important and must-known subjects deep enough; shows applications well, has enough examples, right to the point; and prepare them to tackle real engineering problems as well as make them ready for the next step in nonlinear vibrations and vibrations of continuous systems. This book is prepared to address those requests. It is a modified, improved, and updated version of the Advanced Vibrations: A Modern Approach that is published in 2013. The present book has an educational division of chapters, sections, and subsections that make it to be more appropriate for traditional classroom teaching.

This book is for designers, practicing engineers, and students of engineering. It introduces the fundamental knowledge used in *mechanical vibrations*. This knowledge can be utilized to develop computer programs for analyzing, designing, and optimizing vibration problems in industrial systems. The main problem in the study of vibrations is to determine the motion of a system subjected to a given excitation. The excitation and the response can be expressed by either kinematic quantities such as displacements, velocities, and accelerations, or by forces.

The subject of vibrations has been in the engineering curriculum for more than hundred years. I consider the 1896 Lord Rayleigh's (John William Strutt, 3rd Baron Rayleigh, a British scientist) "Theory of Sound" as the first modern book on the subject of sound, wave propagation, and vibrations. However, the first books on vibrations with a mechanical engineering viewpoint appeared as "Vibration Problems in Engineering" (1928) by the father of modern engineering mechanics, the Russian-American engineer, Stephen P. Timoshenko (1878–1972), and then as "Mechanical Vibrations" (1934) by the Dutch-American engineer, Jacob P. Den Hartog. Although Timoshenko was the architect of modern structure of engineering education and his various books were used for a long time in educating mechanical and civil engineers, it was Den Hartog's Mechanical Vibrations that was globally accepted as a classical educational book. Almost all mechanical vibration books that appeared after 1940 follow the structure of Den Hartog's, starting from time response and ending with frequency response analysis and optimization. Sometimes a glance at random vibrations, nonlinear vibrations, continuous systems, vibrations control, or modal analysis may also be seen in various books. The current book also follows the same order as Den Hartog's. It begins with a review of fundamental concepts of mechanical vibrations and continues with time response, and ends with frequency responses. A glance on music from vibrations viewpoint, first-order vibrating systems, vibration absorber design, and root mean square optimization method are some topics that solely appear in this book.

Level of the Book

This book has evolved from nearly three decades of research and working on real engineering vibrating systems, and teaching courses in fundamental and advanced vibrations. It is primarily designed to teach the last year of an undergraduate study and/or the first year of graduate study in engineering. Hence, it is an intermediate textbook. It provides the reader with both fundamental and advanced topics. The whole book can be covered in two successive courses; however, it is possible to jump over some sections and cover the book in one course. Students are required to know the fundamentals of kinematics and dynamics, as well as a basic knowledge of linear algebra, differential equations, and numerical methods.

The contents of the book have been kept at a fairly theoretical-practical level. Many concepts are deeply explained and their application emphasized, and most of the related theories and formal proofs have been explained. The book places a strong emphasis on the physical meaning and applications of the concepts. Topics that have been selected are of high interest in the field. An attempt has been made to expose students to a broad range of topics and approaches.

An asterisk ★ indicates a more advanced topic which is not designed for undergraduate teaching and can be dropped in the first reading.

Organization of the Book

The text is organized such that it can be used for teaching or for self-study. It is organized in 3 parts:

Part I, "Vibration Fundamentals," introduces the vibrations as a cause of trans-formation of energy. It covers kinematics of vibrations and develops practical skills to derive the equations of motion of vibrating systems. The concepts of the Newton-Euler dynamics and Lagrangian method are both used equally for the derivation of equations of motion. Part I also covers all mathematical analysis that the reader needs to absorb the advanced topics throughout the book.

Part II, "Time Response," covers the time and transient responses of vibrating systems and non-harmonic excitations, as well as free vibrations.

Part III, "Frequency Response," covers the methods of developing the steady-state frequency response of vibrating systems to harmonic excitations. The principles of vibration optimization will be introduced in this part. The root mean square optimization technique for suspension design of mechanical systems is introduced and applied. The outcome of the optimization technique is the optimal stiffness and damping for suspended equipment.

Method of Presentation

The structure of presentation is in a "*fact-reason-application*" fashion. The "fact" is the main subject we introduce in each section. Then the reason is given as a "proof." Finally, the application of the fact is examined in some "examples." The "examples" are a very important part of the book. They show how to implement the knowledge introduced in "facts." They also cover some other facts that are needed to expand the subject.

Prerequisites

The book is written for senior undergraduate and first-year graduate-level students in engineering. The assumption is that users are familiar with matrix algebra as well as basic dynamics and differential equations. Prerequisites are the fundamentals of kinematics, dynamics, calculus, fundamentals of differential equations, and matrix theory. These topics are usually taught in the first three undergraduate years.

Unit System

The system of units adopted in this book is, unless otherwise stated, the International System of Units (*SI*). The units of degree (deg) or radian (rad) are utilized for variables representing angular quantities.

Melbourne, VIC, Australia   Reza N. Jazar

How to Use This Book

This book is suitable for the first course in vibrations and is written for a full semester of 16 weeks long in graduate level. If the level of the course is undergraduate or mixed, and if the length of the semester is shorter than 16 weeks, then the following pattern can be suggested to fit the contents to different classes. Teaching Chap. 3, Vibration Dynamic, is optional in all cases depending on the background of students. The main topic of this chapter is to review dynamics and remind students of the Lagrangean method of deriving equations of motion. Chapter 2 may also be left for students to read by themselves.

Undergraduate level, 10 or 12 weeks semester: Skip all asterisk sections and teach the following chapters.
I Vibration Fundamentals
    1 Vibration Kinematics
II Time Response
    3 One Degree of Freedom
III Frequency Response
    6 One Degree of Freedom Systems
    7 Multi Degrees of Freedom Systems

Undergraduate level, 16 weeks semester: Skip all asterisk sections and teach the following chapters.
I Vibration Fundamentals
    1 Vibration Kinematics
    2 Vibration Dynamics
II Time Response
    3 One Degree of Freedom
    4 Multi Degrees of Freedom
    5 First-Order Systems
III Frequency Response

6 One Degree of Freedom Systems
7 Multi Degrees of Freedom Systems

Graduate level, 10 or 12 weeks semester: Skip asterisk sections and teach the following chapters.
I Vibration Fundamentals
    1 Vibration Kinematics
    2 Vibration Dynamics
II Time Response
    3 One Degree of Freedom
    4 Multi Degrees of Freedom
    5 First-Order Systems
III Frequency Response
    6 One Degree of Freedom Systems
    7 Multi Degrees of Freedom Systems

Graduate level, 16 weeks semester: Skip asterisk sections based on your preference and teach the following chapters.
I Vibration Fundamentals
    1 Vibration Kinematics
    2 Vibration Dynamics
II Time Response
    3 One Degree of Freedom
    4 Multi Degrees of Freedom
    5 First-Order Systems
III Frequency Response
    6 One Degree of Freedom Systems
    7 Multi Degrees of Freedom Systems
    8 Two Degrees of Freedom Systems

Contents

About the Author

Reza N. Jazar is a professor of Mechanical Engineering. Reza received his PhD degree from the Sharif University of Technology in Nonlinear Vibrations and Applied Mathematics; MSc in Mechanical Engineering and Robotics; and BSc in Mechanical Engineering and Internal Combustion Engines from Tehran Polytechnic. His areas of expertise include Nonlinear Dynamic Systems and Applied Mathematics. He obtained original results in nonsmooth dynamic systems, applied nonlinear vibrating problems, time optimal control, and mathematical modeling of vehicle dynamics stability. He authored several monographs on vehicle dynamics, robotics, dynamics, vibrations, and mathematics, and published numerous professional articles, as well as book chapters in research volumes. Most of his books have been adopted by many universities for teaching and research, and by many research agencies as a standard model for research results.

Dr. Jazar had the pleasure to work in several Canadian, American, Asian, Middle Eastern, and Australian universities, as well as several years in automotive industries all around the world. Working in different engineering firms and educational systems provide him with vast experience and knowledge to publish his research on important topics in engineering and science. His unique revolutionized style of writing helps readers to learn the topics deeply in the easiest possible way.

Part I
Vibration Fundamentals

Vibrations is a science that requires several tools to be understood. Linear algebra, trigonometric functions, principles of dynamics, and differential equations are the most important tools to study vibrations. We will review those preliminary topics that are required to understand vibrations. In the first part, we introduce mechanical elements that every vibrating system has. The mechanical vibrations will be introduced as a phenomenon caused by transformation of kinetic and potential energies to each other repeatedly. In the first chapter of Vibration Kinematics, we will learn that mechanical vibrations will be expressed by mathematical functions. In the second chapter, Vibration Dynamics, we will learn how to derive the equations of motion of vibrating dynamic systems. The Newton-Euler and Lagrange methods are being applied in deriving the equations of motion. The fundamental knowledge on matrix calculus and differential equations are presented in the appendices at the end of the book.

Chapter 1
Vibration Kinematics

In this chapter, we study (1) mechanical elements of vibrating systems, (2) physical causes of mechanical vibrations, (3) kinematics of vibrations, and (4) simplification methods of complex vibrating systems. We reduce mechanical systems to an equivalent standard mass-spring-damper system, whenever possible. We will mathematically solve a general mass-spring-damper system and then will make other vibrating systems to be reduced to an equivalent system. Such equivalence yields a generalization in mathematical treatment of vibrating systems. The first section defines all fundamental terms we need to study vibrations.

1.1 Mechanical Vibration Elements

Mechanical vibration is a result of transformation of kinetic energy K to potential energy P, back and forth. When the potential energy is at its maximum, the kinetic energy is zero, and when the kinetic energy is at its maximum, the potential energy is minimum. Because in mechanical engineering a fluctuation in kinetic energy appears as a periodic motion of a massive body, we call such energy transformations *mechanical vibrations*.

The kinetic and potential energies are stored in physical elements. Any element that stores kinetic energy is called the *mass* or *inertia*, and any element that stores potential energy is called the *spring* or *restoring element*. If the total value of *mechanical energy* $E = K + P$ decreases during vibrations, there is also a phenomenon or an element that dissipates energy. The element that causes energy to dissipate is called a *damper*. The symbolic illustration of a mass, spring, and damper is shown in Fig. 1.1.

© The Author(s), under exclusive license to Springer Nature Switzerland AG 2022

R. N. Jazar, *Advanced Vibrations*, https://doi.org/10.1007/978-3-031-16356-2_1

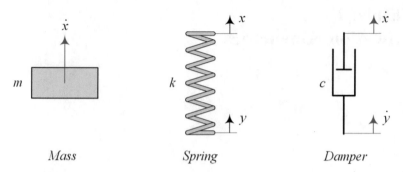

Fig. 1.1 A mass m, spring k, and damper c

1.1.1 Mass

In Newtonian mechanics, a mass m is the proportionality coefficient between an applied force on an object and the acceleration of the object. The mass of the object is the most important characteristic in Newtonian mechanics, and that is why we refer to an object only by its mass instead of an object with mass m. The required force f to move a mass m is proportional to its acceleration $a \equiv \ddot{x}$. The mass m is the proportionality coefficient.

$$f = ma \tag{1.1}$$

In mechanical vibrations, the mass m is an element that acts as a container to store kinetic energy. The amount of stored kinetic energy K in a mass m is proportional to the square of its velocity, v^2, $v \equiv \dot{x} = dx/dt$.

$$K = \frac{1}{2}mv^2 \tag{1.2}$$

The mass m is the resistance of an object to acceleration. The higher the mass, the lower the acceleration of the object under a constant force. The linearity and proportionality properties of the Newton equation of motion make the mass to be an invariant characteristic of the object.

$$f_1 = ma_1 \qquad f_2 = ma_2 \tag{1.3}$$
$$f_1 + f_2 = m(a_1 + a_2) \tag{1.4}$$

So, an object has a constant mass regardless of motion or rest or the amount of the applied force. This fact suggested to accept mass as a fundamental unit instead of force and, hence, define dimension of force in terms of mass and acceleration.

$$[f] = [ma] = MLT^{-2} \tag{1.5}$$

1.1.2 Spring

In Newtonian mechanics, a spring with stiffness k is an element to make a force f proportional to the relative displacement of its ends $x - y$ in opposite direction. The stiffness k is the constant of proportionality. The curve of the force f versus relative displacement $z = x - y$ is called the *spring characteristic curve*. If the stiffness of a spring, k, is constant, it is called a *linear spring*.

$$f = -kz = -k(x - y) \tag{1.6}$$

$$z = x - y \tag{1.7}$$

In mechanical vibrations, a spring is any element that acts as a container to store potential energy. The amount of stored potential energy P in a spring is equal to the work done by the spring force f during the deflection of the spring.

$$P = -\int f \, dz = -\int -kz \, dz \tag{1.8}$$

The spring potential energy is then a function of displacement. For linear springs we have:

$$P = \frac{1}{2}kz^2 \tag{1.9}$$

1.1.3 Damper

In mechanical vibrations, a damper is any element, device, or mechanism that makes the system to lose mechanical energy $E = P + K$.

The *viscous damping model* is the most adopted model of damping in mechanical vibrations. In Newtonian mechanics, a viscose damper with damping c is an element to make a resistance force f proportional to the relative velocity of its ends $\dot{x} - \dot{y}$. The damping c is the constant of proportionality. The curve of the force f versus relative velocity $\dot{z} = \dot{x} - \dot{y}$ is called the *damper characteristic curve*. If the damping of a damper, c, is constant, it is called a *linear damper*.

$$f = -c\dot{z} = -c(\dot{x} - \dot{y}) \tag{1.10}$$

Dampers may also be called *shock absorber*. Viscous damping model is simple, and under certain conditions it brings decoupling to the equations of motion in multi degree-of-freedom systems. Lose of energy can be measured by the average of dissipation-power in one cycle. Assuming a harmonic relative displacement,

$$z = A \cos \omega t \tag{1.11}$$

the dissipated energy W_D in one cycle equals the work done by the damping force.

$$W_D = \int_0^T f \, dz = \int_0^T c\dot{z} \, dz = \int_0^{2\pi/\omega} c\dot{z}^2 \, dt$$

$$= cA^2\omega^2 \int_0^{2\pi/\omega} \sin^2 \omega t \, dt = \pi c \omega A^2 \tag{1.12}$$

$$c = \frac{W_D}{\pi \omega A^2} \tag{1.13}$$

The absolute value of damping force f_d for a harmonic oscillator is:

$$f_d = c\dot{x} = c\omega A \sin \omega t \tag{1.14}$$

$$= c\omega A \sqrt{1 - \cos^2 \omega t} = c\omega \sqrt{A^2 - z^2} \tag{1.15}$$

This equation can be rearranged in the form of an ellipse in the coordinate (x, f_d).

$$\left(\frac{x}{A}\right)^2 + \left(\frac{f_d}{c\omega A}\right)^2 = 1 \tag{1.16}$$

The ellipse is called the hysteresis loop. The area enclosed by the ellipse equals the dissipated energy during one cycle, $\pi c\omega A^2$. The average of dissipation power P_D of viscus damping in one cycle will be:

$$P_D = \frac{1}{T} \int_0^T f \dot{z} \, dt = \frac{\omega}{2\pi} \int_0^{2\pi/\omega} c\dot{z} \, \dot{z} \, dt$$

$$= \frac{c\omega^3}{2\pi} A^2 \int_0^{2\pi/\omega} \sin^2 \omega t \, dt = \frac{1}{2} cA^2\omega^2 \tag{1.17}$$

A more general model of damping is governed by the following mathematical equation.

$$f = -c\dot{z} \, |\dot{z}|^{s-1} \tag{1.18}$$

For $s = 1$, we have the linear viscous damping model (1.10). For $s = 0$, we have Coulomb damping to represent contact friction losses.

$$f = -\mu N \frac{\dot{z}}{|\dot{z}|} = -\mu N \, \text{sgn} \, (\dot{z}) \tag{1.19}$$

For $s = 2$, we have quadratic damping to represent hydrodynamic damping.

$$f = -c\dot{z} \, |\dot{z}| \tag{1.20}$$

These models are all functions of velocity only. It may define other damping models to be functions of velocity as well as displacement.

1.1.4 Period

Any periodic motion x is characterized by a *period* T, which is the required time for one complete cycle of vibration, starting from and ending at the same conditions such as $(\dot{x} = 0, 0 < \ddot{x})$. The *frequency* f is the number of cycles in one T.

$$f = \frac{1}{T} \tag{1.21}$$

$$x\,(t) = x\,(t + T) \tag{1.22}$$

$$\dot{x}\,(t) = \dot{x}\,(t + T) \tag{1.23}$$

In theoretical vibrations, we usually work with *angular frequency* ω [rad/s], and in applied vibrations, we use the *cyclic frequency* f [Hz]:

$$\omega = 2\pi f \tag{1.24}$$

If a variable x is periodic in T, it is also periodic in $2T, 3T, \cdots$. So, the period of a periodic variable, x, is the least value of required time to make a repetition.

1.1.5 Vibration Modes

When there is no applied external force or *excitation* on a vibrating system, any possible oscillation of the system is called a *free vibration*. A free vibrating system will oscillate if any one of the kinematic states of position, or speed, x, $\dot{x}$, is not zero. When we apply an excitation then, any possible motion of the system is called a *forced vibration*. There are four types of applied excitation: *harmonic*, *periodic*, *transient* , and *random*. The harmonic and transient excitations are more applied and more predictable than the periodic and random types. When the excitation is a sinusoidal function of time, it is called a *harmonic excitation*, and when the excitation force disappears after a while or stays steady, it is a *transient excitation*. Any periodic excitation can theoretically be decomposed into a series of harmonic excitations with different coefficients and multiple frequencies. Therefore, a *periodic excitation* is a combination of infinite number of harmonic excitations with decreasing weight factor coefficients. Fourier analysis will reveal the harmonic functions and their coefficients within the periodic excitation. A *random excitation* has no short-term pattern and is not predictable; however, we may define some long-term averages to characterize a random excitation.

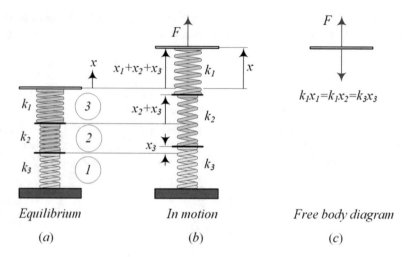

Fig. 1.2 Three serial springs

We use f to indicate a force. If f is a harmonically variable force, we show its amplitude by F, to be consistent with a harmonic motion x with amplitude X. We also use f for cyclic frequency; however, f is a force unless it is indicated that it is a frequency.

Example 1 Serial springs and dampers. It mathematically defines series of springs and shows how we replace a series of springs with a single equivalent spring.

By serial springs we refer to their head to tail geometrical attachments as is shown in Fig. 1.2. Serial springs have the same force and a resultant displacement equal to the sum of individual displacements. Figure 1.2 illustrates three serial springs attached to a massless plate and the ground.

The equilibrium position of the springs is their unstretched configuration in Fig. 1.2a. Because springs are assumed to be massless, serial attachments of springs will not generate any displacement in them. Applying a displacement x, in Fig. 1.2b, generates the free body diagram of the upper plate as shown in Fig. 1.2c. Each spring makes a force $f_i = -k_i x_i$ where x_i is the length change in the spring number i. The total displacement of the springs, x, is the sum of their individual displacements $\sum x_i$.

$$x = \sum x_i = x_1 + x_2 + x_3 \tag{1.25}$$

We may substitute a set of serial springs with only one equivalent spring of stiffness k_e that produces the same displacement x under the same force $F = f_k$.

$$F = f_k = -k_1 x_1 = -k_2 x_2 = -k_3 x_3 = -k_e x \tag{1.26}$$

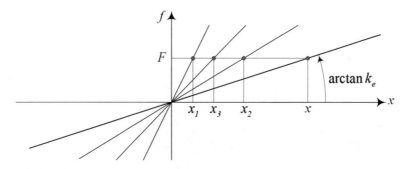

Fig. 1.3 The characteristic curves of serial springs with stiffnesses $k_1 > k_2 > k_3$ and their equivalent stiffness k_{eq}

Substituting (1.26) into (1.25) yields:

$$\frac{F}{k_e} = \frac{F}{k_1} + \frac{F}{k_2} + \frac{F}{k_3} = \left(\frac{1}{k_1} + \frac{1}{k_2} + \frac{1}{k_3}\right) F \tag{1.27}$$

It shows that the inverse of the equivalent stiffness, $1/k_e$, of the serial springs is the sum of their inverse stiffness, as $\sum 1/k_i$:

$$\frac{1}{k_e} = \frac{1}{k_1} + \frac{1}{k_2} + \frac{1}{k_3} \tag{1.28}$$

We assume that the force of a linear spring is not affected by kinematic variables x, $\dot{x}$, $\ddot{x}$, and time t. The characteristic curves of springs k_1, k_2, k_3, and k_e and their force-displacement behaviors are illustrated in Fig. 1.3. The equivalent stiffness of a serial springs is always less than the stiffness of every individual springs.

Similarly, serial dampers have the same force, $F = f_c$, and a resultant velocity $\dot{x}$ that is equal to the sum of individual velocities, $\sum \dot{x}_i$. We may substitute a set of serial dampers with only one equivalent damping c_e that with the same velocity $\dot{x}$ produces the same force f_c. For three parallel dampers, the velocity and force balance equations are:

$$\dot{x} = \dot{x}_1 + \dot{x}_2 + \dot{x}_3 \tag{1.29}$$

$$F = f_c = -c_1\dot{x} = -c_2\dot{x} = -c_3\dot{x} = -c_e\dot{x} \tag{1.30}$$

It show that the inverse of the equivalent damping of the serial dampers, $1/c_e$, is the sum of their inverse dampings, $\sum 1/c_i$:

$$\frac{1}{c_e} = \frac{1}{c_1} + \frac{1}{c_2} + \frac{1}{c_3} \tag{1.31}$$

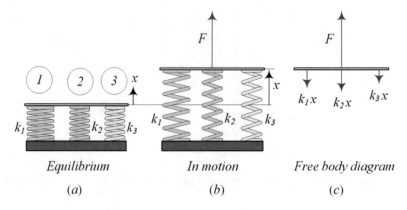

Fig. 1.4 Three parallel springs

We assume that the force of a linear damper is not affected by kinematics x, $\dot{x}$, $\ddot{x}$, and time t. The equivalent damping of a serial damper is always less than the damping of every individual dampers.

Example 2 Parallel springs and dampers. Here we mathematically define parallel springs and show how to replace parallel springs with a single equivalent spring.

By parallel springs we refer to their geometrical attachments as is shown in Fig. 1.4. Parallel springs have the same displacement, x, and a resultant force, F, equal to the sum of the individual forces, $\sum f_i$. Figure 1.4 illustrates three parallel springs between a massless plate and the ground. The equilibrium position of the springs is the unstretched configuration, shown in Fig. 1.4a. Applying a displacement x to all the springs in Fig. 1.4b generates the free body diagram of Fig. 1.4c. Each spring makes a force $-kx$ opposite to the direction of displacement. The resultant force of the springs $F = f_k$ is:

$$F = f_k = -k_1 x - k_2 x - k_3 x = -(k_1 + k_2 + k_3)\, x \qquad (1.32)$$

We may substitute parallel springs with only one equivalent spring of stiffness k_e that produces the same force F under the same displacement x.

$$f_k = -k_e x \qquad k_e = k_1 + k_2 + k_3 \qquad (1.33)$$

Therefore, the equivalent stiffness, k_e, of parallel springs is the sum of their stiffnesses $\sum k_i$. The characteristic curves of springs k_1, k_2, k_3, and k_e and their force-displacement behaviors are illustrated in Fig. 1.5.

Parallel dampers have the same speed $\dot{x}$, and a resultant force $F = f_c$ equal to the sum of the individual forces $\sum f_i$. We may substitute parallel dampers with only one equivalent damping c_e that produces the same force F under the same velocity $\dot{x}$. Consider three parallel dampers such as shown in Fig. 1.6. Their force balance

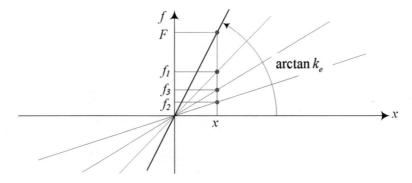

Fig. 1.5 The characteristic curves of parallel springs with stiffnesses $k_1 < k_2 < k_3$ and their equivalent k_{eq}

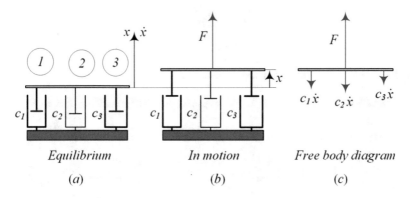

Fig. 1.6 Three parallel dampers

and equivalent damper are:

$$f_c = -c_e \dot{x} = -c_1 \dot{x} - c_2 \dot{x} - c_3 \dot{x} = -(c_1 + c_2 + c_3) \dot{x} \qquad (1.34)$$

$$c_e = c_1 + c_2 + c_3 \qquad (1.35)$$

Example 3 Flexible frame. This is an example to show how we determine equivalent spring for two springs that are not parallel nor in series.

Figure 1.7 depicts a mass m hanging from a frame. The frame is flexible, so it can be modeled by some springs attached to each other as is shown in Fig. 1.8a. If we assume that every beam is simply supported, then the equivalent stiffness for a lateral deflection of each beam at their midspan is:

$$k_3 = \frac{48E_3 I_3}{l_3^3} \qquad k_4 = \frac{48E_4 I_4}{l_4^3} \qquad k_5 = \frac{48E_5 I_5}{l_5^3} \qquad (1.36)$$

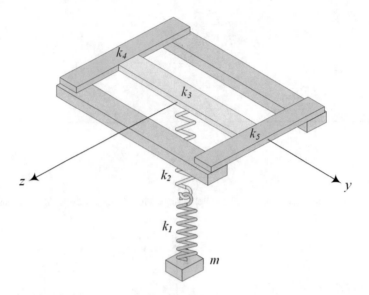

Fig. 1.7 A mass m hanging from a flexible frame

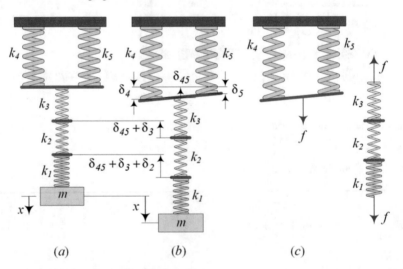

Fig. 1.8 Equivalent springs model for the flexible frame

When the mass is vibrating, the elongation of the springs would be similar to
Fig. 1.8b. Assume that we separate the mass and springs and then apply a force
f at the end of spring k_1 as shown in Fig. 1.8c. Because the springs k_1, k_2, and
k_3 have the same force, and their resultant displacement is the sum of individual
displacements, they are in series.

The springs k_4 and k_5 are neither in series nor parallel. To find their equivalent, we need a geometric analysis. Let us assume that each of the springs k_4 and k_5 support a force equal to $f/2$. Therefore their deflections are:

$$\delta_4 = \frac{f}{2k_4} \qquad \delta_5 = \frac{f}{2k_5} \tag{1.37}$$

The displacement at midspan of the lateral beam is δ_{45}.

$$\delta_{45} = \frac{\delta_4 + \delta_5}{2} \tag{1.38}$$

Assuming

$$\delta_{45} = \frac{f}{k_{45}} \tag{1.39}$$

we can define an equivalent stiffness k_{45} for k_4 and k_5 as:

$$\frac{1}{k_{45}} = \frac{1}{2}\left(\frac{1}{2k_4} + \frac{1}{2k_5}\right) = \frac{1}{4}\left(\frac{1}{k_4} + \frac{1}{k_5}\right) \tag{1.40}$$

Now the equivalent spring k_{45} is in series with the series of k_1, k_2, and k_3. Hence the overall equivalent spring k_e can be calculated.

$$\begin{aligned}
\frac{1}{k_e} &= \frac{1}{k_1} + \frac{1}{k_2} + \frac{1}{k_3} + \frac{1}{k_{45}} \\
&= \frac{1}{k_1} + \frac{1}{k_2} + \frac{1}{k_3} + \frac{1}{4k_4} + \frac{1}{4k_5}
\end{aligned} \tag{1.41}$$

Example 4 Different length parallel springs. Assembling springs with different length happens frequently in engineering application. How to determine equivalent spring of two parallel springs is important to know.

In practical situations, it frequently happens that we should adjust and fit a shorter or longer spring in a physical space. Figure 1.9 illustrates a mass m that is supposed to sit on two unequal springs. To examine the system, let us assume that m can only move in the x direction and the length of spring k_1 is shorter than k_2 by δ_0. When we attach k_1 to m, the equilibrium position of m will move to have a distance of δ from its original position. At this equilibrium position, the spring k_1 is elongated by δ_1, and k_2 is shortened by δ_2. There are two equations to determine δ_1 and δ_2. Force balance on m indicates that

$$-k_1\delta_1 + k_2\delta_2 = 0 \tag{1.42}$$

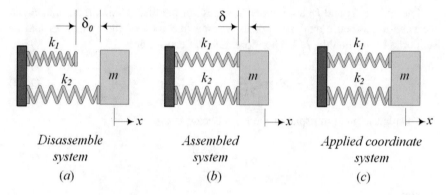

Fig. 1.9 Different length parallel springs assembling

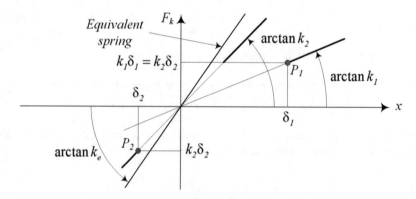

Fig. 1.10 Characteristic curves of k_1 and k_2 and their working points

and geometric compatibility provides us with:

$$\delta_1 = \delta_0 - \delta \qquad \delta_2 = \delta \tag{1.43}$$

Therefore, δ_1 and δ_2 can be calculated.

$$\delta_1 = \frac{\delta_0}{1 + k_1/k_2} \qquad \delta_2 = \frac{\delta_0}{k_2/k_1 + 1} \tag{1.44}$$

The assembled system indicates a mass m that is attached to two springs which are not at their neutral lengths. The working points of k_1 and k_2 are shown by P_1 and P_2 in characteristic curves of k_1 and k_2 of Fig. 1.10. Although the displacements of the springs are equal while m is moving, the resultant force on m needs to be analyzed to make sure $k_e = k_1 + k_2$. To examine the possible equivalent spring, let us apply a displacement x to m and determine the force-displacement equation:

$$F_1 = k_1\delta_1 + k_1 x \tag{1.45}$$

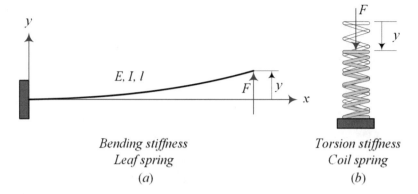

Bending stiffness
Leaf spring
(a)

Torsion stiffness
Coil spring
(b)

Fig. 1.11 Leaf springs work based on banding stiffness, and coil springs work based on torsion stiffness

$$F_2 = -k_2\delta_2 + k_2 x = -k_1\delta_1 + k_2 x \tag{1.46}$$

$$F_k = F_1 + F_2 = (k_1 + k_2)\,x = k_e x \tag{1.47}$$

Therefore, as long as the displacements of the assembled springs are equal, there is an equivalent spring $k_e = k_1 + k_2$, to be substituted for the two springs with the equilibrium point at $x = 0$ of the assembled system.

Example 5 Stiffness and length of springs. The stiffness value of springs is approximate value, and it depends on the type of spring. Here are equations for leaf and coil springs.

The value of stiffness k is a combined property of the material and geometric characteristics of a spring. Figure 1.11 illustrates a leaf and a coil springs. The linear approximated force-displacement equation of the leaf spring, $y = \frac{l^3}{3EI}F$, indicates that the equivalent stiffness k of the spring is:

$$k = k_{leaf} = \frac{3EI}{l^3} \tag{1.48}$$

where E is the Young modulus of the material, I is geometrical moment of the cross sectional area of the beam about lateral neutral axis, and l is the length of the beam. When we cut a leaf spring in half, its stiffness increases *eight* times.

$$k_{l/2} = \frac{3EI}{(l/2)^3} = 8\frac{3EI}{l^3} = 8k_{leaf} \tag{1.49}$$

The approximate equivalent stiffness of a coil spring is:

$$k = k_{coil} = \frac{Gpd^4}{3lD^3} \tag{1.50}$$

where G is the shear modulus of the spring material, D is the mean diameter of the coil spring measured from the centers of the wire cross sections, d is the diameter of the wire, and n is the number of coils. If l is the free length of the spring and p is its pitch, then we have

$$p = \frac{l}{n} \tag{1.51}$$

When we cut a coil spring in half, its stiffness increases *two* times.

$$k_{l/2} = \frac{Gpd^4}{3\,(l/2)\,D^3} = 2\frac{Gpd^4}{3lD^3} = 2k_{coil} \tag{1.52}$$

Let us assume a coil spring of length l is made of a series of two half length similar coil springs. The stiffness of the spring would be half of a short spring.

$$k_{coil} = \frac{1}{\dfrac{1}{k_{l/2}} + \dfrac{1}{k_{l/2}}} = \frac{1}{2}k_{l/2} \tag{1.53}$$

Example 6 ★ Massive spring. Springs are assumed to be massless; however, they are massive element, and hence, they collect some kinetic energy. Here is an approximate method to approximate a massive spring with and equivalent massless spring.

In modeling of vibrating systems, we ignore the mass of springs and dampers. This assumption is valid as long as the masses of springs and dampers are much smaller than the mass of the body they support. However, when the mass of spring m_s or damper m_d is comparable with the mass of body m, we may define a new system with an equivalent mass m_e which is supported by a massless spring and damper.

Consider a vibrating system with a massive spring as shown in Fig. 1.12a. When the system is at equilibrium, the spring has a mass m_s and a length l. The mass of spring is uniformly distributed along its length, so we may define a mass density ρ.

$$\rho = \frac{m}{l} \tag{1.54}$$

To verify that the equivalent mass m_e of an equivalent massless spring system is

$$m_e = m + \frac{1}{3}m_s \tag{1.55}$$

we seek a system which keeps the same amount of kinetic energy as the original system. Figure 1.12b illustrates the original system when the mass m is at position x and has a velocity $\dot{x}$. The spring is extended between the mass m and the ground.

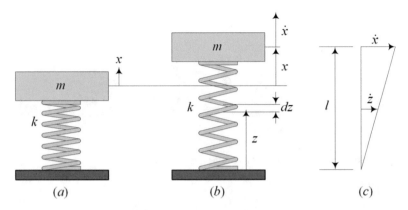

Fig. 1.12 A vibrating system with a massive spring

So, the base of the spring has no velocity, while the other end has the same velocity as m. Let us define a coordinate position z that goes from the grounded base of the spring to the end point at m. An element of spring at z has a length dz and a mass dm.

$$dm = \rho\, dz \tag{1.56}$$

Assuming $x \ll l$ and a linear velocity distribution of the elements of spring, as shown in Fig. 1.12c, we find the velocity $\dot{z}$ of dm.

$$\dot{z} = \frac{z}{l}\dot{x} \tag{1.57}$$

The kinetic energy of the system is a summation of kinetic energy of the mass m plus the kinetic energy of the spring.

$$
\begin{aligned}
K &= \frac{1}{2}m\dot{x}^2 + \frac{1}{2}\int_0^l \left(dm\,\dot{z}^2\right) = \frac{1}{2}m\dot{x}^2 + \frac{1}{2}\int_0^l \rho\left(\frac{z}{l}\dot{x}\right)^2 dz \\
&= \frac{1}{2}m\dot{x}^2 + \frac{1}{2}\frac{\rho}{l^2}\dot{x}^2\int_0^l z^2\, dz = \frac{1}{2}m\dot{x}^2 + \frac{1}{2}\frac{\rho}{l^2}\dot{x}^2\left(\frac{1}{3}l^3\right) \\
&= \frac{1}{2}\left(m + \frac{1}{3}m_s\right)\dot{x}^2 = \frac{1}{2}m_e\dot{x}^2
\end{aligned}
\tag{1.58}
$$

Therefore, an equivalent system should have a massless spring attached to a mass $m_e = m + \frac{1}{3}m_s$ to keep the same amount of kinetic energy. Because the potential energy of a spring is not a function of its mass, the equivalent system will vibrate in similar to the original system.

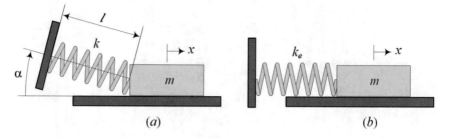

Fig. 1.13 A tilted spring and its equivalent stiffness

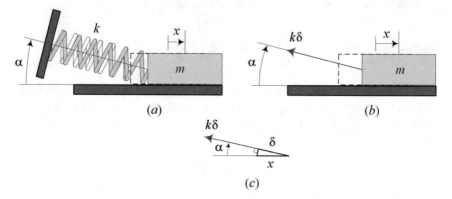

Fig. 1.14 A mass-spring system such that the spring makes and angle α with directing of mass translation

Example 7 Tilted spring. This is to show how we can replace a tilted spring with an equivalent collinear spring.

Consider a mass-spring system such that the spring makes an angle α with the axis of mass translation, as shown in Fig. 1.13a. We may substitute such a tilted spring with an equivalent spring k_e that is on the same axis of mass translation, as shown in Fig. 1.13b. We have:

$$k_e \approx k \cos^2 \alpha \qquad (1.59)$$

When the mass m is in motion, such as is in Fig. 1.14a, its free body diagram is as shown in Fig. 1.14b. If the motion of mass m is much smaller than the length of the spring, $x \ll l$, we ignore any changes in α and calculate the spring elongation δ as shown in Fig. 1.14c.

$$\delta \approx x \cos \alpha \qquad (1.60)$$

Therefore, the spring force f_k is:

$$f_k = k\delta \approx kx \cos \alpha \qquad (1.61)$$

The spring force should be projected on the x-axis to find the x component, f_x, that causes the mass m to move.

$$f_x = f_k \cos \alpha \approx \left(k \cos^2 \alpha \right) x \qquad (1.62)$$

Therefore, the tilted spring can be substituted by an equivalent spring k_e on the x-axis that needs the same force f_x to elongate the same amount x as the mass moves.

$$f_x = k_e x \qquad k_e \approx k \cos^2 \alpha \qquad (1.63)$$

Example 8 Alternative proof for equivalent tilted spring. Equivalent masses can be found by equating kinetic energies K, and equivalent springs can be found by equating potential energies P. Here is to show how we find equivalent spring for a tilted spring by equating potential energies.

Consider a spring that makes an angle α with the direction of motion as shown in Fig. 1.14a. When the mass m translates x, the elongation of the spring would be δ.

$$\delta \approx x \cos \alpha \qquad (1.64)$$

The potential energy of such a spring is:

$$P = \frac{1}{2} k \delta^2 = \frac{1}{2} \left(k \cos^2 \alpha \right) x^2 \qquad (1.65)$$

An equivalent spring with stiffness k_e must collect the same amount of potential energy for the same displacement x.

$$P = \frac{1}{2} k_e x^2 \qquad (1.66)$$

Therefore, the equivalent stiffness k_e is:

$$k_e = k \cos^2 \alpha \qquad (1.67)$$

Example 9 Displaced spring. Springs may be installed offset with respect to the measuring displacement point.

Figure 1.15a illustrates a mass m attached to the tip of a massless bar with length b. The bar is pivoted to a wall, and a spring k is attached to the bar at a distance a from the pivot. When the mass oscillates with displacement $x \ll b$, the elongation δ of the spring is:

$$\delta \approx \frac{a}{b} x \qquad (1.68)$$

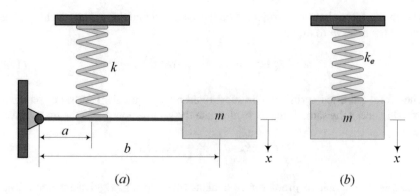

Fig. 1.15 A mass m attached to the tip of a massless bar with length b

We may substitute the system with a translational mass-spring system such as shown in Fig. 1.15b. The new system has the same mass m and an equivalent spring k_e.

$$k_e = \left(\frac{a}{b}\right)^2 k \tag{1.69}$$

The equivalent spring provides us with the same potential energy as the original spring.

$$P = \frac{1}{2}k_e x^2 = \frac{1}{2}k\delta^2 = \frac{1}{2}k\left(\frac{a}{b}x\right)^2 = \frac{1}{2}k\left(\frac{a}{b}\right)^2 x^2 \tag{1.70}$$

Example 10 Equivalent spring and damper of a MacPherson suspension. How to model a vehicle suspension with equivalent mass-spring-damper.

Figure 1.16 illustrates a MacPherson strut mechanism and its equivalent model of vibrating system. We assume the tire is stiff, and therefore, the wheel center gets the same motion y as the tireprint of the wheel. Furthermore, we assume the wheel and the body of the vehicle to move only vertically.

To find the equivalent parameters for the vibrating model, we use m equal to $1/4$ of the vehicle body mass. The spring k and damper c make an angle α with the direction of wheel motion. They are also displaced $b - a$ from the wheel center, laterally. So, the equivalent spring k_e and damper c_e are:

$$k_e = k\left(\frac{a}{b}\cos\alpha\right)^2 \qquad c_e = c\left(\frac{a}{b}\cos\alpha\right)^2 \tag{1.71}$$

As an application assume we have determined the following stiffness and damping as a result of an optimization algorithm.

$$k_e = 9869.6 \text{ N/m} \qquad c_e = 87.965 \text{ Ns/m} \tag{1.72}$$

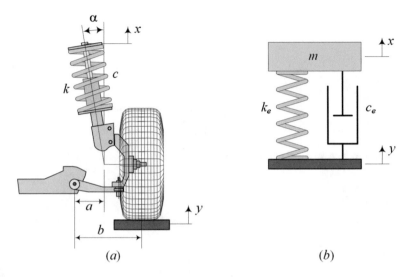

Fig. 1.16 A MacPherson suspension and its equivalent vibrating system

The actual k and c for a MacPherson suspension with

$$a = 19 \text{ cm} \qquad b = 32 \text{ cm} \qquad \alpha = 27 \text{ deg} \tag{1.73}$$

will be:

$$k = 35264 \text{ N/ m} \qquad c = 314.3 \text{ N s/ m} \tag{1.74}$$

Example 11 Equivalent viscous damping and energy loss. It is to show how we approximate damping of real systems.

Consider the system in Fig. 1.16b when $y = 0$ and $x = X \sin \omega t$. The required force to produce this motion is:

$$f = kx + c\dot{x} = k\,(X \sin \omega t) + c\,(X\omega \cos \omega t) \tag{1.75}$$

In one period of vibration, the consumed energy per cycle E_c is proportional to the viscous damping of the system.

$$E_c = \int_0^T f x\, dt = \int_0^{2\pi/\omega} (k\,(X \sin \omega t) + c\,(X\omega \cos \omega t))\,(X\omega \cos \omega t)\, dt$$

$$= c\pi \omega X^2 \tag{1.76}$$

The definition of linear viscous damping is then equal to the dissipated energy in one cycle divided by $\pi \omega X^2$:

$$c = \frac{E_c}{\pi \omega X^2} \tag{1.77}$$

This equation is being used to calculate equivalent damping of other types of damping.

$$c_e = \frac{E_c}{\pi \omega X^2} \tag{1.78}$$

As an example consider a system with

$$m = 2 \text{ kg} \qquad k = 10{,}000 \text{ N/m} \qquad x = 0.01 \sin 100t \ \text{m} \tag{1.79}$$

that consumes $E_c = 3.1416 \text{ J}/cycle$. The equivalent linear damping of the system is:

$$c = \frac{E_c}{\pi \omega X^2} = \frac{3.1416}{\pi \times 100 \times 0.01^2} = 100 \text{ Ns/m} \tag{1.80}$$

For Coulomb friction model with Coulomb force $f_C = \mu N$ that dissipates energy $E_c = 4 f_C X$, the equivalent viscous damper is:

$$c_e = \frac{4 \mu N}{\pi \omega X} \tag{1.81}$$

For hydraulic damping with force $f_H = c_L \, \dot{x}^2 \, \text{sgn} \, \dot{x}$, the dissipates energy E_c and the equivalent viscous damper are:

$$E_c = \int_0^T c_L \dot{x}^2 dx = c_L \int_0^T \dot{x}^3 dt = 4 c_L \omega^3 X^3 \int_0^{\pi/(2\omega)} \cos^3 \omega t \ dt$$
$$= \frac{8}{3} c_L \omega^3 X^3 \tag{1.82}$$

$$c_e = \frac{8}{3} \frac{c_L \omega^2 X}{\pi} \tag{1.83}$$

1.2   Kinematics of Vibrations

The *fundamental equation of vibration* is the free and undamped equation of motion of a mass m attached to a linear spring with stiffness k.

$$m\ddot{x} + kx = 0 \qquad x(0) = x_0 \qquad \dot{x}(0) = \dot{x}_0 \tag{1.84}$$

$$\ddot{x} + \omega_n^2 x = 0 \qquad \omega_n = \sqrt{\frac{k}{m}} \tag{1.85}$$

The solution of the fundamental equation of vibration is:

$$x = A \sin \omega_n t + B \cos \omega_n t \tag{1.86}$$

$$= x_0 \cos \omega_n + \frac{1}{\omega_n} \dot{x}_0 \sin \omega_n \tag{1.87}$$

$$= X \sin (\omega_n t + \varphi) \tag{1.88}$$

$$= X \cos (\omega t + \theta) \tag{1.89}$$

The solution (1.88) or (1.89) is called the *harmonic* or *wave* solution, and the solution (1.86) is called the *weighted harmonic* or *complete harmonic* solution. The X is the *amplitude*, ω is the *angular frequency*, and φ is the *phase* of the harmonic solution, and the coefficients A and B are the *weight factors* of the weighted harmonic solution. The parameters X, φ, θ, A, and B are functions of the initial conditions x_0, $\dot{x}_0$.

$$A = x_0 \qquad B = \frac{1}{\omega} \dot{x}_0 \tag{1.90}$$

$$X = \sqrt{A^2 + B^2} = \sqrt{x_0^2 + \frac{\dot{x}_0^2}{\omega^2}} \tag{1.91}$$

The parameters X, φ, A, and B are related by the following equations.

$$A = X \cos \varphi \qquad B = X \sin \varphi \qquad \tan \varphi = \frac{B}{A} \tag{1.92}$$

$$A = X \sin \theta \qquad B = -X \cos \theta \qquad \tan \theta = \frac{A}{-B} \tag{1.93}$$

Proof The fundamental equation of vibration (1.84) is a linear second-order ordinary deferential equation, and hence its solution will be exponential function. To find the solution, we substitute a general exponential function with unknown coefficients and unknown power and determine the unknowns.

$$x = X e^{st} \tag{1.94}$$

It is done in two steps:

1. We determine the exponent s. They value depend on the parameters of the system, and the number of s is equal to the order of the differential equation. In this case, we will have two values for s, and they will be functions of the

parameters k and m.

$$m\left(Xs^2e^{st}\right) + k\left(Xe^{st}\right) = \left(ms^2 + k\right)Xe^{st} = 0 \qquad (1.95)$$

$$ms^2 + k = 0 \qquad (1.96)$$

$$s_1 = \sqrt{-\frac{k}{m}} = i\sqrt{\frac{k}{m}} = i\omega_n \qquad (1.97)$$

$$s_2 = -\sqrt{-\frac{k}{m}} = -i\sqrt{\frac{k}{m}} = -i\omega_n \qquad (1.98)$$

$$\omega_n = \sqrt{\frac{k}{m}} \qquad (1.99)$$

2. We determine the coefficient X. Every value of s provides us with an exponential function. All these exponential functions are linearly independent, and hence the general solution of the equation would be a linear combination of the exponential functions, each with an unknown coefficients. The coefficients depend on the initial conditions of the system. In this case, we have two exponential functions with coefficients X_1 and X_2.

$$x = X_1e^{s_1t} + X_2e^{s_2t} \qquad (1.100)$$

$$\dot{x} = X_1s_1e^{s_1t} + X_2s_2e^{s_2t} \qquad (1.101)$$

The initial conditions (1.84) provide us with two equations to find X_1 and X_2.

$$x_0 = X_1 + X_2 \qquad (1.102)$$

$$\dot{x}_0 = X_1s_1 + X_2s_2 \qquad (1.103)$$

$$\begin{bmatrix} x_0 \\ \dot{x}_0 \end{bmatrix} = \begin{bmatrix} 1 & 1 \\ s_1 & s_2 \end{bmatrix} \begin{bmatrix} X_1 \\ X_2 \end{bmatrix} \qquad (1.104)$$

$$\begin{bmatrix} X_1 \\ X_2 \end{bmatrix} = \begin{bmatrix} 1 & 1 \\ s_1 & s_2 \end{bmatrix}^{-1} \begin{bmatrix} x_0 \\ \dot{x}_0 \end{bmatrix} = \begin{bmatrix} \frac{1}{s_1-s_2}(\dot{x}_0 - s_2x_0) \\ -\frac{1}{s_1-s_2}(\dot{x}_0 - s_1x_0) \end{bmatrix} \qquad (1.105)$$

The complete solution is achieved.

$$x = X_1 e^{s_1 t} + X_2 e^{s_2 t} = \frac{\dot{x}_0 - s_2 x_0}{s_1 - s_2} e^{s_1 t} - \frac{\dot{x}_0 - s_1 x_0}{s_1 - s_2} e^{s_2 t} \tag{1.106}$$

Although it is the complete solution, it is not in the shape we want to work with. To rewrite it with harmonic functions, we employ the Euler identity.

$$e^{i\theta} = \cos\theta + i\sin\theta \tag{1.107}$$

$$e^{i\omega_n t} = \cos\omega_n t + i\sin\omega_n t \tag{1.108}$$

$$e^{-i\omega_n t} = \cos\omega_n t - i\sin\omega_n t \tag{1.109}$$

$$
\begin{aligned}
x &= X_1 e^{i\omega_n t} + X_2 e^{-i\omega_n t} \\
&= X_1 (\cos\omega_n t + i\sin\omega_n t) + X_2 (\cos\omega_n t - i\sin\omega_n t) \\
&= (X_1 + X_2)\cos\omega_n t + (X_1 - X_2)\,i\sin\omega_n t
\end{aligned}
\tag{1.110}
$$

The coefficients X_1 and X_2 from (1.105) are complex conjugates.

$$
\begin{bmatrix} X_1 \\ X_2 \end{bmatrix} =
\begin{bmatrix} \frac{1}{s_1 - s_2}(\dot{x}_0 - s_2 x_0) \\ -\frac{1}{s_1 - s_2}(\dot{x}_0 - s_1 x_0) \end{bmatrix} =
\frac{1}{2}\begin{bmatrix} x_0 - i\frac{1}{\omega_n}\dot{x}_0 \\ x_0 + i\frac{1}{\omega_n}\dot{x}_0 \end{bmatrix}
\tag{1.111}
$$

Therefore,

$$X_1 + X_2 = x_0 \tag{1.112}$$

$$X_1 - X_2 = -i\frac{1}{\omega_n}\dot{x}_0 \tag{1.113}$$

and it makes the solution x to be a full harmonic function with real coefficients.

$$x = x_0 \cos\omega_n t + \frac{1}{\omega_n}\dot{x}_0 \sin\omega_n t \tag{1.114}$$

The parameter ω_n is the frequency of both functions $\cos\omega_n t$ and $\sin\omega_n t$ and, hence, the frequency of x. This is the frequency of response with no external excitation and no damping. That is the reason for calling this the natural frequency.

This solution suggests to begin with a solution made up of a full harmonic function with unknown coefficients and determine the coefficients. This solution is a full harmonic because it has both $\cos\omega_n t$ and $\sin\omega_n t$. These two functions are orthogonal, and hence, they are linearly independent. Each of them satisfies the differential equation, and therefore, a complete solution would be a linear combination of them.

$$x = A \sin \omega_n t + B \cos \omega_n t \tag{1.115}$$

Employing the initial conditions determines the coefficients and retrieve the same solution (1.115).

$$A = x_0 \qquad B = \frac{1}{\omega_n} \dot{x}_0 \tag{1.116}$$

Any full harmonic function such as the solution (1.115) may be combined into a single cos or sin function with a phase lag.

$$x = X \sin (\omega_n t + \varphi) \tag{1.117}$$

$$x = X \cos (\omega_n t + \theta) \tag{1.118}$$

The relationship between the amplitude X, phase φ, and weight factors A and B can be found by Eqs. (1.86) and (1.88):

$$A \sin \omega_n t + B \cos \omega_n t = X \sin (\omega_n t + \varphi)$$
$$= X \cos \varphi \sin \omega_n t + X \sin \varphi \cos \omega_n t \tag{1.119}$$

$$A = X \cos \varphi \qquad B = X \sin \varphi \tag{1.120}$$

$$X = \sqrt{A^2 + B^2} \qquad \tan \varphi = \frac{B}{A} \tag{1.121}$$

Similarly, the relationship between the amplitude X, phase θ, and weight factors A and B will be found by Eqs. (1.86) and (1.89):

$$A \sin \omega_n t + B \cos \omega_n t = X \cos (\omega_n t + \theta)$$
$$= X \sin \theta \sin \omega_n t - X \cos \theta \cos \omega_n t \tag{1.122}$$

$$A = X \sin \theta \qquad B = -X \cos \theta \tag{1.123}$$

$$X = \sqrt{A^2 + B^2} \qquad \tan \theta = \frac{A}{-B} \tag{1.124}$$

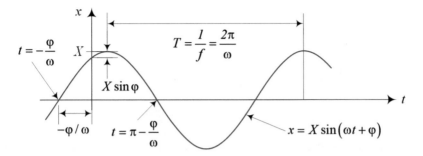

Fig. 1.17 A harmonic wave $x = X \sin(\omega_n t + \varphi)$ and its characteristics

Fig. 1.18 A mass-spring, single degree-of-freedom vibrating system

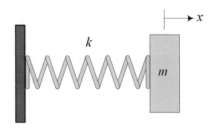

A harmonic wave $x = X \sin(\omega_n t + \varphi)$ and its characteristics are plotted in Fig. 1.17.

There are three characteristics for a harmonic vibration:

1. *Amplitude* X is the maximum deviation of x form the $x = 0$. Amplitude is the size of vibration and is equivalent to *loudness* and *intensity* in music.
2. *Frequency* ω_n is the number of periodic motion in unit of time. Frequency is equivalent to *pitch* in music.
3. *Period* T, *length*, or *duration* is the elapsed time for a complete wave. Period is equivalent to *rhythm* in music.

■

Example 12 Natural frequency. The natural frequency is a unique parameter at which a free vibrating system will oscillates, regardless of initial conditions.

Consider the mass-spring system of Fig. 1.18. The system is undamped and force-free, with a harmonic equation of motion.

$$m\,\ddot{x} + kx = 0 \tag{1.125}$$

To find the solution, let us try a harmonic solution with an unknown frequency Ω.

$$x = A \sin \Omega t + B \cos \Omega t \tag{1.126}$$

Substituting (1.126) in (1.125) provides us with an algebraic equation,

$$- \Omega^2 m \, (A \sin \Omega t + B \cos \Omega t) + k \, (A \sin \Omega t + B \cos \Omega t) = 0 \qquad (1.127)$$

which can be collected as a full harmonic equation.

$$\left(Bk - Bm\Omega^2 \right) \cos \Omega t + \left(Ak - Am\Omega^2 \right) \sin \Omega t = 0 \qquad (1.128)$$

The functions $\sin \Omega t$ and $\cos \Omega t$ are orthogonal, and hence, this equation can be correct only if their coefficients be zero. These conditions provide us with Ω and the possible solution x.

$$\Omega = \sqrt{\frac{k}{m}} \qquad (1.129)$$

$$x = A \sin \sqrt{\frac{k}{m}} t + B \cos \sqrt{\frac{k}{m}} t \qquad (1.130)$$

The parameter $\Omega = \sqrt{k/m}$ is the frequency of vibration of a free and undamped mass-spring system. It is the natural frequency and is shown by a special character ω_n.

$$\omega_n = \sqrt{\frac{k}{m}} \qquad (1.131)$$

Figure 1.19 illustrates how the natural frequency varies as a function of the ratio of stiffness to mass. Natural frequency is very sensitive to the ratio of k/m at low values and loses its sensitivity at higher values of k/m. Doubling the ratio of k/m makes ω_n to increase by a factor of 1.414. As an example, a system with $k/m = 5$ has the natural frequency of $\omega_n = 2.236$ rad/ s. Doubling k/m to $k/m = 10$ will increase the natural frequency to $\omega_n = 3.162$ rad/ s, and another doubling to $k/m = 20$ will make $\omega_n = 4.472$. Engineering point of these analyses is that it is harder to change the natural frequency of a heavy mechanical system than a light system. A system has as many natural frequencies as it has degrees of freedom.

From another viewpoint, we see that a system with no external excitation is a *free system* and, with no damping, is an *undamped system*. A single *DOF* undamped-free system is governed by Eq. (1.125). As long as the mass and stiffness are constant, any nonzero response of the free system is harmonic.

$$x = A \sin \omega_n t + B \cos \omega_n t \qquad (1.132)$$

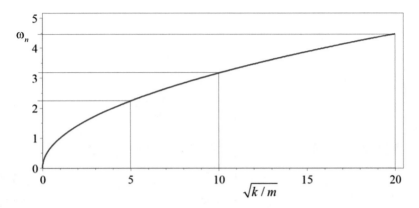

Fig. 1.19 Variation of the natural frequency as a function of the ratio of stiffness to mass

Substituting the solution in the equation provides us with two equations:

$$-\omega_n^2 m \left(A \sin \omega_n t + B \cos \omega_n t\right) + k \left(A \sin \omega_n t + B \cos \omega_n t\right) = 0 \qquad (1.133)$$

$$Ak - Am\omega_n^2 = 0 \qquad (1.134)$$

$$Bk - Bm\omega_n^2 = 0 \qquad (1.135)$$

which shows that A and B can be nonzero if:

$$\omega_n = \sqrt{\frac{k}{m}} \qquad (1.136)$$

The frequency of the harmonic solution (1.132) is called the natural frequency of the system, just because any disturbance of the system puts it in a vibration with this frequency.

Example 13 Boundary conditions. To determine the coefficients of the solution (1.86), we need the value of position x or velocity $\dot{x}$ at two different times, or a combination of them. Having the states x and $\dot{x}$ at any time t_0 other time than $t_0 = 0$ is called boundary conditions.

If the state of the harmonic solution (1.86) is given at a time t_0,

$$x(t_0) = x_0 \qquad \dot{x}(t_0) = \dot{x}_0 \qquad (1.137)$$

then, we must have

$$x_0 = A \sin \omega t_0 + B \cos \omega t_0 \qquad (1.138)$$

$$\dot{x}_0 = A\omega \cos \omega t_0 - B\omega \sin \omega t_0 \qquad (1.139)$$

which yield:

$$\begin{bmatrix} A \\ B \end{bmatrix} = \begin{bmatrix} \sin \omega t_0 & \cos \omega t_0 \\ \omega \cos \omega t_0 & -\omega \sin \omega t_0 \end{bmatrix}^{-1} \begin{bmatrix} x_0 \\ \dot{x}_0 \end{bmatrix}$$

$$= \frac{1}{\omega} \begin{bmatrix} \dot{x}_0 \cos \omega t_0 + \omega x_0 \sin \omega t_0 \\ -\dot{x}_0 \sin \omega t_0 + \omega x_0 \cos \omega t_0 \end{bmatrix} \tag{1.140}$$

Substituting the initial conditions in the harmonic solution, we have

$$x_0 = X \sin (\omega t_0 + \varphi) \tag{1.141}$$

$$\dot{x}_0 = X \omega \cos (\omega t_0 + \varphi) \tag{1.142}$$

that provide us with:

$$X = \sqrt{x_0^2 + \frac{\dot{x}_0^2}{\omega^2}} \qquad \tan (\omega t_0 + \varphi) = \frac{x_0}{\dot{x}_0/\omega} \tag{1.143}$$

Example 14 Libration of harmonic oscillators. The simplest equation in vibrations is the harmonic oscillator. It can be solved by integration method.

Consider a particle of mass m and an attraction force towards a fixed origin O on the line of x-axis. The magnitude of attraction is proportional to the distance from O.

$$F = -kx \tag{1.144}$$

The equation of motion of the particle is a linear differential equation with constant coefficients.

$$\ddot{x} + \omega^2 x = 0 \qquad \omega^2 = \frac{k}{m} \tag{1.145}$$

The dynamic system is called the harmonic oscillator. We may rewrite the equation of motion in integral form,

$$\int \dot{x} d\dot{x} = - \int \omega^2 x dx \tag{1.146}$$

and find:

$$\dot{x}^2 + \omega^2 x^2 = \omega^2 C^2 \tag{1.147}$$

$$C^2 = \frac{\dot{x}_0^2}{\omega^2} + x_0^2 \tag{1.148}$$

This motion may be described by a moving point on an ellipse in the $(x, \dot{x})$-plane with semi-axis ωC on $\dot{x}$ and C on x axes. Such a motion is a libration motion between $x = \pm C$, and the $(x, \dot{x})$-plane is called the phase plane. Integration of Eq. (1.147) provides the solution of the equation of motion:

$$\int \frac{dx}{\omega\sqrt{C^2 - x^2}} = \int dt \tag{1.149}$$

$$x = C \sin(\omega t - \varphi) \qquad \varphi = -\arcsin\frac{x_0}{C} \tag{1.150}$$

The word "harmonic" describes a quality of sound, and harmonic functions get their name from their connection with vibrating strings as a source of sound. The movement of a point on a vibrating string is a harmonic motion. Such motion can be expressed using sine and cosine functions. In classical Fourier analysis, functions on the unit circle are expanded in terms of sines and cosines.

Example 15 Energy of a harmonic oscillator. The energy of a harmonic oscillator is constant and can be changed only by changing the initial conditions.

The force of spring, $F = -kx$, is conservative. The work done by the spring, W, is path-independent and depends only on the initial position x_i and the final position x_f.

$$\int_{x_i}^{x_f} (-kx)\, dx = \frac{1}{2}k\left(x_i^2 - x_f^2\right) \tag{1.151}$$

Let us calculate the mechanical energy of the system.

$$
\begin{aligned}
E = K + P &= \frac{1}{2}m\dot{x}^2 + \frac{1}{2}kx^2 \\
&= \frac{1}{2}m\left(X\omega_n \cos(\omega_n t + \varphi)\right)^2 + \frac{1}{2}k\left(X\sin(\omega_n t + \varphi)\right)^2 \\
&= \frac{1}{2}mX^2\omega_n^2 \cos^2(\omega_n t + \varphi) + \frac{1}{2}kX^2 \sin^2(\omega_n t + \varphi) \\
&= \frac{1}{2}kX^2\left(\cos^2(\omega_n t + \varphi) + \sin^2(\omega_n t + \varphi)\right) = \frac{1}{2}kX^2
\end{aligned}
\tag{1.152}
$$

This is the potential energy of the spring when it is stretched to the maximum and the mass is at rest.

Example 16 Superposition of linear vibrations. Linear systems possess the important property of superposition. This property enables us to break a complicated linear system into several simpler linear systems and determine the solution of the original system by superposing the solution of all the simpler systems.

The fundamental equation of vibration (1.85) is linear. Superposition is the necessary and sufficient condition for linearity. A system is linear if it is superposable, and any superposable system is linear.

We can mathematically define the superposition principle as follows. Consider a linear function f.

$$y = f(x) \tag{1.153}$$

If we have

$$y_1 = f(x_1) \qquad y_2 = f(x_2) \tag{1.154}$$

then it is linear, provided we have:

$$f(ax_1 + bx_2) = ay_1 + by_2 \tag{1.155}$$

Therefore, the sum of two inputs gives rise to the sum of two outputs. Furthermore, magnifying the input by a constant factor multiplies the output by the same factor. Superposing two waves is the same as adding their wave functions.

Consider two wave functions with the same amplitude, phase, and frequency.

$$x = C \cos(\omega t + \varphi) \qquad y = C \sin(\omega t + \varphi) \tag{1.156}$$

If we assume x and y to be components of a moving point in the phase plane, then the identity $\sin^2 \theta + \cos^2 \theta = 1$ provides the path of motion.

$$x^2 + y^2 = c^2 \tag{1.157}$$

It is a circle with radius C and a center at the origin. As t varies, the point (x, y) moves counterclockwise around this circle with frequency $f = \omega/(2\pi)$ times a second. If two of these types of combined wave are moving together,

$$(x_1, y_1) = (C_1 \cos \Omega_1, C_1 \sin \Omega_1) \tag{1.158}$$

$$(x_2, y_2) = (C_2 \cos \Omega_2, C_2 \sin \Omega_2) \tag{1.159}$$

$$\Omega_1 = \omega_1 t + \varphi_1 \qquad \Omega_2 = \omega_2 t + \varphi_2 \tag{1.160}$$

the resultant wave would be the addition of these vectors.

$$(x, y) = (C_1 \cos \Omega_1 + C_2 \cos \Omega_2, C_1 \sin \Omega_1 + C_2 \sin \Omega_2)$$
$$= (C \cos \omega, C \sin \omega) \tag{1.161}$$

Example 17 ★ Two-dimensional harmonic motion. The solutions of two dimensional harmonic equation with equal frequencies are conic sections depending on their initial conditions. Here are classifications of the paths of solutions.

A force $\mathbf{F}$ that is a function of position $\mathbf{r}$ is applied on a particle with mass m that is moving on the (x, y) -plane.

$$\mathbf{F} = -k\mathbf{r} \tag{1.162}$$

The equation of motion of the particle is:

$$m\ddot{\mathbf{r}} = -k\mathbf{r} \tag{1.163}$$

The equation can be decomposed into x and y directions to make two coupled harmonic differential equations.

$$\ddot{x} + \omega^2 x = 0 \qquad \omega = \sqrt{\frac{k}{m}} \tag{1.164}$$

$$\ddot{y} + \omega^2 y = 0 \qquad \omega = \sqrt{\frac{k}{m}} \tag{1.165}$$

The solution of Eqs. (1.164) and (1.165) are:

$$x(t) = A_1 \cos(\omega t) + A_2 \sin(\omega t) = A \sin(\omega t - \alpha) \tag{1.166}$$
$$y(t) = B_1 \cos(\omega t) + B_2 \sin(\omega t) = B \sin(\omega t - \beta) \tag{1.167}$$

where α, β, A, and B are related to the initial conditions $x_0 = x(0)$, $y_0 = y(0)$, $\dot{x}_0 = \dot{x}(0)$, $\dot{y}_0 = \dot{y}(0)$.

$$x_0 = -A \sin\alpha \qquad y_0 = -B \sin\beta \tag{1.168}$$
$$\dot{x}_0 = A\omega \cos\alpha \qquad \dot{y}_0 = B\omega \cos\beta \tag{1.169}$$

To find the path of motion in the (x, y)-plane, we should eliminate time t between x and y. Let us define γ,

$$\gamma = \alpha - \beta \tag{1.170}$$

to rewrite y,

$$\begin{aligned} y(t) &= B \sin((\omega t - \alpha) + (\alpha - \beta)) \\ &= B \sin(\omega t - \alpha) \cos(\alpha - \beta) + B \cos(\omega t - \alpha) \sin(\alpha - \beta) \\ &= B \sin(\omega t - \alpha) \cos\gamma + B \cos(\omega t - \alpha) \sin\gamma \end{aligned} \tag{1.171}$$

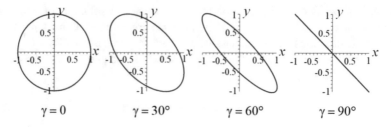

$$\gamma = 0 \qquad\qquad \gamma = 30° \qquad\qquad \gamma = 60° \qquad\qquad \gamma = 90°$$

Fig. 1.20 The path of harmonic motion of a particle in the (x, y)-plane

and substitute x to find the solution in the (x, y)-plane. Solutions in phase plane are called paths.

$$y = \frac{B}{A}x\cos\gamma + B\sqrt{1 - \left(\frac{x}{A}\right)^2}\sin\gamma \tag{1.172}$$

The path (1.172) can be rearranged in the form of conic sections second in analytic geometry.

$$A^2y^2 - 2ABxy\cos\gamma + B^2x^2 = A^2B^2\sin^2\gamma \tag{1.173}$$

A special case of the path happens when $A = B$ and $\gamma = \pm\pi/2$, which indicates a circular motion.

$$x^2 + y^2 = A^2 \tag{1.174}$$

If $A \neq B$, the path is an ellipse:

$$\frac{x^2}{A^2} + \frac{y^2}{B^2} = 1 \tag{1.175}$$

If $\gamma = 0$ or $\gamma = \pm\pi$ the path reduces to a straight line:

$$y = \frac{B}{A}x \tag{1.176}$$

or

$$y = -\frac{B}{A}x \tag{1.177}$$

Figure 1.20 illustrates the path of motion for $A = B = 1$, $\omega = 1$, and different γ.

Example 18 ★ Lissajous curves. The solutions of two-dimensional harmonic equation with unequal frequencies are Lissajous curves. The path of solution depends on initial conditions and frequency ratio. Here are classifications of the paths of solutions.

Consider a particle with mass m that is moving on the (x, y) -plane under the action of the force $\mathbf{F}$.

$$\mathbf{F} = -k_1 x\, \hat{\imath} - k_2 y\, \hat{\jmath} \tag{1.178}$$

The equation of motion of the particle is

$$m\ddot{\mathbf{r}} = \mathbf{F} \tag{1.179}$$

$$m\left(\ddot{x}\, \hat{\imath} + \ddot{y}\, \hat{\jmath}\right) = -k_1 x\, \hat{\imath} - k_2 y\, \hat{\jmath} \tag{1.180}$$

or

$$\ddot{x} + \omega_1 x = 0 \qquad \omega_1 = \sqrt{\frac{k_1}{m}} \tag{1.181}$$

$$\ddot{y} + \omega_2 y = 0 \qquad \omega_2 = \sqrt{\frac{k_2}{m}} \tag{1.182}$$

The solution of Eqs. (1.181) and (1.182) are:

$$x(t) = A_1 \cos(\omega_1 t) + A_2 \sin(\omega_1 t) = A \cos(\omega_1 t - \alpha) \tag{1.183}$$

$$y(t) = B_1 \cos(\omega_2 t) + B_2 \sin(\omega_2 t) = B \cos(\omega_2 t - \beta) \tag{1.184}$$

These equations indicate the parametric path of motion of the particle in the phase plane. If the frequencies ω_1 and ω_2 are commensurable, the path of the motion will be closed. The frequencies ω_1 and ω_2 are commensurable when their ratio is a rational fraction, $\omega_1/\omega_2 = m/n$, $\{m, n \in N\}$. The path of motion in this case is called the Lissajous curve. Figure 1.21 depicts some Lissajous curves. If the frequencies ω_1 and ω_2 are not commensurable, the path of motion will be open, which means the moving particle will never pass twice through the same point with the same velocity. Figure 1.22 shows a few curves for incommensurable cases.

The Lissajous curves can be illustrated better if we assume $A = B$ and write the parametric equations in the form of

$$x(t) = \cos(r\tau + \delta) \qquad y(t) = \cos(\tau) \tag{1.185}$$

where

$$r = \frac{\omega_1}{\omega_2} \qquad \tau = \omega_2 t - \beta \qquad \delta = \beta r - \alpha \tag{1.186}$$

Therefore, the component x is a 2π-periodic, and y is a $(2\pi/r)$ -periodic function of τ.

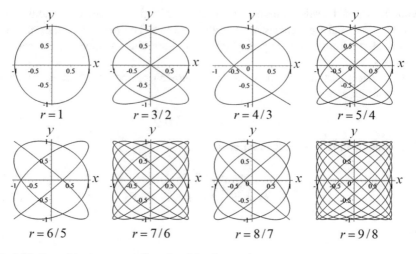

Fig. 1.21 Some Lissajous curves for rational fraction, ω_1/ω_2

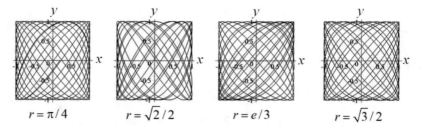

Fig. 1.22 Some 2D harmonic curves for incommensurable ω_1 and ω_2

Example 19 ★ Lissajous motion and period. Period of Lissajous curves depends on frequency ratio of the two harmonic motions. Here is a mathematical description of the period.

A periodic motion as a result of composition of two harmonic motions of different frequencies is a Lissajous motion. It is defined by an equation of the following form.

$$ x = A_1 \cos(\omega_1 t) + A_2 \sin(\omega_2 t + \alpha) \qquad \frac{\omega_1}{\omega_2} \neq 1 \qquad (1.187) $$

Without loss of generality, we may assume $\omega_1/\omega_2 > 0$, $A_1 > 0$, and $A_2 > 0$.

$$ \frac{\omega_1}{\omega_2} = C \qquad (1.188) $$

If the period of x is T, it is necessary and sufficient that both terms of (1.187) have the same period. Therefore, T must be a common multiple of $2\pi/\omega_1$ and $2\pi/\omega_2$.

$$T = C_1 \frac{2\pi}{\omega_1} = C_2 \frac{2\pi}{\omega_2} \qquad C_1, C_2 \in N \tag{1.189}$$

T is the least common multiple of the periods of the components.

Example 20 Combination of harmonic waves. A full harmonic function of a combined cos and sin functions can be expressed by a single cos or single sin function.

The functions $\cos\theta$ and $\sin\theta$ are orthogonal. Hence, a full harmonic function will have both functions with their weight factors.

$$x = A_1 \cos \omega t + A_2 \sin \omega t \tag{1.190}$$

A full harmonic function may always be converted to a cos function,

$$x = X \cos (\omega t + \alpha) \tag{1.191}$$

where their coefficients are calculated by expanding Eq. (1.191) and comparing with (1.190).

$$A_1 = X \cos \alpha \qquad\qquad A_2 = -X \sin \alpha \tag{1.192}$$

$$\alpha = \arctan \frac{-A_2}{A_1} \qquad X = \sqrt{A_1^2 + A_2^2} \tag{1.193}$$

The combination (1.191) may also be converted to a sin function.

$$x = X \sin (\omega t + \beta) \tag{1.194}$$

The coefficients are similarly calculated by expanding Eq. (1.194) and comparing with (1.190).

$$A_1 = X \sin \beta \qquad\qquad A_2 = X \cos \beta \tag{1.195}$$

$$\beta = \arctan \frac{A_1}{A_2} \qquad X = \sqrt{A_1^2 + A_2^2} \tag{1.196}$$

Two harmonic waves at the same frequency with phase difference can be converted to the above cases and combined.

$$x_1 = B_1 \sin \omega t \qquad x_2 = B_2 \sin (\omega t + \varphi) \tag{1.197}$$

To combine x_1 and x_2, we expand the $x_1 + x_2$.

$$x = x_1 + x_2 = B_1 \sin \omega t + B_2 \sin (\omega t + \varphi)$$
$$= B_2 \sin \varphi \cos \omega t + (B_1 + B_2 \cos \varphi) \sin \omega t \tag{1.198}$$

Assuming,

$$A_1 = B_2 \sin \varphi \qquad A_2 = B_1 + B_2 \cos \varphi \qquad (1.199)$$

combination of $x_1 + x_2$ reduces to (1.190). Therefore, combined function for waves (1.197) will be equivalent to (1.194).

$$x = x_1 + x_2 = X \sin(\omega t + \gamma) \qquad (1.200)$$

$$X = \sqrt{(B_2 \sin \varphi)^2 + (B_1 + B_2 \cos \varphi)^2}$$

$$= B_1^2 + B_2^2 + 2B_1 B_2 \cos \varphi \qquad (1.201)$$

$$\gamma = \arctan \frac{B_2 \sin \varphi}{B_1 + B_2 \cos \varphi} \qquad (1.202)$$

Example 21 Beating. When two harmonic waves, x_1, x_2, with close frequencies, ω_1, ω_2, are superposed, their resultant wave x will have two frequencies: a high frequency $\frac{\omega_1 + \omega_2}{2}$, associated to actual fluctuation of x, and a low frequency $\frac{\omega_1 - \omega_2}{2}$, associated to the envelop to the wave function. The envelop frequency sounds a beat for two close notes. Combination of two harmonic waves whose frequencies are very close generates beating phenomenon. Here is the mathematical explanation.

When two waves of same amplitude and slightly different frequency superpose each other, they produce a resultant wave whose amplitude rises and falls periodically. These periodic changes in amplitudes produce rises and falls in the intensity of wave which we call beats.

Consider two vibrations of equal amplitudes but different frequencies to be superposed.

$$x_1(t) = X \cos \omega_1 t \qquad x_2(t) = X \cos \omega_2 t \qquad \omega_1 > \omega_2 \qquad (1.203)$$

$$x(t) = x_1(t) + x_2(t) = X \cos \omega_1 t + X \cos \omega_2 t$$

$$= 2X \cos \frac{\omega_1 - \omega_2}{2} t \sin \frac{\omega_1 + \omega_2}{2} t$$

$$= \left(2X \cos \frac{\omega_1 - \omega_2}{2} t \right) \sin \frac{\omega_1 + \omega_2}{2} t \qquad (1.204)$$

It represents a sinusoidal oscillation at the average frequency $(\omega_1 + \omega_2)/2$ having a displacement amplitude which varies between $2X$ and zero under the influence of the cosine term of a slower frequency $(\omega_1 - \omega_2)/2$. This growth and decay of the amplitude is acoustically known as "beats" that will be heard when two sounds of almost equal frequency are superposed.

Now consider two wave equations $x_1(t)$ and $x_2(t)$ that are played together.

$$x_1(t) = X_1 \cos(\omega_1 t + \varphi_1) \tag{1.205}$$

$$x_2(t) = X_2 \cos(\omega_2 t + \varphi_2) \tag{1.206}$$

The resultant wave would be $x(t) = x_1(t) + x_2(t)$.

$$
\begin{aligned}
x(t) &= x_1(t) + x_2(t) \\
&= X_1 \cos(\omega_1 t + \varphi_1) + X_2 \cos(\omega_2 t + \varphi_2)
\end{aligned} \tag{1.207}
$$

It is convenient to express $x(t)$ in an alternative way

$$
\begin{aligned}
x(t) = {} & \frac{1}{2}(X_1 + X_2)\left(\cos(\omega_1 t + \varphi_1) + \cos(\omega_2 t + \varphi_2)\right) \\
& + \frac{1}{2}(X_1 - X_2)\left(\cos(\omega_1 t + \varphi_1) - \cos(\omega_2 t + \varphi_2)\right)
\end{aligned} \tag{1.208}
$$

and convert the sums to a product:

$$
\begin{aligned}
x(t) = {} & (X_1 + X_2)\cos\left(\frac{\omega_1 + \omega_2}{2}t - \frac{\varphi_1 + \varphi_2}{2}\right) \\
& \times \cos\left(\frac{\omega_1 - \omega_2}{2}t - \frac{\varphi_1 - \varphi_2}{2}\right) \\
& - (X_1 - X_2)\sin\left(\frac{\omega_1 + \omega_2}{2}t - \frac{\varphi_1 + \varphi_2}{2}\right) \\
& \times \sin\left(\frac{\omega_1 - \omega_2}{2}t - \frac{\varphi_1 - \varphi_2}{2}\right)
\end{aligned} \tag{1.209}
$$

This equation may be expressed better as

$$
\begin{aligned}
x(t) = {} & (X_1 + X_2)\cos(\Omega_1 t - \Phi_1)\cos(\Omega_2 t - \Phi_2) \\
& - (X_1 - X_2)\sin(\Omega_1 t - \Phi_1)\sin(\Omega_2 t - \Phi_2)
\end{aligned} \tag{1.210}
$$

by using the following notations.

$$\Omega_1 = \frac{\omega_1 + \omega_2}{2} \qquad \Omega_2 = \frac{\omega_1 - \omega_2}{2} \tag{1.211}$$

$$\Phi_1 = \frac{\varphi_1 + \varphi_2}{2} \qquad \Phi_2 = \frac{\varphi_1 - \varphi_2}{2} \tag{1.212}$$

The resultant wave $x(t)$ shows a periodic function of a fast and a slow frequencies Ω_1, Ω_2. The best way to comprehend the combined wave function is to plot the

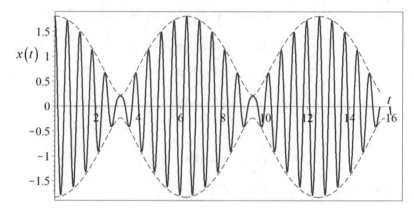

Fig. 1.23 Beating phenomena

function. Figure 1.23 illustrates a sample plot of $x(t)$ for the following data.

$$X_1 = 1 \qquad X_2 = 0.8 \qquad \omega_1 = 10$$
$$\omega_2 = 11 \qquad \phi_1 = 0 \qquad \phi_2 = 0 \tag{1.213}$$

The wave $x(t)$ indicates an oscillation between $X_1 + X_2$ and $X_1 - X_2$, with the higher frequency Ω_1 inside an envelop that oscillates at the lower frequency Ω_2. This behavior is called beating. When $X_1 = X_2 = X$, then

$$x(t) = 2X \cos(\Omega_1 t - \Phi_1) \cos(\Omega_2 t - \Phi_2) \tag{1.214}$$

which becomes zero at every half period $T = 2\pi / \Omega_2$.

Assume we tune a stretched string of A_4 of a piano to $f = 440$ Hz. The notes before and after A_4 are G_4 and B_4 at $f = 392$ Hz and $f = 493.88$ Hz, respectively. When we hit the key of A_4, a wave at 440 Hz will sound, and because the other strings are tuned at other frequencies, they will not sound. If we hit both keys A_4 and B_4 at the same time, the following sound will be heard.

$$x = \sin(880\pi t) + \sin(987.76\pi t) \tag{1.215}$$

We may write this as:

$$x = 2\cos(53.88\pi t) \sin(933.9\pi t) \tag{1.216}$$

It indicates the combined wave as a sine wave of frequency 466.95 Hz , the average of the frequencies of the two keys, and an amplitude modulation by a slow cosine wave with frequency 26.94 Hz, half of the difference of the frequencies of the two keys.

If we add two waves $x_1 = X_1 \cos \omega t$ and $x_2 = X_2 \cos \omega t$ whose frequencies, ω, are exactly equal and there is no phase difference, the result would be another similar wave.

$$x_3 = x_1 + x_2 = (X_1 + X_2) \cos \omega t \tag{1.217}$$

Now suppose we add two waves x_1 and x_2,

$$x_1 = X_1 \cos \omega t \tag{1.218}$$

$$x_2 = X_2 \cos \left(\omega' t + \alpha\right) = X_2 \cos \left(\omega t + \beta\right) \tag{1.219}$$

$$\beta = \varepsilon \omega t + \alpha \tag{1.220}$$

whose frequencies are ω and ω' differ only by a small amount ε.

$$\frac{\omega'}{\omega} = 1 + \varepsilon \qquad 0 < \varepsilon << 1 \tag{1.221}$$

If ω'/ω is not a rational number, then ω' and ω are not commensurable, and the resultant motion of $x_1 + x_2$ is not periodic.

Let us observe the resultant motion during a period $T = 2\pi/\omega$ of x_1, from a time t_1 to $t_1 + T$. During this interval, β increases by a small value of $\varepsilon \omega t = 2\pi \varepsilon$ and increases continuously when time goes by. At the times $t_k = (2k\pi - \alpha)/(\varepsilon \omega)$ when $\beta = \varepsilon \omega t_k + \alpha = 2k\pi$, the resultant motion $x_3 = x_1 + x_2$ is:

$$x_3 = x_1 + x_2 = (X_1 + X_2) \cos \omega t_k \tag{1.222}$$

$$x_1 = X_1 \cos \omega t_k \qquad x_2 = X_2 \cos \omega t_k \tag{1.223}$$

At the times $t_k = ((2k + 1)\pi - \alpha)/(\varepsilon \omega)$, when $\beta = \varepsilon \omega t_k + \alpha = (2k + 1)\pi$, the resultant motion $x_3 = x_1 + x_2$ is:

$$x_3 = x_1 + x_2 = (X_1 - X_2) \cos \omega t_k \tag{1.224}$$

$$x_1 = X_1 \cos \omega t_k \qquad x_2 = -X_2 \cos \omega t_k \tag{1.225}$$

So, the resultant motion of $x_3 = x_1 + x_2$ can be assumed as a harmonic motion with a periodic variable amplitude between $X_1 + X_2$ and $X_1 - X_2$. This phenomenon is a beat. The period T_B of beating is:

$$T_B = t_{k+1} - t_k = 2\frac{\pi}{\varepsilon \omega} = 2\frac{\pi}{\omega' - \omega} \tag{1.226}$$

and therefore, the frequency f_B of beating is:

$$f_B = \frac{\varepsilon \omega}{2\pi} = \frac{\omega' - \omega}{2\pi} = f_2 - f_1 \tag{1.227}$$

which is the difference of the frequencies of x_1 and x_2.

Superposition of the notes A_4 at $f = 440$ Hz and B_5 at $f = 493.88$ Hz provides us with a beat with the following characteristics.

$$T_B = 2\frac{\pi}{\varepsilon\omega} = 2\frac{\pi}{\omega' - \omega} = 1.856 \times 10^{-2} \text{ s} \tag{1.228}$$

$$f_B = \frac{\varepsilon\omega}{2\pi} = \frac{\omega' - \omega}{2\pi} = 53.88 \text{ Hz} \tag{1.229}$$

$$\varepsilon = \frac{\omega'}{\omega} - 1 = \frac{987.76\pi}{880\pi} - 1 = 0.12245 \tag{1.230}$$

From another viewpoint, the sum of two sine waves of closed but distinct frequencies is perceived as a single sound whose intensity slowly oscillates from large to small values. These vibrations are called beats. When the amplitude is large, the interference is said to be constructive, and when the amplitude shrinks, the interference is destructive. If the frequencies differ by an amount Δf, the resulting sound can be viewed as a sine wave of frequency $f + \Delta f/2$ with a slowly varying envelop $A = 2\cos(\pi \Delta f t)$ of frequency Δf.

$$\sin(2\pi f t) + \sin(2\pi (f + \Delta f) t) = 2\cos(\pi \Delta f t) \sin(2\pi (f + \Delta f/2) t) \tag{1.231}$$

The number of beats per second is the difference of the two frequencies. For example, two sounds of 440 Hz and 444 Hz create 4 beats per second. Beats are also considered between partials. The third partial of A_4 at 440 Hz and the second partial of E at 659 Hz create *two beats*/ sec.

$$3f_1 - 2f_2 = 3 \times 440 - 2 \times 659 = 2 \text{ beats}/\sec \tag{1.232}$$

Example 22 Wavy road and excitation frequency. Motion of a wheel on a wavy road is a practical example of harmonic motion. Here is the to show the parameters of a wavy road to be compared with a harmonic function.

Figure 1.24 illustrates a 1/8 car model moving with speed v on a wavy road with length d_1 and peak-to-peak height d_2. Assuming a stiff tire with a small radius compared to the road waves, we may consider y as the fluctuation of the road. The required time T to pass one wave length d_1 is the period of the excitation.

$$T = \frac{d_1}{v} \tag{1.233}$$

Therefore the frequency of excitation is:

$$\omega = \frac{2\pi}{T} = \frac{2\pi v}{d_1} \tag{1.234}$$

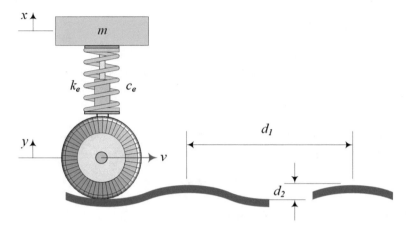

Fig. 1.24 A 1/8 car model moving with speed v on a wavy road

and the excitation $y = Y \sin \omega t$ is:

$$y = \frac{d_2}{2} \sin \frac{2\pi v}{d_1} t \tag{1.235}$$

Example 23 Work done by a harmonic force. In mechanical vibrations, applying a harmonic force on a linear system will produce a harmonic response. Here is the required energy to make such motion by calculating the work done by the force.

The work done by a harmonic force

$$f(t) = F \sin(\omega t + \varphi) \tag{1.236}$$

acting on a body with a harmonic displacement

$$x(t) = X \sin(\omega t + \varphi) \tag{1.237}$$

during one period T

$$T = \frac{2\pi}{\omega} \tag{1.238}$$

is equal to:

$$W = \int_0^{2\pi/\omega} f(t)dx = \int_0^{2\pi/\omega} f(t)\frac{dx}{dt}dt$$

$$= FX\omega \int_0^{2\pi/\omega} \sin(\omega t + \varphi)\cos(\omega t)\, dt$$

$$= FX \int_0^{2\pi} \sin(\omega t + \varphi) \cos(\omega t) \, d(\omega t)$$

$$= FX \int_0^{2\pi} \left(\sin\varphi \cos^2 \omega t + \cos\varphi \sin\omega t \cos\omega t \right) d(\omega t)$$

$$= \pi F X \sin\varphi \tag{1.239}$$

The work W is a function of the phase angle φ between f and x . If $\varphi = \frac{\pi}{2}$, then the work is maximum

$$W_{Max} = \pi F X \tag{1.240}$$

and when $\varphi = 0$, the work is minimum.

$$W_{min} = 0 \tag{1.241}$$

Example 24 ★ Orthogonality of wave functions $\sin\omega t$ and $\cos\omega t$. Orthogonal functions are linearly independent, and hence, they can be two solutions of a second-order differential equations.

Two functions $f(t)$ and $g(t)$ are orthogonal in domain $t \in [a, b]$ if the finite integral of their product in the domain is zero.

$$\int_a^b f(t) \, g(t) \, dt = 0 \tag{1.242}$$

The functions $\sin\omega t$ and $\cos\omega t$ are orthogonal in a period $T = [0, 2\pi/\omega]$.

$$\int_0^{2\pi/\omega} \sin\omega t \cos\omega t \, dt = 0 \tag{1.243}$$

The orthogonality allows us to treat the coefficients of the functions sin and cos in an equation independently.

1.3   ★ Fourier Series

The purpose of Fourier series is to decompose a periodic function $f(t)$ of period $2\pi\lambda$

$$f(t) = f(t + 2\pi\lambda) \tag{1.244}$$

into its harmonic components of $\sin(jt)$ and $\cos(jt)$,

$$f(t) = \frac{1}{2}a_0 + \sum_{j=1}^{\infty} \left(a_j \cos \frac{jt}{\lambda} + b_j \sin \frac{jt}{\lambda} \right) \tag{1.245}$$

where a_j and b_j are weight factors of the $\cos(jt/\lambda)$ and $\sin(jt/\lambda)$ harmonic functions.

$$a_0 = \frac{1}{\pi} \int_{-\pi\lambda}^{\pi\lambda} f(s) \, ds \tag{1.246}$$

$$a_j = \frac{1}{\pi\lambda} \int_{-\pi\lambda}^{\pi\lambda} f(s) \cos \frac{js}{\lambda} \, ds \tag{1.247}$$

$$b_j = \frac{1}{\pi\lambda} \int_{-\pi\lambda}^{\pi\lambda} f(s) \sin \frac{js}{\lambda} \, ds \tag{1.248}$$

The number $a_0/2$ is the mean value of the periodic function. The coefficients are obtained by multiplying $f(t)$ by $\cos(jt/\lambda)$ and $\sin(jt/\lambda)$ and integrating term by term over $(-\pi\lambda, -\pi\lambda)$.

In case the function $f(t)$ is a periodic function of T,

$$f(t) = f(t+T) \tag{1.249}$$

we can define a new 2π-periodic function $g(\theta)$,

$$g(\theta) = f\left(\frac{2\pi}{T}t\right) \tag{1.250}$$

and then expand $g(\theta)$ into a 2π-periodic series.

$$g(\theta) = \frac{1}{2}a_0 + \sum_{j=1}^{\infty} \left(a_j \cos \left(\frac{2\pi}{T} j\theta \right) + b_j \sin \left(\frac{2\pi}{T} j\theta \right) \right) \tag{1.251}$$

$$a_0 = \frac{1}{\pi} \int_{-\pi}^{\pi} g(\theta) \, d\theta \tag{1.252}$$

$$a_j = \frac{1}{\pi} \int_{-\pi}^{\pi} g(\theta) \cos \left(\frac{2\pi}{T} j\theta \right) d\theta \tag{1.253}$$

$$b_j = \frac{1}{\pi} \int_{-\pi}^{\pi} g(\theta) \sin \left(\frac{2\pi}{T} j\theta \right) d\theta \tag{1.254}$$

It is also possible to expand $f(t)$ directly over $T = 2\pi/\omega$.

$$f(t) = f(t + T) = f\left(t + \frac{2\pi}{\omega}\right) \tag{1.255}$$

$$f(t) = \frac{1}{2}a_0 + \sum_{j=1}^{\infty} \left(a_j \cos(j\omega t) + b_j \sin(j\omega t)\right) \tag{1.256}$$

$$a_0 = \frac{2}{T} \int_0^T f(t)\, dt \tag{1.257}$$

$$a_j = \frac{2}{T} \int_0^T f(t)\ \cos(j\omega t)\ dt \tag{1.258}$$

$$b_j = \frac{2}{T} \int_0^T f(t)\ \sin(j\omega t)\ dt \tag{1.259}$$

Proof A polynomial of the form

$$\frac{1}{2}a_0 + a_1 \cos t + a_2 \cos 2t + \cdots + a_n \cos nt$$

$$+ b_1 \sin t + b_2 \sin 2t + \cdots + b_n \sin nt \tag{1.260}$$

where $a_j^2 + b_j^2 \neq 0$ is a trigonometric polynomial of order n . The Fourier series to approximate a function $f(t)$ is a trigonometric series whose coefficients are calculated by the integrals (1.246)–(1.248) or (1.252)–(1.254). Among all trigonometric polynomials of order n, the polynomial (1.260) whose coefficients are given by Fourier rule is the polynomial which is at the minimum distance from $f(t)$.

Let us assume that Eq. (1.245) is given. Taking an integral of both sides over the interval $(-\pi\lambda, \pi\lambda)$ provides us with

$$\int_{-\pi\lambda}^{\pi\lambda} f(s)\, ds = \int_{-\pi\lambda}^{\pi\lambda} \left(\frac{1}{2}a_0 + \sum_{j=1}^{\infty} \left(a_j \cos\frac{js}{\lambda} + b_j \sin\frac{js}{\lambda}\right)\right) ds$$

$$= \frac{1}{2}a_0 \int_{-\pi\lambda}^{\pi\lambda} ds + \sum_{j=1}^{\infty} a_j \int_{-\pi\lambda}^{\pi\lambda} \cos\frac{js}{\lambda}\, ds + \sum_{j=1}^{\infty} b_j \int_{-\pi\lambda}^{\pi\lambda} \sin\frac{js}{\lambda}\, ds$$

$$= \pi\lambda a_0 \tag{1.261}$$

and hence, $a_0/2$ is the average value of $f(t)$ over the interval $(-\pi\lambda, -\pi\lambda)$.

$$a_0 = \frac{1}{\pi\lambda} \int_{-\pi\lambda}^{\pi\lambda} f(s)\, ds \tag{1.262}$$

Multiplying both sides of (1.245) by $\cos{(ks/\lambda)}$ and taking an integral yields:

$$\int_{-\pi\lambda}^{\pi\lambda} \cos{\frac{ks}{\lambda}} f(s)\,ds$$

$$= \int_{-\pi\lambda}^{\pi\lambda} \cos{\frac{ks}{\lambda}} \left(\frac{1}{2}a_0 + \sum_{j=1}^{\infty} \left(a_j \cos{\frac{js}{\lambda}} + b_j \sin{\frac{js}{\lambda}} \right) \right) ds$$

$$= \frac{1}{2}a_0 \int_{-\pi\lambda}^{\pi\lambda} \cos{\frac{ks}{\lambda}}\,ds + \sum_{j=1}^{\infty} a_j \int_{-\pi\lambda}^{\pi\lambda} \cos{\frac{ks}{\lambda}} \cos{\frac{js}{\lambda}}\,ds$$

$$+ \sum_{j=1}^{\infty} b_j \int_{-\pi\lambda}^{\pi\lambda} \cos{\frac{ks}{\lambda}} \sin{\frac{js}{\lambda}}\,ds = \pi a_j \tag{1.263}$$

and therefore, we have:

$$a_j = \frac{1}{\pi} \int_{-\pi\lambda}^{\pi\lambda} \cos{\frac{ks}{\lambda}} f(t)\,dt \tag{1.264}$$

Multiplying both sides of (1.245) by $\sin{(ks/\lambda)}$ and taking an integral shows that:

$$\int_{-\pi\lambda}^{\pi\lambda} \sin{\frac{ks}{\lambda}} f(s)\,ds$$

$$= \int_{-\pi\lambda}^{\pi\lambda} \sin{\frac{ks}{\lambda}} \left(\frac{1}{2}a_0 + \sum_{j=1}^{\infty} \left(a_j \cos{\frac{js}{\lambda}} + b_j \sin{\frac{js}{\lambda}} \right) \right) ds$$

$$= \frac{1}{2}a_0 \int_{-\pi\lambda}^{\pi\lambda} \sin{\frac{ks}{\lambda}}\,ds + \sum_{j=1}^{\infty} a_j \int_{-\pi\lambda}^{\pi\lambda} \sin{\frac{ks}{\lambda}} \cos{\frac{js}{\lambda}}\,ds$$

$$+ \sum_{j=1}^{\infty} b_j \int_{-\pi\lambda}^{\pi\lambda} \sin{\frac{ks}{\lambda}} \sin{\frac{js}{\lambda}}\,ds = \pi b_j \tag{1.265}$$

and therefore, we have:

$$b_j = \frac{1}{\pi} \int_{-\pi\lambda}^{\pi\lambda} \sin{\frac{js}{\lambda}} f(s)\,ds \tag{1.266}$$

Therefore, a periodic function $f(t)$ can be decomposed and substituted by a series of its harmonic components. Working with a truncated series of harmonic functions is much easier than working with an arbitrary periodic function. Because $\sin{(jt)}$ are odd functions and $\cos{(jt)}$ are even functions, the Fourier series breaks a periodic function into the sum of an even and an odd series, associated with an even and an odd function, respectively. If $f(t)$ is itself an even function, then $f(-t) = f(t)$ and $b_j = 0$, and if $f(t)$ is itself an odd function, then $f(-t) = -f(t)$ and $a_j = 0$.

An alternative representation of Eq. (1.245) is:

$$f(t) = \frac{1}{2}a_0 + \sum_{j=1}^{\infty} C_j \sin\left(jt - \varphi_j\right) \tag{1.267}$$

$$C_j = \sqrt{a_j^2 + b_j^2} \tag{1.268}$$

$$\varphi_j = \arctan \frac{a_j}{b_j} \tag{1.269}$$

Equating real and imaginary parts on the two sides of De Moivre's formula

$$(\cos t + i \sin t)^k = \cos kt + i \sin kt \qquad i^2 = -1 \tag{1.270}$$

we have

$$\cos kt = \binom{k}{0} \cos^k t - \binom{k}{2} \cos^{k-2} \sin^2 t + \cdots \tag{1.271}$$

$$\sin kt = \binom{k}{1} \cos^{k-1} t \sin t - \binom{k}{3} \cos^{k-3} \sin^3 t + \cdots \tag{1.272}$$

where

$$\binom{n}{k} = \frac{n!}{k!\,(n-k)!} \tag{1.273}$$

Therefore, every trigonometric polynomial of order n is also a polynomial of degree n in $\cos t$ $\sin t$. Also, by equating real and imaginary parts of the two sides of the exponential expressions of $\cos^k t$ and $\sin^k t$,

$$\cos^k t = \frac{e^{it} + e^{-it}}{2} \qquad \sin^k t = \frac{e^{it} - e^{-it}}{2} \tag{1.274}$$

we have:

$$2^{2k-1} \cos^{2k} t = \binom{2k}{0} \cos 2kt + \binom{2k}{1} \cos(2k-2)t + \cdots$$

$$+ \binom{2k}{k-1} \cos 2t + \frac{1}{2}\binom{2k}{k} \tag{1.275}$$

$$2^{2k} \cos^{2k+1} t = \binom{2k+1}{0} \cos (2k+1) t$$

$$+ \binom{2k+1}{1} \cos (2k-1) t + \cdots + \binom{2k+1}{k} \cos t \qquad (1.276)$$

$$(-1)^k 2^{2k-1} \sin^{2k} t = \binom{2k}{0} \sin 2kt - \binom{2k}{1} \sin (2k-2) t$$

$$+ \binom{2k}{2} \sin (2k-4) t + \cdots + (-1)^k \frac{1}{2} \binom{2k}{k} \qquad (1.277)$$

$$(-1)^k 2^{2k} \sin^{2k+1} t = \binom{2k+1}{0} \sin (2k+1) t - \binom{2k+1}{1} \sin (2k-1) t$$

$$+ \binom{2k+1}{2} \sin (2k-3) t + \cdots$$

$$+ (-1)^k \frac{1}{2} \binom{2k+1}{k} \sin t \qquad (1.278)$$

Assume $f_1 (t)$ is a given periodic function and $f_2 (t)$ is a trigonometric polynomial to approximate $f_1 (t)$.

$$f_1 (t) = f (t) \qquad f (t) - f (t + 2\pi) \qquad (1.279)$$

$$f_2 (t) = \frac{1}{2} a_0 + a_1 \cos t + b_1 \sin t + a_2 \cos 2t + b_2 \sin 2t + \cdots$$

$$= \frac{1}{2} a_0 + \sum_{j=1}^{\infty} \left(a_j \cos (j\omega t) + b_j \sin (j\omega t) \right) \qquad (1.280)$$

The distance between two functions $f_1 (t)$ and $f_2 (t)$ in $[-\pi, \pi]$ is:

$$d^2 = \int_{-\pi}^{\pi} (f_1 - f_2)^2 = \int_{-\pi}^{\pi} f_1^2 + \int_{-\pi}^{\pi} f_2^2 - 2 \int_{-\pi}^{\pi} f_1 f_2 \qquad (1.281)$$

To minimize the distance of $f_2 (t)$ from $f_1 (t)$, we must have

$$\frac{\partial d^2}{\partial a_k} = \frac{\partial d^2}{\partial b_k} = 0 \qquad (1.282)$$

and therefore, the coefficients of the trigonometric polynomial should be the Fourier coefficients.

$$a_0 = \frac{1}{\pi} \int_{-\pi}^{\pi} f(s) \, ds \tag{1.283}$$

$$a_j = \frac{1}{\pi} \int_{-\pi}^{\pi} f(s) \, \cos js \, ds \tag{1.284}$$

$$b_j = \frac{1}{\pi} \int_{-\pi}^{\pi} f(s) \, \sin js \, ds \tag{1.285}$$

Every generalized periodic function $f(s)$ is a sum of a trigonometric series. The trigonometric series converges, in the generalized sense, if and only if for a $k \geq 0$ we have:

$$\frac{a_j}{j^k} \to 0 \qquad \frac{b_j}{j^k} \to 0 \tag{1.286}$$

Joseph Fourier (1768–1830) introduced his method to answer the question of how a string can vibrate with a number of different frequencies at the same time. Fourier showed how we can decompose a periodic wave into a sum of sine and cosine waves with different frequencies. The frequencies are integer multiples of the fundamental frequency of the periodic wave.

■

Example 25 Square wave. Square wave is a good model for switching on-off function similar to a series of step functions with a gap in between. This periodic function is very common in vibration analysis with periodic inputs.

A periodic square wave $f(t)$ is defined by

$$f(t) = f(t + 2\pi) = \begin{cases} 1 & 0 \leq t < \pi \\ -1 & \pi \leq t \leq 2\pi \end{cases} \tag{1.287}$$

To express the square wave function with a Fourier series, we need to determine the Fourier coefficients.

$$a_0 = \frac{1}{\pi} \int_{-\pi}^{\pi} f(t) \, dt = 0 \tag{1.288}$$

$$a_j = \frac{1}{\pi} \int_0^{2\pi} \cos(jt) \, f(t) \, dt = \frac{1}{\pi} \left(\int_0^{\pi} \cos(jt) \, dt + \int_{\pi}^{2\pi} \cos(jt) \, dt \right)$$

$$= \frac{1}{\pi} \left(\left(\frac{1}{j} \sin(jt) \right)_0^{\pi} - \left(\frac{1}{j} \sin(jt) \right)_{\pi}^{2\pi} \right) = 0 \tag{1.289}$$

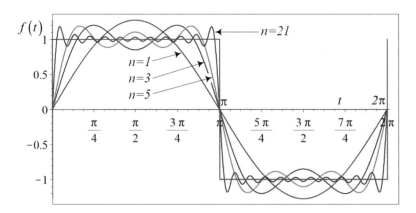

Fig. 1.25 Fourier series of the square wave $f(t)$ for $n = 1, 3, 5, 21$

$$b_j = \frac{1}{\pi} \int_0^{2\pi} \sin(jt) \, f(t) \, dt = \frac{1}{\pi} \left(\int_0^\pi \sin(jt) \, dt + \int_\pi^{2\pi} \sin(jt) \, dt \right)$$

$$= \frac{1}{\pi} \left(\left(\frac{-\cos(jt)}{j} \right)_0^\pi - \left(\frac{-\cos(jt)}{j} \right)_\pi^{2\pi} \right)$$

$$= \frac{2}{\pi} \left(\frac{1}{j} - \frac{(-1)^j}{j} \right) = \begin{cases} \dfrac{4}{j\pi} & j \text{ odd} \\ 0 & j \text{ even} \end{cases} \qquad (1.290)$$

Therefore, the Fourier series of the square wave function is:

$$f(t) = \frac{4}{\pi} \left(\sin t + \frac{1}{3} \sin 3t + \frac{1}{5} \sin 5t + \cdots \right) \qquad (1.291)$$

Figure 1.25 illustrates the Fourier series of $f(t)$ for $j = 1, 3, 5, 21$ and the square wave function.

Example 26 An even periodic function. Even and odd functions will have simpler Fourier series calculation because either a_j or b_j would be zero. Here is calculation of an even periodic function.

A function $f(t)$ is called **even** if $f(-t) = f(t)$ and is called **odd** if $f(-t) = -f(t)$. The harmonic function $\cos t$ is even, and $\sin t$ is odd. Any arbitrary function $f(t)$ can be split into an even plus and odd function.

$$f(t) = \frac{f(t) + f(-t)}{2} + \frac{f(t) - f(-t)}{2} \qquad (1.292)$$

Consider the periodic function in Fig. 1.26. Because the function is even, we have:

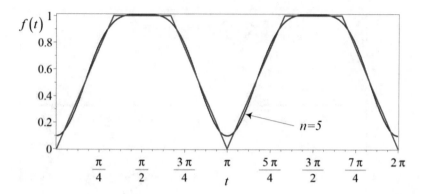

Fig. 1.26   An even periodic function

$$b_j = 0 \tag{1.293}$$

To determine a_0 and a_j, we use Eqs. (1.252) and (1.253).

$$
\begin{aligned}
a_0 &= \frac{2}{T} \int_0^T f(t)\, dt \\
&= \frac{2}{T} \left(\int_0^{T/3} \frac{3F}{T} t\, dt + \int_{T/3}^{2T/3} F\, dt + \int_{2T/3}^T 3F \left(1 - \frac{t}{T}\right) dt \right) \\
&= \frac{4}{3} F
\end{aligned}
\tag{1.294}
$$

$$
\begin{aligned}
a_j &= \frac{2}{T} \int_0^T \cos\left(\frac{2\pi}{T} jt\right) f(t)\, dt \\
&= \frac{2}{T} \int_0^{T/3} \frac{3F}{T} t \cos\left(\frac{2\pi}{T} jt\right) dt + \frac{2}{T} \int_{T/3}^{2T/3} F \cos\left(\frac{2\pi}{T} jt\right) dt \\
&\quad + \frac{2}{T} \int_{2T/3}^T 3F \left(1 - \frac{t}{T}\right) \cos\left(\frac{2\pi}{T} jt\right) dt \\
&= \frac{3F}{\pi^2 j^2} \left(\frac{1}{2} \cos\frac{2\pi j}{3} + \frac{1}{2} \cos\frac{4\pi j}{3} - 1 \right) \\
&= \begin{cases} -\dfrac{9F}{2\pi^2 j^2} & j = 1, 2, 4, 5, 7, 8, \cdots \\[2mm] 0 & j = 3, 6, 9, 12, \cdots \end{cases}
\end{aligned}
\tag{1.295}
$$

Therefore, the Fourier series of $f(t)$ is:

$$f(t) = \frac{4}{3}F - \frac{9F}{2\pi^2} \sum_{j=1,2,4,5,7,8,\cdots}^{\infty} \frac{1}{j^2} \cos\left(\frac{2\pi}{T}jt\right) \tag{1.296}$$

Example 27 ★ Bounded integrals of sin and cos. The bounded integrals of sin and cos happens frequently in Fourier series calculations. Here is a description and proof.

Bounded integrals of the multiplications of sin and cos functions are:

$$\int_{-\pi}^{\pi} \cos(jt)\cos(kt)\,dt = \begin{cases} \pi & j = k \neq 1 \\ 2\pi & j = k = 0 \\ 0 & j \neq k \end{cases} \tag{1.297}$$

$$\int_{-\pi}^{\pi} \sin(jt)\sin(kt)\,dt = \begin{cases} \pi & j = k \neq 1 \\ 0 & j \neq k \end{cases} \tag{1.298}$$

$$\int_{-\pi}^{\pi} \cos(jt)\sin(kt)\,dt = 0 \qquad \forall j, \forall k \tag{1.299}$$

Proof Let us add the identities

$$\cos(j+k)t = \cos jt \cos kt - \sin jt \sin kt \tag{1.300}$$

$$\cos(j-k)t = \cos jt \cos kt + \sin jt \sin kt \tag{1.301}$$

and integrate for $-\pi \leq t \leq \pi$:

$$\int_{-\pi}^{\pi} \cos(jt)\cos(kt)\,dt = \frac{1}{2}\int_{-\pi}^{\pi} (\cos(j+k)t + \cos(j-k)t)\,dt$$

$$= \frac{1}{2}\left(\frac{\sin(j+k)t}{j+k} + \frac{\sin(j-k)t}{j-k}\right)_{-\pi}^{\pi} \tag{1.302}$$

This integral is zero if $j \neq k$. However, if $j = k \geq 1$ then the integral simplifies to:

$$\int_{-\pi}^{\pi} \cos^2(jt)\,dt = \frac{1}{2}\int_{-\pi}^{\pi} (1 + \cos(2jt))\,dt = \pi \tag{1.303}$$

If $j = k = 0$, then the integral reduces to:

$$\int_{-\pi}^{\pi} dt = 2\pi \tag{1.304}$$

This completes the proof of Eq. (1.297). ■

Example 28 Periodic functions. Some of useful properties of period functions are summarized here.

A function $x = f(t)$ is called a periodic function if there exists a period $T > 0$ such that for all t we have:

$$f(t) = f(t + T) = f(t - T) \tag{1.305}$$

If $f(t)$ is periodic of period T, then $f(t)$ is also periodic for kT, $k \in \mathbb{N}$. Usually we take the minimum value of T as the period for which we have (1.305). The function $f(t) = \cos t$ has period $T = 2k\pi$.

If $f(t)$ is periodic of period T, then $f(\omega t)$ is of period $T_1 = T/\omega$. The function $f(t) = \cos 5t$ has period $T = 2\pi/5$, and the function $f(t) = \cos \omega t$ has period $T = 2\pi/\omega$.

If $f(t)$ is periodic of period T and $g(t)$ is periodic of period T, then $f(t) g(t)$ is also of period T, but T may not be the minimum period. The function $\cos t$ and $\sin t$ both have period 2π, and the function $\cos t \sin t$ has also period 2π, but its minimum period is π.

If $f(t)$ is periodic of period T, then for any number $0 < a < T$, we have $\int_a^{a+T} f(x)\, dx = \int_0^T f(x)\, dx$.

If a function $f(t)$ is the sum of two periodic functions with different periods, then $f(t)$ may not be periodic. However if two functions have equal periods, then their sum will also have the same period.

Example 29 Sawtooth wave function. Sawtooth wave function is another example to illustrate how Fourier coefficients are being calculated.

Let us approximate a full sawtooth function $f(x)$ by Fourier series.

$$f(x) = x \qquad (2k - 1)\pi < x - 2k\pi < (2k + 1)\pi \tag{1.306}$$

$$k = 0, 1, 2, 3, \cdots$$

The Fourier coefficients are:

$$a_j = \frac{1}{\pi} \int_{-\pi}^{\pi} f(s) \cos(js)\, ds = \frac{1}{\pi} \int_{-\pi}^{\pi} s \cos(js)\, ds = 0 \tag{1.307}$$

$$b_j = \frac{1}{\pi} \int_{-\pi}^{\pi} f(s) \sin(js)\, ds = \frac{1}{\pi} \int_{-\pi}^{\pi} s \sin(js)\, ds = \frac{2(-1)^{j+1}}{j} \tag{1.308}$$

Hence, the Fourier series of the sawtooth function (1.306) is:

$$f(x) = \frac{1}{2}a_0 + \sum_{j=1}^{\infty} (a_j \cos jx + b_j \sin jx) = \sum_{j=1}^{\infty} \frac{2(-1)^{j+1}}{j} \sin jx \tag{1.309}$$

Its expanded form for the first few harmonics is:

$$f(x) = 2\left(\sin x - \frac{1}{2}\sin 2x + \frac{1}{3}\sin 3x - \cdots\right) \tag{1.310}$$

Example 30 Complex form of the Fourier series. Fourier series can be expressed by a series of exponential functions. Here is the proof.

By using the Euler relations between trigonometric functions and exponential functions

$$\cos j\omega t = \frac{e^{ij\omega t} + e^{-ij\omega t}}{2} \qquad \sin j\omega t = \frac{e^{ij\omega t} - e^{-ij\omega t}}{2i} \tag{1.311}$$

$$i^2 = -1 \tag{1.312}$$

the Fourier series of the periodic function $f(t) = f(t + T)$ will be:

$$f(t) = \frac{1}{2}a_0 + \frac{1}{2}\sum_{j=1}^{\infty}\left(a_j\left(e^{ij\omega t} + e^{-ij\omega t}\right) - ib_j\left(e^{ij\omega t} - e^{-ij\omega t}\right)\right)$$

$$= \frac{1}{2}a_0 + \frac{1}{2}\sum_{j=1}^{\infty}\left((a_j - ib_j)e^{ij\omega t} + (a_j + ib_j)e^{-ij\omega t}\right) \tag{1.313}$$

We can define

$$C_0 = \frac{1}{2}a_0 \tag{1.314}$$

$$C_j = \frac{1}{2}(a_j - ib_j) \tag{1.315}$$

$$C_j^* = \frac{1}{2}(a_j + ib_j) \tag{1.316}$$

to rewrite the series as

$$f(t) = C_0 + \sum_{j=1}^{\infty}\left(C_j e^{ij\omega t} + C_j^* e^{-ij\omega t}\right) = \sum_{j=-\infty}^{\infty} C_j e^{ij\omega t} \tag{1.317}$$

The complex coefficients C_j can be calculated from (1.257) to (1.259):

$$C_j = \frac{1}{2}(a_j - ib_j) = \frac{1}{T}\int_0^T f(t)\left(\cos j\omega t - i\sin j\omega t\right)dt$$

$$= \frac{1}{T}\int_0^T f(t)\,e^{-ij\omega t}\,dt \qquad i = 1, 2, 3, \cdots \tag{1.318}$$

Example 31 Non-periodic functions. Non-periodic functions can also be expanded to Fourier series. Here is to explain how it works.

Consider a non-periodic function $f(t)$ that is defined only on the interval $[0, L]$ and we want to have a Fourier series to approximate the function. To develop the Fourier series, we can make $f(t)$ a periodic function by extending it as an even or odd periodic function over $[-L, L]$ with period $2L$. As an example, assume a non-periodic function $f(t)$ such as

$$f(t) = t \qquad 0 < t < L \tag{1.319}$$

which may be extended over $[-L, L]$ as an odd function $f_1(t)$

$$f_1(t) = t \qquad -L < t < L \tag{1.320}$$

or as an even function $f_2(t)$.

$$f_2(t) = \begin{cases} -t & -L < t < 0 \\ t & 0 < t < L \end{cases} \tag{1.321}$$

A function $f(t)$ is called even if $f(-t) = f(t)$ and is called odd if $f(-t) = -f(t)$. The harmonic function $\cos t$ is even, and $\sin t$ is odd. Any arbitrary function $f(t)$ can be split into an even plus and odd function:

$$f(t) = \frac{f(t) + f(-t)}{2} + \frac{f(t) - f(-t)}{2} \tag{1.322}$$

In case of extension $f(t)$ to an odd function, we use $\lambda L = \pi$ to have:

$$b_j = \frac{2}{L} \int_0^L f_1(s) \sin(j\lambda s)\, ds$$
$$= \frac{2}{L} \int_0^L s \sin(j\lambda s)\, ds = \frac{2L}{\pi} \frac{(-1)^{j+1}}{j} \tag{1.323}$$

$$f(t) = \frac{2L}{\pi} \sum_{j=1}^{\infty} \frac{(-1)^{j+1}}{j} \sin(j\lambda t) \tag{1.324}$$

and in case of extension $f(t)$ to an even function, we have:

$$a_0 = \frac{2}{L} \int_0^L f_2(s)\, ds = \frac{2}{L} \int_0^L s\, ds = L \tag{1.325}$$

$$a_n = \frac{2}{L} \int_0^L f_2(s) \cos(j\lambda s) \, ds$$

$$= \frac{2}{L} \int_0^L s \cos(j\lambda s) \, ds = \frac{2L}{\pi^2} \frac{(-1)^j - 1}{j^2} \tag{1.326}$$

$$f(t) = \frac{L}{2} - \frac{4L}{\pi^2} \sum_{j=1}^{\infty} \frac{1}{(2j-1)^2} \cos \frac{(2j-1)\pi}{L} t \tag{1.327}$$

Although both odd and even extension of $f(t)$ are acceptable and there would be an associated Fourier series, we may compare the series (1.324) and (1.327) to understand their differences. The odd series (1.324) goes with $1/j$, while the even series (1.327) goes with $1/(2j-1)^2$. Therefore, the even series (1.327) converges quicker than the odd series. Fourier coefficients of discontinuous functions decay as $1/j$, and those of continuous functions decay $1/j^2$ or quicker.

1.4 ★ Music of Vibrations

Music is made of harmonized sounds. Sound is anything could be heard by ears created by differences in air pressure. Ears and brains are equipped with biological hardware and software tools to detect and sense sound vibrations. Musical sound deals with specific quantum periodic sounds called notes. Starting with a fundamental frequency f_0 at 440 Hz for note A, all musical pitches are encoded sine waves using a logarithmic formula between pitches and frequencies.

$$f_0 = 440 \, \text{Hz} \tag{1.328}$$

$$f_n = 2^{n/12} \times 440 \, \text{Hz} \tag{1.329}$$

Any note is characterized by a pitch number n, a playing period of t, and a loudness X. A melody then becomes a sequence of notes. Pitch is characterized and measured by frequency f. Loudness is characterized and measured by amplitude X. There is a third characteristic for musical sound that is called timber and identified with the quality of the sound produced by the particular instrument. Timbre is a factor that enables us to distinguish a note played on the violin from an equally played note on a piano or any other instrument.

Proof Keyboard is the best musical instrument to study notes and frequency. Every key is associated to a note and makes a musical sound as a harmonic function. Naming notes are easier on keyboard as well. Figure 1.27 illustrates a keyboard and their notes. Every note is associated to a specific frequency. So, the frequency of the wave equation of each note is known.

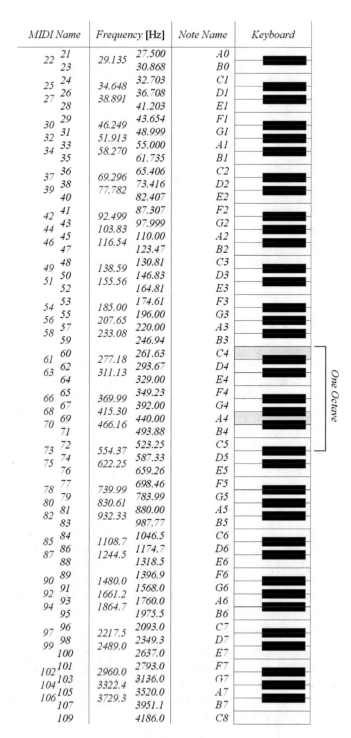

MIDI Name	Frequency [Hz]		Note Name	Keyboard
22 21 23	29.135	27.500 30.868	A0 B0	
25 24 26 27 28	34.648 38.891	32.703 36.708 41.203	C1 D1 E1	
30 29 31 32 33 34 35	46.249 51.913 58.270	43.654 48.999 55.000 61.735	F1 G1 A1 B1	
37 36 38 39 40	69.296 77.782	65.406 73.416 82.407	C2 D2 E2	
42 41 43 44 45 46 47	92.499 103.83 116.54	87.307 97.999 110.00 123.47	F2 G2 A2 B2	
49 48 50 51 52	138.59 155.56	130.81 146.83 164.81	C3 D3 E3	
54 53 55 56 57 58 59	185.00 207.65 233.08	174.61 196.00 220.00 246.94	F3 G3 A3 B3	
61 60 62 63 64	277.18 311.13	261.63 293.67 329.00	C4 D4 E4	
66 65 67 68 69 70 71	369.99 415.30 466.16	349.23 392.00 440.00 493.88	F4 G4 A4 B4	
73 72 74 75 76	554.37 622.25	523.25 587.33 659.26	C5 D5 E5	
78 77 79 80 81 82 83	739.99 830.61 932.33	698.46 783.99 880.00 987.77	F5 G5 A5 B5	
85 84 86 87 88	1108.7 1244.5	1046.5 1174.7 1318.5	C6 D6 E6	
90 89 91 92 93 94 95	1480.0 1661.2 1864.7	1396.9 1568.0 1760.0 1975.5	F6 G6 A6 B6	
97 96 98 99 100	2217.5 2489.0	2093.0 2349.3 2637.0	C7 D7 E7	
102 101 104 103 106 105 107 109	2960.0 3322.4 3729.3	2793.0 3136.0 3520.0 3951.1 4186.0	F7 G7 A7 B7 C8	

Fig. 1.27 A keyboard and their associated notes and frequency

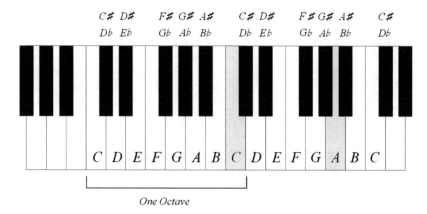

Fig. 1.28 The flat, ♯, and sharp, ♭, keys

The black keys are sharp and flat. The black key to the left (above in Fig. 1.27) of a given white key, say A_4, is A_4-flat and is shown by $A\flat$. The black key to the right (below in Fig. 1.27) of the note A_4 is A_4-sharp and is shown by $A\sharp$. A sharp note is a half tone up from a given white key note, and a flat note is a half tone down from a given white key note. Therefore, any black or white key is the sharp or flat of its adjacent key, as shown in Fig. 1.28.

To write notes, we use a set of five equidistant lines, called a stave. Each of the lines on a stave and the blank spaces between them is used to indicate notes. Figure 1.29 illustrates the whole notes of the middle octave. The convention of note-naming specifies a letter and an octave number. Any note is an integer away from middle A_4 at frequency $f = 440$ Hz. Let us denote the distance by n. If the note is above A_4, then n is positive, and if it is below A_4, then n is negative. The frequency f_n of the note n is calculated mathematically.

$$f_n = 2^{n/12} \times 440 \text{ Hz} \tag{1.330}$$

As an example, let us use this equation to calculate the frequency of B_4, the first B above A_4. There are 2 half-steps between A_4 and B_4 (A_4, $A_4\sharp$, B_4), and the note is above A_4, so $n = +2$. Therefore, the B note's frequency is $f_2 = 493.88$ Hz.

$$B_4 = 2^{2/12} \times 440 = 493.88 \text{ Hz} \tag{1.331}$$

The frequency of F_4 below A_4 is $f_{-4} = 349.23$ Hz, because there are 4 half-steps between A_4 and F_4 (A_4, $A_4\flat$, G_4, $G_4\flat$, F_4), and the note F_4 is below A_4, so $n = -4$.

$$F_4 = 2^{-4/12} \times 440 = 349.23 \text{ Hz} \tag{1.332}$$

Frequency of notes in octave number m is double the frequency of notes in the previous octave $m - 1$ and is half the frequency of notes in the next octave $m + 1$.

Fig. 1.29 Illustration of the whole notes of the middle octave

Octaves are factors of 2 times the original frequency, because in this case n is a multiple of 12. So, the exponent for octave m would be $12m$, where m is the number of octaves up or down.

$$f_m = 2^{12m/12} \times 440 = 2^m \times 440 \text{ Hz} \qquad (1.333)$$

Therefore, the frequency of a note would be double in a higher octave and half in a lower octave. For example, if frequency of $A4$ is 440 Hz, then frequency of $A3$ is 220 Hz, and frequency of $A5$ is 880 Hz.

Fourier series can be utilized to model musical sound waves. In graphing functions of sound waves generated by Fourier series, the vertical axis is the amplitude, and the horizontal axis represents time or period. Therefore, volume, or amplitude, is measured on the y-axis, and frequency, or cycles per second (Hz), is measured on the x-axis. Each Fourier-generated function represents one tone, or one quantum of musical sound. Each tone is made up of infinitely many frequencies combined together. Each frequency is mathematically represented by a type of harmonic wave function. The lowest frequency of a Fourier series is the fundamental frequency and is the most important frequency of the tone. It is the fundamental frequency of the tone that an ear perceives to be the pitch of the tone. Hence, if the lowest frequency is what the human ear perceives to be pitch, then the tone's remaining infinite frequencies are referred to as overtones. They are sound waves whose frequency is an integer multiple of that of the fundamental. In terms of Fourier theory, the fundamental is a harmonic way such as $\sin \omega t$, while the overtones are the $\sin 2\omega t$, $\sin 3\omega t$, and so on. The fundamental frequency has the largest coefficient, and hence, it is the loudest frequency and is the particular frequency which the ear recognized as the pitch. Thus, each tone that is represented by a Fourier function is comprised of a fundamental plus infinitely many overtones. While the fundamental frequency of a tone determines its pitch, it is the series of overtones that determine its timbre. Hence, the human ear's ability to determine two sounds as different is in the different series of overtones which the tone possesses above the fundamental.

∎

Example 32 Wave, sound, and musical notes. The main characteristic of a musical note is its frequency. Here is to show harmonic function associated to a note and what would be mathematical equation for combination of notes.

Fig. 1.30 Note A_4 is at 440 Hz

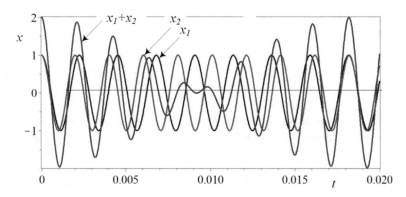

Fig. 1.31 The wave of notes A_4, A_5, and $A_4 + A_5$

We mathematically show a sine wave x with frequency $\omega = 2\pi f$, amplitude X, and phase φ.

$$x = X \sin(2\pi f t + \varphi) \tag{1.334}$$

The frequency ω is associated to notes that indicates the vibration speed of the wave. The higher ω is, the faster the wave vibrates and the more oscillation in time we have.

The note A_4 is the fundamental note that all other notes are calculated from. The frequency of A_4 is $f = 440$ Hz, and it sits above middle C as is shown in Fig. 1.30. The wave equation for note A_4 is:

$$x = X \sin(880\pi t + \varphi) \tag{1.335}$$

When two notes, say A_4 at $f = 440$ Hz and B_5 at $f = 493.88$ Hz, are hit, the sound would be a linear combination of the two notes.

$$x = X_1 \sin(880\pi t + \varphi_1) + X_2 \sin(987.76\pi t + \varphi_2) \tag{1.336}$$

Assuming

$$X_1 = X_2 = 1 \tag{1.337}$$

Figure 1.31 illustrates A_4, A_5, and $A_4 + A_5$ for $\varphi_1 = \varphi_2 = 0$.

Example 33 Musical sounds. Mathematical expression of musical sound is reviewed here.

A fundamental sound is a harmonic wave,

$$x = X \sin(\omega t + \varphi) \qquad \omega = 2\pi f \tag{1.338}$$

and a complicated periodic sound is a collection of fundamental waves, called spectrum.

$$x = \sum_{i=1}^{n} X_i \sin(\omega_i t + \varphi_i) \tag{1.339}$$

A fundamental wave is characterized by its frequency ω, amplitude X, and phase φ. A wave is produced when a medium is disturbed. The medium can be air, water, steel, or any other material in any phase. The disturbance produces a fluctuation in the ambient condition of the medium. The fluctuation propagates as a wave from the source of the disturbance. If we use one second as a time span, the number of fluctuations above and below the ambient condition per second is the frequency of the wave. Therefore the frequency f is the number of vibrations per second. It is measured in *Hertz* (Hz) or cycle per second. Low frequencies are heard as bass notes, and high frequencies sound high. Human sensitive ears hear sounds from about 20 to 20,000 Hz. Concert pitch is defined as 440 Hz. The different frequencies present in a sound are called partials. The lowest frequency is called the fundamental, and the frequencies above the fundamental are the overtones. Overtones are harmonic if their frequency is an integer multiple of the fundamental frequency $f, 2f, 3f$, etc.; otherwise they are inharmonic. The harmonic overtones are also called the harmonic series, harmonic partials, or just harmonics. Pitch is a subjective quantity related to frequency and to the overtone series.

The Fechner law states that the perceived pitch is proportional to the logarithm of the frequency. The amplitude of the sound wave evolves over time and determines the envelop of a wave. The sound envelop is the collection of the envelops of all partials. It is characterized by four segments: attack, sustain, decay, and release. A sine wave emits at a point O of the space will be perceive by a sound source located at a point M with a decay $\theta = d/c$ where d is the distance OM and c the speed of the sound. The period of the sound is the inverse of its frequency, $T = 1/f$. The wavelength $\lambda = cT = c/f$ is the distance covered by the sound during period T seconds at the speed c. The speed of the sound depends on the medium characteristics. At the atmospheric pressure, the speed of sound in air varies with temperature: at $0°C$, $c = 330$ m/s; at $10°C$, $c = 337$ m/s; at $15°C$, $c = 343$ m/ s; at $20°C$, $c = 334$ m/s; and at $30°C$, $c = 349$ m/s.

The acoustic power I is the total energy due to the movement of air molecules. For a sine wave of amplitude A, the acoustic power I for an acoustic pressure P is given by

$$I = \frac{P^2}{\rho c} = 4\pi^2 \rho c A^2 f^2 \tag{1.340}$$

The acoustic power is measured in decibels (dB). The level is defined from a threshold of hearing $I_0 = 1012 \text{ W/m}^2$ and $P_0 = 2 \times 10^{-5}$ Pa.

$$dB = 10 \log \frac{I}{I_0} = 20 \log \frac{P}{P_0} \tag{1.341}$$

The limit of audibility is 0 dB. A low voice is about 20 dB, a normal voice around 40 dB, and a tutti of a symphonic orchestra about 100 dB.

If we start at the lowest audible frequency of 20 Hz and increase it by a 2 to 1 ratio, the result is 40 Hz , and it is an interval of one octave. Doubling 40 Hz yields 80 Hz. This is also a one-octave span that contains twice the frequencies of the previous octave. Each successive frequency doubling makes another octave increase, and each higher octave has twice the spectral content of the one below it. This way makes the logarithmic scale suitable for displaying frequency.

Example 34 Sound intensity, decibel, and noise measurement unit. Loudness of sound is measured by decibels (dB) according to a nonlinear mathematical equation. Here is the expression.

Intensity of sounds is measured by decibels (dB) which is a very useful tool created for audio practitioners. It combines the changes in system parameters such as power, voltage, or distance to be reflected to level changes that a listener hears. The decibel is a measure to express how loud is a sound to the human perception.

A value of n decibels is equal to a power intensity of $10^{(n/10)-12}$ Watts per square meter.

$$n \text{ dB} = 10^{(n/10)-12} \text{ W/m}^2 \tag{1.342}$$

Therefore, decibel is a logarithmic function of power density. Decibel n increases ten units when the power density increases *ten* times. Figure 1.32 illustrates the power density and dB logarithmic relationship. To have a power density of 1 W/ m^2, we need a sound with 120 dB loudness.

Zero decibel is equal to a power density of 10^{-12} W/ m^2, and ten decibels (one bel) is 10^{-11} W/ m^2:

$$0 \text{ dB} = 10^{-12} \text{ W/m}^2 \qquad 10 \text{ dB} = 10^{-11} \text{ W/m}^2 \tag{1.343}$$

Decibel is also used for the ratio L_P of a physical quantity, usually power or acceleration, relative to a specified or implied reference level. Decibel is dimensionless when we use it to express the ratio of a physical quantity in the same units.

$$L_P = 10 \log \frac{P}{P_0} \text{ dB}$$

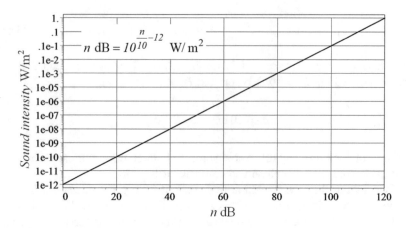

Fig. 1.32 Sound intensity and dB relationship

The beauty of dB is its nonlinear character. We usually consider physical variables to be linear such that twice as much of a quantity produces double the end result. For example, twice as much flour produces twice as much croissants. However, the linear relationship does not hold for the human sense of hearing. Hence, it is not true that twice the amplifier power will sound twice as loud. Expected changes in the loudness and frequency of sound are based on the percentage change from an initial condition. Therefore, the loudness of an audio is defined by ratio.

Experiment shows that the applied power to a loudspeaker must be increased by at least 25% to produce an audible change. Hence, a ratio of 1.25 to 1 produces the minimum audible change, regardless of the initial power quantity. If the initial amount of power is 1 W, then an increase to 1.25 W will produce a just audible increase. If the initial quantity is 10 W, then 12.5 W will be required to produce a just audible increase. Such scale that is calibrated proportionally is called a logarithmic scale. Employing base 10 logarithms for audio work, the resultant number is the level change between the two wattages expressed in Bels. Then we scale the Bel quantity by a factor of ten to convert Bels to decibels.

A decibel conversion requires two quantities that are in the same unit, such as watts, current, volts, and meters. Hence, the decibel is a dimensionless quantity. If a standard reference quantity is used in the denominator of the ratio, the result is an absolute level, and the unit is dB relative to the original unit.

Example 35 ★ Musical intervals. The range of notes in an octave is divided to 1200 cents, and hence, any semitone is divided into 100 cents. This is how the continuous line on notes is discretized. Subdivision of notes are expressed here.

By the Fechner law, perceived pitches are approximately proportional to the logarithm of the frequency. The ratio of the frequencies are measured in cents. The musical interval of two sounds of frequencies f_1 and f_0 is given by

$$\frac{1200}{\ln 2} \ln \frac{f_1}{f_0} \text{ cents} \tag{1.344}$$

There are 100 cents in the tempered semitone and 1200 cents in an octave. In the n-tone equal temperament (n-tet), the octave is divided into n tones. Two consecutive tones are separated by the interval of $2^{1/n}$ or $1200/n$ cents. The ratios of the just intonation scale are unison (1), minor tone (10/9), major tone (9/8), minor third (6/5), major third (5/4), fourth (4/3), and fifth (3/2). The following intervals will be used in the chapters about tuning and temperaments.

The Pythagorean comma is the difference between 12 just fifths and 7 octaves.

$$C_P = \frac{(3/2)^{12}}{2^7} = \frac{3^{12}}{2^{19}} = \frac{531\,441}{524\,288} \equiv 23\,cents \tag{1.345}$$

The syntonic comma is the difference between a major Pythagorean third (81/64) and a just third (5/4).

$$C_s = \frac{81/64}{5/4} = \frac{3^4}{2^4 \times 5} = \frac{81}{80} \equiv 22\,cents \tag{1.346}$$

The Holderian comma is the semitone in the 53-tet.

$$C_H = 2^{1/53} \equiv 23 \text{ cents} \tag{1.347}$$

The Fokker diesis is the 1/31 part of one octave.

$$D_F = 2^{1/31} \equiv 39 \text{ cents} \tag{1.348}$$

The septimal comma is the difference between the Pythagorean minor seventh (16/9) and the just minor seventh (7/4).

$$C_7 = \frac{16/9}{7/4} = \frac{2^6}{3^2 \times 7} = \frac{64}{63} \equiv 27 \text{ cents} \tag{1.349}$$

The leimma is the difference between three octaves and five just fifths.

$$L = \frac{2^3}{(3/2)^5} = \frac{2^8}{3^5} = \frac{256}{243} \equiv 90 \text{ cents} \tag{1.350}$$

The apotome is the difference between seven just fifths and four octaves.

$$A = \frac{(3/2)^7}{2^4} = \frac{3^7}{2^{11}} = \frac{2187}{2048} \equiv 114 \text{ cents} \tag{1.351}$$

1.5 Summary

In this chapter, we study (1) mechanical elements of vibrating systems, (2) physical causes of mechanical vibrations, (3) kinematics of vibrations, and (4) simplification methods of complex vibrating systems.

Mechanical vibration is a result of periodical transformation of kinetic energy K to potential energy P. When the potential energy is at its maximum, the kinetic energy is zero, and when the kinetic energy is at its maximum, the potential energy is minimum. A fluctuation in kinetic energy appears as a periodic motion of a massive body, and this is the reason we call such energy transformations the mechanical vibrations. The kinetic and potential energies are stored in physical elements. Any element that stores kinetic energy is called the *mass* or *inertia*, and any element that stores potential energy is called the *spring* or *restoring element*. If the total value of *mechanical energy* $E = K + P$ decreases during vibrations, there is also a phenomenon or an element that dissipates energy. The element that causes energy to dissipate is called *damper*.

Any periodic motion $x(t) = x(t + T)$ is characterized by a *period* T, which is the required time for one complete cycle of vibration, starting from and ending at the same conditions such as ($\dot{x} = 0, 0 < \ddot{x}$). The *frequency* f is the number of cycles in one T.

$$f = \frac{1}{T}$$

In theoretical vibrations, we usually work with *angular frequency* ω [rad/s], and in applied vibrations, we use the *cyclic frequency* f [Hz].

$$\omega = 2\pi f \tag{1.352}$$

The inverse of the equivalent stiffness, $1/k_e$, of a set of serial springs is the sum of their inverse stiffness, as $\sum 1/k_i$:

$$\frac{1}{k_e} = \frac{1}{k_1} + \frac{1}{k_2} + \frac{1}{k_3} \tag{1.353}$$

The equivalent stiffness, k_e, of parallel springs is the sum of their stiffnesses $\sum k_i$.

The *fundamental equation of vibration* is the free and undamped equation of motion of a mass m attached to a linear spring with stiffness k.

$$m\ddot{x} + kx = 0 \qquad x(0) = x_0 \qquad \dot{x}(0) = \dot{x}_0 \tag{1.354}$$

$$\ddot{x} + \omega_n^2 x = 0 \qquad \omega_n = \sqrt{\frac{k}{m}} \tag{1.355}$$

The solution of the fundamental equation of vibration can be expressed in either of the following forms:

$$x = A \sin \omega_n t + B \cos \omega_n t$$

$$= x_0 \cos \omega_n + \frac{1}{\omega_n} \dot{x}_0 \sin \omega_n \tag{1.356}$$

$$x = X \sin (\omega_n t + \varphi)$$

$$= \sqrt{x_0^2 + \frac{\dot{x}_0^2}{\omega^2}} \sin \left(\omega_n t + \arctan \frac{\dot{x}_0 / \omega_n}{x_0} \right) \tag{1.357}$$

The purpose of Fourier series is to decompose a periodic function $f(t) = f(t + T)$ of period T into its harmonic components of $\sin(jt)$ and $\cos(jt)$,

$$f(t) = f(t + T) = f\left(t + \frac{2\pi}{\omega} \right) \tag{1.358}$$

$$f(t) = \frac{1}{2} a_0 + \sum_{j=1}^{\infty} (a_j \cos (j\omega t) + b_j \sin (j\omega t)) \tag{1.359}$$

where a_j and b_j are weight factors of the $\cos(j\omega t)$ and $\sin(j\omega t)$ harmonic functions.

$$a_0 = \frac{2}{T} \int_0^T f(t)\, dt \tag{1.360}$$

$$a_j = \frac{2}{T} \int_0^T f(t) \, \cos (j\omega t) \, dt \tag{1.361}$$

$$b_j = \frac{2}{T} \int_0^T f(t) \, \sin (j\omega t) \, dt \tag{1.362}$$

The number $a_0/2$ is the mean value of the periodic function.

1.6   Key Symbols

$a \equiv \ddot{x}$	Acceleration
a, b	Distance, Fourier series coefficients
A	Amplitude
A, B	Weight factor, coefficients for frequency responses
$A, B, \cdots, G$	Musical notes
c	Damping
c_e	Equivalent damping
c_L	Drag coefficient
C	Coefficient, amplitude
C	Complex Fourier series coefficient
d	Diameter of wire
d_1	Wavelength
d_2	Peak-to-peak height of a wave
D	Mean diameter of coil, dissipation function
e	Eccentricity arm, exponential function
E	Mechanical energy
E	Young modulus of elasticity
E_c	Consumed energy of a damper per cycle
$f = 1/T$	Cyclic frequency [Hz]
$f, F, \mathbf{F}$	Force
f, g	Function, periodic function
f_B	Beating frequency
f_H	Hydraulic damping force
f_c	Damper force
f_C	Coulomb friction force
f_e	Equivalent force
f_k	Spring force
f_m	Required force to move a mass m
f_x	x-component of a force f
F	Amplitude of a harmonic force $f = F \sin \omega t$, total force
F_0	Constant force
g	Gravitational acceleration, function
G	Shear modulus of elasticity
i	$i^2 = -1$, imaginary unit number
I	Area moment, mass moment
$\mathbf{I}$	Identity matrix
j	Dummy index
J	Polar moment of cross-sectional area
k	Stiffness
k_e	Equivalent stiffness
k_{ij}	Element of row i and column j of a stiffness matrix
k_R	Antiroll bar torsional stiffness

$[k]$	Stiffness matrix
K	Kinetic energy
l	Length
L	Length dimension symbol
m	Mass
m_b	Device mass
m_d	Damper mass
m_e	Eccentric mass, equivalent mass
m_s	Spring mass
m_{ij}	Element of row i and column j of a mass matrix
m_k	Spring mass
$[m]$	Mass matrix
M	Mass dimension symbol
n	Number of coils, number of decibels, number of note
N	Natural numbers
N	Normal force
O	Fixed point, origin
p	Pitch of a coil
P	Potential energy
P	Power
P_D	Dissipated power in one cycle
r	Frequency ratio
$\mathbf{r}$	Position vector
s	Exponent, eigenvalue
t	Time
t_0	Initial time
T	Period
T	Time dimension symbol
T_B	Beating period
T_n	Natural period
$v \equiv \dot{x}$	Velocity
W	Work
W_D	Dissipated energy in one cycle
$x, y, z, \mathbf{x}$	Displacement
x_0	Initial displacement
x_i	Initial position
x_f	Final position
$\dot{x}_0$	Initial velocity
$\dot{x}, \dot{y}, \dot{z}$	Velocity, time derivative of x, y, z
$\ddot{x}$	Acceleration
X	Amplitude of x
z	Relative displacement
Z	Amplitude of z

Greek

α, β, γ	Angle, angle of spring with respect to displacement
δ	Deflection
δ	Angle
δ_s	Static deflection
ε	Mass ratio
ε	Small coefficient
ρ	Length mass density
θ	Angular motion, phase angle
λ	Eigenvalue
π	3.141592653589793...
ω, Ω	Angular frequency
ω_n	Natural frequency
φ, Φ	Phase angle
τ	Time varying angle
μ	Amplitude frequency response

Symbols

$\forall$	For all
[ ]	Dimension
[ ]	Matrix
$\int$	Integral
$\vert\,\vert$	Absolute value
Hz	Hertz
d	Differential
∂	Partial derivative
!	Factorial
Δ	Small amount

Exercises

1. Wave determination.

   Determine the wave equation if we know:

   (a) $x_{max} = 0.022$ mm, $\ddot{x}_{max} = 5.632$ mm/ s^2
   (b) $\dot{x}_{max} = 0.352$ mm/ s, $\ddot{x}_{max} = 5.632$ mm/ s^2
   (c) $x(t_1) = x_1$, $x(t_2) = x_2$
   (d) $x(t_1) = x_1$, $\ddot{x}(t_2) = \ddot{x}_2$
   (e) $\ddot{x}(t_1) = x_1$, $\ddot{x}(t_2) = \ddot{x}_2$
   (f) $x = x_1 + x_2$, $x = 2\sin(\omega t - \pi/6)$, $x_1 = \sin(\omega t - \pi/3)$

2. Wave combination.

   Determine the combined wave of $x = x_1 + x_2 = X\sin(\omega t + \varphi)$ if

   (a) $x_1 = 3\sin\left(\omega t + \dfrac{\pi}{3}\right)$ and $x_2 = 4\sin\left(\omega t + \dfrac{\pi}{4}\right)$
   (b) $x_1 = 3\cos\left(\omega t + \dfrac{\pi}{3}\right)$ and $x_2 = 4\sin\left(\omega t + \dfrac{\pi}{4}\right)$
   (c) $x_1 = 3\cos\left(\omega t + \dfrac{\pi}{3}\right)$ and $x_2 = 4\cos\left(\omega t + \dfrac{\pi}{4}\right)$

3. Wave decomposition.

   The combination of the waves x_1 and x_2 is $x = x_1 + x_2 = 4\sin\left(\omega t + \frac{\pi}{4}\right)$.
   What is x_2 if

   (a) $x_1 = 3\sin\left(\omega t + \dfrac{\pi}{3}\right)$
   (b) $x_1 = 3\cos\left(\omega t + \dfrac{\pi}{3}\right)$
   (c) $x_1 = 3\sin\left(\omega t - \dfrac{\pi}{3}\right)$

4. Average and root mean square of waves.

   The absolute average, x_{av}, and root mean square (RMS) of a vibrating spectrum are good values to determine the effect of changing parameters and compare different oscillating systems.

   $$x_{av} = \frac{1}{T}\int_0^T |x(t)|\ dt \tag{1.363}$$

   $$x_{RMS} = \sqrt{\frac{1}{T}\int_0^T x^2(t)\ dt} \tag{1.364}$$

   Calculate x_{av} and x_{RMS} for a harmonic wave, $x = A\sin\omega t$, and show the following.

   $$x_{av} = \frac{2}{\pi}X = 0.9x_{RMS} \tag{1.365}$$

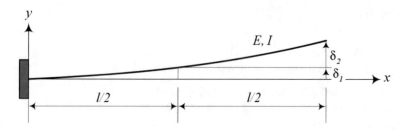

Fig. 1.33 A uniform leaf spring of length l may be assumed that is made of two similar leaf springs of lengths $l/2$

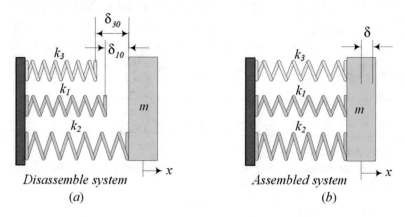

Fig. 1.34 Characteristic curves of k_1, k_2, and k_3 and their working points

$$x_{RMS} = \frac{\sqrt{2}}{2}X = 1.11x_{av} \tag{1.366}$$

5. Half-length leaf spring.

Figure 1.33 illustrates a uniform leaf spring of length l with the assumption that it is made of two similar leaf springs of lengths $l/2$. Calculate $\delta = \delta_1 + \delta_2$ to prove that:

$$k_l = 8k_{l/2} \tag{1.367}$$

6. Dimensional analysis.

   (a) Show that dimension of c^2 is the same as mk.
   (b) What combination of m, c, k make dimension of work?
   (c) What combination of m, c, k make dimension of power?

7. Different length springs in parallel.

Examine the system of Fig. 1.34 and determine the length change of the springs when they are assembled to m.

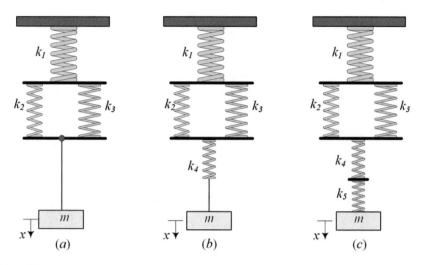

Fig. 1.35 Combinations of linear springs between a mass m and ceiling

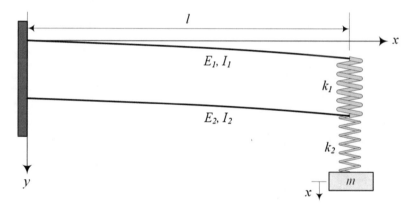

Fig. 1.36 A combination of two cantilever beams and two coil springs

8. ★ Stiffness of elastic systems.

Show that the equivalent spring constant of a bar in longitudinal direction is $k_e = EA/l$, a cantilever beam in lateral direction is $k_e = 3EI/l^3$, and a bar in torsional vibration is $k_e = GJ/l$.

9. Equivalent springs.

(a) Determine the equivalent spring for the systems shown in Fig. 1.35a–c.
(b) Determine the equivalent spring for the systems shown in Fig. 1.36.
(c) Determine the equivalent spring for the systems shown in Fig. 1.37a and b.

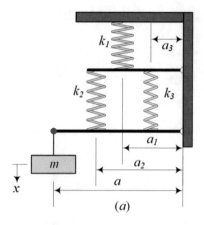
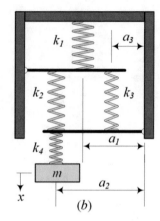

Fig. 1.37 Two combinations of lever and springs

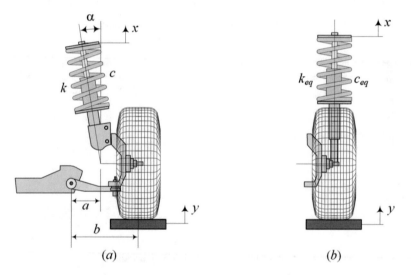

Fig. 1.38 A MacPherson suspension and its equivalent vibrating system

10. Equivalent MacPherson suspension parameters.

Figure 1.38a illustrates a MacPherson suspension. Its equivalent vibrating system is shown in Fig. 1.38b.

(a) Determine k_e and c_e if

$$a = 22 \text{ cm} \qquad b = 45 \text{ cm} \qquad k = 10{,}000 \text{ N/ m}$$

$$c = 1000 \text{ N s/ m} \qquad \alpha = 12 \text{ deg} \tag{1.368}$$

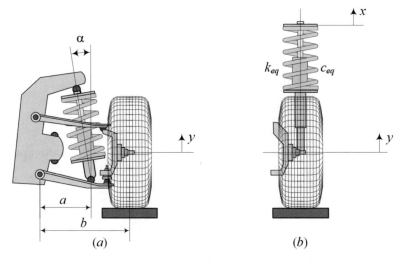

Fig. 1.39 A double A-arm suspension and its equivalent vibrating system

(b) Determine the stiffness k such that the natural frequency of the vibrating system is $f_n = 1$ Hz, if

$$a = 22 \text{ cm} \qquad b = 45 \text{ cm} \qquad m = 1000/4 \text{ kg} \qquad \alpha = 12 \text{ deg} \qquad (1.369)$$

(c) Determine the damping c such that the damping ratio of the vibrating system is $\xi = c/\left(2\sqrt{km}\right) = 0.4$, if

$$a = 22 \text{ cm} \qquad b = 45 \text{ cm} \qquad m = 1000/4 \text{ kg}$$

$$\alpha = 12 \text{ deg} \qquad f_n = 1 \text{ Hz} \qquad (1.370)$$

11. Equivalent double A-arm suspension parameters.

    Figure 1.39a illustrates a double A-arm suspension. Its equivalent vibrating system is shown in Fig. 1.39b.

    (a) Determine k_e and c_e if

    $$a = 32 \text{ cm} \qquad b = 45 \text{ cm} \qquad c = 1000 \text{ N s/ m}$$

    $$k = 8000 \text{ N/ m} \qquad \alpha = 10 \text{ deg} \qquad (1.371)$$

    (b) Determine the stiffness k such that the natural frequency of the vibrating system is $f_n = 1$ Hz, if

    $$a = 32 \text{ cm} \qquad b = 45 \text{ cm} \qquad m = 1000/4 \text{ kg} \qquad \alpha = 10 \text{ deg} \qquad (1.372)$$

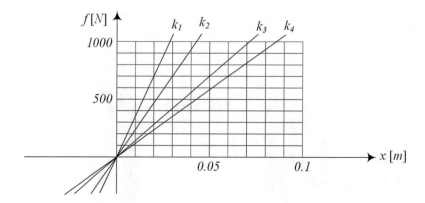

Fig. 1.40 Characteristic curves of four serial springs

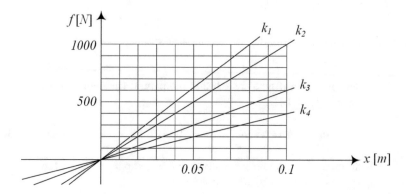

Fig. 1.41 Characteristic curves of four parallel springs

(c) Determine the damping c such that the damping ratio of the vibrating system is $\xi = 0.4$, if

$$a = 32 \text{ cm} \qquad b = 45 \text{ cm} \qquad m = 1000/4 \text{ kg}$$

$$\alpha = 10 \text{ deg} \qquad f_n = 1 \text{ Hz} \tag{1.373}$$

12. Serial springs.
    Figure 1.40 illustrates the characteristic curves of four serial springs.

    (a) Determine the stiffness of the equivalent spring, graphically.
    (b) Prove that the equivalent stiffness of a set of serial springs is less than the stiffness of every one of them.

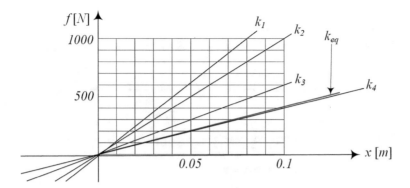

Fig. 1.42 Characteristic curves of four springs and their equivalent spring

13. Parallel springs.
    Figure 1.41 illustrates the characteristic curves of four parallel springs.

    (a) Determine the stiffness of the equivalent spring, graphically.
    (b) Prove that the equivalent stiffness of a set of parallel springs is more than
        the stiffness of every one of them.

14. ★ Parallel and series springs.
    Figure 1.41 illustrates the characteristic curves of four springs.

    (a) Suppose k_1 is parallel to k_2 and k_3 is parallel to k_4. Determine graphically
        the equivalent stiffness of all springs if the combination of k_1 and k_2 is in
        series with the combination of k_3 and k_4.
    (b) Suppose k_1 is in series with k_2 and k_3 is in series with k_4. Determine
        graphically the equivalent stiffness of all springs if the combination of k_1
        and k_2 is parallel to the combination of k_3 and k_4.

15. ★ Unknown spring combination.
    In Fig. 1.42, the stiffness of the equivalent spring k_{eq} and four individual
    springs is illustrated.

    (a) Determine all possible combination of k_1, k_2, k_3, k_4 if they can only be
        parallel or in series.
    (b) Determine the situation of the combination of k_1, k_2, k_3 , k_4 for the given
        equivalent.

16. Road excitation frequency.
    A car is moving on a wavy road. What is the wavelength d_1 if the excitation
    frequency is $f_n = 5$ Hz and

    (a) $v = 30$ km/ h
    (b) $v = 60$ km/ h
    (c) $v = 100$ km/ h

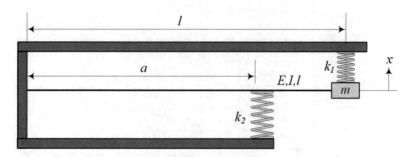

Fig. 1.43 Spring connected cantilever beam

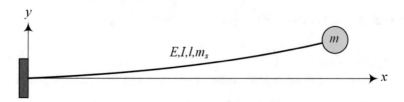

Fig. 1.44 An elastic and massive cantilever beam with a tip mass m

17. Road excitation frequency and wheelbase.
    A car is moving on a wavy road.

    (a) What is the wavelength d_1 if the excitation frequency is $f_n = 8$ Hz and $v = 60$ km/ h?
    (b) What is the phase difference between the front and rear wheel excitations if the car's wheelbase is $l = 2.82$ m?
    (c) At what speed the front and rear wheel excitations have no phase difference?

18. ★ Road excitation amplitude.
    A car is moving on a wavy road with a wavelength $d_1 = 25$ m. What is the damping ratio ξ if $S_2 = Z/Y = 1.02$ when the car is moving with $v = 120$ km/ h?

$$k = 10,000 \text{ N/ m} \qquad m = 1000/4 \text{ kg} \qquad (1.374)$$

19. Natural frequency and damping ratio.
    A one DOF mass-spring-damper has $m = 1$ kg, $k = 1000$ N/ m and $c = 100$ N s/ m. Determine the natural frequency and the damping ratio of the system.
20. Equivalent spring.
    Determine the equivalent spring for the vibrating system that is shown in Fig. 1.43.

21. ★ Equivalent mass for massive spring.

Figure 1.44 illustrates an elastic cantilever beam with a tip mass m. The beam has characteristics: elasticity E, area moment I, and mass m_s. Assume that when the tip mass m oscillates laterally, the beam gets a harmonic shape:

$$y = Y \sin \frac{\pi x}{2l}$$

If the mass of the beam is m_s, what is the new tip mass of an equivalent massless spring system?

22. Wave determination.

Determine the weight factors A and B of the wave

$$x = A \sin \omega t + B \cos \omega t = X \sin (\omega t + \varphi)$$

if $\varphi = \frac{\pi}{3}$ and

(a) $x(0.1) = 1.7773$, and $x(0.5) = -0.46761$
(b) $x(0.1) = 1.7773$, and $\dot{x}(0.1) = -9.1717$
(c) $\dot{x}(0.1) = -9.1717$, and $\dot{x}(0.5) = 19.446$

23. Beating.

(a) Add the waves $x_1 = X_1 \sin (\omega t)$ and $x_1 = X_1 \sin (\omega + \varepsilon) t$, where $\varepsilon \ll \omega$, and determine the amplitude and the beat frequency.
(b) Determine the wave x_2 that when is added to $x_1 = \sin (40\pi t)$ generates a beating frequency of 2 Hz with maximum amplitude of 1.8.

24. Free contact.

Consider a table in a vertical harmonic vibration $y = Y \sin \omega t$ with a constant frequency. What is the largest amplitude of the table if an object on the table remain in contact?

25. Trigonometric identities.

Prove the following equations.

$$\sum_{k=1}^{n} \sin kt = \frac{\cos \frac{1}{2}t - \cos \left(n + \frac{1}{2}\right)t}{2 \sin \frac{1}{2}t} = \frac{\sin \frac{1}{2}nt \, \sin \frac{1}{2}(n+1)t}{\sin \frac{1}{2}t}$$

$$\sum_{k=1}^{n} \cos kt = \frac{\sin \left(n + \frac{1}{2}\right)t - \sin \frac{1}{2}t}{2 \sin \frac{1}{2}t} = \frac{\cos \frac{1}{2}(n+1)t \, \sin \frac{1}{2}nt}{\sin \frac{1}{2}t}$$

26. Periodicity.

Determine the period of

(a) $x = 2 \cos 2t + 4 \cos 4t$
(b) $x = 2 \cos^2 2t$
(c) $x = 2 \cos 2t + 4 \sin 4t$

(d) ★ $x = 2\cos^2 2t + 4\sin^2 4t$

(e) ★ Is the motion $x = \cos 2t + \cos(2 + \pi)t$ periodic?

27. Fourier series and interesting numerical series.

   Fourier series of functions may be converted to numerical series for a fixed value of the variable. Many of those series show interesting numerical series.

   (a) Expand the function $f(x) = x$ in the range $-\pi < x \le \pi$ and show that:

$$1 - \frac{1}{3} + \frac{1}{5} - \frac{1}{7} + \frac{1}{9} - \cdots = \frac{\pi}{4} \tag{1.375}$$

   (b) Expand the function $f(x) = x^2$ in the range $-\pi < x \le \pi$ and show that:

$$1 + \frac{1}{2^2} + \frac{1}{3^2} + \frac{1}{4^2} + \frac{1}{5^2} + \cdots = \frac{\pi^2}{6} \tag{1.376}$$

28. Fourier series of $f(x) = c$.

   (a) Expand $f(x)$ to Fourier series,

$$f(x) = \begin{cases} 0 & -\pi < x < 0 \\ \pi & 0 < x < \pi \end{cases} \tag{1.377}$$

   (b) Expand $f(x)$ to Fourier series,

$$f(x) = \begin{cases} 1 & x_0 - c < x < x_0 + c \\ 0 & elsewhere \end{cases} \tag{1.378}$$

   (c) Expand $f(x)$ to Fourier series,

$$f(x) = \begin{cases} 1 & 0 < x < L \\ -1 & L < x < 2L \end{cases} \tag{1.379}$$

29. Fourier series of $f(x) = x$.

   (a) Show that Fourier series expansion of the given function $f(x)$

$$f(x) = |x| \qquad -\pi < x < \pi \tag{1.380}$$

   is

$$a_0 = \pi \qquad a_n = \frac{2}{\pi j^2}\left((-1)^j - 1\right) \tag{1.381}$$

$$f(x) = |x| = \frac{\pi}{2} - \frac{4}{\pi} \sum_{j=1}^{\infty} \frac{\cos(2j-1)x}{(2j-1)^2} \tag{1.382}$$

(b) Show that Fourier series expansion of the given function $f(x)$

$$f(x) = x \qquad -\pi < x < \pi \tag{1.383}$$

is

$$f(x) = x = 2 \sum_{j=1}^{\infty} \frac{(-1)^{j+1}}{j} \sin jx \tag{1.384}$$

(c) Show that Fourier series expansion of the given function $f(x)$

$$f(x) = \begin{cases} (\pi - x)/2 & 0 < x \le \pi \\ f(x) = -f(-x) & -\pi \le x < 0 \end{cases} \tag{1.385}$$

is a sine series

$$f(x) = \sin x + \frac{1}{2} \sin 2x + \frac{1}{3} \sin 3x + \cdots \qquad 0 < x \le \pi \tag{1.386}$$

30. Fourier series of $f(x) = x^2$.

    (a) Expand $f(x)$ to Fourier series,

$$f(x) = 1 - x^2 \qquad -\pi < x < \pi \tag{1.387}$$

    (b) Expand $f(x)$ to Fourier series,

$$f(x) = 1 + x^2 \qquad -\pi < x < \pi \tag{1.388}$$

    (c) Expand $f(x)$ to Fourier series,

$$f(x) = \frac{1}{2}x^2 \qquad -L < x < L \tag{1.389}$$

31. Fourier series of $f(x) = x^3$.
    Expand $f(x)$ to Fourier series,

$$f(x) = x^3 \qquad -\pi < x < \pi \tag{1.390}$$

32. Half-sine wave.

Expand $f(x)$ in a Fourier series

$$f(x) = \begin{cases} \sin x & 0 < x \leq \pi \\ -\sin x & -\pi \leq x < 0 \end{cases} \tag{1.391}$$

33. ★ Integral by Fourier series.

Show that if

$$f(t) = \frac{1}{2}a_0 + \sum_{j=1}^{\infty} (a_j \cos jt + b_j \sin jt) \tag{1.392}$$

then

$$\int_0^t f(s)\,ds = \frac{1}{2}a_0 t + \sum_{j=1}^{\infty} \left(\frac{a_j}{j} \sin jt - \frac{b_j}{j} (\cos jt - 1) \right) \tag{1.393}$$

34. Wave determination.

Determine the wave equation if we know:

(a) $x_{\max} = 0.022$ mm, $\ddot{x}_{\max} = 5.632$ mm/ s^2
(b) $\dot{x}_{\max} = 0.352$ mm/ s, $\ddot{x}_{\max} = 5.632$ mm/ s^2
(c) $x(t_1) = x_1$, $x(t_2) = x_2$
(d) $x(t_1) = x_1$, $\ddot{x}(t_2) = \ddot{x}_2$
(e) $\ddot{x}(t_1) = x_1$, $\ddot{x}(t_2) = \ddot{x}_2$
(f) $x = x_1 + x_2$, $x = 2\sin(\omega t - \pi/6)$, $x_1 = \sin(\omega t - \pi/3)$

35. Wave combination.

Determine the combined wave of $x = x_1 + x_2 = X \sin(\omega t + \varphi)$ if

(a) $x_1 = 3\sin\left(\omega t + \frac{\pi}{3}\right)$ and $x_2 = 4\sin\left(\omega t + \frac{\pi}{4}\right)$
(b) $x_1 = 3\cos\left(\omega t + \frac{\pi}{3}\right)$ and $x_2 = 4\sin\left(\omega t + \frac{\pi}{4}\right)$
(c) $x_1 = 3\cos\left(\omega t + \frac{\pi}{3}\right)$ and $x_2 = 4\cos\left(\omega t + \frac{\pi}{4}\right)$

36. Note addition.

Assume we play the note A_4 along with the following notes. Determine and draw the resultant wave. Examine if there is any beating.

(a) G_4
(b) F_4
(c) G_4 and E_4
(d) ★ G_4 and E_4 and D_4
(e) ★ G_4 and E_4 and $D_4$and C_4
(f) A_3
(g) A_3 and A_2
(h) ★ A_3 and A_2 and A_1

Chapter 2
Vibration Dynamics

In this chapter, we review the dynamics of vibrations and the methods of deriving the equations of motion of vibrating systems. The Newton-Euler and Lagrange methods are the most applied methods of deriving the equations of motion. Having symmetric coefficient matrices for multi-degree-of-freedom linear vibrating systems is the main advantage of using the Lagrange method in mechanical vibrations.

2.1 Newton-Euler Method

When a vibrating system is modeled as a combination of masses m_i, dampers c_i, and springs k_i, it is called a *discrete* or *lumped* model of the system. To derive the equations of motion of a low degree-of-freedom (DOF) discrete model of a vibrating system, the Newton-Euler method works very well. We move all the masses m_i out of their equilibria at positions x_i with velocities $\dot{x}_i$. Then, a free body diagram (FBD) of the lumped masses indicates the total forces $\mathbf{F}_i$ on mass m_i. Employing the momentum $\mathbf{p}_i = m_i \mathbf{v}_i$ of the mass m_i, the Newton equation provides us with the equation of motion of the system:

$$\mathbf{F}_i = \frac{d}{dt} \mathbf{p}_i = \frac{d}{dt} (m_i \mathbf{v}_i) \tag{2.1}$$

When the motion of a massive body with mass moment I_i is rotational, then its equation of motion will be determined by the Euler equation, in which we employ the moment of momentum $\mathbf{L}_i = I_i \boldsymbol{\omega}$ of the mass m_i:

$$\mathbf{M}_i = \frac{d}{dt} \mathbf{L}_i = \frac{d}{dt} (I_i \boldsymbol{\omega}) \tag{2.2}$$

For example, Fig. 2.1 illustrates a one degree-of-freedom vibrating system. Figure 2.1b depicts the system when m is out of the equilibrium position at x and

© The Author(s), under exclusive license to Springer Nature Switzerland AG 2022
R. N. Jazar, *Advanced Vibrations*, https://doi.org/10.1007/978-3-031-16356-2_2

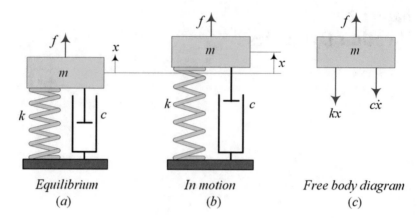

Fig. 2.1   A one DOF vibrating systems and its FBD. (a) The mathematical model, (b) The system in motion, (c) The free-body-diagram of the system in motion

moving with velocity $\dot{x}$, both in positive direction. The free body diagram of the system is as shown in Fig. 2.1c. The Newton equation generates the equations of motion:

$$f(x, v, t) - c\dot{x} - kx = m\ddot{x} \tag{2.3}$$

We measure displacement of a vibrating system from its equilibrium configuration. The *equilibrium position* of a vibrating system is where the potential energy of the system, P, is extremum:

$$\frac{\partial P}{\partial x} = 0 \tag{2.4}$$

We set $P = 0$ at the equilibrium position. Linear systems have only one equilibrium configuration or infinity equilibria, while nonlinear systems may have multiple equilibria. An equilibrium is *stable* if the second spatial derivative of the potential energy P is positive

$$\frac{\partial^2 P}{\partial x^2} > 0 \tag{2.5}$$

and is *unstable* if P is negative.

$$\frac{\partial^2 P}{\partial x^2} < 0 \tag{2.6}$$

The geometric arrangement and the number of employed mechanical elements can be used to classify discrete vibrating systems. The number of masses times the

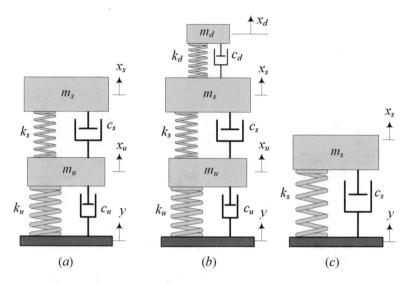

Fig. 2.2 Two, three, and one DOF models for vertical vibrations of vehicles

number of DOF of each mass makes the total DOF of the vibrating system n. Each independent DOF is indicated by an independent variable, called the *generalized coordinate*. The final set of equations of motion would be n coupled second-order differential equations to be solved for n generalized coordinates. When every moving mass has one DOF, the system's DOF is equal to the number of masses. The DOF of a system is the minimum number of independent coordinates that defines the configuration of a system.

The equation of motion of an n DOF linear vibrating system can always be arranged in matrix form of a set of second-order differential equations

$$[m]\ddot{\mathbf{x}} + [c]\dot{\mathbf{x}} + [k]\mathbf{x} = \mathbf{F} \tag{2.7}$$

where $\mathbf{x}$ is a column array of generalized coordinates of the system and $\mathbf{F}$ is a column array of the associated applied forces. The *square matrices* of $[m]$, $[c]$, $[k]$ are the mass, damping, and stiffness matrices, respectively.

Example 36 The one, two, and three DOF model of vehicles. Vertical vibration of vehicles may be modeled by base vibrating system with one or two or three DOF systems by increasing complexity and accuracy.

The one, two, and three DOF models for analysis of vertical vibrations of a vehicle are shown in Fig. 2.2a–c. The system in Fig. 2.2a is called the quarter car model, in which m_s represents a quarter mass of the body and m_u represents a wheel. The parameters k_u and c_u are models for tire stiffness and damping. Similarly, k_s and c_s are models for the stiffness and damping of the main suspension vehicle. Figure 2.2c is called the 1/8 car model, which does not show the wheel of the car.

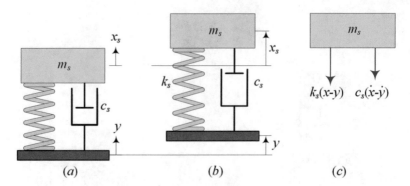

Fig. 2.3 A 1/8 car model and its free body diagram

Figure 2.2b is a quarter car with a driver m_d. The stiffness and damping of the driver's seat are modeled by k_d and c_d.

The 1/8 car model is the simplest vibrating model of vehicles to study and optimize the main characteristics of vehicle suspensions. To derive the equation of motion of the model, Fig. 2.3a illustrates the mathematical model in which the mass m_s represents one quarter of the car's body. The mass m_s is mounted on a suspension made of a spring k_s and a damper c_s. When m_s is at an off-equilibrium position such as shown in Fig. 2.3b, its free body diagram is as shown in Fig. 2.3c.

Applying Newton's method on the free body diagram, the equation of motion would be

$$m_s \ddot{x} = -k_s (x_s - y) - c_s (\dot{x}_s - \dot{y}) \qquad (2.8)$$

which can be simplified to:

$$m_s \ddot{x} + c_s \dot{x}_s + k_s x_s = k_s y + c_s \dot{y} \qquad (2.9)$$

The coordinate y indicates the input from the road, and x indicates the absolute displacement of the body. Absolute displacement refers to displacement with respect to the motionless background.

Figure 2.4a–c illustrates the equilibrium, motion, and FBD of a two DOF system for quarter car model. The FBD is plotted based on the following assumption.

$$x_s > x_u > y \qquad (2.10)$$

Applying Newton's method on the free body diagram of Fig. 2.4c provides us with two equations of motion:

$$m_s \ddot{x}_s = -k_s (x_s - x_u) - c_s (\dot{x}_s - \dot{x}_u) \qquad (2.11)$$

$$m_u \ddot{x}_u = k_s (x_s - x_u) + c_s (\dot{x}_s - \dot{x}_u)$$

$$-k_u (x_u - y) - c_u (\dot{x}_u - \dot{y}) \qquad (2.12)$$

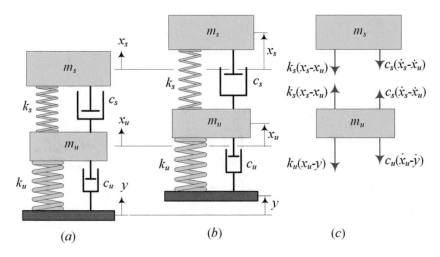

Fig. 2.4 A 1/4 car model and its free body diagram

An assumption for relative order of displacements, such as (2.10), is not necessary to develop equations of motion correctly. We can find the same Eqs. (2.11) and (2.12) using any other assumption, such as $x_s < x_u > y$, $x_s > x_u < y$, or $x_s < x_u < y$. However, having an assumption helps to make a consistent free body diagram.

We usually arrange equations of motion of linear systems in matrix form to take advantage of matrix calculus.

$$[M]\ddot{\mathbf{x}} + [C]\dot{\mathbf{x}} + [K]\mathbf{x} = \mathbf{F} \tag{2.13}$$

Rearrangement of Eqs. (2.11) and (2.12) yields:

$$\begin{bmatrix} m_s & 0 \\ 0 & m_u \end{bmatrix} \begin{bmatrix} \ddot{x}_s \\ \ddot{x}_u \end{bmatrix} + \begin{bmatrix} c_s & -c_s \\ -c_s & c_s + c_u \end{bmatrix} \begin{bmatrix} \dot{x}_s \\ \dot{x}_u \end{bmatrix} +$$
$$\begin{bmatrix} k_s & -k_s \\ -k_s & k_s + k_u \end{bmatrix} \begin{bmatrix} x_s \\ x_u \end{bmatrix} = \begin{bmatrix} 0 \\ k_u y + c_u \dot{y} \end{bmatrix} \tag{2.14}$$

Example 37 Equivalent mass and spring. Mass-spring-damper analogy is a traditional method in mechanical engineering to simulate all vibrating systems. Here is to determine mass-spring equivalent to a oscillating pendulum in linear approximation.

Figure 2.5a illustrates a pendulum made by a point mass m attached to a massless bar with length l. The coordinate θ shows the angular position of the bar. The equation of motion for the pendulum can be found by using the Euler equation and employing the FBD shown in Fig. 2.5b:

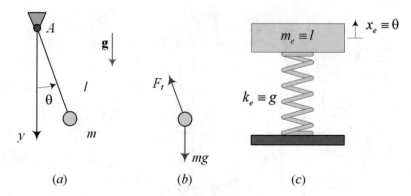

Fig. 2.5 Equivalent mass-spring vibrator for a pendulum

$$I_A \ddot{\theta} = \sum \mathbf{M}_A \tag{2.15}$$

$$ml^2 \ddot{\theta} = -mgl \sin \theta \tag{2.16}$$

Equation of motion of pendulum is a nonlinear differential equation. The equilibrium positions of the pendulum will be determined by Eq. (2.4).

$$\frac{\partial P}{\partial \theta} = 0 \tag{2.17}$$

Setting $P = 0$ at the horizontal level of the fulcrum A at $y = 0$, the potential energy of m will be:

$$P = -mgy = -mgl \cos \theta \tag{2.18}$$

There are two equilibria for the pendulum.

$$\frac{\partial P}{\partial \theta} = mgl \sin \theta = 0 \tag{2.19}$$

$$\theta = 0 \qquad \theta = \pi \tag{2.20}$$

$$\frac{\partial^2 P}{\partial \theta^2} = mg \cos \theta \tag{2.21}$$

The equilibrium position $\theta = 0$ is stable because,

$$\frac{\partial^2 P}{\partial \theta^2}(0) = mg \cos 0 = mg > 0 \tag{2.22}$$

and equilibrium position $\theta = \pi$ is unstable.

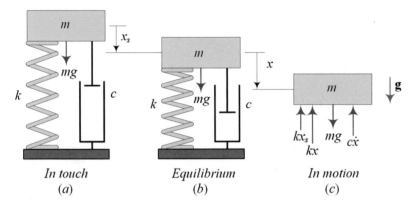

Fig. 2.6 A mass-spring-damper system indicating that the gravitational force in rectilinear vibrations provides a static deflection

$$\frac{\partial^2 P}{\partial \theta^2}(\pi) = mg \cos \pi = -mg < 0 \qquad (2.23)$$

Simplifying the equation of motion and assuming a very small swing angle yields:

$$l\ddot{\theta} + g\theta = 0 \qquad (2.24)$$

This equation is equivalent to an equation of motion of a mass-spring system made by a mass $m_e \equiv l$ and a spring with stiffness $k_e \equiv g$. The displacement of the mass would be measured by $x_e \equiv \theta$. Figure 2.5c depicts such an equivalent mass-spring system.

Example 38 Gravitational force in rectilinear vibrations. Gravity and static force in spring are two opposite and equal forces that may be cancelled out from equation of motion and ignored from the beginning of modeling.

When the direction of the gravitational force on a mass m is not varied with respect to the direction of motion of m, the effect of the weight force can be ignored in deriving the equation of motion. In such a case, the equilibrium position of the system is at a point where the gravity is in balance with the elastic force because of static deflection in the elastic member. This force-balance equation will not be altered during vibration. Consequently we may cancel out and ignore both forces: the gravitational force and the static elastic force. It may also be interpreted as an energy balance situation where the work of gravitational force is always equal to the extra stored energy in the elastic member.

Consider a spring k and damper c as is shown in Fig. 2.6a. A mass m is put on the force free spring and damper. The weight of m compresses the spring a static length x_s to bring the system at equilibrium in Fig. 2.6b. When m is at equilibrium, it is under the balance of two forces, mg and $-kx_s$:

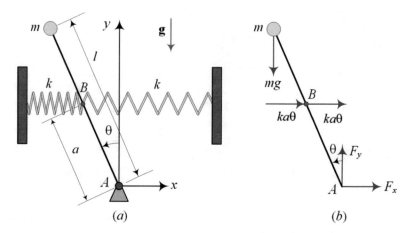

Fig. 2.7 An inverted pendulum with a tip mass m and two supportive springs

$$mg - kx_s = 0 \tag{2.25}$$

While the mass is in motion, its FBD is as shown in Fig. 2.6c, and its equation of motion is:

$$m\ddot{x} = -kx - c\dot{x} + mg - kx_s$$
$$= -kx - c\dot{x} \tag{2.26}$$

It shows that if we examine the motion of the system from equilibrium, we can ignore both the gravitational force and the initial compression of the elastic member of the system.

The equation of motion for a vibrating system is a balance between four different forces: a force proportional to displacement, $-kx$; a force proportional to speed, $-c\dot{x}$; a force proportional to acceleration, ma; and an applied external force $f(x, \dot{x}, t)$. The external force may be a function of displacement, velocity, and time. Based on Newton method, the force proportional to acceleration, ma, is always equal to the sum of all the other forces.

$$ma = -c\dot{x} - kx + f(x, \dot{x}, t) \tag{2.27}$$

Example 39 ★ Inverted pendulum and negative stiffness. Inverted pendulum is a good system to study linearization, negative stiffness, and determining equilibrium configurations from minimization of potential energy.

Figure 2.7a illustrates an inverted pendulum with a tip mass m and a length l. The pendulum is supported by two identical springs attached to point B at a distance $a < l$ from the pivot A. A free body diagram of the pendulum is shown in Fig. 2.7b. The equation of motion may be found by writing Euler equation about A:

$$\sum M_A = I_A \ddot{\theta} \tag{2.28}$$

$$mg \, (l \sin \theta) - 2ka\theta \, (a \cos \theta) = ml^2 \ddot{\theta} \tag{2.29}$$

To derive Eq. (2.29), we assumed that the springs are long enough to remain almost straight when the pendulum oscillates. Rearrangement and assuming a very small θ show that the nonlinear equation of motion (2.29) can be approximated by a linear equation,

$$ml^2 \ddot{\theta} + \left(mgl - 2ka^2 \right) \theta = 0 \tag{2.30}$$

which is equivalent to a linear oscillator with an equivalent mass m_e and equivalent stiffness k_e.

$$m_e \ddot{\theta} + k_e \theta = 0 \tag{2.31}$$

$$m_e = ml^2 \qquad k_e = mgl - 2ka^2 \tag{2.32}$$

The potential energy P of the inverted pendulum has a zero value at $\theta = 0$.

$$P = -mgl \, (1 - \cos \theta) + ka^2 \theta^2 \tag{2.33}$$

If θ is very small, the potential energy is approximately equal to the following equation,

$$P \approx -\frac{1}{2} mgl\theta^2 + ka^2 \theta^2 \tag{2.34}$$

because

$$\cos \theta \approx 1 - \frac{1}{2}\theta^2 + O\left(\theta^4\right) \tag{2.35}$$

To find the equilibrium positions of the system, we solve the equation $\partial P / \partial \theta = 0$ for any possible θ.

$$\frac{\partial P}{\partial \theta} = -2mgl\theta + 2ka^2 \theta = 0 \tag{2.36}$$

The solution of the equation is:

$$\theta = 0 \tag{2.37}$$

which shows that the upright vertical position is the only equilibrium of the inverted pendulum as long as θ is very small. However, if

$$mgl = ka^2 \tag{2.38}$$

then any θ around $\theta = 0$ would be an equilibrium position, and hence, the inverted pendulum would have an infinity of equilibria.

The second derivative of the potential energy

$$\frac{\partial^2 P}{\partial x^2} = -2mgl + 2ka^2 \tag{2.39}$$

indicates that the equilibrium position $\theta = 0$ is stable if:

$$ka^2 > mgl \tag{2.40}$$

A stable equilibrium pulls the system back if it deviates from the equilibrium, while an unstable equilibrium repels the system. Vibration happens around stable equilibria.

This example also indicates the fact that having a negative equivalent stiffness $k_e = mgl - 2ka^2$ is possible by geometric arrangement of mechanical components of a vibrating system.

Example 40 ★ Force function in equation of motion. In Newtonian mechanics, force may only be a function of time t, position $\mathbf{r}$, and velocity $\dot{\mathbf{r}}$. Moreover, forces and Newton equation of motion are superposable. Here is a mathematical explanation.

Qualitatively, force is whatever that changes the motion, and quantitatively, force is equal to mass times acceleration. Mathematically, the equation of motion provides us with a vectorial second-order differential equation.

$$m\ddot{\mathbf{r}} = \mathbf{F}\left(\dot{\mathbf{r}}, \mathbf{r}, t\right) \tag{2.41}$$

We assume that the force function may generally be a function of time t, position $\mathbf{r}$, and velocity $\dot{\mathbf{r}}$. In other words, the Newton equation of motion is correct as long as we can show that the force is only a function of $\dot{\mathbf{r}}, \mathbf{r}, t$.

If there is a force that depends on the acceleration, jerk, or other variables that cannot be reduced to $\dot{\mathbf{r}}, \mathbf{r}, t$, the system is not Newtonian, and we do not know the equation of motion.

$$\mathbf{F}\left(\mathbf{r}, \dot{\mathbf{r}}, \ddot{\mathbf{r}}, \dddot{\mathbf{r}}, \cdots, t\right) \neq m\ddot{\mathbf{r}} \tag{2.42}$$

In Newtonian mechanics, we assume that force can only be a function of $\dot{\mathbf{r}}, \mathbf{r}, t$ and nothing else. In real world, however, force may be a function of everything: however, we always ignore any other variables than $\dot{\mathbf{r}}, \mathbf{r}, t$.

Because Eq. (2.41) is a linear equation for force $\mathbf{F}$, it accepts the superposition principle. When a mass m is affected by several forces $\mathbf{F}_1$, $\mathbf{F}_2$, $\mathbf{F}_3$, $\cdots$, we may calculate their summation vectorially and apply the resultant force on m.

$$\mathbf{F} = \mathbf{F}_1 + \mathbf{F}_2 + \mathbf{F}_3 + \cdots \tag{2.43}$$

So, if a force $\mathbf{F}_1$ provides us with acceleration $\ddot{\mathbf{r}}_1$, and $\mathbf{F}_2$ provides us with $\ddot{\mathbf{r}}_2$,

$$m\ddot{\mathbf{r}}_1 = \mathbf{F}_1 \qquad m\ddot{\mathbf{r}}_2 = \mathbf{F}_2 \tag{2.44}$$

then the resultant force $\mathbf{F}_3 = \mathbf{F}_1 + \mathbf{F}_2$ provides an acceleration $\ddot{\mathbf{r}}_3$ such that:

$$\ddot{\mathbf{r}}_3 = \ddot{\mathbf{r}}_1 + \ddot{\mathbf{r}}_2 \tag{2.45}$$

To examine if the Newton equation of motion is not correct when the force is not only a function of $\dot{\mathbf{r}}, \mathbf{r}, t$, let us assume that a particle with mass m is under two acceleration-dependent forces $F_1(\ddot{x})$ and $F_2(\ddot{x})$ on x-axis.

$$m\ddot{x}_1 = F_1(\ddot{x}_1) \qquad m\ddot{x}_2 = F_2(\ddot{x}_2) \tag{2.46}$$

The acceleration of m under the action of both forces would be $\ddot{x}_3$

$$m\ddot{x}_3 = F_1(\ddot{x}_3) + F_2(\ddot{x}_3) \tag{2.47}$$

however, we must have

$$\ddot{x}_3 = \ddot{x}_1 + \ddot{x}_2 \tag{2.48}$$

but we do have a violation of superposition of Newton equation of motion.

$$m(\ddot{x}_1 + \ddot{x}_2) = F_1(\ddot{x}_1 + \ddot{x}_2) + F_2(\ddot{x}_1 + \ddot{x}_2)$$
$$\neq F_1(\ddot{x}_1) + F_2(\ddot{x}_2) \tag{2.49}$$

2.2 Energy Method

In Newtonian mechanics, the acting forces on a system of bodies can be divided into *internal* and *external forces*. Internal forces are acting between bodies of the system, and external forces are acting from outside of the system. External forces and moments are called *load*. Principle of work and energy is a strong method to derive equations of motion of energy-conserved dynamic system. The equations we find for such systems are first-order differential equations, much simpler to solve.

2.2.1   Force System

The resultant of acting forces and moments on a rigid body is called a *force system*. The *resultant* or *total force* **F** is the vectorial sum of all the external forces acting on the body, and the *resultant* or *total moment* **M** is the vectorial sum of all the moments of the external forces about a point, such as the origin of a coordinate frame.

$$\mathbf{F} = \sum_i \mathbf{F}_i \qquad \mathbf{M} = \sum_i \mathbf{M}_i \tag{2.50}$$

The moment of a force **F** an acting on a point P at $\mathbf{r}_P$ is:

$$\mathbf{M} = \mathbf{r}_P \times \mathbf{F} \tag{2.51}$$

The moment of the force **F**, about a point Q at $\mathbf{r}_Q$, is:

$$\mathbf{M}_Q = (\mathbf{r}_P - \mathbf{r}_Q) \times \mathbf{F} \tag{2.52}$$

The *moment of the force* about a directional line l with unit vector $\hat{u}$ passing through the origin is:

$$\mathbf{M}_l = \hat{u} \cdot (\mathbf{r}_P \times \mathbf{F}) \tag{2.53}$$

The moment of a force may also be called *torque* or *moment* although they are conceptually different.

The effect of a force system is equivalent to the effect of the resultant force and resultant moment of the force system. Any two force systems are equivalent if their resultant forces and resultant moments are equal. If the resultant force of a force system is zero, the resultant moment of the force system will be independent of the origin of the coordinate frame. Such a resultant moment is called a *couple*.

When a force system is reduced to a resultant $\mathbf{F}_P$ and $\mathbf{M}_P$ with respect to a reference point P, we may change the reference point to another point Q and find the new resultants as:

$$\mathbf{F}_Q = \mathbf{F}_P \tag{2.54}$$

$$\mathbf{M}_Q = \mathbf{M}_P + (\mathbf{r}_P - \mathbf{r}_Q) \times \mathbf{F}_P = \mathbf{M}_P + {}_Q\mathbf{r}_P \times \mathbf{F}_P \tag{2.55}$$

The application of a force system is by *Newton's second* and *third laws of motion*. The second law of motion, also called the *Newton's equation of motion*, states that the global rate of change of *linear momentum* is proportional to the global *applied force*.

$$^G\mathbf{F} = \frac{^Gd}{dt} {}^G\mathbf{p} = \frac{^Gd}{dt}\left(m\,{}^G\mathbf{v}\right) \tag{2.56}$$

The third Newton's law of motion states that the action and reaction forces acting between two bodies are equal and opposite.

The second law of motion can be expanded to include rotational motions. Hence, the second law of motion also states that the global rate of change of *angular momentum* is proportional to the global *applied moment*.

$$^G\mathbf{M} = \frac{^Gd}{dt} {}^G\mathbf{L} \tag{2.57}$$

Proof Differentiating the angular momentum (2.61) shows that

$$\frac{^Gd}{dt} {}^G\mathbf{L} = \frac{^Gd}{dt}(\mathbf{r}_C \times \mathbf{p}) = \left(\frac{^Gd\mathbf{r}_C}{dt} \times \mathbf{p} + \mathbf{r}_C \times \frac{^Gd\mathbf{p}}{dt}\right)$$

$$= {}^G\mathbf{r}_C \times \frac{^Gd\mathbf{p}}{dt} = {}^G\mathbf{r}_C \times {}^G\mathbf{F} = {}^G\mathbf{M} \tag{2.58}$$

∎

2.2.2 Momentum

The *momentum* $\mathbf{p}$ of a moving body is a vector quantity equal to the total mass of the body times the translational velocity of the mass center of the body.

$$\mathbf{p} = m\mathbf{v} \tag{2.59}$$

The momentum $\mathbf{p}$ may also be called the *translational momentum* or *linear momentum*.

Consider a rigid body with momentum $\mathbf{p}$. The *moment of momentum*, $\mathbf{L}$, about a directional line l passing through the origin, is:

$$\mathbf{L}_l = \hat{u} \cdot (\mathbf{r}_C \times \mathbf{p}) \tag{2.60}$$

where $\hat{u}$ is a unit vector indicating the direction of the line and $\mathbf{r}_C$ is the position vector of the mass center C. The moment of momentum about the origin is

$$\mathbf{L} = \mathbf{r}_C \times \mathbf{p} \tag{2.61}$$

The moment of momentum $\mathbf{L}$ is also called *angular momentum*.

2.2.3   Mechanical Energy

Kinetic energy K of a moving body point with mass m at a position $^G\mathbf{r}$, and having a velocity $^G\mathbf{v}$, in the global coordinate frame G is

$$K = \frac{1}{2}m \, ^G\mathbf{v} \cdot \, ^G\mathbf{v} = \frac{1}{2}m \, ^G v^2 \tag{2.62}$$

where G indicates the global coordinate frame in which the velocity vector $\mathbf{v}$ is expressed.

The work $_1W_2$ done by the applied force $^G\mathbf{F}$ on m in moving from spatial point 1 to point 2 on a path, indicated by a vector $^G\mathbf{r}$, is:

$$_1W_2 = \int_1^2 \, ^G\mathbf{F} \cdot d \, ^G\mathbf{r} \tag{2.63}$$

However,

$$\int_1^2 \, ^G\mathbf{F} \cdot d \, ^G\mathbf{r} = m \int_1^2 \, ^G\frac{d}{dt} \, ^G\mathbf{v} \cdot \, ^G\mathbf{v} dt = \frac{1}{2}m \int_1^2 \frac{d}{dt} v^2 dt$$

$$= \frac{1}{2}m \left(v_2^2 - v_1^2 \right) = K_2 - K_1 \tag{2.64}$$

which shows that $_1W_2$ is equal to the difference of the kinetic energy of terminal and initial points.

$$_1W_2 = K_2 - K_1 \tag{2.65}$$

Equation (2.65) is called the *principle of work and energy*. If there is a scalar *potential field function* $P = P(x, y, z)$ such that

$$\mathbf{F} = -\nabla P = -\frac{dP}{d\mathbf{r}} = -\left(\frac{\partial P}{\partial x}\hat{\imath} + \frac{\partial P}{\partial y}\hat{\jmath} + \frac{\partial P}{\partial z}\hat{k} \right) \tag{2.66}$$

then the principle of work and energy (2.65) simplifies to the principle of *conservation of energy*,

$$K_1 + P_1 = K_2 + P_2 \tag{2.67}$$

The value of the potential field function $P = P(x, y, z)$ is the *potential energy* of the system.

The sum of kinetic and potential energies, $E = K + P$, is called the mechanical energy. Mechanical energy of an energy conserved system is a constant of motion, and hence, its time derivative is zero, $\dot{E} = 0$.

Proof The spatial integral of Newton equation of motion is:

$$\int_1^2 \mathbf{F} \cdot d\mathbf{r} = m \int_1^2 \mathbf{a} \cdot d\mathbf{r} \tag{2.68}$$

We can simplify the right-hand side of the integral (2.68) by the change of variable

$$\int_{\mathbf{r}_1}^{\mathbf{r}_2} \mathbf{F} \cdot d\mathbf{r} = m \int_{\mathbf{r}_1}^{\mathbf{r}_2} \mathbf{a} \cdot d\mathbf{r} = m \int_{t_1}^{t_2} \frac{d\mathbf{v}}{dt} \cdot \mathbf{v} dt$$

$$= m \int_{\mathbf{v}_1}^{\mathbf{v}_2} \mathbf{v} \cdot d\mathbf{v} = \frac{1}{2} m \left(\mathbf{v}_2^2 - \mathbf{v}_1^2 \right) = K_2 - K_1 \tag{2.69}$$

The kinetic energy of a point mass m at position $^G\mathbf{r}$ and having a velocity $^G\mathbf{v}$ is defined by (2.62). Whenever the global coordinate frame G is the only involved frame, we may drop the superscript G for simplicity. The work done by the applied force $^G\mathbf{F}$ on m in going from point $\mathbf{r}_1$ to $\mathbf{r}_2$ is defined by (2.63). Hence the spatial integral of equation of motion (2.68) reduces to the principle of work and energy (2.65) which says that the work $_1W_2$ done by the applied force $^G\mathbf{F}$ on m during the displacement $\mathbf{r}_2 - \mathbf{r}_1$ is equal to the difference of the kinetic energy of m.

$$_1W_2 = K_2 - K_1 \tag{2.70}$$

If the force $\mathbf{F}$ is the gradient of a potential function P,

$$\mathbf{F} = -\nabla P \tag{2.71}$$

then $\mathbf{F} \cdot d\mathbf{r}$ in Eq. (2.68) is an exact differential.

$$\int_1^2 \mathbf{F} \cdot d\mathbf{r} = \int_1^2 dP = -(P_2 - P_1) \tag{2.72}$$

$$E = K_1 + P_1 = K_2 + P_2 \tag{2.73}$$

Therefore, the work done by a potential force $\mathbf{F} = -\nabla P$ is independent of the path of motion between $\mathbf{r}_1$ and $\mathbf{r}_2$ and depends only on the value of the potential P at start and end points of the path. The function P is the potential energy, Eq. (2.73) is called the principle of conservation of energy, and the force $\mathbf{F} = -\nabla P$ is called a potential, or a conservative force. The kinetic plus potential energy of a dynamic system is called the mechanical energy of the system and is denoted by $E = K + P$. The mechanical energy E is a constant of motion if all the applied forces are conservative. Hence, $\dot{E} = 0$ will make the equations of motion.

A force $\mathbf{F}$ is conservative only if it is the gradient of a stationary scalar function. The components of a conservative force will only be functions of space coordinates.

$$\mathbf{F} = F_x\,(x,\,y,\,z)\,\hat{\imath} + F_y\,(x,\,y,\,z)\,\hat{\jmath} + F_z\,(x,\,y,\,z)\,\hat{k} \qquad (2.74)$$

∎

Example 41 Energy and equation of motion. If a vibrating system is energy conserved, derivative of mechanical energy will provide us with the equation of motion.

Whenever there is no loss of energy in a mechanical vibrating system, the sum of kinetic and potential energies is a constant of motion.

$$E = K + P = const. \qquad (2.75)$$

A system with constant energy is called a conservative system. The time derivative of a constant of motion must be zero at all time.

Consider a mass-spring system which is a conservative system with the total mechanical energy E.

$$E = \frac{1}{2}m\dot{x}^2 + \frac{1}{2}kx^2 \qquad (2.76)$$

Having a zero rate of energy,

$$\dot{E} = m\dot{x}\ddot{x} + kx\dot{x} = \dot{x}\,(m\ddot{x} + kx) = 0 \qquad (2.77)$$

and knowing that $\dot{x}$ cannot be zero at all times, provides us with the equation of motion as:

$$m\ddot{x} + kx = 0 \qquad (2.78)$$

Example 42 Energy and multi DOF systems. Here is a conservative system whose equation of motion can be found by principle of conservation of energy.

We may use energy method and determine the equations of motion of multi DOF conservative systems. Consider the 3 DOF system in Fig. 2.8 whose mechanical energy is:

$$E = K + P = \frac{1}{2}m_1\dot{x}_1^2 + \frac{1}{2}m_2\dot{x}_2^2 + \frac{1}{2}m_3\dot{x}_3^2$$
$$+ \frac{1}{2}k_1x_1^2 + \frac{1}{2}k_2\,(x_1 - x_2)^2 + \frac{1}{2}k_3\,(x_2 - x_3)^2 + \frac{1}{2}k_4x_3^2 \qquad (2.79)$$

Because the system has 3 degrees of freedom, we must find three equations of motions. To find the first equation of motion associated to x_1, we assume x_2 and x_3 and their time rates are constant and take a time directive from E.

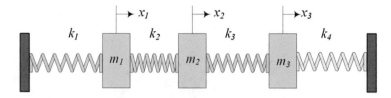

Fig. 2.8 A multi DOF conservative vibrating system

$$\dot{E} = m_1 \dot{x}_1 \ddot{x}_1 + k_1 x_1 \dot{x}_1 + k_2 (x_1 - x_2) \dot{x}_1 = 0 \tag{2.80}$$

Because $\dot{x}_1$ cannot be zero at all times, we may factor $\dot{x}_1$ out and find the first equation of motion.

$$m_1 \ddot{x}_1 + k_1 x_1 + k_2 (x_1 - x_2) = 0 \tag{2.81}$$

To find the second equation of motion associated to x_2, we assume that x_1 and x_3 and their time rates are constant and take a time directive of E,

$$\dot{E} = m_2 \dot{x}_2 \ddot{x}_2 - k_2 (x_1 - x_2) \dot{x}_2 + k_3 (x_2 - x_3) \dot{x}_2 = 0 \tag{2.82}$$

which provides us with the second equation of motion.

$$m_2 \ddot{x}_2 - k_2 (x_1 - x_2) + k_3 (x_2 - x_3) = 0 \tag{2.83}$$

To find the third equation of motion associated to x_2, we assume that x_1 and x_3 and their time rates are constant and we take the time directive of E,

$$\dot{E} = m_3 \dot{x}_3 \ddot{x}_3 - k_3 (x_2 - x_3) \dot{x}_3 + k_4 x_3 \dot{x}_3 = 0 \tag{2.84}$$

to find the third equation of motion.

$$m_3 \ddot{x}_3 - k_3 (x_2 - x_3) + k_4 x_3 = 0 \tag{2.85}$$

We may set up the equations in a matrix form.

$$\begin{bmatrix} m_1 & 0 & 0 \\ 0 & m_2 & 0 \\ 0 & 0 & m_3 \end{bmatrix} \begin{bmatrix} \ddot{x}_1 \\ \ddot{x}_2 \\ \ddot{x}_3 \end{bmatrix} + \begin{bmatrix} k_1 + k_2 & -k_2 & 0 \\ -k_2 & k_2 + k_3 & -k_3 \\ 0 & -k_3 & k_3 + k_4 \end{bmatrix} \begin{bmatrix} x_1 \\ x_2 \\ x_3 \end{bmatrix} = 0 \tag{2.86}$$

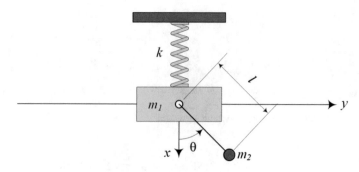

Fig. 2.9 A two DOF conservative nonlinear vibrating system

Alternatively we may take a time derivative of the total mechanical energy E.

$$\dot{E} = m_1\dot{x}_1\ddot{x}_1 + m_2\dot{x}_2\ddot{x}_2 + m_3\dot{x}_3\ddot{x}_3 + k_1x_1\dot{x}_1 + k_2(x_1 - x_2)\dot{x}_1$$
$$\qquad - k_2(x_1 - x_2)\dot{x}_2 + k_3(x_2 - x_3)\dot{x}_2 - k_3(x_2 - x_3)\dot{x}_3 + k_4x_3\dot{x}_3$$
$$= (m_1\dot{x}_1 + k_1x_1 + k_2(x_1 - x_2))\dot{x}_1$$
$$\qquad + (m_2\ddot{x}_2 - k_2(x_1 - x_2) + k_3(x_2 - x_3))\dot{x}_2$$
$$\qquad + (m_3\dot{x}_3 - k_3(x_2 - x_3) + k_4x_3)\dot{x}_3 \tag{2.87}$$

The variables $\dot{x}_1$, $\dot{x}_2$, $\dot{x}_3$ are independent, and because $\dot{x}_1$, $\dot{x}_2$, $\dot{x}_3$ cannot always be zero, their coefficients must be zero independently. Those coefficients will provide us with the same equations of motion.

$$m_1\ddot{x}_1 + k_1x_1 + k_2(x_1 - x_2) = 0 \tag{2.88}$$
$$m_2\ddot{x}_2 - k_2(x_1 - x_2) + k_3(x_2 - x_3) = 0 \tag{2.89}$$
$$m_3\ddot{x}_3 - k_3(x_2 - x_3) + k_4x_3 = 0 \tag{2.90}$$

Example 43 Energy and nonlinear multi DOF systems. The energy method to derive the equation of motion work on all conservative systems regardless of linearity or nonlinearity.

Figure 2.9 illustrates a two DOF nonlinear conservative system. The coordinates of m_1 and m_2 are:

$$x_1 = x \qquad\qquad y_1 = 0$$
$$x_2 = x + l\cos\theta \qquad y_2 = l\sin\theta \tag{2.91}$$

Their time rate are:

$$\dot{x}_1 = \dot{x} \qquad\qquad\qquad \dot{y}_1 = 0$$

$$\dot{x}_2 = \dot{x} - l\dot{\theta} \sin\theta \qquad y_2 = l\dot{\theta} \cos\theta \tag{2.92}$$

Hence, their velocity square, which will contribute in kinetic energy of the system, is:

$$v_1^2 = \dot{x}_1^2 + \dot{y}_1^2 = \dot{x}^2$$
$$v_2^2 = \dot{x}_2^2 + \dot{y}_2^2 = \dot{x}^2 + l^2\dot{\theta}^2 - 2l\dot{x}\dot{\theta}\sin\theta$$

The potential energies of the system are due to length change in spring and height change of m_2. Therefore, the kinetic and potential energies of the system are:

$$K = \frac{1}{2}m_1 v_1^2 + \frac{1}{2}m_2 v_2^2$$
$$= \frac{1}{2}m_1\dot{x}^2 + \frac{1}{2}m_2\left(\dot{x}^2 + l^2\dot{\theta}^2 - 2l\dot{x}\dot{\theta}\sin\theta\right) \tag{2.93}$$

$$P = \frac{1}{2}kx^2 - m_2 g x_2 = \frac{1}{2}kx^2 - m_2 g\left(x + l\cos\theta\right) \tag{2.94}$$

$$E = K + P = \frac{1}{2}m_1\dot{x}^2 + \frac{1}{2}m_2\left(\dot{x}^2 + l^2\dot{\theta}^2 - 2l\dot{x}\dot{\theta}\sin\theta\right)$$
$$+\frac{1}{2}kx^2 - m_2 g\left(x + l\cos\theta\right) \tag{2.95}$$

We assumed that the motionless hanging down position is the equilibrium configuration of the system and that the gravitational energy is zero at the level of m_1 at the equilibrium.

The system has two DOF with generalized coordinates x and θ. Let us take a time derivative of mechanical energy $E = K + P$.

$$\frac{dE}{dt} = (m_1 + m_2)\dot{x}\ddot{x} + m_2 l^2\dot{\theta}\ddot{\theta} - m_2 l\ddot{x}\dot{\theta}\sin\theta - m_2 l\dot{x}\ddot{\theta}\sin\theta$$
$$-m_2 l\dot{x}\dot{\theta}^2\cos\theta + kx\dot{x} - m_2 g\dot{x} - m_2 g l\dot{\theta}\sin\theta = 0 \tag{2.96}$$

Factoring out $\dot{x}$ and $\dot{\theta}$ makes $\dot{E}$ to be of two terms.

$$\left((m_1 + m_2)\ddot{x} - m_2 l\ddot{\theta}\sin\theta + kx - m_2 g - m_2 l\dot{\theta}^2\cos\theta\right)\dot{x}$$
$$+ \left(m_2 l^2\ddot{\theta} - m_2 l\ddot{x}\sin\theta - m_2 g l\sin\theta\right)\dot{\theta} = 0 \tag{2.97}$$

The variables $\dot{x}$ and $\dot{\theta}$ are independent, and $\dot{x}$ and $\dot{\theta}$ cannot always be zero; their coefficients must be zero independently. The coefficients are the equations of motion of the system.

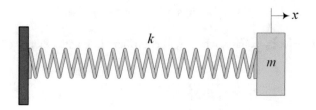

Fig. 2.10 A mass-spring system

$$(m_1 + m_2)\ddot{x} - m_2 l\ddot{\theta}\sin\theta + kx - m_2 g - m_2 l\dot{\theta}^2\cos\theta = 0 \qquad (2.98)$$

$$m_2 l^2\ddot{\theta} - m_2 l\ddot{x}\sin\theta - m_2 gl\sin\theta = 0 \qquad (2.99)$$

Example 44 Maximum energy and frequency of vibrations. The maximum of potential and kinetic energies of a conserved system is equal. This information may be used to calculate the time rate of their exchange which is called natural frequency of the system.

Mechanical vibrations is a continuous exchange of energy between kinetic and potential. If there is no waste of energy, their maximum values must be equal. Consider the mass-spring system of Fig. 2.10. The harmonic motion, kinetic energy K, and potential energy P of the system are:

$$x = X\sin\omega t \qquad (2.100)$$

$$K = \frac{1}{2}m\dot{x}^2 = \frac{1}{2}mX^2\omega^2\cos^2\omega t \qquad (2.101)$$

$$P = \frac{1}{2}kx^2 = \frac{1}{2}kX^2\sin^2\omega t \qquad (2.102)$$

Equating the maximum K and P

$$\frac{1}{2}mX^2\omega^2 = \frac{1}{2}kX^2 \qquad (2.103)$$

provides us with the frequency of vibrations ω.

$$\omega^2 = \frac{k}{m} \qquad (2.104)$$

Example 45 ★ Falling wheel. Work-energy principle is a very useful rule to solve many dynamic problems. Here is an example how to use the principle.

Figure 2.11 illustrates a turning wheel over a cylindrical hill. We may use the conservation of mechanical energy to find the angle at which the wheel leaves the hill. Initially, the wheel is at point A. We assume the initial kinetic energy K and potential energy P are zero. When the wheel is turning over the hill, its angular velocity, ω, is:

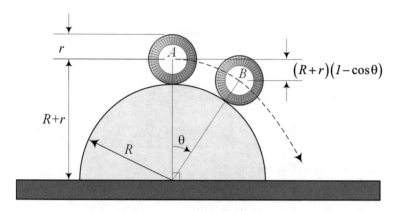

Fig. 2.11 A wheel turning, without slip, over a cylindrical hill

$$\omega = \frac{v}{r} \tag{2.105}$$

where v is the speed at the center of the wheel. At any other point B, the wheel gains some kinetic energy and loses some potential energy. At a certain angle, where the normal component of the weight cannot provide more centripetal force, the wheel separates from the surface.

$$mg \cos \theta = \frac{mv^2}{R+r} \tag{2.106}$$

Employing the conservation of energy yields:

$$K_A + P_A = K_B + P_B \tag{2.107}$$

Using I_C for the mass moment for the wheel about its center, the kinetic and potential energies at the separation point B are:

$$K_B = \frac{1}{2}mv^2 + \frac{1}{2}I_C\omega^2 \tag{2.108}$$

$$P_B = -mg\,(R+r)\,(1 - \cos\theta) \tag{2.109}$$

Therefore, the principle of work-energy will be:

$$\frac{1}{2}mv^2 + \frac{1}{2}I_C\omega^2 = mg\,(R+r)\,(1 - \cos\theta) \tag{2.110}$$

Substituting (2.105) and (2.106) yields:

$$\left(1 + \frac{I_C}{mr^2}\right)(R+r)\,g\cos\theta = 2g\,(R+r)\,(1 - \cos\theta) \tag{2.111}$$

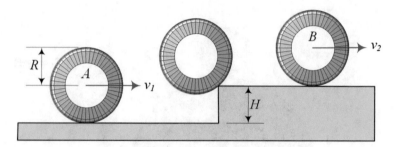

Fig. 2.12 A turning wheel moving up a step

and therefore, the separation angle will be:

$$\theta = \cos^{-1}\frac{2mr^2}{I_C + 3mr^2} \tag{2.112}$$

Let us examine the equation for a disc with a known I_C,

$$I_C = \frac{1}{2}mr^2 \tag{2.113}$$

and find the separation angle.

$$\theta = \arccos\frac{4}{7} \approx 0.96\,\text{rad} \approx 55.15\,\text{deg} \tag{2.114}$$

Example 46 ★ Turning wheel over a step. An application of conservation of energy in determining the speed of a wheel to go over a step.

Figure 2.12 illustrates a wheel of radius R rolling with central speed v to go over a step with height $H < R$. We may use the principle of energy conservation and find the speed of the wheel after getting over the step. Employing the conservation of energy, we have:

$$K_A + P_A = K_B + P_B \tag{2.115}$$

$$\frac{1}{2}mv_1^2 + \frac{1}{2}I_C\omega_1^2 + 0 = \frac{1}{2}mv_2^2 + \frac{1}{2}I_C\omega_2^2 + mgH \tag{2.116}$$

$$\left(m + \frac{I_C}{R^2}\right)v_1^2 = \left(m + \frac{I_C}{R^2}\right)v_2^2 + 2mgH \tag{2.117}$$

and therefore,

$$v_2 = \sqrt{v_1^2 - \frac{2gH}{1 + \dfrac{I_C}{mR^2}}} \tag{2.118}$$

The condition for having a real v_2 is:

$$v_1 > \sqrt{\frac{2gH}{1 + \dfrac{I_C}{mR^2}}} \tag{2.119}$$

The second speed (2.118) and the condition (2.119) for a solid disc with $I_C = mR^2/2$ are:

$$v_2 = \sqrt{v_1^2 - \frac{4}{3}Hg} \qquad v_1 > \sqrt{\frac{4}{3}Hg} \tag{2.120}$$

Example 47 ★ Integral and constant of motion. There are three conservation laws. They will be expressed by equations of the form $f\,(\mathbf{q}, \dot{\mathbf{q}}, t) = c$ and called integral or constant of motion. All of those equations may also be derived by temporal or spatial integrals of equations of motion. Having enough integrals of motion, we will not need to solve the differential equations of motion to determine the solution.

Any equation of the form

$$f\,(\mathbf{q}, \dot{\mathbf{q}}, t) = c \tag{2.121}$$

$$c = f\,(\mathbf{q}_0, \dot{\mathbf{q}}_0, t_0) \tag{2.122}$$

$$q = \begin{bmatrix} q_1 & q_2 & \cdots & q_n \end{bmatrix}^T \tag{2.123}$$

with generalized positions $\mathbf{q}$ and velocities $\dot{\mathbf{q}}$ of a dynamic system, such that its total differential df/dt is zero, is called an integral of motion.

$$\frac{df}{dt} = \sum_{i=1}^{n} \left(\frac{\partial f}{\partial q_i} \dot{q}_i + \frac{\partial f}{\partial \dot{q}_i} \ddot{q}_i \right) + \frac{\partial f}{\partial t} = 0 \tag{2.124}$$

The parameter c, whose value depends on the initial conditions, is called a constant of motion. The maximum number of independent integrals of motion for a dynamic system with n degrees of freedom is $2n$. A constant of motion is a quantity of which the value remains constant during the motion.

Any integral of motion is a result of a conservation principle or a combination of them. There are only three conservation principles for a dynamic system: energy, momentum, and moment of momentum. Every conservation principle is the result of a symmetry in position and time. The conservation of energy indicates the homogeneity of time, the conservation of momentum indicates the homogeneity in position space, and the conservation of moment of momentum indicates the isotropy in position space. In the science of vibrations, the conservation of energy is the most useful and applied integral of motion.

Proof Consider a mechanical system with f_C number of degrees of freedom. Mathematically, the dynamics of the system is expressed by a set of $n = f_C$ second-order differential equations of n unknown generalized coordinates $q_i(t)$, $i = 1, 2, \cdots, n$:

$$\ddot{q}_i = F_i(q_i, \dot{q}_i, t) \qquad i = 1, 2, \cdots, n \tag{2.125}$$

The general solution of the equations contains $2n$ constants of integrals.

$$\dot{q}_i = \dot{q}_i(c_1, c_2, \cdots c_n, t) \qquad i = 1, 2, \cdots, n \tag{2.126}$$

$$q_i = q_i(c_1, c_2, \cdots c_{2n}, t) \qquad i = 1, 2, \cdots, n \tag{2.127}$$

To determine these constants to uniquely identify the motion of the system, it is necessary to know the initial conditions $q_i(t_0)$, $\dot{q}_i(t_0)$, which specify the state of the system at some given instant t_0.

$$c_j = c_j(\mathbf{q}(t_0), \dot{\mathbf{q}}(t_0), t_0) \qquad j = 1, 2, \cdots, 2n \tag{2.128}$$

$$f_j(\mathbf{q}(t), \dot{\mathbf{q}}(t), t) = c_j(\mathbf{q}(t_0), \dot{\mathbf{q}}(t_0), t_0) \tag{2.129}$$

Every one of these functions f_j is an integral of the motion, and every c_i is a constant of the motion. An integral of motion may also be called a *first integral*, and a constant of motion may also be called a *constant of integral*.

When an integral of motion is given,

$$f_1(\mathbf{q}, \dot{\mathbf{q}}, t) = c_1 \tag{2.130}$$

we can substitute one of the equations of motion (2.125) with a first-order equation

$$\dot{q}_1 = f(c_1, q_i, \dot{q}_{i+1}, t) \qquad i = 1, 2, \cdots, n \tag{2.131}$$

and solve a set of $n - 1$ second-order and one first-order differential equations:

$$\begin{cases} \ddot{q}_{i+1} = F_{i+1}(q_i, \dot{q}_i, t) \\ \dot{q}_1 = f(c_1, q_i, \dot{q}_{i+1}, t) \end{cases} \qquad i = 1, 2, \cdots, n \tag{2.132}$$

If there exist $2n$ independent first integrals f_j, $j = 1, 2, \cdots, 2n$, then instead of solving n second-order equations of motion (2.125), we can solve a set of $2n$ algebraic equations

$$f_j(\mathbf{q}, \dot{\mathbf{q}}) = c_j(\mathbf{q}(t_0), \dot{\mathbf{q}}(t_0), t_0) \qquad j = 1, 2, \cdots, 2n \tag{2.133}$$

and determine the n generalized coordinates q_i, $i = 1, 2, \cdots, n$.

$$q_i = q_i(c_1, c_2, \cdots c_{2n}, t) \qquad i = 1, 2, \cdots, n \tag{2.134}$$

Generally speaking, an integral of motion f is a function of generalized coordinates $\mathbf{q}$ and velocities $\dot{\mathbf{q}}$ such that its value remains constant. The value of an integral of motion is the constant of motion c, which can be calculated by substituting the given value of the variables $\mathbf{q}(t_0)$, $\dot{\mathbf{q}}(t_0)$ at the associated time t_0.

∎

Example 48 ★ A mass-spring-damper vibrator. It is a good example to see how constants of motion can be determined and used.

Consider a mass m attached to a spring with stiffness k and a damper with damping c. The equation of motion of the system and its initial conditions are:

$$m\ddot{x} + c\dot{x} + kx = 0 \tag{2.135}$$

$$x(0) = x_0 \qquad \dot{x}(0) = \dot{x}_0 \tag{2.136}$$

Its solution $x = x(t)$ is:

$$x = c_1 \exp(s_1 t) + c_2 \exp(s_2 t) \tag{2.137}$$

$$s_1 = \frac{c - \sqrt{c^2 - 4km}}{-2m} \qquad s_2 = \frac{c + \sqrt{c^2 - 4km}}{-2m} \tag{2.138}$$

Taking a time derivative, we find $\dot{x}$.

$$\dot{x} = c_1 s_1 \exp(s_1 t) + c_2 s_2 \exp(s_2 t) \tag{2.139}$$

Using x and $\dot{x}$, we determine the integrals of motion f_1 and f_2.

$$f_1 = \frac{\dot{x} - x s_2}{(s_1 - s_2)\exp(s_1 t)} = c_1 \tag{2.140}$$

$$f_2 = \frac{\dot{x} - x s_1}{(s_2 - s_1)\exp(s_2 t)} = c_2 \tag{2.141}$$

Because the constants of integral remain constant during the motion, we can calculate their values at any particular time such as $t = 0$.

$$c_1 = \frac{\dot{x}_0 - x_0 s_2}{(s_1 - s_2)} \qquad c_2 = \frac{\dot{x}_0 - x_0 s_1}{(s_2 - s_1)} \tag{2.142}$$

Substituting s_1 and s_2 provides the constants of motion c_1 and c_2 as functions of the initial conditions.

$$c_1 = \frac{\sqrt{c^2 - 4km}\left(cx_0 + x_0\sqrt{c^2 - 4km} + 2m\dot{x}_0\right)}{2\left(c^2 - 4km\right)} \tag{2.143}$$

$$c_2 = \frac{\sqrt{c^2 - 4km}\left(cx_0 - x_0\sqrt{c^2 - 4km} + 2m\dot{x}_0\right)}{2\left(c^2 - 4km\right)} \tag{2.144}$$

In this example, the constants of motion c_1 and c_2 are calculated based on the solution of the differential equation of motion. The best application of constants of motion is to calculate them from conservation laws to avoid solving equations of motion. For example, assume we determined Eqs. (2.140) and (2.141) from some conservation laws. Because the system is one DOF, its equation of motion would be a second-order ordinary differential equations. Therefore, two constants of motion are enough to determine position and velocity of m without any need to derive and solve its equation of motion. Using Eqs. (2.140) and (2.141), we are able to calculate $x = x(t)$ and $\dot{x} = \dot{x}(t)$.

$$x = c_1 \exp(s_1 t) + c_2 \exp(s_2 t) \tag{2.145}$$

$$\dot{x} = c_1 s_1 \exp(s_1 t) + c_2 s_2 \exp(s_2 t) \tag{2.146}$$

The value of c_1 and c_2 must be substituted from (2.143) and (2.144). The value of s_1 and s_2 must also be substituted from (2.138). Then, we will have the solutions $x = x(t)$ and $\dot{x} = \dot{x}(t)$.

Example 49 ★ Constraint and first integral of a pendulum. A pendulum with no friction is a good example of an energy-conserved dynamic system to study kinematic constraints, constants of motion, and work with first-order equation instead of second-order equation of motion.

Figure 2.13a illustrates a planar pendulum. The free body diagram of Fig. 2.13b provides us with two equations of motion:

$$m\ddot{x} = -T\frac{x}{l} \qquad m\ddot{y} = -mg + T\frac{y}{l} \tag{2.147}$$

Eliminating the tension force T, we have one second-order equation of two variables:

$$\ddot{y}x + \ddot{x}y + gx = 0 \tag{2.148}$$

Because the constant length of the connecting bar, we have a constraint equation between x and y.

$$x^2 + y^2 - l^2 = 0 \tag{2.149}$$

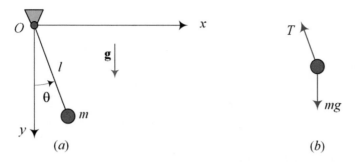

Fig. 2.13 A planar pendulum

Having one constraint indicates that we can express the dynamic of the system by only one generalized coordinate. Choosing θ as the generalized coordinate, we express x and y by θ, writing the equation of motion (2.148) as

$$\ddot{\theta} + \frac{g}{l} \sin \theta = 0 \qquad (2.150)$$

Multiplying the equation by $\dot{\theta}$ and integrating provide us with the integral of energy.

$$f\left(\theta, \dot{\theta}\right) = \frac{1}{2}\dot{\theta}^2 - \frac{g}{l} \cos \theta = E \qquad (2.151)$$

$$E = \frac{1}{2}\dot{\theta}_0^2 - \frac{g}{l} \cos \theta_0 \qquad (2.152)$$

The integral of motion (2.151) is a first-order differential equation:

$$\dot{\theta} = \sqrt{2E + 2\frac{g}{l} \cos \theta} \qquad (2.153)$$

This equation expresses the dynamic of the pendulum upon solution.

Let us assume that θ is too small to approximate the equation of motion as:

$$\ddot{\theta} + \frac{g}{l}\theta = 0 \qquad (2.154)$$

The first integral of this equation is:

$$f\left(\theta, \dot{\theta}\right) = \frac{1}{2}\dot{\theta}^2 - \frac{g}{l}\theta = E \qquad (2.155)$$

$$E = \frac{1}{2}\dot{\theta}_0^2 - \frac{g}{l}\theta_0 \qquad (2.156)$$

that provides us with a separated first-order differential equation.

$$\dot{\theta} = \sqrt{2E + 2\frac{g}{l}\theta} \tag{2.157}$$

Its solution is:

$$t = \int \frac{d\theta}{\sqrt{2E + 2\frac{g}{l}\theta}} = \sqrt{2}\frac{l}{g}\sqrt{\frac{g}{l}\theta + E} - p \tag{2.158}$$

where p is the second constant of motion.

$$p = \frac{l}{g}\dot{\theta}_0 \tag{2.159}$$

Now, let us ignore the energy integral and solve the second-order equation of motion (2.154).

$$\theta = c_1 \cos\sqrt{\frac{g}{l}}t + c_2 \sin\sqrt{\frac{g}{l}}t \tag{2.160}$$

The time derivative of the solution

$$\sqrt{\frac{l}{g}}\dot{\theta} = -c_1 \sin\sqrt{\frac{g}{l}}t + c_2 \cos\sqrt{\frac{g}{l}}t \tag{2.161}$$

can be used to determine the integrals and constants of motion.

$$f_1 = \theta \cos\sqrt{\frac{g}{l}}t - \sqrt{\frac{l}{g}}\dot{\theta} \sin\sqrt{\frac{g}{l}}t \tag{2.162}$$

$$f_2 = \theta \sin\sqrt{\frac{g}{l}}t + \sqrt{\frac{l}{g}}\dot{\theta} \cos\sqrt{\frac{g}{l}}t \tag{2.163}$$

Using the initial conditions $\theta(0) = \theta_0$, $\dot{\theta}(0) = \dot{\theta}_0$, we have:

$$c_1 = \theta_0 \qquad c_2 = \sqrt{\frac{l}{g}}\dot{\theta}_0 \tag{2.164}$$

A second-order equation has two constants of integrals. Therefore, we should be able to express E and p in terms of c_1 and c_2 or vice versa.

$$E = \frac{1}{2}\dot{\theta}_0^2 - \frac{g}{l}\theta_0 = \frac{1}{2}\frac{g}{l}c_2^2 - \frac{g}{l}c_1 \tag{2.165}$$

Fig. 2.14 A falling object on a spring

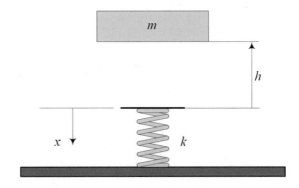

$$p = \frac{l}{g}\dot{\theta}_0 = \frac{l}{g}\sqrt{\frac{g}{l}}c_2 \tag{2.166}$$

$$c_2 = \sqrt{\frac{l}{g}}\dot{\theta}_0 = \frac{g}{l}\sqrt{\frac{l}{g}}p \tag{2.167}$$

$$c_1 = \theta_0 = \frac{1}{2}\frac{g}{l}p^2 - \frac{l}{g}E \tag{2.168}$$

E is the mechanical energy of the pendulum, and p is proportional to its moment of momentum.

Example 50 A falling object on a spring. A falling object experiment can be used to evaluate the value of the gravitational acceleration.

An object with mass m falls from a height h on a linear spring with stiffness k, as is shown in Fig. 2.14. We can determine the maximum compression of the spring using the work-energy principle.

The gravity force mg and the spring force $-kx$ are the acting forces on m. If x_M is the maximum compression of the spring, then $K_2 = K_1 = 0$, and we have

$$_1W_2 = mg(h + x_M) - \int_0^{x_M} kx \, dx = 0 \tag{2.169}$$

$$x_M = \frac{mg}{k} + \sqrt{\left(\frac{mg}{k}\right)^2 + \frac{2mgh}{k}} \tag{2.170}$$

If we put m on the spring, it will deflect statically to

$$x_0 = \frac{mg}{k} \tag{2.171}$$

so we may compare x_M to x_0 and write x_M as

$$x_M = x_0 \left(1 + \sqrt{1 + \frac{2h}{x_0}} \right) \tag{2.172}$$

Considering

$$h = 1\,\text{m} \qquad m = 1\,\text{kg} \qquad k = 1000\,\text{N}\,/\,\text{m} \qquad g = 9.81\,\text{m}\,/\,\text{s}^2 \tag{2.173}$$

we have:

$$x_0 = \frac{mg}{k} = \frac{9.81}{1000} = 0.00981\,\text{m} = 9.81\,\text{mm} \tag{2.174}$$

$$x_M = x_0 \left(1 + \sqrt{1 + \frac{2h}{x_0}} \right) = 0.15022\,\text{m} = 150.22\,\text{mm} \tag{2.175}$$

However, if this experiment is performed on the moon surface, then $g = 1.6\,\text{m}\,/\,\text{s}^2$, and we would have:

$$x_0 = \frac{mg}{k} = \frac{1.6}{1000} = 0.0016\,\text{m} = 1.6\,\text{mm} \tag{2.176}$$

$$x_M = x_0 \left(1 + \sqrt{1 + \frac{2h}{x_0}} \right) = 0.058191\,\text{m} = 58.191\,\text{mm} \tag{2.177}$$

It is theoretically possible to measure g based on such an experiment:

$$g = \frac{2hk}{m \left(\left(\frac{x_M}{x_0} - 1 \right)^2 - 1 \right)} \tag{2.178}$$

The mass of the spring and friction are the main sources of error.

2.3 ★ Rigid Body Dynamics

A rigid body may have up to three translational and three rotational degrees of freedom (DOF). The translational and rotational equations of motion of the rigid body are determined by the Newton-Euler equations. Majority of mechanical vibrating systems are modeled by lumped masses such that every mass has a translational or a rotational degree of freedom. Masses in translational motion may be treated as a point mass, and rotational rigid bodies must be analyzed by one

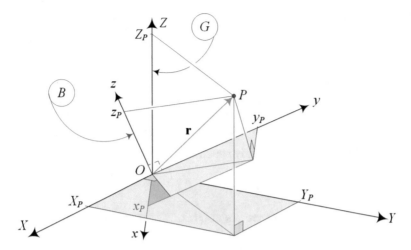

Fig. 2.15 A globally fixed G-frame and a body B-frame with a fixed common origin at O

dimensional Euler equations. However, in this section, kinematics and dynamics of rigid bodies in three-dimensional space will be reviewed. The number of vibrational systems in which we need to apply $3D$ Euler equations are very limited and usually complicated.

Kinematics of rigid bodies is equivalent to analyze relative coordinate frames. Every rigid body will be substituted by a Cartesian orthogonal coordinate frame B_i attached to its mass center C_i. All body frames will be moving relatively in a motionless global coordinate frame G.

2.3.1 ★ Coordinate Frame Transformation

Consider a rotating body coordinate frame $B(Oxyz)$ with respect to a global frame $G(OXYZ)$ about their common origin O as illustrated in Fig. 2.15. The components of any vector $\mathbf{r}$ may be expressed in either frame. There is always a *transformation matrix* $^{G}R_B$ to map the components of $\mathbf{r}$ from the body frame $B(Oxyz)$ to the frames $G(OXYZ)$.

$$^{G}\mathbf{r} = {}^{G}R_B \, {}^{B}\mathbf{r} \tag{2.179}$$

$$^{G}R_B = \begin{bmatrix} \cos(\hat{I}, \hat{\imath}) & \cos(\hat{I}, \hat{\jmath}) & \cos(\hat{I}, \hat{k}) \\ \cos(\hat{J}, \hat{\imath}) & \cos(\hat{J}, \hat{\jmath}) & \cos(\hat{J}, \hat{k}) \\ \cos(\hat{K}, \hat{\imath}) & \cos(\hat{K}, \hat{\jmath}) & \cos(\hat{K}, \hat{k}) \end{bmatrix} \tag{2.180}$$

$$^{B}\mathbf{r} = \begin{bmatrix} x & y & z \end{bmatrix}^{T} \tag{2.181}$$

$$^{G}\mathbf{r} = \begin{bmatrix} X & Y & Z \end{bmatrix}^{T} \tag{2.182}$$

The inverse map from G-frame to B-frame will be performed by an inverse transformation matrix $^B R_G$.

$$^B\mathbf{r} = {}^G R_B^{-1}\, {}^B\mathbf{r} = {}^B R_G\, {}^G\mathbf{r} \tag{2.183}$$

All rotational coordinate transformations are orthogonal; that means their inverse is equal to their transpose and their determinant is equal to one. Because of the matrix orthogonality condition, only three of the nine elements of $^G R_B$ are independent.

Proof Employing the orthogonality condition

$$\mathbf{r} = (\mathbf{r} \cdot \hat{\imath})\hat{\imath} + (\mathbf{r} \cdot \hat{\jmath})\hat{\jmath} + (\mathbf{r} \cdot \hat{k})\hat{k} \tag{2.184}$$

and decomposition of the unit vectors of $G(OXYZ)$ along the axes of $B(Oxyz)$,

$$\hat{I} = (\hat{I} \cdot \hat{\imath})\hat{\imath} + (\hat{I} \cdot \hat{\jmath})\hat{\jmath} + (\hat{I} \cdot \hat{k})\hat{k} \tag{2.185}$$

$$\hat{J} = (\hat{J} \cdot \hat{\imath})\hat{\imath} + (\hat{J} \cdot \hat{\jmath})\hat{\jmath} + (\hat{J} \cdot \hat{k})\hat{k} \tag{2.186}$$

$$\hat{K} = (\hat{K} \cdot \hat{\imath})\hat{\imath} + (\hat{K} \cdot \hat{\jmath})\hat{\jmath} + (\hat{K} \cdot \hat{k})\hat{k} \tag{2.187}$$

introduces the transformation matrix $^G R_B$ to map the local axes to the global axes.

$$\begin{bmatrix} \hat{I} \\ \hat{J} \\ \hat{K} \end{bmatrix} = \begin{bmatrix} \hat{I}\cdot\hat{\imath} & \hat{I}\cdot\hat{\jmath} & \hat{I}\cdot\hat{k} \\ \hat{J}\cdot\hat{\imath} & \hat{J}\cdot\hat{\jmath} & \hat{J}\cdot\hat{k} \\ \hat{K}\cdot\hat{\imath} & \hat{K}\cdot\hat{\jmath} & \hat{K}\cdot\hat{k} \end{bmatrix} \begin{bmatrix} \hat{\imath} \\ \hat{\jmath} \\ \hat{k} \end{bmatrix} = {}^G R_B \begin{bmatrix} \hat{\imath} \\ \hat{\jmath} \\ \hat{k} \end{bmatrix} \tag{2.188}$$

$$^G R_B = \begin{bmatrix} \cos(\hat{I},\hat{\imath}) & \cos(\hat{I},\hat{\jmath}) & \cos(\hat{I},\hat{k}) \\ \cos(\hat{J},\hat{\imath}) & \cos(\hat{J},\hat{\jmath}) & \cos(\hat{J},\hat{k}) \\ \cos(\hat{K},\hat{\imath}) & \cos(\hat{K},\hat{\jmath}) & \cos(\hat{K},\hat{k}) \end{bmatrix} \tag{2.189}$$

Each column of $^G R_B$ is the decomposition of a unit vector of the local frame $B(Oxyz)$ in the global frame $G(OXYZ)$.

$$^G R_B = \begin{bmatrix} {}^G\hat{\imath} & {}^G\hat{\jmath} & {}^G\hat{k} \end{bmatrix} \tag{2.190}$$

Similarly, each row of $^G R_B$ is decomposition of a unit vector of the global frame $G(OXYZ)$ in the local frame $B(Oxyz)$.

$$^G R_B = \begin{bmatrix} {}^B\hat{I}^T \\ {}^B\hat{J}^T \\ {}^B\hat{K}^T \end{bmatrix} \tag{2.191}$$

Therefore, the elements of $^G R_B$ are directional cosines of the axes of $G(OXYZ)$ in $B(Oxyz)$ or B in G. This set of nine directional cosines completely specifies the orientation of $B(Oxyz)$ in $G(OXYZ)$ and can be used to map the coordinates of any point (x, y, z) to its corresponding global coordinates (X, Y, Z).

Alternatively, using the method of unit-vector decomposition to develop the matrix $^B R_G$ yields:

$$^B \mathbf{r} = {}^B R_G {}^G \mathbf{r} = {}^G R_B^{-1} {}^G \mathbf{r} \tag{2.192}$$

$$^B R_G = \begin{bmatrix} \hat{\imath} \cdot \hat{I} & \hat{\imath} \cdot \hat{J} & \hat{\imath} \cdot \hat{K} \\ \hat{\jmath} \cdot \hat{I} & \hat{\jmath} \cdot \hat{J} & \hat{\jmath} \cdot \hat{K} \\ \hat{k} \cdot \hat{I} & \hat{k} \cdot \hat{J} & \hat{k} \cdot \hat{K} \end{bmatrix}$$

$$= \begin{bmatrix} \cos(\hat{\imath}, \hat{I}) & \cos(\hat{\imath}, \hat{J}) & \cos(\hat{\imath}, \hat{K}) \\ \cos(\hat{\jmath}, \hat{I}) & \cos(\hat{\jmath}, \hat{J}) & \cos(\hat{\jmath}, \hat{K}) \\ \cos(\hat{k}, \hat{I}) & \cos(\hat{k}, \hat{J}) & \cos(\hat{k}, \hat{K}) \end{bmatrix} \tag{2.193}$$

It shows that the inverse of a rotation transformation matrix is equal to the transpose of the matrix.

$$^G R_B^{-1} = {}^G R_B^T \tag{2.194}$$

$$^G R_B \cdot {}^G R_B^T = \mathbf{I} \tag{2.195}$$

A matrix with condition (2.194) is called an *orthogonal matrix*. Orthogonality of $^G R_B$ comes from the fact that it maps an orthogonal coordinate frame to another orthogonal coordinate frame.

An orthogonal transformation matrix $^G R_B$ has only three *independent* elements. The constraint equations among the elements of $^G R_B$ will be found by applying the matrix orthogonality condition (2.194):

$$\begin{bmatrix} r_{11} & r_{12} & r_{13} \\ r_{21} & r_{22} & r_{23} \\ r_{31} & r_{32} & r_{33} \end{bmatrix} \begin{bmatrix} r_{11} & r_{21} & r_{31} \\ r_{12} & r_{22} & r_{32} \\ r_{13} & r_{23} & r_{33} \end{bmatrix} = \begin{bmatrix} 1 & 0 & 0 \\ 0 & 1 & 0 \\ 0 & 0 & 1 \end{bmatrix} \tag{2.196}$$

Therefore, the inner product of any two different rows of $^G R_B$ is zero, and the inner product of any row of $^G R_B$ by itself is unity.

$$r_{11}^2 + r_{12}^2 + r_{13}^2 = 1$$

$$r_{21}^2 + r_{22}^2 + r_{23}^2 = 1$$

$$r_{31}^2 + r_{32}^2 + r_{33}^2 = 1$$

$$r_{11}r_{21} + r_{12}r_{22} + r_{13}r_{23} = 0 \tag{2.197}$$

$$r_{11}r_{31} + r_{12}r_{32} + r_{13}r_{33} = 0$$

$$r_{21}r_{31} + r_{22}r_{32} + r_{23}r_{33} = 0$$

These relations are also true for columns of $^G R_B$ and, evidently, for rows and columns of $^B R_G$. The orthogonality condition can be summarized by the equation,

$$\sum_{i=1}^{3} r_{ij} r_{ik} = \delta_{jk} \qquad j, k = 1, 2, 3 \tag{2.198}$$

where r_{ij} is the element of row i and column j of the transformation matrix $^G R_B$ and δ_{jk} is the Kronecker delta δ_{ij}.

$$\delta_{ij} = \delta_{ji} = \begin{cases} 1 & i = j \\ 0 & i \neq j \end{cases} \tag{2.199}$$

Equation (2.198) provides us with six independent relations that must be satisfied by the nine directional cosines of $^G R_B$. Therefore, there are only three independent directional cosines. The independent elements of the matrix $^G R_B$ cannot be in the same row or column or any diagonal.

The determinant of a transformation matrix is equal to unity,

$$\left| ^G R_B \right| = 1 \tag{2.200}$$

because of Eq. (2.195).

$$\left| ^G R_B \cdot {^G R_B^T} \right| = \left| ^G R_B \right| \cdot \left| ^G R_B^T \right| = \left| ^G R_B \right| \cdot \left| ^G R_B \right| = \left| ^G R_B \right|^2 = 1 \tag{2.201}$$

Using linear algebra and column vectors $^G\hat{\imath}$, $^G\hat{\jmath}$, $^G\hat{k}$ of $^G R_B$, we know that

$$\left| ^G R_B \right| = {^G\hat{\imath}} \cdot \left({^G\hat{\jmath}} \times {^G\hat{k}} \right) \tag{2.202}$$

and because the coordinate system is right handed, we have $^G\hat{\jmath} \times {^G\hat{k}} = {^G\hat{\imath}}$ and therefore

$$\left| ^G R_B \right| = {^G\hat{\imath}^T} \cdot {^G\hat{\imath}} = +1 \tag{2.203}$$

∎

Example 51 Global position using $^B\mathbf{r}$ and $^B R_G$. A numerical example to show how to use a rotation transformation matrix.

The position vector $\mathbf{r}$ of a point P may be expressed in either $G\,(OXYZ)$ or $B\,(Oxyz)$-frame. If

$$^B\mathbf{r} = 10\hat{\imath} - 5\hat{\jmath} + 15\hat{k}$$

and the transformation matrix $^B R_G$ to map $^G \mathbf{r}$ to $^B \mathbf{r}$ is:

$$^B R_G = \begin{bmatrix} 0.866 & 0 & 0.5 \\ -0.353 & 0.707 & 0.612 \\ 0.353 & 0.707 & -0.612 \end{bmatrix} \tag{2.204}$$

then the components of $^G \mathbf{r}$ in $G\,(OXYZ)$ would be:

$$^G \mathbf{r} = {}^G R_B \, {}^B \mathbf{r} = {}^B R_G^T \, {}^B \mathbf{r} = \begin{bmatrix} 15.72 \\ 7.07 \\ -7.24 \end{bmatrix} \tag{2.205}$$

Expressing a vector in different frames utilizing rotation matrices does not affect the length and direction properties of the vector. Therefore, the length of a vector is an invariant property.

$$|\mathbf{r}| = \left|{}^G \mathbf{r}\right| = \left|{}^B \mathbf{r}\right| \tag{2.206}$$

The length invariant property can be shown as:

$$\begin{aligned} |\mathbf{r}|^2 &= {}^G \mathbf{r}^T \, {}^G \mathbf{r} = \left[{}^G R_B \, {}^B \mathbf{r}\right]^T \, {}^G R_B \, {}^B \mathbf{r} = {}^B \mathbf{r}^T \, {}^G R_B^T \, {}^G R_B \, {}^B \mathbf{r} \\ &= {}^B \mathbf{r}^T \, {}^B \mathbf{r} \end{aligned} \tag{2.207}$$

Example 52 Two-point transformation matrix. How to discover $^G R_B$ from information of two body points.

The global position vectors of two points P_1 and P_2 of a rigid body B are given.

$$^G \mathbf{r}_{P_1} = \begin{bmatrix} 1.077 \\ 1.365 \\ 2.666 \end{bmatrix} \qquad {}^G \mathbf{r}_{P_2} = \begin{bmatrix} -0.473 \\ 2.239 \\ -0.959 \end{bmatrix} \tag{2.208}$$

The origin of the body $B\,(Oxyz)$ is fixed on the origin of $G\,(OXYZ)$, and the points P_1 and P_2 are lying on the local x and y-axis, respectively.

To find $^G R_B$, we use the local unit vectors $^G \hat{\imath}$ and $^G \hat{\jmath}$,

$$^G \hat{\imath} = \frac{{}^G \mathbf{r}_{P_1}}{\left|{}^G \mathbf{r}_{P_1}\right|} = \begin{bmatrix} 0.338 \\ 0.429 \\ 0.838 \end{bmatrix} \qquad {}^G \hat{\jmath} = \frac{{}^G \mathbf{r}_{P_2}}{\left|{}^G \mathbf{r}_{P_2}\right|} = \begin{bmatrix} -0.191 \\ 0.902 \\ -0.387 \end{bmatrix} \tag{2.209}$$

to obtain $^G \hat{k}$

$$
{}^G \hat{k} = \hat{i} \times \hat{j} = \begin{bmatrix} -0.922 \\ -0.029 \\ 0.387 \end{bmatrix} \tag{2.210}
$$

Hence, the transformation matrix ${}^G R_B$ would be

$$
{}^G R_B = \begin{bmatrix} {}^G \hat{i} & {}^G \hat{j} & {}^G \hat{k} \end{bmatrix} = \begin{bmatrix} 0.338 & -0.191 & -0.922 \\ 0.429 & 0.902 & -0.029 \\ 0.838 & -0.387 & 0.387 \end{bmatrix} \tag{2.211}
$$

Example 53 Multiple rotations about global axes. Multiple rotations are equivalent to matrix multiplication. Here are examples for multiple rotations about global and body axes.

Consider a globally fixed point P at ${}^G \mathbf{r}$.

$$
{}^G \mathbf{r} = \begin{bmatrix} 1 & 2 & 3 \end{bmatrix}^T \tag{2.212}
$$

The body B will turn 45 deg about the X-axis and then 45 deg about the Y-axis. An observer in B will see the point P at ${}^B \mathbf{r}$.

$$
{}^B \mathbf{r} = R_{y,-45} \, R_{x,-45} \, {}^G \mathbf{r} \tag{2.213}
$$

$$
= \begin{bmatrix} \cos\dfrac{-\pi}{4} & 0 & -\sin\dfrac{-\pi}{4} \\ 0 & 1 & 0 \\ \sin\dfrac{-\pi}{4} & 0 & \cos\dfrac{-\pi}{4} \end{bmatrix} \begin{bmatrix} 1 & 0 & 0 \\ 0 & \cos\dfrac{-\pi}{4} & \sin\dfrac{-\pi}{4} \\ 0 & -\sin\dfrac{-\pi}{4} & \cos\dfrac{-\pi}{4} \end{bmatrix} \begin{bmatrix} 1 \\ 2 \\ 3 \end{bmatrix}
$$

$$
= \begin{bmatrix} 0.707 & 0.5 & 0.5 \\ 0 & 0.707 & -0.707 \\ -0.707 & 0.5 & 0.5 \end{bmatrix} \begin{bmatrix} 1 \\ 2 \\ 3 \end{bmatrix} = \begin{bmatrix} 3.207 \\ -0.707 \\ 1.793 \end{bmatrix}
$$

To check the result, let us change the role of B and G. So, the body point at

$$
{}^B \mathbf{r} = \begin{bmatrix} 1 & 2 & 3 \end{bmatrix}^T \tag{2.214}
$$

undergoes a rotation of 45 deg about the x-axis followed by 45 deg about the y-axis. The global coordinates of the point would be

$$
{}^B \mathbf{r} = R_{y,45} \, R_{x,45} \, {}^G \mathbf{r} \tag{2.215}
$$

so

$$
{}^G \mathbf{r} = \begin{bmatrix} R_{y,45} \, R_{x,45} \end{bmatrix}^T \, {}^B \mathbf{r} = R_{x,45}^T \, R_{y,45}^T \, {}^B \mathbf{r} \tag{2.216}
$$

Now consider a globally fixed point P at $^G\mathbf{r}$.

$$^G\mathbf{r} = \begin{bmatrix} 1 & 2 & 3 \end{bmatrix}^T \tag{2.217}$$

The body B will turn 45 deg about the x-axis and then 45 deg about the y-axis. An observer in B will see P at $^B\mathbf{r}$.

$$^B\mathbf{r} = R_{Y,-45}\, R_{X,-45}\, ^G\mathbf{r} \tag{2.218}$$

$$= \begin{bmatrix} \cos\dfrac{-\pi}{4} & 0 & \sin\dfrac{-\pi}{4} \\ 0 & 1 & 0 \\ -\sin\dfrac{-\pi}{4} & 0 & \cos\dfrac{-\pi}{4} \end{bmatrix} \begin{bmatrix} 1 & 0 & 0 \\ 0 & \cos\dfrac{-\pi}{4} & -\sin\dfrac{-\pi}{4} \\ 0 & \sin\dfrac{-\pi}{4} & \cos\dfrac{-\pi}{4} \end{bmatrix} \begin{bmatrix} 1 \\ 2 \\ 3 \end{bmatrix}$$

$$= \begin{bmatrix} 0.707 & 0.5 & -0.5 \\ 0 & 0.707 & 0.707 \\ 0.707 & -0.5 & 0.5 \end{bmatrix} \begin{bmatrix} 1 \\ 2 \\ 3 \end{bmatrix} = \begin{bmatrix} 0.20711 \\ 3.5356 \\ 1.2071 \end{bmatrix}$$

Example 54 Successive rotations about global or body axes. Here is the theory of multiple rotations to be equivalent to one resultant transformation.

After a series of sequential rotations R_1, R_2, R_3, ..., R_n about the global axes, the final global position of a body point P can be found by a single rotation,

$$^G\mathbf{r} = \,^G R_B \,^B\mathbf{r} \tag{2.219}$$

where

$$^G R_B = R_n \cdots R_3\, R_2\, R_1 \tag{2.220}$$

The vectors $^G\mathbf{r}$ and $^B\mathbf{r}$ indicate the position vectors of the point P in the global and local coordinate frames, respectively. The matrix $^G R_B$, which transforms the local coordinates to their corresponding global coordinates, is called the *global rotation matrix*.

Because matrix multiplications do not commute, the sequence of performing rotations is important and indicates the order of rotations.

Proof Consider a body frame B that undergoes two sequential rotations R_1 and R_2 about the global axes. Assume the body coordinate frame B is initially coincident with the global coordinate frame G. The rigid body rotates about a global axis, and the global rotation matrix R_1 gives us the new global coordinate $^G\mathbf{r}_1$ of the body point.

$$^G\mathbf{r}_1 = R_1\, ^B\mathbf{r} \tag{2.221}$$

Before the second rotation, the situation is similar to the one before the first rotation. We put the B-frame aside and assume that a new body coordinate frame B_1 that is coincident with the global frame. Therefore, the new body coordinate would be $^{B_1}\mathbf{r} \equiv {}^{G}\mathbf{r}_1$. The second global rotation matrix R_2 provides us with the new global position $^{G}\mathbf{r}_2$ of the body vector $^{B_1}\mathbf{r}$:

$$^{B_1}\mathbf{r} = R_2 \, {}^{B_1}\mathbf{r} \tag{2.222}$$

Substituting (2.221) into (2.222) shows that

$$^{G}\mathbf{r} = R_2 \, R_1 \, {}^{B}\mathbf{r} \tag{2.223}$$

Following the same procedure we can determine the final global position of a body point after a series of sequential rotations $R_1, R_2, R_3, ..., R_n$ as (2.220). ∎

Now consider a rigid body B with body coordinate frame $B(Oxyz)$ that undergoes a series of sequential rotations $R_1, R_2, R_3, ..., R_n$ about the body axes. Having the final global position vector $^{G}\mathbf{r}$ of a body point P, we can determine its body position vector $^{B}\mathbf{r}$ by

$$^{B}\mathbf{r} = {}^{B}R_G \, {}^{G}\mathbf{r} \tag{2.224}$$

where

$$^{B}R_G = R_n \cdots R_3 R_2 R_1 \tag{2.225}$$

The matrix $^{B}R_G$ is called the *body rotation matrix*, and it maps the global coordinates of body points to their body coordinates.
Proof Assume that the body coordinate frame B was initially coincident with the global coordinate frame G. The rigid body rotates about a body axis, and a body rotation matrix R_1 relates the global coordinates of a body point to the associated body coordinates.

$$^{B}\mathbf{r} = R_1 \, {}^{G}\mathbf{r} \tag{2.226}$$

If we introduce an intermediate space-fixed frame G_1 coincident with the new position of the body coordinate frame, then

$$^{G_1}\mathbf{r} \equiv {}^{B}\mathbf{r} \tag{2.227}$$

and we may give the rigid body a second rotation about a body coordinate axis. Now another proper body rotation matrix R_2 relates the coordinates in the intermediate fixed frame to the corresponding body coordinates.

$$^{B}\mathbf{r} = R_2 \, {}^{G_1}\mathbf{r} \tag{2.228}$$

Hence, to relate the final coordinates of the point, we must first transform its global coordinates to the intermediate fixed frame and then transform to the original body frame. Substituting (2.226) in (2.228) shows that

$$^B\mathbf{r} = R_2\, R_1\, {}^G\mathbf{r} \tag{2.229}$$

Following the same procedure, we can determine the final global position of a body point after a series of sequential rotations $R_1, R_2, R_3, ..., R_n$ as (2.225).

Rotation about the body coordinate axes is conceptually interesting. This is because in a sequence of rotations each rotation is about one of the axes of the local coordinate frame, which has been moved to its new global position during the previous rotation. ■

2.3.2 ★ Velocity Kinematics

Consider a rotating rigid body $B(Oxyz)$ with a fixed point O in a reference frame $G(OXYZ)$, as shown in Fig. 2.16. We express the motion of the body by a time-varying rotation transformation matrix between B and G to transform the instantaneous coordinates of body points to their coordinates in the global frame:

$$^G\mathbf{r}(t) = {}^G R_B(t)\, {}^B\mathbf{r} \tag{2.230}$$

The velocity of a body point in the global frame is

$$
\begin{aligned}
^G\mathbf{v}(t) = {}^G\dot{\mathbf{r}}(t) &= {}^G\dot{R}_B(t)\, {}^B\mathbf{r} \\
&= {}^G\dot{R}_B\, {}^G R_B^T\, {}^G\mathbf{r}(t) = {}_G\tilde{\omega}_B\, {}^G\mathbf{r}(t) = {}_G\boldsymbol{\omega}_B \times {}^G\mathbf{r}(t)
\end{aligned} \tag{2.231}
$$

where ${}_G\boldsymbol{\omega}_B$ is the *angular velocity vector* of B with respect to G. It is equal to a rotation with *angular speed* $\dot{\phi}$ about an *instantaneous axis of rotation* $\hat{u}$:

$$\boldsymbol{\omega} = \begin{bmatrix} \omega_1 \\ \omega_2 \\ \omega_3 \end{bmatrix} = \dot{\phi}\,\hat{u} \tag{2.232}$$

The angular velocity vector is associated with the skew-symmetric *angular velocity matrix* ${}_G\tilde{\omega}_B$,

$$_G\tilde{\omega}_B = \begin{bmatrix} 0 & -\omega_3 & \omega_2 \\ \omega_3 & 0 & -\omega_1 \\ -\omega_2 & \omega_1 & 0 \end{bmatrix} \tag{2.233}$$

Fig. 2.16 A rotating rigid
body $B(Oxyz)$ with a fixed
point O in a global frame
$G(OXYZ)$

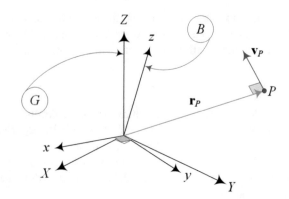

which is equivalent to:

$$_G\tilde{\omega}_B = {}^G\dot{R}_B \, {}^G R_B^T = \dot{\phi}\,\tilde{u} \tag{2.234}$$

A matrix $\tilde{\omega}$ is skew-symmetric if its transpose is equal to its negative.

$$\tilde{\omega}^T = -\tilde{\omega} \tag{2.235}$$

The B-expression of the angular velocity is similarly defined:

$$_G^B\tilde{\omega}_B = {}^G R_B^T \, {}^G\dot{R}_B \tag{2.236}$$

Employing the global and body expressions of the angular velocity of the body
relative to the global coordinate frame, $_G\tilde{\omega}_B$ and $_G^B\tilde{\omega}_B$, we determine the global and
body expressions of the velocity of a body point as

$$_G^G v_P = {}_G^G\omega_B \times {}^G r_P \tag{2.237}$$

$$_G^B v_P = {}_G^B\omega_B \times {}^B r_P \tag{2.238}$$

The G-expression $_G\tilde{\omega}_B$ and B-expression $_G^B\tilde{\omega}_B$ of the angular velocity matrix
can be transformed to each other using the rotation matrix ${}^G R_B$:

$$_G\tilde{\omega}_B = {}^G R_B \, {}_G^B\tilde{\omega}_B \, {}^G R_B^T \tag{2.239}$$

$$_G^B\tilde{\omega}_B = {}^G R_B^T \, {}_G\tilde{\omega}_B \, {}^G R_B \tag{2.240}$$

The relative angular velocity vectors of relatively moving rigid bodies can be done
only if all the angular velocities are expressed in one coordinate frame.

$$_0\omega_n = {}_0\omega_1 + {}_1^0\omega_2 + {}_2^0\omega_3 + \cdots + {}_{n-1}^0\omega_n = \sum_{i=1}^{n} {}_{i-1}^0\omega_i \tag{2.241}$$

Proof Consider a rigid body with a fixed point O and an attached frame $B(Oxyz)$ as shown in Fig. 2.16. The body frame B is initially coincident with the global frame G. Therefore, the position vector of a body point P at the initial time $t = t_0$ is:

$$G_{\mathbf{r}}(t_0) = {}^B\mathbf{r} \tag{2.242}$$

The position vector of P at any other time is found by the associated transformation matrix ${}^G R_B(t)$:

$$G_{\mathbf{r}}(t) = {}^G R_B(t)\, {}^B\mathbf{r} = {}^G R_B(t)\, {}^G\mathbf{r}(t_0) \tag{2.243}$$

The global time derivative of ${}^G\mathbf{r}$ is:

$$
{}^G\mathbf{v} = {}^G\dot{\mathbf{r}} = \frac{{}^G d}{dt}\, {}^G\mathbf{r}(t) = \frac{{}^G d}{dt}\left[{}^G R_B(t)\, {}^B\mathbf{r}\right] = \frac{{}^G d}{dt}\left[{}^G R_B(t)\, {}^G\mathbf{r}(t_0)\right]
$$
$$
= {}^G\dot{R}_B(t)\, {}^G\mathbf{r}(t_0) = {}^G\dot{R}_B(t)\, {}^B\mathbf{r} \tag{2.244}
$$

Eliminating ${}^B\mathbf{r}$ between (2.243) and (2.244) determines the velocity of the point in the global frame.

$$G_{\mathbf{v}} = {}^G\dot{R}_B(t)\, {}^G R_B^T(t)\, {}^G\mathbf{r}(t) \tag{2.245}$$

We denote the coefficient of ${}^G\mathbf{r}(t)$ by ${}_G\tilde{\omega}_B$

$$_G\tilde{\omega}_B = {}^G\dot{R}_B\, {}^G R_B^T \tag{2.246}$$

and rewrite Eq. (2.245) as

$$G_{\mathbf{v}} = {}_G\tilde{\omega}_B\, {}^G\mathbf{r}(t) = {}_G\boldsymbol{\omega}_B \times {}^G\mathbf{r}(t) \tag{2.247}$$

where ${}_G\boldsymbol{\omega}_B$ is the *instantaneous angular velocity* of the body B relative to the global frame G expressed the G-frame.

Transforming ${}^G\mathbf{v}$ to the body frame provides us with the body expression of the velocity vector:

$$
{}^B_G\mathbf{v}_P = {}^G R_B^T\, {}^G\mathbf{v} = {}^G R_B^T\, {}_G\tilde{\omega}_B\, {}^G\mathbf{r} = {}^G R_B^T\, {}^G\dot{R}_B\, {}^G R_B^T\, {}^G\mathbf{r}
$$
$$
= {}^G R_B^T\, {}^G\dot{R}_B\, {}^B\mathbf{r} \tag{2.248}
$$

We denote the coefficient of ${}^B\mathbf{r}$ by ${}^B_G\tilde{\omega}_B$

$$^B_G\tilde{\omega}_B = {}^G R_B^T\, {}^G\dot{R}_B \tag{2.249}$$

and rewrite Eq. (2.248) as

$$
{}_{G}^{B}\mathbf{v}_P = {}_{G}^{B}\tilde{\omega}_B \, {}^{B}\mathbf{r}_P \tag{2.250}
$$

or equivalently as

$$
{}_{G}^{B}\mathbf{v}_P = {}_{G}^{B}\boldsymbol{\omega}_B \times {}^{B}\mathbf{r}_P \tag{2.251}
$$

where ${}_{G}^{B}\boldsymbol{\omega}_B$ is the *instantaneous angular velocity* of B relative to the global frame G expressed in the B-frame.

The time derivative of the orthogonality condition, ${}^{G}R_B \, {}^{G}R_B^T = \mathbf{I}$, introduces an important identity,

$$
{}^{G}\dot{R}_B \, {}^{G}R_B^T + {}^{G}R_B \, {}^{G}\dot{R}_B^T = 0 \tag{2.252}
$$

which can be used to show that the angular velocity matrix ${}_{G}\tilde{\omega}_B = [{}^{G}\dot{R}_B \, {}^{G}R_B^T]$ is skew-symmetric:

$$
{}^{G}R_B \, {}^{G}\dot{R}_B^T = -\left[{}^{G}\dot{R}_B \, {}^{G}R_B^T\right]^T \tag{2.253}
$$

Generally speaking, an angular velocity vector is the instantaneous rotation of a coordinate frame A with respect to another frame B that can be expressed in or seen from a third coordinate frame C. We indicate the first coordinate frame A by a right subscript, the second frame B by a left subscript, and the third frame C by a left superscript, ${}_{B}^{C}\boldsymbol{\omega}_A$, to show the angular velocity of the A-frame with respect to B-frame, expressed in C-frame. If the left super and subscripts are the same, we only show the subscript, ${}_{G}\tilde{\omega}_B = {}_{G}^{G}\tilde{\omega}_B$.

We can transform the G-expression of the global velocity of a body point P, ${}^{G}\mathbf{v}_P$, and the B-expression of the global velocity of the point P, ${}_{G}^{B}\mathbf{v}_P$, to each other using rotation matrix rule:

$$
\begin{aligned}
{}_{G}^{B}\mathbf{v}_P &= {}^{B}R_G \, {}^{G}\mathbf{v}_P = {}^{B}R_G \, {}_{G}\tilde{\omega}_B \, {}^{G}\mathbf{r}_P = {}^{B}R_G \, {}_{G}\tilde{\omega}_B \, {}^{G}R_B \, {}^{B}\mathbf{r}_P \\
&= {}^{B}R_G \, {}^{G}\dot{R}_B \, {}^{G}R_B^T \, {}^{G}R_B \, {}^{B}\mathbf{r}_P = {}^{B}R_G \, {}^{G}\dot{R}_B \, {}^{B}\mathbf{r}_P \\
&= {}^{G}R_B^T \, {}^{G}\dot{R}_B \, {}^{B}\mathbf{r}_P = {}_{G}^{B}\tilde{\omega}_B \, {}^{B}\mathbf{r}_P = {}_{G}^{B}\boldsymbol{\omega}_B \times {}^{B}\mathbf{r}_P
\end{aligned} \tag{2.254}
$$

$$
\begin{aligned}
{}^{G}\mathbf{v}_P &= {}^{G}R_B \, {}_{G}^{B}\mathbf{v}_P = {}^{G}R_B \, {}_{G}^{B}\tilde{\omega}_B \, {}^{B}\mathbf{r}_P = {}^{G}R_B \, {}_{G}^{B}\tilde{\omega}_B \, {}^{G}R_B^T \, {}^{G}\mathbf{r}_P \\
&= {}^{G}R_B \, {}^{G}R_B^T \, {}^{G}\dot{R}_B \, {}^{G}R_B^T \, {}^{G}\mathbf{r}_P = {}^{G}\dot{R}_B \, {}^{G}R_B^T \, {}^{G}\mathbf{r}_P \\
&= {}_{G}\tilde{\omega}_B \, {}^{G}\mathbf{r}_P = {}_{G}\boldsymbol{\omega}_B \times {}^{G}\mathbf{r}_P = {}^{G}R_B \left({}_{G}^{B}\boldsymbol{\omega}_B \times {}^{B}\mathbf{r}_P\right)
\end{aligned} \tag{2.255}
$$

From the definitions of $_G\tilde{\omega}_B$ and $_G^B\tilde{\omega}_B$ in (2.246) and (2.249) and comparing with (2.254) and (2.255), we are able to transform the two angular velocity matrices by the following rules

$$_G\tilde{\omega}_B = {}^G R_B \, {}_G^B\tilde{\omega}_B \, {}^G R_B^T \tag{2.256}$$

$$_G^B\tilde{\omega}_B = {}^G R_B^T \, {}_G\tilde{\omega}_B \, {}^G R_B \tag{2.257}$$

and derive the following useful equations:

$$^G\dot{R}_B = {}_G\tilde{\omega}_B \, {}^G R_B \tag{2.258}$$

$$^G\dot{R}_B = {}^G R_B \, {}_G^B\tilde{\omega}_B \tag{2.259}$$

$$_G\tilde{\omega}_B \, {}^G R_B = {}^G R_B \, {}_G^B\tilde{\omega}_B \tag{2.260}$$

The angular velocity of B in G is negative of the angular velocity of G in B if both are expressed in the same coordinate frame:

$$_G^G\tilde{\omega}_B = -{}_B^G\tilde{\omega}_G \qquad _G^G\omega_B = -{}_B^G\omega_G \tag{2.261}$$

$$_G^B\tilde{\omega}_B = -{}_B^B\tilde{\omega}_G \qquad _G^B\omega_B = -{}_B^B\omega_G \tag{2.262}$$

The vector $_G\omega_B$ can always be expressed in the natural form

$$_G\omega_B = \omega\hat{u} \tag{2.263}$$

with a magnitude ω and a unit vector $\hat{u}$ parallel to $_G\omega_B$ that indicates the *instantaneous axis of rotation*.

To show the addition of relative angular velocities in Eq. (2.241), we start from a combination of rotations

$$^0R_2 = {}^0R_1 \, {}^1R_2 \tag{2.264}$$

and take the time derivative:

$$^0\dot{R}_2 = {}^0\dot{R}_1 \, {}^1R_2 + {}^0R_1 \, {}^1\dot{R}_2 \tag{2.265}$$

Substituting the derivative of the rotation matrices with

$$^0\dot{R}_2 = {}_0\tilde{\omega}_2 \, {}^0R_2 \tag{2.266}$$

$$^0\dot{R}_1 = {}_0\tilde{\omega}_1 \, {}^0R_1 \tag{2.267}$$

$$^1\dot{R}_2 = {}_1\tilde{\omega}_2 \, {}^1R_2 \tag{2.268}$$

results in

$$\begin{aligned}
{}_0\tilde{\omega}_2\,{}^0R_2 &= {}_0\tilde{\omega}_1\,{}^0R_1\,{}^1R_2 + {}^0R_1\,{}_1\tilde{\omega}_2\,{}^1R_2 \\
&= {}_0\tilde{\omega}_1\,{}^0R_2 + {}^0R_1\,{}_1\tilde{\omega}_2\,{}^0R_1^T\,{}^0R_1\,{}^1R_2 \\
&= {}_0\tilde{\omega}_1\,{}^0R_2 + {}_1^0\tilde{\omega}_2\,{}^0R_2
\end{aligned}$$

(2.269)

where

$$^0R_1\,{}_1\tilde{\omega}_2\,{}^0R_1^T = {}_1^0\tilde{\omega}_2$$

(2.270)

Therefore, we find

$$_0\tilde{\omega}_2 = {}_0\tilde{\omega}_1 + {}_1^0\tilde{\omega}_2$$

(2.271)

which indicates that two angular velocities may be added when they are expressed in the same frame:

$$_0\omega_2 = {}_0\omega_1 + {}_1^0\omega_2$$

(2.272)

The expansion of this equation for any number of angular velocities would be Eq. (2.241).

Employing the relative angular velocity formula (2.272), we can find the relative velocity formula of a point P in B_2 at $^0\mathbf{r}_P$:

$$\begin{aligned}
_0\mathbf{v}_2 = {}_0\omega_2\,{}^0\mathbf{r}_P &= \left({}_0\omega_1 + {}_1^0\omega_2\right){}^0\mathbf{r}_P = {}_0\omega_1\,{}^0\mathbf{r}_P + {}_1^0\omega_2\,{}^0\mathbf{r}_P \\
&= {}_0\mathbf{v}_1 + {}_1^0\mathbf{v}_2
\end{aligned}$$

(2.273)

∎

Example 55 ★ Rotation of a body point about a global axis. A continuous rotation about a stationary axis is the simplest rotational velocity.

Consider a rigid body is turning about the Z-axis with a constant angular speed $\dot{\alpha} = 10\,\mathrm{deg}/s$. The global velocity of a body point at $P(5, 30, 10)$ when the body is at $\alpha = 30\,\mathrm{deg}$ is $^G\mathbf{v}_P$.

$$^G\mathbf{v}_P = {}^G\dot{R}_B(t)\,{}^B\mathbf{r}_P$$

(2.274)

$$= \frac{^Gd}{dt}\left(\begin{bmatrix} \cos\alpha & -\sin\alpha & 0 \\ \sin\alpha & \cos\alpha & 0 \\ 0 & 0 & 1 \end{bmatrix}\right)\begin{bmatrix} 5 \\ 30 \\ 10 \end{bmatrix}$$

$$= \dot{\alpha} \begin{bmatrix} -\sin\alpha & -\cos\alpha & 0 \\ \cos\alpha & -\sin\alpha & 0 \\ 0 & 0 & 0 \end{bmatrix} \begin{bmatrix} 5 \\ 30 \\ 10 \end{bmatrix}$$

$$= \frac{10\pi}{180} \begin{bmatrix} -\sin\frac{\pi}{6} & -\cos\frac{\pi}{6} & 0 \\ \cos\frac{\pi}{6} & -\sin\frac{\pi}{6} & 0 \\ 0 & 0 & 0 \end{bmatrix} \begin{bmatrix} 5 \\ 30 \\ 10 \end{bmatrix} = \begin{bmatrix} -4.97 \\ -1.86 \\ 0 \end{bmatrix}$$

The point P is now at ${}^G\mathbf{r}_P$.

$${}^G\mathbf{r}_P = {}^G R_B {}^B\mathbf{r}_P \tag{2.275}$$

$$= \begin{bmatrix} \cos\frac{\pi}{6} & -\sin\frac{\pi}{6} & 0 \\ \sin\frac{\pi}{6} & \cos\frac{\pi}{6} & 0 \\ 0 & 0 & 1 \end{bmatrix} \begin{bmatrix} 5 \\ 30 \\ 10 \end{bmatrix} = \begin{bmatrix} -10.67 \\ 28.48 \\ 10 \end{bmatrix}$$

Example 56 ★ Simple derivative transformation formula. Differentiating a coordinate frame-dependent operation. Here is the theory of how to relate differentiation between a rotating and stationary frames.

Consider a point P that can move in a body coordinate frame B $(Oxyz)$. The body position vector ${}^B\mathbf{r}_P$ is not constant, and, therefore, the B-expression of the G-velocity of such a point is:

$$\frac{{}^G d}{dt}{}^B\mathbf{r}_P = {}^B_G\dot{\mathbf{r}}_P = \frac{{}^B d}{dt}{}^B\mathbf{r}_P + {}^B_G\boldsymbol{\omega}_B \times {}^B\mathbf{r}_P \tag{2.276}$$

The result of Eq. (2.276) defines the transformation of the differential operator on a B-vector ${}^B\square$ from the body to the global coordinate frame:

$$\frac{{}^G d}{dt}{}^B\square = {}^B_G\dot{\square} = \frac{{}^B d}{dt}{}^B\square + {}^B_G\boldsymbol{\omega}_B \times {}^B\square \tag{2.277}$$

However, special attention must be paid to the coordinate frame in which the vector ${}^B\square$ and the final result are expressed. The final result is ${}^B_G\dot{\square}$, showing the global time derivative expressed in the body frame. In other words it is the B-expression of the G -derivative of the vector ${}^B\square$. The vector ${}^B\square$ may be any vector quantity such as position, velocity, angular velocity, momentum, angular momentum, and a time-varying force vector.

Equation (2.277) is called a *simple derivative transformation formula* and relates the derivative of a B-vector as it would be seen from the G-frame to its derivative as seen from the B -frame. The derivative transformation formula (2.277) is more general and can be applied to every vector for derivative transformation between every two relatively moving coordinate frames.

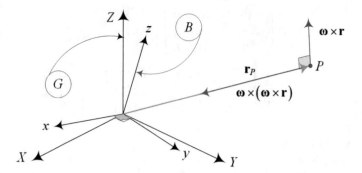

Fig. 2.17 A rotating rigid body $B(Oxyz)$ with a fixed point O in a reference frame $G(OXYZ)$

2.3.3 ★ Acceleration Kinematics

Consider a rotating rigid body $B(Oxyz)$ with a fixed point O in a reference frame $G(OXYZ)$ such as shown in Fig. 2.17. When the body rotates in G, the global acceleration of a body point P is given as $^G\mathbf{a}$

$$^G\mathbf{a} = {}^G\dot{\mathbf{v}} = {}^G\ddot{\mathbf{r}} = {}_G S_B\,{}^G\mathbf{r} \tag{2.278}$$

$$= {}_G\boldsymbol{\alpha}_B \times {}^G\mathbf{r} + {}_G\boldsymbol{\omega}_B \times \left({}_G\boldsymbol{\omega}_B \times {}^G\mathbf{r}\right) \tag{2.279}$$

$$= \left({}_G\tilde{\alpha}_B + {}_G\tilde{\omega}_B^2\right){}^G\mathbf{r} \tag{2.280}$$

$$= {}^G\ddot{R}_B\,{}^G R_B^T\,{}^G\mathbf{r} \tag{2.281}$$

where ${}_G\boldsymbol{\alpha}_B$ is the *angular acceleration vector* of B relative to G

$$_G\boldsymbol{\alpha}_B = \frac{{}^G d}{dt}\,{}_G\boldsymbol{\omega}_B \tag{2.282}$$

and ${}_G\tilde{\alpha}_B$ is the *angular acceleration matrix*.

$$_G\tilde{\alpha}_B = {}_G\dot{\tilde{\omega}}_B = {}^G\ddot{R}_B\,{}^G R_B^T + {}^G\dot{R}_B\,{}^G\dot{R}_B^T \tag{2.283}$$

The relative angular acceleration of two bodies B_1, B_2 in the global frame G can be combined as

$$_G\boldsymbol{\alpha}_2 = \frac{{}^G d}{dt}\,{}_G\boldsymbol{\omega}_2 = {}_G\boldsymbol{\alpha}_1 + {}_1^G\boldsymbol{\alpha}_2 \tag{2.284}$$

The B-expressions of ${}^G\mathbf{a}$ is:

$$\frac{B}{G}\mathbf{a} = \frac{B}{G}\boldsymbol{\alpha}_B \times {}^B\mathbf{r} + \frac{B}{G}\boldsymbol{\omega}_B \times \left(\frac{B}{G}\boldsymbol{\omega}_B \times {}^B\mathbf{r}\right) \tag{2.285}$$

Proof The global position and velocity vectors of the body point P are:

$$ {}^G\mathbf{r} = {}^GR_B\,{}^B\mathbf{r} \tag{2.286}$$

$$ {}^G\mathbf{v} = {}^G\dot{\mathbf{r}} = {}^G\dot{R}_B\,{}^B\mathbf{r} = {}_G\tilde{\omega}_B\,{}^G\mathbf{r} = {}_G\boldsymbol{\omega}_B \times {}^G\mathbf{r} \tag{2.287}$$

Differentiating Equation (2.287) and using the notation ${}_G\boldsymbol{\alpha}_B = \frac{{}^Gd}{dt}\,{}_G\boldsymbol{\omega}_B$ for angular acceleration yields:

$$ {}^G\mathbf{a} = {}^G\ddot{\mathbf{r}} = {}_G\dot{\boldsymbol{\omega}}_B \times {}^G\mathbf{r} + {}_G\boldsymbol{\omega}_B \times {}^G\dot{\mathbf{r}}$$

$$ = {}_G\boldsymbol{\alpha}_B \times {}^G\mathbf{r} + {}_G\boldsymbol{\omega}_B \times \left({}_G\boldsymbol{\omega}_B \times {}^G\mathbf{r}\right) \tag{2.288}$$

We may substitute the matrix expressions of angular velocity and acceleration in (2.288) to derive Eqs. (2.280):

$$ {}^G\ddot{\mathbf{r}} = {}_G\boldsymbol{\alpha}_B \times {}^G\mathbf{r} + {}_G\boldsymbol{\omega}_B \times \left({}_G\boldsymbol{\omega}_B \times {}^G\mathbf{r}\right)$$

$$ = {}_G\tilde{\alpha}_B\,{}^G\mathbf{r} + {}_G\tilde{\omega}_B\,{}_G\tilde{\omega}_B\,{}^G\mathbf{r}$$

$$ = \left({}_G\tilde{\alpha}_B + {}_G\tilde{\omega}_B^2\right){}^G\mathbf{r} \tag{2.289}$$

Recalling that

$$ {}_G\tilde{\omega}_B = {}^G\dot{R}_B\,{}^GR_B^T \tag{2.290}$$

$$ {}^G\dot{\mathbf{r}}(t) = {}_G\tilde{\omega}_B\,{}^G\mathbf{r}(t) \tag{2.291}$$

we find Eqs. (2.281) and (2.283):

$$ {}^G\ddot{\mathbf{r}} = \frac{{}^Gd}{dt}\left({}^G\dot{R}_B\,{}^GR_B^T\,{}^G\mathbf{r}\right)$$

$$ = {}^G\ddot{R}_B\,{}^GR_B^T\,{}^G\mathbf{r} + {}^G\dot{R}_B\,{}^G\dot{R}_B^T\,{}^G\mathbf{r} + \left[{}^G\dot{R}_B\,{}^GR_B^T\right]\left[{}^G\dot{R}_B\,{}^GR_B^T\right]{}^G\mathbf{r}$$

$$ = \left[{}^G\ddot{R}_B\,{}^GR_B^T + {}^G\dot{R}_B\,{}^G\dot{R}_B^T + \left[{}^G\dot{R}_B\,{}^GR_B^T\right]^2\right]{}^G\mathbf{r}$$

$$ = \left[{}^G\ddot{R}_B\,{}^GR_B^T - \left[{}^G\dot{R}_B\,{}^GR_B^T\right]^2 + \left[{}^G\dot{R}_B\,{}^GR_B^T\right]^2\right]{}^G\mathbf{r}$$

$$ = {}^G\ddot{R}_B\,{}^GR_B^T\,{}^G\mathbf{r} \tag{2.292}$$

$$
\begin{aligned}
{}_{G}\tilde{\alpha}_B = {}_{G}\dot{\tilde{\omega}}_B &= {}^{G}\ddot{R}_B \, {}^{G}R_B^T + {}^{G}\dot{R}_B \, {}^{G}\dot{R}_B^T \\
&= {}^{G}\ddot{R}_B \, {}^{G}R_B^T + {}^{G}\dot{R}_B \, {}^{G}R_B^T \, {}^{G}R_B \, {}^{G}\dot{R}_B^T \\
&= {}^{G}\ddot{R}_B \, {}^{G}R_B^T + \left[{}^{G}\dot{R}_B \, {}^{G}R_B^T \right] \left[{}^{G}\dot{R}_B \, {}^{G}R_B^T \right]^T \\
&= {}^{G}\ddot{R}_B \, {}^{G}R_B^T + {}_{G}\tilde{\omega}_B \, {}_{G}\tilde{\omega}_B^T = {}^{G}\ddot{R}_B \, {}^{G}R_B^T - {}_{G}\tilde{\omega}_B^2
\end{aligned}
\tag{2.293}
$$

The expanded forms of the angular accelerations ${}_{G}\alpha_B$, ${}_{G}\tilde{\alpha}_B$ are:

$$
\begin{aligned}
{}_{G}\tilde{\alpha}_B = {}_{G}\dot{\tilde{\omega}}_B = \ddot{\phi}\tilde{u} + \dot{\phi}\dot{\tilde{u}} &=
\begin{bmatrix}
0 & -\dot{\omega}_3 & \dot{\omega}_2 \\
\dot{\omega}_3 & 0 & -\dot{\omega}_1 \\
-\dot{\omega}_2 & \dot{\omega}_1 & 0
\end{bmatrix} \\
&=
\begin{bmatrix}
0 & -\dot{u}_3\dot{\phi} - u_3\ddot{\phi} & \dot{u}_2\dot{\phi} + u_2\ddot{\phi} \\
\dot{u}_3\dot{\phi} + u_3\ddot{\phi} & 0 & -\dot{u}_1\dot{\phi} - u_1\ddot{\phi} \\
-\dot{u}_2\dot{\phi} - u_2\ddot{\phi} & \dot{u}_1\dot{\phi} + u_1\ddot{\phi} & 0
\end{bmatrix}
\end{aligned}
\tag{2.294}
$$

$$
{}_{G}\alpha_B =
\begin{bmatrix}
\dot{\omega}_1 \\
\dot{\omega}_2 \\
\dot{\omega}_3
\end{bmatrix}
=
\begin{bmatrix}
\dot{u}_1\dot{\phi} + u_1\ddot{\phi} \\
\dot{u}_2\dot{\phi} + u_2\ddot{\phi} \\
\dot{u}_3\dot{\phi} + u_3\ddot{\phi}
\end{bmatrix}
\tag{2.295}
$$

The angular accelerations of several bodies rotating relative to each other can be related according to:

$$
{}_{0}\alpha_n = {}_{0}\alpha_1 + {}_{1}^{0}\alpha_2 + {}_{2}^{0}\alpha_3 + \cdots + {}_{n-1}^{0}\alpha_n
\tag{2.296}
$$

To show this fact and develop the relative acceleration formula, we consider a pair of relatively rotating rigid bodies in a base coordinate frame B_0 with a fixed point at O. The angular velocities of the links are related as:

$$
{}_{0}\omega_2 = {}_{0}\omega_1 + {}_{1}^{0}\omega_2
\tag{2.297}
$$

So, their angular accelerations are:

$$
{}_{0}\alpha_1 = \frac{{}^{0}d}{dt}\,{}_{0}\omega_1
\tag{2.298}
$$

$$
{}_{0}\alpha_2 = \frac{{}^{0}d}{dt}\,{}_{0}\omega_2 = {}_{0}\alpha_1 + {}_{1}^{0}\alpha_2
\tag{2.299}
$$

We can transform the G and B-expressions of the global acceleration of a body point P to each other using a rotation matrix:

$$
\begin{aligned}
{}_G^B \mathbf{a}_P &= {}^B R_G \, {}^G \mathbf{a}_P = {}^B R_G \, {}^G \ddot{R}_B \, {}^G R_B^T \, {}^G R_B \, {}^B \mathbf{r}_P \\
&= {}^B R_G \, {}^G \ddot{R}_B \, {}^B \mathbf{r}_P = {}^G R_B^T \, {}^G \ddot{R}_B \, {}^B \mathbf{r}_P \\
&= {}_G^B \boldsymbol{\alpha}_B \times {}^B \mathbf{r} + {}_G^B \boldsymbol{\omega}_B \times \left({}_G^B \boldsymbol{\omega}_B \times {}^B \mathbf{r} \right)
\end{aligned}
\tag{2.300}
$$

$$
\begin{aligned}
{}^G \mathbf{a}_P &= {}^G R_B \, {}_G^B \mathbf{a}_P = {}^G R_B \, {}^G R_B^T \, {}^G \ddot{R}_B \, {}^G R_B^T \, {}^G \mathbf{r}_P \\
&= {}^G \ddot{R}_B \, {}^G R_B^T \, {}^G \mathbf{r}_P \\
&= {}_G \boldsymbol{\alpha}_B \times {}^G \mathbf{r} + {}_G \boldsymbol{\omega}_B \times \left({}_G \boldsymbol{\omega}_B \times {}^G \mathbf{r} \right)
\end{aligned}
\tag{2.301}
$$

The angular acceleration of B in G is negative of the angular acceleration of G in B if both are expressed in the same coordinate frame:

$$
{}_G \tilde{\alpha}_B = - {}_B^G \tilde{\alpha}_G \qquad {}_G \boldsymbol{\alpha}_B = - {}_B^G \boldsymbol{\alpha}_G
\tag{2.302}
$$

$$
{}_G^B \tilde{\alpha}_B = - {}_B \tilde{\alpha}_G \qquad {}_G^B \boldsymbol{\alpha}_B = - {}_B \boldsymbol{\alpha}_G
\tag{2.303}
$$

The term ${}_G \boldsymbol{\alpha}_B \times {}^G \mathbf{r}$ in (2.288) is called the *tangential acceleration*, which is a function of the angular acceleration of B in G. The term ${}_G \boldsymbol{\omega}_B \times \left({}_G \boldsymbol{\omega}_B \times {}^G \mathbf{r} \right)$ in ${}^G \mathbf{a}$ is called the *centripetal acceleration* and is a function of the angular velocity of B in G. ∎

Example 57 ★ Rotation of a body point about a global axis. Rotation of a body frame B about an axis of the global frame G with constant acceleration is the simplest accelerated motion of a rigid body with a fixed point.

Consider a rigid body that is turning about the Z-axis with a constant angular acceleration $\ddot{\alpha} = 2 \, \text{rad}/\text{s}^2$. The global acceleration of a body point at $P(5, 30, 10)$ cm when the body is at angular speed $\dot{\alpha} = 10 \, \text{rad}/\text{s}$ and angular position $\alpha = 30 \, \text{deg}$ is:

$$
{}^G \mathbf{a}_P = {}^G \ddot{R}_B(t) \, {}^B \mathbf{r}_P
\tag{2.304}
$$

$$
= \begin{bmatrix} -87.6 & 48.27 & 0 \\ -48.27 & -87.6 & 0 \\ 0 & 0 & 0 \end{bmatrix} \begin{bmatrix} 5 \\ 30 \\ 10 \end{bmatrix} = \begin{bmatrix} 1010 \\ -2869.4 \\ 0 \end{bmatrix} \text{cm}/\text{s}
$$

where

$$
{}^G \ddot{R}_B = \frac{{}^G d^2}{dt^2} \, {}^G R_B = \dot{\alpha} \frac{{}^G d}{d\alpha} \, {}^G R_B = \ddot{\alpha} \frac{{}^G d}{d\alpha} \, {}^G R_B + \dot{\alpha}^2 \frac{{}^G d^2}{d\alpha^2} \, {}^G R_B
\tag{2.305}
$$

$$
= \ddot{\alpha} \begin{bmatrix} -\sin\alpha & -\cos\alpha & 0 \\ \cos\alpha & -\sin\alpha & 0 \\ 0 & 0 & 0 \end{bmatrix} + \dot{\alpha}^2 \begin{bmatrix} -\cos\alpha & \sin\alpha & 0 \\ -\sin\alpha & -\cos\alpha & 0 \\ 0 & 0 & 0 \end{bmatrix}
$$

At this moment, the point P is at $^G\mathbf{r}_P$.

$$^G\mathbf{r}_P = {}^G R_B \, {}^B\mathbf{r}_P \tag{2.306}$$

$$= \begin{bmatrix} \cos\dfrac{\pi}{6} & -\sin\dfrac{\pi}{6} & 0 \\ \sin\dfrac{\pi}{6} & \cos\dfrac{\pi}{6} & 0 \\ 0 & 0 & 1 \end{bmatrix} \begin{bmatrix} 5 \\ 30 \\ 10 \end{bmatrix} = \begin{bmatrix} -10.67 \\ 28.48 \\ 10 \end{bmatrix} \text{cm}$$

Example 58 ★ *B-expression of angular acceleration.* Expression of an absolute acceleration in G-frame in the moving B-frame must be done by derivative transformation formula.

The angular acceleration expressed in the body frame is the body derivative of the angular velocity vector. To show this, we use the derivative transport formula (2.277):

$$\begin{aligned} {}^B_G\boldsymbol{\alpha}_B = {}^B_G\dot{\boldsymbol{\omega}}_B &= \frac{{}^Gd}{dt}\, {}^B_G\boldsymbol{\omega}_B \\ &= \frac{{}^Bd}{dt}\, {}^B_G\boldsymbol{\omega}_B + {}^B_G\boldsymbol{\omega}_B \times {}^B_G\boldsymbol{\omega}_B = \frac{{}^Bd}{dt}\, {}^B_G\boldsymbol{\omega}_B \end{aligned} \tag{2.307}$$

Interestingly, the global and body derivatives of $^B_G\omega_B$ are equal:

$$\frac{{}^Gd}{dt}\, {}^B_G\boldsymbol{\omega}_B = \frac{{}^Bd}{dt}\, {}^B_G\boldsymbol{\omega}_B = {}^B_G\boldsymbol{\alpha}_B \tag{2.308}$$

This is because $_G\omega_B$ is about an axis $\hat{u}$ that is instantaneously fixed in both B and G.

A vector $\boldsymbol{\alpha}$ can generally indicate the angular acceleration of a coordinate frame A with respect to another frame B. It can be expressed in a third coordinate frame C. We indicate the first coordinate frame A by a right subscript, the second frame B by a left subscript, and the third frame C by a left superscript, $^C_B\boldsymbol{\alpha}_A$. If the left super and subscripts are the same, we only show the subscript. So, the angular acceleration of A with respect to B as seen from C is the C-expression of $_B\boldsymbol{\alpha}_A$:

$$^C_B\boldsymbol{\alpha}_A = {}^C R_B \, {}_B\boldsymbol{\alpha}_A \tag{2.309}$$

Example 59 Velocity and acceleration of a simple pendulum. Simple pendulum, a point mass attached to a massless bar, is an excellent example to practice on a variable acceleration rotating system.

A point mass attached to a massless rod hanging from a revolute joint is what we call a *simple pendulum*. Figure 2.18 illustrates a sample. A local coordinate frame B is attached to the pendulum, which rotates in a global frame G about the Z-axis. The kinematic informations of the mass are:

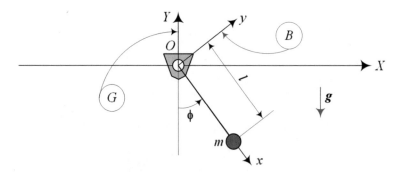

Fig. 2.18 A simple pendulum

$$^B\mathbf{r} = l\hat{\imath} \qquad ^G\mathbf{r} = {}^G R_B \, {}^B\mathbf{r} = \begin{bmatrix} l\sin\phi \\ -l\cos\phi \\ 0 \end{bmatrix} \tag{2.310}$$

$$^B_G\boldsymbol{\omega}_B = \dot{\phi}\hat{k} \qquad _G\boldsymbol{\omega}_B = {}^G R_B^T \, {}^B_G\boldsymbol{\omega}_B = \dot{\phi}\,\hat{K} \tag{2.311}$$

$$^G R_B = \begin{bmatrix} \cos\left(\frac{3}{2}\pi + \phi\right) & -\sin\left(\frac{3}{2}\pi + \phi\right) & 0 \\ \sin\left(\frac{3}{2}\pi + \phi\right) & \cos\left(\frac{3}{2}\pi + \phi\right) & 0 \\ 0 & 0 & 1 \end{bmatrix}$$

$$= \begin{bmatrix} \sin\phi & \cos\phi & 0 \\ -\cos\phi & \sin\phi & 0 \\ 0 & 0 & 1 \end{bmatrix} \tag{2.312}$$

Therefore, the velocity and acceleration of m are:

$$^B_G\mathbf{v} = {}^B\dot{\mathbf{r}} + {}^B_G\boldsymbol{\omega}_B \times {}^B_G\mathbf{r} = 0 + \dot{\phi}\hat{k} \times l\hat{\imath} = l\,\dot{\phi}\hat{\jmath} \tag{2.313}$$

$$^G\mathbf{v} = {}^G R_B \, {}^B\mathbf{v} = \begin{bmatrix} l\,\dot{\phi}\cos\phi \\ l\,\dot{\phi}\sin\phi \\ 0 \end{bmatrix} \tag{2.314}$$

$$^B_G\mathbf{a} = {}^B_G\dot{\mathbf{v}} + {}^B_G\boldsymbol{\omega}_B \times {}^B_G\mathbf{v} = l\,\ddot{\phi}\hat{\jmath} + \dot{\phi}\hat{k} \times l\,\dot{\phi}\hat{\jmath} = l\,\ddot{\phi}\hat{\jmath} - l\,\dot{\phi}^2\hat{\imath} \tag{2.315}$$

$$^G\mathbf{a} = {}^G R_B \, {}^B\mathbf{a} = \begin{bmatrix} l\,\ddot{\phi}\cos\phi - l\,\dot{\phi}^2\sin\phi \\ l\,\ddot{\phi}\sin\phi + l\,\dot{\phi}^2\cos\phi \\ 0 \end{bmatrix} \tag{2.316}$$

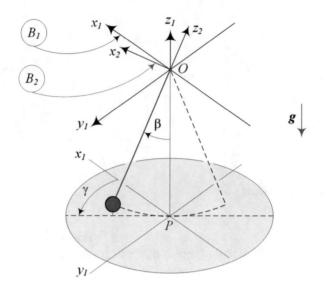

Fig. 2.19 A spherical pendulum

Example 60 Spherical pendulum. Spherical pendulum is a two DOF oscillating system with complicated equations of motion and kinematics. Here is a detailed analysis.

A pendulum that is free to oscillate in any plane is called a spherical pendulum. This name comes from the codominants that we use to locate the tip mass. Consider a pendulum with a point mass m at the tip point of a long, massless, and straight rod with length l. The pendulum is hanging from a point O $(0, 0, 0)$ in the global coordinate frame B_1 (x_1, y_1, z_1) as is shown in Fig. 2.19.

To indicate the mass m, we attach a coordinate frame B_2 (x_2, y_2, z_2) to the pendulum at point O. The pendulum makes an angle β with the vertical z_1-axis. The pendulum swings at the moment in the plane (x_2, z_2) and makes an angle γ with the plane (x_1, z_1). Therefore, the transformation matrix between B_2 and B_1 is a combination of rotation γ about the z_2-axis followed by a rotation $-\beta$ about the y_2-axis.

$$
\begin{aligned}
{}^2R_1 &= R_{y_2,-\beta}\, R_{z_2,\gamma} \\
&= \begin{bmatrix} \cos\gamma\cos\beta & \cos\beta\sin\gamma & \sin\beta \\ -\sin\gamma & \cos\gamma & 0 \\ -\cos\gamma\sin\beta & -\sin\gamma\sin\beta & \cos\beta \end{bmatrix}
\end{aligned}
\qquad (2.317)
$$

The position vectors of m in B_1 and B_2 are:

$$
{}^{2}\mathbf{r} = \begin{bmatrix} 0 \\ 0 \\ -l \end{bmatrix} \qquad {}^{1}\mathbf{r} = {}^{1}R_{2}\,{}^{2}\mathbf{r} = \begin{bmatrix} l\cos\gamma\sin\beta \\ l\sin\beta\sin\gamma \\ -l\cos\beta \end{bmatrix} \tag{2.318}
$$

The equation of motion of m will be a unidirectional Euler equation.

$$
{}^{1}\mathbf{M} = I\,{}_{1}\boldsymbol{\alpha}_{2} \tag{2.319}
$$

$$
{}^{1}\mathbf{r} \times m\,{}^{1}\mathbf{g} = ml^{2}\,{}_{1}\boldsymbol{\alpha}_{2} \tag{2.320}
$$

$$
\begin{bmatrix} l\cos\gamma\sin\beta \\ l\sin\beta\sin\gamma \\ -l\cos\beta \end{bmatrix} \times m \begin{bmatrix} 0 \\ 0 \\ -g_{0} \end{bmatrix} = ml^{2}\,{}_{1}\boldsymbol{\alpha}_{2} \tag{2.321}
$$

It simplifies to:

$$
{}_{1}\boldsymbol{\alpha}_{2} = \frac{g_{0}}{l} \begin{bmatrix} -\sin\beta\sin\gamma \\ \cos\gamma\sin\beta \\ 0 \end{bmatrix} \tag{2.322}
$$

To find the angular acceleration ${}_{1}\boldsymbol{\alpha}_{2}$ of B_{2} in B_{1}, we use ${}^{1}R_{2}$:

$$
\begin{aligned}
{}^{1}R_{2} &= {}^{2}R_{1}^{T} \\
&= \begin{bmatrix} \cos\beta\cos\gamma & -\sin\gamma & -\cos\gamma\sin\beta \\ \cos\beta\sin\gamma & \cos\gamma & -\sin\beta\sin\gamma \\ \sin\beta & 0 & \cos\beta \end{bmatrix}
\end{aligned} \tag{2.323}
$$

$$
\begin{aligned}
{}^{1}\dot{R}_{2} &= \dot{\beta}\frac{d}{d\beta}{}^{1}R_{2} + \dot{\gamma}\frac{d}{d\gamma}{}^{1}R_{2} \\
&= \begin{bmatrix} -\dot{\beta}c\gamma s\beta - \dot{\gamma}c\beta s\gamma & -\dot{\gamma}c\gamma & \dot{\gamma}s\beta s\gamma - \dot{\beta}c\beta c\gamma \\ \dot{\gamma}c\beta c\gamma - \dot{\beta}s\beta s\gamma & -\dot{\gamma}s\gamma & -\dot{\beta}c\beta s\gamma - \dot{\gamma}c\gamma s\beta \\ \dot{\beta}c\beta & 0 & -\dot{\beta}s\beta \end{bmatrix}
\end{aligned} \tag{2.324}
$$

$$
\begin{aligned}
{}^{1}\dot{R}_{2} &= \dot{\beta}\frac{d}{d\beta}{}^{1}R_{2} + \dot{\gamma}\frac{d}{d\gamma}{}^{1}R_{2} \\
&= \begin{bmatrix} -\dot{\beta}c\gamma s\beta - \dot{\gamma}c\beta s\gamma & -\dot{\gamma}c\gamma & \dot{\gamma}s\beta s\gamma - \dot{\beta}c\beta c\gamma \\ \dot{\gamma}c\beta c\gamma - \dot{\beta}s\beta s\gamma & -\dot{\gamma}s\gamma & -\dot{\beta}c\beta s\gamma - \dot{\gamma}c\gamma s\beta \\ \dot{\beta}c\beta & 0 & -\dot{\beta}s\beta \end{bmatrix}
\end{aligned} \tag{2.325}
$$

$$
{}_1\tilde{\omega}_2 = {}^1\dot{R}_2\,{}^1R_2^T = \begin{bmatrix} 0 & -\dot{\gamma} & -\dot{\beta}\cos\gamma \\ \dot{\gamma} & 0 & -\dot{\beta}\sin\gamma \\ \dot{\beta}\cos\gamma & \dot{\beta}\sin\gamma & 0 \end{bmatrix} \tag{2.326}
$$

$$
{}^1\ddot{R}_2 = \ddot{\beta}\frac{d}{d\beta}\,{}^1R_2 + \dot{\beta}^2\frac{d^2}{d\beta^2}\,{}^1R_2 + \dot{\beta}\dot{\gamma}\frac{d^2}{d\gamma\,d\beta}\,{}^1R_2
$$
$$
+\ddot{\gamma}\frac{d}{d\gamma}\,{}^1R_2 + \dot{\gamma}\dot{\beta}\frac{d^2}{d\beta\,d\gamma}\,{}^1R_2 + \dot{\gamma}^2\frac{d^2}{d\gamma^2}\,{}^1R_2 \tag{2.327}
$$

$$
{}_1\tilde{\alpha}_2 = {}^1\ddot{R}_2\,{}^1R_2^T - {}_1\tilde{\omega}_2^2
$$
$$
= \begin{bmatrix} 0 & -\ddot{\gamma} & -\ddot{\beta}c\gamma + \dot{\beta}\dot{\gamma}s\gamma \\ \ddot{\gamma} & 0 & -\ddot{\beta}s\gamma - \dot{\beta}\dot{\gamma}c\gamma \\ \ddot{\beta}c\gamma - \dot{\beta}\dot{\gamma}s\gamma & \ddot{\beta}s\gamma + \dot{\beta}\dot{\gamma}c\gamma & 0 \end{bmatrix} \tag{2.328}
$$

Therefore, the equation of motion of the pendulum would be:

$$
\frac{g_0}{l}\begin{bmatrix} -\sin\beta\sin\gamma \\ \cos\gamma\sin\beta \\ 0 \end{bmatrix} = \begin{bmatrix} \ddot{\beta}\sin\gamma + \dot{\beta}\dot{\gamma}\cos\gamma \\ -\ddot{\beta}\cos\gamma + \dot{\beta}\dot{\gamma}\sin\gamma \\ \ddot{\gamma} \end{bmatrix} \tag{2.329}
$$

The third equation is easy to solve.

$$
\dot{\gamma} = \dot{\gamma}_0 \qquad \gamma = \dot{\gamma}_0 t + \gamma_0 \tag{2.330}
$$

The second and first equations can be combined to the form

$$
\ddot{\beta} = -\sqrt{\frac{g_0^2}{l^2}\sin^2\beta + \dot{\beta}^2\dot{\gamma}_0^2} \tag{2.331}
$$

which reduces to the equation of a simple pendulum if $\dot{\gamma}_0 = 0$.

Example 61 Equation of motion of a spherical pendulum. We may derive the equations of motions of a spherical pendulum from another view point by starting with Newton equation of motion. Here is the analysis.

Consider a particle P of mass m that is suspended by a massless bar of length l from a point O, as shown in Fig. 2.19. If we show the tension force of the bar by $\mathbf{T}$, then the equation of motion of m is:

$$
{}^1\mathbf{T} + m\,{}^1\mathbf{g} = m\,{}^1\ddot{\mathbf{r}} \tag{2.332}
$$

or

$$- T \, {}^1\mathbf{r} + m \, {}^1\mathbf{g} = m \, {}^1\ddot{\mathbf{r}} \tag{2.333}$$

To eliminate ${}^1\mathbf{T}$, we multiply the equation by ${}^1\mathbf{r}$,

$$ {}^1\mathbf{r} \times {}^1\mathbf{g} = {}^1\mathbf{r} \times {}^1\ddot{\mathbf{r}} \tag{2.334}$$

$$\begin{bmatrix} l \cos \gamma \sin \beta \\ l \sin \beta \sin \gamma \\ -l \cos \beta \end{bmatrix} \times \begin{bmatrix} 0 \\ 0 \\ -g_0 \end{bmatrix} = \begin{bmatrix} l \cos \gamma \sin \beta \\ l \sin \beta \sin \gamma \\ -l \cos \beta \end{bmatrix} \times \begin{bmatrix} \ddot{x} \\ \ddot{y} \\ \ddot{z} \end{bmatrix}$$

and find

$$\begin{bmatrix} -l g_0 \sin \beta \sin \gamma \\ l g_0 \cos \gamma \sin \beta \\ 0 \end{bmatrix} = \begin{bmatrix} l\ddot{y} \cos \beta + l\ddot{z} \sin \beta \sin \gamma \\ -l\ddot{x} \cos \beta - l\ddot{z} \cos \gamma \sin \beta \\ l\ddot{y} \cos \gamma - l\ddot{x} \sin \gamma \end{bmatrix} \tag{2.335}$$

These are the equations of motion of m in terms of x, y, z. The system has only two *DOF*, and hence, it is better to express the equations only in terms of two variables γ and β. To do so, we may either take time derivatives of ${}^1\mathbf{r}$ or employ ${}_1\boldsymbol{\alpha}_2$ from Example 60 to find ${}^1\ddot{\mathbf{r}}$:

$$ {}^1\ddot{\mathbf{r}} = {}_1\boldsymbol{\alpha}_2 \times {}^1\mathbf{r} \tag{2.336}$$

In either case, Eq. (2.329) would be the equations of motion in terms of γ and β.

Example 62 ★ Foucault pendulum. A spherical pendulum on the Earth cannot oscillate in a stationary plane because of Earth's rotation. It is an interesting vibrating system. Here is a mathematical tackling the system to determine how the plane of oscillation changes.

Consider a pendulum with a point mass m at the tip of a long, massless, and straight bar with length l. The pendulum is hanging from a point O $(0, 0, l)$ in a local coordinate frame B_1 (x_1, y_1, z_1) at a point P on the Earth surface. Point P at longitude φ and latitude λ is indicated by ${}^E\mathbf{d}$ in the Earth frame E $(Oxyz)$. The E-frame is turning in a motionless global frame G $(OXYZ)$ about the Z-axis.

To indicate the mass m, we attach a coordinate frame B_1 (x_1, y_1, z_1) to the pendulum at point A as shown in Fig. 2.20. The pendulum makes an angle β with the local vertical z_1-axis at point P. The pendulum swings at the moment in the plane (x_2, z_2) and makes an angle γ with the plane (x_1, z_1). Therefore, the kinematic equation between B_2 and B_1 is:

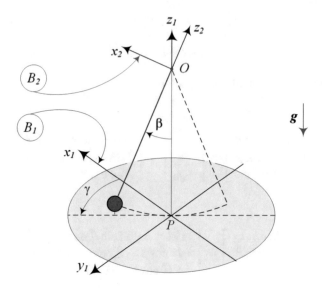

Fig. 2.20 Foucault pendulum is a simple pendulum hanging from a point A above a point P on Earth's surface

$$
{}^1\mathbf{r} = {}^1R_2\,{}^2\mathbf{r} + {}^1_1\mathbf{d}_2
$$

$$
= \begin{bmatrix} \cos\gamma\cos\beta & -\sin\gamma & -\cos\gamma\sin\beta \\ \cos\beta\sin\gamma & \cos\gamma & -\sin\gamma\sin\beta \\ \sin\beta & 0 & \cos\beta \end{bmatrix} \begin{bmatrix} 0 \\ 0 \\ -l \end{bmatrix} + \begin{bmatrix} 0 \\ 0 \\ l \end{bmatrix}
$$

$$
= \begin{bmatrix} l\cos\gamma\sin\beta \\ l\sin\beta\sin\gamma \\ l - l\cos\beta \end{bmatrix} \tag{2.337}
$$

The position vector of m in B_2 and B_1 are:

$$
{}^2\mathbf{r} = \begin{bmatrix} 0 \\ 0 \\ -l \end{bmatrix} \qquad {}^1\mathbf{r} = \begin{bmatrix} x_1 \\ y_1 \\ z_1 \end{bmatrix} = \begin{bmatrix} l\cos\gamma\sin\beta \\ l\sin\beta\sin\gamma \\ l - l\cos\beta \end{bmatrix} \tag{2.338}
$$

Employing the acceleration equation,

$$
{}^1_G\mathbf{a} = {}^1\mathbf{a} + {}^1_G\boldsymbol{\alpha}_1 \times {}^1\mathbf{r} + 2\,{}^1_G\boldsymbol{\omega}_1 \times {}^1\mathbf{v} + {}^1_G\boldsymbol{\omega}_1 \times \left({}^1_G\boldsymbol{\omega}_1 \times {}^1\mathbf{r} \right) \tag{2.339}
$$

we can write the Newton equation of motion of m,

$$
{}^1_G\mathbf{F} - m\,{}^1_G\mathbf{g} = m\,{}^1_G\mathbf{a} \tag{2.340}
$$

where $^1\mathbf{F}$ is the applied non-gravitational force on m.

Recalling that

$$\frac{1}{G}\boldsymbol{\alpha}_1 = 0 \tag{2.341}$$

we find the general equation of motion of a particle in frame B_1 as:

$$\frac{1}{G}\mathbf{F} + m\,\frac{1}{G}\mathbf{g} = m\left(^1\mathbf{a} + 2\,\frac{1}{G}\boldsymbol{\omega}_1 \times {}^1\mathbf{v} + \frac{1}{G}\boldsymbol{\omega}_1 \times \left(\frac{1}{G}\boldsymbol{\omega}_1 \times {}^1\mathbf{r}\right)\right) \tag{2.342}$$

The individual vectors in this equation are:

$$^1\mathbf{g} = \begin{bmatrix} 0 \\ 0 \\ -g_0 \end{bmatrix} \qquad \frac{1}{G}\mathbf{F} = \begin{bmatrix} F_x \\ F_y \\ F_z \end{bmatrix} \qquad \frac{1}{G}\boldsymbol{\omega}_1 = \begin{bmatrix} \omega_E \cos\lambda \\ 0 \\ \omega_E \sin\lambda \end{bmatrix} \tag{2.343}$$

$$^1\mathbf{v} = \begin{bmatrix} \dot{x}_1 \\ \dot{y}_1 \\ \dot{z}_1 \end{bmatrix} = \begin{bmatrix} l\dot{\beta}\cos\beta\cos\gamma - l\dot{\gamma}\sin\beta\sin\gamma \\ l\dot{\beta}\cos\beta\sin\gamma + l\dot{\gamma}\cos\gamma\sin\beta \\ l\dot{\beta}\sin\beta \end{bmatrix} \tag{2.344}$$

$$^1\mathbf{a} = \begin{bmatrix} \ddot{x} \\ \ddot{y} \\ \ddot{z} \end{bmatrix} = \begin{bmatrix} l\left(\ddot{\beta}\cos\gamma - \dot{\beta}^2\sin\gamma - \dot{\beta}\dot{\gamma}\sin\gamma\right)\cos\beta \\ \quad -l\left(\ddot{\gamma}\sin\gamma + \dot{\gamma}^2\cos\gamma + \dot{\beta}\dot{\gamma}\cos\gamma\right)\sin\beta \\ l\left(\ddot{\beta}\sin\gamma + \dot{\beta}^2\cos\gamma + \dot{\beta}\dot{\gamma}\cos\gamma\right)\cos\beta \\ \quad +l\left(\ddot{\gamma}\cos\gamma - \dot{\gamma}^2\sin\gamma - \dot{\beta}\dot{\gamma}\sin\gamma\right)\sin\beta \\ l\ddot{\beta}\sin\beta \end{bmatrix} \tag{2.345}$$

For a spherical pendulum, the external force $^1\mathbf{F}$ is the tension of the string.

$$\frac{1}{G}\mathbf{F} = -\frac{F}{l}\,^1\mathbf{r} \tag{2.346}$$

Substituting the above vectors in (2.342) provides us with three coupled ordinary differential equations for two angular variables γ and β. One of the equations is not independent, and the others may theoretically be integrated to determine $\gamma = \gamma(t)$ and $\beta = \beta(t)$.

For example, let us use

$$\omega_E \approx 7.292\,1 \times 10^{-5}\,\text{rad}/\text{s} \qquad g_0 \approx 9.81\,\text{m}/\text{s}^2$$

$$l = 100\,\text{m} \qquad x_0 = l\cos 10 = 17.365\,\text{m}$$

$$\lambda = 28°58'30''\ N \approx 28.975\,\text{deg}\ N$$

$$\varphi = 50°50'17''\ E \approx 50.838\,\text{deg}\ E \tag{2.347}$$

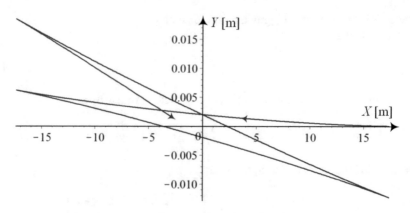

Fig. 2.21 The projection of the path of a pendulum with length $l = 100\,\text{m}$ at latitude $\lambda \approx 28.975\,\text{deg}\ N$ on Earth for a few oscillations (not to scale)

and find

$$x = 8.6839 \cos{(0.313\,16t)} + 8.6811 \cos{(-0.313\,26t)} \tag{2.348}$$

$$y = 8.6839 \sin{(0.313\,16t)} + 8.6811 \sin{(-0.313\,26t)} \tag{2.349}$$

At the given latitude, which approximately corresponds to several large cities of the world such as New Delhi, India; Fujian, China; Chihuahua, Mexico; Island of Tenerife, Spain; Orlando, Florida; and Bushehr, Iran, on the shore of Persian Gulf, the plane of oscillation turns about the local **g**-axis with an angular speed $\omega = -3.5325 \times 10^{-5}\,\text{rad}\,/\,\text{s} \approx -87.437\,\text{deg}\,/\,\text{d}$. These results are independent of longitude. Therefore, the same phenomena will be seen at all places at the same latitude. Figure 2.21 depicts the projection of m on the (x, y)-plane for a few oscillations. The (x, y)-plane is the tangential plan to the Earth at p. The oscillation takes $T \approx 49.4\,\text{h}$ for the pendulum to turn 2π:

$$T = \frac{2\pi}{3.5325 \times 10^{-5}} = 1.7787 \times 10^5\,\text{s} = 49.408\,\text{h} \tag{2.350}$$

However, the pendulum gets back to the (y, x)-plane after $t = T/2 = 24.704\,\text{h}$. By that time, the pendulum must have oscillated about $n \approx 4433$ times:

$$n = \frac{\omega_n}{2\pi} \frac{T}{2} = \frac{0.313\,21}{2\pi} \frac{1.7787 \times 10^5}{2} = 4433.3 \tag{2.351}$$

$$\omega_n = \sqrt{\frac{g}{l}} \tag{2.352}$$

By shortening the length of the pendulum, say $l = 1\,\text{m}$, the rotation speed remains the same while the number of oscillations increases to $n \approx 44333$.

2.3.4 ★ *Rotational Dynamics*

The rigid body rotational equations of motion come from the *Euler equation*,

$$
{}^B\mathbf{M} = \frac{{}^G d}{dt} {}^B\mathbf{L} = {}^B\dot{\mathbf{L}} + {}^B_G\boldsymbol{\omega}_B \times {}^B\mathbf{L}
$$

$$
= {}^BI \, {}^B_G\dot{\boldsymbol{\omega}}_B + {}^B_G\boldsymbol{\omega}_B \times \left({}^BI \, {}^B_G\boldsymbol{\omega}_B \right) \tag{2.353}
$$

where $\mathbf{L}$ is the *angular momentum* and I is the *mass moment* of the rigid body.

$$
{}^B\mathbf{L} = {}^BI \, {}^B_G\boldsymbol{\omega}_B \tag{2.354}
$$

$$
I = \begin{bmatrix} I_{xx} & I_{xy} & I_{xz} \\ I_{yx} & I_{yy} & I_{yz} \\ I_{zx} & I_{zy} & I_{zz} \end{bmatrix} \tag{2.355}
$$

The expanded form of the Euler equation (2.353) is:

$$
M_x = I_{xx}\dot{\omega}_x + I_{xy}\dot{\omega}_y + I_{xz}\dot{\omega}_z - \left(I_{yy} - I_{zz} \right)\omega_y\omega_z
$$

$$
- I_{yz}\left(\omega_z^2 - \omega_y^2 \right) - \omega_x\left(\omega_z I_{xy} - \omega_y I_{xz} \right) \tag{2.356}
$$

$$
M_y = I_{yx}\dot{\omega}_x + I_{yy}\dot{\omega}_y + I_{yz}\dot{\omega}_z - \left(I_{zz} - I_{xx} \right)\omega_z\omega_x
$$

$$
- I_{xz}\left(\omega_x^2 - \omega_z^2 \right) - \omega_y\left(\omega_x I_{yz} - \omega_z I_{xy} \right) \tag{2.357}
$$

$$
M_z = I_{zx}\dot{\omega}_x + I_{zy}\dot{\omega}_y + I_{zz}\dot{\omega}_z - \left(I_{xx} - I_{yy} \right)\omega_x\omega_y
$$

$$
- I_{xy}\left(\omega_y^2 - \omega_x^2 \right) - \omega_z\left(\omega_y I_{xz} - \omega_x I_{yz} \right) \tag{2.358}
$$

When the body coordinate is the *principal coordinate frame*, Euler equations reduce to:

$$
M_1 = I_1\dot{\omega}_1 - (I_2 - I_2)\,\omega_2\omega_3
$$

$$
M_2 = I_2\dot{\omega}_2 - (I_3 - I_1)\,\omega_3\omega_1 \tag{2.359}
$$

$$
M_3 = I_3\dot{\omega}_3 - (I_1 - I_2)\,\omega_1\omega_2
$$

The principal coordinate frame is denoted by numbers 123, instead of xyz, to indicate the first, second, and third *principal axes*. The parameters I_{ij}, $i \neq j$, are zero in the principal frame. The principal body coordinate frame sits at the mass center C.

The kinetic energy K of a rotating rigid body is:

$$K = \frac{1}{2}\left(I_{xx}\omega_x^2 + I_{yy}\omega_y^2 + I_{zz}\omega_z^2\right)$$
$$-I_{xy}\omega_x\omega_y - I_{yz}\omega_y\omega_z - I_{zx}\omega_z\omega_x \qquad (2.360)$$
$$= \frac{1}{2}\boldsymbol{\omega}\cdot\mathbf{L} = \frac{1}{2}\boldsymbol{\omega}^T I \boldsymbol{\omega} \qquad (2.361)$$

which in the principal coordinate frame reduces to:

$$K = \frac{1}{2}\left(I_1\omega_1^2 + I_2\omega_2^2 + I_3\omega_3^2\right) \qquad (2.362)$$

Proof Let m_i be the mass of the ith particle of a rigid body B, which is made of n particles. Let $\mathbf{r}_i$ be the Cartesian position vector of m_i in a body coordinate frame $B\,(Oxyz)$.

$$\mathbf{r}_i = {}^B\mathbf{r}_i = \begin{bmatrix} x_i & y_i & z_i \end{bmatrix}^T \qquad (2.363)$$

Assume that $\boldsymbol{\omega}$ is the angular velocity of the rigid body with respect to the global coordinate frame $G\,(OXYZ)$, expressed in the body coordinate frame.

$$\boldsymbol{\omega} = {}^B_G\boldsymbol{\omega}_B = \begin{bmatrix} \omega_x & \omega_y & \omega_z \end{bmatrix}^T \qquad (2.364)$$

The angular momentum of m_i is $\mathbf{L}_i$.

$$\mathbf{L}_i = \mathbf{r}_i \times m_i\dot{\mathbf{r}}_i = m_i\,[\mathbf{r}_i \times (\boldsymbol{\omega} \times \mathbf{r}_i)]$$
$$= m_i\,[(\mathbf{r}_i \cdot \mathbf{r}_i)\,\boldsymbol{\omega} - (\mathbf{r}_i \cdot \boldsymbol{\omega})\,\mathbf{r}_i]$$
$$= m_i r_i^2\boldsymbol{\omega} - m_i\,(\mathbf{r}_i \cdot \boldsymbol{\omega})\,\mathbf{r}_i \qquad (2.365)$$

Hence, the angular momentum of the whole rigid body would be:

$$\mathbf{L} = \boldsymbol{\omega}\sum_{i=1}^n m_i r_i^2 - \sum_{i=1}^n m_i\,(\mathbf{r}_i \cdot \boldsymbol{\omega})\,\mathbf{r}_i \qquad (2.366)$$

Substitution for $\mathbf{r}_i$ and $\boldsymbol{\omega}$ in $\mathbf{L}$ yields:

$$\mathbf{L} = \left(\omega_x\hat{i} + \omega_y\hat{j} + \omega_z\hat{k}\right)\sum_{i=1}^n m_i\left(x_i^2 + y_i^2 + z_i^2\right)$$
$$-\sum_{i=1}^n m_i\left(x_i\omega_x + y_i\omega_y + z_i\omega_z\right)\cdot\left(x_i\hat{i} + y_i\hat{j} + z_i\hat{k}\right) \qquad (2.367)$$

which can be rearranged as:

$$\mathbf{L} = \sum_{i=1}^{n} m_i \left(y_i^2 + z_i^2 \right) \omega_x \hat{i} + \sum_{i=1}^{n} m_i \left(z_i^2 + x_i^2 \right) \omega_y \hat{j} + \sum_{i=1}^{n} m_i \left(x_i^2 + y_i^2 \right) \omega_z \hat{k}$$

$$- \left(\sum_{i=1}^{n} (m_i x_i y_i) \omega_y + \sum_{i=1}^{n} (m_i x_i z_i) \omega_z \right) \hat{i}$$

$$- \left(\sum_{i=1}^{n} (m_i y_i z_i) \omega_z + \sum_{i=1}^{n} (m_i y_i x_i) \omega_x \right) \hat{j}$$

$$- \left(\sum_{i=1}^{n} (m_i z_i x_i) \omega_x + \sum_{i=1}^{n} (m_i z_i y_i) \omega_y \right) \hat{k} \qquad (2.368)$$

By introducing the mass moment matrix I with the following elements,

$$I_{xx} = \sum_{i=1}^{n} \left[m_i \left(y_i^2 + z_i^2 \right) \right] \qquad (2.369)$$

$$I_{yy} = \sum_{i=1}^{n} \left[m_i \left(z_i^2 + x_i^2 \right) \right] \qquad (2.370)$$

$$I_{zz} = \sum_{i=1}^{n} \left[m_i \left(x_i^2 + y_i^2 \right) \right] \qquad (2.371)$$

$$I_{xy} = I_{yx} = - \sum_{i=1}^{n} (m_i x_i y_i) \qquad (2.372)$$

$$I_{yz} = I_{zy} = - \sum_{i=1}^{n} (m_i y_i z_i) \qquad (2.373)$$

$$I_{zx} = I_{xz} = - \sum_{i=1}^{n} (m_i z_i x_i) . \qquad (2.374)$$

we may rewrite the angular momentum $\mathbf{L}$ in concise form.

$$L_x = I_{xx} \omega_x + I_{xy} \omega_y + I_{xz} \omega_z \qquad (2.375)$$

$$L_y = I_{yx} \omega_x + I_{yy} \omega_y + I_{yz} \omega_z \qquad (2.376)$$

$$L_z = I_{zx} \omega_x + I_{zy} \omega_y + I_{zz} \omega_z \qquad (2.377)$$

$$\mathbf{L} = I \cdot \boldsymbol{\omega} \tag{2.378}$$

$$\begin{bmatrix} L_x \\ L_y \\ L_z \end{bmatrix} = \begin{bmatrix} I_{xx} & I_{xy} & I_{xz} \\ I_{yx} & I_{yy} & I_{yz} \\ I_{zx} & I_{zy} & I_{zz} \end{bmatrix} \begin{bmatrix} \omega_x \\ \omega_y \\ \omega_z \end{bmatrix} \tag{2.379}$$

For a rigid body that is a continuous solid, the summations must be replaced by integrations over the volume of the body.

If $^B\mathbf{M}$ denotes the resultant of the external moments applied on the rigid body, the Euler equation of motion for a rigid body will be:

$$^B\mathbf{M} = \frac{^G d}{dt} \, {}^B\mathbf{L} \tag{2.380}$$

The angular momentum $^B\mathbf{L}$ is a vector quantity defined in the body coordinate frame. Hence, its time derivative in the global coordinate frame is:

$$\frac{^G d \, {}^B\mathbf{L}}{dt} = {}^B\dot{\mathbf{L}} + {}^B_G\boldsymbol{\omega}_B \times {}^B\mathbf{L}. \tag{2.381}$$

Therefore, the vectorial form of the Euler equation of motion is

$$^B\mathbf{M} = \frac{d\mathbf{L}}{dt} = \dot{\mathbf{L}} + \boldsymbol{\omega} \times \mathbf{L} = I\dot{\boldsymbol{\omega}} + \boldsymbol{\omega} \times (I\boldsymbol{\omega}) \tag{2.382}$$

and in expanded form is:

$$\begin{aligned} ^B\mathbf{M} = &\left(I_{xx}\dot{\omega}_x + I_{xy}\dot{\omega}_y + I_{xz}\dot{\omega}_z\right)\hat{\imath} + \omega_y\left(I_{xz}\omega_x + I_{yz}\omega_y + I_{zz}\omega_z\right)\hat{\imath} \\ &-\omega_z\left(I_{xy}\omega_x + I_{yy}\omega_y + I_{yz}\omega_z\right)\hat{\imath} \\ &+\left(I_{yx}\dot{\omega}_x + I_{yy}\dot{\omega}_y + I_{yz}\dot{\omega}_z\right)\hat{\jmath} + \omega_z\left(I_{xx}\omega_x + I_{xy}\omega_y + I_{xz}\omega_z\right)\hat{\jmath} \\ &-\omega_x\left(I_{xz}\omega_x + I_{yz}\omega_y + I_{zz}\omega_z\right)\hat{\jmath} \\ &+\left(I_{zx}\dot{\omega}_x + I_{zy}\dot{\omega}_y + I_{zz}\dot{\omega}_z\right)\hat{k} + \omega_x\left(I_{xy}\omega_x + I_{yy}\omega_y + I_{yz}\omega_z\right)\hat{k} \\ &-\omega_y\left(I_{xx}\omega_x + I_{xy}\omega_y + I_{xz}\omega_z\right)\hat{k} \end{aligned} \tag{2.383}$$

Therefore, the most general form of the Euler equations of motion for a rigid body in a body frame attached to C is:

$$\begin{aligned} M_x = &I_{xx}\dot{\omega}_x + I_{xy}\dot{\omega}_y + I_{xz}\dot{\omega}_z - \left(I_{yy} - I_{zz}\right)\omega_y\omega_z \\ &-I_{yz}\left(\omega_z^2 - \omega_y^2\right) - \omega_x\left(\omega_z I_{xy} - \omega_y I_{xz}\right) \end{aligned} \tag{2.384}$$

$$M_y = I_{yx}\dot{\omega}_x + I_{yy}\dot{\omega}_y + I_{yz}\dot{\omega}_z - (I_{zz} - I_{xx})\,\omega_z\omega_x$$

$$-I_{xz}\left(\omega_x^2 - \omega_z^2\right) - \omega_y\left(\omega_x I_{yz} - \omega_z I_{xy}\right) \tag{2.385}$$

$$M_z = I_{zx}\dot{\omega}_x + I_{zy}\dot{\omega}_y + I_{zz}\dot{\omega}_z - \left(I_{xx} - I_{yy}\right)\omega_x\omega_y$$

$$-I_{xy}\left(\omega_y^2 - \omega_x^2\right) - \omega_z\left(\omega_y I_{xz} - \omega_x I_{yz}\right) \tag{2.386}$$

Assume that we are able to rotate the body frame about its origin to find an orientation that makes. In a *principal coordinate frame*, we have $I_{ij} = 0$, for $i \neq j$, and hence, the equations simplify to:

$$M_1 = I_1\dot{\omega}_1 - (I_2 - I_2)\,\omega_2\omega_3 \tag{2.387}$$

$$M_2 = I_2\dot{\omega}_2 - (I_3 - I_1)\,\omega_3\omega_1 \tag{2.388}$$

$$M_3 = I_3\dot{\omega}_3 - (I_1 - I_2)\,\omega_1\omega_2 \tag{2.389}$$

The kinetic energy of a rigid body may be found by the integral of the kinetic energy of the mass element dm, over the whole body:

$$K = \frac{1}{2}\int_B \dot{\mathbf{v}}^2 dm = \frac{1}{2}\int_B (\boldsymbol{\omega} \times \mathbf{r}) \cdot (\boldsymbol{\omega} \times r)\,dm$$

$$= \frac{\omega_x^2}{2}\int_B \left(y^2 + z^2\right)dm + \frac{\omega_y^2}{2}\int_B \left(z^2 + x^2\right)dm + \frac{\omega_z^2}{2}\int_B \left(x^2 + y^2\right)dm$$

$$-\omega_x\omega_y \int_B xy\,dm - \omega_y\omega_z \int_B yz\,dm - \omega_z\omega_x \int_B zx\,dm$$

$$= \frac{1}{2}\left(I_{xx}\omega_x^2 + I_{yy}\omega_y^2 + I_{zz}\omega_z^2\right)$$

$$-I_{xy}\omega_x\omega_y - I_{yz}\omega_y\omega_z - I_{zx}\omega_z\omega_x \tag{2.390}$$

The kinetic energy can be rearranged to a matrix multiplication form.

$$K = \frac{1}{2}\boldsymbol{\omega}^T I\,\boldsymbol{\omega} = \frac{1}{2}\boldsymbol{\omega} \cdot \mathbf{L} \tag{2.391}$$

When the body frame is principal, the kinetic energy will be simplified.

$$K = \frac{1}{2}\left(I_1\omega_1^2 + I_2\omega_2^2 + I_3\omega_3^2\right) \tag{2.392}$$

Kinetic energy is what we need to apply energy method, as well as developing Lagrange method. ∎

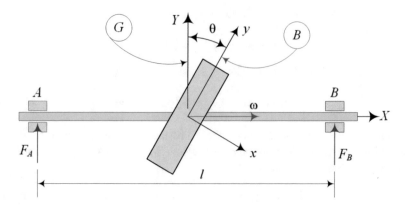

Fig. 2.22 A disc with mass m and radius r, mounted on a massless turning shaft

Example 63 ★ A tilted disc on a massless shaft. Rotation of a rigid body about a fixed axis is the simplest example of a rotating rigid body and is the most common case in industry. Here is to show how to develop equations of motion and determine supporting forces.

Figure 2.22 illustrates a disc with mass m and radius R, mounted on a massless shaft. The shaft is turning with a constant angular speed ω. The disc is attached to the shaft at an angle θ. Because of θ, the bearings at A and B must support a rotating force.

To model the system, we attach a principal body coordinate frame at the disc center as shown in the figure. The G-expression of the angular velocity is a constant vector

$$^G\boldsymbol{\omega}_B = \omega\hat{I} \tag{2.393}$$

and the expression of the angular velocity vector in the body frame is:

$$^B_G\boldsymbol{\omega}_B = \omega\cos\theta\,\hat{\imath} + \omega\sin\theta\,\hat{\jmath} \tag{2.394}$$

The mass moment of inertia matrix of the disc in its body frame is:

$$^B I = \begin{bmatrix} mR^2/2 & 0 & 0 \\ 0 & mR^2/4 & 0 \\ 0 & 0 & mR^2/4 \end{bmatrix} \tag{2.395}$$

Substituting (2.394) and (2.395) in (2.387)–(2.389), with $1 \equiv x$, $2 \equiv y$, $3 \equiv z$, yields

$$M_x = 0 \qquad M_y = 0 \tag{2.396}$$

$$M_z = \frac{mr^2}{4}\omega \cos\theta \sin\theta \tag{2.397}$$

Therefore, the bearing reaction forces F_A and F_B are:

$$F_A = -F_B = -\frac{M_z}{l} = -\frac{mr^2}{4l}\omega \cos\theta \sin\theta \tag{2.398}$$

Example 64 Steady rotation of a freely rotating rigid body. Motion of a particle with no external force is rest or steady-state motion on a straight line with constant velocity. However, motion of a rigid body with no external moment is more complicated. Here is the equations of motion and the conditions to have a steady-state rotation.

Consider a situation in which the resultant applied force and moment on a rigid body are zero.

$$^G\mathbf{F} = \, ^B\mathbf{F} = 0 \tag{2.399}$$

$$^G\mathbf{M} = \, ^B\mathbf{M} = 0 \tag{2.400}$$

Based on the Newton's equation,

$$^G\mathbf{F} = m \, ^G\dot{\mathbf{v}} \tag{2.401}$$

the velocity of the mass center will be constant in the global coordinate frame. However, the Euler equation

$$^B\mathbf{M} = I \, _G^B\dot{\boldsymbol{\omega}}_B + \, _G^B\boldsymbol{\omega}_B \times \, ^B\mathbf{L} \tag{2.402}$$

reduces to

$$\dot{\omega}_1 = \frac{I_2 - I_3}{I_1}\omega_2\omega_3 \tag{2.403}$$

$$\dot{\omega}_2 = \frac{I_3 - I_1}{I_{22}}\omega_3\omega_1 \tag{2.404}$$

$$\dot{\omega}_3 = \frac{I_1 - I_2}{I_3}\omega_1\omega_2 \tag{2.405}$$

which show that the angular velocity can be constant if

$$I_1 = I_2 = I_3 \tag{2.406}$$

or if two principal moments of inertia, say I_1 and I_2, are zero and the third angular velocity, in this case ω_3, is initially zero, or if the angular velocity vector is initially parallel to a principal axis.

2.4 Lagrange Method

The Lagrange method in deriving the equations of motion of vibrating systems has some advantages over Newton-Euler due to its simplicity and generality, especially for multi-degree-of-freedom (DOF) systems. Let us assume that for some forces **F** there is a *potential energy* function P such that the force is derivable from P. Such a force is called a *potential* or *conservative force*.

$$\mathbf{F} = -\nabla P = \frac{\partial P}{\partial x}\hat{\imath} + \frac{\partial P}{\partial y}\hat{\jmath} + \frac{\partial P}{\partial z}\hat{k}$$

$$= \left[F_x \; F_y \; F_z \right]^T \tag{2.407}$$

The Lagrange equation of motion of a dynamic system can be written as

$$\frac{d}{dt}\left(\frac{\partial \mathcal{L}}{\partial \dot{q}_r}\right) - \frac{\partial \mathcal{L}}{\partial q_r} = Q_r \qquad r = 1, 2, \cdots n \tag{2.408}$$

where $\mathcal{L}$ is the *Lagrangean* of the system

$$\mathcal{L} = K - P \tag{2.409}$$

and Q_r is the nonpotential generalized force for which there is no potential function.

$$Q_r = \sum_{i=1}^{n} \left(F_{ix}\frac{\partial f_i}{\partial q_1} + F_{iy}\frac{\partial g_i}{\partial q_2} + F_{iz}\frac{\partial h_i}{\partial q_n} \right) \tag{2.410}$$

Proof Assume that the external forces $\mathbf{F} = \left[F_x \; F_y \; F_z \right]^T$ acting on the system are conservative:

$$\mathbf{F} = -\nabla P \tag{2.411}$$

The work done by these forces in an arbitrary virtual displacement δq_1, δq_2, δq_3, $\cdots$, δq_n is

$$\partial W = -\frac{\partial P}{\partial q_1}\delta q_1 - \frac{\partial P}{\partial q_2}\delta q_2 - \cdots \frac{\partial P}{\partial q_n}\delta q_n \tag{2.412}$$

and then the Lagrange equation becomes:

$$\frac{d}{dt}\left(\frac{\partial K}{\partial \dot{q}_r}\right) - \frac{\partial K}{\partial q_r} = -\frac{\partial P}{\partial q_1} \qquad r = 1, 2, \cdots, n \tag{2.413}$$

Introducing the Lagrangean function $\mathcal{L} = K - P$ converts the Lagrange equation of a conservative system to:

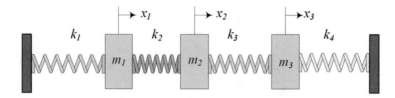

Fig. 2.23 An undamped three-DOF system

$$\frac{d}{dt}\left(\frac{\partial \mathcal{L}}{\partial \dot{q}_r}\right) - \frac{\partial \mathcal{L}}{\partial q_r} = 0 \qquad r = 1, 2, \cdots n \tag{2.414}$$

If there is a nonpotential force, then the virtual work done by the force is:

$$\delta W = \sum_{i=1}^{n}\left(F_{xi}\frac{\partial f_i}{\partial q_r} + F_{yi}\frac{\partial g_i}{\partial q_r} + F_{zi}\frac{\partial h_i}{\partial q_r}\right)\delta q_r$$

$$= Q_r \, \delta q_r \tag{2.415}$$

and the Lagrange equation of motion would be:

$$\frac{d}{dt}\left(\frac{\partial \mathcal{L}}{\partial \dot{q}_r}\right) - \frac{\partial \mathcal{L}}{\partial q_r} = Q_r \qquad r = 1, 2, \cdots n \tag{2.416}$$

where Q_r is the nonpotential generalized force doing work in a virtual displacement of the rth generalized coordinate q_r.

The Lagrangean $\mathcal{L}$ is also called the *kinetic potential*. ∎

Example 65 An undamped three DOF system. An example of a multi DOF vibrating system and showing how to derive the equations of motion by applying Lagrange method.

Figure 2.23 illustrates an undamped three DOF linear vibrating systems. The kinetic and potential energies of the system are:

$$K = \frac{1}{2}m_1\dot{x}_1^2 + \frac{1}{2}m_2\dot{x}_2^2 + \frac{1}{2}m_3\dot{x}_3^2 \tag{2.417}$$

$$P = \frac{1}{2}k_1x_1^2 + \frac{1}{2}k_2(x_1 - x_2)^2 + \frac{1}{2}k_3(x_2 - x_3)^2 + \frac{1}{2}k_4x_3^2 \tag{2.418}$$

Because there is no damping in the system, we may find the Lagrangean $\mathcal{L}$ and use Eq. (2.499) with $Q_r = 0$.

$$\mathcal{L} = K - P \tag{2.419}$$

$$\frac{\partial \mathcal{L}}{\partial x_1} = -k_1 x_1 - k_2 (x_1 - x_2) \tag{2.420}$$

$$\frac{\partial \mathcal{L}}{\partial x_2} = k_2 (x_1 - x_2) - k_3 (x_2 - x_3) \tag{2.421}$$

$$\frac{\partial \mathcal{L}}{\partial x_3} = k_3 (x_2 - x_3) - k_4 x_3 \tag{2.422}$$

$$\frac{\partial \mathcal{L}}{\partial \dot{x}_1} = m_1 \dot{x}_1 \qquad \frac{\partial \mathcal{L}}{\partial \dot{x}_2} = m_2 \dot{x}_2 \qquad \frac{\partial \mathcal{L}}{\partial \dot{x}_3} = m_3 \dot{x}_3 \tag{2.423}$$

The equations of motion are:

$$m_1 \ddot{x}_1 + k_1 x_1 + k_2 (x_1 - x_2) = 0 \tag{2.424}$$

$$m_2 \ddot{x}_2 - k_2 (x_1 - x_2) + k_3 (x_2 - x_3) = 0 \tag{2.425}$$

$$m_3 \ddot{x}_3 - k_3 (x_2 - x_3) + k_4 x_3 = 0 \tag{2.426}$$

These equations can be rewritten in matrix form for simpler calculation.

$$\begin{bmatrix} m_1 & 0 & 0 \\ 0 & m_2 & 0 \\ 0 & 0 & m_3 \end{bmatrix} \begin{bmatrix} \ddot{x}_1 \\ \ddot{x}_2 \\ \ddot{x}_3 \end{bmatrix}$$

$$+ \begin{bmatrix} k_1 + k_2 & -k_2 & 0 \\ -k_2 & k_2 + k_3 & -k_3 \\ 0 & -k_3 & k_3 + k_4 \end{bmatrix} \begin{bmatrix} x_1 \\ x_2 \\ x_3 \end{bmatrix} = 0 \tag{2.427}$$

Example 66 Spherical pendulum. Spherical pendulum is a system with only one point mass and two *DOF*. Here is how to derive its equations of motion.

The pendulum analogy is utilized to model many dynamic problems. Figure 2.24 illustrates a spherical pendulum with mass m and length l. The angles φ and θ may be used as describing coordinates of the system. The Cartesian coordinates of the mass as functions of the generalized coordinates are:

$$\begin{bmatrix} X \\ Y \\ Z \end{bmatrix} = \begin{bmatrix} r \cos \varphi \sin \theta \\ r \sin \theta \sin \varphi \\ -r \cos \theta \end{bmatrix} \tag{2.428}$$

Therefore, the kinetic and potential energies of the pendulum are:

$$K = \frac{1}{2} m \left(l^2 \dot{\theta}^2 + l^2 \dot{\varphi}^2 \sin^2 \theta \right) \tag{2.429}$$

$$P = -mgl \cos \theta \tag{2.430}$$

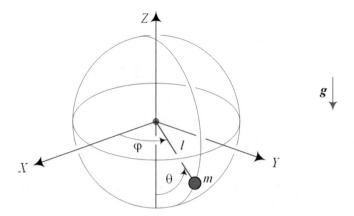

Fig. 2.24 A spherical pendulum

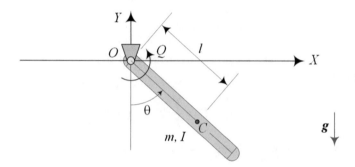

Fig. 2.25 A controlled compound pendulum

The Lagrangean $\mathcal{L}$ of this system is:

$$\mathcal{L} = \frac{1}{2}m\left(l^2\dot{\theta}^2 + l^2\dot{\varphi}^2 \sin^2\theta\right) + mgl\cos\theta \tag{2.431}$$

which leads to the following equations of motion.

$$\ddot{\theta} - \dot{\varphi}^2 \sin\theta \cos\theta + \frac{g}{l}\sin\theta = 0 \tag{2.432}$$

$$\ddot{\varphi}\sin^2\theta + 2\dot{\varphi}\dot{\theta}\sin\theta\cos\theta = 0 \tag{2.433}$$

Example 67 Controlled compound pendulum. A sample of a dynamic system with nonpotential force. The nonpotential force is an external torque to control position of a compound pendulum.

A massive arm is attached to the ceiling at a pin joint O as illustrated in Fig. 2.25. Assume there is viscous friction in the joint where an ideal motor can apply a torque

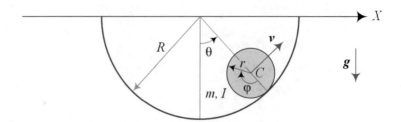

Fig. 2.26 A uniform disc, rolling in a circular path

Q to move the arm. The rotor of an ideal motor has no mass moment by assumption. The kinetic and potential energies of the controlled pendulum are:

$$K = \frac{1}{2} I \dot{\theta}^2 = \frac{1}{2} \left(I_C + m l^2 \right) \dot{\theta}^2 \tag{2.434}$$

$$P = -mg \cos \theta \tag{2.435}$$

where m is the mass and I is the mass moment of the pendulum about O. The Lagrangean of the manipulator is:

$$\mathcal{L} = K - P = \frac{1}{2} I \dot{\theta}^2 + mg \cos \theta \tag{2.436}$$

and therefore, the equation of motion of the pendulum is:

$$M = \frac{d}{dt} \left(\frac{\partial \mathcal{L}}{\partial \dot{\theta}} \right) - \frac{\partial \mathcal{L}}{\partial \theta} = I \ddot{\theta} + mgl \sin \theta \tag{2.437}$$

The generalized force M is the contribution of the motor torque Q and the viscous friction torque $-c\dot{\theta}$. Hence, the equation of motion of the pendulum would be:

$$Q = I \ddot{\theta} + c \dot{\theta} + mgl \sin \theta \tag{2.438}$$

Example 68 A rolling disc in a circular path. It is an example of a rotating rigid body with some kinematic constraints to derive the equation of motion.

Figure 2.26 illustrates a uniform disc with mass m and radius r. The disc is rolling without slip in a circular path of radius R. The disc may have free oscillations around $\theta = 0$. To find the equation of motion, we employ the Lagrange method. The energies of the system are:

$$K = \frac{1}{2}mv_C^2 + \frac{1}{2}I_c\omega^2$$

$$= \frac{1}{2}m(R-r)^2\dot{\theta}^2 + \frac{1}{2}\left(\frac{1}{2}mr^2\right)(\dot{\varphi} - \dot{\theta})^2 \qquad (2.439)$$

$$P = -mg(R-r)\cos\theta \qquad (2.440)$$

When there is no slip, there is a constraint between θ and φ ,

$$R\theta = r\varphi \qquad R\dot{\theta} = r\dot{\varphi} \qquad (2.441)$$

which can be used to eliminate φ from K.

$$K = \frac{3}{4}m(R-r)^2\dot{\theta}^2 \qquad (2.442)$$

Based on the partial derivatives

$$\frac{d}{dt}\left(\frac{\partial\mathcal{L}}{\partial\dot{\theta}}\right) = \frac{3}{2}m(R-r)^2\ddot{\theta} \qquad (2.443)$$

$$\frac{\partial\mathcal{L}}{\partial\theta} = mg(R-r)\sin\theta \qquad (2.444)$$

we find the equation of motion for the oscillating disc.

$$\frac{3}{2}(R-r)\ddot{\theta} + g\sin\theta = 0 \qquad (2.445)$$

When θ is very small, this equation is equivalent to a mass-spring system with $m_e = 3(R-r)$ and $k_e = 2g$.

$$m_e\ddot{\theta} + 2g\theta = 0 \qquad (2.446)$$

Example 69 ★ A double pendulum. Double pendulum is a very good example of nonlinear coupled multi *DOF* vibrating system. A good example to show that Lagrange method is simpler than Newton method to derive equations of motion of multi *DOF* systems.

Figure 2.27 illustrates a double pendulum. There are two massless rods with lengths l_1 and l_2, and two point masses m_1 and m_2, subject to gravity forces and constrained by the hinges in the rods to move in a plane. The variables θ_1 and θ_2 are being used as the generalized coordinates to express the system configuration.

First let us derive the equations of motion by Lagrange method. To calculate the Lagrangean of the system and derive the equations of motion, we start by defining the global position of the masses.

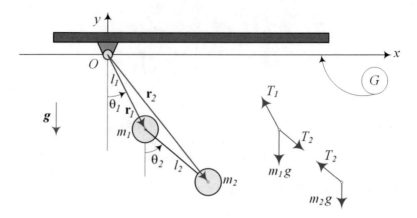

Fig. 2.27 A double pendulum

$$x_1 = l_1 \sin \theta_1 \qquad y_1 = -l_1 \cos \theta_1 \tag{2.447}$$

$$x_2 = l_1 \sin \theta_1 + l_2 \sin \theta_2 \tag{2.448}$$

$$y_2 = -l_1 \cos \theta_1 - l_2 \cos \theta_2 \tag{2.449}$$

The time derivatives of the coordinates are:

$$\dot{x}_1 = l_1 \dot{\theta}_1 \cos \theta_1 \qquad \dot{y}_1 = l_1 \dot{\theta}_1 \sin \theta_1 \tag{2.450}$$

$$\dot{x}_2 = l_1 \dot{\theta}_1 \cos \theta_1 + l_2 \dot{\theta}_2 \cos \theta_2 \tag{2.451}$$

$$\dot{y}_2 = l_1 \dot{\theta}_1 \sin \theta_1 + l_2 \dot{\theta}_2 \sin \theta_2 \tag{2.452}$$

and therefore, the squares of the masses' velocities are:

$$v_1^2 = \dot{x}_1^2 + \dot{y}_1^2 = l_1^2 \dot{\theta}_1^2 \tag{2.453}$$

$$v_2^2 = \dot{x}_2^2 + \dot{y}_2^2 = l_1^2 \dot{\theta}_1^2 + l_2^2 \dot{\theta}_2^2 + 2 l_1 l_2 \dot{\theta}_1 \dot{\theta}_2 \cos (\theta_1 - \theta_2) \tag{2.454}$$

The kinetic energy of the pendulum is then equal to:

$$\begin{aligned} K &= \frac{1}{2} m_1 v_1^2 + \frac{1}{2} m_2 v_2^2 \\ &= \frac{1}{2} m_1 l_1^2 \dot{\theta}_1^2 + \frac{1}{2} m_2 \left(l_1^2 \dot{\theta}_1^2 + l_2^2 \dot{\theta}_2^2 + 2 l_1 l_2 \dot{\theta}_1 \dot{\theta}_2 \cos (\theta_1 - \theta_2) \right) \end{aligned} \tag{2.455}$$

The potential energy of the pendulum is equal to the sum of the potentials of each mass.

$$P = m_1 g y_1 + m_2 g y_2$$
$$= -m_1 g l_1 \cos \theta_1 - m_2 g (l_1 \cos \theta_1 + l_2 \cos \theta_2) \qquad (2.456)$$

The kinetic and potential energies constitute the following Lagrangean:

$$\mathcal{L} = K - P$$
$$= \frac{1}{2} m_1 l_1^2 \dot{\theta}_1^2 + \frac{1}{2} m_2 \left(l_1^2 \dot{\theta}_1^2 + l_2^2 \dot{\theta}_2^2 + 2 l_1 l_2 \dot{\theta}_1 \dot{\theta}_2 \cos (\theta_1 - \theta_2) \right)$$
$$+ m_1 g l_1 \cos \theta_1 + m_2 g (l_1 \cos \theta_1 + l_2 \cos \theta_2) \qquad (2.457)$$

Employing Lagrange method we find the following equations of motion:

$$\frac{d}{dt} \left(\frac{\partial \mathcal{L}}{\partial \dot{\theta}_1} \right) - \frac{\partial \mathcal{L}}{\partial \theta_1} = (m_1 + m_2) l_1^2 \ddot{\theta}_1 + m_2 l_1 l_2 \ddot{\theta}_2 \cos (\theta_1 - \theta_2)$$
$$- m_2 l_1 l_2 \dot{\theta}_2^2 \sin (\theta_1 - \theta_2) + (m_1 + m_2) l_1 g \sin \theta_1 = 0 \quad (2.458)$$

$$\frac{d}{dt} \left(\frac{\partial \mathcal{L}}{\partial \dot{\theta}_2} \right) - \frac{\partial \mathcal{L}}{\partial \theta_2} = m_2 l_2^2 \ddot{\theta}_2 + m_2 l_1 l_2 \ddot{\theta}_1 \cos (\theta_1 - \theta_2)$$
$$+ m_2 l_1 l_2 \dot{\theta}_1^2 \sin (\theta_1 - \theta_2) + m_2 l_2 g \sin \theta_2 = 0 \qquad (2.459)$$

Now let us derive the equations of motion by Newton method. Let us indicate the absolute position of m_1 by vector $\mathbf{r}_1$ and the position of m_2 by vector $\mathbf{r}_2$.

$${}^G \mathbf{r}_1 = l_1 \sin \theta_1 \hat{I} - l_1 \cos \theta_1 \hat{J} \qquad (2.460)$$
$${}^G \mathbf{r}_2 = {}^G \mathbf{r}_1 + l_2 \sin \theta_2 \hat{I} - l_2 \cos \theta_2 \hat{J} \qquad (2.461)$$

$${}^G \mathbf{v}_1 = l_1 \dot{\theta}_1 \left(\cos \theta_1 \hat{I} + \sin \theta_1 \hat{J} \right) \qquad (2.462)$$
$${}^G \mathbf{v}_2 = {}^G \mathbf{v}_1 + l_2 \dot{\theta}_2 \left(\cos \theta_2 \hat{I} + \sin \theta_2 \hat{J} \right) \qquad (2.463)$$

$${}^G \mathbf{a}_1 = l_1 \ddot{\theta}_1 \left(\cos \theta_1 \hat{I} + \sin \theta_1 \hat{J} \right) - l_1 \dot{\theta}_1^2 \left(\sin \theta_1 \hat{I} - \cos \theta_1 \hat{J} \right) \quad (2.464)$$
$${}^G \mathbf{a}_2 = {}^G \mathbf{a}_1 + l_2 \ddot{\theta}_2 \left(\cos \theta_2 \hat{I} + \sin \theta_2 \hat{J} \right)$$
$$- l_2 \dot{\theta}_2^2 \left(\sin \theta_2 \hat{I} - \cos \theta_2 \hat{J} \right) \qquad (2.465)$$

The forces on m_1 are the tension in the two rods and gravity. The tension in the upper rod is along the direction $-\mathbf{r}_1$, and the tension force on m_1 due to the lower rod is along the direction $\mathbf{r}_2 - \mathbf{r}_1$. Therefore, the resultant force on m_1 is:

$$
\begin{aligned}
{}^G\mathbf{F}_1 &= T_1 \frac{-\mathbf{r}_1}{|\mathbf{r}_1|} + T_2 \frac{\mathbf{r}_2 - \mathbf{r}_1}{|\mathbf{r}_2 - \mathbf{r}_1|} + m_1 \mathbf{g} \\
&= -\frac{T_1}{l_1} \mathbf{r}_1 + \frac{T_2}{l_2} (\mathbf{r}_2 - \mathbf{r}_1) - m_1 g \hat{J}
\end{aligned}
\tag{2.466}
$$

The forces on m_2 are the tension in the lower rod and gravity. The tension on m_2 is along the direction of $-(\mathbf{r}_2 - \mathbf{r}_1)$.

$$
{}^G\mathbf{F}_2 = -T_2 \frac{\mathbf{r}_2 - \mathbf{r}_1}{|\mathbf{r}_2 - \mathbf{r}_1|} + m_2 \mathbf{g} = -\frac{T_2}{l_2} (\mathbf{r}_2 - \mathbf{r}_1) - m_2 g \hat{J}
\tag{2.467}
$$

Therefore Newton's second law on each particle makes the following equations of motion.

$$
{}^G\mathbf{F}_1 = m_1 {}^G\mathbf{a}_1
\tag{2.468}
$$

$$
{}^G\mathbf{F}_2 = m_2 {}^G\mathbf{a}_2
\tag{2.469}
$$

Substituting the vectors and setting the coefficients of $\hat{I}$ and $\hat{J}$ zero provide us with four differential equations for four unknowns $\theta_1, \theta_2, T_1, T_2$.

$$
m_1 l_1 \left(\ddot{\theta}_1 \cos \theta_1 - \dot{\theta}_1^2 \sin \theta_1 \right) = T_2 \sin \theta_2 - T_1 \sin \theta_1
\tag{2.470}
$$

$$
m_1 l_1 \left(\ddot{\theta}_1 \sin \theta_1 + \dot{\theta}_1^2 \cos \theta_1 \right) = T_1 \cos \theta_1 - T_2 \cos \theta_2 - m_1 g
\tag{2.471}
$$

$$
\begin{aligned}
&- m_2 l_1 \ddot{\theta}_1 \cos \theta_1 - m_2 l_2 \ddot{\theta}_2 \cos \theta_2 \\
&+ m_2 l_1 \dot{\theta}_1^2 \sin \theta_1 + m_2 l_2 \dot{\theta}_2^2 \sin \theta_2 = T_2 \sin \theta_2
\end{aligned}
\tag{2.472}
$$

$$
\begin{aligned}
&m_2 l_1 \ddot{\theta}_1 \sin \theta_1 + m_2 l_2 \ddot{\theta}_2 \sin \theta_2 \\
&+ m_2 l_1 \dot{\theta}_1^2 \cos \theta_1 + m_2 l_2 \dot{\theta}_2^2 \cos \theta_2 = T_2 \cos \theta_2 - m_2 g
\end{aligned}
\tag{2.473}
$$

Multiplying Eq. (2.470) by $\cos \theta_1$ and (2.471) by $\sin \theta_1$ and then adding them make a new equation for $\ddot{\theta}_1$. Multiplying Eq. (2.470) by $\sin \theta_1$ and (2.471) by $\cos \theta_1$ and then subtracting them make a new equation for $\dot{\theta}_1^2$. We perform similar actions on Eqs. (2.472) and (2.473) to derive the following four equations.

$$
m_1 l_1 \ddot{\theta}_1 = T_2 \sin (\theta_2 - \theta_1) - m_1 g \sin \theta_1
\tag{2.474}
$$

$$
m_1 l_1 \dot{\theta}_1^2 = -T_2 \cos (\theta_2 - \theta_1) + T_1 - m_1 g \cos \theta_1
\tag{2.475}
$$

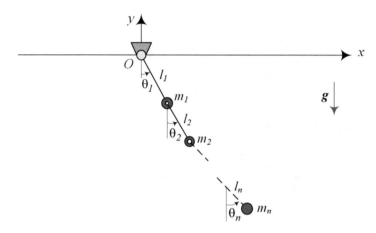

Fig. 2.28 A chain pendulum

$$l_1\ddot{\theta}_1 \cos(\theta_2 - \theta_1) + l_1\dot{\theta}_1^2 \sin(\theta_2 - \theta_1) + l_2\ddot{\theta}_2 = -g\sin\theta_2 \tag{2.476}$$

$$-l_1\ddot{\theta}_1 \sin(\theta_2 - \theta_1) + l_1\dot{\theta}_1^2 \cos(\theta_2 - \theta_1) + l_2\ddot{\theta}_2 = \frac{T_2}{m_2} - g\cos\theta_2 \tag{2.477}$$

Manipulating these equations, we can derive the following two equations of motion, as well as two equations to determine the tension forces T_1, T_2.

$$(m_1 + m_2) l_1\ddot{\theta}_1 + m_2 l_2\ddot{\theta}_2 \cos(\theta_1 - \theta_2)$$

$$-m_2 l_2\dot{\theta}_2^2 \sin(\theta_1 - \theta_2) + (m_1 + m_2) g\sin\theta_1 = 0 \tag{2.478}$$

$$l_2\ddot{\theta}_2 + l_1\ddot{\theta}_1 \cos(\theta_1 - \theta_2)$$

$$+l_1\dot{\theta}_1^2 \sin(\theta_1 - \theta_2) + g\sin\theta_2 = 0 \tag{2.479}$$

$$T_1 = -m_1 \frac{l_2\ddot{\theta}_2}{\sin(\theta_2 - \theta_1)} \tag{2.480}$$

$$T_2 = m_1 \frac{l_1\ddot{\theta}_1 + g\sin\theta_1}{\sin(\theta_2 - \theta_1)} \tag{2.481}$$

Example 70 ★ Chain pendulum. Adding the number of pendulums, we make a multi *DOF* system to model a continuous flexible rope. Here is to derive the equations of motion of a chain of n pendulums.

Consider an n-chain-pendulum as shown in Fig. 2.28. Each pendulum has a massless rod of length l_i with a concentrated point mass m_i and a generalized angular coordinate θ_i measured from the vertical direction. The x_i and y_i components of the mass m_i are:

$$x_i = \sum_{j=1}^{i} l_j \sin \theta_j \qquad y_i = -\sum_{j=1}^{i} l_j \cos \theta_j \qquad (2.482)$$

We find their time derivatives

$$\dot{x}_i = \sum_{j=1}^{i} l_j \dot{\theta}_j \cos \theta_j \qquad \dot{y}_i = \sum_{j=1}^{i} l_j \dot{\theta}_j \sin \theta_j \qquad (2.483)$$

and the square of $\dot{x}_i$ and $\dot{y}_i$

$$\dot{x}_i^2 = \left(\sum_{j=1}^{i} l_j \dot{\theta}_j \cos \theta_j \right) \left(\sum_{k=1}^{i} l_k \dot{\theta}_k \cos \theta_k \right) = \sum_{j=1}^{n} \sum_{k=1}^{n} l_j l_k \dot{\theta}_j \dot{\theta}_k \cos \theta_j \cos \theta_k$$
$$(2.484)$$

$$\dot{y}_i^2 = \left(\sum_{j=1}^{i} l_j \dot{\theta}_j \sin \theta_j \right) \left(\sum_{k=1}^{i} l_k \dot{\theta}_k \sin \theta_k \right) = \sum_{j=1}^{i} \sum_{k=1}^{i} l_j l_k \dot{\theta}_j \dot{\theta}_k \sin \theta_j \sin \theta_k$$
$$(2.485)$$

to calculate the squared of velocity v_i^2 of the mass m_i.

$$v_i^2 = \dot{x}_i^2 + \dot{y}_i^2$$

$$= \sum_{j=1}^{i} \sum_{k=1}^{i} l_j l_k \dot{\theta}_j \dot{\theta}_k \left(\cos \theta_j \cos \theta_k + \sin \theta_j \sin \theta_k \right)$$

$$= \sum_{j=1}^{i} \sum_{k=1}^{i} l_j l_k \dot{\theta}_j \dot{\theta}_k \cos \left(\theta_j - \theta_k \right)$$

$$= \sum_{r=1}^{i} l_r^2 \dot{\theta}_r^2 + 2 \sum_{j=1}^{i} \sum_{k=j+1}^{i} l_j l_k \dot{\theta}_j \dot{\theta}_k \cos \left(\theta_j - \theta_k \right) \qquad (2.486)$$

Now, we calculate the kinetic energy, K, of the chain.

$$K = \frac{1}{2} \sum_{i=1}^{n} m_i v_i^2$$

$$= \frac{1}{2} \sum_{i=1}^{n} m_i \left(\sum_{r=1}^{i} l_r^2 \dot{\theta}_r^2 + 2 \sum_{j=1}^{i} \sum_{k=j+1}^{i} l_j l_k \dot{\theta}_j \dot{\theta}_k \cos \left(\theta_j - \theta_k \right) \right)$$

$$= \frac{1}{2} \sum_{i=1}^{n} \sum_{r=1}^{i} m_i l_r^2 \dot{\theta}_r^2 + \sum_{i=1}^{n} \sum_{j=1}^{i} \sum_{k=j+1}^{i} m_i l_j l_k \dot{\theta}_j \dot{\theta}_k \cos\left(\theta_j - \theta_k\right) \qquad (2.487)$$

The potential energy P_i of the ith pendulum is related to m_i

$$P_i = m_i g y_i = -m_i g \sum_{j=1}^{i} l_j \cos\theta_j \qquad (2.488)$$

and therefore, the potential energy P of the chain is:

$$P = \sum_{i=1}^{n} m_i g y_i = -\sum_{i=1}^{n} \sum_{j=1}^{i} m_i g l_j \cos\theta_j \qquad (2.489)$$

To find the equations of motion for the chain, we use the Lagrangean $\mathcal{L}$

$$\mathcal{L} = K - P \qquad (2.490)$$

and apply the Lagrange equation.

$$\frac{d}{dt}\left(\frac{\partial \mathcal{L}}{\partial \dot{q}_s}\right) - \frac{\partial \mathcal{L}}{\partial q_s} = 0 \qquad s = 1, 2, \cdots n \qquad (2.491)$$

Example 71 Mechanical energy. Conservation of energy appears in a system with only potential forces. Here is a proof.

If a system of masses m_i is moving in a potential force field P

$$\mathbf{F}_{m_i} = -\nabla_i P \qquad (2.492)$$

their Newton equations of motion will be:

$$m_i \ddot{\mathbf{r}}_i = -\nabla_i P \qquad i = 1, 2, \cdots n \qquad (2.493)$$

The inner product of equations of motion with $\dot{\mathbf{r}}_i$ and adding the equations,

$$\sum_{i=1}^{n} m_i \dot{\mathbf{r}}_i \cdot \ddot{\mathbf{r}}_i = -\sum_{i=1}^{n} \dot{\mathbf{r}}_i \cdot \nabla_i P \qquad (2.494)$$

and then integrating over time

$$\frac{1}{2} \sum_{i=1}^{n} m_i \dot{\mathbf{r}}_i \cdot \dot{\mathbf{r}}_i = -\int \sum_{i=1}^{n} \mathbf{r}_i \cdot \nabla_i P \qquad (2.495)$$

show that

$$K = -\int \sum_{i=1}^{n} \left(\frac{\partial P}{\partial x_i} x_i + \frac{\partial P}{\partial y_i} y_i + \frac{\partial P}{\partial z_i} z_i \right) = -P + E \qquad (2.496)$$

where E is the constant of integration. E is called the mechanical energy of the system and is equal to kinetic plus potential energies:

$$E = K + P \qquad (2.497)$$

2.5 Dissipation Function

The Lagrange equation based on kinetic energy K,

$$\frac{d}{dt} \left(\frac{\partial K}{\partial \dot{q}_r} \right) - \frac{\partial K}{\partial q_r} = F_r \qquad r = 1, 2, \cdots n \qquad (2.498)$$

or based on Lagrangean $\mathcal{L}$, as introduced in Eq. (2.408), can both be applied to derive the equations of motion of a vibrating system.

$$\frac{d}{dt} \left(\frac{\partial \mathcal{L}}{\partial \dot{q}_r} \right) - \frac{\partial \mathcal{L}}{\partial q_r} = Q_r \qquad r = 1, 2, \cdots n \qquad (2.499)$$

However, we may use a simpler and more practical Lagrange equation, for linear vibrations:

$$\frac{d}{dt} \left(\frac{\partial K}{\partial \dot{q}_r} \right) - \frac{\partial K}{\partial q_r} + \frac{\partial D}{\partial \dot{q}_r} + \frac{\partial P}{\partial q_r} = f_r \qquad r = 1, 2, \cdots n \qquad (2.500)$$

where K is the kinetic energy, P is the potential energy, and D is the *dissipation function* of the system and f_r is the applied force on the mass m_r.

$$K = \frac{1}{2} \dot{\mathbf{q}}^T [m] \dot{\mathbf{q}} = \frac{1}{2} \sum_{i=1}^{n} \sum_{j=1}^{n} \dot{q}_i m_{ij} \dot{q}_j \qquad (2.501)$$

$$P = \frac{1}{2} \mathbf{q}^T [k] \mathbf{q} = \frac{1}{2} \sum_{i=1}^{n} \sum_{j=1}^{n} q_i k_{ij} q_j \qquad (2.502)$$

$$D = \frac{1}{2} \dot{\mathbf{q}}^T [c] \dot{\mathbf{q}} = \frac{1}{2} \sum_{i=1}^{n} \sum_{j=1}^{n} \dot{q}_i c_{ij} \dot{q}_j \qquad (2.503)$$

Proof Consider a one DOF mass-spring-damper vibrating system. When viscous damping is the only type of damping in the system, we may employ a function known as the *Rayleigh dissipation function D*,

$$D = \frac{1}{2}c\dot{x}^2 \qquad (2.504)$$

to find the damping force f_c by differentiation.

$$f_c = -\frac{\partial D}{\partial \dot{x}} \qquad (2.505)$$

Remembering that the elastic force f_k can be found from a potential energy P,

$$f_k = -\frac{\partial P}{\partial x} \qquad (2.506)$$

the generalized force F can be expressed as a collection of the forces,

$$F = f_c + f_k + f = -\frac{\partial D}{\partial \dot{x}} - \frac{\partial P}{\partial x} + f \qquad (2.507)$$

where f is the non-conservative applied force on mass m. Substituting (2.507) in (2.498)

$$\frac{d}{dt}\left(\frac{\partial K}{\partial \dot{x}}\right) - \frac{\partial K}{\partial x} = -\frac{\partial D}{\partial \dot{x}} - \frac{\partial P}{\partial x} + f \qquad (2.508)$$

gives us the Lagrange equation for a viscous damped vibrating system.

$$\frac{d}{dt}\left(\frac{\partial K}{\partial \dot{x}}\right) - \frac{\partial K}{\partial x} + \frac{\partial D}{\partial \dot{x}} + \frac{\partial P}{\partial x} = f \qquad (2.509)$$

For vibrating systems with n DOF, the kinetic energy K, potential energy P, and dissipating function D are as (2.501)–(2.503). Applying the Lagrange equation to the n DOF system would result in n second-order differential equations (2.500). ∎

Example 72 A one DOF forced mass-spring-damper system. A simple example of employing dissipation function D to derive equation of motion of a damped vibrating system.

Figure 2.29 illustrates a single DOF mass-spring-damper system with an external force f applied on the mass m. The kinetic and potential energies of the system, when it is in motion, are:

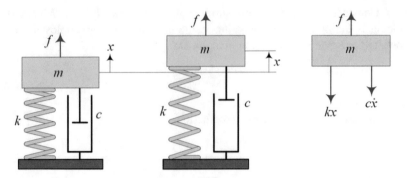

Fig. 2.29 A one DOF forced mass-spring-damper system

$$K = \frac{1}{2}m\dot{x}^2 \qquad P = \frac{1}{2}kx^2 \tag{2.510}$$

and its dissipation function D is:

$$D = \frac{1}{2}c\dot{x}^2 \tag{2.511}$$

Substituting (2.510)–(2.511) into the Lagrange equation (2.500) provides us with the equation of motion:

$$\frac{\partial K}{\partial \dot{x}} = m\dot{x} \qquad \frac{\partial K}{\partial x} = 0 \qquad \frac{\partial D}{\partial \dot{x}} = c\dot{x} \qquad \frac{\partial P}{\partial x} = kx \tag{2.512}$$

$$\frac{d}{dt}(m\dot{x}) + c\dot{x} + kx = f \tag{2.513}$$

Example 73 An eccentric excited one DOF system. When a device with a rotating shaft is mounted on a base, there would be vibrations due to eccentric excitations. Here is to derive equation of motion.

An eccentric excited one DOF system is shown in Fig. 2.30 with mass m supported by a suspension of a spring k and a damper c. There is also a mass m_e at a distance e that is rotating with an angular velocity ω. We may find the equation of motion of the system by applying the Lagrange method. The kinetic energy K of the system is:

$$K = \frac{1}{2}(m - m_e)\dot{x}^2 + \frac{1}{2}m_e(\dot{x} + e\omega\cos\omega t)^2 + \frac{1}{2}m_e(-e\omega\sin\omega t)^2 \tag{2.514}$$

The velocity of the main vibrating mass $m - m_e$ is $\dot{x}$, and the velocity of the eccentric mass m_e has two components $\dot{x} + e\omega\cos\omega t$ and $-e\omega\sin\omega t$. The potential energy P and dissipation function of the system are:

Fig. 2.30 An eccentric
excitated single DOF system

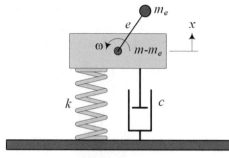

Fig. 2.31 A one DOF
eccentric base excited
vibrating system

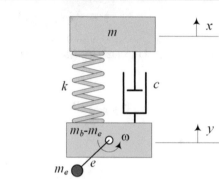

$$P = \frac{1}{2}kx^2 \qquad D = \frac{1}{2}c\dot{x}^2 \tag{2.515}$$

Applying the Lagrange equation (2.500)

$$\frac{\partial K}{\partial \dot{x}} = m\dot{x} + m_e e\omega \cos \omega t \tag{2.516}$$

$$\frac{d}{dt}\left(\frac{\partial K}{\partial \dot{x}}\right) = m\ddot{x} - m_e e\omega^2 \sin \omega t \tag{2.517}$$

$$\frac{\partial D}{\partial \dot{x}} = c\dot{x} \qquad \frac{\partial P}{\partial x} = kx \tag{2.518}$$

provides us with the equation of motion.

$$m\ddot{x} + c\dot{x} + kx = m_e e\omega^2 \sin \omega t \tag{2.519}$$

Example 74 An eccentric base excited vibrating system. When a device is mounted
on another device with a rotating shaft, there would be vibrations due to eccentric
excitations. Here is how to derive equation of motion by Lagrange method.

Figure 2.31 illustrates a one DOF eccentric base excited vibrating system. A
mass m is mounted on an eccentric excited base by a spring k and a damper c. The

base has a mass m_b with an attached unbalance mass m_e at a distance e. The mass m_e is rotating with an angular velocity ω. We may derive the equation of motion of the system by applying Lagrange method. The required functions are:

$$K = \frac{1}{2} m \dot{x}^2 + \frac{1}{2} (m_b - m_e) \dot{y}^2$$

$$+ \frac{1}{2} m_e (\dot{y} - e\omega \cos \omega t)^2 + \frac{1}{2} m_e (e\omega \sin \omega t)^2 \tag{2.520}$$

$$P = \frac{1}{2} k (x - y)^2 \qquad D = \frac{1}{2} c (\dot{x} - \dot{y})^2 \tag{2.521}$$

Applying the Lagrange method (2.500) provides us with the equations

$$m \ddot{x} + c (\dot{x} - \dot{y}) + k (x - y) = 0 \tag{2.522}$$

$$m_b \ddot{y} + m_e e\omega^2 \sin \omega t - c (\dot{x} - \dot{y}) - k (x - y) = 0 \tag{2.523}$$

because

$$\frac{\partial K}{\partial \dot{x}} = m \dot{x} \qquad \frac{d}{dt} \left(\frac{\partial K}{\partial \dot{x}} \right) = m \ddot{x} \tag{2.524}$$

$$\frac{\partial D}{\partial \dot{x}} = c (\dot{x} - \dot{y}) \qquad \frac{\partial P}{\partial x} = k (x - y) \tag{2.525}$$

$$\frac{\partial K}{\partial \dot{y}} = m_b \dot{y} - m_e e\omega \cos \omega t \tag{2.526}$$

$$\frac{d}{dt} \left(\frac{\partial K}{\partial \dot{y}} \right) = m_b \ddot{y} + m_e e\omega^2 \sin \omega t \tag{2.527}$$

$$\frac{\partial D}{\partial \dot{y}} = -c (\dot{x} - \dot{y}) \qquad \frac{\partial P}{\partial y} = -k (x - y) \tag{2.528}$$

Using a relative variable $z = x - y$, we may combine Eqs. (2.522) and (2.523) to find the equation of relative motion.

$$\frac{m m_b}{m_b + m} \ddot{z} + c \dot{z} + k z = \frac{m m_e}{m_b + m} e\omega^2 \sin \omega t \tag{2.529}$$

Example 75 ★ Generalized forces. In mechanical vibrations, the generalized force is usually one of two kinds of elastic and dissipative.

1. Elastic force:

   An elastic force is a recoverable force from an elastic body, such as a spring. An elastic body is the one for which any produced work is stored in the body in the form of internal energy and is recoverable. Therefore, the variation of the internal potential energy of the body, $P = P(q, t)$, would be

$$\delta P = -\delta W = \sum_{i=1}^{n} Q_i \delta q_i \tag{2.530}$$

   where q_i is the generalized coordinate of the particle i of the body and δW is the virtual work of the generalized elastic force Q.

$$Q_i = -\frac{\partial P}{\partial q_i} \tag{2.531}$$

2. Dissipation force:

   A dissipative force between two bodies is proportional to and in opposite direction of the relative velocity vector $\mathbf{v}$ between two bodies.

$$Q_i = -c_i \, f_i(v_i) \, \frac{\mathbf{v}_i}{v_i} \tag{2.532}$$

   The coefficient c_i is assumed constant, $f_i(v_i)$ is the velocity function of the force, and v_i is the magnitude of the relative velocity:

$$v_i = \sqrt{\sum_{j=1}^{3} v_{ij}^2} \tag{2.533}$$

   The virtual work of the dissipation force is:

$$\delta W = \sum_{i=1}^{n_1} Q_i \, \delta q_i \tag{2.534}$$

$$Q_i = -\sum_{k=1}^{n_1} c_k \, f_k(v_k) \, \frac{\partial v_k}{\partial \dot{q}_i} \tag{2.535}$$

   where n_1 is the total number of dissipation forces. By introducing the dissipation function D as

$$D = \sum_{i=1}^{n_1} \int_0^{v_i} c_k \, f_k(z_k) \, dz \tag{2.536}$$

we have

$$Q_i = -\frac{\partial D}{\partial \dot{q}_i} \tag{2.537}$$

The dissipation power P of the dissipation force Q_i is:

$$P = \sum_{i=1}^{n} Q_i \, \dot{q}_i = \sum_{i=1}^{n} \dot{q}_i \frac{\partial D}{\partial \dot{q}_i} \tag{2.538}$$

2.6 Quadratures

If $[m]$ is an $n \times n$ square matrix and $\mathbf{x}$ is an $n \times 1$ vector, then S is a scalar function called *quadrature*.

$$S = \mathbf{x}^T [m] \mathbf{x} \tag{2.539}$$

The derivative of the quadrature S with respect to the vector $\mathbf{x}$ is:

$$\frac{\partial S}{\partial \mathbf{x}} = \left([m] + [m]^T \right) \mathbf{x} \tag{2.540}$$

Kinetic energy K, potential energy P, and dissipation function D are quadratures.

$$K = \frac{1}{2} \dot{\mathbf{x}}^T [m] \dot{\mathbf{x}} \tag{2.541}$$

$$P = \frac{1}{2} \mathbf{x}^T [k] \mathbf{x} \tag{2.542}$$

$$D = \frac{1}{2} \dot{\mathbf{x}}^T [c] \dot{\mathbf{x}} \tag{2.543}$$

Therefore,

$$\frac{\partial K}{\partial \dot{\mathbf{x}}} = \frac{1}{2} \left([m] + [m]^T \right) \dot{\mathbf{x}} \tag{2.544}$$

$$\frac{\partial P}{\partial \mathbf{x}} = \frac{1}{2} \left([k] + [k]^T \right) \mathbf{x} \tag{2.545}$$

$$\frac{\partial D}{\partial \dot{\mathbf{x}}} = \frac{1}{2} \left([c] + [c]^T \right) \dot{\mathbf{x}} \tag{2.546}$$

Employing quadrature derivatives and the Lagrange method,

$$\frac{d}{dt}\frac{\partial K}{\partial \dot{\mathbf{x}}} + \frac{\partial K}{\partial \mathbf{x}} + \frac{\partial D}{\partial \dot{\mathbf{x}}} + \frac{\partial P}{\partial \mathbf{x}} = \mathbf{F} \tag{2.547}$$

$$\delta W = \mathbf{F}^T \partial \mathbf{x} \tag{2.548}$$

the equation of motion for a linear n degree-of-freedom vibrating system is:

$$[\underline{m}]\ddot{\mathbf{x}} + [\underline{c}]\dot{\mathbf{x}} + [\underline{k}]\mathbf{x} = \mathbf{F} \tag{2.549}$$

where $[\underline{m}], [\underline{c}], [\underline{k}]$ are symmetric matrices.

$$[\underline{m}] = \frac{1}{2}\left([m] + [m]^T\right) \tag{2.550}$$

$$[\underline{c}] = \frac{1}{2}\left([c] + [c]^T\right) \tag{2.551}$$

$$[\underline{k}] = \frac{1}{2}\left([k] + [k]^T\right) \tag{2.552}$$

Quadratures are also called *Hermitian forms*.
Proof Let us define a general asymmetric quadrature S as

$$S = \mathbf{x}^T[a]\mathbf{y} = \sum_i \sum_j x_i a_{ij} y_j \tag{2.553}$$

If the quadrature is symmetric, then $\mathbf{x} = \mathbf{y}$.

$$S = \mathbf{x}^T[a]\mathbf{x} = \sum_i \sum_j x_i a_{ij} x_j \tag{2.554}$$

The vectors $\mathbf{x}$ and $\mathbf{y}$ may be functions of n generalized coordinates q_i and time t:

$$\mathbf{x} = \mathbf{x}(q_1, q_2, \cdots, q_n, t) \tag{2.555}$$

$$\mathbf{y} = \mathbf{y}(q_1, q_2, \cdots, q_n, t) \tag{2.556}$$

$$\mathbf{q} = [q_1 \ q_2 \ \cdots \ q_n]^T \tag{2.557}$$

The derivative of $\mathbf{x}$ with respect to $\mathbf{q}$ is a square matrix

$$\frac{\partial \mathbf{x}}{\partial \mathbf{q}} = \begin{bmatrix} \dfrac{\partial x_1}{\partial q_1} & \dfrac{\partial x_2}{\partial q_1} & \cdots & \dfrac{\partial x_n}{\partial q_1} \\ \dfrac{\partial x_1}{\partial q_2} & \dfrac{\partial x_2}{\partial q_2} & \cdots & \cdots \\ \cdots & \cdots & \cdots & \cdots \\ \dfrac{\partial x_1}{\partial q_n} & & \cdots & \dfrac{\partial x_n}{\partial q_n} \end{bmatrix} \tag{2.558}$$

which can also be expressed by

$$\frac{\partial \mathbf{x}}{\partial \mathbf{q}} = \begin{bmatrix} \dfrac{\partial \mathbf{x}}{\partial q_1} \\ \dfrac{\partial \mathbf{x}}{\partial q_2} \\ \cdots \\ \dfrac{\partial \mathbf{x}}{\partial q_n} \end{bmatrix} \tag{2.559}$$

or

$$\frac{\partial \mathbf{x}}{\partial \mathbf{q}} = \begin{bmatrix} \dfrac{\partial x_1}{\partial \mathbf{q}} & \dfrac{\partial x_2}{\partial \mathbf{q}} & \cdots & \dfrac{\partial x_n}{\partial \mathbf{q}} \end{bmatrix} \tag{2.560}$$

The derivative of S with respect to an element of q_k is

$$\begin{aligned} \frac{\partial S}{\partial q_k} &= \frac{\partial}{\partial q_k} \sum_i \sum_j x_i a_{ij} y_j \\ &= \sum_i \sum_j \frac{\partial x_i}{\partial q_k} a_{ij} y_j + \sum_i \sum_j x_i a_{ij} \frac{\partial y_j}{\partial q_k} \\ &= \sum_j \sum_i \frac{\partial x_i}{\partial q_k} a_{ij} y_j + \sum_i \sum_j \frac{\partial y_j}{\partial q_k} a_{ij} x_i \\ &= \sum_j \sum_i \frac{\partial x_i}{\partial q_k} a_{ij} y_j + \sum_j \sum_i \frac{\partial y_i}{\partial q_k} a_{ji} x_j \end{aligned} \tag{2.561}$$

and hence, the derivative of S with respect to $\mathbf{q}$ is:

$$\frac{\partial S}{\partial \mathbf{q}} = \frac{\partial \mathbf{x}}{\partial \mathbf{q}} [a] \mathbf{y} + \frac{\partial \mathbf{y}}{\partial \mathbf{q}} [a]^T \mathbf{x} \tag{2.562}$$

If S is a symmetric quadrature, then

$$\frac{\partial S}{\partial \mathbf{q}} = \frac{\partial}{\partial \mathbf{q}} \left(\mathbf{x}^T [a] \mathbf{x} \right) = \frac{\partial \mathbf{x}}{\partial \mathbf{q}} [a] \mathbf{x} + \frac{\partial \mathbf{x}}{\partial \mathbf{q}} [a]^T \mathbf{x} \tag{2.563}$$

and if $\mathbf{q} = \mathbf{x}$, then the derivative of a symmetric S with respect to $\mathbf{x}$ is

$$\begin{aligned} \frac{\partial S}{\partial \mathbf{x}} &= \frac{\partial}{\partial \mathbf{x}} \left(\mathbf{x}^T [a] \mathbf{x} \right) = \frac{\partial \mathbf{x}}{\partial \mathbf{x}} [a] \mathbf{x} + \frac{\partial \mathbf{x}}{\partial \mathbf{x}} [a]^T \mathbf{x} \\ &= [a] \mathbf{x} + [a]^T \mathbf{x} = \left([a] + [a]^T \right) \mathbf{x} \end{aligned} \tag{2.564}$$

If $[a]$ is a symmetric matrix, then

$$[a] + [a]^T = 2[a] \tag{2.565}$$

and if $[a]$ is not a symmetric matrix, then $\left[\underline{a}\right] = [a] + [a]^T$ is a symmetric matrix because

$$\underline{a}_{ij} = a_{ij} + a_{ji} = a_{ji} + a_{ij} = \underline{a}_{ji} \tag{2.566}$$

and therefore,

$$\left[\underline{a}\right] = \left[\underline{a}\right]^T \tag{2.567}$$

Kinetic energy K, potential energy P, and dissipation function D can be expressed by quadratures.

$$K = \frac{1}{2}\dot{\mathbf{x}}^T [m]\dot{\mathbf{x}} \tag{2.568}$$

$$P = \frac{1}{2}\mathbf{x}^T [k]\mathbf{x} \tag{2.569}$$

$$D = \frac{1}{2}\dot{\mathbf{x}}^T [c]\dot{\mathbf{x}} \tag{2.570}$$

Substituting K, P, D in the Lagrange equation provides us with the equations of motion:

$$
\begin{aligned}
\mathbf{F} &= \frac{d}{dt}\frac{\partial K}{\partial \dot{\mathbf{x}}} + \frac{\partial K}{\partial \mathbf{x}} + \frac{\partial D}{\partial \dot{\mathbf{x}}} + \frac{\partial P}{\partial \mathbf{x}} \\
&= \frac{1}{2}\frac{d}{dt}\frac{\partial}{\partial \dot{\mathbf{x}}}\left(\dot{\mathbf{x}}^T [m]\dot{\mathbf{x}}\right) + \frac{1}{2}\frac{\partial}{\partial \dot{\mathbf{x}}}\left(\dot{\mathbf{x}}^T [c]\dot{\mathbf{x}}\right) + \frac{1}{2}\frac{\partial}{\partial \mathbf{x}}\left(\mathbf{x}^T [k]\mathbf{x}\right) \\
&= \frac{1}{2}\left[\frac{d}{dt}\left(\left([m] + [m]^T\right)\dot{\mathbf{x}}\right) + \left([c] + [c]^T\right)\dot{\mathbf{x}} + \left([k] + [k]^T\right)\mathbf{x}\right] \\
&= \frac{1}{2}\left([m] + [m]^T\right)\ddot{\mathbf{x}} + \frac{1}{2}\left([c] + [c]^T\right)\dot{\mathbf{x}} + \frac{1}{2}\left([k] + [k]^T\right)\mathbf{x} \\
&= [\underline{m}]\ddot{\mathbf{x}} + [\underline{c}]\dot{\mathbf{x}} + [\underline{k}]\mathbf{x} \tag{2.571}
\end{aligned}
$$

where

$$[\underline{m}] = \frac{1}{2}\left([m] + [m]^T\right) \tag{2.572}$$

$$[\underline{c}] = \frac{1}{2}\left([k] + [k]^T\right) \tag{2.573}$$

$$[\underline{k}] = \frac{1}{2}\left([c] + [c]^T\right) \tag{2.574}$$

Fig. 2.32 A quarter car
model with driver

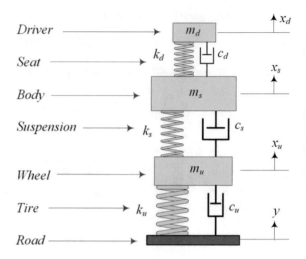

From now on, we assume that every equation of motion is found from the
Lagrange method to have symmetric coefficient matrices. Hence, we show the
equations of motion in the form of

$$[m]\ddot{\mathbf{x}} + [c]\dot{\mathbf{x}} + [k]\mathbf{x} = \mathbf{F} \qquad (2.575)$$

and use $[m]$, $[c]$, $[k]$ as a substitute for $[\underline{m}]$, $[\underline{c}]$, $[\underline{k}]$:

$$[m] \equiv [\underline{m}] \qquad (2.576)$$

$$[c] \equiv [\underline{c}] \qquad (2.577)$$

$$[k] \equiv [\underline{k}] \qquad (2.578)$$

Symmetric matrices are equal to their transpose:

$$[m] \equiv [m]^T \qquad (2.579)$$

$$[c] \equiv [c]^T \qquad (2.580)$$

$$[k] \equiv [k]^T \qquad (2.581)$$

■

Example 76 ★ A quarter car model with driver's chair. Quarter car is the simplest
model of a vehicle to study its vertical vibrations. Adding a chair and driver makes
the model to have three *DOF*. Here is to show how to derive the equations of motion
by Lagrange method.

Figure 2.32 illustrates a quarter car model plus a driver, which is modeled by a
mass m_d over a linear cushion above the sprung mass m_s . Assuming

$$y = 0 \qquad (2.582)$$

we find the free vibration equations of motion. The kinetic energy K of the system

$$
\begin{aligned}
K &= \frac{1}{2} m_u \dot{x}_u^2 + \frac{1}{2} m_s \dot{x}_s^2 + \frac{1}{2} m_d \dot{x}_d^2 \\
&= \frac{1}{2} \begin{bmatrix} \dot{x}_u & \dot{x}_s & \dot{x}_d \end{bmatrix} \begin{bmatrix} m_u & 0 & 0 \\ 0 & m_s & 0 \\ 0 & 0 & m_d \end{bmatrix} \begin{bmatrix} \dot{x}_u \\ \dot{x}_s \\ \dot{x}_d \end{bmatrix} \\
&= \frac{1}{2} \dot{\mathbf{x}}^T [m] \dot{\mathbf{x}}
\end{aligned}
\qquad (2.583)
$$

and the potential energy P are:

$$
\begin{aligned}
P &= \frac{1}{2} k_u (x_u)^2 + \frac{1}{2} k_s (x_s - x_u)^2 + \frac{1}{2} k_d (x_d - x_s)^2 \\
&= \frac{1}{2} \begin{bmatrix} x_u & x_s & x_d \end{bmatrix} \begin{bmatrix} k_u + k_s & -k_s & 0 \\ -k_s & k_s + k_d & -k_d \\ 0 & -k_d & k_d \end{bmatrix} \begin{bmatrix} x_u \\ x_s \\ x_d \end{bmatrix} \\
&= \frac{1}{2} \mathbf{x}^T [k] \mathbf{x}
\end{aligned}
\qquad (2.584)
$$

Similarly, the dissipation function D can be expressed as:

$$
\begin{aligned}
D &= \frac{1}{2} c_u (\dot{x}_u)^2 + \frac{1}{2} c_s (\dot{x}_s - \dot{x}_u)^2 + \frac{1}{2} c_d (\dot{x}_d - \dot{x}_s)^2 \\
&= \frac{1}{2} \begin{bmatrix} \dot{x}_u & \dot{x}_s & \dot{x}_d \end{bmatrix} \begin{bmatrix} c_u + c_s & -c_s & 0 \\ -c_s & c_s + c_d & -c_d \\ 0 & -c_d & c_d \end{bmatrix} \begin{bmatrix} \dot{x}_u \\ \dot{x}_s \\ \dot{x}_d \end{bmatrix} \\
&= \frac{1}{2} \dot{\mathbf{x}}^T [c] \dot{\mathbf{x}}
\end{aligned}
\qquad (2.585)
$$

Employing the quadrature derivative method, we may find the derivatives of K, P, D with respect to their variable vectors:

$$
\begin{aligned}
\frac{\partial K}{\partial \dot{\mathbf{x}}} &= \frac{1}{2} \left([m] + [m]^T \right) \dot{\mathbf{x}} = \frac{1}{2} \left([k] + [k]^T \right) \begin{bmatrix} \dot{x}_u \\ \dot{x}_s \\ \dot{x}_d \end{bmatrix} \\
&= \begin{bmatrix} m_u & 0 & 0 \\ 0 & m_s & 0 \\ 0 & 0 & m_d \end{bmatrix} \begin{bmatrix} \dot{x}_u \\ \dot{x}_s \\ \dot{x}_d \end{bmatrix}
\end{aligned}
\qquad (2.586)
$$

$$\frac{\partial P}{\partial \mathbf{x}} = \frac{1}{2}\left([k]+[k]^T\right)\mathbf{x} = \frac{1}{2}\left([k]+[k]^T\right)\begin{bmatrix} x_u \\ x_s \\ x_d \end{bmatrix}$$

$$= \begin{bmatrix} k_u + k_s & -k_s & 0 \\ -k_s & k_s + k_d & -k_d \\ 0 & -k_d & k_d \end{bmatrix}\begin{bmatrix} x_u \\ x_s \\ x_d \end{bmatrix} \tag{2.587}$$

$$\frac{\partial D}{\partial \dot{\mathbf{x}}} = \frac{1}{2}\left([c]+[c]^T\right)\dot{\mathbf{x}} = \frac{1}{2}\left([c]+[c]^T\right)\begin{bmatrix} \dot{x}_u \\ \dot{x}_s \\ \dot{x}_d \end{bmatrix}$$

$$= \begin{bmatrix} c_u + c_s & -c_s & 0 \\ -c_s & c_s + c_d & -c_d \\ 0 & -c_d & c_d \end{bmatrix}\begin{bmatrix} \dot{x}_u \\ \dot{x}_s \\ \dot{x}_d \end{bmatrix} \tag{2.588}$$

Therefore, we find the system's free vibration equations of motion.

$$[m]\,\ddot{\mathbf{x}} + [c]\,\dot{\mathbf{x}} + [k]\,\mathbf{x} = 0 \tag{2.589}$$

$$\begin{bmatrix} m_u & 0 & 0 \\ 0 & m_s & 0 \\ 0 & 0 & m_d \end{bmatrix}\begin{bmatrix} \ddot{x}_u \\ \ddot{x}_s \\ \ddot{x}_d \end{bmatrix} + \begin{bmatrix} c_u + c_s & -c_s & 0 \\ -c_s & c_s + c_d & -c_d \\ 0 & -c_d & c_d \end{bmatrix}\begin{bmatrix} \dot{x}_u \\ \dot{x}_s \\ \dot{x}_d \end{bmatrix}$$

$$+ \begin{bmatrix} k_u + k_s & -k_s & 0 \\ -k_s & k_s + k_d & -k_d \\ 0 & -k_d & k_d \end{bmatrix}\begin{bmatrix} x_u \\ x_s \\ x_d \end{bmatrix} = 0 \tag{2.590}$$

Example 77 ★ Discrete multiple vibrating systems. Employing dissipation function and using Lagrange method are the best methods to derive the equations of motion of multi DOF systems with linear damping.

Figure 2.33 indicates a two DOF with absolute coordinates x_1 and x_2. The kinetic energy K, potential energy P, and dissipation function D can be expressed as:

$$K = \frac{1}{2}m_1\dot{x}_1^2 + \frac{1}{2}m_2\dot{x}_2^2 = \frac{1}{2}\dot{\mathbf{x}}^T[m]\dot{\mathbf{x}}$$

$$= \frac{1}{2}\begin{bmatrix} \dot{x}_1 & \dot{x}_2 \end{bmatrix}\begin{bmatrix} m_1 & 0 \\ 0 & m_2 \end{bmatrix}\begin{bmatrix} \dot{x}_1 \\ \dot{x}_2 \end{bmatrix} \tag{2.591}$$

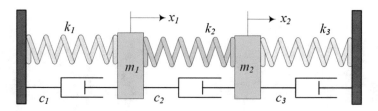

Fig. 2.33 A two DOF with absolute coordinates x_1 and x_2

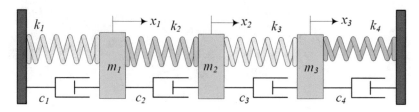

Fig. 2.34 A three DOF system with absolute coordinates x_1, x_2 and x_3

$$P = \frac{1}{2}k_1x_1^2 + \frac{1}{2}k_2(x_1 - x_2)^2 + \frac{1}{2}k_3x_2^2 = \frac{1}{2}\mathbf{x}^T[k]\mathbf{x}$$

$$= \frac{1}{2}\begin{bmatrix} x_1 & x_2 \end{bmatrix}\begin{bmatrix} k_1 + k_2 & -k_2 \\ -k_2 & k_2 + k_3 \end{bmatrix}\begin{bmatrix} x_1 \\ x_2 \end{bmatrix} \tag{2.592}$$

$$D = \frac{1}{2}c_1\dot{x}_1^2 + \frac{1}{2}c_2(\dot{x}_1 - \dot{x}_2)^2 + \frac{1}{2}c_3\dot{x}_2^2 = \frac{1}{2}\dot{\mathbf{x}}^T[c]\dot{\mathbf{x}}$$

$$= \frac{1}{2}\begin{bmatrix} \dot{x}_1 & \dot{x}_2 \end{bmatrix}\begin{bmatrix} c_1 + c_2 & -c_2 \\ -c_2 & c_2 + c_3 \end{bmatrix}\begin{bmatrix} \dot{x}_1 \\ \dot{x}_2 \end{bmatrix} \tag{2.593}$$

Now consider the three DOF system of Fig. 2.34 with absolute coordinates x_1, x_2, and x_3. The kinetic energy K, potential energy P, and dissipation function D of the system are:

$$K = \frac{1}{2}m_1\dot{x}_1^2 + \frac{1}{2}m_2\dot{x}_2^2 + \frac{1}{2}m_3\dot{x}_3^2 = \frac{1}{2}\dot{\mathbf{x}}^T[m]\dot{\mathbf{x}}$$

$$= \frac{1}{2}\begin{bmatrix} \dot{x}_1 & \dot{x}_2 & \dot{x}_3 \end{bmatrix}\begin{bmatrix} m_1 & 0 & 0 \\ 0 & m_2 & 0 \\ 0 & 0 & m_3 \end{bmatrix}\begin{bmatrix} \dot{x}_1 \\ \dot{x}_2 \\ \dot{x}_3 \end{bmatrix} \tag{2.594}$$

$$P = \frac{1}{2}k_1x_1^2 + \frac{1}{2}k_2(x_1 - x_2)^2 + \frac{1}{2}k_3(x_2 - x_3)^2 + \frac{1}{2}k_4x_3^2$$

$$= \frac{1}{2}\mathbf{x}^T[k]\mathbf{x} \tag{2.595}$$

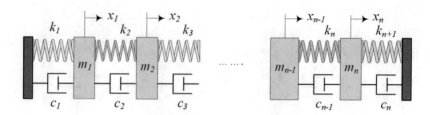

Fig. 2.35 A discrete n DOF system with absolute coordinates of $x_1, x_2, ..., x_n$

$$[k] = \begin{bmatrix} k_1 + k_2 & -k_2 & 0 \\ -k_2 & k_2 + k_3 & -k_3 \\ 0 & -k_3 & k_3 + k_4 \end{bmatrix} \tag{2.596}$$

$$D = \frac{1}{2}c_1\dot{x}_1^2 + \frac{1}{2}c_2\left(\dot{x}_1 - \dot{x}_2\right)^2 + \frac{1}{2}c_3\left(\dot{x}_2 - \dot{x}_3\right)^2 + \frac{1}{2}c_4\dot{x}_3^2$$

$$= \frac{1}{2}\dot{\mathbf{x}}^T[c]\,\dot{\mathbf{x}} \tag{2.597}$$

$$[c] = \begin{bmatrix} c_1 + c_2 & -c_2 & 0 \\ -c_2 & c_2 + c_3 & -c_3 \\ 0 & -c_3 & c_3 + c_4 \end{bmatrix} \tag{2.598}$$

Similarly, if we have a discrete n DOF system with absolute coordinates of x_1, $x_2, ..., x_n$ such as shown in Fig. 2.35, then the kinetic energy K, potential energy P, and dissipation function D of the system are:

$$[k] = \begin{bmatrix} m_1 & 0 & 0 & \cdots & 0 \\ 0 & m_2 & 0 & \cdots & 0 \\ 0 & 0 & m_3 & \vdots & \vdots \\ \vdots & \vdots & \vdots & \ddots & \vdots \\ 0 & 0 & \cdots\cdots & & m_n \end{bmatrix} \tag{2.599}$$

$$[k] = \begin{bmatrix} k_1 + k_2 & -k_2 & 0 & \cdots & & 0 \\ -k_2 & k_2 + k_3 & -k_3 & \cdots & & \vdots \\ 0 & -k_3 & \ddots & \ddots & & \vdots \\ \vdots & \vdots & \ddots & \ddots & & -k_n \\ 0 & 0 & \cdots & & -k_n & k_n + k_{n+1} \end{bmatrix} \tag{2.600}$$

$$[c] = \begin{bmatrix} c_1 + c_2 & -c_2 & 0 & \cdots & & 0 \\ -c_2 & c_2 + c_3 & -c_3 & \cdots & & \vdots \\ 0 & -c_3 & \ddots & \ddots & & \vdots \\ \vdots & \vdots & \ddots & \ddots & -c_n & \\ 0 & 0 & \cdots & -c_n & c_n + c_{n+1} \end{bmatrix} \qquad (2.601)$$

Employing the Lagrange equation, we find the general equations of motion.

$$m_i \ddot{x}_i + c_1 \dot{x}_1 + c_2 (\dot{x}_1 - \dot{x}_2) + c_3 (\dot{x}_2 - \dot{x}_3) + \cdots$$
$$+ c_n (\dot{x}_{n-1} - \dot{x}_n) + c_{n+1} \dot{x}_n$$
$$+ k_1 x_1 + k_2 (x_1 - x_2) + k_3 (x_2 - x_3) + \cdots$$
$$+ k_n (x_{n-1} - x_n) k_{n+1} x_n = 0 \qquad (2.602)$$

Example 78 ★ Different $[m]$, $[c]$, and $[k]$ arrangements. The coefficient matrices of quadratures in mechanical energy may be arranged differently at the beginning; however, after applying Lagrange method, all coefficient matrices will be changed to symmetric.

Mass, damping, and stiffness matrices $[m]$, $[c]$, $[k]$ for a vibrating system may be arranged in different forms with the same overall kinetic energy K, potential energy P, and dissipation function D. For example, the potential energy P for the quarter car model that is shown in Fig. 2.32 may be expressed by different stiffness matrix $[k]$:

$$P = \frac{1}{2} k_u (x_u)^2 + \frac{1}{2} k_s (x_s - x_u)^2 + \frac{1}{2} k_d (x_d - x_s)^2 \qquad (2.603)$$

$$P = \frac{1}{2} \mathbf{x}^T \begin{bmatrix} k_u + k_s & -k_s & 0 \\ -k_s & k_s + k_d & -k_d \\ 0 & -k_d & k_d \end{bmatrix} \mathbf{x} \qquad (2.604)$$

$$P = \frac{1}{2} \mathbf{x}^T \begin{bmatrix} k_u + k_s & -2k_s & 0 \\ 0 & k_s + k_d & -2k_d \\ 0 & 0 & k_d \end{bmatrix} \mathbf{x} \qquad (2.605)$$

$$P = \frac{1}{2} \mathbf{x}^T \begin{bmatrix} k_u + k_s & 0 & 0 \\ -2k_s & k_s + k_d & 0 \\ 0 & -2k_d & k_d \end{bmatrix} \mathbf{x} \qquad (2.606)$$

The matrices $[m]$, $[c]$, and $[k]$ in K, D, and P may not be symmetric; however, the matrices $[\underline{m}]$, $[\underline{c}]$, and $[\underline{k}]$ in $\partial K/\partial \dot{\mathbf{x}}$, $\partial D/\partial \dot{\mathbf{x}}$, and $\partial P/\partial \mathbf{x}$ are always symmetric.

When a matrix $[a]$ is diagonal, it is symmetric and

$$[a] = [\underline{a}] \tag{2.607}$$

A diagonal matrix cannot be written in different forms. The mass matrix $[m]$ in Example 76 is diagonal, and, hence, K has only one form, (2.583).

Example 79 Quadratic form and sum of squares. Any physical quantity that its terms are made of squares of a variable can be written in a quadrature form. Here is the proof.

We can write the sum of x_i^2 in a quadratic form

$$\sum_{i=1}^{n} x_i^2 = \mathbf{x}^T \mathbf{x} = \mathbf{x}^T \mathbf{I} \mathbf{x} \tag{2.608}$$

where

$$\mathbf{x}^T = \begin{bmatrix} x_1 & x_2 & x_3 & \cdots & x_n \end{bmatrix} \tag{2.609}$$

and $\mathbf{I}$ is an $n \times n$ identity matrix. If we are looking for the sum of squares around a mean value x_0, then

$$\sum_{i=1}^{n} (x_i - x_0)^2 = \sum_{i=1}^{n} x_i^2 - nx_0^2 = \mathbf{x}^T \mathbf{x} - \frac{1}{n}\left(\sum_{i=1}^{n} x_i\right)\left(\sum_{i=1}^{n} x_i\right)$$

$$= \mathbf{x}^T \mathbf{x} - \frac{1}{n}\left(\mathbf{x}^T \mathbf{1}_n\right)\left(\mathbf{1}_n^T \mathbf{x}\right)$$

$$= \mathbf{x}^T \left(\mathbf{I} - \frac{1}{n}\mathbf{1}_n\mathbf{1}_n^T\right)\mathbf{x} \tag{2.610}$$

where

$$\mathbf{1}_n^T = \begin{bmatrix} 1 & 1 & \cdots & 1 \end{bmatrix} \tag{2.611}$$

Example 80 ★ Positive definite matrix. The kinetic energy of a mechanical vibrating system cannot be negative, and this reflects as a condition on the mass matrix $[m]$. Here is the mathematical definitions.

A matrix $[a]$ is called positive definite if $\mathbf{x}^T [a] \mathbf{x} > 0$ for all $\mathbf{x} \neq 0$. A matrix $[a]$ is called positive semidefinite if $\mathbf{x}^T [a] \mathbf{x} \geq 0$ for all $\mathbf{x}$.

The kinetic energy is positive definite, and this means we cannot have $K = 0$ unless $\dot{\mathbf{x}} = 0$. The potential energy is positive semidefinite, and this means we have

$P \geq 0$ as long as $\mathbf{x} > 0$; however, it is possible to have a especial $\mathbf{x}_0 > 0$ at which $P = 0$. A positive definite matrix, such as the mass matrix $[m]$, satisfies Sylvester's criterion, which is that the determinant of $[m]$ and determinant of all the diagonal minors must be positive:

$$\Delta_n = \begin{vmatrix} m_{11} & m_{12} & \cdots & m_{1n} \\ m_{21} & m_{22} & \cdots & m_{2n} \\ \vdots & \ddots & \ddots & \vdots \\ m_{n1} & m_{n2} & \cdots & m_{nn} \end{vmatrix} > 0 \tag{2.612}$$

$$\Delta_{n-1} = \begin{vmatrix} m_{11} & m_{12} & \cdots & m_{1,n-1} \\ m_{21} & m_{22} & \cdots & m_{2,n-1} \\ \vdots & \ddots & \ddots & \vdots \\ m_{n1} & m_{n2} & \cdots & m_{n-1,n-1} \end{vmatrix} > 0 \tag{2.613}$$

$$\cdots \Delta_2 = \begin{vmatrix} m_{11} & m_{12} \\ m_{21} & m_{22} \end{vmatrix} > 0 \qquad \Delta_1 = m_{11} > 0 \tag{2.614}$$

Example 81 ★ Symmetric matrices. Working with symmetric matrices is critical in linear vibrations. There are many simplicities with symmetric matrices such as easy to inverse, real eigenvalues, etc.

Employing the Lagrange method guarantees that the coefficient matrices of equations of motion of linear vibrating systems are symmetric. A matrix $[A]$ is symmetric if the columns and rows of $[A]$ are interchangeable, so $[A]$ is equal to its transpose:

$$[A] = [A]^T \tag{2.615}$$

The characteristic equation of a symmetric matrix $[A]$ is a polynomial for which all the roots are real. Therefore, the eigenvalues of $[A]$ are real and distinct, and $[A]$ is diagonalizable. Any two eigenvectors that come from distinct eigenvalues of a symmetric matrix $[A]$ are orthogonal.

Example 82 Discretization of energies. Kinetic and potential energies are the most important quadratures in development of equations of motion. Here is a more detailed view in their structures from a generalized coordinate viewpoint.

The kinetic energy of a system with n particles is

$$K = \frac{1}{2} \sum_{i=1}^{n} m_i \left(\dot{x}_i^2 + \dot{y}_i^2 + \dot{z}_i^2 \right) = \frac{1}{2} \sum_{i=1}^{3n} m_i \dot{u}_i^2 \tag{2.616}$$

Expressing the configuration coordinate u_i in terms of generalized coordinates q_j, we have:

$$\dot{u}_i = \sum_{s=1}^{n} \frac{\partial u_i}{\partial q_s} \dot{q}_s + \frac{\partial u_i}{\partial t} \qquad s = 1, 2, \cdots, N \qquad (2.617)$$

Therefore, the kinetic energy in terms of generalized coordinates is

$$K = \frac{1}{2} \sum_{i=1}^{N} m_i \left(\sum_{s=1}^{n} \frac{\partial u_i}{\partial q_s} \dot{q}_s + \frac{\partial u_i}{\partial t} \right)^2$$

$$= \frac{1}{2} \sum_{j=1}^{n} \sum_{k=1}^{n} a_{jk} \dot{q}_j \dot{q}_k + \sum_{j=1}^{n} b_j \dot{q}_j + c \qquad (2.618)$$

where

$$a_{jk} = \sum_{i=1}^{N} m_i \frac{\partial u_i}{\partial q_j} \frac{\partial u_i}{\partial q_k} \qquad (2.619)$$

$$b_j = \sum_{i=1}^{N} m_i \frac{\partial u_i}{\partial q_j} \frac{\partial u_i}{\partial t} \qquad (2.620)$$

$$c = \frac{1}{2} \sum_{i=1}^{N} m_i \left(\frac{\partial u_i}{\partial t} \right)^2 \qquad (2.621)$$

and

$$\left(\sum_{s=1}^{n} \frac{\partial u_i}{\partial q_s} \dot{q}_s + \frac{\partial u_i}{\partial t} \right)^2 = \left(\sum_{j=1}^{n} \frac{\partial u_i}{\partial q_j} \dot{q}_j + \frac{\partial u_i}{\partial t} \right) \left(\sum_{k=1}^{n} \frac{\partial u_i}{\partial q_k} \dot{q}_k + \frac{\partial u_i}{\partial t} \right)$$

$$= \sum_{j=1}^{n} \sum_{k=1}^{n} \left(\frac{\partial u_i}{\partial q_j} \frac{\partial u_i}{\partial q_k} \right) \dot{q}_j \dot{q}_k$$

$$+ 2 \sum_{j=1}^{n} \frac{\partial u_i}{\partial q_j} \frac{\partial u_i}{\partial t} \dot{q}_j + \left(\frac{\partial u_i}{\partial t} \right)^2 \qquad (2.622)$$

Using these expressions, we may show the kinetic energy of the dynamic system in the following form,

$$K = K_0 + K_1 + K_2 \qquad (2.623)$$

where

$$K_0 = \frac{1}{2} \sum_{i=1}^{N} m_i \left(\frac{\partial u_i}{\partial t} \right)^2 \tag{2.624}$$

$$K_1 = \sum_{j=1}^{n} \sum_{i=1}^{N} m_i \frac{\partial u_i}{\partial q_j} \frac{\partial u_i}{\partial t} \dot{q}_j \tag{2.625}$$

$$K_2 = \frac{1}{2} \sum_{j=1}^{n} \sum_{k=1}^{n} \sum_{i=1}^{N} m_i \frac{\partial u_i}{\partial q_j} \frac{\partial u_i}{\partial q_k} \dot{q}_j \dot{q}_k \tag{2.626}$$

$$a_{kj} = a_{jk} = \sum_{i=1}^{N} m_i \frac{\partial u_i}{\partial q_k} \frac{\partial u_i}{\partial q_j} = \sum_{i=1}^{N} m_i \frac{\partial \dot{u}_i}{\partial \dot{q}_k} \frac{\partial \dot{u}_i}{\partial \dot{q}_j} \tag{2.627}$$

If the coordinates u_i do not depend explicitly on time t, then $\partial u_i / \partial t = 0$, and we have:

$$K = K_2 = \frac{1}{2} \sum_{j=1}^{n} \sum_{k=1}^{n} a_{jk} \dot{q}_j \dot{q}_k = \frac{1}{2} \sum_{j=1}^{n} \sum_{k=1}^{n} \frac{\partial^2 K(0)}{\partial q_j \partial q_i} \dot{q}_j \dot{q}_k$$

$$= \frac{1}{2} \sum_{j=1}^{n} \sum_{k=1}^{n} m_{ij} \dot{q}_j \dot{q}_k \tag{2.628}$$

The kinetic energy is a scalar quantity and, because of (2.616), must be positive definite. The first term of (2.618) is a positive quadratic form. The third term of (2.618) is also a nonnegative quantity, as indicated by (2.621). The second term of (2.616) can be negative for some $\dot{q}_j$ and t. However, because of (2.616), the sum of all three terms of (2.618) must be positive.

$$K = \frac{1}{2} \dot{\mathbf{q}}^T [m] \dot{\mathbf{q}} \tag{2.629}$$

The generalized coordinate q_i represents deviations from equilibrium. The potential energy P is a continuous function of generalized coordinates q_i, and hence, its expansion is:

$$P(\mathbf{q}) = P(0) + \sum_{i=1}^{n} \frac{\partial P(0)}{\partial q_i} q_i + \frac{1}{2} \sum_{j=1}^{n} \sum_{i=1}^{n} \frac{\partial^2 P(0)}{\partial q_j \partial q_i} q_i q_j + \cdots \tag{2.630}$$

where $\partial P (0) / \partial q_i$ and $\partial^2 P (0) / (\partial q_j \, \partial q_i)$ are the values of $\partial P / \partial q_i$ and $\partial^2 P / (\partial q_j \, \partial q_i)$ at $\mathbf{q} = 0$, respectively. By assuming $P (0) = 0$, and knowing that the first derivative of P is zero at equilibrium,

$$\frac{\partial P}{\partial q_i} = 0 \qquad i = 1, 2, 3, \cdots, n \tag{2.631}$$

we have the second-order approximation as

$$P (\mathbf{q}) = \frac{1}{2} \sum_{j=1}^{n} \sum_{i=1}^{n} k_{ij} q_i q_j \qquad k_{ij} = \frac{\partial^2 P (0)}{\partial q_j \, \partial q_i} \tag{2.632}$$

where k_{ij} are the elastic coefficients. The second-order approximation of P is zero only at $\mathbf{q} = 0$. The expression (2.632) can also be written as a quadrature.

$$P = \frac{1}{2} \mathbf{q}^T [k] \, \mathbf{q} \tag{2.633}$$

2.7   Summary

In this chapter, we review the dynamics of vibrations and the methods of deriving the equations of motion of vibrating systems. The Newton-Euler and Lagrange methods are the most applied methods of deriving the equations of motion.

The translational and rotational equations of motion for a rigid body, expressed in the global coordinate frame G, are

$$^G\mathbf{F} = \frac{^G d}{dt} \, ^G\mathbf{p} = \frac{^G d}{dt} \left(m \, ^G\mathbf{v} \right) \tag{2.634}$$

$$^G\mathbf{M} = \frac{^G d}{dt} \, ^G\mathbf{L} = \frac{^G d}{dt} \left(I_i \, _G \omega_B \right) \tag{2.635}$$

where $^G\mathbf{F}$ and $^G\mathbf{M}$ indicate the resultant of the external forces and moments applied on the rigid body, measured at the mass center C.

The equation of motion of an $n \, DOF$ linear vibrating system can always be arranged in matrix form of a set of second-order differential equations

$$[m] \ddot{\mathbf{x}} + [c] \dot{\mathbf{x}} + [k] \mathbf{x} = \mathbf{F} \tag{2.636}$$

where $\mathbf{x}$ is a column array of generalized coordinates of the system and $\mathbf{F}$ is a column array of the associated applied forces. The *square matrices* of $[m]$, $[c]$, $[k]$ are the mass, damping, and stiffness matrices, respectively.

The work $_1W_2$ done by an applied force $^G\mathbf{F}$ on m in moving from spatial point 1 to point 2 on a path, indicated by a vector $^G\mathbf{r}$, is:

$$_1W_2 = \int_1^2 {}^G\mathbf{F} \cdot d\,{}^G\mathbf{r} \tag{2.637}$$

However, $_1W_2$ is equal to the difference of the kinetic energy of terminal and initial points. This is called the *principle of work and energy*.

$$\int_1^2 {}^G\mathbf{F} \cdot d\,{}^G\mathbf{r} = m \int_1^2 \frac{{}^G d}{dt}\,{}^G\mathbf{v} \cdot {}^G\mathbf{v}dt = \frac{1}{2}m \int_1^2 \frac{d}{dt}v^2 dt$$

$$= \frac{1}{2}m\left(v_2^2 - v_1^2\right) = K_2 - K_1 \tag{2.638}$$

$$_1W_2 = K_2 - K_1 \tag{2.639}$$

If there is a scalar *potential field function* $P = P(x, y, z)$ such that

$$\mathbf{F} = -\nabla P = -\frac{dP}{d\mathbf{r}} = -\left(\frac{\partial P}{\partial x}\hat{\imath} + \frac{\partial P}{\partial y}\hat{\jmath} + \frac{\partial P}{\partial z}\hat{k}\right) \tag{2.640}$$

then the principle of work and energy simplifies to the principle of *conservation of energy*.

$$K_1 + P_1 = K_2 + P_2 \tag{2.641}$$

The sum of kinetic and potential energies, $E = K + P$, is called the mechanical energy. Mechanical energy of an energy conserved system is a constant of motion, and hence, its time derivative is zero, $\dot{E} = 0$.

The expression of the equations of motion in the body coordinate frame is

$$^B\mathbf{F} = {}^G\dot{\mathbf{p}} + {}^B_G\omega_B \times {}^B\mathbf{p} = m\,{}^B\mathbf{a}_B + m\,{}^B_G\omega_B \times {}^B\mathbf{v}_B \tag{2.642}$$

$$^B\mathbf{M} = {}^B\dot{\mathbf{L}} + {}^B_G\omega_B \times {}^B\mathbf{L} = {}^B I\,{}^B_G\dot{\omega}_B + {}^B_G\omega_B \times \left({}^B I\,{}^B_G\omega_B\right) \tag{2.643}$$

where I is the mass moment of the rigid body.

$$I = \begin{bmatrix} I_{xx} & I_{xy} & I_{xz} \\ I_{yx} & I_{yy} & I_{yz} \\ I_{zx} & I_{zy} & I_{zz} \end{bmatrix} \tag{2.644}$$

The elements of I are functions of the mass distribution of the rigid body and are defined by

$$I_{ij} = \int_B \left(r_i^2 \delta_{mn} - x_{im}x_{jn}\right) dm \qquad i, j = 1, 2, 3 \tag{2.645}$$

where δ_{ij} is Kronecker's delta.

Every rigid body has a principal body coordinate frame in which the mass moment matrix is diagonal

$$^BI = \begin{bmatrix} I_1 & 0 & 0 \\ 0 & I_2 & 0 \\ 0 & 0 & I_3 \end{bmatrix} \tag{2.646}$$

The rotational equation of motion in the principal coordinate frame simplifies to:

$$\begin{aligned} M_1 &= I_1\dot{\omega}_1 - (I_2 - I_2)\,\omega_2\omega_3 \\ M_2 &= I_2\dot{\omega}_2 - (I_3 - I_1)\,\omega_3\omega_1 \\ M_3 &= I_3\dot{\omega}_3 - (I_1 - I_2)\,\omega_1\omega_2 \end{aligned} \tag{2.647}$$

The equations of motion for a mechanical system having n DOF can also be found by the Lagrange equation, for linear vibrations:

$$\frac{d}{dt}\left(\frac{\partial K}{\partial \dot{q}_r}\right) - \frac{\partial K}{\partial q_r} + \frac{\partial D}{\partial \dot{q}_r} + \frac{\partial P}{\partial q_r} = f_r \qquad r = 1, 2, \cdots n \tag{2.648}$$

where K is the kinetic energy, P is the potential energy, and D is the *dissipation function* of the system and f_r is the applied force on the mass m_r.

$$K = \frac{1}{2}\dot{\mathbf{q}}^T [m]\,\dot{\mathbf{q}} = \frac{1}{2}\sum_{i=1}^{n}\sum_{j=1}^{n}\dot{q}_i m_{ij}\dot{q}_j \tag{2.649}$$

$$P = \frac{1}{2}\mathbf{q}^T [k]\,\mathbf{q} = \frac{1}{2}\sum_{i=1}^{n}\sum_{j=1}^{n}q_i k_{ij} q_j \tag{2.650}$$

$$D = \frac{1}{2}\dot{\mathbf{q}}^T [c]\,\dot{\mathbf{q}} = \frac{1}{2}\sum_{i=1}^{n}\sum_{j=1}^{n}\dot{q}_i c_{ij}\dot{q}_j \tag{2.651}$$

When (x_i, y_i, z_i) are Cartesian coordinates in a globally fixed coordinate frame for the particle m_i, then its coordinates may be functions of another set of generalized coordinates $q_1, q_2, q_3, \cdots, q_n$ and possibly time t.

$$x_i = f_i(q_1, q_2, q_3, \cdots, q_n, t) \tag{2.652}$$

$$y_i = g_i(q_1, q_2, q_3, \cdots, q_n, t) \tag{2.653}$$

$$z_i = h_i(q_1, q_2, q_3, \cdots, q_n, t) \tag{2.654}$$

2.8   Key Symbols

1	Identity vector, identity matrix
$a \equiv \ddot{x}$	Acceleration
a, b	Distance, length
a, b, w, h	Length
$\mathbf{a}, \ddot{\mathbf{r}}$	Acceleration
A, B	Weight factors
A, B	Coefficients for frequency responses
B	Body coordinator
c	Damping
c	Constant of integral, integral of motion
$[c], [C]$	Damping matrix
c_e	Equivalent damping
C	Mass center
$\mathbf{d}$	Position vector of the body coordinate frame
D	Dissipation function
DOF	Degree of freedom
$d\mathbf{f}$	Infinitesimal force
dm	Infinitesimal mass
$d\mathbf{m}$	Infinitesimal moment
E	Mechanical energy
E	Young modulus of elasticity
$f = 1/T$	Cyclic frequency [Hz]
f	Function
f	Integral of motion
$f, F, \mathbf{f}, \mathbf{F}$	Force
F_C	Coriolis force
FBD	Free body diagram
$g, \mathbf{g}$	Gravitational acceleration
G	Global coordinate
H	Height
$I, \mathbf{I}, [I]$	Identity matrix
$[I]$	Mass moment matrix
I_1, I_2, I_3	Principal moment of inertia
k	Stiffness
k_e	Equivalent stiffness
k_s	Sprung spring stiffness
k_u	Unsprung spring stiffness
k_{ij}	Element of row i and column j of a stiffness matrix
$[k], [K]$	Stiffness matrix
K	Kinetic energy
l	Directional line
$L, \mathbf{L}$	Moment of momentum

$\mathcal{L} = K - P$	Lagrangean
m	Mass
m_e	Eccentric mass, equivalent mass
m_{ij}	Element of row i and column j of a mass matrix
m_k	Spring mass
m_s	Sprung mass
m_u	Unsprung mass
$[m] , [M]$	Mass matrix
$M, \mathbf{M}$	Moment
n	Number of coils, number of decibels, number of note
N	Natural numbers
O	Order of magnitude
$\mathbf{p}$	Momentum, translational momentum
P	Potential energy
P	Power
P, Q	Fixed points on a rigid body
$q, \mathbf{q}$	Generalized coordinate
Q	Generalized force
Q	Torque
r	Frequency ratio
$\mathbf{r}$	Position vector
r, R	Radius
R	Rotation transformation matrix
s	Eigenvalue, characteristic value
S	Quadrature
t	Time
t_0	Initial time
T	Period
T	Tension
T	Transpose
T_n	Natural period
$\hat{u}$	Unit vector
$\mathbf{u}, \mathbf{v}$	Velocity
$v \equiv \dot{x}, \; v, \dot{\mathbf{r}}$	Velocity
W	Work
$x, y, z, \mathbf{x}$	Displacement
$\mathbf{x}$	Vector of variables
x_0	Initial displacement
$\dot{x}_0$	Initial velocity
$\dot{x}, \dot{y}, \dot{z}$	Velocity, time derivative of x, y, z
$\ddot{x}$	Acceleration
X	Amplitude of x
z	Relative displacement

Greek

α, β, γ	Angle, angle of spring with respect to displacement
$\boldsymbol{\alpha}$	Angular acceleration
δ	Deflection, angle
δ	Kronecker delta function
δ_s	Static deflection
λ	Longitude
ε	Mass ratio
θ	Angular motion coordinate
$\omega, \boldsymbol{\omega}, \Omega$	Angular frequency
φ, Φ	Phase angle
φ	Latitude
ϕ	Angle of coordinate frame rotation
λ	Eigenvalue
π	3.141592653589793...

Symbols

[ ]	Matrix
$\int$	Integral
$\mid \, \mid$	Absolute value
Hz	Hertz
d	Differential
∂	Partial derivative
$\mathcal{L}$	Lagrangean
O	Order of magnitude
∇	Gradient
$\square$	Vector quantity
$\tilde{\square}$	Skew-symmetric matrix for vector $\square$

Exercises

1. Equation of motion of vibrating systems.
     Determine the equation of motion of the systems in Fig. 2.36a and b by

   (a) Energy method
   (b) Newton method
   (c) Lagrange method

2. Friction coefficient measurement.
     Figure 2.37 illustrates two turning rollers in opposite directions with equal angular speed ω. A slab of size $t \times w \times l$ and mass m is put on the rollers. A small disturbance and misplacement will cause the slab to oscillate about the

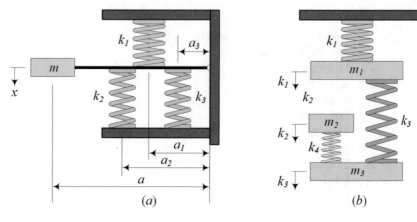

Fig. 2.36 Two undamped discrete vibrating systems

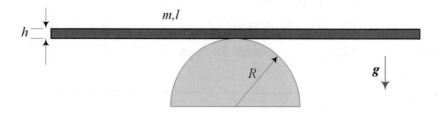

Fig. 2.37 A friction coefficient measurement device

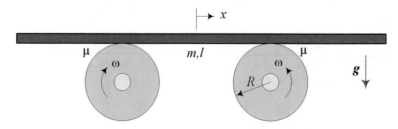

Fig. 2.38 A slab on a semicylinder

equilibrium position. If the coefficient of friction between the roller and the slab is μ, determine

(a) The equation of motion of the slab
(b) The natural frequency of the slab's oscillation for a given μ
(c) The value of μ for a measured frequency of oscillation ω

3. A slab on a semicylinder.
   Determine the equation of motion of the slab $h \times w \times l$ in Fig. 2.38 if

(a) The thickness of the slab is ignorable, $h = 0$
(b) ★ The thickness of the slab is not ignorable, $h \neq 0$

4. ★ Velocity-dependent friction.

The device in problem 2 is to measure the friction coefficient between the slab and the rollers.

Assume the friction force is $f = f_0 - Cv$ where v is the relative velocity of slab and rollers.

(a) Determine the equation of motion of the slab if x and $\dot{x}$ are assumed small.
(b) Determine the equation of motion of the slab if $\dot{x}$ is comparable with $R\omega$.
(c) Determine the equation of motion of the slab if one roller is turning with angular speed of 2ω, assuming $\dot{x}$ is very small.

5. Oscillating motion of a particle on the x-axis.

The displacement of a particle moving along the x-axis is given by

$$x = 0.01t^4 - t^3 + 4.5t^2 - 10 \qquad t \geq 0 \tag{2.655}$$

(a) Determine t_1 at which x becomes positive.
(b) For how long does x remain positive after $t = t_1$?
(c) How long does it take for x to become positive for the second time?
(d) When and where does the particle reach its maximum acceleration?
(e) Derive an equation to calculate its acceleration when its speed is given.

6. ★ Kinetic energy of a rigid link.

Consider a straight and uniform bar as a rigid link. The link has a mass m . Show that the kinetic energy of the bar can be expressed as

$$K = \frac{1}{6}m \left(\mathbf{v}_1 \cdot \mathbf{v}_1 + \mathbf{v}_1 \cdot \mathbf{v}_2 + \mathbf{v}_2 \cdot \mathbf{v}_2 \right) \tag{2.656}$$

where $\mathbf{v}_1$ and $\mathbf{v}_2$ are the velocity vectors of the end points of the link.

7. Ideal spring connected pendulum.

(a) Determine the kinetic and potential energy of the pendulum in Fig. 2.39.
(b) Determine the potential energy of the pendulum in Fig. 2.39, at an angle θ, if:
(c) The free length of the spring is $l = a - b$
(d) The free length of the spring is $l = a - 1.2b$
(e) The free length of the spring is $l = a - 0.8b$

8. ★ Moving a particle on a cycloid path.

A particle is moving on a planar curve with the following parametric expression:

$$x = r\left(\omega t - \sin \omega t\right) \qquad y = r\left(1 - \cos \omega t\right) \tag{2.657}$$

Fig. 2.39 Spring connected
pendulum

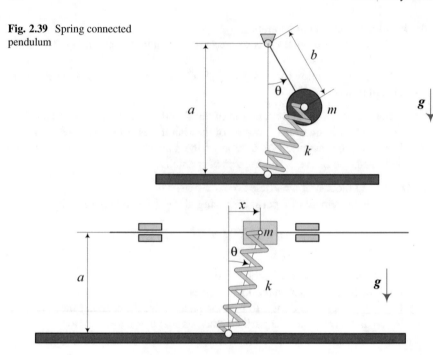

Fig. 2.40 Spring connected rectilinear oscillator

(a) Determine the speed of the particle at time t.
(b) Show that the magnitude of acceleration of the particle is constant.
(c) ★ Determine the tangential and normal accelerations of the particle.
(d) ★ Using $ds = v\,dt$, determine the length of the path that the particle travels up to time t.
(e) Examine if the magnitude of acceleration of the particle is constant for the following path:

$$x = a\,(\omega t - \sin \omega t) \qquad y = b\,(1 - \cos \omega t) \qquad (2.658)$$

9. ★ Spring connected rectilinear oscillator.
   Determine the kinetic and potential energies of the oscillator shown in Fig. 2.40. The free length of the spring is a.

(a) Express your answers in terms of the variable angle θ.
(b) Express your answers in terms of the variable distance x.
(c) Determine the equation of motion for large and small θ.
(d) Determine the equation of motion for large and small x.

Fig. 2.41 Mathematical
model for cushion suspension

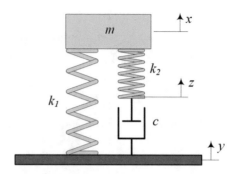

10. ★ Cushion mathematical model.

Figure 2.41 illustrates a mathematical model for cushion suspension. Such a model can be used to analyze the driver's seat, or a rubbery pad suspension.

(a) Derive the equations of motion for the variables x and z and using y as a known input function of time.

(b) Eliminate z and derive a third-order equation for x.

11. ★ Relative frequency.

Consider a body B that is moving along the x-axis with a constant velocity u and every T seconds emits small particles which move with a constant velocity c along the x-axis. If f denotes the frequency and λ the distance between two successively emitted particles, then we have

$$f = \frac{1}{T} = \frac{c - u}{\lambda} \qquad (2.659)$$

Now suppose that an observer moves along the x-axis with velocity v. Let us show the number of particles per second that the observer meets by the relative frequency f' and the time between meeting the two successive particles by the relative period T', where

$$f' = \frac{c - v}{\lambda} \qquad (2.660)$$

Show that

$$f' \approx f \left(1 - \frac{v - u}{c} \right) \qquad (2.661)$$

12. Absolute and relative coordinates and equations of motion of double pendulum.

Figure 2.42 illustrates two similar double pendulums. We express the motion of the left one using absolute coordinates θ_1 and θ_2 and express the motion of the right one with absolute coordinate θ_1 and relative coordinate φ.

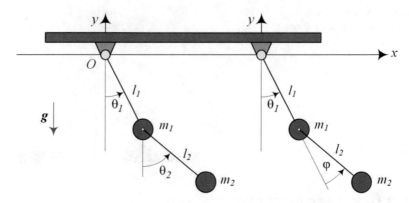

Fig. 2.42 Two similar double pendulums, expressed by absolute and relative coordinates

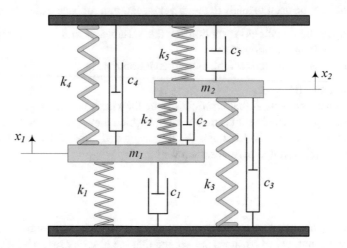

Fig. 2.43 A two DOF vibrating system

(a) Determine the equation of motion of the absolute coordinate double pendulum.
(b) Determine the equation of motion of the relative coordinate double pendulum.
(c) Compare their mass and stiffness matrices.

13. Equation of motion of a multiple DOF system.
    Figure 2.43 illustrates a two DOF vibrating system.

(a) Determine the kinetic and potential energies K and P and dissipation function
   D.
(b) Determine the equations of motion using the Lagrange method.
(c) ★ Rewrite K, P, and D in quadrature form.

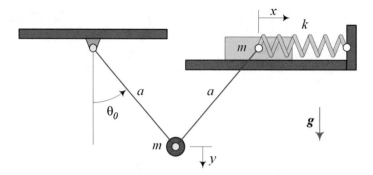

Fig. 2.44 A combination of pendulums and springs

Fig. 2.45 An elastic
pendulum

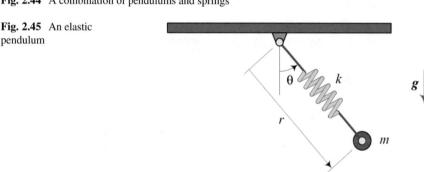

14. Static equilibrium position.

   Figure 2.44 illustrates a combination of pendulums and springs.

   (a) Determine the value of equilibrium θ_0 and static stretch of spring δ_0 if we assemble the system and let it go slowly.
   (b) Determine the equation of motion of the system in terms of x measured from the shown equilibrium position.
   (c) Determine the equation of motion of the system in terms of y.

15. Elastic pendulum.

   Figure 2.45 illustrates an elastic pendulum. Such a pendulum has two DOF.

   (a) Determine the equations of motion using the energy method.
   (b) Determine the equations of motion using the Lagrange method.
   (c) Determine the equations of motion using the Newton-Euler method.
   (d) Linearize the equations of motion and rewrite them in matrix form.

16. Compound pendulum and variable density.

   Figure 2.46a illustrates a slender as a compound pendulum with variable density, and Fig. 2.46b illustrates a simple pendulum with the same length. Determine the equivalent mass m_e if the mass density $\rho = m/l$ is:

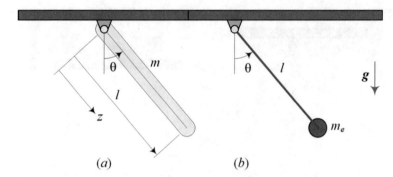

Fig. 2.46 (**a**) A slender as a pendulum with variable density. (**b**) A simple pendulum with the same length

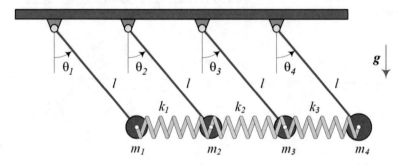

Fig. 2.47 Four connected pendulums

(a) $\rho = C_1 z$

(b) $\rho = C_2 (l - z)$

(c) $\rho = C_3 \left(z - \dfrac{l}{2} \right)^2$

(d) $\rho = C_4 \left(\dfrac{l}{2} - \left(z - \dfrac{l}{2} \right) \right)$

17. A four DOF nonlinear connected vibrating system.

   Four pendulums are connected as shown in Fig. 2.47.

   (a) Determine the kinetic energy K, linearize the equation, and find the mass matrix $[m]$.
   (b) Determine the potential energy P, linearize the equation, and find the stiffness matrix $[k]$.
   (c) Determine the equations of motion using K and P and determine the symmetric matrices $[\underline{m}]$ and $[\underline{k}]$ of the linearize equations.

18. Two spring connected to two heavy discs.

   The two spring connected disc system of Fig. 2.48 is linear for small θ_1 and θ_2. Find the equations of motion by energy and Lagrange methods.

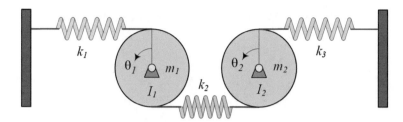

Fig. 2.48 Two spring connected heavy discs

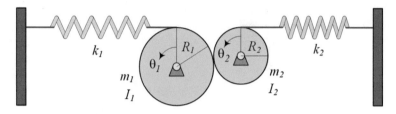

Fig. 2.49 Two connected disc system

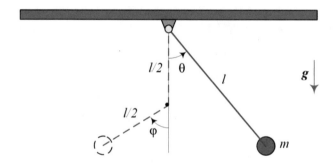

Fig. 2.50 A pendulum and a peg

19. Two connected heavy discs.

    Determine the equations of motion of the two connected disc system of Fig. 2.49 by energy and Lagrange methods.

20. ★ A pendulum and a peg.

    Determine the equation of motion of the pendulum in Fig. 2.50 using only one variable θ or φ.

21. ★ A pendulum on circular wall.

    Determine the equation of motion of the pendulum in Fig. 2.51 using the variable θ.

22. ★ A wire of the shape $y = f(x)$.

    Consider a wire in an arbitrary shape given by $y = f(x)$ as is shown in Fig. 2.52. Determine the equation of motion of a bid with mass m that is sliding frictionless on the wire.

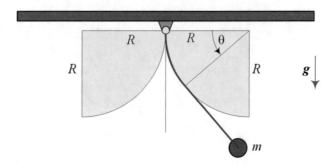

Fig. 2.51 A pendulum on circular wall

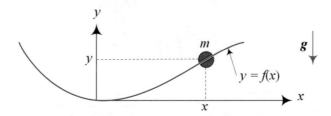

Fig. 2.52 A wire in shape of $y = f(x)$

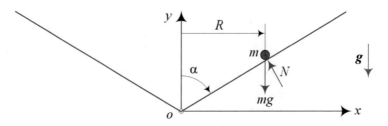

Fig. 2.53 A particle of mass m slides inside a conical shell

23. ★ A particle moving in a cone deep ground.

A particle of mass m slides without friction inside an upside down conical shell of semi-vertical angle α, as is shown in Fig. 2.53.

(a) Determine the equations of motion of the particle, using Euler equation.
(b) Show that it is possible for the particle to move such that it is at a constant R with the cone axis.
(c) Determine the angular speed of the particle for a uniform motion of part (b).

24. ★ A particle on a circular surface.

Draw *FBD* of the particle in Fig. 2.54a and b for $a = cR$ and $c < 1$ and determine their equation of motion. The spring is linear and applies a tangential force on m.

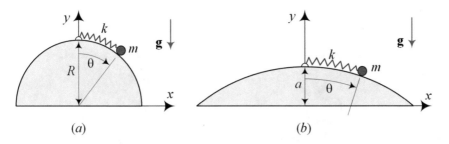

Fig. 2.54 A particle on a circular surface

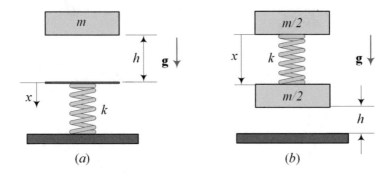

Fig. 2.55 A solid mass m falls on or with a spring

25. ★ Falling on a spring.

A solid mass m falls on a spring as shown in Fig. 2.55a or with a spring as shown Fig. 2.55b. The spring exerts a stiffness force F_s. Determine the maximum compression x_{Max} of the springs if:

(a) The restitution coefficient $e = 0$ and $F_s = kx$
(b) The restitution coefficient $e = 0$ and $F_s = kx^3$
(c) The restitution coefficient $e = 1$ and $F_s = kx^3$

We define a *restitution coefficient e* by

$$v_2' - v_1' = e\,(v_1 - v_2) \qquad 0 \le e \le 1 \qquad (2.662)$$

where v_1 and v_2 are the speed of the two particles before impact and v_1' and v_2' are their speeds after impact. The case $e = 1$ indicates an *inelastic collision* in which the particles stick to each other after impact, and the case $e = 0$ is called the *elastic collision* in which the energy conserves in impact.

26. ★ Pendulum with flexible support.

Figure 2.56a and b illustrate two pendulums with flexible supports in directions of x and y, respectively. Determine the equations of motion for:

(a) A pendulum with a flexible support in the x-direction of Fig. 2.56a

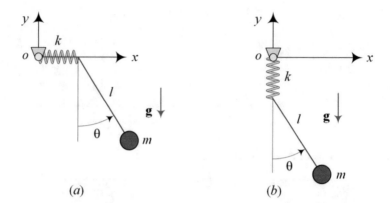

Fig. 2.56 Pendulums with flexible support

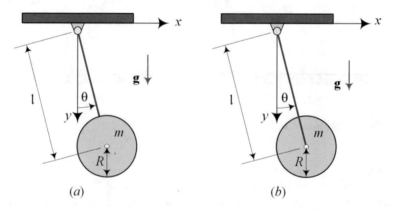

Fig. 2.57 Heavy pendulums, with and without a revolute joint

(b) A pendulum with a flexible support in the y-direction of Fig. 2.56b.

27. ★ Heavy pendulum.

Figure 2.57a illustrates a heavy disc with mass m and radius R suspended by a massless rod of length l. Figure 2.57b illustrates another heavy disc with mass m and radius R that is attached to a massless rod of length l by a frictionless revolute joint at its center.

(a) Derive the equations of motion for the pendulums in (a) and (b).
(b) Linearize the equations of motion. Is it possible to compare the periods of the oscillations?
(c) ★ Assume the disc of Fig. 2.57b has an angular velocity of ω when $\theta = 0$. Determine the equation of motion, linearize the equation, and determine the period of oscillation.

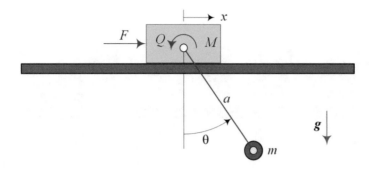

Fig. 2.58 A cart on a horizontal surface with a hanging pendulum

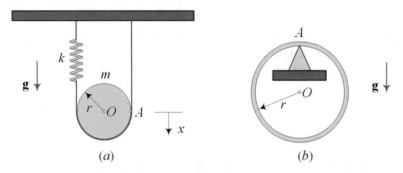

Fig. 2.59 A roller hung on an unstretchable rope and a spring

28. Controlled cart with hanging pendulum.

Figure 2.58 illustrates a cart on a horizontal surface with a hanging pendulum. Both the cart and the pendulum are controlled by the force F and torque Q, respectively. Employing the Lagrange method shows that the equations of motion are:

$$(M + m)\, l\ddot{x} - \frac{1}{2}ml\left(\ddot{\theta}\cos\theta - \dot{\theta}^2\sin\theta\right) = F(t) \tag{2.663}$$

$$\frac{1}{3}ml^2\ddot{\theta} - \frac{1}{2}ml^2\ddot{x}\cos\theta + \frac{1}{2}mgl\sin\theta = Q(t) \tag{2.664}$$

29. Equation of motion of a roller hung on an unstretchable rope and a spring and a thin circular ring.

   (a) Figure 2.59a illustrates a roller of radius r and mass m, and mass moment $I = mr^2/2$ is hung on an unstretchable rope and a spring. Derive the equation of motion by taking a moment about point A or by Lagrange method and show that its natural frequency is ω_n.

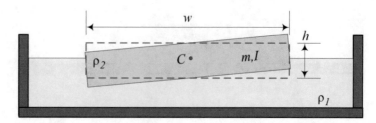

Fig. 2.60 A floating wooden slab

$$\omega_n = \sqrt{\frac{8k}{3m}} \qquad\qquad (2.665)$$

(b) Derive the equation of motion of the thin circular ring of Fig. 2.59b and show that its natural frequency is ω_n.

$$\omega_n = \sqrt{\frac{8}{2r}} \qquad\qquad (2.666)$$

30. A floating wooden slab.

   Figure 2.60 illustrates a floating wooden slab on water. The slab has dimensions of $l \times w \times h$, with mass of m and mass moment about the axis normal to the figure outward of I. Assume the density of the liquid is ρ_1 and density of the wood is $\rho_2 < \rho_1$. Let us consider two modes of vibrations of the slab, vertical (up and down) and rotational.

   (a) Derive the equation of motion of the vertical oscillation of the slab when it oscillates up and down. What is the frequency of oscillation?
   (b) Derive the equation of motion of rotational oscillation of the slab when two sides of the slab oscillate up and down opposite to each other. Does such rotational oscillation make the mass center C to move up and down as well? Linearize the equation of motion. What is the frequency of oscillation?
   (c) Assume two degrees of freedom for the slab and derive the equations of motion. Are they dynamically or statically coupled? Linearize the equations and determine the natural frequencies. How different are these frequencies with part a and b.

31. A beam with enharmonic displacement excitation.

   Figure 2.61 illustrates a pivoted beam supported by a linear damper. There is displacement excitation at the tip point through a linear spring. The displacement is not a simple harmonic; even the angular velocity ω is constant. Derive the equation of motion of the system.

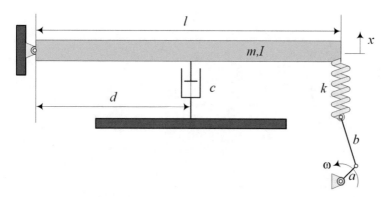

Fig. 2.61 A pivoted beam with a displacement excitation

Part II
Time Response

Time response is the reaction of a vibrating system due to nonzero initial conditions and/or time forcing function, usually transient excitations. Therefore, time response may also be called transient response.

The linear discrete vibrating systems are expressed by linear ordinary differential equations. If the parameters of the system are constant, the equations of motion are differential equations with constant coefficients with known general solutions. If there are transient excitations, then such systems have the following properties for their solutions:

1. Superposition of time responses is applied.

$$x(C_1 f_1 + C_2 f_2) = C_1 x(f_1) + C_2 x(f_2)$$

2. The total time response is a combination of two parts: *natural* and *forced*, or *general* and *particular*, or *homogeneous* and *inhomogenous*.

$$x(t) = x_h(t) + x_p(t)$$

3. The natural solution of linear equations is always exponential time functions.

$$x_h(t) = \sum_{i=1}^{n} C_i x(0) e^{-st}$$

4. The forced solution always has the same form as the force function with proportional magnitude.

$$x_p(t) = Cf(t)$$

In this part we review the natural, transient, and nonharmonic forced vibrations of vibrating systems in time domain.

Chapter 3
One Degree of Freedom

Consider a one degree of freedom (DOF) vibrating system as is shown in Fig. 3.1. The equation of motion of the system is a forced linear second-order differential equation along with a set of given initial conditions.

$$m\ddot{x} + c\dot{x} + kx = f(x, \dot{x}, t) \tag{3.1}$$

$$x(0) = x_0 \qquad \dot{x}(0) = \dot{x}_0 \tag{3.2}$$

The coefficients of mass m, stuffiness k, and damping c, are assumed constant, although they may be functions of time in more general problems. The solution of such problem, $x = x(t)$, $t > 0$, will be studied in this chapter.

When there is no excitation, $f = 0$, the equation is called *homogeneous*,

$$m\ddot{x} + c\dot{x} + kx = 0 \tag{3.3}$$

otherwise it is *nonhomogeneous*. The solution of the general equation of motion (3.1) is made up of two parts: $x_h(t)$ which is the *homogeneous solution* and $x_p(t)$ which is the *particular solution*.

$$x(t) = x_h(t) + x_p(t) \tag{3.4}$$

In mechanical vibrations, the no force case $f = 0$ is called a *free vibration*, and its solution $x_h(t)$ is called the *free vibration response*. The nonhomogeneous case is called *forced vibration* and its solution is called *forced vibration response*.

The *order* of an equation is the highest number of derivatives in the equation. In this chapter we study all systems that can be modeled by the second-order differential equation (3.1). A second-order differential equation will have two

© The Author(s), under exclusive license to Springer Nature Switzerland AG 2022
R. N. Jazar, *Advanced Vibrations*, https://doi.org/10.1007/978-3-031-16356-2_3

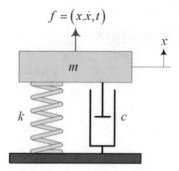

Fig. 3.1 The model of a one DOF vibrating system

independent solutions. If $x_1(t)$, and $x_2(t)$ are the two independent solutions of a second-order equation, then its general solution is a linear combination of them with unknown coefficients to be determined by initial conditions of the system:

$$x(t) = a_1 x_1(t) + a_2 x_2(t) \tag{3.5}$$

Determination of homogeneous solution x_h and particular solution x_p is the subject of this chapter.

Proof Assuming $x_h(t)$ is the solution of Eq. (3.3) and $x_p(t)$ is the particular solution of Eq. (3.1) associated to the excitation f, we observe that $x = x_h + x_p$ satisfies $m\ddot{x} + c\dot{x} + kx = f$ and hence is the solution:

$$\begin{aligned}
m\ddot{x} + c\dot{x} + kx &= m\left(\ddot{x}_h + \ddot{x}_p\right) + c\left(\dot{x}_h + \dot{x}_p\right) + k\left(x_h + x_p\right) \\
&= \left(m\ddot{x}_h + c\dot{x}_h + kx_h\right) + \left(m\ddot{x}_p + c\dot{x}_p + kx_p\right) \\
&= 0 + f = f
\end{aligned} \tag{3.6}$$

Therefore, if x is any solution of the nonhomogeneous equation (3.1), then $x - x_p$ is a solution of the homogeneous equation:

$$\begin{aligned}
m\frac{d^2}{dt^2}\left(x - x_p\right) &+ c\frac{d}{dt}\left(x - x_p\right) + k\left(x - x_p\right) \\
&= m\left(\ddot{x} - \ddot{x}_p\right) + c\left(\dot{x} - \dot{x}_p\right) + k\left(x - x_p\right) \\
&= \left(m\ddot{x} + c\dot{x} + kx\right) - \left(m\ddot{x}_p + c\dot{x}_p + kx_p\right) \\
&= f - f = 0
\end{aligned} \tag{3.7}$$

Hence, we find the solution of the nonhomogeneous equation (3.1) by determining the solution x_h of the associated homogeneous equation (3.3) and add the particular solution x_p of the nonhomogeneous equation associated to the forcing term f. ∎

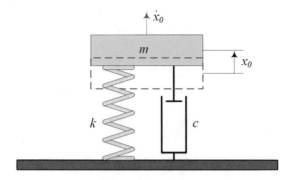

Fig. 3.2 The model of a forced-free one degree of freedom vibrating system with initial conditions

3.1 Free Vibrations

Consider a forced-free mass-spring-damper system (m, k, c) as is shown in Fig. 3.2. The equation of motion of the system and the general initial conditions are:

$$m\ddot{x} + c\dot{x} + kx = 0 \tag{3.8}$$

$$x(0) = x_0 \qquad \dot{x}(0) = \dot{x}_0 \tag{3.9}$$

Using the definition of the *natural frequency* ω_n and *damping ratio* ξ,

$$\omega_n = \sqrt{\frac{k}{m}} \tag{3.10}$$

$$\xi = \frac{c}{2\sqrt{km}} = \frac{c}{2m\omega_n} = \frac{c\omega_n}{2k} \tag{3.11}$$

we can rewrite the equation of motion to be dependent on only two parameters ω_n and ξ, instead of three parameters $m, k,$ and c.

$$\ddot{x} + 2\xi\omega_n\dot{x} + \omega_n^2 x = 0 \tag{3.12}$$

The time response of the free vibration of the system (3.8) is:

$$x(t) = X_1 e^{s_1 t} + X_2 e^{s_2 t} \tag{3.13}$$

where the *characteristic parameters* s_1 and s_2 are the solutions of the *characteristic equation* of the system:

$$s^2 + 2\xi\omega_n s + \omega_n^2 = 0 \tag{3.14}$$

$$s_1 = -\xi\omega_n + \omega_n\sqrt{\xi^2 - 1} \tag{3.15}$$

$$s_2 = -\xi\omega_n - \omega_n\sqrt{\xi^2 - 1} \tag{3.16}$$

The constants of integrations X_1 and X_2 depend on the initial conditions:

$$X_1 = \frac{\dot{x}_0 - s_2 x_0}{s_1 - s_2} \tag{3.17}$$

$$X_2 = -\frac{\dot{x}_0 - s_1 x_0}{s_1 - s_2} \tag{3.18}$$

Depending on the value of the damping ratio ξ, we will have five types of time response for free vibrations:

1. Underdamped: $0 < \xi < 1$. The system shows vibrations with a decaying amplitude:

$$x = e^{-\xi\omega_n t} \left(A \cos \omega_d t + B \sin \omega_d t\right) \tag{3.19}$$

$$= X e^{-\xi\omega_n t} \sin \left(\omega_d t + \varphi\right) \tag{3.20}$$

$$= X e^{-\xi\omega_n t} \cos \left(\omega_d t + \theta\right) \tag{3.21}$$

$$A = x_0 \qquad B = \frac{\dot{x}_0 + \xi\omega_n x_0}{\omega_d} \tag{3.22}$$

$$X = \sqrt{A^2 + B^2} \tag{3.23}$$

$$\omega_d = \omega_n\sqrt{1 - \xi^2} \tag{3.24}$$

2. Critically damped: $\xi = 1$. The system approaches its equilibrium in fastest time with no vibrations:

$$x = x_0 e^{-t\omega_n} \left(1 + \omega_n t\right) + \dot{x}_0 t e^{-t\omega_n} \tag{3.25}$$

3. Overdamped: $\xi > 1$. The system approaches its equilibrium asymptotically and shows no vibrations:

$$x(t) = X_1 e^{s_1 t} + X_2 e^{s_2 t} \tag{3.26}$$

$$X_1 = \frac{\dot{x}_0 - s_2 x_0}{s_1 - s_2} \qquad X_2 = -\frac{\dot{x}_0 - s_1 x_0}{s_1 - s_2} \tag{3.27}$$

$$s_1 = \left(-\xi + \sqrt{\xi^2 - 1}\right)\omega_n < 0 \tag{3.28}$$

$$s_2 = \left(-\xi - \sqrt{\xi^2 - 1}\right)\omega_n < 0 \qquad (3.29)$$

4. Undamped: $\xi = 0$. The system vibrates harmonically with a constant amplitude:

$$x = A \cos \omega_n t + B \sin \omega_n t \qquad (3.30)$$

$$= X \sin (\omega_n t + \varphi) \qquad (3.31)$$

$$= X \cos (\omega_n t + \theta) \qquad (3.32)$$

$$A = x_0 \qquad B = \frac{\dot{x}_0}{\omega_d} \qquad (3.33)$$

$$X = \sqrt{A^2 + B^2} \qquad (3.34)$$

5. Negative damping: $\xi < 0$. The system is unstable and vibrates with an increasing amplitude unboundedly:

$$x = \frac{\dot{x}_0 + \left(\xi + \sqrt{\xi^2 - 1}\right)\omega_n x_0}{2\omega_n \sqrt{\xi^2 - 1}} e^{-\left(\xi - \sqrt{\xi^2 - 1}\right)\omega_n t}$$

$$- \frac{\dot{x}_0 + \left(\xi - \sqrt{\xi^2 - 1}\right)\omega_n x_0}{2\omega_n \sqrt{\xi^2 - 1}} e^{-\left(\xi + \sqrt{\xi^2 - 1}\right)\omega_n t} \qquad (3.35)$$

Proof Dividing the free vibration equation of motion (3.8) by m:

$$\ddot{x} + \frac{c}{m}\dot{x} + \frac{k}{m}x = 0 \qquad (3.36)$$

$$x(0) = x_0 \qquad \dot{x}(0) = \dot{x}_0 \qquad (3.37)$$

yields:

$$\ddot{x} + 2\xi\omega_n \dot{x} + \omega_n^2 x = 0 \qquad (3.38)$$

where:

$$2\xi\omega_n = \frac{c}{m} \qquad (3.39)$$

$$\omega_n^2 = \frac{k}{m} \qquad (3.40)$$

The parameter ω_n is the natural frequency of the system, which is the frequency of vibration of the undamped system, and the dimensionless parameter ξ is a measure of the damping of the system.

The free vibration of a system is a *transient response* that for a set of parameters ω_n and ξ depends solely on the initial conditions $x_0 = x(0)$ and $\dot{x}_0 = \dot{x}(0)$. The natural frequency ω_n is a positive number that indicates the fluctuations of the x in a unit of time. Variation of ω_n will not change the category of x. The value of the positive damping ratio ξ indicates how quick the effects of initial conditions $x_0 = x(0)$ and $\dot{x}_0 = \dot{x}(0)$ will disappear and x stops at equilibrium point. However, variation of ξ will provide different category of responses x.

The general solution of a second-order homogeneous linear equation:

$$\ddot{x} + p(t)\,\dot{x} + q(t)\,x = 0 \tag{3.41}$$

is

$$x = X_1 x_1(t) + X_2 x_2(t) \tag{3.42}$$

where the X_i are constants of integration and $x_1(t)$ and $x_2(t)$ are linearly independent functions, each satisfying Eq. (3.41). There are always exactly *two* linearly independent solutions to Eq. (3.41) in any region in which the coefficient functions $p(t)$ and $q(t)$ are continuous. For linear differential equation, the exponential functions $x = X\exp(st)$ are the solution. Substituting the exponential function into the differential equation will determine two values for s that makes the two solutions $x_1 = X_1\exp(s_1 t)$ and $x_2 = X_2\exp(s_2 t)$. In special case when $s_1 = s_2 = s$, the two solutions would be $x_1 = X_1\exp(st)$ and $x_2 = X_2 t\exp(st)$.

The equation of vibration (3.38) is a linear second-order ordinary deferential equation, and hence its solution will be exponential function. To find the solution, we substitute a general exponential function with unknown coefficients and unknown power and determine the unknowns.

$$x = X\,e^{st} \tag{3.43}$$

Determination of the solution is perform in two steps:

1. We determine the exponent s. Their value depends on the parameters of the system, and the number of s is equal to the order of the differential equation. In this case, we will have two values for s, and they will be functions of the parameters ω_n and ξ.

   Substituting (3.43) in (3.38)

$$\left(Xs^2 e^{st}\right) + 2\xi\omega_n\left(Xse^{st}\right) + \omega_n^2\left(Xe^{st}\right) = \left(s^2 + 2\xi\omega_n s + \omega_n^2\right)Xe^{st} = 0 \tag{3.44}$$

provides us with an second-degree algebraic equation, called the characteristic equation:

$$s^2 + 2\xi\omega_n s + \omega_n^2 = 0 \tag{3.45}$$

The solution of the characteristic equation is the characteristic values s_1 and s_2:

$$s_1 = -\xi\omega_n + \sqrt{\xi^2 - 1}\,\omega_n \tag{3.46}$$

$$s_2 = -\xi\omega_n - \sqrt{\xi^2 - 1}\,\omega_n \tag{3.47}$$

2. We determine the coefficient X. Every value of s provides us with an exponential function. All of these exponential functions are linearly independent, and hence the general solution of the equation would be a linear combination of the exponential functions, each with an unknown coefficient. The coefficients depend on the initial conditions of the system. In this case, we have two exponential functions with coefficients X_1 and X_2:

$$x = X_1 e^{s_1 t} + X_2 e^{s_2 t} \tag{3.48}$$

$$\dot{x} = X_1 s_1 e^{s_1 t} + X_2 s_2 e^{s_2 t} \tag{3.49}$$

The initial condition (3.37) provides us with two equations to find X_1 and X_2:

$$x_0 = X_1 + X_2 \tag{3.50}$$

$$\dot{x}_0 = X_1 s_1 + X_2 s_2 \tag{3.51}$$

$$\begin{bmatrix} x_0 \\ \dot{x}_0 \end{bmatrix} = \begin{bmatrix} 1 & 1 \\ s_1 & s_2 \end{bmatrix} \begin{bmatrix} X_1 \\ X_2 \end{bmatrix} \tag{3.52}$$

$$\begin{bmatrix} X_1 \\ X_2 \end{bmatrix} = \begin{bmatrix} 1 & 1 \\ s_1 & s_2 \end{bmatrix}^{-1} \begin{bmatrix} x_0 \\ \dot{x}_0 \end{bmatrix} = \begin{bmatrix} \frac{1}{s_1 - s_2}(\dot{x}_0 - s_2 x_0) \\ -\frac{1}{s_1 - s_2}(\dot{x}_0 - s_1 x_0) \end{bmatrix} \tag{3.53}$$

Therefore, the complete solution is achieved:

$$x = X_1 e^{s_1 t} + X_2 e^{s_2 t} = \frac{\dot{x}_0 - s_2 x_0}{s_1 - s_2} e^{s_1 t} - \frac{\dot{x}_0 - s_1 x_0}{s_1 - s_2} e^{s_2 t} \tag{3.54}$$

The complete solution will have different categories of appearance depending on the value of the damping ration ξ as classified in the following example. ∎

Example 83 The characteristic values. This is to show how to determine the characteristic values of a second-order equation of motion and classifying the solutions according to the value of damping ratio ξ. The two characteristic values provide the two independent solutions of the equation.

The exponential function is the general solution of every linear differential equation. The equation of motion of the free vibrations of a one DOF is a second-order linear equation:

$$m\ddot{x} + c\dot{x} + kx = 0 \tag{3.55}$$

Substituting an exponential function with unknown coefficient to satisfy the equation:

$$x = Ce^{st} \tag{3.56}$$

provides us with:

$$\left(ms^2 + cs + k\right) Ce^{st} = 0 \tag{3.57}$$

Because $Ce^{st} \neq 0$ at any time, we must have:

$$ms^2 + cs + k = 0 \tag{3.58}$$

This is a second-degree algebraic equation, called the characteristic equation of Eq. (3.55). The solutions of the characteristic equation are all possible values of the exponent s of the exponential solution (3.56):

$$s_1 = -\frac{1}{2m}\left(c + \sqrt{-4km + c^2}\right) \tag{3.59}$$

$$s_2 = -\frac{1}{2m}\left(c - \sqrt{-4km + c^2}\right) \tag{3.60}$$

Employing the parameters ω_n and ξ from (3.10) and (3.11), we usually rewrite the equation of motion (3.55) in the following form to generalize the equation and work with the minimum number of parameters:

$$\ddot{x} + 2\xi\omega_n\dot{x} + \omega_n^2 x = 0 \tag{3.61}$$

Changing the parameters does not change the linearity of the equation, and therefore the exponential function is still the solution. The characteristic equation will be changed to:

$$s^2 + 2\xi\omega_n s + \omega_n^2 = 0 \tag{3.62}$$

with equivalent solutions:

$$s_1 = -\xi\omega_n + \omega_n\sqrt{\xi^2 - 1} = \omega_n\left(-\xi + \sqrt{\xi^2 - 1}\right)$$

$$= -\xi\omega_n\left(1 - \sqrt{1 - \frac{1}{\xi^2}}\right) \tag{3.63}$$

$$s_2 = -\xi\omega_n - \omega_n\sqrt{\xi^2 - 1} = \omega_n\left(-\xi - \sqrt{\xi^2 - 1}\right)$$

$$= -\xi\omega_n\left(1 + \sqrt{1 - \frac{1}{\xi^2}}\right) \tag{3.64}$$

The values of s_1 and s_2 depend on the natural frequency $\omega_n > 0$ and the damping ratio ξ of the system. The natural frequency ω_n appears as a coefficient and hence will not change the character of s_1 and s_2. If $\xi < 1$, then s_1 and s_2 will be two complex conjugate numbers. We will have only one characteristic value $s_1 = s_2$ if $\xi = 1$. The s_1 and s_2 provide two distinct real values as long as $1 < \xi$. The behavior of the system is different for different ranges of ξ. The ranges and special cases for different ξ are:

1. Underdamped: $0 < \xi < 1$. The system shows vibrations with a decaying amplitude.
2. Critically damped: $\xi = 1$. The system approaches its equilibrium in fastest time with no vibrations.
3. Overdamped: $1 < \xi$. The system approaches its equilibrium asymptotically and shows no vibrations.
4. Undamped: $\xi = 0$. The system vibrates harmonically with a constant amplitude.
5. Oscillatory unstable: $-1 < \xi < 0$. The system vibrates with an increasing amplitude unbounded.
6. Monotonically unstable: $\xi < -1$. The system goes to infinity and shows no vibrations.

The characteristic equation of an n-order real system is an n-degree algebraic equation with n characteristic values. The solution of the equation of motion of the system would be a linear combination of the exponential time functions with the characteristic values as exponents. Therefore, there are n coefficients in the linear combined solutions that should be calculated based on n initial conditions of the system. The solution for a one DOF with a second-order equation of motion will be:

$$x = C_1 e^{s_1 t} + C_2 e^{s_2 t} \tag{3.65}$$

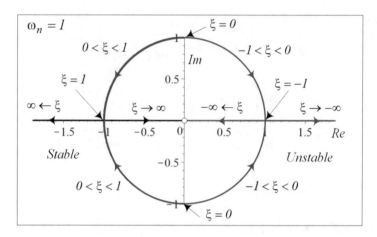

Fig. 3.3 Root locus of the characteristic values for $-\infty < \xi < \infty$

Example 84 ★ Root locus. Locating the position of the characteristic values on the imaginary plane predicts the behavior of a vibrating system.

The location of the characteristic values on the imaginary plane is a good indication of the behavior of a vibrating system. Figure 3.3 illustrates the location of the characteristic values s_1 and s_2 for $-\infty < \xi < \infty$.

The characteristic values s are expressed by complex numbers for $0 < \xi < 1$ and are on the left half of the complex plane:

$$s_1 = -\xi\omega_n + \omega_n\sqrt{\xi^2 - 1} = -\xi\omega_n + i\omega_n\sqrt{1 - \xi^2} \tag{3.66}$$

$$s_2 = -\xi\omega_n - \omega_n\sqrt{\xi^2 - 1} = -\xi\omega_n - i\omega_n\sqrt{1 - \xi^2} \tag{3.67}$$

The real parts of s_1 and s_2 are equal and negative that indicate the decay rate $-\xi\omega_n$ at which the amplitude of the oscillations shrinks in time. For stability, the real part of s must be negative. The imaginary part of s is the damped frequency $\omega_d = \omega_n\sqrt{1 - \xi^2}$ of oscillations. Almost all engineering applications of vibrating systems belong to this category.

The characteristic values s are real numbers for $\xi < 0$ and $1 < \xi$. When $1 < \xi$ then the system is stable, and s_1 moves from the point $s_1 = -1$ toward $s_1 = 0$ for $\xi \to \infty$, while s_2 moves from point $s_2 = -1$ toward $s_2 = -\infty$ for $\xi \to \infty$. When $\xi < 0$ then the system is unstable, and s_1 moves from $s_1 = -1$ toward $s_1 = 0$ for $\xi \to \infty$, while s_2 moves from $s_2 = -1$ toward $s_2 = -\infty$ for $\xi \to \infty$.

The characteristic values s are also complex for $-1 < \xi < 0$ and are on the right half of the complex plane:

$$s_1 = \xi\omega_n + \omega_n\sqrt{\xi^2 - 1} = \xi\omega_n + i\omega_n\sqrt{1 - \xi^2} \tag{3.68}$$

$$s_2 = \xi \omega_n - \omega_n \sqrt{\xi^2 - 1} = \xi \omega_n - i \omega_n \sqrt{1 - \xi^2} \tag{3.69}$$

The real parts of s_1 and s_2 are equal and positive that indicate the exponential rate $\xi \omega_n$ at which the amplitude of the oscillations grows in time. The imaginary part of s is the frequency $\omega_d = \omega_n \sqrt{1 - \xi^2}$ of unstably growing oscillations of the system.

Example 85 A homogeneous solution of a second-order linear equation. A numerical example to determine the solution of a second-order differential equation.

Consider a system with equation of motion:

$$\ddot{x} + \dot{x} - 2x = 0 \qquad x_0 = 1 \qquad \dot{x}_0 = 7 \tag{3.70}$$

To find the solution, we substitute an exponential solution $x = Ce^{st}$ in the equation of motion and find the characteristic equation:

$$s^2 + s - 2 = 0 \tag{3.71}$$

The characteristic values and the solution are:

$$s_{1,2} = 1, -2 \tag{3.72}$$

$$x = C_1 e^t + C_2 e^{-2t} \tag{3.73}$$

Taking a derivative and employing the initial conditions:

$$\dot{x} = C_1 e^t - 2C_2 e^{-2t} \tag{3.74}$$

$$1 = C_1 + C_2 \qquad 7 = C_1 - 2C_2 \tag{3.75}$$

provide us with the coefficients C_1 and C_2 and the solution $x = x(t)$

$$C_1 = 3 \qquad C_2 = -2 \tag{3.76}$$

$$x = 3e^t - 2e^{-2t} \tag{3.77}$$

Example 86 A fourth-order equation. All linear ordinary differential equations are solved similarly. Here is to show how to determine general solution of a fourth-order differential equation.

The characteristic equation of the fourth-order equation:

$$x^{(4)} - 4\dddot{x} + 12\ddot{x} + 4\dot{x} - 13x = 0 \tag{3.78}$$

is:

$$s^4 - 4s^3 + 12s^2 + 4s - 13 = 0 \tag{3.79}$$

The roots of the characteristic equation are:

$$s = \pm 1, 2 \pm 3i \tag{3.80}$$

Therefore the solution of the differential equation (3.78) is any linear combination of the fundamental solutions e^{st}:

$$
\begin{aligned}
x(t) &= C_1 e^t + C_2 e^{-t} + C_3 e^{(2+3i)t} + C_4 e^{(2-3i)t} \\
&= C_1 e^t + C_2 e^{-t} + e^{2t} (C_5 \cos 3t + C_6 \sin 3t)
\end{aligned} \tag{3.81}
$$

The value of the four coefficients $C_1, C_2, C_5, and C_6$ will be determined if the initial conditions of the equation are given.

Example 87 ★ Repeating characteristic values. The characteristic equation may have repeated roots. Here is an example to show how the solution of a differential equation with repeated characteristic roots will be derived.

If a differential equation has an n times repeating characteristic value s, then e^{st}, $te^{st}, t^2 e^{st}, \cdots,$ and $t^{n-1} e^{st}$ are the fundamental solutions of the equation associated to the n times repeating characteristic value. Any linear combination of fundamental solutions is a solution of the differential equation. Consider the following fifth-order differential equation:

$$x^{(5)} + 3x^{(4)} + 3\dddot{x} + \ddot{x} = 0 \tag{3.82}$$

Its characteristic equation is:

$$s^2 (s + 1)^3 = 0 \tag{3.83}$$

The solutions of the equation are:

$$1, t, e^{-t}, te^{-t}, t^2 e^{-t} \tag{3.84}$$

and, hence, any solution of the equation is of the following form:

$$x(t) = C_1 + C_2 t + \left(C_3 + C_4 t + C_5 t^2 \right) e^{-t} \tag{3.85}$$

Example 88 ★ State space representation of the equation of motion. Any differential equation of order n can be expressed by n first-order differential equations by introducing a set of new variables called state variables.

Consider a damped vibrating system:

$$\ddot{x} + 2\xi\omega_n\dot{x} + \omega_n^2 x = 0 \qquad x(0) = x_0 \qquad \dot{x}(0) = \dot{x}_0 \tag{3.86}$$

We can rewrite the equation of the system in state-variable form, by transforming it into a system of two first-order, linear, constant-coefficient ordinary differential equations. Let us define the following new variables:

$$x_1 = x \qquad x_2 = \dot{x} \tag{3.87}$$

Therefore:

$$\dot{x}_1 = x_2 \qquad \dot{x}_2 = -2\xi\omega_n x_2 - \omega_n^2 x_1 \tag{3.88}$$

$$x_1(0) = x_0 \qquad x_2(0) = \dot{x}_0 \tag{3.89}$$

Introducing a variable state vector $\mathbf{x}$, the equation of motion (3.86) will be defined by a set first-order equations in matrix form:

$$\mathbf{x} = \begin{bmatrix} x_1 \\ x_2 \end{bmatrix} \qquad \mathbf{x}_0 = \begin{bmatrix} x_1(0) \\ x_2(0) \end{bmatrix} = \begin{bmatrix} x_0 \\ \dot{x}_0 \end{bmatrix} \tag{3.90}$$

$$\dot{\mathbf{x}} = \frac{d\mathbf{x}}{dt} = \mathbf{A}\,\mathbf{x} = \begin{bmatrix} 0 & 1 \\ -\omega_n^2 & -2\xi\omega_n \end{bmatrix} \begin{bmatrix} x_1 \\ x_2 \end{bmatrix} \tag{3.91}$$

The matrix $\mathbf{A}$ is called the coefficient matrix and the solution of the first-order equation is:

$$\mathbf{x} = \mathbf{x}_0\, e^{\mathbf{A}} \qquad 0 < t \tag{3.92}$$

The exponential functions of the coefficient matrix, $e^{\mathbf{A}}$ for underdamped, critically damped, and overdamped are as below.

Underdamped: $\xi < 1$

$$e^{\mathbf{A}} = e^{-\xi\omega_n t} \begin{bmatrix} \cos\omega_d t + \dfrac{\xi\omega_n}{\omega_d}\sin\omega_d t & \dfrac{\sin\omega_d t}{\omega_d} \\[2ex] -\dfrac{\omega_n^2}{\omega_d}\sin\omega_d t & \cos\omega_d t - \dfrac{\xi\omega_n}{\omega_d}\sin\omega_d t \end{bmatrix} \tag{3.93}$$

$$\omega_d = \omega_n\sqrt{1-\xi^2}$$

$$x = e^{-\xi\omega_n t}\left(\cos\omega_d t + \frac{\xi\omega_n}{\omega_d}\sin\omega_d t\right)x_0 + \frac{e^{-\xi\omega_n t}\sin\omega_d t}{\omega_d}\dot{x}_0 \tag{3.94}$$

$$\dot{x} = -\frac{\omega_n^2 e^{-\xi\omega_n t} \sin\omega_d t}{\omega_d} x_0 + e^{-\xi\omega_n t}\left(\cos\omega_d t - \frac{\xi\omega_n}{\omega_d}\sin\omega_d t\right)\dot{x}_0 \qquad (3.95)$$

Critically damped: $\xi = 1$

$$e^{\mathbf{A}} = e^{-\omega_n t}\begin{bmatrix} 1+\omega_n t & t \\ -\omega_n^2 t & 1-\omega_n t \end{bmatrix} \qquad (3.96)$$

$$x = e^{-\omega_n t}\left((1+\omega_n t)\,x_0 + t\,\dot{x}_0\right) \qquad (3.97)$$

$$\dot{x} = e^{-\omega_n t}\left(-\omega_n^2\,t\,x_0 + (1-\omega_n t)\,\dot{x}_0\right) \qquad (3.98)$$

Overdamped: $1 < \xi$

$$e^{\mathbf{A}} = e^{-\xi\omega_n t}\begin{bmatrix} \cosh\omega_s t + \dfrac{\xi\omega_n}{\omega_s}\sinh\omega_s t & \dfrac{\sinh\omega_s t}{\omega_s} \\ -\dfrac{\omega_n^2}{\omega_s}\sinh\omega_s t & \cosh\omega_s t - \dfrac{\xi\omega_n}{\omega_s}\sinh\omega_s t \end{bmatrix} \qquad (3.99)$$

$$\omega_s = \omega_n\sqrt{\xi^2 - 1} \qquad (3.100)$$

$$x = e^{-\xi\omega_n t}\left(\cosh\omega_s t + \frac{\xi\omega_n}{\omega_s}\sin\omega_s t\right)x_0 + \frac{e^{-\xi\omega_n t}\sinh\omega_s t}{\omega_s}\dot{x}_0 \qquad (3.101)$$

$$\dot{x} = -\frac{\omega_n^2 e^{-\xi\omega_n t}\sinh\omega_s t}{\omega_s} x_0 + e^{-\xi\omega_n t}\left(\cosh\omega_s t - \frac{\xi\omega_n}{\omega_s}\sinh\omega_s t\right)\dot{x}_0 \qquad (3.102)$$

In all three cases, the response is a linear combination of the two initial conditions x_0 and $\dot{x}_0$.

Example 89 Stability conditions of homogenous linear differential equations. If the root of characteristic equation are real or complex, the real part must be negative to have stable solutions. It is equivalent to eigenvalues with negative real part of coefficient matrix.

Any homogeneous linear system of differential equations of order n can be written in the form of a set of first-order equations:

$$\dot{\mathbf{x}} = \mathbf{A}\,\mathbf{x} \qquad (3.103)$$

To have stable solutions, all eigenvalues of the coefficient matrix $\mathbf{A}$ must have negative real parts. Let us examine the coefficient matrix of a one DOF vibrating system:

$$\ddot{x} + 2\xi\omega_n\,\dot{x} + \omega_n^2 x = 0 \tag{3.104}$$

$$\dot{\mathbf{x}} = \mathbf{A}\,\mathbf{x} \tag{3.105}$$

$$\begin{bmatrix} \dot{x}_1 \\ \dot{x}_2 \end{bmatrix} = \begin{bmatrix} 0 & 1 \\ -\omega_n^2 & -2\xi\omega_n \end{bmatrix}\begin{bmatrix} x_1 \\ x_2 \end{bmatrix} \tag{3.106}$$

To determine the eigenvalues of $\mathbf{A}$, we need to calculate the characteristic equation of $\mathbf{A}$:

$$\begin{vmatrix} -\lambda & 1 \\ -\omega_n^2 & -2\xi\omega_n - \lambda \end{vmatrix} = 0 \tag{3.107}$$

$$\lambda^2 + 2\xi\omega_n\lambda + \omega_n^2 = 0 \tag{3.108}$$

The solutions of the characteristic equation are the eigenvalues. Assuming $\xi < 1$, we have:

$$\lambda_1 = -\xi\omega_n + i\sqrt{1 - \xi^2}\,\omega_n \tag{3.109}$$

$$\lambda_2 = -\xi\omega_n - i\sqrt{1 - \xi^2}\,\omega_n \tag{3.110}$$

Therefore, a one DOF vibrating system is stable as long as $0 < \xi$, because natural frequency is always positive, $0 < \omega_n$.

3.1.1 Underdamped: $0 < \xi < 1$

When $0 < \xi < 1$, the two roots of the characteristic equation (3.14) are complex conjugate numbers with negative real parts. The underdamped system is the vibrating class of responses of free vibrating systems with the most important engineering applications. The response of a system with parameters of natural frequency ω_n and damping ratio ξ from initial conditions $x(0) = x_0$ and $\dot{x}(0) = \dot{x}_0$ is an oscillating function with damped frequency $\omega_d = \omega_n\sqrt{1 - \xi^2}$ and an exponentially decreasing amplitude $Xe^{-\xi\omega_n t}$:

$$x = e^{-\xi\omega_n t}\left(x_0 \cos\omega_d t + \frac{\dot{x}_0 + \xi\omega_n x_0}{\omega_d}\sin\omega_d t\right) \tag{3.111}$$

$$= Xe^{-\xi\omega_n t}\sin(\omega_d t + \varphi) \tag{3.112}$$

$$= Xe^{-\xi\omega_n t}\cos(\omega_d t + \theta) \tag{3.113}$$

$$X = \sqrt{x_0^2 + \left(\frac{\dot{x}_0 + \xi\omega_n x_0}{\omega_d}\right)^2} \tag{3.114}$$

$$\tan\varphi = \frac{x_0\omega_d}{\dot{x}_0 + \xi\omega_n x_0} \tag{3.115}$$

$$\tan\theta = -\frac{\dot{x}_0 + \xi\omega_n x_0}{x_0\omega_d} \tag{3.116}$$

The solutions (3.111)–(3.113) are equivalent, which depending on the given problem and preference of the engineer, one of the expressions will work better than the other. This fact will be clarified by studying some examples.

Proof Having $0 < \xi < 1$ makes the characteristics s_1 and s_2 to be complex conjugate:

$$s_1 = -\xi\omega_n + \sqrt{\xi^2 - 1}\omega_n = -\xi\omega_n + i\sqrt{1 - \xi^2}\omega_n$$
$$= -\xi\omega_n + i\omega_d \tag{3.117}$$

$$s_2 = -\xi\omega_n - \sqrt{\xi^2 - 1}\omega_n = -\xi\omega_n - i\sqrt{1 - \xi^2}\omega_n$$
$$= -\xi\omega_n - i\omega_d \tag{3.118}$$

$$\omega_d = \sqrt{1 - \xi^2}\omega_n \qquad 0 < \omega_d < \omega_n \tag{3.119}$$

To rewrite it with harmonic functions, we employ the Euler identity:

$$e^{i\omega_d t} = \cos\omega_d t + i\sin\omega_d t \qquad i^2 = -1 \tag{3.120}$$

$$e^{-i\omega_d t} = \cos\omega_d t - i\sin\omega_d t \tag{3.121}$$

$$x = X_1 e^{(-\xi\omega_n + i\omega_d)t} + X_2 e^{(-\xi\omega_n - i\omega_d)t}$$
$$= e^{-\xi\omega_n t}(X_1(\cos\omega_d t + i\sin\omega_d t) + X_2(\cos\omega_d t - i\sin\omega_d t))$$
$$= (X_1 + X_2)e^{-\xi\omega_n t}\cos\omega_d t + (X_1 - X_2)ie^{-\xi\omega_n t}\sin\omega_d t \tag{3.122}$$

The coefficients X_1 and X_2 from (3.53) will be complex conjugates for $0 < \xi < 1$:

$$\begin{bmatrix} X_1 \\ X_2 \end{bmatrix} = \begin{bmatrix} \frac{1}{s_1 - s_2} (\dot{x}_0 - s_2 x_0) \\ -\frac{1}{s_1 - s_2} (\dot{x}_0 - s_1 x_0) \end{bmatrix}$$

$$= \begin{bmatrix} \frac{1}{2\omega_d} (\omega_d x_0 - i (\dot{x}_0 + \xi \omega_n x_0)) \\ \frac{1}{2\omega_d} (\omega_d x_0 + i (\dot{x}_0 + \xi \omega_n x_0)) \end{bmatrix} \tag{3.123}$$

Therefore:

$$X_1 + X_2 = x_0 \tag{3.124}$$

$$X_1 - X_2 = -\frac{i}{\omega_d} (\dot{x}_0 + \xi \omega_n x_0) \tag{3.125}$$

and it makes the solution x to be a full harmonic function with real coefficients and exponentially decreasing amplitude:

$$x = x_0 e^{-\xi \omega_n t} \cos \omega_d t - \frac{i}{\omega_d} (\dot{x}_0 + \xi \omega_n x_0) i e^{-\xi \omega_n t} \sin \omega_d t \tag{3.126}$$

$$= e^{-\xi \omega_n t} \left(x_0 \cos \omega_d t + \frac{\dot{x}_0 + \xi \omega_n x_0}{\omega_d} \sin \omega_d t \right) \tag{3.127}$$

The parameter ω_d is the frequency of both functions $\cos \omega_d t$ and $\sin \omega_d t$ and, hence, is the frequency of x. This is the frequency of response with damping which is called the damped frequency. The damping ratio ξ reduces the angular frequency from ω_n and increases the period of the oscillations from T_n:

$$T_d = \frac{T_n}{\sqrt{1 - \xi^2}} = \frac{2\pi}{\sqrt{1 - \xi^2} \omega_n} = \frac{2\pi}{\omega_d} \tag{3.128}$$

The solution (3.127) includes a full harmonic function which may be combined into a single cos or sin function with a phase lag:

$$x = X e^{-\xi \omega_n t} \sin (\omega_d t + \varphi) \tag{3.129}$$

$$x = X e^{-\xi \omega_n t} \cos (\omega_d t + \theta) \tag{3.130}$$

The relationship between the amplitude X, phase φ, and weight factors A and B can be found by equating Eqs. (3.127) and (3.129):

$$e^{-\xi \omega_n t} (A \sin \omega_d t + B \cos \omega_d t) = X e^{-\xi \omega_n t} \sin (\omega_d t + \varphi)$$

$$= e^{-\xi \omega_n t} (X \cos \varphi \sin \omega_d t + X \sin \varphi \cos \omega_d t) \tag{3.131}$$

$$A = \frac{\dot{x}_0 + \xi \omega_n x_0}{\omega_d} \qquad B = x_0 \tag{3.132}$$

$$A = X \cos \varphi \qquad B = X \sin \varphi \tag{3.133}$$

$$X = \sqrt{A^2 + B^2} = \sqrt{\left(\frac{\dot{x}_0 + \xi \omega_n x_0}{\omega_d}\right)^2 + x_0^2} \tag{3.134}$$

$$\tan \varphi = \frac{B}{A} = \frac{x_0 \omega_d}{\dot{x}_0 + \xi \omega_n x_0} \tag{3.135}$$

Similarly, the relationship between the amplitude X, phase θ, and weight factors A and B will be found by equating Eqs. (3.127) and (3.130):

$$e^{-\xi \omega_n t} (A \sin \omega_d t + B \cos \omega_d t) = X e^{-\xi \omega_n t} \cos (\omega_d t + \theta)$$

$$= (X \sin \theta \sin \omega_d t - X \cos \theta \cos \omega_d t) \tag{3.136}$$

$$A = X \sin \theta \qquad P = -X \cos \theta \tag{3.137}$$

$$X = \sqrt{A^2 + B^2} = \sqrt{x_0^2 + \left(\frac{\dot{x}_0 + \xi \omega_n x_0}{\omega_d}\right)^2} \tag{3.138}$$

$$\tan \theta = \frac{A}{-B} = -\frac{\dot{x}_0 + \xi \omega_n x_0}{x_0 \omega_d} \tag{3.139}$$

$$\tan \varphi \tan \theta = -1 \tag{3.140}$$

The solution (3.127) suggests to begin with a solution made up of a full harmonic function with unknown coefficients and exponentially decreasing amplitude and then determine the coefficients by initial conditions:

$$x = e^{-\xi \omega_n t} (A \sin \omega_d t + B \cos \omega_d t) \tag{3.141}$$

When $\xi = 0$ the response is sinusoidal. Damping is a frictional force that generates heat and dissipates energy. When the damping of the system is small, it still oscillates but with decreasing amplitude, and hence it goes to rest at equilibrium over time. ∎

Example 90 Underdamped system $\xi < 1$. Graphical illustration of underdamped vibrating systems.

The underdamped case of a vibrating systems is the most practical engineering situation. By "damped system," we usually refer to the class of underdamped vibrating systems. It may also be called "subcritically damped." An underdamped system has an oscillatory time response with a decaying amplitude as shown in Fig. 3.4 for $\xi - 0.15$, $\omega_n - 20\pi$ rad / s, $x_0 = 1$, and $\dot{x}_0 = 0$. The exponential functions $\pm e^{-\xi \omega_n t}$ are the envelops of the response. Envelops are curves to be tangent to a family of curves differing by family parameter.

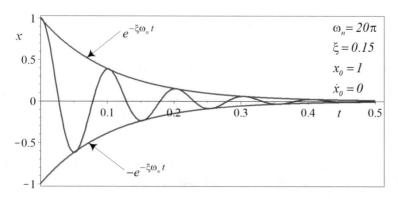

Fig. 3.4 A sample time response for an underdamped system

The time response of free vibrations of an underdamped one *DOF* system is expressed by:

$$x = e^{-\xi\omega_n t}\left(\frac{\dot{x}_0 + \xi\omega_n x_0}{\omega_d}\sin\omega_d t + x_0\cos\omega_d t\right) \tag{3.142}$$

which can also be written in the following equivalent forms:

$$x = e^{-\xi\omega_n t}\left(x_0\left(\cos\omega_d t + \frac{\xi}{\omega_d}\omega_n\sin\omega_d t\right) + \frac{\dot{x}_0}{\omega_d}\sin\omega_d t\right) \tag{3.143}$$

$$x = \frac{x_0}{\sin\varphi}e^{-\xi\omega_n t}\sin\left(\omega_d t + \arctan\frac{\omega_d x_0}{\xi\omega_n x_0 + \dot{x}_0}\right) \tag{3.144}$$

$$\varphi = \arctan\frac{\omega_d x_0}{\xi\omega_n x_0 + \dot{x}_0} \tag{3.145}$$

$$x = \frac{e^{-\xi\omega_n t}}{\omega_d}\sqrt{(\omega_d x_0)^2 + (\dot{x}_0 + \xi\omega_n x_0)^2}$$

$$\times\sin\left(\omega_d t + \arctan\frac{\omega_d x_0}{\xi\omega_n x_0 + \dot{x}_0}\right) \tag{3.146}$$

The response of the damped system is always bounded by the exponential functions x_B:

$$x_B = \pm e^{-\xi\omega_n t} \tag{3.147}$$

Therefore, if we are only interested in the maximum possible values of x, we may use x_B. It is possible to reach x_B at any desired time t by adjusting ω_d.

Example 91 ★ Envelop. Mathematical determination of equations of envelops of a family function.

Consider a function $y = f(x)$ that depends on a parameter c:

$$y = f(x, c) \tag{3.148}$$

Changing c will produce other curves for y as a function of x. If it is possible to eliminate c between (3.148) and the equation of:

$$\frac{dy}{dc} = 0 \tag{3.149}$$

we find an envelop equation:

$$y = g(x) \tag{3.150}$$

which is tangent to all members of the family $y = f(x, c)$ for all possible c. Such an equation, $y = g(x)$, if it exists, is called the envelop to the family of curves $y = f(x, c)$.

As an example, let us examine the family of the paths of motion of a projectile for a constant initial velocity v_0 and different shooting angle θ:

$$z = -\frac{1}{2} g \frac{x^2}{v_0^2 \cos^2 \theta} + x \tan \theta \tag{3.151}$$

Regardless of the shooting angle, the projectile cannot get out of a closed area. The reachable boundary of space is denoted by an envelop curve that is tangent to all paths. To find the envelop of the family, we should eliminate the family parameter θ between the equation of the family and the equation made up of its derivative with respect to the parameter θ. The derivative of z with respect to θ is:

$$\frac{dz}{d\theta} = -\frac{1}{v_0^2 \cos^2 \theta} \left(g x^2 \tan \theta - x v_0^2 \right) = 0 \tag{3.152}$$

which yields

$$\tan \theta = \frac{v_0^2}{gx} \tag{3.153}$$

$$\frac{1}{\cos^2 \theta} = 1 + \left(\frac{v_0^2}{gx} \right)^2 \tag{3.154}$$

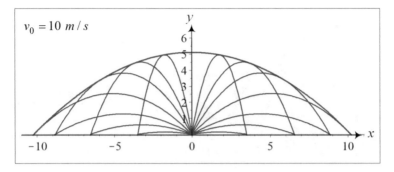

Fig. 3.5 The path of motion of a projectile with a constant initial speed v_0 and different angles θ

Substituting (3.153) and (3.154) in (3.151), we eliminate the parameter c and find the equation of the envelop:

$$z = \frac{1}{2}\left(\frac{v_0^2}{g} - \frac{g}{v_0^2}x^2\right) \tag{3.155}$$

When the shooting device is similar to an antiaircraft gun that can turn about the z-axis, the envelop of the reachable space is a circular paraboloid called a projectile umbrella:

$$z = \frac{v_0^2}{2g} - \frac{g}{2v_0^2}\left(x^2 + y^2\right) \tag{3.156}$$

Such a paraboloid is called the forbidden umbrella for military aircraft. The reachable space under the umbrella is:

$$0 \le z \le \frac{v_0^2}{2g} - \frac{g}{2v_0^2}\left(x^2 + y^2\right) \tag{3.157}$$

These results are independent of the mass of the projectile. Figure 3.5 illustrates paths of motion of a projectile of Eq. (3.151), with a constant initial speed v_0 and different angles θ as well as the envelop (3.155).

Example 92 The envelop of underdamped time responses. To use the envelop theory and determine the envelop curves of underdamped responses.

The time response x of underdamped systems, $\xi < 1$, for different $\omega_d = \omega_n\sqrt{1 - \xi^2}$ is always bounded between two envelop exponential functions x_e:

$$x = Xe^{-\xi\omega_n t}\sin(\omega_d t + \varphi) \tag{3.158}$$

$$x_e = \pm Xe^{-\xi\omega_n t} \tag{3.159}$$

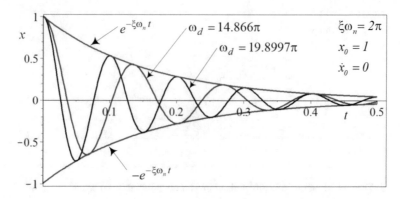

Fig. 3.6 The graph of x for $\omega_d = 19.8997\pi$ and $\omega_d = 14.866\pi$; at a constant $\xi\omega_n = 2\pi$

Figure 3.6 illustrates x for $\omega_d = 19.8997\pi$ and $\omega_d = 14.866\pi$ and a constant $\xi\omega_n = 2\pi$. To have the condition $\xi\omega_n = const$ and different ω_d, we select two values of ω_d and calculate ξ and ω_n:

$$\omega_d = 19.8997\pi \qquad \xi\omega_n = 2\pi \qquad \xi = 0.1 \qquad \omega_n = 20\pi \qquad (3.160)$$

$$\omega_d = 14.866\pi \qquad \xi\omega_n = 2\pi \qquad \xi = 0.1333 \qquad \omega_n = 15\pi \qquad (3.161)$$

To determine the equation of envelop curves that are tangent to the solution curves $x = x\,(t, \omega_d)$, we need to eliminate ω_d between (3.158) and $dx/d\omega_d$:

$$\frac{dx}{d\omega_d} = Xte^{-\xi\omega_n t} \cos\left(\varphi + \omega_d t\right) = 0 \qquad (3.162)$$

The equation $dx/d\omega_d = 0$ yields:

$$\varphi + \omega_d t = \pm\frac{\pi}{2} \qquad (3.163)$$

which after substituting in (3.158) provides us with the envelop curves that are tangent to the family of solution curves $x = x\,(t, \omega_d)$:

$$x = \pm Xe^{-\xi\omega_n t} \qquad (3.164)$$

Example 93 Damped frequency ω_d. To illustrate how the damping ratio ξ affects the amplitude and frequency of oscillations.

When $\xi = 0$, the system is undamped and the solution of an (m, k, c) system is simplified to an undamped harmonic function:

$$x = \frac{\dot{x}_0}{\omega_d} \sin\omega_n t + x_0 \cos\omega_n t \qquad (3.165)$$

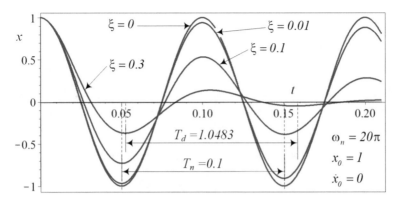

Fig. 3.7 The response of a damped one DOF vibrating system for $\omega_n = 20\pi$ and $\xi = 0, 0.01,$ 0.1, 0.3

The frequency of the undamped solution is $\omega_n = 2\pi f_n = 2\pi / T_n$ where, T_n is the natural period of the response. Introducing damping to the system yields the solution of a damped harmonic function x:

$$x = e^{-\xi \omega_n t} \left(\frac{\dot{x}_0 + \xi \omega_n x_0}{\omega_d} \sin \omega_d t + x_0 \cos \omega_d t \right) \tag{3.166}$$

The frequency of the damped solution is $\omega_d = 2\pi f_d = 2\pi / T_d$ where ω_d is the frequency of the vibration and T_d is the damping period of the response. Figure 3.7 illustrates x for $\omega_n = 20\pi$ and $\xi = 0, 0.01, 0.1, 0.3$. The periods of vibrations for $\xi = 0$ and $\xi = 0.3$ are compared.

The damped frequency of vibration $\omega_d = \omega_n \sqrt{1 - \xi^2}$ decreases by increasing ξ as is shown in Fig. 3.8. So, a damped system vibrates slower than its undamped system. Damping makes the system sluggish and decreases the frequency of oscillation and, hence, increases the period of oscillation.

Example 94 Real solution from complex characteristic values. A detailed discussion on how a real solution will be achieved from complex characteristic values and coefficients.

When $\xi < 1$, the roots of the characteristic equation $s^2 + 2\xi \omega_n s + \omega_n^2 = 0$ are complex conjugates:

$$s_1 = -\xi \omega_n - \omega_n i \sqrt{1 - \xi^2} = a - ib \tag{3.167}$$

$$s_2 = -\xi \omega_n + \omega_n i \sqrt{1 - \xi^2} = a + ib \tag{3.168}$$

$$a = -\xi \omega_n \qquad b = \omega_n \sqrt{1 - \xi^2} \qquad \xi < 1 \tag{3.169}$$

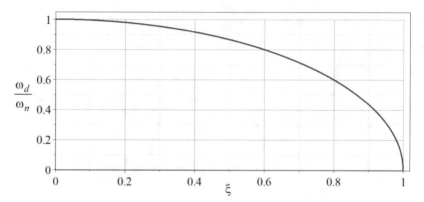

Fig. 3.8 The effect of increasing damping ξ on the frequency ratio $\omega_d/\omega_n = T_n/T_d = \sqrt{1-\xi^2}$

and the solution of the equation of motion x is:

$$x = C_1 e^{(a-ib)t} + C_2 e^{(a+ib)t} \tag{3.170}$$

Employing the Euler identity $e^{i\theta} = \cos\theta + i\sin\theta$ yields:

$$x = e^{at} C_1 (\cos bt - i \sin bt) + e^{at} C_2 (\cos bt + i \sin bt)$$
$$= e^{at} ((C_1 + C_2) \cos bt + i (C_1 - C_2) \sin bt) \tag{3.171}$$

The coefficients C_1 and C_2 must be complex conjugate to provide a real x:

$$C_1 = A_1 - i B_1 \qquad C_2 = A_1 + i B_1 \tag{3.172}$$

Therefore,

$$C_1 + C_2 = 2A_1 \qquad C_1 - C_2 = -2i B_1 \tag{3.173}$$

and

$$x = e^{at} C_1 (\cos bt - i \sin bt) + e^{at} C_2 (\cos bt + i \sin bt)$$
$$= 2e^{at} (A_1 \cos bt - B_1 \sin bt) = e^{at} (A \cos bt + B \sin bt) \tag{3.174}$$

This is how to write the solution of the equation all with real parameters and coefficients:

$$x = e^{at} (A \cos bt + B \sin bt)$$
$$= e^{-\xi\omega_n t} (A \cos \omega_d t + B \sin \omega_d t) \tag{3.175}$$

$$\omega_d = b = \omega_n \sqrt{1-\xi^2} \tag{3.176}$$

Example 95 Mechanical energy of a damped oscillator. The damping mechanism in a vibrating system dissipates energy from the system. Here is to determine the energy of a damped vibrating system mathematically.

The total mechanical energy $E = T + P$ of a damped oscillator is the sum of the kinetic K and potential P energies:

$$K = \frac{1}{2}m\dot{x}^2 \qquad P = \frac{1}{2}kx^2 \tag{3.177}$$

$$E = \frac{1}{2}m\dot{x}^2 + \frac{1}{2}kx^2 \tag{3.178}$$

The mechanical energy of a vibrating system will be wasted when there is a damping component. Having the mathematical expression of dishevelment x, we can evaluate E:

$$x = Xe^{-\xi\omega_n t}\sin(\omega_d t + \varphi) \tag{3.179}$$

$$X = \sqrt{x_0^2 + \left(\frac{\dot{x}_0 + \xi\omega_n x_0}{\omega_d}\right)^2} \qquad \tan\varphi = \frac{x_0\omega_d}{\dot{x}_0 + \xi\omega_n x_0} \tag{3.180}$$

$$E = \frac{1}{2}kX^2 e^{-2\xi\omega_n t}\sin^2(\omega_d t + \varphi)$$

$$+\frac{1}{2}mX^2 e^{-2\xi\omega_n t}(\omega_d\cos(\omega_d t + \varphi) - \xi\omega_n\sin(\omega_d t + \varphi))^2$$

$$= \frac{1}{4}X^2 e^{-2\xi\omega_n t}\left(k + m\omega_d^2 + m\xi^2\omega_n^2\right)$$

$$-\frac{1}{2}X^2 e^{-2\xi\omega_n t}m\xi\omega_d\omega_n\sin 2(\omega_d t + \varphi)$$

$$-\frac{1}{4}X^2 e^{-2\xi\omega_n t}\left(\left(k - m\omega_d^2 + m\xi^2\omega_n^2\right)\cos 2(\omega_d t + \varphi)\right) \tag{3.181}$$

Employing these relationships

$$\omega_d = \sqrt{1 - \xi^2}\,\omega_n \qquad m\omega_n^2 = k \tag{3.182}$$

we may simplify E:

$$\frac{E}{\frac{1}{2}kX^2} = e^{-2\xi\omega_n t} - e^{-2\xi\omega_n t}\xi\sqrt{1 - \xi^2}\sin 2(\omega_d t + \varphi)$$

$$-e^{-2\xi\omega_n t}\xi^2\cos 2(\omega_d t + \varphi) \tag{3.183}$$

$$= e^{-2\xi\omega_n t} - e^{-2\xi\omega_n t}\sin 2(\omega_d t + \varphi + \alpha)$$

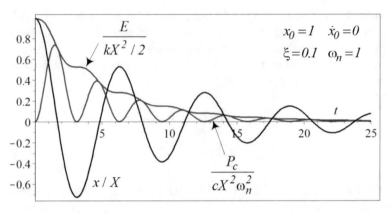

Fig. 3.9 Plot of $E/\left(\frac{1}{2}kX^2\right)$ and x/X as function of time for a set of given data and initial conditions

$$= e^{-2\xi\omega_n t}\left(1 - \sin 2\left(\omega_d t + \varphi + \alpha\right)\right)$$

$$= e^{-2\xi\omega_n t}\left(1 - \sin\left(2\omega_d t + \beta\right)\right) \tag{3.184}$$

$$\sin 2\alpha = \xi^2 \qquad \cos 2\alpha = \sqrt{1 - \xi^4} \qquad \tan 2\alpha = \frac{\xi^2}{\sqrt{1 - \xi^4}} \tag{3.185}$$

$$\beta = 2\varphi + 2\alpha = 2\arctan\frac{x_0\omega_d}{\dot{x}_0 + \xi\omega_n x_0} + \arctan\frac{\xi^2}{\sqrt{1 - \xi^4}} \tag{3.186}$$

Therefore, the dimensionless energy parameter $E/\left(\frac{1}{2}kX^2\right)$ will be equal to 1 plus a fluctuating harmonic function with frequency $2\omega_d$, all multiplied by an exponentially amplitude $e^{-2\xi\omega_n t}$. Figure 3.9 depicts $E/\left(\frac{1}{2}kX^2\right)$ and x/X as function of time for a set of given data:

$$x_0 = 1 \qquad \dot{x}_0 = 0 \tag{3.187}$$

$$\omega_n = r \qquad \xi = 0.1 \tag{3.188}$$

The energy lost by the oscillator can be calculated as the difference between the initial energy E_0 and the energy E at the time t:

$$\frac{E_0}{\frac{1}{2}kX^2} = 1 - \xi\sqrt{1 - \xi^2}\sin 2\varphi - \xi^2\cos 2\varphi \tag{3.189}$$

The dissipated power P_c in the damper is equal to the product of the dissipative force and the velocity:

$$P_c = c\dot{x} \cdot \dot{x} = c\dot{x}^2$$

$$= cX^2 e^{-2\xi\omega_n t} (\omega_d \cos(\omega_d t + \varphi) - \xi\omega_n \sin(\omega_d t + \varphi))^2$$

$$= c\omega_n^2 X^2 e^{-2\xi\omega_n t} \left(\sqrt{1 - \xi^2} \cos(\omega_d t + \varphi) - \xi \sin(\omega_d t + \varphi)\right)^2 \quad (3.190)$$

$$\frac{P_c}{c\omega_n^2 X^2} = e^{-2\xi\omega_n t} \left(\sqrt{1 - \xi^2} \cos(\omega_d t + \varphi) - \xi \sin(\omega_d t + \varphi)\right)^2 \quad (3.191)$$

3.1.2 Critically Damped: $\xi = 1$

When $\xi = 1$, the damping c would be equal to the critical damping c_c and the equation of motion will only be dependent of one parameter, ω_n:

$$\ddot{x} + 2\omega_n \dot{x} + \omega_n^2 x = 0 \quad (3.192)$$

$$x(0) = x_0 \qquad \dot{x}(0) = \dot{x}_0$$

$$\xi = \frac{c}{2m\omega_n} = \frac{c}{c_c} = 1 \quad (3.193)$$

$$c_c = 2m\omega_n = 2\sqrt{km} \quad (3.194)$$

The two characteristic values s_1 and s_2 of Eq. (3.14) are equal and negative, indicating the response to be approaching equilibrium when time goes $t \to \infty$:

$$s_1 = s_2 = -\omega_n \quad (3.195)$$

The critically damped is a nonvibrating class of responses of free vibrating systems. The two linearly independent solutions of the differential Equation (3.192) would be $e^{-\omega_n t}$ and $te^{-\omega_n t}$. These solutions make the general solution of the critically damped oscillator:

$$x = x_0 e^{-\omega_n t} + (\dot{x}_0 + \omega_n x_0) te^{-\omega_n t} \quad (3.196)$$

Proof Substituting a general exponential Equation Xe^{st} into the linear differential equation makes the characteristic equation of the critically damped oscillator to have only one solution:

$$\left(s^2 + 2\omega_n s + \omega_n^2\right) X e^{st} = 0 \tag{3.197}$$

$$s^2 + 2\omega_n s + \omega_n^2 = 0 \tag{3.198}$$

$$s = -\omega_n \tag{3.199}$$

Therefore, $x_1 = X_1 e^{-\omega_n t}$ is one of the solutions of Eq. (3.192). The second solution of the equation would be $x_2 = Xt e^{-\omega_n t}$. Let us examine x_2 as a solution:

$$\ddot{x} + 2\omega_n \dot{x} + \omega_n^2 x = \left((t\omega_n - 2)\omega_n - 2\omega_n (t\omega_n - 1) + \omega_n^2 t\right) X e^{-t\omega_n} = 0 \tag{3.200}$$

Therefore, the general solution of the equation would be:

$$x = X_1 e^{-\omega_n t} + X_2 t e^{-\omega_n t} \tag{3.201}$$

The coefficients X_1 and X_2 will be determined by initial conditions:

$$x(0) = X_1 = x_0 \tag{3.202}$$

$$\dot{x}(0) = -\omega_n x_0 - X_2 (t\omega_n - 1) = \dot{x}_0 \tag{3.203}$$

$$X_1 = x_0 \qquad X_2 = \dot{x}_0 + \omega_n x_0 \tag{3.204}$$

The solution of critically damped of free vibration may alternatively be proven by approaching $\xi \to 1$ from higher or lower values. Let us assume

$$\xi = 1 + \varepsilon \qquad 0 < \varepsilon << 1 \tag{3.205}$$

and calculate the solution of the characteristic Equation (3.14):

$$s_1 \simeq -\omega_n (1 + \varepsilon) \tag{3.206}$$

$$s_2 \simeq -\omega_n (1 - \varepsilon) \tag{3.207}$$

Therefore, the solution of the equation of motion (3.192) will be:

$$
\begin{aligned}
x &= X_1 e^{s_1 t} + X_2 e^{s_2 t} = \frac{\dot{x}_0 - s_2 x_0}{s_1 - s_2} e^{s_1 t} - \frac{\dot{x}_0 - s_1 x_0}{s_1 - s_2} e^{s_2 t} \\
&= \frac{s_1 e^{s_2 t} - s_2 e^{s_1 t}}{s_1 - s_2} x_0 + \frac{e^{s_1 t} - e^{s_2 t}}{s_1 - s_2} \dot{x}_0 \\
&= \frac{-\omega_n (1 + \varepsilon) e^{-\omega_n (1 - \varepsilon)t} + \omega_n (1 - \varepsilon) e^{e^{-\omega_n (1+\varepsilon)t}}}{-2\varepsilon \omega_n} x_0 \\
&\quad + \frac{e^{e^{-\omega_n (1+\varepsilon)t}} - e^{e^{-\omega_n (1-\varepsilon)t}}}{-2\varepsilon \omega_n} \dot{x}_0
\end{aligned}
\tag{3.208}
$$

Expanding $e^{-\varepsilon t}$ in a power series of ε,

$$e^{-\varepsilon t} = 1 - t\varepsilon + \frac{1}{2}t^2\varepsilon^2 + O\left(\varepsilon^3\right) \tag{3.209}$$

yields:

$$e^{-\omega_n(1+\varepsilon)t} = e^{-\omega_n t} - \varepsilon\omega_n t e^{-\omega_n t} + \frac{1}{2}\varepsilon^2\omega_n^2 t^2 e^{-\omega_n t} + O\left(\varepsilon^3\right) \tag{3.210}$$

$$e^{-\omega_n(1-\varepsilon)t} = e^{-\omega_n t} + \varepsilon\omega_n t e^{-\omega_n t} + \frac{1}{2}\varepsilon^2\omega_n^2 t^2 e^{-\omega_n t} + O\left(\varepsilon^3\right) \tag{3.211}$$

and we will have:

$$\begin{aligned}
x = {} & \frac{x_0}{2\varepsilon}\left((1+\varepsilon)\left(e^{-\omega_n t} + t\varepsilon\omega_n e^{-\omega_n t} + \frac{1}{2}t^2\varepsilon^2\omega_n^2 e^{-\omega_n t} + \cdots\right)\right. \\
& \left. - (1-\varepsilon)\left(e^{-\omega_n t} - t\varepsilon\omega_n e^{-\omega_n t} + \frac{1}{2}t^2\varepsilon^2\omega_n^2 e^{-\omega_n t} + \cdots\right)\right) \\
& + \frac{\dot{x}_0}{-2\varepsilon\omega_n}\left(\left(e^{-\omega_n t} - t\varepsilon\omega_n e^{-\omega_n t} + \frac{1}{2}t^2\varepsilon^2\omega_n^2 e^{-\omega_n t} + \cdots\right)\right. \\
& \left. - \left(e^{-\omega_n t} + t\varepsilon\omega_n e^{-\omega_n t} + \frac{1}{2}t^2\varepsilon^2\omega_n^2 e^{-\omega_n t} + \cdots\right)\right)
\end{aligned} \tag{3.212}$$

or:

$$\begin{aligned}
x = {} & x_0\frac{e^{-\omega_n t}}{2\varepsilon}\left(1 + \varepsilon + \varepsilon\omega_n t + \varepsilon^2\omega_n t + \frac{1}{2}\varepsilon^2\omega_n^2 t^2 + \frac{1}{2}\varepsilon^3\omega_n^2 t^2 + \cdots\right. \\
& \left. - \left(1 - \varepsilon - \varepsilon\omega_n t + \varepsilon^2\omega_n t + \frac{1}{2}t^2\varepsilon^2\omega_n^2 - \frac{1}{2}\varepsilon^3\omega_n^2 t^2 + \cdots\right)\right) \\
& + \frac{\dot{x}_0 e^{-\omega_n t}}{-2\varepsilon\omega_n}\left(\left(1 - t\varepsilon\omega_n + \frac{1}{2}t^2\varepsilon^2\omega_n^2 + \cdots\right)\right. \\
& \left. - \left(1 + t\varepsilon\omega_n + \frac{1}{2}t^2\varepsilon^2\omega_n^2 + \cdots\right)\right) \\
= {} & x_0\frac{e^{-\omega_n t}}{2\varepsilon}\left(2\varepsilon + 2\varepsilon\omega_n t + \varepsilon^3\omega_n^2 t^2 + \cdots\right) \\
& + \frac{\dot{x}_0 e^{-\omega_n t}}{-2\varepsilon\omega_n}\left(-2t\varepsilon\omega_n - \frac{1}{3}t^3\varepsilon^3\omega_n^3 + \cdots\right)
\end{aligned} \tag{3.213}$$

Taking the limit $\varepsilon \to 0$, the solution x will be simplified to (3.196):

$$x = x_0 e^{-\omega_n t}(1 + \omega_n t) + \dot{x}_0 t e^{-\omega_n t} \tag{3.214}$$

$$x = x_0 e^{-\omega_n t} + (\dot{x}_0 + \omega_n x_0)\, t e^{-\omega_n t} \tag{3.215}$$

Critically damping is practically not a significant case in engineering applications. There is not practically much difference between the responses of a vibrating system for $\xi = 1$, $\xi = 1.01$, and $\xi = 0.99$. It is a mathematical issue where the eigenvalues of the system become equal. The solution will then be $X_1 e^{st}$ plus $X_1 t e^{st}$. The critically damped response does not have any envelop to cover of x. ∎

Example 96 Critically damped system $\xi = 1$. Graphical illustration of critically damped vibrating systems.

If the system is critically damped, $\xi = 1$, then the time response to free vibrations is:

$$x = e^{-\xi \omega_n t}\,(X_1 + X_2 t) \tag{3.216}$$

Using the initial conditions, $x(0) = x_0$, $\dot{x}(0) = \dot{x}_0$, we find the coefficients X_1 and X_2:

$$X_1 = x_0 \qquad X_2 = \dot{x}_0 + \omega_n x_0 \tag{3.217}$$

and therefore the general response of critically damped system is calculated.

$$x = e^{-\omega_n t}\,(x_0 + (\dot{x}_0 + \omega_n x_0)\, t) \tag{3.218}$$

Figure 3.10 shows a critically damped response for $\xi = 1$, $\omega_n = 3\pi\ \mathrm{rad}$, $x_0 = 1$, and $\dot{x}_0 = 0$.

Example 97 Reduction of order. When we have one solution of a second-order differential equation, we are able to use the solution and simplify the equation to determine the second solution.

When a free vibrating system is critically damped, the characteristic values will equal, and only one solution will be determined for the equation:

$$\ddot{x} + 2\omega_n \dot{x} + \omega_n^2 x = 0 \tag{3.219}$$

$$s^2 + 2\omega_n s + \omega_n^2 = 0 \qquad s = -\omega_n \tag{3.220}$$

$$x_1 = X_1 e^{-\omega_n t} \tag{3.221}$$

Let introduce a new variable, u, based on the first solution, x_1, and substitute it in the differential equation:

$$x_2 = x_1 u = u e^{-\omega_n t} \tag{3.222}$$

$$\dot{x}_2 = \dot{u} e^{-\omega_n t} - \omega_n u e^{-\omega_n t} \tag{3.223}$$

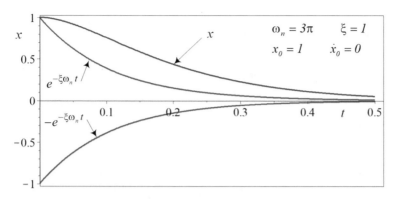

Fig. 3.10 A sample time response of a critically damped vibrating system.

$$\ddot{x}_2 = \ddot{u}e^{-\omega_n t} - 2\omega_n \dot{u}e^{-\omega_n t} + \omega_n^2 u e^{-\omega_n t} \tag{3.224}$$

$$\ddot{x} + 2\omega_n \dot{x} + \omega_n^2 x = \ddot{u}e^{-\omega_n t} - 2\omega_n \dot{u}e^{-\omega_n t} + \omega_n^2 u e^{-\omega_n t}$$
$$+ 2\omega_n \left(\dot{u}e^{-\omega_n t} - \omega_n u e^{-\omega_n t}\right) + \omega_n^2 u e^{-\omega_n t} = \ddot{u}e^{-\omega_n t} = 0 \tag{3.225}$$

The resultant equation can be solved:

$$u = C_1 t + C_2 \tag{3.226}$$

to determine the second solution of the original differential equation:

$$x_2 = (C_1 t + C_2) e^{-\omega_n t} \tag{3.227}$$

Therefore the general solution of the equation would be:

$$x = A_1 x_1 + A_2 x_2 = A_1 e^{-\omega_n t} + A_2 (C_1 t + C_2) e^{-\omega_n t}$$
$$= X_1 e^{-\omega_n t} + X_2 t e^{-\omega_n t} \tag{3.228}$$

Example 98 ★Crossing the equilibrium. The critically damped system will approach equilibrium position from any initial conditions. However, the curve of x may pass the line $x = 0$ once if the initial condition is set properly.

Let us assume the following numerical values for a critically damped system:

$$x_0 = 1 \qquad \omega_n = 1 \tag{3.229}$$

The response x of the system will be:

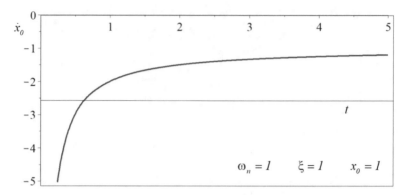

Fig. 3.11 Relationship between $\dot{x}_0$ and t for a critically damped system to cross $x = 0$ for $\omega_n = 1$, $x_0 = 1$

$$x = e^{-\omega_n t} (x_0 + (\dot{x}_0 + \omega_n x_0) t)$$
$$= e^{-t} (1 + (\dot{x}_0 + 1) t) \tag{3.230}$$

Setting $x = 0$ provides us with a relationship between $\dot{x}_0$ and t to cross the equilibrium $x = 0$:

$$\dot{x}_0 = -\frac{1}{t} (t + 1) \tag{3.231}$$

This relationship is plotted in Fig. 3.11 indicating a design chart. We are able to select the time at which we wish x to cross the equilibrium position, and this chart will determine what should be the initial velocity $\dot{x}_0$. On the other hand, we are able to determine the time at which we wish x to cross the equilibrium position for a given initial velocity, $\dot{x}_0$. The position variable, x, will cross the equilibrium position, $x = 0$, only for negative initial velocity if $x_0 > 0$.

To make the design chart to be general, let us begin with the solution and make it nondimensionalized:

$$x = e^{-\omega_n t} (x_0 + (\dot{x}_0 + \omega_n x_0) t) = 0 \tag{3.232}$$

Because $e^{-\omega_n t}$ can ot be zero, the equation reduces to an algebraic equation:

$$x_0 + (\dot{x}_0 + \omega_n x_0) t = 0 \tag{3.233}$$

$$\frac{x_0}{\dot{x}_0 t} = \frac{-1}{1 + \omega_n t} \tag{3.234}$$

The design plot of $\frac{x_0}{\dot{x}_0 t}$ against $\omega_n t$ is shown in Fig. 3.12. This design chart is nondimensionalized, and, hence, it is valid for all critically damped free vibrating systems. Setting a value for $\omega_n t$ on horizontal axis will provide us with the

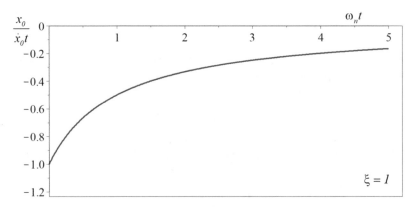

Fig. 3.12 Nondimensionalized relationship between $\frac{x_0}{\dot{x}_0 t}$ and $\omega_n t$ for a critically damped vibrating system to cross $x = 0$ for $\omega_n = 1$

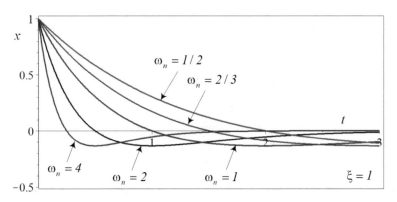

Fig. 3.13 Time response of a critically damped system for some ω_n and their associate $\dot{x}_0$ to cross $x = 0$ at $t = 1/\omega_n$

associated value for $\frac{x_0}{\dot{x}_0 t}$ on vertical axis to have the initial conditions to cross the equilibrium position. There would be no crossing the $x = 0$ axis for $0 < \frac{x_0}{\dot{x}_0 t} < -1$.

As an example, a point on horizontal axis at $\omega_n t = 1$ gives $\frac{x_0}{\dot{x}_0 t} = -\frac{1}{2}$ on vertical axis. A system with $\omega_n = 1$ from $x_0 = 1$ will need $\dot{x}_0 = -2x_0/t = -2$ to pass $x = 0$ at $t = 1$. Also a point on horizontal axis at $\omega_n t = 1$ gives $\frac{x_0}{\dot{x}_0 t} = -\frac{1}{2}$ on vertical axis, and a system with $\omega_n = 2$ from $x_0 = 1$ will need $\dot{x}_0 = -2x_0/t = -4$ to pass $x = 0$ at $t = 1/2$. Keeping $x_0 = 1$ and $\omega_n t = 1$, we will need $\dot{x}_0 = -2/t = -4/3$ for $t = 1.5$, and $\dot{x}_0 = -1$ for $t = 2$, and $\dot{x}_0 = -8$ for $t = 1/4$. Figure 3.13 depicts these time response.

3.1.3   Overdamped: $1 < \xi$

When $1 < \xi$ the vibrating system is called overdamped. The characteristic values s_1 and s_2 of Eq. (3.14) are both real and negative for $1 < \xi$:

$$s_1 = \left(-\xi + \sqrt{\xi^2 - 1}\right)\omega_n < 0 \tag{3.235}$$

$$s_2 = \left(-\xi - \sqrt{\xi^2 - 1}\right)\omega_n < 0 \tag{3.236}$$

$$x(t) = X_1 e^{s_1 t} + X_2 e^{s_2 t} \tag{3.237}$$

$$X_1 = \frac{\dot{x}_0 - s_2 x_0}{s_1 - s_2} \qquad X_2 = -\frac{\dot{x}_0 - s_1 x_0}{s_1 - s_2} \tag{3.238}$$

The negative characteristic values make the response of the system to have no oscillatory motion and go to zero for large t:

$$x = \frac{\dot{x}_0 + \left(\xi + \sqrt{\xi^2 - 1}\right)\omega_n x_0}{2\omega_n\sqrt{\xi^2 - 1}} e^{-\left(\xi - \sqrt{\xi^2 - 1}\right)\omega_n t}$$

$$- \frac{\dot{x}_0 + \left(\xi - \sqrt{\xi^2 - 1}\right)\omega_n x_0}{2\omega_n\sqrt{\xi^2 - 1}} e^{-\left(\xi + \sqrt{\xi^2 - 1}\right)\omega_n t} \tag{3.239}$$

The overdamped case is mathematically the simplest with least engineering applications.

Example 99 Overdamped systems, $1 < \xi$. Graphical illustration of overdamped vibrating systems and comparison with critically and underdamped cases.

An overdamped system has damping c greater than the critical damping, $c > 2\xi\omega_2$ or $1 < \xi$. The time response of the system will be exponentially decaying to approach the equilibrium position, starting from any set of initial conditions. Figure 3.14 shows an overdamped response for $\xi = 2$, $\omega_n = 20\pi$, $x_0 = 1$, and $\dot{x}_0 = 0$.

Using the initial conditions, $x(0) = x_0$, $\dot{x}(0) = \dot{x}_0$, and the characteristic values (3.235)–(3.236), the general overdamped response is:

$$x = \frac{\dot{x}_0 - s_2 x_0}{s_1 - s_2} e^{s_1 t} + \frac{s_1 x_0 - \dot{x}_0}{s_1 - s_2} e^{s_2 t}$$

$$= \left(\frac{s_1 e^{s_2 t}}{s_1 - s_2} - \frac{s_2 e^{s_1 t}}{s_1 - s_2}\right) x_0 + \left(\frac{e^{s_1 t}}{s_1 - s_2} - \frac{e^{s_2 t}}{s_1 - s_2}\right) \dot{x}_0$$

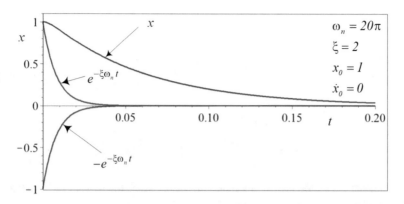

Fig. 3.14 A sample time response for an overdamped system

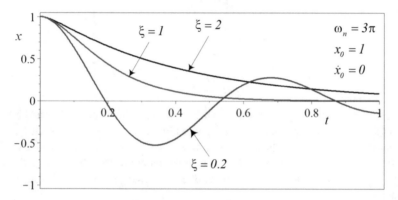

Fig. 3.15 Comparison of an underdamped, critically damped, and overdamped vibrating systems with $\omega_n = 3\pi$ and initial conditions of $x_0 = 1$ and $\dot{x}_0 = 0$

$$= \frac{e^{-\xi\omega_n t}}{2\sqrt{\xi^2 - 1}} \left(\left(\frac{\dot{x}_0}{\omega_n} + x_0 \left(\xi + \sqrt{\xi^2 - 1} \right) \right) e^{\omega_n \sqrt{\xi^2 - 1}t} \right.$$

$$\left. + \left(-\frac{\dot{x}_0}{\omega_n} + x_0 \left(-\xi - \sqrt{\xi^2 - 1} \right) \right) e^{-\omega_n \sqrt{\xi^2 - 1}t} \right) \qquad (3.240)$$

Figure 3.15 compares an underdamped, critically damped, and overdamped vibrating systems with $\omega_n = 3\pi$ and initial conditions of $x_0 = 1$ and $\dot{x}_0 = 0$.

3.1.4 Negative Damping: $\xi < 0$

When $\xi < 0$ the vibrating system is called unstable. The characteristic values s_1 and s_2 of Eq. (3.14) are both real and positive for negative damping $\xi < 0$:

$$s_1 = \left(\xi + \sqrt{\xi^2 - 1} \right) \omega_n > 0 \qquad (3.241)$$

$$s_2 = \left(\xi - \sqrt{\xi^2 - 1} \right) \omega_n > 0 \qquad (3.242)$$

$$x(t) = X_1 e^{s_1 t} + X_2 e^{s_2 t} \qquad (3.243)$$

$$X_1 = \frac{\dot{x}_0 - s_2 x_0}{s_1 - s_2} \qquad X_2 = -\frac{\dot{x}_0 - s_1 x_0}{s_1 - s_2} \qquad (3.244)$$

The positive characteristic values make the response of the system to have oscillatory motion with exponentially growing amplitude:

$$x = \frac{\dot{x}_0 + \left(\xi + \sqrt{\xi^2 - 1} \right) \omega_n x_0}{2\omega_n \sqrt{\xi^2 - 1}} e^{-\left(\xi - \sqrt{\xi^2 - 1} \right) \omega_n t}$$

$$- \frac{\dot{x}_0 + \left(\xi - \sqrt{\xi^2 - 1} \right) \omega_n x_0}{2\omega_n \sqrt{\xi^2 - 1}} e^{-\left(\xi + \sqrt{\xi^2 - 1} \right) \omega_n t} \qquad (3.245)$$

The negative damping is always due to a condition at which energy inserted into the vibrating system. An example could be wind excitation on structures. In practical vibration engineering applications, negative damping will be provided by semi-active system to control damping rate of a computer controlled shock absorber. Such damping is not constant and will be changed according to constraints and control strategy.

Example 100 ★ Negative damping and instability. Graphical illustration of overdamped vibrating systems and comparison with critically and underdamped cases.

When there is a mechanism that pumps energy into a vibrating system during the oscillation, the damping is negative. The equation of motion of such a system is:

$$m\ddot{x} = -kc + c\dot{x} + f (x, \dot{x}, t) \qquad (3.246)$$

$$m > 0 \qquad k > 0 \qquad c > 0 \qquad (3.247)$$

If we move all terms but the forcing term to the other side, the coefficient of $\dot{x}$ becomes negative, and this justifies the name of negative damping:

$$m\ddot{x} - c\dot{x} + kx = f (x, \dot{x}, t) \qquad (3.248)$$

The free vibration of a system with negative damping is divergent and, hence, unstable. It means that the amplitude of vibration increases periodically until the system practically breaks down:

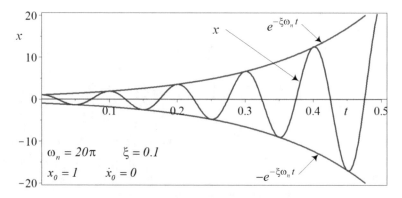

Fig. 3.16 A sample time response of a negative damped system

$$\ddot{x} - 2\xi\omega_n\dot{x} + \omega_n^2 x = 0 \tag{3.249}$$

$$2\xi\omega_n = \frac{c}{m} \qquad \omega_n^2 = \frac{k}{m} \tag{3.250}$$

$$x(0) = x_0 \qquad \dot{x}(0) = \dot{x}_0 \tag{3.251}$$

When $\xi > 1$, both characteristic values $s_{1,2}$ are real and positive, $s_{1,2} \in \mathbb{R}$ and $s_{1,2} > 0$:

$$s^2 - 2\xi\omega_n s + \omega_n^2 = 0 \tag{3.252}$$

$$s_{1,2} = \xi\omega_n \pm \omega_n\sqrt{\xi^2 - 1} \tag{3.253}$$

Therefore, both terms of the general solution of Eq. (3.249) $X_1 e^{s_1 t}$ and $X_2 e^{s_2 t}$ exponentially grow with time, with different weights and rates:

$$\begin{aligned} x &= X_1 e^{s_1 t} + X_2 e^{s_2 t} \\ &= X_1 e^{\omega_n\left(\xi+\sqrt{\xi^2-1}\right)t} + X_2 e^{\omega_n\left(\xi-\sqrt{\xi^2-1}\right)t} \end{aligned} \tag{3.254}$$

Figure 3.16 depicts the response of a negative damping system.

Example 101 ★Temporary negative damping. Negative damping is usually a temporary effect and may disappear after certain conditions.

Let us consider a vibrating system whose damping ratio is negative for a while and change to be zero according to a safety control system. Such control system usually shut down the damping mechanism when the displacement exceeds certain value. Here let us assume a system that is equipped with a control system that shuts down the damping of the vibrating system when it is in negative range for a period of time:

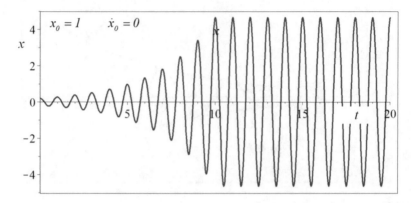

Fig. 3.17 Vibration of a system with negative damping with a control system to shut down the damping after 10 s

$$\ddot{x} + 2\xi\omega_n\dot{x} + \omega_n^2 x = 0 \tag{3.255}$$

$$\xi = \begin{cases} -0.5 & 0 < t \le 10\,\text{s} \\ 0 & 10\,\text{s} < t \end{cases} \tag{3.256}$$

To define such damping ratio mathematically, we may use Heaviside, H, function:

$$H(t - \tau) = \begin{cases} 1 & t \ge \tau \\ 0 & t < \tau \end{cases} \tag{3.257}$$

$$\ddot{x} + 2\left(\xi H(10 - t)\right)\omega_n\dot{x} + \omega_n^2 x = 0 \tag{3.258}$$

This will make a system to have complex dynamics. A complex dynamic system will have more than one different equations of motion depending on certain conditions. In this example we will have:

$$\ddot{x} + 2\xi\omega_n\dot{x} + \omega_n^2 x = 0 \qquad 0 < t \le 10\,\text{s} \tag{3.259}$$

$$\ddot{x} + \omega_n^2 x = 0 \qquad 10\,\text{s} < t \tag{3.260}$$

Figure 3.17 illustrates the time history of the system starting from an initial unit displacement:

$$x(0) = 1 \qquad \dot{x}(0) = 0 \tag{3.261}$$

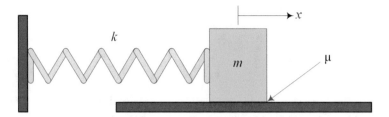

Fig. 3.18 A mass-spring oscillator with a resisting dry friction force

3.1.5 Coulomb Friction Damping

Consider a particle with mass m moving on a rough surface. Then there would exist a friction force $\mathbf{F}_f$ that resists the relative motion between the surfaces. The friction force $\mathbf{F}_f$ is proportional to the normal force F_N that presses the two surfaces together. Consider an object with mass m is moving on a rough surface as is shown in Fig. 3.18; then:

$$\mathbf{F}_f = -\mu N \frac{\mathbf{v}}{v} \tag{3.262}$$

where μ is the coefficient of friction, N is the normal force exerted by the surfaces on the particle, and $\mathbf{v}/v$ is the unit vector in the direction of the object's velocity relative to the surface. This is a simplified Coulomb theory of friction that states: the friction force is proportional to the force pressing two physical surfaces together and is in the tangent plane to the surfaces at the contact point.

If Coulomb friction is the only damping mechanism in a one DOF vibrating system, then its equation of motion will be:

$$m\ddot{x} + \mu N \, \mathrm{sgn} \, \dot{x} + kx = \ddot{x} + \omega_n^2 x + \mu \frac{N}{m} g \frac{\dot{x}}{|\dot{x}|}$$

$$= \ddot{x} + \omega^2 x + \mu \frac{N}{m} g \, \mathrm{sgn} \, \dot{x} = 0 \tag{3.263}$$

$$x(0) = x_0 \qquad \dot{x}(0) = 0 \qquad \omega_n^2 = \frac{k}{m} \tag{3.264}$$

The equation of motion is piecewise including two equations depending on the direction of motion:

$$\ddot{x} + \omega_n^2 x = \begin{cases} \mu \dfrac{N}{m} & \dot{x} < 0 \\[2mm] -\mu \dfrac{N}{m} & 0 < \dot{x} \end{cases} \tag{3.265}$$

The system will have an equilibrium zone $x < |\mu N/k|$ in which the spring force kx is less than the maximum friction force μN:

$$-\frac{\mu N}{k} < x < \frac{\mu N}{k} \tag{3.266}$$

As long as the mass m is within the equilibrium zone with zero velocity, it remains at rest. Let us count the motion starting from the stop position $x(0) = x_0 > \mu N/k$, $\dot{x}(0) = 0$ on the right side of to the equilibrium zone to the point where the mass stops again on the left side of equilibrium zone as the part 1 of motion after half a cycle of oscillation. Then count the second half of the first cycle of motion from the left stop to the right stop of the equilibrium zone as part 2 and so on. The solution of the equation is as below, where j is the number of motions:

$$x = \left(x_0 - (2j-1)\frac{N}{k}\mu\right)\cos\omega_n t - (-1)^j \mu\frac{N}{k} \qquad t_j < t < t_{j+1} \tag{3.267}$$

$$\dot{x} = -\left(x_0 - (2j-1)\frac{N}{k}\mu\right)\omega_n \sin\omega_n t \qquad t_j = \frac{j\pi}{\omega_n} \tag{3.268}$$

The extreme displacements x_j on both sides of the equilibrium zone where the mass stops before returning to oscillation are:

$$x_j = (-1)^j\left(x_0 - 2j\,\mu\frac{N}{k}\right) \qquad j = 1, 2, 3, \cdots, s \tag{3.269}$$

$$x_{j+1} = -x_j + 2\mu\frac{N}{k} \qquad x_0 = x(0) \tag{3.270}$$

The displacement of m in every half a cycle will be shortened by $2\mu\frac{N}{k}$. The value of $2\mu\frac{N}{k}$ is equal to the total length of the equilibrium zone.

The initial energy of the system is the stored potential energy in the spring:

$$P_0 = \frac{1}{2}kx_0^2 \tag{3.271}$$

The potential energy drops continuously after every half a cycle:

$$P_i = \frac{1}{2}k\left(x_0 - 2i\,\mu\frac{N}{k}\right)^2 \tag{3.272}$$

The mass finally stops at some point in the equilibrium zone when the stop x_i is within the zone.

Proof The equation of motion of a mass-spring system with Coulomb friction can be written as two piecewise linear equations:

$$m\ddot{x} + kx = \begin{cases} -\mu N & 0 < \dot{x} \\ \mu N & \dot{x} < 0 \end{cases} \tag{3.273}$$

or as a single equation using signum function sgn $\dot{x}$:

$$\ddot{x} + \mu \frac{N}{m} \operatorname{sgn} \dot{x} + \omega_n^2 x = 0 \qquad x(0) = x_0 > 0 \qquad \dot{x}(0) = \dot{x}_0 = 0 \tag{3.274}$$

$$\operatorname{sgn} \dot{x} = \begin{cases} -1 & \dot{x} < 0 \\ 0 & 0 \\ 1 & 0 < \dot{x} \end{cases} \tag{3.275}$$

As the velocity changes sign, the sign of friction force that is always opposing the motion will also change. Therefore, the friction force will slow down the mass, which will eventually come to a stop. The mass will be in motion as long as the supplied force by the spring is larger than the maximum friction force, μN. Motion will eventually stop when the spring force is not larger than the friction force.

To find the time response of the system, we assume the mass starts from the initial conditions $x(0) = x_0 > 0$, $x_0 > \mu \frac{N}{m}$, and $\dot{x}(0) = 0$.

The equation may equivalently be transformed to the following form:

$$\ddot{x} + \omega_n^2 x = -\mu \frac{N}{m} \operatorname{sgn} \dot{x} = \begin{cases} \mu \frac{N}{m} & \dot{x} < 0 \\ -\mu \frac{N}{m} & 0 < \dot{x} \end{cases} \tag{3.276}$$

Assuming $x_0 > 0$ and $\dot{x}(0) = 0$, we expect the motion to start in negative direction, $\dot{x} < 0$ until the mass goes to stop. The equation for this part of motion will be:

$$\ddot{x} + \omega_n^2 x = \mu \frac{N}{m} \qquad x_0 > 0 \qquad \dot{x}_0 = 0 \tag{3.277}$$

The solution of the equation will have a homogenous solution x_h for $\ddot{x} + \omega_n^2 x = 0$, plus a particular solution for the external force function $f = \mu \frac{N}{m}$. The homogenous solution is the free vibrations (3.19) for $\xi = 0$:

$$x_h = A \sin \omega_n t + B \cos \omega_n t \tag{3.278}$$

Because the forcing function is constant, the particular solution x_p will be a constant function that satisfies Eq. (3.277):

$$x_p = \mu \frac{N}{m\omega_n^2} = \mu \frac{N}{k} \tag{3.279}$$

Therefore, the total solution of Eq. (3.277) will be:

$$x = x_h + x_p = A \sin \omega_n t + B \cos \omega_n t + \mu \frac{N}{k} \qquad (3.280)$$

$$\dot{x} = \omega_n (A \cos \omega_n t - B \sin \omega_n t) \qquad (3.281)$$

Applying the initial conditions:

$$x_0 = B + \mu \frac{N}{k} \qquad (3.282)$$

$$\dot{x}_0 = \omega_n A = 0 \qquad (3.283)$$

we find the solution:

$$x = \left(x_0 - \mu \frac{N}{k} \right) \cos \omega_n t + \mu \frac{N}{k} \qquad 0 < t < t_1 \qquad (3.284)$$

$$\dot{x} = - \left(x_0 - \mu \frac{N}{k} \right) \omega_n \sin \omega_n t \qquad t_1 = \frac{\pi}{\omega_n} \qquad (3.285)$$

The mass will come to the first stop at $t = t_1 = \pi/\omega_n$, when m is at x_1. The mass stops at a point which is $2\mu \frac{N}{k}$ closer to $x = 0$:

$$x_1 = x(t_1) = \mu \frac{N}{k} - \left(x_0 - \mu \frac{N}{k} \right) = -x_0 + 2\mu \frac{N}{k} \qquad (3.286)$$

Now the problem reduces to:

$$\ddot{x} + \omega_n^2 x = -\mu \frac{N}{m} \qquad x(t_1) = x_1 < 0 \qquad \dot{x}_1 = 0 \qquad (3.287)$$

whose solution will be found similar to (3.284):

$$x = \left(x_0 - 3 \frac{N}{k} \mu \right) \cos \omega_n t - \mu \frac{N}{k} \qquad t_1 < t < t_2 \qquad (3.288)$$

$$\dot{x} = - \left(x_0 - 3 \frac{N}{k} \mu \right) \omega_n \sin \omega_n t \qquad t_2 = \frac{2\pi}{\omega_n} \qquad (3.289)$$

The mass will come to the second stop at $t = t_2 = 2\pi/\omega_n$, when m is at x_2. In one cycle of oscillation, the mass will lose its amplitude by $4\mu \frac{N}{k}$:

$$x_2 = x(t_2) = x_0 - 4\mu \frac{N}{k} \qquad (3.290)$$

The new problem will be similar to (3.277) with a new initial position as x_2. The next step will show that x_3 at t_3 is:

$$x_3 = x(t_3) = -x_0 + 6\mu\frac{N}{k} \qquad t_3 = \frac{3\pi}{\omega_n} \qquad (3.291)$$

Similarly we derive the time response of m and the stop position of the mass at both sides of $x = 0$ at $x_i, i = 1, 2, 3, \cdots, s$:

$$x = \left(x_0 - (2i-1)\frac{N}{k}\mu\right)\cos\omega_n t - (-1)^i \mu\frac{N}{k} \qquad t_i < t < t_{i+1} \qquad (3.292)$$

$$\dot{x} = -\left(x_0 - (2i-1)\frac{N}{k}\mu\right)\omega_n \sin\omega_n t \qquad t_i = \frac{i\pi}{\omega_n} \qquad (3.293)$$

$$x_i = (-1)^i \left(x_0 - 2i\,\mu\frac{N}{k}\right) \qquad i = 1, 2, 3, \cdots, s \qquad (3.294)$$

$$x_{i+1} = -x_i + 2\mu\frac{N}{k} \qquad x_0 = x(0) \qquad (3.295)$$

The final position x_s will be reached when the friction force is greater than the spring force while the mass is at rest. The mass then stops forever. The value s is the lowest integer for which we have:

$$k\,|x_s| < \mu N < k\,|x_{s-1}| \qquad (3.296)$$

or

$$|x_s| < \frac{\mu N}{k} < |x_{s-1}| \qquad (3.297)$$

The rest or equilibrium zone of the system is $x < |\mu N/k|$ in which the spring force is less than the maximum friction force:

$$-\frac{\mu N}{k} < x < \frac{\mu N}{k} \qquad (3.298)$$

Friction force is an example of forces that arise from the presence of holonomic constraints. It is a function of constraint force, but its line of action lies in the tangent plane and hence does virtual work.

Phenomena related to dry friction appear in poorly lubricated or non-lubricated contact surfaces of bodies in relative motion.

French scientist Charles-Augustin de Coulomb (1736 − −1806) developed a series of two-term equations to model the friction force. The first term is a constant, and the second term varies with time, normal force, velocity, or other parameters.

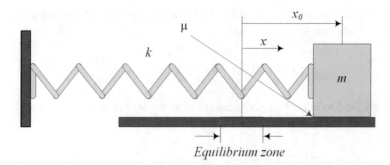

Fig. 3.19 A mass-spring oscillator on a rough surface with Coulomb friction

Leonardo Da Vinci (1452–1519) was one of the first scholars to systematically study friction. Friction force $\mathbf{F}_f$ is a given force that exists because of a constraint and hence is a function of constraint force N. ∎

Example 102 Oscillation with Coulomb friction. A step-by-step detailed example is needed to learn vibrations with friction, numerically.

Consider a mass and spring system as shown in Fig. 3.19 where the mass is sliding on a rough surface with friction coefficient μ. Let us work with the following data:

$$\mu = 0.5 \qquad N = 10\,\mathrm{N} \qquad k = 100\,\mathrm{N}\,/\,\mathrm{m}$$

$$\omega_n = 4\,\mathrm{s}^{-1} \qquad x_0 = 1\,\mathrm{m} \qquad \dot{x}_0 = 0 \tag{3.299}$$

The mass is released from $x_0 = 1\,\mathrm{m}$ on the right side of $x = 0$. The equation of motion and its solution will be:

$$\ddot{x} + \omega_n^2 x = \mu \frac{N}{m} \qquad x_0 = 1\,\mathrm{m} \qquad \dot{x}_0 = 0 \tag{3.300}$$

$$\ddot{x} + 16x = \mu \frac{N}{k/\omega_n^2} = 0.5 \frac{10}{100/16} = 0.8 \tag{3.301}$$

$$x = \left(x_0 - \mu \frac{N}{k} \right) \cos \omega_n t + \mu \frac{N}{k} \qquad 0 < t < t_1 \tag{3.302}$$

$$= 0.95 \cos 4t + 0.05 \qquad t_1 = \frac{\pi}{\omega_n} = \frac{\pi}{4} \tag{3.303}$$

$$\dot{x} = -3.8 \sin 4t \tag{3.304}$$

The mass will come to stop at the left-hand side of $x = 0$ at x_1:

$$x_1 = x(t_1) = -x_0 + 2\mu \frac{N}{k} = -0.9 \, \text{m} \qquad (3.305)$$

The second and other parts of oscillations are:

$$x = 0.85 \cos 4t - 0.05 \qquad t_1 < t < t_2 \qquad t_2 = \frac{2\pi}{4} \qquad x_2 = 0.8 \, \text{m} \qquad (3.306)$$

$$x = 0.75 \cos 4t + 0.05 \qquad t_2 < t < t_3 \qquad t_3 = \frac{3\pi}{4} \qquad x_3 = -0.7 \, \text{m} \qquad (3.307)$$

$$x = 0.65 \cos 4t - 0.05 \qquad t_3 < t < t_4 \qquad t_4 = \frac{4\pi}{4} \qquad x_4 = 0.6 \, \text{m} \qquad (3.308)$$

$$x = 0.55 \cos 4t + 0.05 \qquad t_4 < t < t_5 \qquad t_5 = \frac{5\pi}{4} \qquad x_5 = -0.5 \, \text{m} \qquad (3.309)$$

$$x = 0.45 \cos 4t - 0.05 \qquad t_5 < t < t_6 \qquad t_6 = \frac{6\pi}{4} \qquad x_6 = 0.4 \, \text{m} \qquad (3.310)$$

$$x = 0.35 \cos 4t + 0.05 \qquad t_6 < t < t_7 \qquad t_7 = \frac{7\pi}{4} \qquad x_7 = -0.3 \, \text{m} \qquad (3.311)$$

$$x = 0.25 \cos 4t - 0.05 \qquad t_7 < t < t_8 \qquad t_8 = \frac{8\pi}{4} \qquad x_8 = 0.2 \, \text{m} \qquad (3.312)$$

$$x = 0.15 \cos 4t + 0.05 \qquad t_8 < t < t_9 \qquad t_9 = \frac{9\pi}{4} \qquad x_9 = -0.1 \, \text{m} \qquad (3.313)$$

$$x = 0.05 \cos 4t - 0.05 \qquad t_9 < t < t_{10} \qquad t_{10} = \frac{10\pi}{4} \qquad x_{10} = 0.0 \, \text{m}$$
$$(3.314)$$

Figure 3.20 illustrates m at the initial position x_0 and at the first stop at x_1, as well as the location of x_i, $i = 1, 2, 3, \cdots, x_{10}$. The time response of the system is shown in Fig. 3.21. Because the amplitude drops with a constant value, the response curve stays within a pair of linear envelops. The oscillation ends up at rest position $x_{10} = 0$, in the middle of equilibrium zone:

$$-0.05 < x < 0.05 \qquad (3.315)$$

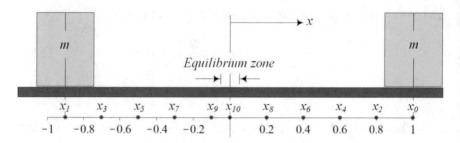

Fig. 3.20 Illustration of m at the initial position x_0 and at the first stop at x_1, as well as the location of x_i, $i = 1, 2, 3, \cdots, x_{10}$

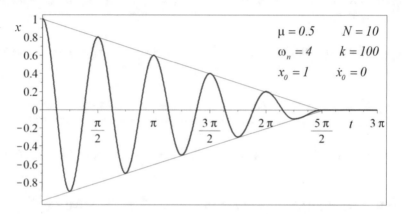

Fig. 3.21 The time response of the system

The equilibrium zone is an area in which the spring force is less than friction force. In this example, the mass will stop exactly at the middle of the equilibrium point; however, other data may make m to stop at other point of the zone. For example, if we increase the friction coefficient to $\mu = 0.51$, the mass will stop at $x_1 = -0.898$, $x_2 = 0.796$, $x_3 = -0.694$, $x_4 = 0.592$, $x_5 = -0.49$, $x_6 = 0.388$, $x_7 = -0.286$, $x_8 = 0.184$, $x_9 = -0.082$, and finally $x_{10} = -0.02$ entering the zone from the left.

Example 103 ★ Approximate solution for friction and mass-spring oscillator. Perturbation methods are good to provide approximate continuous solution for these types of piecewise problems. Here is an approximate solution for frictional oscillator using averaging method.

Friction is a natural phenomenon that exists in every type of mechanical oscillators. Consider a mass-spring oscillator with a resisting dry friction force as is shown in Fig. 3.18. Assuming the coefficient of friction to be indicated by μ, the equation of motion of the system will be:

$$\ddot{x} + \omega^2 x + 2\varepsilon\omega\frac{\dot{x}}{|\dot{x}|} = \ddot{x} + \omega^2 x + 2\varepsilon\omega \, \text{sgn} \, \dot{x} = 0 \tag{3.316}$$

$$x\,(0) = x_0 \qquad \dot{x}\,(0) = 0 \qquad \omega^2 = \frac{k}{m} \qquad \varepsilon = \frac{\mu N}{2m\omega} \tag{3.317}$$

To determine the motion of m starting from the given initial conditions, we apply the averaging method which is assuming a harmonic solution with time-dependent amplitude $A\,(t)$ and argument $\varphi\,(t)$:

$$x = A\,(t) \sin\varphi\,(t) \qquad \dot{x} = A\,(t)\,\omega \cos\varphi\,(t) \tag{3.318}$$

$$\varphi\,(t) = \omega t + \beta\,(t) \qquad \dot{\varphi}\,(t) = \omega + \dot{\beta}\,(t) \tag{3.319}$$

$$\ddot{x} = \dot{A}\omega \cos\varphi - A\omega^2 \sin\varphi - A\omega\dot{\beta} \sin\varphi \tag{3.320}$$

Therefore, we have two equations for $\dot{A}$ and $\dot{\beta}$:

$$\dot{A} \sin\varphi + A\dot{\beta} \cos\varphi = 0 \tag{3.321}$$

$$\dot{A}\omega \cos\varphi - A\omega^2 \sin\varphi - A\omega\dot{\beta} \sin\varphi + \omega^2 x + 2\varepsilon\omega \operatorname{sgn} \dot{x}$$
$$= \dot{A}\omega \cos\varphi - A\omega\dot{\beta} \sin\varphi + 2\varepsilon\omega \operatorname{sgn} \dot{x} = 0 \tag{3.322}$$

These equations may be arranged as a set of algebraic equations for $\dot{A}$ and $\dot{\varphi}$:

$$\begin{bmatrix} \sin\varphi & A \cos\varphi \\ \omega \cos\varphi & -A\omega \sin\varphi \end{bmatrix} \begin{bmatrix} \dot{A} \\ \dot{\beta} \end{bmatrix} = \begin{bmatrix} 0 \\ -2\varepsilon\omega \operatorname{sgn} \dot{x} \end{bmatrix} \tag{3.323}$$

$$\dot{A} = -2\varepsilon \cos\varphi \operatorname{sgn} \dot{x} = -2\varepsilon \cos\varphi \operatorname{sgn}\,(A \cos\varphi) \tag{3.324}$$

$$\dot{\varphi} = \frac{2}{A}\varepsilon \sin\varphi \operatorname{sgn} \dot{x} = \frac{2}{A}\varepsilon \sin\varphi \operatorname{sgn}\,(A \cos\varphi) \tag{3.325}$$

In averaging method, we replace $\dot{A}$ and $\dot{\beta}$ by the average of their integrals of over a period, assuming their average remains constant. The average integrals of $\dot{A}$ and $\dot{\beta}$ over one period are:

$$\dot{A} = \frac{1}{2\pi} \int_0^{2\pi} \dot{A}d\varphi = \frac{1}{2\pi} \int_{-\frac{\pi}{2}}^{\frac{\pi}{2}} -2\varepsilon \cos\varphi d\varphi$$

$$+ \frac{1}{2\pi} \int_{\frac{\pi}{2}}^{\frac{3\pi}{2}} 2\varepsilon \cos\varphi d\varphi = -\frac{4\varepsilon}{\pi} \tag{3.326}$$

$$\dot{\beta} = \frac{1}{2\pi} \int_0^{2\pi} \dot{\varphi}d\varphi = \frac{1}{2\pi} \int_{-\frac{\pi}{2}}^{\frac{\pi}{2}} \frac{2}{A}\varepsilon \sin\varphi d\varphi$$

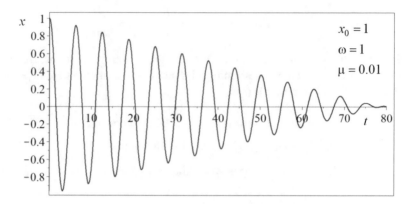

Fig. 3.22 Response of a mass-spring oscillator with a resisting dry friction

$$+\frac{1}{2\pi}\int_{\frac{\pi}{2}}^{\frac{3\pi}{2}}\frac{-2}{A}\varepsilon\sin\varphi d\varphi = 0 \tag{3.327}$$

Therefore, we have:

$$A = -\frac{4\varepsilon}{\pi}t + x_0 \qquad \beta = \beta_0 \tag{3.328}$$

To satisfy the initial conditions, we must have:

$$\beta_0 = \frac{\pi}{2} \tag{3.329}$$

and hence:

$$x = \left(-\frac{4\varepsilon}{\pi}t + x_0\right)\sin\frac{\pi}{2} \tag{3.330}$$

Figure 3.22 illustrates the response of the system to a low dry friction. Although the solution (3.330) and its graphical illustration agree with expected linear amplitude reduction, it is only correct for the time $0 < t \leq \pi x_0/(4\varepsilon)$. The amplitude at $t = \pi x_0/(4\varepsilon)$ becomes zero, and there would be no more displacement after that. However, the solution (3.330) is not capable to keep x at zero after $t = \pi x_0/(4\varepsilon)$.

3.2 Forced Vibrations

The equation of motion of forced linear vibrations of a single degree of freedom is Eq. (3.331). The solution of equation has two parts: the homogenous solution x_h

plus the particular solution x_p. The homogenous solution x_h is the solution of free vibrations for $f = 0$ and the particular solution is the special solution of the equation associated to the forcing function $f(x, \dot{x}, t)$:

$$m\ddot{x} + c\dot{x} + kx = f(x, \dot{x}, t) \tag{3.331}$$

$$x(t) = x_h(t) + x_p(t) \tag{3.332}$$

The particular solution x_p of a forced vibrating system is not possible to be found for a general force function $f = f(x, \dot{x}, t)$. However, if the force function is a continuous function of time $f = f(t)$ and is a combination of the following functions:

1. A constant, such as $f = F$
2. A polynomial in t, such as $f = a_0 + a_1 t + a_2 t^2 + \cdots + a_n t^n$
3. An exponential function, such as $f = Fe^{at}$
4. A harmonic function, such as $f = F_1 \sin \omega t + F_2 \cos \omega t$;

then the particular solution $x_p(t)$ of the linear equation of motion has the same form as the forcing term:

1. $x_p(t) = $ a constant, such as $x_p(t) = X$
2. $x_p(t) = $ a polynomial of the same degree, such as $x_p(t) = C_0 + C_1 t + C_2 t^2 + \cdots + C_n t^n$
3. $x_p(t) = $ an exponential function, such as $x_p(t) = Ce^{at}$
4. $x_p(t) = $ a harmonic function, such as $x_p(t) = A \sin \omega t + B \cos \omega t$

The coefficients of particular solution are calculated by substituting the solution in the differential equation and solving the resultant algebraic equations to satisfy the differential equation. The coefficients will be functions of the differential equation known coefficients. The coefficients of homogenous solution are functions of the initial conditions. The initial conditions must be applied after the exact particular solution x_p and the general homogenous solution x_h are found and added together to have the general solution $x = x_h + x_p$.

Proof Any solution that satisfies the equation of motion (3.331) is a particular solution x_p, and it completes the solution of the equation by adding to the homogenous solution, $x = x_h + x_p$. When the equation is linear, the particular solution would be similar to the forcing term. To find a particular solution x_p of the linear differential equation (3.331), we use the *method of undetermined coefficients* and assume a trial solution x_p of the same form as $f = f(t)$ with unknown coefficients. Substituting x_p in the equation, we determine the coefficients of x_p to satisfy the equation.

1. Constant forcing functions:

$$m\ddot{x} + c\dot{x} + kx = f = F \tag{3.333}$$

Assuming a constant solution, $x = X$, and substituting the assumed solution into the equation provide us with an algebraic equation to determine the solution:

$$x = X \qquad \dot{x} = 0 \qquad \ddot{x} = 0 \tag{3.334}$$

$$X = \frac{F}{k} \tag{3.335}$$

The case of a constant force $f = F$ is a special case of the polynomial forcing function.

2. Polynomial forcing functions:

$$f = a_0 + a_1 t + a_2 t^2 + a_3 t^3 + \cdots + a_n t^n \tag{3.336}$$

Assuming a trial polynomial solution:

$$x_p = C_0 + C_1 t + C_2 t^2 + C_3 t^3 + \cdots + C_n t^n \tag{3.337}$$

$$\dot{x}_p = C_1 + 2C_2 t + 3C_3 t^2 + \cdots + nC_n t^{n-1} \tag{3.338}$$

$$\ddot{x}_p = 2C_2 + 6t C_3 + 12C_4 t^2 + \cdots + n(n-1) C_n t^{n-2} \tag{3.339}$$

and substituting it in Eq. (3.331):

$$m \left(2C_2 + 6C_3 t + 12C_4 t^2 + 20C_5 t^3 + \cdots + n(n-1) C_n t^{n-2} \right)$$
$$+ c \left(C_1 + 2C_2 t + 3C_3 t^2 + 4C_4 t^3 + \cdots + nC_n t^{n-1} \right)$$
$$+ k \left(C_0 + C_1 t + C_2 t^2 + C_3 t^3 + \cdots + C_{n-1} t^{n-1} + C_n t^n \right)$$
$$= a_0 + a_1 t + a_2 t^2 + a_3 t^3 + \cdots + a_{n-1} t^{n-1} + a_n t^n \tag{3.340}$$

provide us with a set of algebraic equations to calculate the coefficient C_i:

$$kC_0 + cC_1 + 2mC_2 = a_0 \tag{3.341}$$

$$kC_1 + 2cC_2 + 6mC_3 = a_1 \tag{3.342}$$

$$kC_2 + 3cC_3 + 12mC_4 = a_2 \tag{3.343}$$

$$\vdots = \vdots$$

$$kC_{n-1} + ncC_n = a_{n-1} \tag{3.344}$$

$$kC_n = a_n \tag{3.345}$$

There are $n + 1$ known constants $a_0, \cdots, a_n$, and $n + 1$ unknowns $C_0, \cdots, C_n$, and we are able to set the equations in matrix form:

$$
\begin{bmatrix}
k & c & 2m & 0 & \cdots & 0 \\
0 & k & 2c & 6m & \cdots & 0 \\
0 & 0 & k & 3c & \cdots & 0 \\
\vdots & \vdots & \vdots & \vdots & \ddots & \vdots \\
0 & 0 & 0 & \cdots & k & nc \\
0 & 0 & 0 & \cdots & 0 & k
\end{bmatrix}
\begin{bmatrix}
C_0 \\
C_1 \\
C_2 \\
\vdots \\
C_{n-1} \\
C_n
\end{bmatrix}
=
\begin{bmatrix}
a_0 \\
a_1 \\
a_2 \\
\vdots \\
a_{n-1} \\
a_n
\end{bmatrix}
\tag{3.346}
$$

The solution of the set of equation indicates that C_i, $i = 0, \cdots, n$ will be determined as functions of $a_0, \cdots, a_n$. Therefore, the trial function (3.337) is a solution of the equation of motion.

3. Exponential forcing function:

$$
f = Fe^{at}
\tag{3.347}
$$

The trial solution would also be an exponential function:

$$
x_p = Ce^{at} \qquad \dot{x}_p = Cae^{at} \qquad \ddot{x}_p = Ca^2 e^{at}
\tag{3.348}
$$

Substituting the solution into Eq. (3.331),

$$
mCa^2 e^{at} + cCae^{at} + kCe^{at} = Fe^{at}
\tag{3.349}
$$

yields:

$$
C = \frac{F}{ma^2 + ca + k}
\tag{3.350}
$$

$$
x_p = \frac{F}{ma^2 + ca + k} e^{at}
\tag{3.351}
$$

4. Harmonic forcing functions:

$$
f = F_1 \sin \omega t + F_2 \cos \omega t
\tag{3.352}
$$

The particular solution for a harmonic excitation is also a harmonic function:

$$
x_p = A \sin \omega t + B \cos \omega t
\tag{3.353}
$$

$$
\dot{x}_p = \omega A \cos \omega t - \omega B \sin \omega t
\tag{3.354}
$$

$$
\ddot{x}_p = -\omega^2 A \sin \omega t - \omega^2 B \cos \omega t
\tag{3.355}
$$

Substituting the harmonic solution in the equation of motion makes an algebraic equation:

$$-m\omega^2 (A \sin \omega t + B \cos \omega t) + c\omega (A \cos \omega t - B \sin \omega t)$$

$$+k (A \sin \omega t + B \cos \omega t) = F_1 \sin \omega t + F_2 \cos \omega t \qquad (3.356)$$

The harmonic functions $\sin \omega t$ and $\cos \omega t$ are orthogonal; therefore, their coefficients must be balanced on both sides of the equal sign. Balancing the coefficients of $\sin \omega t$ and $\cos \omega t$ provides us with a set of two equations for A_1 and B_1:

$$\begin{bmatrix} k - m\omega^2 & -c\omega \\ c\omega & k - m\omega^2 \end{bmatrix} \begin{bmatrix} A \\ B \end{bmatrix} = \begin{bmatrix} F_1 \\ F_2 \end{bmatrix} \qquad (3.357)$$

Solving for coefficients A and B:

$$\begin{bmatrix} A \\ B \end{bmatrix} = \begin{bmatrix} k - m\omega^2 & -c\omega \\ c\omega & k - m\omega^2 \end{bmatrix}^{-1} \begin{bmatrix} F_1 \\ F_2 \end{bmatrix}$$

$$= \begin{bmatrix} \dfrac{\left(k - m\omega^2\right) F_1 + c\omega F_2}{\left(k - m\omega^2\right)^2 + c^2\omega^2} \\[2ex] \dfrac{\left(k - m\omega^2\right) F_2 - c\omega F_1}{\left(k - m\omega^2\right)^2 + c^2\omega^2} \end{bmatrix} \qquad (3.358)$$

determines the particular solution for the harmonic force function.

$$x_p = \frac{\left(k - m\omega^2\right) F_1 + c\omega F_2}{\left(k - m\omega^2\right)^2 + c^2\omega^2} \sin \omega t + \frac{\left(k - m\omega^2\right) F_2 - c\omega F_1}{\left(k - m\omega^2\right)^2 + c^2\omega^2} \cos \omega t \qquad (3.359)$$

∎

Example 104 Response to a step input. The response of dynamic systems to step input is the most important response to identify the characteristics of the system. Here is the analytical calculation of the response of a vibrating system to the unit step excitation.

The response of a damped system to a step input is a standard and the most important response by which we examine and compare all vibrating systems. Consider a linear second-order system with zero initial conditions:

$$m\ddot{x} + c\dot{x} + kx = f(t) \qquad (3.360)$$

$$x(0) = 0 \qquad \dot{x}(0) = 0 \qquad (3.361)$$

The equation may equivalently be transformed to the following form:

$$\ddot{x} + 2\xi\omega_n\dot{x} + \omega_n^2 x = \frac{1}{m}f(t) \qquad \xi < 1 \qquad (3.362)$$

A step input is a sudden change of the forcing function $f(t)$ from zero to a constant and steady value. If the value is unity, then the excitation is called unit step input:

$$f(t) = \begin{cases} 1\,\text{N} & t > 0 \\ 0 & t \le 0 \end{cases} \qquad (3.363)$$

The response of the system to a unit step input is called the unit step response. Linearity of the equation of motion guarantees that the response to a non-unit step input is proportional to the unit step response. A general step function will have a constant force of magnitude F:

$$f(t) = \begin{cases} F\,\text{N} & t > 0 \\ 0 & t \le 0 \end{cases} \qquad (3.364)$$

The general solution of Eq. (3.360) along with sudden force (3.364) is equal to the sum of the homogeneous and particular solutions of the equation, $x = x_h + x_p$. The homogeneous solution is given by Eq. (3.141):

$$x_h = e^{-\xi\omega_n t}(A\sin\omega_d t + B\cos\omega_d t) \qquad (3.365)$$

The particular solution would be a constant function, $x_p = X$, because the forcing term is constant $f(t) = F$. Substituting $x_p = X$ in Eq. (3.360) provides us with x_p:

$$x_p = X = \frac{F}{m\omega_n^2} = \frac{F}{k} \qquad (3.366)$$

Therefore, the general solution of Eq. (3.360) is:

$$x = x_h + x_p$$

$$= e^{-\xi\omega_n t}(A\sin\omega_d t + B\cos\omega_d t) + \frac{F}{m\omega_n^2} \qquad 0 \le t \qquad (3.367)$$

$$\omega_d = \omega_n\sqrt{1 - \xi^2} \qquad (3.368)$$

To determine the unknown coefficients of the homogenous solution, the initial conditions of the system must be applied after summation of the homogenous and particular solutions. The zero initial conditions are the best to explore the natural behavior of systems to step inputs. Applying a set of zero initial conditions provides us with two equations for A and B:

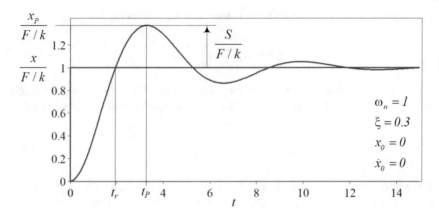

Fig. 3.23 Response of a one DOF vibrating system to a unit step input

$$B + \frac{F}{m\omega_n^2} = 0 \qquad \omega_d A - \xi\omega_n B = 0 \qquad (3.369)$$

with the following solutions:

$$A = -\frac{\xi F}{m\omega_d\omega_n} = -\frac{\xi\omega_n F}{\omega_d k} \qquad B = -\frac{F}{m\omega_n^2} = -\frac{F}{k} \qquad (3.370)$$

Therefore, the step response is:

$$\frac{x}{F/k} = 1 - e^{-\xi\omega_n t}\left(\frac{\xi\omega_n}{\omega_d}\sin\omega_d t + \cos\omega_d t\right) \qquad (3.371)$$

or equivalently is:

$$\frac{x}{F/k} = 1 - \frac{e^{-\xi\omega_n t}}{\sqrt{1-\xi^2}}\sin\left(\omega_d t + \arctan\frac{\sqrt{1-\xi^2}}{\xi}\right) \qquad (3.372)$$

When $t \to \infty$, the homogenous equation approaches zero, and the step response approaches $x/(F/k) \to 1$. Therefore the response x approaches the particular solution $x_p = F/k$, which is the static displacement of the system under the constant force $f = F$. Figure 3.23 depicts a step input response for the following numerical values:

$$\xi = 0.3 \qquad \omega_n = 1 \qquad F = 1 \qquad (3.373)$$

Figure 3.24 illustrates the effect of the damping ratio ξ on the step input response, and Fig. 3.25 illustrates the effect of the natural frequency ω_n.

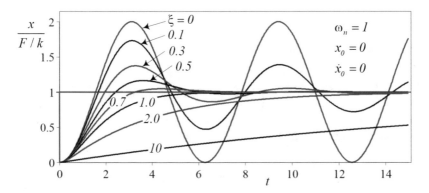

Fig. 3.24 The effect of damping ratio ξ on step input response

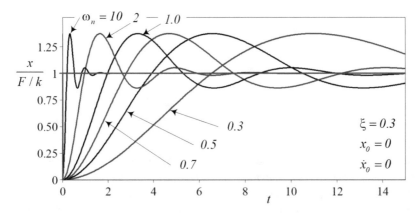

Fig. 3.25 The effect of natural frequency ω_n on step input response

Example 105 Rise time, peak time, overshoot. The characteristics of response of dynamic systems to step input. These characteristics are measurable, and an equivalent second-order dynamic system can be identified based on them.

There are some measurable characteristics for a step response that are being used to identify the system. They are rise time t_r, peak time t_P, peak value x_P, overshoot $S = x_P - F/k$, and settling time t_s. The response of a second-order system with natural frequency ω_n and damping ratio ξ to a step input with magnitude F from zero initial conditions is:

$$\frac{x}{F/k} = 1 - \frac{e^{-\xi\omega_n t}}{\sqrt{1-\xi^2}} \sin\left(\omega_d t + \arctan\frac{\sqrt{1-\xi^2}}{\xi}\right) \tag{3.374}$$

$$= 1 - e^{-\xi\omega_n t}\left(\frac{\xi}{\sqrt{1-\xi^2}} \sin\omega_d t + \cos\omega_d t\right) \tag{3.375}$$

A graphical illustration of this function is shown in Fig. 3.25.

The rise time t_r is the first time that the response $x(t)$ reaches the value of the steady-state response of the step input, F/k:

$$\frac{F}{k} = \frac{F}{k}\left(1 - e^{-\xi\omega_n t}\left(\frac{\xi}{\sqrt{1-\xi^2}}\sin\omega_d t + \cos\omega_d t\right)\right) \tag{3.376}$$

Because $e^{-\xi\omega_n t} \neq 0$, Eq. (3.376) yields:

$$\frac{\xi}{\sqrt{1-\xi^2}}\sin\omega_d t + \cos\omega_d t = 0 \tag{3.377}$$

or

$$\tan\omega_d t = -\frac{\sqrt{1-\xi^2}}{\xi} \qquad n = 0, 1, 2, 3, \cdots \tag{3.378}$$

Therefore, the rise time t_r can be calculated:

$$t_r = \frac{1}{\omega_d}\left(\pi - \arctan\frac{\sqrt{1-\xi^2}}{\xi}\right) \tag{3.379}$$

There are also other definitions for rise time. It may also be defined as the inverse of the largest slope of the step response, or as the time it takes to pass from 10% to 90% of the steady-state value. Such alternative definitions include the cases of critically or overdamped as well as underdamped systems. However, the definition of rise time as the first time that the response reaches the value of the steady-state value is more applied. In fact majority of dynamic systems are designed underdamped.

The peak time t_P is the first time that the response $x(t)$ reaches its maximum value. The times at which $x(t)$ is maximum or minimum are the solutions of the equation $\dot{x} = 0$:

$$\frac{dx}{dt} = \frac{F}{k}\frac{\omega_n}{\sqrt{1-\xi^2}}e^{-\xi\omega_n t}\sin\omega_d t = 0 \tag{3.380}$$

which simplifies to:

$$\sin\omega_d t = 0 \tag{3.381}$$

The time of the first maximum is the peak time t_P:

$$t_P = \frac{\pi}{\omega_d} \tag{3.382}$$

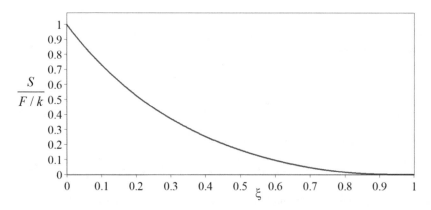

Fig. 3.26 The variation of overshoot $\frac{S}{F/k}$ which is only a function of damping ratio ξ

The value of $x\,(t)$ at t_P is the peak value x_P.

$$\frac{x_P}{F/k} = 1 + e^{-\xi\omega_n \frac{\pi}{\omega_d}} = 1 + e^{-\xi\pi/\sqrt{1-\xi^2}} \qquad (3.383)$$

The overshoot S indicates how much the response $x(t)$ exceeds the steady state response of the step input, F/k:

$$S = x_P - \frac{F}{k} = \frac{F}{k} e^{-\xi\pi/\sqrt{1-\xi^2}} \qquad (3.384)$$

$$\frac{S}{F/k} = e^{-\xi\pi/\sqrt{1-\xi^2}} \qquad (3.385)$$

The value of the overshoot $\frac{S}{F/k}$ is only a function of damping ratio ξ and is always positive $0 < \frac{S}{F/k} < 1$ for underdamped system, $0 < \xi < 1$. It exponentially decreases from $\frac{S}{F/k} = 1$ to $\frac{S}{F/k} = 0$ when ξ increases from zero to one. Figure 3.26 depicts how $\frac{S}{F/k}$ varies by ξ:

$$\lim_{\xi \to 0} \frac{S}{F/k} = 1 \qquad \lim_{\xi \to 1} \frac{S}{F/k} = 0 \qquad (3.386)$$

The settling time t_s is four times of the time constant $\tau = 1/(\xi\omega_n)$ of the exponential function $e^{-\xi\omega_n t}$:

$$t_s = \frac{4}{\xi\omega_n} \qquad (3.387)$$

There are other definitions for settling time as well. The settling time may also be defined as the required time that the step response $x(t)$ needs to settle within a $\pm n\%$ window of the steady-state value, F/k. The value $n = 2$ is commonly used:

$$t_s \approx \frac{\ln\left(n\sqrt{1-\xi^2}\right)}{\xi\omega_n} \tag{3.388}$$

As an example, for a set of sample data:

$$\xi = 0.3 \qquad \omega_n = 1 \qquad F = 1 \tag{3.389}$$

we find the following characteristic values:

$$t_r = 1.966 \tag{3.390}$$

$$t_P = 3.2933 \qquad \frac{x_P}{F/k} = 1.3723 \qquad \frac{S}{F/k} = 0.3723 \tag{3.391}$$

$$t_s = 13.333 \tag{3.392}$$

On the other hand, measuring only two of the four characteristics, t_r, t_P, x_P, S, and t_s would be enough to evaluate ξ, ω_n.

Example 106 ★ Boundary of step response. The response curve to step input is bounded between two exponentially decaying curves. The value of $\frac{x}{F/k}$ at time constant $\tau = \frac{1}{\xi\omega_n}$ is only a function of ξ.

It is shown in Fig. 3.27 that the step response curve is bounded by two functions:

$$x = 1 \pm \frac{e^{-\xi\omega_n t}}{\sqrt{1-\xi^2}} \tag{3.393}$$

The time constant τ of the boundary functions is $\tau = \frac{1}{\xi\omega_n}$ which is four times of the settling time of the system $t_s = \frac{4}{\xi\omega_n}$:

$$\tau = \frac{1}{\xi\omega_n} \tag{3.394}$$

The value of the response of the system $x_\tau/(F/k)$ at $t = \tau$ is only a function of the damping ratio ξ:

$$\frac{x_\tau}{F/k} = 1 - \frac{1}{e}\left(\cos\frac{\sqrt{1-\xi^2}}{\xi} + \frac{\xi}{\sqrt{1-\xi^2}}\sin\frac{\sqrt{1-\xi^2}}{\xi}\right) \tag{3.395}$$

$$0 < \xi < 1 \tag{3.396}$$

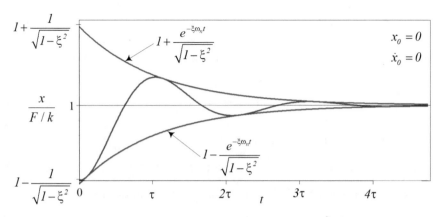

Fig. 3.27 Any step response curve $\frac{x}{F/k}$ has two envelops $x = 1 \pm \frac{e^{-\xi\omega_n t}}{\sqrt{1-\xi^2}}$

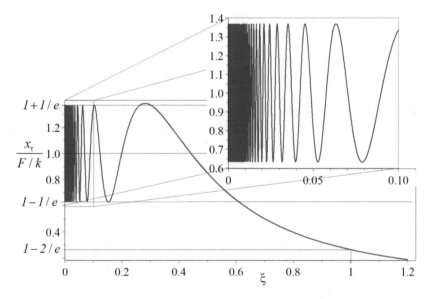

Fig. 3.28 The value of the function x_τ is highly sensitive to ξ for very low values of ξ

Figure 3.28 illustrates $\frac{x_\tau}{F/k}$ as a function of ξ. The value of the function $\frac{x_\tau}{F/k}$ is highly sensitive to ξ for very low values of ξ, and, therefore, the value of the function at $t = \tau$ is hard to calculate at low values of ξ. However, $\frac{x_\tau}{F/k}$ is bounded between two values, and they can be determined by moving $\xi \to 0$:

$$\lim_{\xi \to 0} \frac{x_\tau}{F/k} = 1 \pm \frac{1}{e} = 1.3679,\ \ 0.63212 \tag{3.397}$$

Increasing ξ from zero makes $\frac{x_\tau}{F/k}$ to fluctuate between $1 - 1/e$ and $1 + 1/e$ by a decreasing frequency $\omega = \sqrt{1 - \xi^2}/\xi$. By approaching to the limit of underdamped case, $\xi = 1$, the value of $\frac{x_\tau}{F/k}$ approaches $1 - 2/e$:

$$\lim_{\xi \to 1} \frac{x_\tau}{F/k} = 1 - \frac{2}{e} \tag{3.398}$$

Employing the hyperbolic functions:

$$\sin(ix) = i \sinh(x) \qquad \cos(ix) = \cosh(x) \tag{3.399}$$

we are able to mathematically define $\frac{x_\tau}{F/k}$ for overdamped situations:

$$\frac{x_\tau}{F/k} = 1 - \frac{1}{e} \left(\cos \frac{i\sqrt{\xi^2 - 1}}{\xi} + \frac{\xi}{i\sqrt{\xi^2 - 1}} \sin \frac{i\sqrt{\xi^2 - 1}}{\xi} \right)$$

$$= 1 - \frac{1}{e} \left(\cosh \frac{\sqrt{\xi^2 - 1}}{\xi} + \frac{\xi}{\sqrt{\xi^2 - 1}} \sinh \frac{\sqrt{\xi^2 - 1}}{\xi} \right) \tag{3.400}$$

$$1 < \xi$$

Example 107 ★ Forced vibration of undamped systems. Applying a general force function $f(t)$ on an undamped mass-spring vibrating system will have a closed form solution. Here is the mathematical analysis.

An undamped one DOF vibrating system under an external force $f(t)$ has a solution x in integral form:

$$m\ddot{x} + kx = f(t) \tag{3.401}$$

$$x = -\cos \omega_n t \int_0^t \frac{f(t)}{m\omega_n} \sin \omega_n t \, dt + \sin \omega_n t \int_0^t \frac{f(t)}{m\omega_n} \cos \omega_n t \, dt$$

$$+ x_0 \cos \omega_n t + \frac{\dot{x}_0}{\omega_n} \sin \omega_n t \tag{3.402}$$

Proof Consider a forced vibration with a second-order differential equation:

$$\ddot{x} + \omega_n^2 x = \frac{1}{m} f(t) \tag{3.403}$$

We rewrite the equation by introducing a complex variable z:

$$\frac{d}{dt} (\dot{x} + i\omega_n x) - i\omega_n (\dot{x} + i\omega_n x) = \frac{1}{m} f(t) \tag{3.404}$$

$$z = \dot{x} + i\omega_n x \qquad i^2 = -1 \tag{3.405}$$

to transform the equation to a solvable first-order equation in terms of the complex variable z:

$$\frac{d}{dt}z - i\omega_n z = \frac{1}{m}f(t) \tag{3.406}$$

$$z = e^{i\omega_n t}\left(\int_0^t \frac{1}{m}f(t)e^{-i\omega_n t}dt + z_0\right) \tag{3.407}$$

$$z_0 = z(0) \tag{3.408}$$

Expanding the solution into real and imaginary parts:

$$\begin{aligned}
\dot{x} + i\omega_n x &= (\cos\omega_n t + i\sin\omega_n t)\left(\int_0^t \frac{f(t)}{m}(\cos\omega_n t - i\sin\omega_n t)\,dt + z_0\right)\\
&= \cos\omega_n t\left(\int_0^t \frac{f(t)}{m}\cos\omega_n t\,dt - i\int_0^t \frac{f(t)}{m}\sin\omega_n t\,dt + z_0\right)\\
&\quad + i\sin\omega_n t\left(\int_0^t \frac{f(t)}{m}\cos\omega_n t\,dt - i\int_0^t \frac{f(t)}{m}\sin\omega_n t\,dt + z_0\right)\\
&= \cos\omega_n t\int_0^t \frac{f(t)}{m}\cos\omega_n t\,dt + \sin\omega_n t\int_0^t \frac{f(t)}{m}\sin\omega_n t\,dt\\
&\quad + \dot{x}_0\cos\omega_n t - \omega_n x_0\sin\omega_n t\\
&\quad - i\cos\omega_n t\int_0^t \frac{f(t)}{m}\sin\omega_n t\,dt + i\sin\omega_n t\int_0^t \frac{f(t)}{m}\cos\omega_n t\,dt\\
&\quad + i\omega_n x_0\cos\omega_n t + i\dot{x}_0\sin\omega_n t \tag{3.409}
\end{aligned}$$

$$\begin{aligned}
\dot{x} &= \cos\omega_n t\int_0^t \frac{f(t)}{m}\cos\omega_n t\,dt + \sin\omega_n t\int_0^t \frac{f(t)}{m}\sin\omega_n t\,dt\\
&\quad + \dot{x}_0\cos\omega_n t - \omega_n x_0\sin\omega_n t \tag{3.410}
\end{aligned}$$

and equating the real and imaginary parts of both sides provide us with the following real solutions:

$$\begin{aligned}
x &= \frac{-\cos\omega_n t}{m\omega_n}\int_0^t f(t)\sin\omega_n t\,dt + \frac{\sin\omega_n t}{m\omega_n}\int_0^t f(t)\cos\omega_n t\,dt\\
&\quad + x_0\cos\omega_n t + \frac{\dot{x}_0}{\omega_n}\sin\omega_n t \tag{3.411}
\end{aligned}$$

$$\dot{x} = \frac{\cos \omega_n t}{m} \int_0^t f(t) \cos \omega_n t \, dt + \frac{\sin \omega_n t}{m} \int_0^t f(t) \sin \omega_n t \, dt$$

$$+ \dot{x}_0 \cos \omega_n t - \omega_n x_0 \sin \omega_n t \tag{3.412}$$

∎

As an example let us solve the undamped system with a constant force $f(t) = F$:

$$\ddot{x} + \omega_n^2 x = \frac{1}{m} F \tag{3.413}$$

The solution (3.407) provides:

$$\dot{x} + i\omega_n x = e^{i\omega_n t} \left(\int_0^t \frac{1}{m} F e^{-i\omega_n t} \, dt + \dot{x}_0 + i\omega_n x_0 \right)$$

$$= \frac{iF}{m\omega_n} + \left(\dot{x}_0 + i\omega_n x_0 - \frac{iF}{m\omega_n} \right) e^{i\omega_n t}$$

$$= \dot{x}_0 \cos \omega_n t + \frac{F \sin \omega_n t}{m\omega_n} - \omega_n x_0 \sin \omega_n t$$

$$+ i \left(\frac{F(1 - \cos \omega_n t)}{m\omega_n} + \dot{x}_0 \sin \omega_n t + \omega_n x_0 \cos \omega_n t \right) \tag{3.414}$$

and, therefore, the response x is calculated:

$$x = \frac{F}{k}(1 - \cos \omega_n t) + \frac{\dot{x}_0}{\omega_n} \sin \omega_n t + x_0 \cos \omega_n t \tag{3.415}$$

It is also possible to separate the real and imaginary parts before integration and use Eqs. (3.411) and (3.412) and find the solution:

$$x = -\frac{F}{m\omega_n} \cos \omega_n t \int_0^t \sin \omega_n t \, dt + \frac{F}{m\omega_n} \sin \omega_n t \int_0^t \cos \omega_n t \, dt$$

$$+ x_0 \cos \omega_n t + \frac{\dot{x}_0}{\omega_n} \sin \omega_n t$$

$$= \frac{F}{k} \cos \omega_n t (\cos \omega_n t - 1) + \frac{F}{k} \sin^2 \omega_n t + x_0 \cos \omega_n t + \frac{\dot{x}_0}{\omega_n} \sin \omega_n t$$

$$= \frac{F}{k}(1 - \cos t\omega_n) + x_0 \cos \omega_n t + \frac{\dot{x}_0}{\omega_n} \sin \omega_n t \tag{3.416}$$

$$\dot{x} = \frac{F}{m}\cos\omega_n t \int_0^t \cos\omega_n t\, dt + \frac{F}{m}\sin\omega_n t \int_0^t \sin\omega_n t\, dt$$

$$+\dot{x}_0\cos\omega_n t - \omega_n x_0\sin\omega_n t$$

$$= \frac{F}{m\omega_n}\cos\omega_n t\,\sin\omega_n t - \frac{F}{m\omega_n}\sin\omega_n t\,(\cos\omega_n t - 1)$$

$$+\dot{x}_0\cos\omega_n t - \omega_n x_0\sin\omega_n t$$

$$= \frac{F}{m\omega_n}\sin\omega_n t + \dot{x}_0\cos\omega_n t - \omega_n x_0\sin\omega_n t \qquad (3.417)$$

Example 108 Response to a polynomial and ramp input. Response of a vibrating system to a polynomial force function is an easy problem to solve. Ramp input is an important standard polynomial input to determine laziness of a system.

Let us examine the response of a second-order system to a second-degree polynomial force function:

$$\ddot{x} + 2\xi\omega_n\dot{x} + \omega_n^2 x = \frac{1}{m}f(t) \qquad \xi < 1 \qquad (3.418)$$

$$f(t) = a_0 + a_1 t + a_2 t^2 \qquad (3.419)$$

$$[a_0] = MLT^{-2} \qquad [a_1] = MLT^{-3} \qquad [a_2] = MLT^{-4} \qquad (3.420)$$

The dimension of the coefficients of the forcing term is shown in Eq. (3.420).

Substituting a trial polynomial particular solution x_p into the equation:

$$2b_2 + 2\xi\omega_n (b_1 + 2t b_2) + \omega_n^2 \left(b_0 + b_1 t + b_2 t^2\right) = \frac{a_0 + a_1 t + a_2 t^2}{m} \qquad (3.421)$$

$$x_p = b_0 + b_1 t + b_2 t^2 \qquad (3.422)$$

$$[b_0] = L \qquad [b_1] = LT^{-1} \qquad [b_2] = LT^{-2} \qquad (3.423)$$

yields:

$$\left(\omega_n^2 b_2 - \frac{a_2}{m}\right) t^2 + \left(b_1\omega_n^2 + 4\xi b_2\omega_n - \frac{a_1}{m}\right) t \qquad (3.424)$$

$$+ \left(b_0\omega_n^2 + 2\xi b_1\omega_n - \frac{a_0}{m} + 2b_2\right) = 0 \qquad (3.425)$$

To have this equation at all time, all coefficients must be zero. Therefore, we will have three equations to determine the coefficients b_0, and b_1, b_2 of the guessed solution:

$$\omega_n^2 b_2 - \frac{a_2}{m} = 0 \qquad (3.426)$$

$$b_1 \omega_n^2 + 4\xi b_2 \omega_n - \frac{a_1}{m} = 0 \tag{3.427}$$

$$b_0 \omega_n^2 + 2\xi b_1 \omega_n - \frac{a_0}{m} + 2b_2 = 0 \tag{3.428}$$

The coefficients are:

$$b_2 = \frac{a_2}{m\omega_n^2} \tag{3.429}$$

$$b_1 = \frac{1}{m\omega_n^3} (\omega_n a_1 - 4\xi a_2) \tag{3.430}$$

$$b_0 = -\frac{1}{m\omega_n^4} \left(2a_2 - 8\xi^2 a_2 - \omega_n^2 a_0 + 2\xi \omega_n a_1 \right) \tag{3.431}$$

Therefore, the general solution of Eq. (3.360) is:

$$\begin{aligned}
x = x_h + x_p &= e^{-\xi \omega_n t} (A \sin \omega_d t + B \cos \omega_d t) + b_0 + b_1 t + b_2 t^2 \\
&= e^{-\xi \omega_n t} (A \sin \omega_d t + B \cos \omega_d t) \\
&\quad - \frac{1}{m\omega_n^4} \left(2a_2 - 8\xi^2 a_2 - \omega_n^2 a_0 + 2\xi \omega_n a_1 \right) \\
&\quad + \frac{1}{m\omega_n^3} (\omega_n a_1 - 4\xi a_2) t + \frac{a_2}{m\omega_n^2} t^2
\end{aligned} \tag{3.432}$$

$$\omega_d = \omega_n \sqrt{1 - \xi^2} \tag{3.433}$$

If $a_1 = 0$, $a_2 = 0$, and $a_0 = F$, the solution (3.432) simplifies to the step input response (3.367).

When $a_0 = 0$, $a_2 = 0$, the force is proportional to time t:

$$f(t) = a_1 t \tag{3.434}$$

Such a forcing term is called the ramp input, and a response to the force is called the ramp response. The ramp input analysis is another standard excitation to identify some characteristics of a dynamic system. The ramp response would be:

$$x = x_h + x_p = e^{-\xi \omega_n t} (A \sin \omega_d t + B \cos \omega_d t) - \frac{2\xi a_1}{m\omega_n^3} + \frac{a_1}{m\omega_n^2} t \tag{3.435}$$

Applying a set of zero initial conditions:

$$x(0) = 0 \qquad \dot{x}(0) = 0 \tag{3.436}$$

provides us with two equations to find the coefficients A and B:

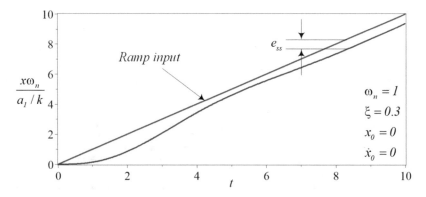

Fig. 3.29 Response of a one DOF vibrating system to a ramp force input

$$B - \frac{2\xi a_1}{m\omega_n^3} = 0 \qquad \frac{a_1}{m\omega_n^2} - \xi\omega_n B + \omega_d A = 0 \qquad (3.437)$$

$$A = \frac{(2\xi^2 - 1)\, a_1}{m\omega_n^2 \omega_d} = \frac{(2\xi^2 - 1)\, a_1}{k\omega_d} \qquad B = \frac{2\xi a_1}{m\omega_n^3} = \frac{2\xi a_1}{k\omega_n} \qquad (3.438)$$

Therefore, the step response to zero initial conditions is calculated:

$$\frac{x\omega_n}{a_1/k} = e^{-\xi\omega_n t}\left(\frac{2\xi^2 - 1}{\sqrt{1 - \xi^2}}\sin\omega_d t + 2\xi\cos\omega_d t\right) - 2\xi + \omega_n t \qquad (3.439)$$

Figure 3.29 depicts a ramp input response for the following numerical values:

$$\xi = 0.3 \qquad \omega_n = 1 \qquad (3.440)$$

The difference between $x\omega_n/\,(a_1/k)$ and $\omega_n t$ at $t \to \infty$ is called the steady-state error e_{ss}:

$$e_{ss} = \lim_{t\to\infty}\left(\frac{x\omega_n}{a_1/k} - \omega_n t\right) = -2\xi \qquad (3.441)$$

The steady-state error has a constant value for ramp input, which means $x\omega_n/\,(a_1/k)$ follows $\omega_n t$ with a constant lag. Figure 3.30 illustrates the effect of the damping ratio ξ on the step input response.

Example 109 Response to a sinusoidal force. The particular solution x_p of a vibrating system to a harmonic forcing term will be a harmonic displacement which must be added to the homogenous solution x_h to make the general solution. The homogenous part is the transient part of the solution which will disappear after a

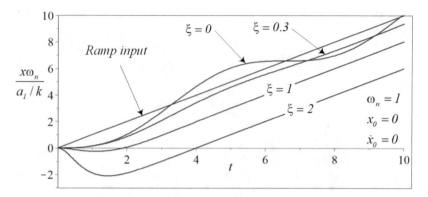

Fig. 3.30 The effect of damping ratio ξ on ramp input response

while. Here we study that transient part before the steady-state harmonic response overcomes.

The steady-state response to a harmonic force is a subject of frequency response. However, the system goes through unsteady vibrations before showing its frequency response. Let us discover the unsteady part of the motion of the system. Consider a linear second-order system with the equation of motion:

$$m\ddot{x} + c\dot{x} + kx = F\cos\omega t \tag{3.442}$$

$$\ddot{x} + 2\xi\omega_n\dot{x} + \omega_n^2 x = \frac{F}{m}\cos\omega t \qquad \xi < 1 \tag{3.443}$$

The trial particular solution must be a full harmonic function of the same frequency as the excitation with both $\sin\omega t$ and $\cos\omega t$ terms:

$$x_p = C_1\sin\omega t + C_2\cos\omega t \tag{3.444}$$

We substitute the trial solution (3.444) in Eq. (3.443), and balancing the harmonic functions on both sides yields:

$$C_1\omega_n^2 - 2\xi C_2\omega\omega_n - C_1\omega^2 = 0 \tag{3.445}$$

$$\omega_n^2 C_2 - \omega^2 C_2 - \frac{F}{m} + 2\xi\omega\omega_n C_1 = 0 \tag{3.446}$$

$$\begin{bmatrix} \omega_n^2 - \omega^2 & -2\xi\omega\omega_n \\ 2\xi\omega\omega_n & \omega_n^2 - \omega^2 \end{bmatrix}\begin{bmatrix} C_1 \\ C_2 \end{bmatrix} = \begin{bmatrix} 0 \\ \dfrac{F}{m} \end{bmatrix} \tag{3.447}$$

$$C_1 = \frac{F}{m} \frac{2\xi\omega\omega_n}{\left(\omega^2 - \omega_n^2\right)^2 + 4\xi^2\omega^2\omega_n^2} \tag{3.448}$$

$$C_2 = -\frac{F}{m} \frac{\omega^2 - \omega_n^2}{\left(\omega^2 - \omega_n^2\right)^2 + 4\xi^2\omega^2\omega_n^2} \tag{3.449}$$

Employing the homogenous solution x_h from free vibration in Eq. (3.19), the general solution of Eq. (3.360) would be:

$$x = x_h + x_p$$
$$= e^{-\xi\omega_n t} \left(A \sin\omega_d t + B \cos\omega_d t\right) + C_1 \sin\omega t + C_2 \cos\omega t \tag{3.450}$$

$$\omega_d = \omega_n\sqrt{1 - \xi^2} \tag{3.451}$$

Assuming a set of zero initial conditions provides us with two equations to determine A and B:

$$x(0) = 0 \qquad \dot{x}(0) = 0 \tag{3.452}$$

$$B + C_2 = 0 \qquad \omega C_1 + A\omega_d - B\xi\omega_n = 0 \tag{3.453}$$

$$A = \frac{\xi\omega_n C_2 - \omega C_1}{\omega_d} \qquad B = C_2 \tag{3.454}$$

Therefore, the response to a harmonic force from zero initial condition is as follows:

$$x = e^{-\xi\omega_n t} \left(\frac{\xi\omega_n C_2 - \omega C_1}{\omega_d} \sin\omega_d t + C_2 \cos\omega_d t\right)$$
$$+ C_1 \sin\omega t + C_2 \cos\omega t \tag{3.455}$$

When $t \to \infty$, the homogeneous solution disappears, and the particular harmonic solution remains. Figure 3.31 illustrates the transient response of the system for the following numerical values:

$$\omega_n = 1 \qquad \omega = 0.2 \qquad \xi = 0.05 \qquad F = 1 \qquad m = 1 \tag{3.456}$$

Example 110 ★ Response to two harmonic forces. In analysis of linear systems, the principal method of solution for multi-terms forcing function is superposition. Particular solution can be found for every forcing term individually and add all of them to have the particular solution for the whole forcing function. However, we are

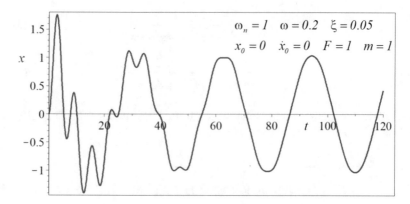

Fig. 3.31 The transient response of a second-order system to a harmonic excitation

able to search for particular solution of similar forcing terms in one try. Here is an example for two harmonic forcing terms.

Let us examine the transient response of a system to two harmonic excitations with two different frequencies:

$$m\ddot{x} + c\dot{x} + kx = F_1 \cos \omega_1 t + (F_2 \cos \omega_2 t + F_3 \sin \omega_2 t) \tag{3.457}$$

The first force function is a cosine function of time, $F_1 \cos \omega_1 t$, and the second force function is a full harmonic function with both sine and cosine terms, $F_2 \cos \omega_2 t + F_3 \sin \omega_2 t$. The equation of motion can also be written in more suitable form by dividing over m:

$$\ddot{x} + 2\xi \omega_n \dot{x} + \omega_n^2 x = \frac{F_1}{m} \cos \omega_1 t + \left(\frac{F_2}{m} \cos \omega_2 t + \frac{F_3}{m} \sin \omega_2 t \right) \tag{3.458}$$

$$\xi < 1 \tag{3.459}$$

The trial particular solution must be two full harmonic functions of both forcing frequencies, ω_1 and ω_2:

$$x_p = C_1 \sin \omega_1 t + C_2 \cos \omega_1 t + C_3 \sin \omega_2 t + C_4 \cos \omega_2 t \tag{3.460}$$

Substituting the trial solution in the equation of motion yields:

$$- \omega_1^2 \left(C_1 \sin t\omega_1 + C_2 \cos t\omega_1 \right) - \omega_2^2 \left(C_3 \sin t\omega_2 + C_4 \cos t\omega_2 \right)$$
$$+ 2\xi \omega_n \left(\omega_1 C_1 \cos t\omega_1 + \omega_2 C_3 \cos t\omega_2 - \omega_1 C_2 \sin t\omega_1 - \omega_2 C_4 \sin t\omega_2 \right)$$
$$+ \omega_n^2 \left(C_1 \sin \omega_1 t + C_2 \cos \omega_1 t + C_3 \sin \omega_2 t + C_4 \cos \omega_2 t \right)$$
$$= \frac{F_1}{m} \cos \omega_1 t + \frac{F_2}{m} \cos \omega_2 t + \frac{F_3}{m} \sin \omega_2 t \tag{3.461}$$

Balancing the harmonic functions $\cos\omega_1 t$, $\cos\omega_2 t$, $\sin\omega_1 t$, and $\sin\omega_2 t$ on both sides provides us with four algebraic equations to find the coefficients of the solution (3.460):

$$C_1\omega_n^2 - 2\xi C_2\omega_1\omega_n - C_1\omega_1^2 = 0 \tag{3.462}$$

$$\omega_n^2 C_2 - \omega_1^2 C_2 - \frac{1}{m}F_1 + 2\xi\omega_1\omega_n C_1 = 0 \tag{3.463}$$

$$\frac{1}{m}F_3 - \omega_2^2 C_3 + \omega_n^2 C_3 - 2\xi\omega_2\omega_n C_4 = 0 \tag{3.464}$$

$$\frac{1}{m}F_2 - \omega_2^2 C_4 + \omega_n^2 C_4 + 2\xi\omega_2\omega_n C_3 = 0 \tag{3.465}$$

$$\begin{bmatrix} \omega_n^2 - \omega_1^2 & -2\xi\omega_1\omega_n & 0 & 0 \\ 2\xi\omega_1\omega_n & \omega_n^2 - \omega_1^2 & 0 & 0 \\ 0 & 0 & -\omega_2^2 + \omega_n^2 & -2\xi\omega_2\omega_n \\ 0 & 0 & 2\xi\omega_2\omega_n & -\omega_2^2 + \omega_n^2 \end{bmatrix} \begin{bmatrix} C_1 \\ C_2 \\ C_3 \\ C_4 \end{bmatrix} = \begin{bmatrix} 0 \\ F_1/m \\ -F_3/m \\ -F_2/m \end{bmatrix} \tag{3.466}$$

The coefficients C_1, C_2, C_3, and C_4 are:

$$C_1 = \frac{2}{m}\xi\omega_1\omega_n\frac{F_1}{4\xi^2\omega_1^2\omega_n^2 + \omega_1^4 - 2\omega_1^2\omega_n^2 + \omega_n^4} \tag{3.467}$$

$$C_2 = -\frac{1}{m}F_1\frac{\omega_1^2 - \omega_n^2}{4\xi^2\omega_1^2\omega_n^2 + \omega_1^4 - 2\omega_1^2\omega_n^2 + \omega_n^4} \tag{3.468}$$

$$\begin{aligned} C_3 &= \frac{1}{m}F_3\frac{\omega_2^2 - \omega_n^2}{4\xi^2\omega_2^2\omega_n^2 + \omega_2^4 - 2\omega_2^2\omega_n^2 + \omega_n^4} \\ &\quad - \frac{2}{m}\xi\omega_2\omega_n\frac{F_2}{4\xi^2\omega_2^2\omega_n^2 + \omega_2^4 - 2\omega_2^2\omega_n^2 + \omega_n^4} \end{aligned} \tag{3.469}$$

$$\begin{aligned} C_4 &= \frac{1}{m}F_2\frac{\omega_2^2 - \omega_n^2}{4\xi^2\omega_2^2\omega_n^2 + \omega_2^4 - 2\omega_2^2\omega_n^2 + \omega_n^4} \\ &\quad + \frac{2}{m}\xi\omega_2\omega_n\frac{F_3}{4\xi^2\omega_2^2\omega_n^2 + \omega_2^4 - 2\omega_2^2\omega_n^2 + \omega_n^4} \end{aligned} \tag{3.470}$$

Therefore, the general solution of Eq. (3.457) is:

$$
\begin{aligned}
x &= x_h + x_p \\
&= e^{-\xi \omega_n t} \left(A \sin \omega_d t + B \cos \omega_d t \right) \\
&\quad + C_1 \sin \omega_1 t + C_2 \cos \omega_1 t + C_3 \sin \omega_2 t + C_4 \cos \omega_2 t \quad (3.471)
\end{aligned}
$$

$$
\omega_d = \omega_n \sqrt{1 - \xi^2} \qquad\qquad\qquad\qquad\qquad\qquad\qquad (3.472)
$$

It is now to apply initial conditions to determine the coefficient of the homogenous part of the solution. Assuming a set of zero initial conditions provides us with two equations for A and B:

$$
x(0) = 0 \qquad \dot{x}(0) = 0 \qquad\qquad\qquad\qquad (3.473)
$$

$$
B + C_2 + C_4 = 0 \qquad \omega_1 C_1 + \omega_2 C_3 + A \omega_d - B \xi \omega_n = 0 \qquad (3.474)
$$

$$
A = -\frac{\omega_1 C_1 + \omega_2 C_3 + \xi \omega_n (C_2 + C_4)}{\omega_d} \qquad B = -C_2 - C_4 \qquad (3.475)
$$

Hence, the response to the two harmonic forces with zero initial conditions is:

$$
\begin{aligned}
x &= e^{-\xi \omega_n t} \left(-\frac{\omega_1 C_1 + \omega_2 C_3 + \xi \omega_n (C_2 + C_4)}{\omega_d} \sin \omega_d t - C_2 - C_4 \cos \omega_d t \right) \\
&\quad + C_1 \sin \omega_1 t + C_2 \cos \omega_1 t + C_3 \sin \omega_2 t + C_4 \cos \omega_2 t \quad (3.476)
\end{aligned}
$$

When $t \to \infty$, the homogeneous solution disappears, and the particular biharmonic solution remains. The steady-state solution is the sum of two harmonic functions $C_1 \sin \omega_1 t + C_2 \cos \omega_1 t$ and $C_3 \sin \omega_2 t + C_4 \cos \omega_2 t$ with frequencies ω_1 and ω_2. The actual values of the excitation frequencies ω_1 and ω_2 as well as the natural frequency ω_n determine the pattern and the amplitude of the steady-state response. Figure 3.32 illustrates the transient response of the system to the combined harmonic excitations for the following numerical values:

$$
\omega_n = 1 \qquad \omega_1 = 0.2 \qquad \omega_2 = 0.3 \qquad \xi = 0.05
$$

$$
F_1 = 1 \qquad F_2 = 1 \qquad F_3 = 1 \qquad m = 1 \qquad\qquad (3.477)
$$

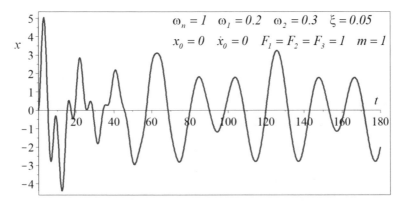

Fig. 3.32 A sample of the transient response of a second-order system to a combined harmonic excitations

3.3 Transient Vibrations

When the forcing function $f(t)$ disappears after a while, the system is under *transient excitation*. The solution of the equation of motion under a transient excitation is called *transient response*. The general solution method of a mass-spring-damper vibrating system under a transient force is the same method of determining and adding two parts: the homogenous solution x_h plus the particular solution x_p.

$$m\ddot{x} + c\dot{x} + kx = f(t) \tag{3.478}$$

$$x(t) = x_h(t) + x_p(t) \tag{3.479}$$

The homogenous part x_h is the solution of free vibrations for $f = 0$, and the particular part x_p is the solution associated to the transient forcing function $f(t)$. A transient excitation is assumed to be weak, so the displacement remains small to keep the system linear.

However, there is an alternative and easier method to determine the response of vibrating systems to a transient forcing term $f(t)$ by using the *convolution integral*:

$$x(t) = \int_0^t f(t-\tau) h(\tau) d\tau = \int_0^t h(t-\tau) f(\tau) d\tau \tag{3.480}$$

where:

$$h(t) = \frac{1}{m\omega_n} e^{-\xi\omega_n t} \sin \omega_d t \tag{3.481}$$

$$\omega_d = \omega_n \sqrt{1 - \xi^2} \qquad \xi < 1 \tag{3.482}$$

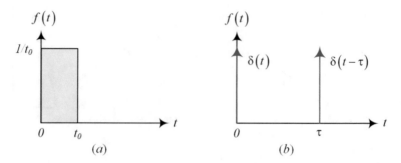

Fig. 3.33 Definition of the unit impulse or Dirac delta function $\delta(t)$

Proof The time integral of a force is called impulse, I:

$$I = \int f(t)\, dt \ \text{N m}/\text{s} \tag{3.483}$$

Consider a constant function $f(t) = 1/t_0$ that is applied for a period of time $0 \le t \le t_0$ as shown in Fig. 3.33a. The time integral and, hence, the impulse of the function $1/t_0$ during its application are unity:

$$I = \int_{-\infty}^{\infty} f(t)\, dt = \int_{-\infty}^{\infty} \frac{1}{t_0} dt = 1\,\text{N m}/\text{s} \tag{3.484}$$

When $t_0 \to 0$, the height of the function approaches infinity, and the integral of the function remains unity. Such function with unit impulse is denoted as *delta function* $\delta(t)$ and is illustrated in Fig. 3.33b. It is zero everywhere except at $t = 0$ at which Eq. (3.485) holds. The delta function is a mathematical concept which is easier to be introduced graphically as is shown in Fig. 3.33. In mechanical vibrations, the function $\delta(t)$ is assumed to have unit impulse, and hence it is called *the unit impulse force* and is defined by an integral:

$$\int_{-\infty}^{\infty} \delta(t)\, dt = \int_{0^-}^{0^+} \delta(t)\, dt = 1 \tag{3.485}$$

$$\delta(t) = \begin{cases} \infty & t = 0 \\ 0 & t \ne 0 \end{cases} \tag{3.486}$$

In case $\delta(t)$ is applied at another time, $t = \tau$, its definition would be the same with a shift in time:

$$\int_{-\infty}^{\infty} \delta(t - \tau)\, dt = \int_{\tau^-}^{\tau^+} \delta(t - \tau)\, dt = 1 \tag{3.487}$$

$$\delta(t) = 0 \qquad t \ne \tau \tag{3.488}$$

The unit impulse force function is also called the *Dirac delta function*. The Dirac function is not similar to traditional functions, and its integrals do not exist in the classical mathematics.

Multiplying any function $g(t)$ by $\delta(t)$ makes the integral (3.485) to provide the value of $g(t)$ at the time $\delta(t)$ applies:

$$\int_{-\infty}^{\infty} g(t)\,\delta(t-\tau)\,dt = g(\tau) \tag{3.489}$$

It should be noted that the unit impulse function is not necessarily a force function. It could be defined for any variable, such as displacement and velocity.

Assume a unit impulse function is applied on (m, c, k) vibrating system:

$$m\ddot{x} + c\dot{x} + kx = \begin{cases} \delta(t) & t = 0 \\ 0 & t \neq 0 \end{cases} \tag{3.490}$$

To determine the response of a vibrating system to a unit impulse force $\delta(t)$, we assume the system has zero initial conditions before applying $\delta(t)$:

$$x\left(0^-\right) = 0 \qquad \dot{x}\left(0^-\right) = 0 \tag{3.491}$$

Because the unit impulse force $\delta(t)$ is applied at a point with zero duration of time, we introduce the notations 0^- and 0^+ to denote an infinitesimal time before and after $t = 0$. The unit impulse force $\delta(t)$ is applied at $t = 0$, and it is over after $0^+ \leq t$. The system would be unforced after $t = 0^+$ performing a free vibrations. Practically there is no difference between 0^- and 0^+. Therefore, the zero initial condition and Eq. (3.490) are equivalent to the free vibration equation $m\ddot{x} + c\dot{x} + kx = 0$ along with new initial conditions. Hence, the unit impulse force function $\delta(t)$ will only change the initial conditions of the system. To find the initial conditions at $t = 0^+$, we integrate the equation of motion and use the definition (3.485).

Twice taking the integrals of Eq. (3.490) shows that:

$$m\left(x\left(0^+\right) - x\left(0^-\right)\right)$$
$$+ \int_{0^-}^{0^+} cx\,dt + \int_{0^-}^{0^+}\int_{0^-}^{0^+} kx\,dt\,dt = \int_{0^-}^{0^+}\int_{0^-}^{0^+} \delta(t)\,dt\,dt \tag{3.492}$$

All integrals on the left-hand side are normal integrals at a point and hence are equal to zero. The first integral of $\delta(t)$ on the right-hand side of the equality sign gives a constant. Then the next integral would be an integral of regular constant functions at a point which is equal to zero. Therefore Eq. (3.492) simplifies to an equation for initial position x before and after the $\delta(t)$.

$$x\left(0^+\right) = x\left(0^-\right) = 0 \tag{3.493}$$

One integral of Eq. (3.490) shows that:

$$m \left(\dot{x} \left(0^+ \right) - \dot{x} \left(0^- \right) \right) + c \left(x \left(0^+ \right) - x \left(0^- \right) \right)$$

$$+ \int_{0^-}^{0^+} kx \, dt = \int_{0^-}^{0^+} \delta \left(t \right) dt \tag{3.494}$$

Therefore, the zero initial condition at $t = 0$ before and after $\delta \left(t \right)$ is:

$$\dot{x} \left(0^- \right) = 0 \qquad \dot{x} \left(0^+ \right) = \frac{1}{m} \tag{3.495}$$

Hence, $\delta \left(t \right)$ changes the problem of:

$$m\ddot{x} + c\dot{x} + kx = \delta \left(t \right) \qquad x \left(0^- \right) = 0 \qquad \dot{x} \left(0^- \right) = 0 \tag{3.496}$$

into the new problem of free vibration with nonzero initial conditions:

$$m\ddot{x} + c\dot{x} + kx = 0 \qquad x \left(0^+ \right) = 0 \qquad \dot{x} \left(0^+ \right) = \frac{1}{m} \tag{3.497}$$

Substituting the initial conditions in the solution (3.20) of free vibrations provides us with the response to a unit impulse force:

$$x = X e^{-\xi \omega_n t} \sin \left(\omega_d t + \varphi \right) \tag{3.498}$$

$$X = \sqrt{x_0^2 + \left(\frac{\dot{x}_0 + \xi \omega_n x_0}{\omega_d} \right)^2} = \frac{1}{m \omega_d} \tag{3.499}$$

To distinguish this special solution associated to the special function $\delta \left(t \right)$, we denote it by $h \left(t \right)$:

$$h \left(t \right) = \frac{1}{m \omega_d} e^{-\xi \omega_n t} \sin \omega_d t \tag{3.500}$$

If the unit impulse force applied at time $t = \tau$, then the response to a unit impulse force will have a time shift, $h \left(t - \tau \right)$, where $h \left(t - \tau \right) = 0$ for $t < \tau$:

$$h \left(t - \tau \right) = \frac{1}{m \omega_d} e^{-\xi \omega_n \left(t - \tau \right)} \sin \omega_d \left(t - \tau \right) \tag{3.501}$$

To determine the response of a mass-spring-damper system to a given force $f \left(t \right)$, we assume the force function to be replaced by a series of pulses as is shown in Fig. 3.34. If the response of the system to a pulse is known, then the response to $f \left(t \right)$ will be obtained by summation of the responses to each pulse in the sequence. Increasing the number of pulses approximates them by impulses and converts the

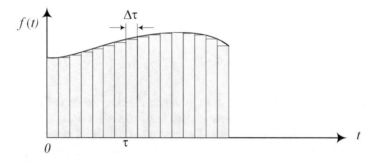

Fig. 3.34 Discretization of a force function $f(t)$ by a series of pulses

summation to an integral. The strength of a pulse is defined by the pulse area. So, the strength of a pulse at $t = \tau$ is $I = f(\tau)\Delta\tau$ in which $f(\tau)$ is assumed to remain constant during $\Delta\tau$. The response to a pulse is the product of the response to a unit impulse force multiplied by the pulse strength, $h(t - \tau)f(\tau)\Delta\tau$:

$$x(t) = h(t - \tau) f(\tau) \Delta\tau \tag{3.502}$$

Superposing the responses of the series of pulses is the response of the system to $f(\tau)$:

$$x(t) = \sum h(t - \tau) f(\tau) \Delta\tau \tag{3.503}$$

As $\Delta\tau \to 0$, the summation becomes integral:

$$x(t) = \int_0^t h(t - \tau) f(\tau) d\tau$$

$$= \frac{1}{m\omega_d} \int_0^t e^{-\xi\omega_n(t-\tau)} f(\tau) \sin \omega_d(t - \tau) \, d\tau \tag{3.504}$$

This is called the *convolution integral* that provides us with the response of the system to $f(\tau)$ with zero initial conditions. The convolution integral can also be equivalently written as:

$$x(t) = \int_0^t h(\tau) f(t - \tau) d\tau$$

$$= \frac{1}{m\omega_d} \int_0^t e^{-\xi\omega_n\tau} f(t - \tau) \sin \omega_d\tau \, d\tau \tag{3.505}$$

When the initial conditions are not zero, we add the homogeneous response of the system to the convolution integral:

$$x = e^{-\xi\omega_n t}\left(\frac{\dot{x}_0 + \xi\omega_n x_0}{\omega_d}\sin\omega_d t + x_0\cos\omega_d t\right)$$

$$+\frac{1}{m\omega_d}\int_0^t e^{-\xi\omega_n(t-\tau)}f(\tau)\sin\omega_d(t-\tau)\,d\tau \tag{3.506}$$

To adapt Equation (3.506) for nonzero initial conditions, we do not need to add a homogeneous solution with unknown coefficient, $e^{-\xi\omega_n t}(A\sin\omega_d t + B\cos\omega_d t)$, and then apply the initial conditions. The convolution integral always provides us with zero initial conditions, as shown below, and will not affect the coefficients A and B. Therefore (3.506) is valid for any function $f(t)$ and for any initial conditions $x(0) = x_0$, and $\dot{x}(0) = \dot{x}_0$.

The initial position of the convolution integral is zero:

$$x(0) = \int_0^0 h(t-\tau)f(\tau)\,d\tau = 0 \tag{3.507}$$

Employing the Leibniz formula:

$$\frac{d}{dt}\int_{a(t)}^{b(t)} f(t,\tau)\,d\tau = f(t,b(t))\frac{d}{dt}b(t) - f(t,a(t))\frac{d}{dt}a(t)$$

$$+\int_{a(t)}^{b(t)}\frac{\partial}{\partial t}f(t,\tau)\,d\tau \tag{3.508}$$

the initial velocity of the convolution integral is also zero:

$$\dot{x}(0) = \frac{d}{dt}\left(\int_0^t h(t-\tau)f(\tau)\,d\tau\right)$$

$$= h(0)f(0) - h(0)f(0) + \int_0^0 f(\tau)\frac{d}{dt}h(t-\tau)\,d\tau = 0 \tag{3.509}$$

The solutions (3.501) and (3.506) are only valid for an underdamped system and for the time t during the application of $f(t)$. As soon as $f(t)$ disappears, a new solution of free vibration occurs. The initial conditions of the free vibration would be the state of the system at the end of duration of application of $f(t)$.

The function $h(t)$ for critically and overdamped systems are:

$$h(t) = \frac{t}{m}e^{-\omega_n t}\qquad \xi = 1 \tag{3.510}$$

$$h(t) - \frac{e^{-\xi\omega_n t}}{m\omega_n\sqrt{\xi^2-1}}\sinh\left(\omega_d\sqrt{\xi^2-1}t\right)\qquad \xi > 1 \tag{3.511}$$

A dimensional analysis is needed to clarify correctness of dimensional balance of the outcomes. Equation (3.489) indicates that the dimension of the function $\delta(t)$ is T^{-1}. Therefore, the function $\delta(t)$ in Eq. (3.490) must have a constant coefficient I of dimension impulse=force×time:

$$m\ddot{x} + c\dot{x} + kx = I\delta(t) \tag{3.512}$$

$$I = 1 \qquad [I] = MLT^{-1} \tag{3.513}$$

That makes Eq. (3.494) to be equal to $I = 1$ with dimension MLT^{-1}. When an impulse, I, is applied to a one degree of freedom system, Newton's second law gives:

$$I = F\Delta t = m\Delta\dot{x} \tag{3.514}$$

$$m\left(\dot{x}\left(0^{+}\right) - \dot{x}\left(0^{-}\right)\right) = I \tag{3.515}$$

Therefore, the dimension of the solution $h(t)$ to impulse in Eq. (3.500) would be displacement:

$$h(t) = \frac{I}{m\omega_d}e^{-\xi\omega_n t}\sin\omega_d t \tag{3.516}$$

$$[h(t)] = \frac{MLT^{-1}}{MT^{-1}} = L \tag{3.517}$$

Therefore, $[I\delta]$ is force, and $[Ih]$ is displacement. These make $[\delta] = T^{-1}$ and $[h] = L$, where $[I] = MLT^{-1}$ is and impulse of a force function $I = \int f(t)\,dt$.

The convolution integral method can also be used on first-order system to determine the transient response easier. Consider a set of coupled first-order systems with a transient forcing function $f(t)$:

$$\dot{\mathbf{x}} + \mathbf{A}\,\mathbf{x} = \mathbf{b}\,f(t) \qquad \mathbf{x}(0) = \mathbf{x}_0 \tag{3.518}$$

The solution of the set of equations will be:

$$\mathbf{x}(t) = e^{-\mathbf{A}t}\,\mathbf{x}_0 + \int_0^t e^{-\mathbf{A}t}\,\mathbf{b}\,f(t - \tau)\,d\tau \tag{3.519}$$

$$= e^{-\mathbf{A}t}\,\mathbf{x}_0 + \int_0^t e^{-\mathbf{A}(t-\tau)}\,\mathbf{b}\,f(\tau)\,d\tau \tag{3.520}$$

The unit impulse function was first introduced by the German physicist Gustav Robert Kirchhoff (1824–1887) and then by the British physicist Paul Adrien Maurice Dirac (1902–1984). Any function which satisfies the definition (3.485) is a Dirac delta function. It does not necessarily need to be the limit of a rectangular pulse.

Convolution integral is also called Duhamel integral. Jean Marie Constant Duhamel (1797–1872) was a French mathematician and physicist who proposed a theory related to the transmission of heat in crystal structures.

■

Example 111 Step input. The convolution integral is a strong method of calculating response of vibrating system to transient forces. It gives us the flexibility to determine the solution to complicated force functions by superposition of solutions to few simple forces. Here is to introduce the Heaviside function and show how the solution to a step input will be calculated and shifted in time.

Assume an (m, k, c)-system is under the step force input of Fig. 3.35a:

$$m\ddot{x} + c\dot{x} + kx = f(t) \tag{3.521}$$

$$f(t) = F \tag{3.522}$$

A step input is a continuos function, and hence the method of homogenous plus particular solutions, $x = x_h + x_p$, works well. That method has been used to determine the general solution (3.367) and:

$$
\begin{aligned}
x &= x_h + x_p \\
&= e^{-\xi \omega_n t}(A \sin \omega_d t + B \cos \omega_d t) + \frac{F}{m\omega_n^2} \qquad 0 \le t
\end{aligned}
\tag{3.523}
$$

$$\omega_d = \omega_n \sqrt{1 - \xi^2} \tag{3.524}$$

or Eq. (3.371) for zero initial conditions:

$$\frac{x}{F/k} = 1 - e^{-\xi \omega_n t}\left(\frac{\xi \omega_n}{\omega_d} \sin \omega_d t + \cos \omega_d t\right) \tag{3.525}$$

To determine the response of the system, we may also use the convolution integral (3.504) and derive the same solution (3.525):

$$
\begin{aligned}
x(t) &= \frac{1}{m\omega_d} \int_0^t e^{-\xi \omega_n \tau} f(t - \tau) \sin \omega_d \tau \, d\tau \\
&= \frac{F}{m\omega_d} \int_0^t e^{-\xi \omega_n \tau} \sin \omega_d \tau \, d\tau \\
&= \frac{F}{k}\left(1 - e^{-\xi \omega_n t}\left(\frac{\xi}{\sqrt{1 - \xi^2}} \sin \omega_d t + \cos \omega_d t\right)\right) \\
& \qquad\qquad 0 < t
\end{aligned}
\tag{3.526}
$$

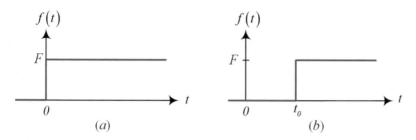

Fig. 3.35 A step force function: (a) at $t = 0$, (b) at $t = t_0$

The response of the system to the lagged step input of Fig. 3.35b is the same as (3.526) with a shift t_0 in time and valid for $t_0 < t$:

$$x(t) = \frac{1}{m\omega_d} \int_0^t e^{-\xi\omega_n(t-\tau)} f(\tau) \sin\omega_d(t-\tau) \, d\tau$$

$$= \int_0^t 0 \, d\tau + \frac{F}{m\omega_d} \int_{t_0}^t e^{-\xi\omega_n(t-\tau)} \sin\omega_d(t-\tau) \, d\tau$$

$$= \frac{F}{m} \frac{1}{\omega_d^2 + \xi^2\omega_n^2} - \frac{Fe^{-\xi\omega_n(t-t_0)}}{m\omega_d} \frac{\omega_d \cos\omega_d(t-t_0) + \xi\omega_n \sin\omega_d(t-t_0)}{\omega_d^2 + \xi^2\omega_n^2}$$

$$= \frac{F}{k} \left(1 - e^{-\xi\omega_n(t-t_0)} \left(\frac{\xi\omega_n}{\omega_d} \sin\omega_d(t-t_0) + \cos\omega_d(t-t_0) \right) \right) \qquad (3.527)$$

$$t_0 < t$$

However, to express the response by a continuous function starting from $t = 0$, we employ the Heaviside function $H(t - t_0)$ to write the solution to the step input of Fig. 3.35b:

$$x(t) = \frac{F}{k} \left(1 - e^{-\xi\omega_n(t-t_0)} \left(\frac{\xi \sin\omega_d(t-t_0)}{\sqrt{1-\xi^2}} + \cos\omega_d(t-t_0) \right) \right) H(t-t_0)$$

$$(3.528)$$

$H(t - t_0)$ is the Heaviside function or the unit step function that acts as an on-off switch in mathematical equations:

$$H(t - t_0) = \begin{cases} 1 & t_0 \le t \\ 0 & t < t_0 \end{cases} \qquad (3.529)$$

The response to a step input at $t_0 = 1$ is shown in Fig. 3.36.

As a more complicated step input, let us determine the response of the system to a double step force input as follows:

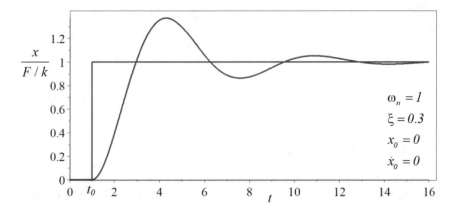

Fig. 3.36 The response of a second-order system to a step input at $\tau = 1$

$$f(t) = \begin{cases} 0 & 0 < t < t_0 \\ F_0 & t_0 < t < t_1 \\ F_1 & t_1 < t \end{cases} \tag{3.530}$$

The response of an underdamped (m, c, k) would be:

$$x = \int_0^t h(\tau) f(t - \tau) \, d\tau = \int_0^t h(\tau) 0 \, d\tau = 0 \qquad 0 < t < t_0 \tag{3.531}$$

$$x = \int_0^t h(\tau) 0 \, d\tau + \int_{t_0}^t h(\tau) F_0 \, d\tau$$

$$= 0 + F_0 \int_{t_0}^t h(\tau) \, d\tau \qquad t_0 < t < t_1 \tag{3.532}$$

$$x = 0 + F_0 \int_{t_0}^t h(\tau) \, d\tau + (F_1 - F_0) \int_{t_1}^t h(\tau) \, d\tau \qquad t_1 < t \tag{3.533}$$

Employing Heaviside functions, all these solutions can be combined and expressed by one equation:

$$x = \int_0^t h(\tau) f(t - \tau) \, d\tau + H(t - t_0) \int_{t_0}^t h(\tau) f(t - \tau) \, d\tau$$

$$+ H(t - t_1) \int_{t_1}^t h(\tau) f(t - \tau) \, d\tau \tag{3.534}$$

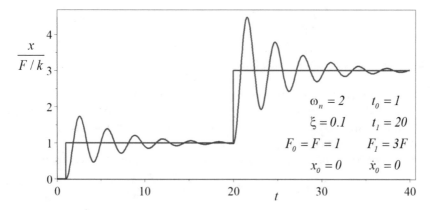

Fig. 3.37 The response of an underdamped (m, c, k) system to two-step force function

$$= 0 + H(t - t_0) F_0 \int_{t_0}^{t} h(\tau) f(t - \tau) \, d\tau$$

$$+ H(t - t_1)(F_1 - F_0) \int_{t_1}^{t} h(\tau) f(t - \tau) \, d\tau \qquad (3.535)$$

Figure 3.37 illustrates the response of the system to two-step force function (3.530) for a set of data.

The response to step input for critically damped and overdamped systems will be:

$$x(t) = \frac{F}{k} \left(1 - (1 - \omega_n t) e^{-\omega_n t} \right) \qquad 0 < t \qquad \xi = 1 \qquad (3.536)$$

$$x(t) = \frac{F}{k} \left(1 - \frac{1}{2} e^{-\xi \omega_n t} \left(e^{\sqrt{1 - \xi^2} \omega_n t} - e^{-\sqrt{1 - \xi^2} \omega_n t} \right) \right) \qquad 0 < t \qquad 1 < \xi \qquad (3.537)$$

The Dirac delta function is the derivative of the Heaviside function:

$$\delta(t - t_0) = \frac{d}{dt} H(t - t_0) \qquad (3.538)$$

The derivative of Dirac delta function is called the doublet function:

$$\dot{\delta}(t - t_0) = \frac{d}{dt} \delta(t - t_0) \qquad (3.539)$$

The unit of the doublet is, s^{-2}, frequency-squared. The doublet can be used to express inputs that cause the position of the mass of a (m, c, k) system to change instantaneously without any change in its velocity. The triplet, the quadruplet, etc.

are similarly defined. In general, the $(n-1)$-derivative of the unit impulse, denoted by $\delta^{(n-1)}(t)$, is termed the n-tuplet function.

The British mathematician and physicist Oliver Heaviside (1850–1925) introduced and applied the Heaviside function in telegraphic communications.

Example 112 A negative ramp triangular force. This is an example to use the general method of analysis of $x = x_h + x_p$ to determine the response of a vibrating system to a complicated transient force. Solving the same problem by convolution integral illustrates how more practical convolution integral is.

Consider a vibrating (m, c, k) system under a negative ramp triangular force:

$$m\ddot{x} + c\dot{x} + kx = f(t) \qquad x(0) = 0 \qquad \dot{x}(0) = 0 \qquad (3.540)$$

$$f(t) = \begin{cases} F\left(1 - \frac{t}{t_0}\right) & 0 < t < t_0 \\ 0 & t_0 \leq t \end{cases} \qquad (3.541)$$

The force varies linearly from a value of F at $t = 0$ to zero at $t = t_0$. Let us transform the equation of motion to a more practical form, by dividing over m:

$$\ddot{x} + 2\xi\omega_n\dot{x} + \omega_n^2 x = \frac{1}{m}f(t) \qquad (3.542)$$

Assuming the system to be underdamped, the homogenous solution x_h for $f = 0$ is given in (3.19). The particular solution x_p will be a first-order polynomial as the forcing function is:

$$x_p = C_1 + C_2 t \qquad (3.543)$$

We substitute the guessed function of x_p into the differential equation (3.542). Sorting the equation for powers of t will provide us with two equations to determine the coefficients C_1 and C_2:

$$cC_2 + k(C_1 + C_2 t) = F\left(1 - \frac{t}{t_0}\right) \qquad (3.544)$$

$$\left(kC_2 + \frac{F}{t_0}\right)t + (cC_2 - F + kC_1) = 0 \qquad (3.545)$$

$$C_2 = -\frac{F}{kt_0} \qquad C_1 = \frac{F}{k}\left(1 + \frac{c}{kt_0}\right) = \frac{F}{k}\left(1 + \frac{2\xi}{\omega_n t_0}\right) \qquad (3.546)$$

Therefore the particular solution x_p and the general solution x of the differential equation will be found:

$$x_p = \frac{F}{k}\left(1 + \frac{2\xi}{\omega_n t_0} - \frac{t}{t_0}\right) \qquad (3.547)$$

$$x_1 = x = x_h + x_p$$

$$= e^{-\xi \omega_n t} (A \sin \omega_d t + B \cos \omega_d t) + \frac{F}{k} \left(1 + \frac{2\xi}{\omega_n t_0} - \frac{t}{t_0} \right) \qquad (3.548)$$

$$0 < t < t_0$$

Let us name this solution x_1 to indicate the first part of the solution for the period that the forcing term applies. It is the time to use the initial conditions and determine the coefficients A and B:

$$x(0) = 0 \qquad B = -\frac{F}{k} \left(1 + \frac{2\xi}{\omega_n t_0} \right) \qquad (3.549)$$

$$\dot{x}(0) = 0 \qquad A = -\frac{F}{k \omega_d t_0} \left(\xi \omega_n t_0 + 2\xi^2 - 1 \right) \qquad (3.550)$$

The solution (3.548) with coefficients (3.549) and (3.550) is valid only for the period of time $0 < t < t_0$ that the force is applied. After $t_0 \leq t$ there will be no external force, and the system will follow free vibrations with similar equation as x_h, with a shift of t_0 in time. We show this solution by x_2 to indicate the second part of the response after the forcing terms vanishes:

$$x_2 = e^{-\xi \omega_n (t - t_0)} (C \sin \omega_d (t - t_0) + D \cos \omega_d (t - t_0)) \qquad t_0 \leq t \qquad (3.551)$$

$$\dot{x}_2 = e^{-\xi \omega_n (t - t_0)} (C \omega_d - \xi \omega_n D) \cos \omega_d (t - t_0)$$

$$- e^{-\xi \omega_n (t - t_0)} (\omega_d D + C \xi \omega_n) \sin \omega_d (t - t_0) \qquad (3.552)$$

The coefficients C and D will be determined by the conditions of $x_2 = x_1$ at t_0:

$$D = x_2(t_0) \qquad C = \frac{\dot{x}_2(t_0) + \xi \omega_n D}{\omega_d} \qquad (3.553)$$

The conditions of $x_2(t_0)$, $\dot{x}_2(t_0)$ must be equal to $x_1(t_0)$ and $\dot{x}_1(t_0)$ which can be found from x_1 at $t = t_0$:

$$x_1(t_0) = x_2(t_0) = e^{-\xi \omega_n t_0} (A \sin \omega_d t_0 + B \cos \omega_d t_0) + \frac{F}{k} \frac{2\xi}{\omega_n t_0} \qquad (3.554)$$

$$\dot{x}_1(t_0) = \dot{x}_2(t_0) = \frac{e^{-\xi \omega_n t_0}}{k t_0} (A k \omega_d t_0 - B k \xi \omega_n t_0) \cos \omega_d t_0$$

$$- \frac{e^{-\xi \omega_n t_0}}{k t_0} (B k \omega_d t_0 + A k \xi \omega_n t_0) \sin \omega_d t_0 - \frac{F}{k t_0} \qquad (3.555)$$

The value of A and B from (3.549) and (3.550) will be substituted in (3.554) and (3.555). They will go into (3.553) to calculate C and D, to be substituted

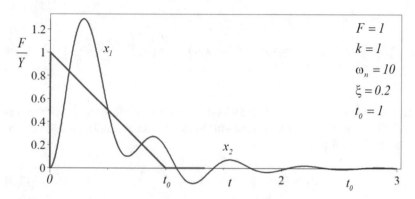

Fig. 3.38 A sample of responses of a vibrating system to a transient force $f(t) = F(1 - t/t_0)$ for a set of data

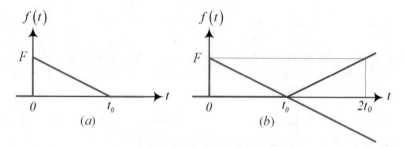

Fig. 3.39 (a) A negative ramp triangular force. (b) Equivalent superposition functions

in (3.551) to determine x_2. Figure 3.38 illustrates a sample of responses x_1 and x_2 of the system for a set of data.

Let us also solve this problem by convolution integral. Using superposition, we can simulate the transient force of (3.541) by addition of two continuous functions as are shown in Fig. 3.39b. Therefore, the response to the transient force of Fig. 3.39a can be calculated by using the convolution integral as follows:

$$
x = \int_0^{t_0} h(\tau) f(t - \tau) \, d\tau = \frac{1}{m\omega_d} \int_0^t e^{-\xi\omega_n \tau} f(t - \tau) \sin \omega_d \tau \, d\tau
$$

$$
= F \int_0^t h(\tau) \left(1 - \frac{t}{t_0}\right) d\tau
$$

$$
+ H(t - t_0) F \int_{t_0}^t h(\tau) \left(\frac{t}{t_0} - 1\right) d\tau \tag{3.556}
$$

Integrating and combining them by a Heaviside function yields:

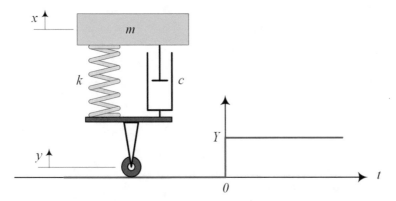

Fig. 3.40 A vibrating model of a vehicle to have a step displacement under its tire

$$x(t) = \frac{-Fe^{-\xi\omega_n t}}{m\,\omega_d^2\,\omega_n^4 t_0}\left(\left(\omega_n^3 t_0 + \omega_n^2\left(2\xi^2 - 1\right)\right)\sin\omega_d t\right.$$

$$\left. +\omega_d\left(\xi^2\omega_n^2 t_0 + \omega_d^2 t_0 + 2\xi\omega_n\right)\cos\omega_d t + \omega_d\left(\omega_n^2(t - t_0) - 2\xi\omega_n\right)\right)$$

$$-\frac{Fe^{-\xi\omega_n(t - t_0)}H(t - t_0)}{m\,\omega_d^2\,\omega_n^4 t_0}\left(\omega_n^2\left(2\xi^2 - 1\right)\sin\omega_d(t - t_0)\right.$$

$$\left. +2\xi\omega_n\omega_d\cos\omega_d(t - t_0) + \omega_d\left(\omega_n^2(t - t_0) - 2\xi\omega_n\right)\right) \tag{3.557}$$

As an engineering application, we can determine the maximum displacement of the system. Considering the maximum displacement happens during application of the transient force (3.541), we should determine the time at which $\dot{x}_1 = 0$ and calculate x_1 at the first nonzero time of equation $\dot{x}_1 = 0$.

Example 113 A vehicle goes over a step. A practical example of step input for a vehicle climbing a step. The input is a displacement, and we need to show it will be converted to a force input.

Figure 3.40 illustrates a vibrating model of a vehicle to have a step displacement under its tire. The equation of motion of the system is:

$$m\ddot{x} + c\dot{x} + kx = f(t) \qquad f(t) = c\dot{y} + ky \tag{3.558}$$

$$y = YH(t) \qquad \dot{y} = Y\frac{d}{dt}H = Y\delta(t) \tag{3.559}$$

The solution for displacement of the body of the vehicle has two parts due to $\delta(t)$ and $H(t)$:

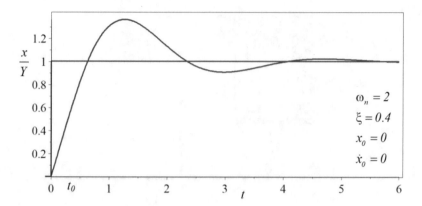

Fig. 3.41 The response of a vehicle going over a step displacement under its tire

$$x = cY h\,(\tau) + kY \int_0^t h\,(\tau)\, f\,(t - \tau)\; d\tau$$

$$= \frac{cY}{m\omega_d} e^{-\xi\omega_n t} \sin \omega_d t$$

$$+ Y \left(1 - e^{-\xi\omega_n t} \left(\frac{\xi}{\sqrt{1 - \xi^2}} \sin \omega_d t + \cos \omega_d t \right) \right) \qquad (3.560)$$

Figure 3.41 illustrates the response for a set of data. The graph looks very much similar to a step input of Fig. 3.36; however, the difference is in the slope of the response at the moment excitation is applied. The slope at t_0 for the step input in Fig. 3.36 is zero. The slope of the vehicle response at $t = 0$ is not zero because the unit impulse force $\delta\,(t)$ term in $c\dot{y} = cY\delta\,(t)$ changes the initial velocity that makes the slope not to be zero.

Example 114 ★ Impulse. It is a review of the concept and application of impulse in dynamics and vibrations.

Assume a constant force F act on a particle of mass m for a time t. In this time the velocity of the particle changes from v_1 to v_2. The constant force will give the particle a constant acceleration $a = (v_2 - v_1)/t$ in the direction of the force. The force is equal the product of the mass and the acceleration:

$$F = m \frac{v_2 - v_1}{t} \qquad (3.561)$$

The product Ft is known as the impulse of the force F acting for time t. The impulse is then equal to change in momentum $p = mv$:

$$Ft = \Delta p = mv_2 - mv_1 \qquad (3.562)$$

The Newton law of motion is a differential equation. However, we may also express the law of motion with an integral equation:

$$mv = \int_{t_1}^{t_2} \mathbf{F}\,(\mathbf{v}, \mathbf{r}, t)\, dt + mv(t_1) \tag{3.563}$$

where $\mathbf{v}(t_1) = \mathbf{v}_1$ is the initial velocity vector. The integral is the impulse of the force $\mathbf{F}$ in that time interval. This equation states that the change of momentum $\Delta\mathbf{p} = m\mathbf{v}_2 - m\mathbf{v}_1$ of a particle during the time interval $[t_1, t_2]$ is equal to the impulse of the resultant force over the same interval. The differential form of the second Newton law of motion or force equals the time rate of momentum has some disadvantages over the integral form (3.563) which is impulse equals momentum change. The position of a particle must be continuous; however, its velocity might have discontinuities. At a velocity discontinuity, the acceleration is not defined, so at such instants, the differential form of the second law is not valid; however, the integral form (3.563) is valid at these instants.

The temporal integral of the Newton equation of motion:

$$\int_1^2 \mathbf{F}\, dt = m \int_1^2 \mathbf{a}\, dt \tag{3.564}$$

reduces to the principle of impulse and momentum:

$$_1\mathbf{I}_2 = \mathbf{p}_2 - \mathbf{p}_1 \tag{3.565}$$

where $\mathbf{p}$ is called the momentum:

$$\mathbf{p} = \frac{1}{2}m\,{}^G\mathbf{v} \tag{3.566}$$

and $_1\mathbf{I}_2$ is the impulse of the force during the time interval $t_2 - t_1$:

$$_1\mathbf{I}_2 = \int_1^2 {}^G\mathbf{F}\, dt \tag{3.567}$$

When, for example, a ball is struck by a bat, a large force acts on the ball for a very short time. It is not easy to measure the force or the very short time, but by measuring the change in momentum of the ball, the impulse of the force can be measured.

Example 115 Pulse input. A pulse input is a step input for a limited period of time. Having the solution for step input enables us to determine the solution any type of pulses inputs by superposition. Here is the to show how we superpose step input solution to determine solution to a pulse input.

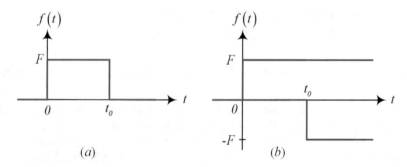

Fig. 3.42 (a) A pulse input, (b) superposition of two-step functions to make the pulse input

Let us determine the response of a mass-spring-damper system to a pulse excitation as is shown in Fig. 3.42a:

$$f(t) = \begin{cases} F & 0 < t < t_0 \\ 0 & t_0 < t \end{cases} \tag{3.568}$$

The pulse is equal to a summation of two-step functions $f(t) = f_1(t) + f_2(t)$ as shown in Fig. 3.42b. Because of linearity of the system, its response to $f(t) = f_1(t) + f_2(t)$ is equal to the sum of its responses to $f_1(t)$ and $f_2(t)$ individually. The response to the step input $f_1(t)$ is given in (3.526), and the response to the lagged step input $f_2(t)$ is given in (3.528). Therefore, the response of an (m, k, c)-system to a pulse excitation will be:

$$x = \int_0^t h(\tau) f(t - \tau) \, d\tau = F \int_0^t h(\tau) \, d\tau \qquad 0 < t < t_0 \tag{3.569}$$

$$x = F \int_0^t h(\tau) \, d\tau - F \int_{t_0}^t h(\tau) \, d\tau \qquad t_0 < t \tag{3.570}$$

Integrating and combining them by a Heaviside function yields:

$$x(t) = \frac{F}{k} \left(1 - e^{-\xi \omega_n t} \left(\frac{\xi \omega_n}{\omega_d} \sin \omega_d t + \cos \omega_d t \right) \right)$$
$$- \frac{F}{k} \left(1 - e^{-\xi \omega_n (t - t_0)} \left(\frac{\xi \sin \omega_d (t - t_0)}{\sqrt{1 - \xi^2}} + \cos \omega_d (t - t_0) \right) \right) H(t - t_0) \tag{3.571}$$

Figure 3.43 illustrates a sample of a time response to a pulse excitation.

The method of superposition helps to develop the solution of a complicated excitation by using the responses to fundamental excitations.

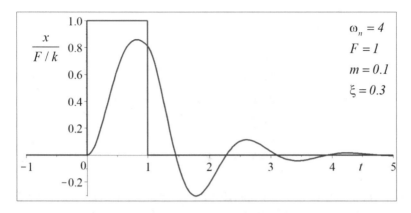

Fig. 3.43 Time response of a one DOF system to a pulse excitation

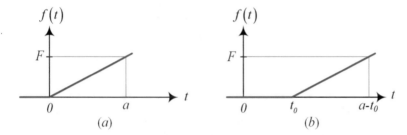

Fig. 3.44 A ramp force function. (**a**) at $t = 0$, (**b**) at $t = t_0$

Example 116 Ramp input. After step input, the ramp input is the second most important excitation to identify dynamic systems' characteristics. Combination of ramp and step input enables us to determine the solutions to a vast domain of transient responses. Here is to use convulsion integral and derive the solution of a (m, c, k)-system to ramp and lagged ramp inputs.

In addition to a step input, the ramp is another fundamental excitation which can be used in superposition methods and breaking down complicated excitations. Figure 3.44a illustrates a ramp excitation:

$$f(t) = \begin{cases} \dfrac{F}{a}t & 0 < t \\ 0 & t < 0 \end{cases} \tag{3.572}$$

The response of a mass-spring-damper system to a ramp input is the result of its convolution integral (3.504) for the ramp function (3.572):

$$x(t) = \frac{1}{m\omega_d} \int_0^t e^{-\xi\omega_n(t-\tau)} f(\tau) \sin \omega_d (t-\tau) \, d\tau$$

$$= \frac{F}{a} \frac{1}{m\omega_d} \int_0^t e^{-\xi\omega_n(t-\tau)} \tau \sin \omega_d (t-\tau) \, d\tau$$

$$= \frac{Fe^{-\xi\omega_n t}}{ka\omega_n^3} \left(\frac{2\xi^2 - 1}{\sqrt{1-\xi^2}} \sin \omega_d t + 2\xi \cos \omega_d t \right)$$

$$+ \frac{F}{ka\omega_n^3} (\omega_n t - 2\xi) \tag{3.573}$$

If there is time lag of $t = a$ in applying the ramp input, as is shown in Fig. 3.44b, the convulsion integral provides similar solution with a shift t_0 in time:

$$f(t) = \begin{cases} \dfrac{F}{a} (t - t_0) & 0 < t \\ 0 & t < 0 \end{cases} \tag{3.574}$$

$$x(t) = \frac{1}{m\omega_d} \int_0^t e^{-\xi\omega_n(t-\tau)} f(\tau) \sin \omega_d (t-\tau) \, d\tau$$

$$= \int_0^{t_0} 0 \, d\tau + \frac{F}{m a\omega_d} \int_{t_0}^t e^{-\xi\omega_n(t-\tau)} (\tau - t_0) \sin \omega_d (t-\tau) \, d\tau$$

$$= \frac{Fe^{-\xi\omega_n(t-t_0)}}{ka\omega_n^3} \left(\frac{2\xi^2 - 1}{\sqrt{1-\xi^2}} \sin \omega_d (t - t_0) + 2\xi \cos \omega_d (t - t_0) \right)$$

$$+ \frac{F}{ka\omega_n^3} (\omega_n (t - t_0) - 2\xi) \qquad t_0 < t \tag{3.575}$$

A sample of lagged ramp force and the response of a system are shown in Fig. 3.45.

Example 117 Ramp-step excitation. Here is to show how the vibrations of a (m, c, k) system to a tapered ramped force can be determined by superposing ramp solutions.

Figure 3.46a shows a tapered ramp excitation:

$$f(t) = \begin{cases} \dfrac{F}{t_0} t & 0 < t < t_0 \\ F & t_0 < t \end{cases} \tag{3.576}$$

Superposition is the simplest way to solve this problem. The ramp-step forcing function of Fig. 3.46a is equivalent to the superposing the responses to the two ramp inputs of Fig. 3.46b. For $0 < t < t_0$, there is a ramp with the solution (3.573) for $a = t_0$. At $t = t_0$, a negative lagged ramp with solution (3.575) for $a = t_0$ will be added to the excitation:

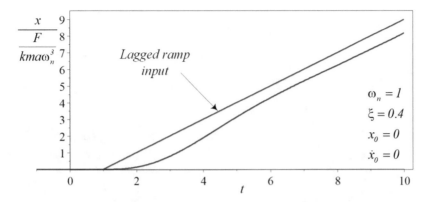

Fig. 3.45 A sample of lagged ramp force and the response of a one DOF system

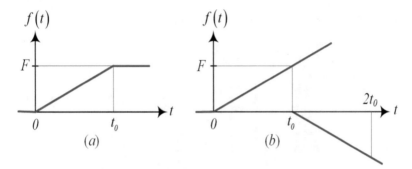

Fig. 3.46 (a) a tapered ramp excitation. (b) superposition of two ramp to make the tapered ramp excitation

$$x(t) = \frac{1}{m\omega_d} \int_0^t e^{-\xi\omega_n(t-\tau)} \frac{F}{t_0} \tau \sin\omega_d (t-\tau) \, d\tau$$

$$-\frac{F}{m\omega_d} H(t-t_0) \int_{t_0}^t e^{-\xi\omega_n(t-\tau)} \frac{F}{t_0} (\tau-t_0) \sin\omega_d (t-\tau) \, d\tau \quad (3.577)$$

Figure 3.47 illustrates a response to the ramp-step input for a set of data.

Example 118 Half sine transient base excitation. Half sine is a reasonable mathematical model for bumps on the roads to make vehicles to drive slowly. Here is the analytic calculation of the response of a vibrating model of vehicle to a bump.

The equation of motion of a base excited system in relative displacement z is:

$$m\ddot{z} + c\dot{z} + kz = -m\ddot{y} \qquad z = x - y \qquad (3.578)$$

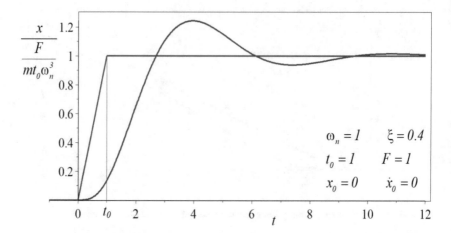

Fig. 3.47 A response to the ramp-step input for a set of data

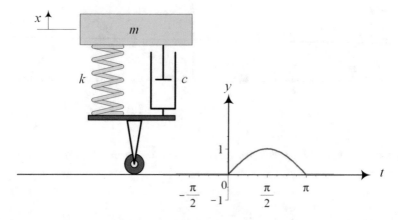

Fig. 3.48 A one DOF system with a half sine transient displacement base excitation

When the base excitation function $y = y(t)$ is given, then $-m\ddot{y}$ acts as an equivalent external force in the equation of motion. We may use the convolution integral to determine the response of the system. Figure 3.48 illustrates a one DOF vibrating model of vehicles along with a half sine transient displacement base excitation:

$$y = \begin{cases} Y \sin \omega t & 0 \leq t \leq t_0 \\ 0 & t_0 < t \end{cases} \qquad t_0 = \frac{\pi}{\omega} \qquad (3.579)$$

The relative displacement response $z(t)$ of the vehicle would be the result of the following convolution integral which is valid only for $0 < t < t_0$:

$$z(t) = \int_0^t h(t - \tau) f(\tau) d\tau \tag{3.580}$$

$$h(t) = \frac{1}{m\omega_n} e^{-\xi\omega_n t} \sin\omega_d t \tag{3.581}$$

$$f(t) = -m\ddot{y} = \begin{cases} mY\omega^2 \sin\omega t & 0 < t < t_0 \\ 0 & t_0 < t \end{cases} \qquad t_0 = \frac{\pi}{\omega} \tag{3.582}$$

Therefore:

$$
\begin{aligned}
z(t) &= \frac{Y\omega^2}{\omega_n} \int_0^t e^{-\xi\omega_n(t-\tau)} \sin\omega_d(t - \tau) \sin\omega\tau \, d\tau \\
&= \frac{Y\omega^2 e^{-\xi\omega_n t} \left(\frac{\omega}{\omega_d} (\xi^2\omega_n^2 + \omega^2 - \omega_d^2) \sin\omega_d t + 2\xi\omega_n\omega \cos\omega_d t \right)}{(\xi^2\omega_n^2 + (\omega - \omega_d)^2)(\xi^2\omega_n^2 + (\omega + \omega_d)^2)} \\
&\quad + \frac{Y\omega^2 \left((\xi^2\omega_n^2 - \omega^2 + \omega_d^2) \sin\omega t - 2\xi\omega_n\omega \cos\omega t \right)}{(\xi^2\omega_n^2 + (\omega - \omega_d)^2)(\xi^2\omega_n^2 + (\omega + \omega_d)^2)}
\end{aligned} \tag{3.583}
$$

$$0 < t < t_0$$

The response of the system for $t_0 < t$ will be a free vibrations with initial conditions at the end of the half sine excitation. The conditions of the system at $t = t_0$ are:

$$z(t_0) = 0 \tag{3.584}$$

$$
\begin{aligned}
\dot{z}(t_0) &= \frac{-Y\omega^3 (\xi^2\omega_n^2 - \omega^2 + \omega_d^2) e^{-\xi\omega_n t_0}}{(\xi^2\omega_n^2 + (\omega - \omega_d)^2)(\xi^2\omega_n^2 + (\omega + \omega_d)^2)} \cos\omega_d t_0 \\
&\quad - \frac{Y\omega^3\xi\omega_n (\xi^2\omega_n^2 + \omega^2 + \omega_d^2) e^{-\xi\omega_n t_0}}{\omega_d (\xi^2\omega_n^2 + (\omega - \omega_d)^2)(\xi^2\omega_n^2 + (\omega + \omega_d)^2)} \sin\omega_d t_0 \\
&\quad + Y\omega^3 \frac{(\xi^2\omega_n^2 - \omega^2 + \omega_d^2) \cos\omega t_0 + 2\xi\omega\omega_n \sin\omega t_0}{(\xi^2\omega_n^2 + (\omega - \omega_d)^2)(\xi^2\omega_n^2 + (\omega + \omega_d)^2)}
\end{aligned} \tag{3.585}
$$

and the response of the system after the bump excitation is over will be free vibrations:

$$z(t) = e^{-\xi\omega_n t} \left(\frac{\dot{z}(t_0) + \xi\omega_n z(t_0)}{\omega_d} \sin\omega_d t + z(t_0) \cos\omega_d t \right) \tag{3.586}$$

$$t_0 < t$$

To define the response by a continuous function, we denote the response in $0 < t < t_0$ by z_1 and for $t_0 < t$ by z_2. Then, using the Heaviside function, we have:

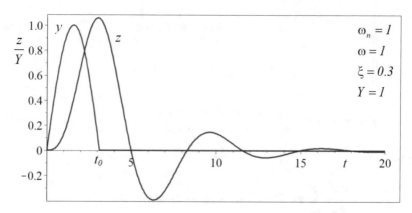

Fig. 3.49 The response of m in relative coordinate $z = x - y$, along with the excitation y

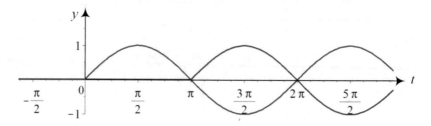

Fig. 3.50 Making a half sine excitation by superposition method

$$z = H(t_0 - t) \, z_1(t) + H(t - t_0) \, z_2(t - t_0) \tag{3.587}$$

Figure 3.49 illustrates the response of m in relative coordinate $z = x - y$, along with the excitation y. The absolute displacement x of m can be calculated by:

$$
\begin{aligned}
x &= z + y \\
&= H(t_0 - t) \, z_1(t) + H(t - t_0) \, z_2(t - t_0) \\
&\quad + H(t - t_0) \, Y \sin \omega t
\end{aligned}
\tag{3.588}
$$

Instead of the general method of solution, we may employ superposition method and model the half sine excitation with two sine functions as is shown in Fig. 3.50:

$$y = \begin{cases} Y \sin \omega t & 0 \le t \\ Y \sin \omega (t - t_0) & t_0 < t \end{cases} \quad t_0 = \frac{\pi}{\omega} \tag{3.589}$$

$$f(t) = -m\ddot{y} = \begin{cases} mY\omega^2 \sin \omega t & 0 \le t \\ mY\omega^2 \sin \omega (t - t_0) & t_0 < t \end{cases} \tag{3.590}$$

Therefore, the response of the vehicle can also be calculated by two-part convolution integrals:

$$
z(t) = \int_0^t h(t - \tau) f(\tau) d\tau
$$

$$
= \frac{Y\omega^2}{\omega_n} \int_0^t e^{-\xi\omega_n(t-\tau)} \sin \omega_d (t - \tau) \sin \omega\tau \, d\tau
$$

$$
+ \frac{Y\omega^2}{\omega_n} \int_{t_0}^t e^{-\xi\omega_n(t-\tau)} \sin \omega_d (t - \tau) \sin \omega (\tau - t_0) \, d\tau \qquad (3.591)
$$

Example 119 Exponentially decaying excitation. A rapidly disappearing step input is modeled by an Exponentially decaying excitation. Here is the response of a (m, c, k) system to such excitation.

Consider a force that suddenly jumps to F_0 and disappears exponentially:

$$
m\ddot{x} + c\dot{x} + kx = F_0 e^{-\alpha t} \qquad (3.592)
$$

The response of a one DOF system to the force is:

$$
x(t) = \int_0^t h(t - \tau) f(\tau) d\tau
$$

$$
= \frac{F_0}{m\omega_n} \int_0^t e^{-\alpha\tau} e^{-\xi\omega_n(t-\tau)} \sin \omega_d (t - \tau) \, d\tau
$$

$$
= \frac{F_0\omega_n^2}{k} \frac{e^{-\xi\omega_n t} \left(\dfrac{\alpha - \xi\omega_n}{\omega_d} \sin \omega_d t - \cos \omega_d t \right) + e^{-\alpha t}}{(\alpha - \xi\omega_n)^2 + \omega_d^2} \qquad (3.593)
$$

Figure 3.51 illustrates a sample of response to a decaying exponential excitation.

Example 120 Excitation with nonzero initial conditions. Any nonzero initial conditions will appear in the homogenous solution because the convolution integral is always calculated based on zero initial conditions. Here is to show how to tackle with nonzero initial conditions.

Consider a mass-spring-damper system with nonzero initial conditions where suddenly an exponentially decaying force is applied on the system:

$$
m\ddot{x} + c\dot{x} + kx = F_0 e^{-\alpha t} \qquad (3.594)
$$

$$
x(0) = 1 \qquad \dot{x}(0) = 0 \qquad (3.595)
$$

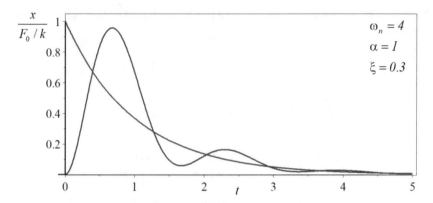

Fig. 3.51 A sample of response to a decaying exponential excitation

Following Eq. (3.506), the response of the system to the force would be:

$$
\begin{aligned}
x(t) &= \int_0^t h(t-\tau) f(\tau) d\tau + e^{-\xi\omega_n t} \left(\frac{\dot{x}_0 + \xi\omega_n x_0}{\omega_d} \sin\omega_d t + x_0 \cos\omega_d t \right) \\
&= \frac{F_0\omega_n^2}{k} \frac{e^{-\xi\omega_n t} \left(\dfrac{\alpha - \xi\omega_n}{\omega_d} \sin\omega_d t - \cos\omega_d t \right) + e^{-\alpha t}}{(\alpha - \xi\omega_n)^2 + \omega_d^2} \\
&\quad + e^{-\xi\omega_n t} \left(\frac{\dot{x}_0 + \xi\omega_n x_0}{\omega_d} \sin\omega_d t + x_0 \cos\omega_d t \right)
\end{aligned} \tag{3.596}
$$

Substituting the initial conditions (3.595) provides us with the time response of the system:

$$
\begin{aligned}
x(t) &= \frac{F_0\omega_n^2}{k} \frac{e^{-\xi\omega_n t} \left(\dfrac{\alpha - \xi\omega_n}{\omega_d} \sin\omega_d t - \cos\omega_d t \right) + e^{-\alpha t}}{(\alpha - \xi\omega_n)^2 + \omega_d^2} \\
&\quad + e^{-\xi\omega_n t} \left(\frac{\xi\omega_n}{\omega_d} \sin\omega_d t + \cos\omega_d t \right)
\end{aligned} \tag{3.597}
$$

Figure 3.52 illustrates a sample of response to a decaying exponential excitation with nonzero initial conditions.

Example 121 ★ A dropped packet. Shock and impact is easy to be analyzed by convolution integral method. Here is an example.

The packet in Fig. 3.53 is dropped with zero initial velocity from the height h and hits the ground. Assuming the box stick to the ground after strike, we may use the convolution integral to determine the motion of m.

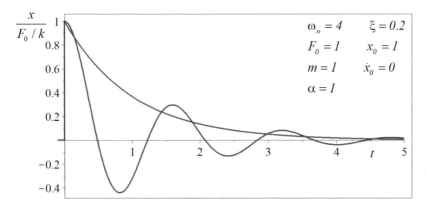

Fig. 3.52 A sample of response to a decaying exponential excitation with nonzero initial conditions

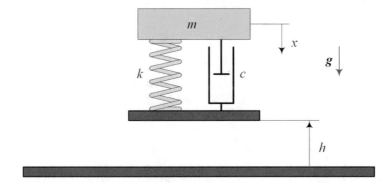

Fig. 3.53 A packet is dropped from the height h and hits the ground

If x is the displacement of the mass m relative to the box and y is the displacement of the box with respect to the ground, then the box hits the ground at $t_0 = \sqrt{2h/g}$. During the fall, the displacement of the m relative to the ground is $(x + y)$. Therefore, the equation of motion of m during the fall is:

$$\ddot{x} + 2\xi\omega_n\dot{x} + \omega_n^2 x = -g \tag{3.598}$$

The solution of the equation is:

$$x(t) = \int_0^t h(t - \tau) f(\tau) d\tau = -g \int_0^t h(t - \tau) d\tau$$

$$= \frac{-g}{m\omega_d} \int_0^t e^{-\xi\omega_n(t-\tau)} \sin\omega_d(t - \tau) \, d\tau$$

$$= \frac{-g}{k} \left(1 - e^{-\xi \omega_n t} \left(\frac{\xi \omega_n}{\omega_d} \sin \omega_d t + \cos \omega_d t \right) \right) \qquad (3.599)$$

$$0 < t < t_0$$

When the box hits the ground at $t = t_0$, the system becomes unforced. The conditions for m at $t = t_0$ are:

$$x(0) = x(t_0) = \frac{-g}{k} \left(1 - e^{-\xi \omega_n t_0} \left(\frac{\xi \omega_n}{\omega_d} \sin \omega_d t_0 + \cos \omega_d t_0 \right) \right) \qquad (3.600)$$

$$\dot{x}(0) = \dot{x}(t_0) + g t_0$$

$$= g t_0 - \frac{g}{k \omega_d} e^{-\xi \omega_n t_0} \left(\xi^2 \omega_n^2 + \omega_d^2 \right) \sin \omega_d t_0 \qquad (3.601)$$

Employing (3.506) with $f(t) = 0$ yields:

$$x = e^{-\xi \omega_n t} \left(\frac{\dot{x}_0 + \xi \omega_n x_0}{\omega_d} \sin \omega_d t + x_0 \cos \omega_d t \right) \qquad t_0 < t \qquad (3.602)$$

Example 122 First-order systems. A practice on determination of the response to unit impulse force on a first-order differential equation.

Consider a first-order system with the following differential equation:

$$\dot{x} + a x = \delta(t) \qquad x\left(0^-\right) = 0 \qquad (3.603)$$

This system is equivalent to an equation with zero input and new initial conditions. To determine the new initial condition at 0^+, we take integral of the equation between 0^- and 0^+:

$$\int_{0^-}^{0^+} \dot{x} \, dt + \int_{0^-}^{0^+} a x \, dt = \int_{0^-}^{0^+} \delta(t) \, dt \qquad (3.604)$$

$$x\left(0^+\right) + 0 = 1 \qquad (3.605)$$

Therefore, the problem is reduced to a new system with exponential solution:

$$\dot{x} + a x = 0 \qquad x\left(0^+\right) = x_0 = 1 \qquad 0 \le t \qquad (3.606)$$

If we show the solution for $\delta(t)$:

$$h(t) = x_0 e^{-at} = e^{-at} \qquad (3.607)$$

then the solution for a forcing term $f(t)$ is:

$$x(t) = \int_0^t f(t - \tau) h(\tau) d\tau = \int_0^t h(t - \tau) f(\tau) d\tau$$

$$= \int_0^t e^{-a(t-\tau)} f(\tau) d\tau \tag{3.608}$$

As an example, the response of a first-order system to a step input $f(t) = F$ is:

$$x(t) = \int_0^t e^{-a(t-\tau)} F \, d\tau = F \int_0^t e^{-a\tau} d\tau = \frac{F}{a}\left(1 - e^{-a\tau}\right) \tag{3.609}$$

The response to a ramp input $f(t) = Ft$ will be:

$$x(t) = \int_0^t e^{-a(t-\tau)} F\tau \, d\tau = F \int_0^t e^{-a(t-\tau)} \tau \, d\tau = \frac{F}{a^2}\left(e^{-at} + at - 1\right) \tag{3.610}$$

Example 123 ★ Doublet response. Doublet $\dot{\delta}(t)$ is time derivative of delta function $\delta(t)$. Here we determine the response of a (m, c, k) system to a doublet.

Doublet has a similar effect to delta function when it is multiplied by a continuous function:

$$x(t)\,\dot{\delta}(\tau) = \dot{x}(\tau) \tag{3.611}$$

Let us determine the response of a second-order (m, c, k) system to a doublet with the following differential equation:

$$\ddot{x} + 2\xi\omega_n\dot{x} + \omega_n^2 x = \frac{1}{m}\dot{\delta}(t) \qquad \dot{x}(0^-) = 0 \qquad x(0^-) = 0 \tag{3.612}$$

Two integral of the equation between $t = 0^-$ and $t = 0^+$ makes a new initial condition at $t = 0^+$:

$$\left(x(0^+) - x(0^-)\right) + 2\xi\omega_n \int_{0^-}^{0^+} x \, dt$$

$$+\omega_n^2 \int_{0^-}^{0^+}\int_{0^-}^{0^+} x \, dt dt = \frac{1}{m}\int_{0^-}^{0^+}\int_{0^-}^{0^+} \dot{\delta}(t) \, dt dt \tag{3.613}$$

$$x(0^+) = \frac{1}{m} \tag{3.614}$$

One integral of Eq. (3.612) between 0^- and 0^+ shows that:

$$\left(\dot{x}\left(0^{+}\right) - \dot{x}\left(0^{-}\right)\right) + 2\xi\omega_n\left(x\left(0^{+}\right) - x\left(0^{-}\right)\right)$$

$$+\omega_n^2 \int_{0^-}^{0^+} x \, dt = \frac{1}{m}\int_{0^-}^{0^+} \dot{\delta}\left(t\right) dt \qquad (3.615)$$

$$\dot{x}\left(0^{+}\right) = -2\xi\omega_n\frac{1}{m} \qquad (3.616)$$

Therefore, the zero initial conditions at $t = 0$ after $\dot{\delta}\left(t\right)$ are:

$$x\left(0^{+}\right) = \frac{1}{m} \qquad \dot{x}\left(0^{+}\right) = -2\xi\omega_n\frac{1}{m} \qquad (3.617)$$

Hence, $\delta\left(t\right)$ changes the problem of:

$$m\ddot{x} + c\dot{x} + kx = \dot{\delta}\left(t\right) \qquad x\left(0^{-}\right) = 0 \qquad \dot{x}\left(0^{-}\right) = 0 \qquad (3.618)$$

into the new problem of free vibration with nonzero initial conditions:

$$m\ddot{x} + c\dot{x} + kx = 0 \qquad x\left(0^{+}\right) = \frac{1}{m} \qquad \dot{x}\left(0^{+}\right) = -2\xi\omega_n\frac{1}{m} \qquad (3.619)$$

The solution for an underdamped system will be:

$$x = e^{-\xi\omega_n t}\left(x_0 \cos\omega_d t + \frac{\dot{x}_0 + \xi\omega_n x_0}{\omega_d}\sin\omega_d t\right)$$

$$= \frac{1}{m}e^{-\xi\omega_n t}\left(\cos\omega_d t - \frac{\xi\omega_n}{\omega_d}\sin\omega_d t\right) \qquad (3.620)$$

Example 124 Alternative proof for convulsion integral method. We may estimate any for function $f\left(t\right)$ by a series of step functions and derive the response of the system to the force $f\left(t\right)$ by adding the responses of the to a series of step inputs.

Figure 3.54 illustrates how a function $f\left(t\right)$ can be estimated by a series of step functions. Let $f\left(t\right) = F_0$ be a step function for $0 < t$ of amplitude as shown in Fig. 3.54. If initial displacement and velocity are both equal to zero and step force in unite $F = 1$, then we indicated the response of a $(m.c, k)$ system to the unit step input by x_u:

$$x_u\left(t\right) = \frac{1}{k}\left(1 - e^{-\xi\omega_n t}\left(\frac{\xi\omega_n}{\omega_d}\sin\omega_d t + \cos\omega_d t\right)\right) \qquad 0 < t \qquad (3.621)$$

Therefore, the response to a step input of magnitude F_0 will be:

$$x = F_0 x_u\left(t\right) \qquad 0 < t \qquad (3.622)$$

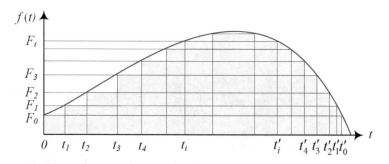

Fig. 3.54 Estimation of a function $f(t)$ by a series of step functions

and if the step input with magnitude F_1 happens at a time $t = t_0$ after the step input F, then the response of the system is the same as with a shift t_0 in time and valid for $t_0 < t$:

$$x = \frac{\Delta F_1}{k} \left(1 - e^{-\xi \omega_n (t - t_0)} \left(\frac{\xi \omega_n}{\omega_d} \sin \omega_d (t - t_0) + \cos \omega_d (t - t_0)\right)\right)$$

$$= \Delta F_1 \, x_u (t - t_0) \qquad\qquad t_0 < t \qquad\qquad \Delta F_1 = F_1 - F_0 \qquad (3.623)$$

Now if we replace a continuous force function $f(t)$ with a series of step inputs, the response of the system to the total force can be evaluated as the cumulative action of the individual step increases:

$$x = F_0 \, x_u (t) + \sum_{i=1}^{n} \Delta F_i \, x_u (t - t_i) \qquad\qquad (3.624)$$

$$= F_0 \, x_u (t) + \sum_{i=1}^{n} \frac{\Delta F_i}{\Delta t_i} x_u (t - t_i) \Delta t_i \qquad\qquad (3.625)$$

hence, in the limit, leads to the continuous solution for x:

$$x = F_0 \, x_u (t) + \int_{t_1}^{t} \frac{df}{d\tau} x_u (t - \tau) \, d\tau \qquad\qquad (3.626)$$

If the force function is transient and exists for a limited period of time, such as shown in Fig. 3.54, then the step inputs are applied for a limited time whose responses are calculated in Eq. (3.571). Therefore Eqs. (3.621)–(3.626) will get new forms:

$$x_u (t) = \frac{1}{k} \left(1 - e^{-\xi \omega_n t} \left(\frac{\xi \omega_n}{\omega_d} \sin \omega_d t + \cos \omega_d t\right)\right) \qquad\qquad (3.627)$$

$$- \frac{1}{k} \left(1 - e^{-\xi \omega_n (t - t'_0)} \left(\frac{\xi \omega_n}{\omega_d} \sin \omega_d (t - t'_0) + \cos \omega_d (t - t'_0)\right)\right) H (t - t'_0)$$

$$x = F_0 \, x_u \, (t) + \sum_{i=1}^{n} \frac{\Delta F_i}{\Delta t_i} \, x_u \, (t - t_i) \, \Delta t_i \tag{3.628}$$

$$x = F_0 \, x_u \, (t) + \int_{t_1}^{t_1'} \frac{df}{d\tau} \, x_u \, (t - \tau) \, d\tau \tag{3.629}$$

Example 125 ★ The transient force is only time-dependent.

Let us assume that besides the linear term of $P_1 = -kx^2/2$, the potential function of a system has a potential term of $P_2 \, (x, t)$, which is a function of position and time. Expanding this potential term as a powers series of small x yields:

$$P_2 \, (x, t) = -P_2 \, (0, t) - x \frac{\partial P_2 \, (0, t)}{\partial x} - \cdots \tag{3.630}$$

where $\partial P_2 \, (0, t) \, / \partial x$ means $\partial P_2 \, (x, t) \, / \partial x$ at $x = 0$. The first term of (3.630) is only a function of time and, therefore, may be omitted from the Lagrangean as being the total time derivative of another function of time. In the second term, $\partial P_2 \, (0, t) \, / \partial x$ is the external force acting on the system at the equilibrium position. It is a function of time and we refer by $f \, (t)$:

$$f \, (t) = \frac{\partial P_2 \, (0, t)}{\partial x} \tag{3.631}$$

Therefore, the Lagrangean of the system is:

$$\mathcal{L} = K - P = K - P_1 - P_2 = \frac{1}{2} m \dot{x}^2 - \frac{1}{2} k x^2 + x \, f \, (t) \tag{3.632}$$

which provides us with the equation of motion with only a time-dependent force function:

$$\ddot{x} + \omega_n^2 x = \frac{1}{m} f \, (t) \tag{3.633}$$

Example 126 ★t and τ are replaceable.

To prove the equality (3.480):

$$\int_0^t f \, (t - \tau) \, h \, (\tau) \, d\tau = \int_0^t h \, (t - \tau) \, f \, (\tau) \, d\tau \tag{3.634}$$

we can define a new variable η by keeping t constant:

$$\eta = t - \tau \qquad d\eta = -d\tau \tag{3.635}$$

and replace the variable in the first integral:

$$
\int_0^t f(t-\tau) h(\tau) d\tau = -\int_t^0 f(\eta) h(t-\eta) d\eta = \int_0^t f(\eta) h(t-\eta) d\eta
$$

$$
= \int_0^t f(\tau) h(t-\tau) d\tau \qquad (3.636)
$$

Example 127 ★ Convolution integral and Laplace transform. Convolution integral or Duhamel integral is a method to derive the inverse transformation of multiplication of two functions when the inverse transform of the individual functions are known. Convolution integral works for any linear transformation. Here is an example based on Laplace transform. trapezoidal

Let's assume $f(t)$ and $g(t)$ denote two sectionally continuous functions, and their Laplace transforms, to be $F(s)$ and $G(s)$:

$$
F(s) = L(f(t)) \qquad G(s) = L(g(t)) \qquad (3.637)
$$

$$
F(s) = \int_0^\infty f(t) e^{-st} ds \qquad (3.638)
$$

$$
e^{-\tau s} F(s) = L(f(t-\tau)) = \int_0^\infty e^{-ts} f(t-\tau) ds \qquad (3.639)
$$

If $F(s)$ and $G(s)$ are the Laplace transforms of two functions $f(\tau)$ and $g(t)$, then the transform of the convolution $f(t) * g(t)$ is $F(s) G(s)$:

$$
f(t) * g(t) = \int_0^t f(\tau) \, g(t-\tau) \, d\tau \qquad (3.640)
$$

$$
F(s) G(s) = L(f(t) * g(t)) \qquad (3.641)
$$

$$
L^{-1}(F(s) G(s)) = f(t) * g(t) \qquad (3.642)
$$

Example 128 Linear excitation. Here is to show how the vibrations of a (m, c, k) system to a linear force can be determined by superposing homogenous and particular solutions.

Figure 3.55 illustrates a linearly varied force between F_1 and F_2 during $t_1 < t < t_2$:

$$
f(t) = F_1 + \frac{F_2 - F_1}{t_2 - t_1}(t - t_1) = F_1 + \frac{\Delta F}{\Delta t}(t - t_1)
$$

$$
= a_0 + a_1 t \qquad a_0 = F_1 - t_1 \frac{\Delta F}{\Delta t} \qquad a_1 = \frac{\Delta F}{\Delta t} \qquad (3.643)
$$

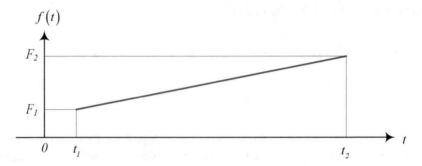

Fig. 3.55 A linearly varying force between F_1 and F_2 during $t_1 < t < t_2$

The equation of motion of an underdamped (m, c, k) system to the linear forcing function will be:

$$\ddot{x} + 2\xi\omega_n\dot{x} + \omega_n^2 x = \frac{a_0 + a_1 t}{m} \qquad \xi < 1 \qquad (3.644)$$

Substituting a trial linear polynomial into the equation: $x_p = b_0 + b_1 t$,

$$2\xi\omega_n b_1 + \omega_n^2 (b_0 + b_1 t) = \frac{a_0 + a_1 t}{m} \qquad (3.645)$$

yields:

$$\left(b_1\omega_n^2 - \frac{a_1}{m}\right) t + b_0\omega_n^2 + 2\xi\omega_n b_1 - \frac{a_0}{m} = 0 \qquad (3.646)$$

To have this equation at all time, all coefficients must be zero. Therefore, we will have two equations to determine the coefficients b_0 and b_1 of the particular solution:

$$b_1\omega_n^2 - \frac{a_1}{m} = 0 \qquad (3.647)$$

$$b_0\omega_n^2 + 2\xi\omega_n b_1 - \frac{a_0}{m} = 0 \qquad (3.648)$$

The coefficients of the solution are:

$$b_1 = \frac{a_1}{m\omega_n^2} \qquad b_0 = \frac{1}{m\omega_n^3} (a_0\omega_n - 2\xi a_1) \qquad (3.649)$$

Therefore, the general solution of Eq. (3.644) for $t_1 < t$ is:

$$x = x_h + x_p = e^{-\xi\omega_n t}(A\sin\omega_d t + B\cos\omega_d t) + b_0 + b_1 t$$

$$= e^{-\xi\omega_n t}(A\sin\omega_d t + B\cos\omega_d t) + \frac{a_0\omega_n - 2\xi a_1}{m\omega_n^3} + \frac{a_1}{m\omega_n^2}t \tag{3.650}$$

$$\omega_d = \omega_n\sqrt{1-\xi^2} \qquad t_1 < t < t_2 \tag{3.651}$$

Substituting for $a_1 = \frac{\Delta F}{\Delta t}$ and $a_0 = F_1 - t_1\frac{\Delta F}{\Delta t}$, the solution simplifies to the linear force input response:

$$x = e^{-\xi\omega_n t}(A\sin\omega_d t + B\cos\omega_d t)$$

$$+\frac{1}{m\omega_n^3}\left(\omega_n F_1 - (\omega_n t_1 + 2\xi)\frac{\Delta F}{\Delta t}\right) + \frac{1}{m\omega_n^2}\frac{\Delta F}{\Delta t}t \tag{3.652}$$

Applying a set of general initial conditions at $t = t_1$ provides us with two equations to find the coefficients A and B and determine the solution:

$$x(t_1) = x_1 \qquad \dot{x}(t_1) = \dot{x}_1 \tag{3.653}$$

$$e^{-\xi\omega_n t_1}(A\sin\omega_d t_1 + B\cos\omega_d t_1) + \frac{1}{m\omega_n^2}\frac{\Delta F}{\Delta t}t_1$$

$$+\frac{1}{m\omega_n^3}\left(\omega_n F_1 - 2\xi\frac{\Delta F}{\Delta t}\right) = x_1 \tag{3.654}$$

$$e^{-\xi\omega_n t_1}(-B\xi\omega_n + A\omega_d)\cos\omega_d t_1$$

$$-e^{-\xi\omega_n t_1}(A\xi\omega_n + B\omega_d)\sin\omega_d t_1 + \frac{1}{m\omega_n^2}\frac{\Delta F}{\Delta t} = \dot{x}_1 \tag{3.655}$$

$$A = \frac{-e^{\xi\omega_n t_1}}{m\omega_d\omega_n^2}\left(\xi F_1 - m\xi\omega_n^3 x_1 - m\omega_n^2\dot{x}_1 + \left(1-2\xi^2\right)\frac{\Delta F}{\Delta t}\right)\cos\omega_d t_1$$

$$-\frac{e^{\xi\omega_n t_1}}{m\omega_n^3}\left(\omega_n F_1 - m\omega_n^3 x_1 - 2\xi\frac{\Delta F}{\Delta t}\right)\sin\omega_d t_1 \tag{3.656}$$

$$B = \frac{-e^{\xi\omega_n t_1}}{m\omega_d\omega_n^2}\left(-\xi\omega_n F_1 + m\xi\omega_n^3 x_1 + m\omega_n^2\dot{x}_1 + \left(1-2\xi^2\right)\frac{\Delta F}{\Delta t}\right)\sin\omega_d t_1$$

$$-\frac{e^{\xi\omega_n t_1}}{m\omega_n^3}\left(\omega_n F_1 - m\omega_n^3 x_1 - 2\xi\frac{\Delta F}{\Delta t}\right)\cos\omega_d t_1 \tag{3.657}$$

Therefore the solution (3.652) has been completed, and it is valid for the time $t_1 \leq t \leq t_2$. The system will have a free vibrations after $t = t_2$ with the conditions at $t = t_2$:

$$x = e^{-\xi \omega_n (t - t_2)} \left(x_2 \cos \omega_d (t - t_2) + \frac{\dot{x}_2 + \xi \omega_n x_2}{\omega_d} \sin \omega_d (t - t_2) \right) \qquad (3.658)$$

$$x_2 = x(t_2) \qquad \dot{x}_2 = \dot{x}(t_2) \qquad t_2 < t \qquad (3.659)$$

$$x_2 = e^{-\xi \omega_n t_2} (A \sin \omega_d t_2 + B \cos \omega_d t_2)$$
$$+ \frac{1}{m \omega_n^3} \left(\omega_n F_1 - (\omega_n t_1 + 2\xi) \frac{\Delta F}{\Delta t} \right) + \frac{1}{m \omega_n^2} \frac{\Delta F}{\Delta t} t_2 \qquad (3.660)$$

$$\dot{x}_2 = -e^{-\xi \omega_n t_2} (B \xi \omega_n - A \omega_d) \cos \omega_d t_2$$
$$- e^{-\xi \omega_n t_2} (A \xi \omega_n - B \omega_d) \sin \omega_d t_2 + \frac{1}{m \omega_n^2} \frac{\Delta F}{\Delta t} \qquad (3.661)$$

The solution has been completed. As a summary, the system starts at $t = t_1$ with conditions (3.653). The linear force (3.643) applies in the system for duration $t_1 \leq t \leq t_2$. The response of the system as long as the force is applied will be (3.652), where A and B are calculated in (3.656) and (3.657). The response of the system after $t = t_2$ will be the free vibration (3.658). All equations may have a simpler expression by substituting $m \omega_n^2 = k$.

Figure 3.56 illustrates a sample response of an underdamped vibrating system to a linear force function that is applied for a short duration of time $1 \leq t \leq 3$.

Example 129 ★ Irregular force. If it is not possible to represent the transient force by a mathematical function due to irregularity, we may model the force by piecewise linear functions and determine the solution step by step.

If the transient force is given as an irregular function of time, we may approximate the function by a series of connected lines as shown in Fig. 3.57. Then convolution integral and unit impulse force method are well applicable to determine the response of a vibrating system. Let us experimentally determine the function by a list of numerical values $f_1, f_2, f_3, \cdots, f_n$ at n points $t_1, t_2, t_3, \cdots, t_n$:

$$f(t) = f_i(t_i) \qquad i = 1, 2, 3, \cdots, n \qquad (3.662)$$

$$\Delta t_i = t_i - t_{i-1} \qquad \Delta F_i = F_i - F_{i-1} \qquad (3.663)$$

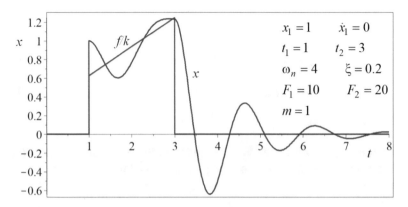

Fig. 3.56 Response of a vibrating system to a linear force function

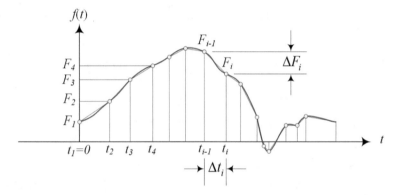

Fig. 3.57 Approximation of an irregular function by a series of connected lines

The forcing function is approximated by a piecewise linear function connecting the n selected points. In the piecewise linear interpolation, the variation of the force $f(t)$ in any time interval is linear. The response of the system in the time interval $t_{i-1} < t < t_i$ will be found by calculating the response of the system to the applied linear force function during the interval.

The response of an underdamped vibrating system to a linear forcing term starting at $f(t_1) = F_1$ to $f(t_2) = F_2$ is calculated and presented in Eq. (3.652) with A and B from Eqs. (3.656) and (3.657). Let us revise the solution and present it in a more useful form:

$$x = \frac{\Delta F_i}{k \Delta t_i} \left(t - t_1 - \frac{2\xi}{\omega_n} \right)$$

$$+ \frac{\Delta F_i}{k \Delta t_i} \left(\frac{2\xi}{\omega_n} \cos \omega_d (t - t_1) - \frac{1 - 2\xi^2}{\omega_d} \sin \omega_d (t - t_1) \right) e^{-\xi \omega_n (t - t_1)}$$

$$+\frac{F_1}{k}\left(1-\left(\cos\omega_d\,(t-t_1)+\frac{\xi\omega_n}{\omega_d}\sin\omega_d\,(t-t_1)\right)e^{-\xi\omega_n(t-t_1)}\right)$$

$$+x_1\cos\omega_d\,(t-t_1)\,e^{-\xi\omega_n(t-t_1)}$$

$$+\frac{\dot{x}_1+\xi\omega_n x_1}{\omega_d}\sin\omega_d\,(t-t_1)\,e^{-\xi\omega_n(t-t_1)}\tag{3.664}$$

$$t_1<t<t_2$$

By substituting t_{i-1} for t_1, we fit this solution to the response to a linear forcing function between times t_{i-1} to t_i of Fig. 3.57 to be valid for $t_{i-1}<t<t_i$:

$$x=\frac{\Delta F_i}{k\Delta t_i}\left(t-t_{i-1}-\frac{2\xi}{\omega_n}\right)$$

$$+\frac{\Delta F_i}{k\Delta t_i}\left(\left(\frac{2\xi}{\omega_n}\cos\omega_d\,(t-t_{i-1})-\frac{1-2\xi^2}{\omega_d}\sin\omega_d\,(t-t_{i-1})\right)e^{-\xi\omega_n(t-t_{i-1})}\right)$$

$$+\frac{F_{i-1}}{k}\left(1-\left(\cos\omega_d\,(t-t_{i-1})+\frac{\xi\omega_n}{\omega_d}\sin\omega_d\,(t-t_{i-1})\right)e^{-\xi\omega_n(t-t_{i-1})}\right)$$

$$+x_{i-1}\cos\omega_d\,(t-t_{i-1})\,e^{-\xi\omega_n(t-t_{i-1})}$$

$$+\frac{\dot{x}_{i-1}+\xi\omega_n x_{i-1}}{\omega_d}\sin\omega_d\,(t-t_{i-1})\,e^{-\xi\omega_n(t-t_{i-1})}\tag{3.665}$$

$$t_{i-1}<t<t_i$$

The solution in the next time interval of $t_i<t<t_{i+1}$ would be similar to (3.665) by replacing the initial conditions x_{i-1} and $\dot{x}_{i-1}$ with x_i and $\dot{x}_i$ and t_{i-1} to t_i. The conditions x_i and $\dot{x}_i$ will be found from (3.665) by setting $t=t_i$ in the equation and its derivative:

$$x_i=\frac{\Delta F_i}{k\Delta t_i}\left(\Delta t_i-\frac{2\xi}{\omega_n}\right)$$

$$+\frac{\Delta F_i}{k\Delta t_i}\left(\left(\frac{2\xi}{\omega_n}\cos\omega_d\Delta t_i-\frac{1-2\xi^2}{\omega_d}\sin\omega_d\Delta t_i\right)e^{-\xi\omega_n\Delta t_i}\right)$$

$$+\frac{F_{i-1}}{k}\left(1-e^{-\xi\omega_n\Delta t_i}\left(\cos\omega_d\Delta t_i+\frac{\xi\omega_n}{\omega_d}\sin\omega_d\Delta t_i\right)\right)$$

$$+e^{-\xi\omega_n\Delta t_i}\left(x_{i-1}\cos\omega_d\Delta t_i+\frac{\dot{x}_{i-1}+\xi\omega_n x_{i-1}}{\omega_d}\sin\omega_d\Delta t_i\right)\tag{3.666}$$

Table 3.1 Piecewise linear approximation of the force function $f(t)$

Point, i	Time, t	Force, $f(t_i)$
1	0	100
2	0.6283185308	128.9734652
3	1.256637062	138.5840227
4	1.884955592	114.7627540
5	2.513274123	57.32511719
6	3.141592654	−20.02935412

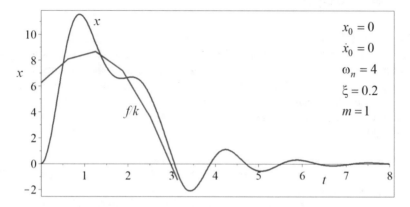

Fig. 3.58 The linearized force function and the response of a vibrating system

$$\dot{x}_i = \frac{\Delta F_i}{k\Delta t_i}\left(1 - e^{-\xi\omega_n\Delta t_i}\left(\cos\omega_d\Delta t_i + \frac{\xi\omega_n}{\omega_d}\sin\omega_d\Delta t_i\right)\right)$$

$$+\frac{F_{i-1}}{k}e^{-\xi\omega_n\Delta t_i}\frac{\omega_n^2}{\omega_d}\sin\omega_d\Delta t_i + \dot{x}_{i-1}e^{-\xi\omega_n\Delta t_i}\cos\omega_d\Delta t_i$$

$$+\left(\frac{\xi\omega_n}{\omega_d}\left(\dot{x}_{i-1} + \frac{\omega_n}{\xi}x_{i-1}\right)\sin\omega_d\Delta t_i\right)e^{-\xi\omega_n\Delta t_i} \tag{3.667}$$

Equations (3.665) and (3.667) make the recurrence relations to determine the response of a vibrating system to a piecewise linear forcing function simulated by lines connecting $f(t) = f_i(t_i)$, $i = 1, 2, 3, \cdots, n$.

As an example, let us determine the response of a vibrating system to $f(t)$ which is approximated by 5 lines as is shown in Table 3.1:

$$f(t) = 100\cos\sqrt{t} + 100\sin t \qquad 0 < t < \pi \tag{3.668}$$

The linearized force function and the response of the system are plotted in Fig. 3.58 for a sample data.

Example 130 ★ Shock. Definition of shock compared to usual excitations.

When the time duration of a transient excitation is small compared to the natural period of the system, the excitation is called a shock. The severity of a shock is measured by the maximum peak response. The maximum peak is also called maximax, which is the point at the maximum deviation from rest. It is possible that two different shocks have equal maximax. A plot of maximax over a parameter is used in the design of the parameter for a better shock response.

3.4   Measurement

The measurable vibration parameters, such as period T and amplitude X, may be used to identify mechanical characteristics of the vibrating system. In most vibration measurement methods, a transient or harmonically steady-state vibration will be applied to the system, and the parameter will be discovered by examination of the response. Using time and kinematic measurement devices, we usually measure amplitude and period of response and employ the analytic equations to find the required data.

Example 131 Evaluating the damping ratio. Damping of a system reduces the amplitude of free vibrations. The rate of reduction may be used to evaluate the damping of the system.

Damping ratio ξ of an underdamped one DOF system can be approximately found by the following equation which is based on a plot of $x = x(t)$ and peak amplitudes x_i:

$$\xi = \frac{1}{\sqrt{4(n-1)^2 \pi^2 + \ln^2 \frac{x_1}{x_n}}} \ln \frac{x_1}{x_n} \approx \frac{1}{2(n-1)\pi} \ln \frac{x_1}{x_n} \qquad (3.669)$$

To prove this equation, consider the free vibration of an underdamped one DOF system:

$$\ddot{x} + 2\xi \omega_n \dot{x} + \omega_n^2 x = 0 \qquad (3.670)$$

The time response of the system is given in Eq. (3.113) as:

$$x = Xe^{-\xi \omega_n t} \cos(\omega_d t + \theta) \qquad (3.671)$$

$$X = \sqrt{x_0^2 + \left(\frac{\dot{x}_0 + \xi \omega_n x_0}{\omega_d}\right)^2} \qquad (3.672)$$

$$\tan \theta = -\frac{\dot{x}_0 + \xi \omega_n x_0}{x_0 \omega_d} \qquad (3.673)$$

where the parameters X and θ are dependent on initial conditions.

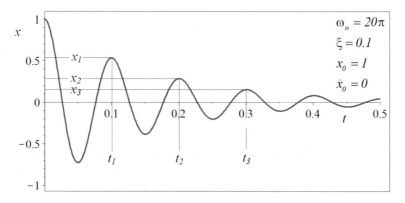

Fig. 3.59 A sample of response for the free vibration of an underdamped one DOF vibrating system

Figure 3.59 illustrates a sample of the free vibration response of an underdamped one DOF system. The peak amplitudes x_i are:

$$x_1 = e^{-\xi \omega_n t_1} \left(X \cos (\omega_d t_1 + \theta) \right) \tag{3.674}$$

$$x_2 = e^{-\xi \omega_n t_2} \left(X \cos (\omega_d t_2 + \theta) \right) \tag{3.675}$$

$$\vdots$$

$$x_n = e^{-\xi \omega_n t_n} \left(X \cos (\omega_d t_n + \theta) \right) \tag{3.676}$$

The ratio of the first two peaks is:

$$\frac{x_1}{x_2} = e^{-\xi \omega_n (t_1 - t_2)} \frac{\cos (\omega_d t_1 + \theta)}{\cos (\omega_d t_2 + \theta)} \tag{3.677}$$

Because the time difference between t_1 and t_2 is the damped period of oscillations:

$$T_d = t_2 - t_1 = \frac{2\pi}{\omega_d} = \frac{2\pi}{\omega_n \sqrt{1 - \xi^2}} \tag{3.678}$$

we may simplify Eq. (3.677):

$$\frac{x_1}{x_2} = \frac{x_k}{x_{k+1}} = e^{\xi \omega_n T_d} \frac{\cos (\omega_d t_k + \phi)}{\cos (\omega_d (t_k + T_d) + \phi)}$$

$$= e^{\xi \omega_n T_d} \frac{\cos (\omega_d t_k + \phi)}{\cos (\omega_d t_k + 2\pi + \phi)} = e^{\xi \omega_n T_d} \tag{3.679}$$

This equation shows that:

$$\ln \frac{x_1}{x_2} = \ln \frac{x_k}{x_{k+1}} = \xi \omega_n T_d = \frac{2\pi \xi}{\sqrt{1 - \xi^2}} \qquad (3.680)$$

which can be used to evaluate the damping ratio ξ:

$$\xi = \frac{1}{\sqrt{4\pi^2 + \ln^2 \frac{x_k}{x_{k+1}}}} \ln \frac{x_k}{x_{k+1}} \qquad (3.681)$$

For a better evaluation, we can measure the ratio between x_1/x_n and use the following equation:

$$\xi = \frac{1}{\sqrt{4(n-1)^2 \pi^2 + \ln^2 \frac{x_1}{x_n}}} \ln \frac{x_1}{x_n} \qquad (3.682)$$

If $\xi \ll 1$, then $\sqrt{1 - \xi^2} \approx 1$, and we may evaluate ξ from (3.680) with a simpler equation:

$$\xi \approx \frac{1}{2(n-1)\pi} \ln \frac{x_1}{x_n} \qquad (3.683)$$

Example 132 Natural frequency determination. Natural frequency indicates the stiffness of the suspension of a vibrating system for a constant mass. Natural frequency and damping ratio are the two parameters of an one DOF vibrating system. If the system is on the ground, then its natural frequency can be estimated by static deflection of the spring. Here is the proof.

The natural frequency of a mass-spring-damper system can be found by measuring the static deflection of the system under gravity. Consider a one DOF system shown in Fig. 3.60a that barely touches the ground. Assume that the spring is at its free length with no deflection. When the system rests on the ground as shown in Fig. 3.60b, the spring is compressed by a static deflection $\delta_s = mg/k$ because of gravity. We may determine the natural frequency of the system by measuring δ_s:

$$\omega_n = \sqrt{\frac{g}{\delta_s}} \qquad (3.684)$$

because:

$$\delta_s = \frac{mg}{k} = \frac{g}{\omega_n^2} \qquad (3.685)$$

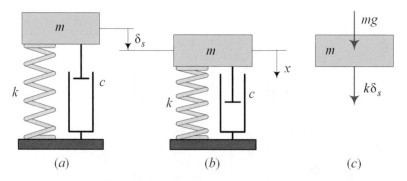

Fig. 3.60 Measurement of natural frequency based on static deflection. (**a**) the system when spring is at free length, (**b**) the system after the mass sits on spring, (**c**) the free body diagram of the system at rest position

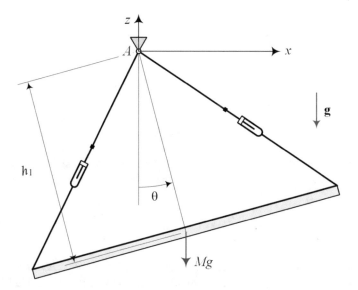

Fig. 3.61 An oscillating platform hung from point A

Example 133 Moments of inertia determination. Mass moment of an object about an axis is easy to be evaluated by measuring period of oscillation of a compound pendulum carrying the object. Here is to show how we can make a pendulum to evaluate mass moment of a vehicle.

Mass moments are important characteristics of a vehicle that affect its dynamic behavior. The main principal mass moments of I_x, I_y, and I_z can be calculated by an oscillating experiment. Figure 3.61 illustrates an oscillating platform hung from point A. Assume that the platform has a mass M and a mass moment I_0 about the pivot point A. Ignoring the mass of the cables, we can write the Euler equation of motion about point A:

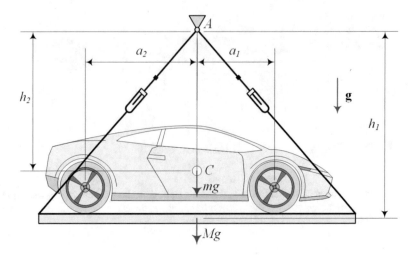

Fig. 3.62 A car with mass m on an oscillating platform hung from point A

$$\sum M_y = I_0 \ddot{\theta} = -Mgh_1 \sin\theta \tag{3.686}$$

$$I_0 \ddot{\theta} + Mgh_1 \sin\theta = 0 \tag{3.687}$$

If the angle of oscillation θ is very small, then $\sin\theta \approx \theta$, and then Eq. (3.687) simplifies to a linear equation where ω_n is the natural frequency of the oscillation:

$$\ddot{\theta} + \omega_n^2 \theta = 0 \qquad \omega_n = \sqrt{\frac{Mgh_1}{I_0}} \tag{3.688}$$

We are able to measure ω_n as the frequency of small oscillation about the point A when the platform is set free after a small deviation from the vertical equilibrium position. The natural period of oscillation $T_n = 2\pi/\omega_n$ is what we can measure, and, therefore, the mass moment I_0 is equally determined:

$$I_0 = \frac{1}{4\pi^2} Mgh_1 T_n^2 \tag{3.689}$$

The natural period T_n may be measured by an average period of a few cycles, or more accurately, by an accelerometer.

Now consider the swing shown in Fig. 3.62. A car with mass m at mass center C is on the platform such that C is exactly above the mass center of the platform. Because the location of the mass center C is known, the distance between C and the fulcrum A is also known, as say, h_2. To find the car's pitch mass moment of inertia I_y about C, we apply the Euler equation about point A, when the oscillator

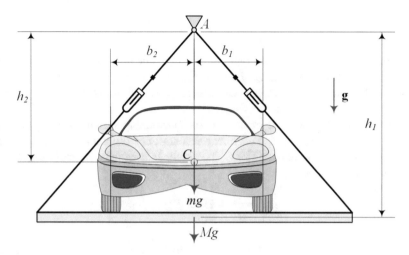

Fig. 3.63 Roll moment of inertia measurement using a swinging platform

is deviated from the equilibrium condition:

$$\sum M_y = I_A \ddot{\theta} \qquad (3.690)$$

$$-Mgh_1 \sin\theta - mgh_2 \sin\theta = I_0 + I_y + mh_2^2 \qquad (3.691)$$

Assuming a very small oscillation, we may use $\sin\theta \approx \theta$, and then Eq. (3.691) reduces to a linear oscillator:

$$\ddot{\theta} + \omega_n^2 \theta = 0 \qquad \omega_n = \sqrt{\frac{(Mh_1 + mh_2)\,g}{I_0 + I_y + mh_2^2}} \qquad (3.692)$$

Therefore, the pitch mass moment I_y can be calculated by measuring the natural period of oscillation $T_n = 2\pi/\omega_n$ from the following equation:

$$I_y = \frac{1}{4\pi^2}(Mh_1 + mh_2)\,g T_n^2 - I_0 - mh_2^2 \qquad (3.693)$$

To determine the roll mass moment I_x, we may put the car on the platform as is shown in Fig. 3.63.

Having I_x and I_y we may put the car on the platform, at an angle α, to find its mass moment about the axis passing through C and parallel to the swing axis. Then the product mass moment I_{xy} can be calculated by transformation calculus.

3.5   Summary

Time response is the reaction of a vibrating system due to nonzero initial conditions and/or time forcing function, usually transient excitations. Therefore, time response may also be called transient response. The response of one DOF system to such conditions will be studied in this chapter.

   Consider a one degree of freedom (DOF) vibrating system. The equation of motion of the system is a forced linear second-order differential equation along with a set of given initial conditions. The coefficients of mass m, stuffiness k, and damping c, are constant:

$$m\ddot{x} + c\dot{x} + kx = f(x, \dot{x}, t) \tag{3.694}$$

$$x(0) = x_0 \qquad \dot{x}(0) = \dot{x}_0 \tag{3.695}$$

When there is no excitation, $f = 0$, the equation is called *homogeneous*:

$$m\ddot{x} + c\dot{x} + kx = 0 \tag{3.696}$$

otherwise it is *nonhomogeneous*. The solution of the general equation of motion is made up of two parts: $x_h(t)$ which is the *homogeneous solution* and $x_p(t)$ which is the *particular solution*:

$$x(t) = x_h(t) + x_p(t) \tag{3.697}$$

The no force case $f = 0$ is called a *free vibration*. and its solution $x_h(t)$ is called the *free vibration response*. The nonhomogeneous case is called *forced vibration*, and its solution is called *forced vibration response*.

   We rewrite the equation of motion of free vibrations based on only two parameters ω_n and ξ, instead of three parameters $m, k, and c$:

$$\ddot{x} + 2\xi\omega_n\dot{x} + \omega_n^2 x = 0 \tag{3.698}$$

$$\omega_n = \sqrt{\frac{k}{m}} \qquad \xi = \frac{c}{2\sqrt{km}} = \frac{c}{2m\omega_n} = \frac{c\omega_n}{2k} \tag{3.699}$$

The time response of the free vibration is:

$$x_h = X_1 e^{s_1 t} + X_2 e^{s_2 t} \tag{3.700}$$

$$= \frac{\dot{x}_0 - s_2 x_0}{s_1 - s_2} e^{s_1 t} - \frac{\dot{x}_0 - s_1 x_0}{s_1 - s_2} e^{s_2 t} \tag{3.701}$$

where the *characteristic parameters* s_1 and s_2 are the solutions of the *characteristic equation* of the system:

$$s^2 + 2\xi\omega_n s + \omega_n^2 = 0 \qquad (3.702)$$

$$s_1 = -\xi\omega_n + \omega_n\sqrt{\xi^2 - 1} \qquad (3.703)$$

$$s_2 = -\xi\omega_n - \omega_n\sqrt{\xi^2 - 1} \qquad (3.704)$$

Depending on the value of the damping ratio ξ, we will have five types of time response for free vibrations. The most practical ones with applications in mechanical vibrations are:

1. Underdamped: $0 < \xi < 1$. The system shows vibrations with a decaying amplitude:

$$x_h = e^{-\xi\omega_n t} (A \cos \omega_d t + B \sin \omega_d t) \qquad (3.705)$$

$$= X e^{-\xi\omega_n t} \sin (\omega_d t + \varphi) \qquad (3.706)$$

$$= X e^{-\xi\omega_n t} \cos (\omega_d t + \theta) \qquad (3.707)$$

$$A = x_0 \qquad B = \frac{\dot{x}_0 + \xi\omega_n x_0}{\omega_d} \qquad (3.708)$$

$$X = \sqrt{A^2 + B^2} \qquad \omega_d = \omega_n\sqrt{1 - \xi^2} \qquad (3.709)$$

2. Critically damped: $\xi = 1$. The system approaches its equilibrium in fastest time with no vibrations:

$$x_h = x_0 e^{-t\omega_n} (1 + \omega_n t) + \dot{x}_0 t e^{-t\omega_n} \qquad (3.710)$$

3. Overdamped: $\xi > 1$. The system approaches its equilibrium asymptotically and shows no vibrations:

$$x_h = \frac{\dot{x}_0 - s_2 x_0}{s_1 - s_2} e^{s_1 t} - \frac{\dot{x}_0 - s_1 x_0}{s_1 - s_2} e^{s_2 t} \qquad (3.711)$$

$$s_1 = \left(-\xi + \sqrt{\xi^2 - 1}\right) \omega_n < 0 \qquad (3.712)$$

$$s_2 = \left(-\xi - \sqrt{\xi^2 - 1}\right) \omega_n < 0 \qquad (3.713)$$

The particular solution of forced equation of motion is the special solution of the equation associated to the forcing function $f(x, \dot{x}, t)$. If the force function is a continuous function of time $f = f(t)$ and is a combination of the following functions:

1. A constant, such as $f = F$
2. A polynomial in t, such as $f = a_0 + a_1 t + a_2 t^2 + \cdots + a_n t^n$
3. An exponential function, such as $f = Fe^{at}$
4. A harmonic function, such as $f = F_1 \sin \omega t + F_2 \cos \omega t$;

then the particular solution $x_p(t)$ of the linear equation of motion has the same form as the forcing term:

1. $x_p(t) = $ a constant, such as $x_p(t) = X$
2. $x_p(t) = $ a polynomial of the same degree, such as $x_p(t) = C_0 + C_1 t + C_2 t^2 + \cdots + C_n t^n$
3. $x_p(t) = $ an exponential function, such as $x_p(t) = Ce^{at}$
4. $x_p(t) = $ a harmonic function, such as $x_p(t) = A \sin \omega t + B \cos \omega t$

The coefficients of particular solution are calculated by substituting the solution in the differential equation and solving the resultant algebraic equations to satisfy the differential equation. The coefficients will be functions of the differential equation known coefficients. The coefficients of homogenous solution are functions of the initial conditions. The initial conditions must be applied after the exact particular solution x_p, and the general homogenous solution x_h is found and added together to have the general solution $x = x_h + x_p$.

When the forcing function $f(t)$ disappears after a while, the system is under *transient excitation*. The particular part x_p of the solution associated to the transient forcing function $f(t)$ will be determined by using the *convolution integral*:

$$x(t) = \int_0^t f(t - \tau) h(\tau) d\tau = \int_0^t h(t - \tau) f(\tau) d\tau \tag{3.714}$$

$$h(t) = \frac{1}{m\omega_n} e^{-\xi \omega_n t} \sin \omega_d t \tag{3.715}$$

$$\omega_d = \omega_n \sqrt{1 - \xi^2} \qquad \xi < 1 \tag{3.716}$$

If it is not possible to represent the transient force by a mathematical function, we model the force by piecewise linear connected functions and determine the solution step-by-step. Then convolution integral and unit impulse force method determine the response. Assume the force function is expressed by a list of numerical values $f_1, f_2, f_3, \cdots, f_n$ at n points $t_1, t_2, t_3, \cdots, t_n$:

$$f(t) = f_i(t_i) \qquad i = 1, 2, 3, \cdots, n \tag{3.717}$$

$$\Delta t_i = t_i - t_{i-1} \qquad \Delta F_i = F_i - F_{i-1} \qquad (3.718)$$

The forcing function is approximated by a piecewise linear function connecting the n selected points. In the piecewise linear interpolation, the variation of the force $f(t)$ in any time interval is linear. The response of the system in the time interval $t_{i-1} < t < t_i$ will be found by calculating the response of the system to the applied linear force function during the interval.

The response of an underdamped vibrating system to a linear forcing term starting at $f(t_{i-1}) = F_{i-1}$ to $f(t_i) = F_i$ is:

$$
\begin{aligned}
x = {} & \frac{\Delta F_i}{k \Delta t_i} \left(t - t_{i-1} - \frac{2\xi}{\omega_n} \right) \\
& + \frac{\Delta F_i}{k \Delta t_i} \left(\left(\frac{2\xi}{\omega_n} \cos \omega_d (t - t_{i-1}) - \frac{1 - 2\xi^2}{\omega_d} \sin \omega_d (t - t_{i-1}) \right) e^{-\xi \omega_n (t - t_{i-1})} \right) \\
& + \frac{F_{i-1}}{k} \left(1 - \left(\cos \omega_d (t - t_{i-1}) + \frac{\xi \omega_n}{\omega_d} \sin \omega_d (t - t_{i-1}) \right) e^{-\xi \omega_n (t - t_{i-1})} \right) \\
& + x_{i-1} \cos \omega_d (t - t_{i-1}) e^{-\xi \omega_n (t - t_{i-1})} \\
& + \frac{\dot{x}_{i-1} + \xi \omega_n x_{i-1}}{\omega_d} \sin \omega_d (t - t_{i-1}) e^{-\xi \omega_n (t - t_{i-1})} \\
& t_{i-1} < t < t_i
\end{aligned}
\qquad (3.719)
$$

The solution in the next time interval of $t_i < t < t_{i+1}$ would be similar by replacing the initial conditions x_{i-1} and $\dot{x}_{i-1}$ with x_i and $\dot{x}_i$ and t_{i-1} to t_i. The conditions x_i and $\dot{x}_i$ will be found by setting $t = t_i$ in the above equation and its derivative:

$$
\begin{aligned}
x_i = {} & \frac{\Delta F_i}{k \Delta t_i} \left(\Delta t_i - \frac{2\xi}{\omega_n} \right) \\
& + \frac{\Delta F_i}{k \Delta t_i} \left(\left(\frac{2\xi}{\omega_n} \cos \omega_d \Delta t_i - \frac{1 - 2\xi^2}{\omega_d} \sin \omega_d \Delta t_i \right) e^{-\xi \omega_n \Delta t_i} \right) \\
& + \frac{F_{i-1}}{k} \left(1 - e^{-\xi \omega_n \Delta t_i} \left(\cos \omega_d \Delta t_i + \frac{\xi \omega_n}{\omega_d} \sin \omega_d \Delta t_i \right) \right) \\
& + e^{-\xi \omega_n \Delta t_i} \left(x_{i-1} \cos \omega_d \Delta t_i + \frac{\dot{x}_{i-1} + \xi \omega_n x_{i-1}}{\omega_d} \sin \omega_d \Delta t_i \right) \qquad (3.720)
\end{aligned}
$$

$$\dot{x}_i = \frac{\Delta F_i}{k \Delta t_i} \left(1 - e^{-\xi \omega_n \Delta t_i} \left(\cos \omega_d \Delta t_i + \frac{\xi \omega_n}{\omega_d} \sin \omega_d \Delta t_i \right) \right)$$

$$+ \frac{F_{i-1}}{k} e^{-\xi \omega_n \Delta t_i} \frac{\omega_n^2}{\omega_d} \sin \omega_d \Delta t_i + \dot{x}_{i-1} e^{-\xi \omega_n \Delta t_i} \cos \omega_d \Delta t_i$$

$$+ \left(\frac{\xi \omega_n}{\omega_d} \left(\dot{x}_{i-1} + \frac{\omega_n}{\xi} x_{i-1} \right) \sin \omega_d \Delta t_i \right) e^{-\xi \omega_n \Delta t_i} \tag{3.721}$$

These three equations make the recurrence relations to determine the response of a vibrating system to any piecewise linear forcing function simulated by lines connecting $f(t) = f_i(t_i)$, $i = 1, 2, 3, \cdots, n$.

3.6 Key Symbols

a, b	Coefficients, real and imaginary parts of complex numbers
$\mathbf{b}$	Coefficient vector
A, B	Coefficients of time response
$\mathbf{A}$	Coefficient matrix
c	Damping
c_c	Critical damping
C	Constant, constant of integration, coefficient, amplitude
e	Exponential function
$E = K + P$	Mechanical energy
f	Force, function
F	Force amplitude, force magnitude, constant force
F_0	Constant force
$\mathbf{F}$	Force vector
$\mathbf{F}_f$	Friction force
g	Functions of displacement, gravitational acceleration
$\mathbf{g}$	Gravitational acceleration vector
h	Height
$h(t)$	Impulse response function
H	Heaviside function
i	Dummy index, unit of imaginary number
I	Mass moment
k	Stiffness
K	Kinetic energy
l	Distance, length
L	Length dimension symbol
m	Mass
M	Mass
M	Mass dimension symbol

M, N	Homogeneous functions of x and t
n	Number of equations, number of coordinates, DOF
N	Normal force
$p(t), q(t)$	Variable coefficients
P	Potential energy
P	Power
P_c	Dissipated power
q	Generalized coordinate
r	Frequency ratio
$\mathbf{r}$	Position vector
s	Eigenvalue, characteristic value
sgn	Signum function
S	Overshoot
t	Time
t_0	Characteristic time
t_P	Peak time
t_r	Rise time
t_s	Settling time
T	Period
T	Time dimension symbol
T_n	Natural period
$v \equiv \dot{x} = \lvert\mathbf{v}\rvert$	Velocity
$v_0 = \dot{x}_0$	Initial velocity
$\mathbf{v}$	Velocity vector
W	Work
$x, \mathbf{x}$	Displacement, $n \times 1$ displacement set
x_0	Initial displacement
x_h	Homogenous solution
x_P	Particular solution
$\mathbf{x}$	Vector of variables
X	Displacement amplitude
X_1, X_2	Constant of integration
y	Displacement excitation, variable, displacement
Y	Amplitude of excitation
z	Relative displacement
z	Height displacement
z	Complex variable

Greek

α	Angular variable
α	Exponent, time constant
δ	Delta function
$\delta(t)$	Unit impulse input
ε	Small number
η	Dummy time variable

θ	Angle, angular variable
ξ	Damping ratio
π	3.1415926535897932...
s	Eigenvalue, characteristic value
τ	Time constant, time variable
μ	Friction coefficient
μ	Integrating factor
ϕ	Phase angle
φ	Phase angle
ω	Frequency
$\omega_d = \omega\sqrt{1-\xi^2}$	Damped frequency
ω_n	Natural frequencyξ
Ω	Characteristic frequency

Subscript

B	Boundary, envelop
d	Damped
D	Damped function
h	Homogeneous solution
min	Minimum
n	Natural
p	Particular solution
P	Peak
r	Rise
R	Real
s	Steady state, settling, static
ss	Steady state
U	Undamped function
y	Lateral

Symbols

[ ]	Matrix
[]	Dimension
$\int$	Integral
| |	Absolute value
H	Heaviside function
Hz	Hertz
d	Differential
Δ	Increment, small change
sgn	Signum function
∂	Partial derivative
$\mathcal{L}$	Lagrangean
∇	Gradient

Exercises

1. Free vibrations of a (m, c, k) system
   Consider a mass-spring-damper system:

   $$m = 1\,\text{kg} \qquad k = 100\,\text{N}/\text{m} \tag{3.722}$$

   (a) Assume $x(0) = 1$, $\dot{x}(0) = 0$ and determine the response of the system if $c = 10\,\text{N s/m}$.
   (b) Assume $x(0) = 1$, $\dot{x}(0) = 0$ and determine the response of the system if $c = 20\,\text{N s/m}$.
   (c) Assume $x(0) = 1$, $\dot{x}(0) = 0$ and determine the response of the system if $c = 30\,\text{N s/m}$.
   (d) Assume $x(0) = 0$, $\dot{x}(0) = 1$ and determine the response of the system if $c = 10\,\text{N s/m}$.
   (e) Assume $x(0) = 0$, $\dot{x}(0) = 1$ and determine the response of the system if $c = 20\,\text{N s/m}$.
   (f) Assume $x(0) = 0$, $\dot{x}(0) = 1$ and determine the response of the system if $c = 30\,\text{N s/m}$.

2. Initial velocity calculation of a (m, c, k) system
   Consider a mass-spring-damper system with given natural frequency and damping ratio:

   $$\omega_n = 20\pi \qquad \xi = 0.15 \tag{3.723}$$

   (a) Determine the response of the system for $x(0) = 1$, $\dot{x}(0) = 0$.
   (b) Determine the response of the system for $x(0) = 0$, $\dot{x}(0) = 1$.
   (c) Determine the initial velocity $\dot{x}(0)$ such that the absolute value of the first peak of the response is equal to the first peak of the case $x(0) = 1$, $\dot{x}(0) = 0$, if possible.

3. Time of maxima and period

   (a) Show that the local maximum for the displacement of an underdamped oscillation does not occur halfway between the times at which the mass passes its equilibrium position.
   (b) Show that the time period between successive local maxima is constant.

4. ★ Equal peak condition
   For a general (m, c, k) system, determine the condition that the absolute value of the first peak would be equal for the cases of $x(0) = x_0$, $\dot{x}(0) = 0$ and $x(0) = 0$, $\dot{x}(0) = \dot{x}_0$ if $\xi\omega_n = a = comst.$ and has a given value.

5. ★ Horizontal peak distance

   For a general (m, c, k) system, determine the horizontal distance between two successive peaks for $\xi \omega_n = a = comst.$, and show that it approaches T_d when $t \rightarrow \infty$.

6. ★ Vertical peak distance

   For a general (m, c, k) system, determine the vertical distance between two successive peaks for $\xi \omega_n = a = comst.$, and show that it approaches zero when $t \rightarrow \infty$.

7. ★ Horizontal peak distance of unstable systems

   For a general (m, c, k) system with negative damping, determine the horizontal distance between two successive peaks for $\xi \omega_n = a = comst.$, and find its limit when $t \rightarrow \infty$.

8. ★ Vertical peak distance of unstable systems

   For a general (m, c, k) system with negative damping, determine the vertical distance between two successive peaks for $\xi \omega_n = a = comst.$, and show that it approaches infinity when $t \rightarrow \infty$.

9. ★ Stable and unstable systems

   Consider a mass-spring-damper system with given $\xi \omega_n = a = comst.$ Determine if the time response of the system in cases of positive and negative damping are mirror. In other words, is it true that when the damping is negative, we can go backward in time and get the same result as positive damping and positive time? Explain and prove.

10. ★ Out of envelop

    Consider a mass-spring-damper system with given $\xi \omega_n = a = comst$:

$$\xi \omega_n = 3\pi \tag{3.724}$$

   (a) Is it possible to reach any point out of the envelop curves $x = \pm e^{-\xi \omega_n t}$? How or why?
   (b) What happens if suddenly the system jumps to a point out of the envelops, such as $x = 0.6, t = 0.2$? Explain.

11. Third-order system

    Determine the time response of a third-order system:

$$\dddot{x} + 2\ddot{x} + \dot{x} + 100x = 0 \tag{3.725}$$

   (a) For the initial condition $x(0) = 1$, $\dot{x}(0) = 0$, $\ddot{x}(0) = 0$.
   (b) For the initial condition $x(0) = 0$, $\dot{x}(0) = 1$, $\ddot{x}(0) = 0$.
   (c) For the initial condition $x(0) = 0$, $\dot{x}(0) = 0$, $\ddot{x}(0) = 1$.

12. ★ Third-order system classification

    Consider a homogeneous third-order system:

$$a\dddot{x} + m\ddot{x} + c\dot{x} + kx = 0 \tag{3.726}$$

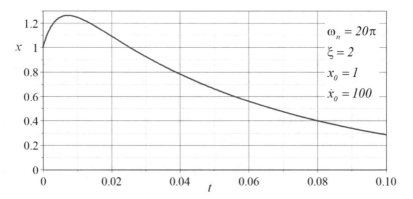

Fig. 3.64 The time response of an overdamped system with nonzero initial conditions

(a) Divide the equation by m and define the three independent coefficients that control the behavior of the system.

(b) Determine the characteristic values of the system, and classify different possible ranges of the parameters.

13. Maximum of overdamped time response

Figure 3.64 illustrates the time response of an overdamped system for the given numerical values.

(a) Use the general solution for the overdamped response of a one DOF system, and show that the maximum of x occurs at $t = 0$ or at:

$$t_M = \frac{\dot{x}_0 - s_2 x_0}{s_1 - s_2} e^{s_1 t} + \frac{s_1 x_0 - \dot{x}_0}{s_1 - s_2} e^{s_2 t}$$

$$= \left(\frac{s_1 e^{s_2 t}}{s_1 - s_2} - \frac{s_2 e^{s_1 t}}{s_1 - s_2} \right) x_0 + \left(\frac{e^{s_1 t}}{s_1 - s_2} - \frac{e^{s_2 t}}{s_1 - s_2} \right) \dot{x}_0$$

$$= \frac{e^{-\xi \omega_n t}}{2\sqrt{\xi^2 - 1}} \left(\left(\frac{\dot{x}_0}{\omega_n} + x_0 \left(\xi + \sqrt{\xi^2 - 1} \right) \right) e^{\omega_n \sqrt{\xi^2 - 1} t} \right.$$

$$\left. + \left(-\frac{\dot{x}_0}{\omega_n} + x_0 \left(-\xi - \sqrt{\xi^2 - 1} \right) \right) e^{-\omega_n \sqrt{\xi^2 - 1} t} \right)$$

(b) Determine the condition for having $1 < t_M$.

(c) ★ Determine the maximum displacement x_M at $t = t_M$.

14. An underdamped and critically damped system

Consider a critically damped (m, c, k) system:

$$\omega_n = 3\pi \qquad x_0 = 1 \qquad \dot{x}_0 = 0 \qquad (3.727)$$

(a) Determine the time at which the slope of the response $x = x(t)$ is maximum.

(b) Determine x at the maximum slope.

(c) ★ Determine ξ such that an underdamped system has the first peak at the same point and time of the critically damped maximum slope. Discuss the options, if there are any.

15. ★ Displacement integral
    Consider a damped (m, c, k) system:

$$\omega_n = 3\pi \qquad x_0 = 1 \qquad \dot{x}_0 = 0 \qquad\qquad (3.728)$$

(a) Determine the area between the response $x = x(t)$ and $x = 0$ for underdamped, critically damped, and overdamped systems.

(b) Determine ξ such that the area is minimized for the underdamped case.

(c) Determine ξ such that the area is minimized for the overdamped case.

(d) Is it true that the area is minimum for $\xi = 1$?

16. Alternative rise time definitions
    In the step input response, the classical definition of the rise time t_r is the first time that the response $x(t)$ reaches the value of the step input F/k. However, the rise time may also be defined as the inverse of the largest slope of the step response, or as the time it takes to pass from 10% to 90% of the steady-state value F/k. Use the step input response and determine:

(a) The time at which the largest slope of the step response occurs

(b) The largest slope of the step response

(c) The time it takes for the response to reach 10% of the final steady-state value

(d) The time it takes for the response to reach 90% of the final steady-state value

(e) The time it takes for the response to pass from 10% to 90% of the steady-state value

(f) ★ What is the slope of the response curve where the first time that the response $x(t)$ reaches the value of the step input F/k

17. Step response
    Consider the response of a second-order system to a step input (3.372):

(a) Determine the horizontal distance between the first and second peaks.

(b) Determine the vertical distance between the first and second peaks.

18. Step response envelop
    Prove that the step response curve is bounded by two envelop functions:

$$x = 1 \pm \frac{e^{-\xi\omega_n t}}{\sqrt{1-\xi^2}} \qquad\qquad (3.729)$$

19. Step response curve and envelop
    The step response curve is bounded by two envelop curves. The envelop is not touching the response curve at the peak points. Determine:

    (a) The time at which the response touches the envelop for the first time after the peak
    (b) The value of the response at the touch point
    (c) The slope of the response at the touch point
    (d) ★ The horizontal distance between the touch and the peak point
    (e) ★ The vertical distance between the touch and the peak points

20. Maximum overshoot
    Is it true that the maximum overshoot is 100% and it occurs for $\xi = 0$?

21. ★ A constant force $f(t) = F$ on an undamped system
    Consider a mass-spring system with:

    $$m = 1\,\text{kg} \qquad k = 100\,\text{N}\,/\,\text{m} \tag{3.730}$$

    which a step force of $f = 10\,\text{N}$ is applied on it:

    (a) Determine and plot the responses x and $\dot{x}$.
    (b) Plot $\dot{x}$ versus x.

22. Initial condition determination
    Determine the time response of an underdamped system:

    $$m\ddot{x} + c\dot{x} + kx = f(t) \tag{3.731}$$

    $$x(0) = x_0 \qquad \dot{x}(0) = \dot{x}_0 \tag{3.732}$$

    for the following forces:

    (a) $f(t) = F_0 + F_1 t$
    (b) $f(t) = F_0 + \cos \omega t$
    (c) ★ $f(t) = F_0 + \cos \omega_n t$
    (d) $f(t) = F_0 + \cos 2\omega_n t$
    (e) $f(t) = F_0 + \cos \dfrac{\omega_n}{2} t$
    (f) $f(t) = F_0 + \cos \omega t$

23. Particular solutions
    Consider a vibrating system, $m\ddot{x} + c\dot{x} + kx = f(t)$, and determine the particular solutions for the following forces:

    (a) $f(t) = F_0$:

    $$x_p = F_0/k \tag{3.733}$$

(b) $f(t) = F_0 t$:

$$x_p = \frac{F_0}{k}\left(t - \frac{c}{k}\right) \tag{3.734}$$

(c) $f(t) = F_0 t^2$:

$$x_p = \frac{2F_0}{k}\left(\frac{t^2}{2} - \frac{ct}{k} - \frac{m}{k} + \frac{c^2}{k^2}\right) \tag{3.735}$$

(d) $f(t) = F_0 t^3$:

$$x_p = \frac{3F_0}{k}\left(\frac{t^3}{3} - \frac{ct^2}{k} - \frac{2mt}{k} + \frac{2c^2 t}{k^2} + \frac{4mc}{k^2} - \frac{2c^3}{k^3}\right) \tag{3.736}$$

(e) $f(t) = F_0\left(e^s + e^{-s}\right)$:

$$x_p = \frac{F_0}{ms^2 + cs + k}\left(e^s + e^{-s}\right) \tag{3.737}$$

24. Step response for a critically damped system
    Determine the step response of a critically damped system:

$$m\ddot{x} + c\dot{x} + kx = F \tag{3.738}$$

$$x(0) = x_0 \qquad \dot{x}(0) = \dot{x}_0 \qquad \xi = 1 \tag{3.739}$$

25. Step response in overdamped case
    Determine the step response of an overdamped system:

$$m\ddot{x} + c\dot{x} + kx = F \tag{3.740}$$

$$x(0) = x_0 \qquad \dot{x}(0) = \dot{x}_0 \qquad \xi > 1 \tag{3.741}$$

26. Transient response 1
    Determine and plot the transient response of a mass-spring-damper system to
    the transient force of Fig. 3.65. Use $F = 1$, $t_0 = 0.1$, $\omega_n = 1$, and $\xi = 0.2$
    when needed.

27. Particular solutions
    Show that particular solution of a (m, c, k) system to the following forcing
    functions:

$$m\ddot{x} + c\dot{x} + kx = f(t) \tag{3.742}$$

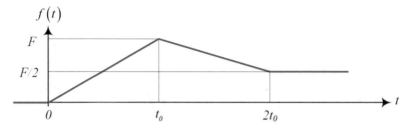

Fig. 3.65 A linearly dropping transient force

(a) $f(t) = Ft$
(b) $f(t) = Ft^2$

are:

(a) $x_p = F(t - c/k)/k$
(b) $f(t) = 2F(t^2/2 - ct/k - m/k + c^2/k^2)$

Determine the particular solutions for the following forcing functions:

(c) $f(t) = Ft^3$
(d) $f(t) = Fe^{-st}$
(e) $f(t) = Fe^{st}$
(f) $f(t) = F \sin \omega t$
(g) $f(t) = F \cos \omega t$

28. Transient response 2
    Determine and plot the transient responses of a mass-spring-damper system to the transient forces of Fig. 3.66a–d. Use $F = 1$, $t_0 = 0.1$, $\omega_n = 1$, and $\xi = 0.2$ when needed.

29. Transient response 3
    Determine and plot the transient responses of a mass-spring-damper system to the transient forces of Fig. 3.67. Use $F = 1$, $t_0 = 1$, and:

    (a) $\omega_n = 1$ and $\xi = 0.2$
    (b) $\omega_n = 1$ and $\xi = 0.8$
    (c) $\omega_n = 1$ and $\xi = 1$
    (d) $\omega_n = 1$ and $\xi = 1.2$
    (e) $T_n = 0.1$ and $\xi = 0.2$
    (f) $T_n = 0.8$ and $\xi = 0.2$
    (g) $T_n = 1$ and $\xi = 0.2$
    (h) $T_n = 1.2$ and $\xi = 0.2$

30. Transient response 4
    Determine and plot the transient responses of the base excited system of Fig. 3.68:

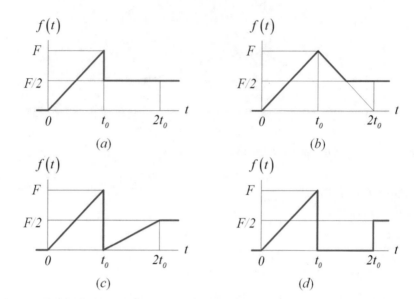

Fig. 3.66 Different transient response with initial ramp

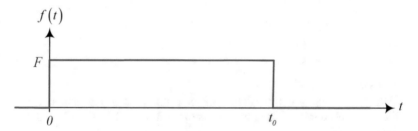

Fig. 3.67 A short duration constant force input

(a) Use $Y = 1$, $\omega = 1$, $\omega_n = 1$ and $\xi = 0.2$ when needed.
(b) Use $Y = 1$, $\omega = 1$, $\omega_n = 0.1$ and $\xi = 0.2$ when needed.
(c) Use $Y = 1$, $\omega = 1$, $\omega_n = 10$ and $\xi = 0.2$ when needed.

31. **Transient response 5**

Determine and plot the transient responses of the base excited system of Fig. 3.68 for $Y = 1$, $\omega = 1$, $\omega_n = 1$ and:

(a) An underdamped system of $\xi = 0.2$
(b) ★ A critically damped system of $\xi = 1$
(c) ★ An overdamped system of $\xi = 1.2$

32. **Transient response 6**

The base excited system of Fig. 3.69 has one period and half period excitations as shown in Fig. 3.69a and b:

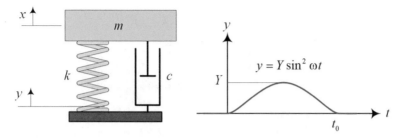

Fig. 3.68 A transient base excited system

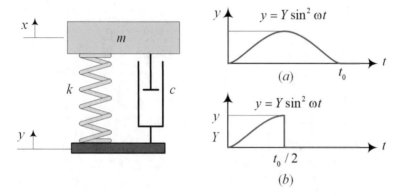

Fig. 3.69 A transient base excited system with two excitations

(a) Determine and plot the transient responses of the system to excitations (a) for $Y = 1, \xi = 0.2, \omega = 1, \omega_n = 1$.

(b) Determine and plot the transient responses of the system to excitations (b) for $Y = 1, \xi = 0.2, \omega = 1, \omega_n = 1$.

(c) Compare the responses of the previous cases on the same graph.

(d) Explain the reason of the difference between the responses. Can we say that the input energy of the case (b) is less than the case (a)?

33. **Maximum amplitude of a base excitation**

The system of Fig. 3.70 is attached to a massless plate and sitting on the ground. If the ground is moving up and down as shown in the figure, what would be the maximum Y before separation of the system from the ground?

$$\xi = 0.2 \qquad \omega = 1 \qquad \omega_n = 1 \tag{3.743}$$

34. **Damping ratio determination**

Figures 3.71, 3.72, 3.73, and 3.74 illustrates the time response of a mass-spring-damper system with $\omega_n = 10 \, \text{Hz}$ and initial condition $x(0) = 1, \dot{x}(0) = 0$. Determine the damping ratio in each case.

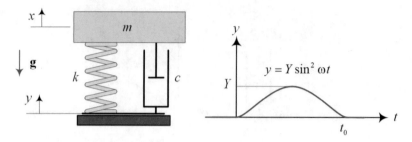

Fig. 3.70 A base excited system that is not attached to the ground

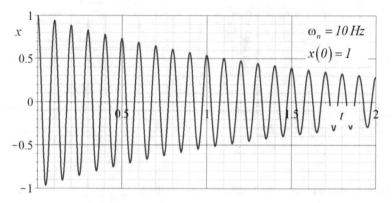

Fig. 3.71 The time response of a mass-spring-damper to the initial condition $x(0) = 1$, $\dot{x}(0) = 0$

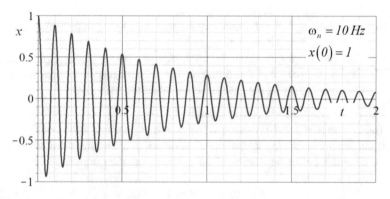

Fig. 3.72 The time response of a mass-spring-damper to initial condition $x(0) = 1$, $\dot{x}(0) = 0$

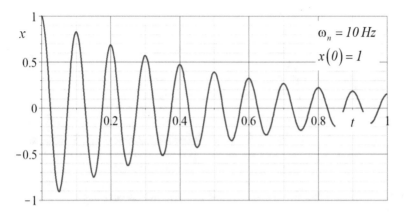

Fig. 3.73 The time response of a mass-spring-damper to initial condition $x(0) = 1$, $\dot{x}(0) = 0$

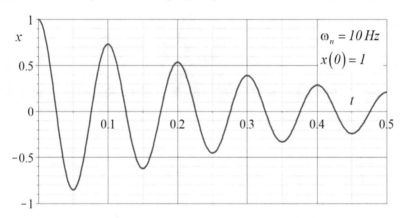

Fig. 3.74 The time response of a mass-spring-damper to the initial condition $x(0) = 1$, $\dot{x}(0) = 0$

35. **The car lateral mass moment**
    Consider a car with the following characteristics.

b_1	746 mm
b_2	740 mm
m	1245 kg
a_1	1100 mm
a_2	1323 mm
h	580 mm
I_x	335 kg m^2
I_y	1095 kg m^2

Determine the period of oscillation when the car is on a solid steel platform with dimension 2000 mm $\times$ 3800 mm $\times$ 35 mm:

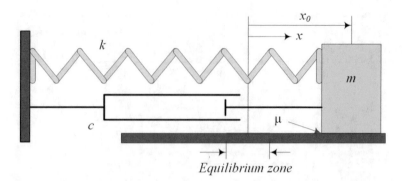

Fig. 3.75 A mass-spring vibrating system with Coulomb and viscous damping

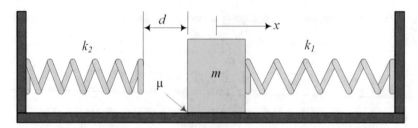

Fig. 3.76 A bilinear frictional mass-spring oscillator

(a) Laterally

(b) Longitudinally

36. Coulomb friction and number of half oscillations

Consider a mass-spring oscillator with Coulomb friction, $m = 4\,\text{kg}$, $k = 100\,\text{N}/\text{m}$, and $\mu = 0.25$. Calculate the amplitude reduction per cycle and the number of half cycles until oscillation stops for the initial conditions, $x\,(0) = 0.3\,\text{m}$ and $\dot{x}\,(0) = 0$.

37. ★ Project. Oscillation with Coulomb friction

Calculate the time solution of the system of Fig. 3.75 from $x\,(0) = x_0$, $\dot{x}\,(0) = 0$ step-by-step, and find the stop positions of the mass. Compare the effect of viscous and Coulomb damping on oscillating behavior of the system.

38. ★ Project. Bilinear frictional oscillator free vibrations

Employ the technic to derive the exact response of a mass-spring with Coulomb friction, and determine the response of the system shown in Fig. 3.76 from $x\,(0) = x_0$, $\dot{x}\,(0) = 0$.

39. Response to pointwise linear force function

Prove that Eq. (3.652) with A and B from Eqs. (3.656) and (3.657) will be simplified to Eq. (3.664).

Table 3.2 Piecewise linear approximation of the force function $f(t)$

Point, i	Time, t	Force, $f(t_i)$
1	0	0.0051723186
2	0.1570796327	0.01353944788
3	1.3141592654	0.03373533816
...	...	...
19	2.984513021	97.56279041
20	3.141592654	100.0

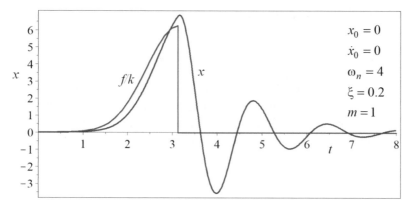

Fig. 3.77 Response of a vibrating system to a half normal distribution force function

40. ★ Project. Piecewise linear approximation force function
    Consider an underdamped (m, c, k) system. A forcing function with normal distribution is applied on the system for $0 < t < \pi$ and disappears when it is at its maximum:

$$f(t) = 100e^{-(t-\pi)^2} \qquad 0 < t < \pi \qquad (3.744)$$

   (a) Approximate the force function with 20 lines connecting the points of Table 3.2.
   (b) Employing the recursive method, determine the response of the system. You may use Fig. 3.77 as a sample answer to check your calculation.
   (c) By trial and error, adjust ξ for the same set of data of Fig. 3.77 such that the maximum displacement of x to be equal to the maximum of $f(t)/k$.
   (d) By trial and error, adjust ξ for the same set of data of Fig. 3.77 such that the second pick of x to be equal to the maximum of $f(t)/2/k$.

41. Friction force between two objects
    In Fig. 3.78, the mass m_1 is connected to the wall by spring k. There is a mass m_2 sitting on m_1, and there is a friction coefficient between the masses. If the initial condition of m_1 is $x(0) = x_0$, $x(0) = 0$, determine:

   (a) The equation/s of motion of the system
   (b) The time response of the system

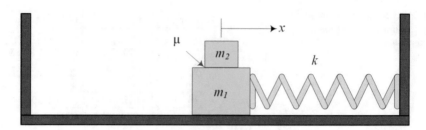

Fig. 3.78 A vibrating system with a mass m_2 sitting on top of a mass m_1 with rough surface

(c) The friction force between m_2 sitting on m_1

(d) The maximum value of x_0 for which, m_2 starts siding on m_1

(e) Assume m_1 is wide enough that m_1 will not fall off. Determine the motion of m_1 and m_2 if x_0 is 1.2 times of its calculated value at part (d).

(f) Assume $m_1 = 10\,\mathrm{kg}$, $m_2 = 1\,\mathrm{kg}$, $\mu = 0.1$, and $k = 100\,\mathrm{N/m}$, and recalculate parts (a) to (e) numerically.

Chapter 4
Multi-degrees of Freedom

Analysis of time response of systems with more than one degree of freedom (DOF) is very similar to analysis of time response of systems with one DOF. It is simple and interesting topic to study. The analysis is very much dependent of natural frequencies and mode shapes of the system. Figure 4.1 illustrates a multi-DOF vibrating system with three oscillating masses.

An n DOF system is expressed by a set of n ordinary differential equations of the second order. The system has as many natural frequencies ω_i as the degrees of freedom. There is a mode of vibration $\mathbf{u}_i$ associated with each natural frequency. The equations of motion are coupled, and, hence, the motion of the masses are combination of the motions of the individual modes.

4.1 Free Vibrations

4.1.1 Undamped Systems

A system shows its natural behavior when it is force-free and undamped. Such system will have its basic responses and expresses its natural behavior which is only dependent on initial conditions. We call a system with no external excitation a *free system* and a system with no damping, an *undamped system*. A linear undamped-free system with constant mass matrix $\mathbf{M}$ and stiffness matrix $\mathbf{K}$ is governed by a set of coupled linear differential equations using $\mathbf{x}$ as the vector of coordinates:

$$\mathbf{M}\,\ddot{\mathbf{x}} + \mathbf{K}\,\mathbf{x} = \mathbf{0} \tag{4.1}$$

$$\mathbf{x} = \begin{bmatrix} x_1 & x_2 & \cdots & x_n \end{bmatrix}^T \tag{4.2}$$

$$\mathbf{x}(0) = \mathbf{x}_0 \qquad \dot{\mathbf{x}}(0) = \dot{\mathbf{x}}_0 \tag{4.3}$$

© The Author(s), under exclusive license to Springer Nature Switzerland AG 2022
R. N. Jazar, *Advanced Vibrations*, https://doi.org/10.1007/978-3-031-16356-2_4

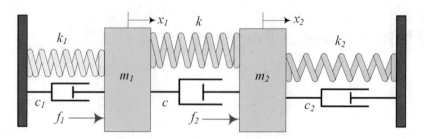

Fig. 4.1 A multi-DOF vibrating system with two oscillating masses

$$\mathbf{M} = \begin{bmatrix} m_{11} & m_{12} & \cdots & m_{1n} \\ m_{21} & m_{22} & \cdots & m_{2n} \\ \cdots & \cdots & \cdots & \cdots \\ m_{n1} & m_{n2} & \cdots & m_{nn} \end{bmatrix} \tag{4.4}$$

$$\mathbf{K} = \begin{bmatrix} k_{11} & k_{12} & \cdots & k_{1n} \\ k_{21} & k_{22} & \cdots & k_{2n} \\ \cdots & \cdots & \cdots & \cdots \\ k_{n1} & k_{n2} & \cdots & k_{nn} \end{bmatrix} \tag{4.5}$$

The time response of the free system is a linear combination of harmonic functions for each coordinate x_i, with ω_i as the *natural frequencies* and $\mathbf{u}_i$ as the *mode shapes* of the system:

$$\mathbf{x} = \sum_{i=1}^{n} \mathbf{u}_i \, (A_i \sin \omega_i t + B_i \cos \omega_i t) \quad i = 1, 2, 3, \cdots, n$$

$$= \sum_{i=1}^{n} X_i \mathbf{u}_i \sin \left(\omega_i t + \varphi_i \right) \quad i = 1, 2, 3, \cdots, n \tag{4.6}$$

$$X_i = \sqrt{A_i^2 + B_i^2} \tag{4.7}$$

$$\tan \varphi_i = \frac{B_i}{A_i} \tag{4.8}$$

The coefficients A_i and B_i, or X_i and φ_i, must be determined from the initial conditions.

The natural frequencies ω_i are solutions of the characteristic equation of the system:

$$\det \left[\mathbf{K} - \omega^2 \, \mathbf{M} \right] = 0 \tag{4.9}$$

and the mode shape $\mathbf{u}_i$, corresponding to ω_i, is the solution of the eigenvector equation:

$$\left[\mathbf{K} - \omega_i^2 \mathbf{M}\right] \mathbf{u}_i = 0 \tag{4.10}$$

Mode shapes are orthogonal with respect to the mass $\mathbf{M}$ and stiffness $\mathbf{K}$ matrices:

$$\mathbf{u}_j^T \mathbf{M} \mathbf{u}_i = 0 \qquad i \neq j \tag{4.11}$$

$$\mathbf{u}_j^T \mathbf{K} \mathbf{u}_i = 0 \qquad i \neq j \tag{4.12}$$

The orthogonality of mode shapes yields:

$$\omega_i^2 = \frac{\mathbf{u}_i^T \mathbf{K} \mathbf{u}_i}{\mathbf{u}_i^T \mathbf{M} \mathbf{u}_i} \tag{4.13}$$

Employing the modal matrix $\mathbf{U}$:

$$\mathbf{U} = \left[\mathbf{u}_1 \; \mathbf{u}_2 \; \mathbf{u}_2 \; \cdots \; \mathbf{u}_n\right] \tag{4.14}$$

we can use a linear transformation to change the variables from the generalized coordinates $\mathbf{x}$ to principal coordinates $\mathbf{p}$ and make the equations decoupled:

$$\mathbf{x} = \mathbf{U} \mathbf{p} \tag{4.15}$$

$$\mathbf{M}' \ddot{\mathbf{p}} + \mathbf{K}' \mathbf{p} = 0 \tag{4.16}$$

$$m_i' \ddot{p}_i + k_i' p_i = 0 \tag{4.17}$$

$$\mathbf{M}' = \begin{bmatrix} m_1' & 0 & 0 & 0 \\ 0 & m_2' & 0 & 0 \\ \vdots & \vdots & \ddots & \vdots \\ 0 & 0 & \cdots & m_n' \end{bmatrix} \qquad \mathbf{K}' = \begin{bmatrix} k_1' & 0 & 0 & 0 \\ 0 & k_2' & 0 & 0 \\ \vdots & \vdots & \ddots & \vdots \\ 0 & 0 & \cdots & k_n' \end{bmatrix} \tag{4.18}$$

The uncoupled equations can be solved individually. The original solution will be found by combining the principal solutions and the transformation (4.15).

Proof If the equations of motion are derived from Lagrange method and the coordinates are all absolute, then all coefficient matrices $\mathbf{M}$ and $\mathbf{K}$ are symmetric, $\mathbf{M}^T = \mathbf{M}$ and $\mathbf{K}^T = \mathbf{K}$. These matrices can also be shown as $[m] \equiv \mathbf{M}$, $[k] \equiv \mathbf{K}$. Simplifying the equations by eliminating the force and damping terms will provide us with the following set of equations:

$$\mathbf{M} \ddot{\mathbf{x}} + \mathbf{K} \mathbf{x} = 0 \tag{4.19}$$

$$\mathbf{x}\left(0\right) = \mathbf{x}_0 \qquad \dot{\mathbf{x}}\left(0\right) = \dot{\mathbf{x}}_0 \tag{4.20}$$

The solution of linear differential equations will be exponential function:

$$\mathbf{x} = \mathbf{u}\,e^{i\omega t} \tag{4.21}$$

$$\dot{\mathbf{x}} = i\omega\,\mathbf{x} = i\omega\,\mathbf{u}\,e^{i\omega t} \tag{4.22}$$

$$\ddot{\mathbf{x}} = -\omega^2 \mathbf{x} = -\omega^2\,\mathbf{u}\,e^{i\omega t} \tag{4.23}$$

Substituting the exponential functions $\mathbf{x}$ and $\ddot{\mathbf{x}}$ in Eq. (4.19) makes a generalized algebraic eigenvalue problem with $\lambda = \omega^2$ as the eigenvalue and $\mathbf{u}$ as the eigenvector:

$$[\mathbf{K} - \lambda\mathbf{M}]\,\mathbf{u} = \mathbf{0} \tag{4.24}$$

$$\lambda = \omega^2 \tag{4.25}$$

Nontrivial solution for $\mathbf{u}$ exists only if determinant of its coefficient is zero:

$$\det{[\mathbf{K} - \lambda\mathbf{M}]} = |\mathbf{K} - \lambda\mathbf{M}| = 0 \tag{4.26}$$

Trivial solution $\mathbf{u} = \mathbf{0}$ indicates the *rest position* of the system and shows no motion.

Determining the constant λ, such that the set of Eqs. (4.24) provide us with a nontrivial solution, is called the *eigenvalue problem*. Expanding (4.26) makes an algebraic *characteristic equation* which is an nth order equation in terms of $\lambda = \omega^2$ and provides us with n natural frequencies ω_i:

$$a_n\lambda^n + a_{n-1}\lambda^{n-1} + a_{n-2}\lambda^{n-2} + \cdots a_1\lambda + a_0 = 0 \tag{4.27}$$

Having n values for ω_i indicates that the solution (4.21) is possible for n different frequencies ω_i, $i = 1, 2, 3, \cdots, n$.

Determination of the vectors $\mathbf{u}_i$ to satisfy Eq. (4.24) is the *eigenvector problem*. To determine $\mathbf{u}_i$, we substitute ω_i into Eq. (4.24) to find n different $\mathbf{u}_i$:

$$\left[\mathbf{K} - \omega_i^2\mathbf{M}\right]\mathbf{u}_i = 0 \tag{4.28}$$

In mechanical vibrations, the eigenvector $\mathbf{u}_i$ corresponding to the eigenvalue ω_i is called the *mode shape*.

Equations (4.28) are homogeneous, so if $\mathbf{u}_i$ is a solution, then $a\mathbf{u}_i$, $a \in \mathbb{R}$, is also a solution. Hence, the eigenvectors may be expressed with any length. However, the ratio of any two elements of an eigenvector is unique, and, therefore, $\mathbf{u}_i$ has a unique shape. In other words, if one of the elements of $\mathbf{u}_i$ is assigned, the remaining $n - 1$ elements are uniquely determined. The shape of an eigenvector $\mathbf{u}_i$ indicates the relative amplitudes of the coordinates of the system at ω_i. Mode shapes are the principal elements of all possible free vibrations of a multi-DOF system.

Having $\omega_i = \pm\sqrt{\lambda}$ and $\mathbf{u}_i$, the solution (4.21) will be determined as:

$$\mathbf{x}_i = \mathbf{u}_i \, e^{i\omega_i t}$$
$$= \mathbf{u}_i \, (A_i \sin \omega_i t + i B_i \cos \omega_i t) \tag{4.29}$$

os as:

$$\mathbf{x}_i = \mathbf{u}_i \, e^{-i\omega_i t}$$
$$= \mathbf{u}_i \, (A_i \sin \omega_i t - i B_i \cos \omega_i t) \tag{4.30}$$

Because both the real and imaginary parts of the solutions (4.29) and (4.30) satisfy Eq. (4.19), the general and real solution of the equation is a linear combination of all solutions:

$$\mathbf{x} = \sum_{i=1}^{n} \mathbf{u}_i \, (A_i \sin \omega_i t + B_i \cos \omega_i t) \quad i = 1, 2, 3, \cdots, n \tag{4.31}$$

$$= \sum_{i=1}^{n} X_i \mathbf{u}_i \sin\left(\omega_i t + \varphi_i\right) \quad i = 1, 2, 3, \cdots, n \tag{4.32}$$

Setting $t = 0$ in these expressions and substituting in the initial conditions of (4.20) provides $2n$ equations to determine the constants of integration:

$$\mathbf{x}\,(0) = \sum_{i=1}^{n} \mathbf{u}_i \, B_i \tag{4.33}$$

$$\dot{\mathbf{x}}\,(0) = \sum_{i=1}^{n} \mathbf{u}_i \omega_i A_i \tag{4.34}$$

Using the form of Eqs. (4.33) and (4.34), the solution of the free vibration problem then becomes:

$$\mathbf{x} = \sum_{i=1}^{n} \frac{\mathbf{u}_i \, \mathbf{u}_i^T \, \mathbf{M}}{\mathbf{u}_i^T \, \mathbf{M} \, \mathbf{u}_i} \left(\frac{\dot{\mathbf{x}}\,(0)}{\omega_i} \sin \omega_i t + \mathbf{x}\,(0) \cos \omega_i t \right) \tag{4.35}$$

We usually set the natural frequencies ω_i in numerical order:

$$\omega_1 \leq \omega_2 \leq \omega_3 \leq \cdots \leq \omega_n \tag{4.36}$$

We can also multiply Eq. (4.19) by $\mathbf{M}^{-1}$ and find the same characteristic equation (4.26):

$$\ddot{\mathbf{x}} + \mathbf{M}^{-1}\mathbf{K}\,\mathbf{x} = \ddot{\mathbf{x}} + \mathbf{A}\,\mathbf{x} = \mathbf{0} \tag{4.37}$$

$$\det\left[\mathbf{A} - \lambda\mathbf{I}\right] = 0 \tag{4.38}$$

$$\mathbf{A} = \mathbf{M}^{-1}\mathbf{K} \tag{4.39}$$

Therefore, determination of the natural frequencies ω_i is equivalent to determining the eigenvalues λ_i of the *characteristic matrix* $\mathbf{A} = \mathbf{M}^{-1}\mathbf{K}$. Alternatively, we can find the eigenvectors $\mathbf{u}_i$ from the matrix $\mathbf{A}$:

$$\left[\mathbf{A} - \lambda\mathbf{I}\right]\mathbf{u}_i = 0 \tag{4.40}$$

Any type of free, transient, or forced response of a linear vibrating system is made by its natural frequencies, mode shapes, and interaction of excitation frequencies. Determination of the natural frequencies ω_i and their associated mode shapes $\mathbf{u}_i$ is the first step in analysis of a multi-DOF vibrating systems. There exists at least one mode shape corresponding to each natural frequency.

Arranging the eigenvalues and mode shapes in the form of the following matrices introduces the modal matrix $\mathbf{U}$, and eigen matrix $\mathbf{\Lambda} \equiv \mathbf{\Omega}^2$:

$$\mathbf{U} = \begin{bmatrix} \mathbf{u}_1 \ \mathbf{u}_2 \ \mathbf{u}_2 \ \cdots \ \mathbf{u}_n \end{bmatrix} \tag{4.41}$$

$$\mathbf{\Lambda} = \begin{bmatrix} \lambda_1 & 0 & \cdots & 0 \\ 0 & \lambda_2 & \cdots & 0 \\ \vdots & \vdots & \ddots & \vdots \\ 0 & 0 & \cdots & \lambda_n \end{bmatrix} = \mathbf{\Omega}^2 = \begin{bmatrix} \omega_1^2 & 0 & \cdots & 0 \\ 0 & \omega_2^2 & \cdots & 0 \\ \vdots & \vdots & \ddots & \vdots \\ 0 & 0 & \cdots & \omega_n^2 \end{bmatrix} \tag{4.42}$$

enables us to collect all equations of eigenvalues:

$$\mathbf{K}\,\mathbf{u}_i = \omega_i^2\,\mathbf{M}\,\mathbf{u}_i \tag{4.43}$$

and rewrite them as a compact expression of the complete solution to the eigenvalue problem:

$$\mathbf{K}\,\mathbf{U} = \mathbf{\Omega}^2\,\mathbf{M}\,\mathbf{U} \tag{4.44}$$

The proof of orthogonality conditions of mode shapes with respect to the mass $\mathbf{M}$ and stiffness $\mathbf{K}$ matrices of Eqs. (4.11) and (4.12) is given in Sect. 8.1.

The orthogonality of the mode shapes with respect to $\mathbf{M}$ and $\mathbf{K}$ implies that the following square matrices are diagonal:

$$\mathbf{U}^T \, \mathbf{K} \, \mathbf{U} = \mathbf{K}' \tag{4.45}$$

$$\mathbf{U}^T \, \mathbf{M} \, \mathbf{U} = \mathbf{M}' \tag{4.46}$$

$$\mathbf{u}_i^T \, \mathbf{K} \, \mathbf{u}_j = \begin{cases} k_i & i = j \\ 0 & i \neq j \end{cases} \tag{4.47}$$

$$\mathbf{u}_i^T \, \mathbf{M} \, \mathbf{u}_j = \begin{cases} m_i & i = j \\ 0 & i \neq j \end{cases} \tag{4.48}$$

$$\mathbf{M}' = \begin{bmatrix} m_1' & 0 & 0 & 0 \\ 0 & m_2' & 0 & 0 \\ \vdots & \vdots & \ddots & \vdots \\ 0 & 0 & \cdots & m_n' \end{bmatrix} \tag{4.49}$$

$$\mathbf{K}' = \begin{bmatrix} k_1' & 0 & 0 & 0 \\ 0 & k_2' & 0 & 0 \\ \vdots & \vdots & \ddots & \vdots \\ 0 & 0 & \cdots & k_n' \end{bmatrix} \tag{4.50}$$

Rearranging the diagonal mass and stiffness matrices make it easier to show the natural frequencies ω_i^2:

$$\mathbf{\Omega}^2 = \frac{\mathbf{K}'}{\mathbf{M}'} = \frac{\mathbf{U}^T \, \mathbf{K} \, \mathbf{U}}{\mathbf{U}^T \, \mathbf{M} \, \mathbf{U}} \tag{4.51}$$

$$\omega_i^2 = \frac{\mathbf{u}_i^T \, \mathbf{K} \, \mathbf{u}_i}{\mathbf{u}_i^T \, \mathbf{M} \, \mathbf{u}_i} \tag{4.52}$$

This is called Rayleigh's quotient. It also may be derived by equating time averages of the maximum value of kinetic energy to the maximum value of potential energy of the system under the assumption that the system is oscillating harmonic at frequency ω with amplitude ratios $\mathbf{u}$.

The diagonal mass and stiffness matrices make the equations of motion independent in principal coordinates $\mathbf{p}$:

$$\mathbf{x} = \mathbf{U} \, \mathbf{p} \tag{4.53}$$

$$\mathbf{M}' \ddot{\mathbf{p}} + \mathbf{K}' \mathbf{p} = \mathbf{0} \tag{4.54}$$

$$
\begin{bmatrix} m'_1 & 0 & 0 & 0 \\ 0 & m'_2 & 0 & 0 \\ \vdots & \vdots & \ddots & \vdots \\ 0 & 0 & \cdots & m'_n \end{bmatrix}
\begin{bmatrix} \ddot{p}_1 \\ \ddot{p}_2 \\ \vdots \\ \ddot{p}_n \end{bmatrix} +
\begin{bmatrix} k'_1 & 0 & 0 & 0 \\ 0 & k'_2 & 0 & 0 \\ \vdots & \vdots & \ddots & \vdots \\ 0 & 0 & \cdots & k'_n \end{bmatrix}
\begin{bmatrix} p_1 \\ p_2 \\ \vdots \\ p_n \end{bmatrix} =
\begin{bmatrix} 0 \\ 0 \\ \vdots \\ 0 \end{bmatrix}
\tag{4.55}
$$

Therefore:

$$
m'_i \ddot{p}_i + k'_i p_i = 0 \tag{4.56}
$$

and, hence, the natural frequency of the ith equation is:

$$
\omega_i^2 = \frac{k_i}{m_i} \tag{4.57}
$$

and, therefore:

$$
\begin{bmatrix} \omega_1^2 & 0 & 0 & 0 \\ 0 & \omega_2^2 & 0 & 0 \\ \vdots & \vdots & \ddots & \vdots \\ 0 & 0 & \cdots & \omega_n^2 \end{bmatrix} = \mathbf{M}'^{-1} \, \mathbf{K}' \tag{4.58}
$$

The principal coordinates and decoupled equations suggest another method to solve time response of the multi-DOF systems. ∎

Example 134 ★ Eigenvector expansion. Eigenvectors of an n dimensional space are orthogonal, and, hence, they make base vectors to decompose any other vectors of the space.

Consider a linear vibrating system with n DOF. Every eigenvector $\mathbf{u}_i$ of the system belongs to an n dimensional configuration space and, hence, will have n elements:

$$
\mathbf{u}_i = \begin{bmatrix} u_{i1} & u_{i2} & u_{i2} & \cdots & u_{in} \end{bmatrix}^T \tag{4.59}
$$

Because eigenvectors are orthogonal:

$$
\mathbf{u}_j \cdot \mathbf{u}_i = \begin{cases} |\mathbf{u}_i| \ |\mathbf{u}_i| \ i = j \\ \quad 0 \qquad i \neq j \end{cases} \tag{4.60}
$$

they can be used as base vectors to express any other vector of the same space. Therefore, any solution $\mathbf{x}$ can be expressed by a linear combination of the eigenvectors. This is called expansion theorem:

$$\mathbf{x} = \sum_{i=1}^{n} C_i \, \mathbf{u}_i \tag{4.61}$$

The initial position $\mathbf{x}_0$ and velocity $\dot{\mathbf{x}}_0$ of the system are vectors of the same space of the eigenvectors of the system. As a result, they can be expressed by the eigenvectors $\mathbf{u}_i$:

$$\mathbf{x}_0 = \sum_{i=1}^{n} a_i \, \mathbf{u}_i \tag{4.62}$$

The coefficients a_i are called the scalar mode multipliers. To calculate a_i, let us multiply both sides of (4.62) by $\mathbf{u}_j^T \, \mathbf{M}$. Because of the orthogonality relations of (4.48), all the terms on the right side vanish except the one for which $j = i$:

$$\mathbf{u}_j^T \, \mathbf{M} \, \mathbf{x}_0 = \sum_{i=1}^{n} a_i \, \mathbf{u}_j^T \, \mathbf{M} \, \mathbf{u}_i \tag{4.63}$$

$$a_i = \frac{\mathbf{u}_i^T \, \mathbf{M} \, \mathbf{x}_0}{\mathbf{u}_i^T \, \mathbf{M} \, \mathbf{u}_i} \tag{4.64}$$

Therefore, the vector $\mathbf{x}_0$ can be written in eigenvector expansion form:

$$\mathbf{x}_0 = \sum_{i=1}^{n} \frac{\mathbf{u}_i \, \mathbf{u}_i^T \, \mathbf{M}}{\mathbf{u}_i^T \, \mathbf{M} \, \mathbf{u}_i} \mathbf{x}_0 \tag{4.65}$$

The time response of undamped-free vibrating system (4.31) at $t = 0$ makes the initial conditions to be:

$$\mathbf{x}(0) = \sum_{i=1}^{n} \mathbf{u}_i \, B_i \qquad \dot{\mathbf{x}}(0) = \sum_{i=1}^{n} \mathbf{u}_i \omega_i A_i \tag{4.66}$$

The eigenvector expansion of $\mathbf{x}(0)$ and $\dot{\mathbf{x}}(0)$ will be:

$$\mathbf{x}(0) = \sum_{i=1}^{n} \frac{\mathbf{u}_i \, \mathbf{u}_i^T \, \mathbf{M}}{\mathbf{u}_i^T \, \mathbf{M} \, \mathbf{u}_i} \mathbf{x}(0) \qquad \dot{\mathbf{x}}(0) = \sum_{i=1}^{n} \frac{\mathbf{u}_i \, \mathbf{u}_i^T \, \mathbf{M} \, \dot{\mathbf{x}}(0)}{\mathbf{u}_i^T \, \mathbf{M} \, \mathbf{u}_i} \frac{}{\omega_i} \tag{4.67}$$

and, hence, the solution for general initial conditions can be expressed as follows:

$$\mathbf{x} = \sum_{i=1}^{n} \frac{\mathbf{u}_i \, \mathbf{u}_i^T \, \mathbf{M}}{\mathbf{u}_i^T \, \mathbf{M} \, \mathbf{u}_i} \left(\frac{\dot{\mathbf{x}}(0)}{\omega_i} \sin \omega_i t + \mathbf{x}(0) \cos \omega_i t \right) \tag{4.68}$$

Example 135 A two *DOF* vibrating system. Study a low *DOF* vibrating system will show all features of analysis of free vibrations of multi-*DOF* systems.

Consider free and undamped case of the system in Fig. 4.1:

$$\mathbf{M} \, \ddot{\mathbf{x}} + \mathbf{K} \, \mathbf{x} = \mathbf{0} \tag{4.69}$$

$$\begin{bmatrix} m_1 & 0 \\ 0 & m_2 \end{bmatrix} \begin{bmatrix} \ddot{x}_1 \\ \ddot{x}_2 \end{bmatrix} + \begin{bmatrix} k + k_1 & -k \\ -k & k + k_2 \end{bmatrix} \begin{bmatrix} x_1 \\ x_2 \end{bmatrix} = \begin{bmatrix} 0 \\ 0 \end{bmatrix} \tag{4.70}$$

$$\mathbf{x}(0) = \mathbf{x}_0 = \begin{bmatrix} x_{10} \\ x_{20} \end{bmatrix} \qquad \dot{\mathbf{x}}(0) = \dot{\mathbf{x}}_0 = \begin{bmatrix} \dot{x}_{10} \\ \dot{x}_{20} \end{bmatrix} \tag{4.71}$$

The natural frequencies of the system are the square root of the eigenvalues of **A**-matrix:

$$\mathbf{A} = \mathbf{M}^{-1}\mathbf{K} = \begin{bmatrix} m_1 & 0 \\ 0 & m_2 \end{bmatrix}^{-1} \begin{bmatrix} k + k_1 & -k \\ -k & k + k_2 \end{bmatrix}$$

$$= \begin{bmatrix} (k + k_1)/m_1 & -k/m_1 \\ -k/m_2 & (k + k_2)/m_2 \end{bmatrix} \tag{4.72}$$

$$\det\left[\mathbf{A} - \lambda \mathbf{I} \right] = 0 \tag{4.73}$$

$$\lambda^2 - \left(\frac{k + k_1}{m_1} + \frac{k + k_2}{m_2} \right) \lambda + \left(\frac{kk_1 + kk_2 + k_1k_2}{m_1m_2} \right) = 0 \tag{4.74}$$

$$\omega_1 = \sqrt{\lambda_1} = \sqrt{\frac{1}{2}Z_1 - \frac{1}{2}\sqrt{Z_1^2 - 4Z_2}} \tag{4.75}$$

$$\omega_2 = \sqrt{\lambda_2} = \sqrt{\frac{1}{2}Z_1 + \frac{1}{2}\sqrt{Z_1^2 - 4Z_2}} \tag{4.76}$$

$$Z_1 = \frac{k + k_1}{m_1} + \frac{k + k_2}{m_2} \tag{4.77}$$

$$Z_2 = \frac{kk_1 + kk_2 + k_1k_2}{m_1m_2} \tag{4.78}$$

The lower-frequency ω_1 is the fundamental, and the others are the harmonics. The mode shapes of the system are:

$$\mathbf{u}_1 = \begin{bmatrix} u_{11} \\ u_{21} \end{bmatrix} = \begin{bmatrix} \lambda_1 - \frac{k+k_2}{m_2} \\ -k/m_2 \end{bmatrix} = \begin{bmatrix} 1 \\ u_1 \end{bmatrix} \tag{4.79}$$

$$\mathbf{u}_2 = \begin{bmatrix} u_{12} \\ A_{22} \end{bmatrix} = \begin{bmatrix} \lambda_2 - \frac{k+k_2}{m_2} \\ -k/m_2 \end{bmatrix} = \begin{bmatrix} 1 \\ u_2 \end{bmatrix} \tag{4.80}$$

Therefore the solution of the equations of motion (4.69) can be expressed in a more informative way:

$$\begin{bmatrix} x_1 \\ x_2 \end{bmatrix} = \begin{bmatrix} A_{11} \\ A_{21} \end{bmatrix} \sin(\omega_1 t + \varphi_1) + \begin{bmatrix} A_{12} \\ A_{22} \end{bmatrix} \sin(\omega_2 t + \varphi_2) \tag{4.81}$$

where the $A's$ and $\varphi's$ are constants depending on initial conditions. A_{ij} is the amplitude of x_i at the frequency ω_j. The relative amplitudes of the harmonic components are:

$$\frac{A_{11}}{A_{21}} = \frac{k}{k+k_1-m_1\omega_1^2} = \frac{k+k_2-m_2\omega_1^2}{k} = \frac{u_{11}}{u_{21}} = \frac{1}{u_1} \tag{4.82}$$

$$\frac{A_{12}}{A_{22}} = \frac{k}{k+k_1-m_1\omega_2^2} = \frac{k+k_2-m_2\omega_2^2}{k} = \frac{u_{12}}{u_{22}} = \frac{1}{u_2} \tag{4.83}$$

where u_i are constants, defining the relative amplitudes of x_1 and x_2 at each of the natural frequencies ω_1 and ω_2. Thus, the solution (4.81) becomes:

$$\begin{bmatrix} x_1 \\ x_2 \end{bmatrix} = \begin{bmatrix} 1 \\ u_1 \end{bmatrix} A_{11} \sin(\omega_1 t + \varphi_1) + \begin{bmatrix} 1 \\ u_2 \end{bmatrix} A_{12} \sin(\omega_2 t + \varphi_2) \tag{4.84}$$

There are four constants of integrations, A_{11}, A_{12}, φ_1, and φ_2 to be determined from initial conditions. Depending on the initial conditions, the system can vibrate in a principal natural mode or any combination of the mode shapes. The first mode occurs when $A_{12} = 0$:

$$\begin{bmatrix} x_1 \\ x_2 \end{bmatrix} = \begin{bmatrix} 1 \\ u_1 \end{bmatrix} A_{11} \sin(\omega_1 t + \varphi_1) \tag{4.85}$$

and the second mode occurs when $A_{11} = 0$:

$$\begin{bmatrix} x_1 \\ x_2 \end{bmatrix} = \begin{bmatrix} 1 \\ u_2 \end{bmatrix} A_{12} \sin(\omega_2 t + \varphi_2) \tag{4.86}$$

Hence a principal mode is specified the modal vector at the given natural frequency.

Example 136 A 2 *DOF* free vibrations. Here is a numerical example of free vibrations of a 2 *DOF* vibrating system to illustrate its solution graphically in detail.

Consider the system of Fig. 4.1 with the following numerical values:

$$m_1 = 1 \qquad m_2 = 2$$
$$k_1 = 10 \qquad k = 20 \qquad k_2 = 30 \tag{4.87}$$

The value of natural frequencies ω_1 and ω_2 are:

$$Z_1 = 55 \qquad Z_2 = 550 \tag{4.88}$$
$$\lambda_1 = 13.1386 \qquad \lambda_2 = 41.861 \tag{4.89}$$

$$\omega_1 = \sqrt{\lambda_1} = 3.62 \qquad \omega_2 = \sqrt{\lambda_2} = 6.47 \tag{4.90}$$

$$u_1 = \frac{A_{21}}{A_{11}} = 0.843 \qquad u_2 = \frac{A_{22}}{A_{12}} = -0.593 \tag{4.91}$$

Therefore, the time function for x_1 and x_2 are:

$$\begin{bmatrix} x_1 \\ x_2 \end{bmatrix} = \begin{bmatrix} 1 \\ u_1 \end{bmatrix} A_{11} \sin(\omega_1 t + \varphi_1) + \begin{bmatrix} 1 \\ u_2 \end{bmatrix} A_{12} \sin(\omega_2 t + \varphi_2)$$
$$= \begin{bmatrix} 1 \\ 0.843 \end{bmatrix} A_{11} \sin(\omega_1 t + \varphi_1) + \begin{bmatrix} 1 \\ -0.593 \end{bmatrix} A_{12} \sin(\omega_2 t + \varphi_2) \tag{4.92}$$

We calculate the constants of integrations for the following initial conditions and calculate the solutions x_1 and x_2:

$$\begin{bmatrix} x_{10} \\ x_{20} \end{bmatrix} = \begin{bmatrix} 0.1 \\ 0 \end{bmatrix} \qquad \begin{bmatrix} \dot{x}_{10} \\ \dot{x}_{20} \end{bmatrix} = \begin{bmatrix} 0 \\ 0 \end{bmatrix} \tag{4.93}$$

$$A_{11} = 0.041 \qquad A_{12} = 0.058 \tag{4.94}$$
$$\varphi_1 = 1.57 \, \text{rad} \qquad \varphi_2 = 1.57 \, \text{rad} \tag{4.95}$$

$$x_1 = 0.041 \sin(3.62t + 1.57) + 0.058 \sin(6.47t + 1.57) \tag{4.96}$$

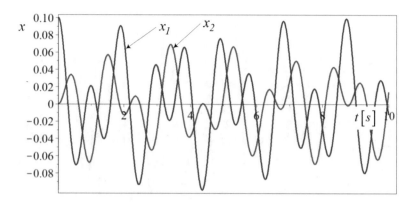

Fig. 4.2 Free vibrations of a 2 DOF vibrating system

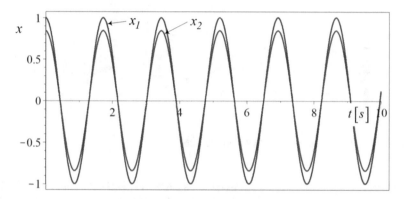

Fig. 4.3 Free vibrations of a 2 DOF vibrating system at first mode of vibrations

$$x_2 = 3.456 \times 10^{-2} \sin(3.62t + 1.57) - 3.439 \times 10^{-2} \sin(6.47t + 1.57) \qquad (4.97)$$

Figure 4.2 illustrates the response functions of x_1 and x_2.

It is more interesting if the initial condition is set equal to mode shapes of the system. If the initial conditions are:

$$\begin{bmatrix} x_{10} \\ x_{20} \end{bmatrix} = \begin{bmatrix} 1 \\ u1 \end{bmatrix} \qquad \begin{bmatrix} \dot{x}_{10} \\ \dot{x}_{20} \end{bmatrix} = \begin{bmatrix} 0 \\ 0 \end{bmatrix} \qquad (4.98)$$

then the system will be released from rest when the initial displacement is proportional to the first mode shape. Figure 4.3 illustrates the oscillation of the system at first mode, and Fig. 4.4 illustrates the oscillation of the system at second mode. The higher frequency and opposite phase of oscillation of the second mode compared to the first mode can be recognized in Fig. 4.4.

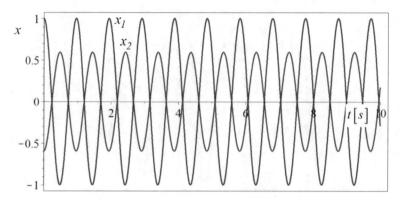

Fig. 4.4 Free vibrations of a 2 DOF vibrating system at second mode of vibrations

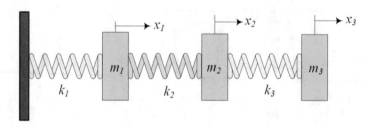

Fig. 4.5 A 3 DOF free undamped vibrating system

Example 137 A three degree of freedom system. A 3 DOF system is a good example to review the method of determining free vibrations of undamped multi-DOF systems.

Consider the 3 DOF vibrating system of Fig. 4.5. The equations of motion of the system are:

$$m_1\ddot{x}_1 = -k_1 x_1 - k_2 (x_1 - x_2) \tag{4.99}$$

$$m_2\ddot{x}_2 = -k_2 (x_2 - x_1) - k_3 (x_2 - x_3) \tag{4.100}$$

$$m_3\ddot{x}_3 = -k_3 (x_3 - x_2) \tag{4.101}$$

or

$$\mathbf{M}\,\ddot{\mathbf{x}} + \mathbf{K}\,\mathbf{x} = \mathbf{0} \tag{4.102}$$

$$\mathbf{M} = \begin{bmatrix} m_1 & 0 & 0 \\ 0 & m_2 & 0 \\ 0 & 0 & m_3 \end{bmatrix} \quad \mathbf{K} = \begin{bmatrix} k_1 + k_2 & -k_2 & 0 \\ -k_2 & k_2 + k_3 & -k_3 \\ 0 & -k_3 & k_3 \end{bmatrix} \tag{4.103}$$

$$\mathbf{x} = \begin{bmatrix} x_1 & x_2 & x_3 \end{bmatrix}^T \tag{4.104}$$

Determinant $[\mathbf{K} - \omega^2\mathbf{M}]$ provides us with the characteristic equation of the system determine the natural frequencies:

$$
\begin{aligned}
&\left| \mathbf{K} - \omega^2 \mathbf{M} \right| \\
&= \begin{vmatrix}
k_1 + k_2 - m_1\omega^2 & -k_2 & 0 \\
-k_2 & k_2 + k_3 - m_2\omega^2 & -k_3 \\
0 & -k_3 & k_3 - m_3\omega^2
\end{vmatrix} = 0
\end{aligned}
\tag{4.105}
$$

Assuming $m_1 = m_2 = m_3 = m$ and $k_1 = k_2 = k_3 = k$ simplifies the analysis and yields:

$$
\left| \mathbf{K} - \omega^2 \mathbf{M} \right| = \begin{vmatrix}
2k - m\omega^2 & -k & 0 \\
-k & 2k - m\omega^2 & -k \\
0 & -k & k - m\omega^2
\end{vmatrix} = 0
\tag{4.106}
$$

$$
\omega^6 - 5\left(\frac{k}{m}\right)\omega^4 + 6\left(\frac{k}{m}\right)^2 \omega^2 - \left(\frac{k}{m}\right)^3 = 0
\tag{4.107}
$$

$$
\omega_1^2 = 0.198\frac{k}{m} \qquad \omega_2^2 = 1.55\frac{k}{m} \qquad \omega_3^2 = 3.25\frac{k}{m}
\tag{4.108}
$$

The mode shapes of the simplified system will be:

$$
\mathbf{U} = \begin{bmatrix} \mathbf{u}_1 & \mathbf{u}_2 & \mathbf{u}_3 \end{bmatrix} = \begin{bmatrix}
1 & 1 & 1 \\
1.802 & 0.445 & -1.247 \\
2.25 & -0.802 & 0.555
\end{bmatrix}
\tag{4.109}
$$

Therefore the time response of the system will be expressed by the following functions:

$$
\begin{aligned}
\mathbf{x} &= \mathbf{u}_1 A_{11} \sin(\omega_1 t + \varphi_1) + \mathbf{u}_2 A_{12} \sin(\omega_2 t + \varphi_2) \\
&\quad + \mathbf{u}_3 A_{13} \sin(\omega_3 t + \varphi_3)
\end{aligned}
\tag{4.110}
$$

Example 138 General method to study free vibrations of multi-DOF systems. Here we consider a general two DOF undamped and free system to learn the standard method of determining the response of such systems.

Consider the equations of motion of any statically coupled two degree of freedom system in the following form:

$$
m_1\ddot{x}_1 + k_{11}x_1 + k_{12}x_2 = 0
\tag{4.111}
$$

$$
m_2\ddot{x}_2 + k_{21}x_1 + k_{22}x_2 = 0
\tag{4.112}
$$

$$x_1(0) = x_{10} \qquad x_2(0) = x_{20} \qquad \dot{x}_1(0) = \dot{x}_{10} \qquad \dot{x}_2(0) = \dot{x}_{20} \qquad (4.113)$$

Let us search for harmonic solutions:

$$x_1 = A_1 \sin(\omega t + \varphi) \qquad x_2 = A_2 \sin(\omega t + \varphi) \qquad (4.114)$$

Equation (4.114) is solutions, provided that:

$$\left(k_{11} - m_1 \omega^2\right) A_1 + k_{12} A_2 = 0 \qquad (4.115)$$

$$k_{21} A_1 + \left(k_{22} - m_2 \omega^2\right) A_2 = 0 \qquad (4.116)$$

Elimination of A_1/A_2 and considering a symmetric **K**-matrix with $k_{12} = k_{21}$ lead to the characteristic equation:

$$m_1 m_2 \omega^4 - (k_{11}m_2 + k_{22}m_1)\, \omega^2 + \left(k_{11}k_{22} - k_{12}^2\right) = 0 \qquad (4.117)$$

The solution of characteristic equation is:

$$\omega^2 = \frac{1}{2}\left(\frac{k_{11}}{m_1} + \frac{k_{22}}{m_2}\right) \pm \frac{1}{2}\sqrt{\left(\frac{k_{11}}{m_1} - \frac{k_{22}}{m_2}\right)^2 + \frac{4k_{12}^2}{m_1 m_2}} \qquad (4.118)$$

This equation gives us two real roots for ω^2 as ω_1^2 and ω_2^2, because the term in the square root is always positive. Rewriting the characteristic equation in the following form:

$$\left(\omega^2 - \omega_1^2\right)\left(\omega^2 - \omega_2^2\right) = 0 \qquad (4.119)$$

we have:

$$\omega_1^2 + \omega_2^2 = \frac{k_{11}m_2 + k_{22}m_1}{m_1 m_2} \qquad (4.120)$$

$$\omega_1^2 \omega_2^2 = \frac{k_{11}k_{22} - k_{12}^2}{m_1 m_2} \qquad (4.121)$$

that indicated ω_1^2 and ω_2^2 will be positive if $k_{11}k_{22} > k_{12}^2$. Assuming $\omega_1^2 < \omega_2^2$, the minus sign of Eq. (4.118) belongs to ω_1^2 and the plus sign indicates ω_2^2.

The ratio A_1/A_2 can be calculated from either equations of (4.115) or (4.116). Thus if u_1 and u_2 are the values of A_1/A_2 in the first and second modes respectively, we have:

$$u_1 = \frac{A_1}{A_2} = \frac{m_1\omega_1^2 - k_{11}}{k_{12}} \qquad u_2 = \frac{A_1}{A_2} = \frac{m_1\omega_2^2 - k_{11}}{k_{12}} \qquad (4.122)$$

Employing Equation (4.122) and (4.121), we can show that:

$$u_1 u_2 = -m_1 m_2 \qquad (4.123)$$

Therefore, one of the ratios u is positive and the other negative. That means the masses are vibrating in phase in the lower mode at the natural frequency ω_1, and they are vibrating with a phase difference of 180 deg in the higher mode at the frequency ω_2.

If A_1 and A_2 are the amplitudes of the mass m_1 and m_2 in the first and second modes respectively, we will have:

$$x_1 = A_1 \sin(\omega_1 t + \varphi_1) \qquad x_2 = u_1 A_1 \sin(\omega_1 t + \varphi_1) \qquad (4.124)$$

$$x_1 = A_2 \sin(\omega_2 t + \varphi_2) \qquad x_2 = u_2 A_2 \sin(\omega_2 t + \varphi_2) \qquad (4.125)$$

Therefore, by superposing the normal modes, the general solution of Eqs. (4.111) and (4.112) are:

$$x_1 = A_1 \sin(\omega_1 t + \varphi_1) + A_2 \sin(\omega_2 t + \varphi_2) \qquad (4.126)$$

$$x_2 = u_1 A_1 \sin(\omega_1 t + \varphi_1) + u_2 A_2 \sin(\omega_2 t + \varphi_2) \qquad (4.127)$$

In the solution functions, ω_1, ω_2, u_1, and u_2 are fully known from the system parameters and are calculated by Eqs. (4.118) and (4.122). The unknown amplitudes A_1 and A_2 and phase angles φ_1 and φ_2 are determined from the initial conditions, x_{10}, x_{20}, $\dot{x}_{10}$, and $\dot{x}_{20}$. In general, the initial conditions will be such that the resulting motion has components of all mode shapes. Practically, there is always some damping in the system, causing a slow decay in the vibrations. Every component of the mode shapes will decay at different rates.

Example 139 Principal coordinate and decoupling the equations of motion. Converting the equations of motion of linear vibrating system to a set of principal coordinates is always possible. The principal coordinate makes the n coupled equations to become n decoupled and independent. The solution of every equation will be simple, and the response of the whole system will be calculated much easier in principal coordinate.

Let us consider the system of Eqs. (4.111)–(4.113). Combining Eqs. (4.123) and (4.122) yields:

$$m_1 A_1 A_2 + m_2 (u_1 A_1)(u_2 A_2) = 0 \qquad (4.128)$$

This is an expression of the orthogonality property of the two mode shapes. We can normalize the mode shapes if the amplitudes A_1 and A_2 in the first and second modes satisfy:

$$m_1 A_1^2 + m_2 (u_1 A_1)^2 = 0 \tag{4.129}$$

$$m_1 A_2^2 + m_2 (u_2 A_2)^2 = 0 \tag{4.130}$$

Let us define the new coordinates, p_1 and p_2, by the following equations by using A_1 and A_2 satisfying Eqs. (4.129) and (4.130):

$$x_1 = A_1 p_1 + A_2 p_2 \tag{4.131}$$

$$x_2 = u_1 A_1 p_1 + u_2 A_2 p_2 \tag{4.132}$$

Substituting x_1 and x_2 in terms of p_1 and p_2 in the equations of motion (4.111)–(4.112) converts them to:

$$m_1 (A_1 \ddot{p}_1 + A_2 \ddot{p}_2) + k_{11} (A_1 p_1 + A_2 p_2)$$
$$+ k_{12} (u_1 A_1 p_1 + u_2 A_2 p_2) = 0 \tag{4.133}$$
$$m_2 (u_1 A_1 \ddot{p}_1 + u_2 A_2 \ddot{p}_2) + k_{21} (A_1 p_1 + A_2 p_2)$$
$$+ k_{22} (u_1 A_1 p_1 + u_2 A_2 p_2) = 0 \tag{4.134}$$

Multiplying the first and second equations by A_1 and $u_1 A_1$, respectively, and adding them up yield:

$$\left(m_2 A_1^2 u_1^2 + m_1 A_1^2 \right) \ddot{p}_1 + (m_1 A_1 A_2 + m_2 A_1 A_2 u_1 u_2) \ddot{p}_2$$
$$+ (A_1 (A_1 k_{11} + A_1 u_1 k_{12}) + A_1 u_1 (A_1 k_{21} + A_1 u_1 k_{22})) p_1$$
$$+ (A_1 (A_2 k_{11} + A_2 u_2 k_{12}) + A_1 u_1 (A_2 k_{21} + A_2 u_2 k_{22})) p_2 = 0 \tag{4.135}$$

Rewriting Eq. (4.115) in terms of the normalized modes (4.129) for the first mode yields:

$$k_{11} A_1 + k_{12} u_1 A_1 = m_1 \omega_1^2 A_1 \tag{4.136}$$

$$k_{21} A_1 + k_{22} u_1 A_1 = m_2 \omega_1^2 u_1 A_1 \tag{4.137}$$

and Eq. (4.116) in terms of the normalized modes (4.130) for the second mode is:

$$k_{11} A_2 + k_{12} u_2 A_2 = m_1 \omega_2^2 A_2 \tag{4.138}$$

$$k_{21} A_2 + k_{22} u_2 A_2 = m_2 \omega_2^2 u_2 A_2 \tag{4.139}$$

Applying Eqs. (4.128), (4.129), (4.130), and (4.136)–(4.139) to the coefficients of Eq. (4.136) reduces to:

$$\ddot{p}_1 + \left(m_1 \omega_1^2 A_1 + m_2 \omega_1^2 u_1 A_1 \right) p_1$$

$$+ \left(m_1 \omega_2^2 A_1 A_2 + m_2 \omega_2^2 u_1 A_1 u_2 A_2 \right) p_2 = 0 \tag{4.140}$$

Using the normalizing condition and the orthogonal property, it simplifies to:

$$\ddot{p}_1 + \omega_1^2 p_1 = 0 \tag{4.141}$$

Similarly, the second equation of motion (4.112):

$$\ddot{p}_2 + \omega_2^2 p_2 = 0 \tag{4.142}$$

Equations (4.141) and (4.142) are uncoupled expressions of the equations of motion in principal coordinates p_1 and p_2. Their solutions are individual simple harmonic functions of time:

$$p_1 = \sin(\omega_1 t + \theta_1) \tag{4.143}$$

$$p_2 = \sin(\omega_2 t + \theta_2) \tag{4.144}$$

The general solution for free vibrations of Eqs. (4.111)–(4.113) in coordinates x_1 and x_2 is obtained by substituting from Eqs. (4.131)–(4.132):

$$x_1 = A_1 \sin(\omega_1 t + \theta_1) + A_2 \sin(\omega_2 t + \theta_2) \tag{4.145}$$

$$x_2 = u_1 A_1 \sin(\omega_1 t + \theta_1) + u_2 A_2 \sin(\omega_2 t + \theta_2) \tag{4.146}$$

Example 140 Principal coordinates from expansion theorem. Employing expansion theorem is an alternative method to show there exists a set of principal coordinates that decoupled equations of motion.

Let us begin with the equations of motion of an undamped-free multi-DOF vibrating system:

$$\mathbf{M}\ddot{\mathbf{x}} + \mathbf{K}\mathbf{x} = \mathbf{0} \tag{4.147}$$

The expansion theorem implies that a general solution can be written as:

$$\mathbf{x} = \sum_{i=1}^{n} p_i(t)\,\mathbf{u}_i \tag{4.148}$$

Substituting this solution in the equations of motion yields:

$$\mathbf{M} \left(\sum_{i=1}^{n} \ddot{p}_i \, (t) \, \mathbf{u}_i \right) + \mathbf{K} \left(\sum_{i=1}^{n} p_i \, (t) \, \mathbf{u}_i \right) = \mathbf{0} \tag{4.149}$$

Inner product of $\mathbf{u}_j$ leads to:

$$\sum_{i=1}^{n} \ddot{p}_i \, (t) \, (\mathbf{u}_j \cdot \mathbf{M} \, \mathbf{u}_i) + \sum_{i=1}^{n} p_i \, (t) \, (\mathbf{u}_j \cdot \mathbf{K} \, \mathbf{u}_i) = \mathbf{0} \tag{4.150}$$

$$\sum_{i=1}^{n} \ddot{p}_i \, (t) \, \left(\mathbf{u}_j^T \, \mathbf{M} \, \mathbf{u}_i \right) + \sum_{i=1}^{n} p_i \, (t) \, \left(\mathbf{u}_j^T \, \mathbf{K} \, \mathbf{u}_i \right) = \mathbf{0} \tag{4.151}$$

$$m'_i \, \ddot{p}_i \, (t) + k'_i \, p_i \, (t) = 0 \qquad i = 1, 2, \cdots, n \tag{4.152}$$

Equation (4.148) can be viewed as a linear transformation between the generalized coordinates, $\mathbf{x}$, and the principal coordinates $\mathbf{p}$:

$$\mathbf{x} = \sum_{i=1}^{n} p_i \, (t) \, \mathbf{u}_i = \mathbf{U} \, \mathbf{p} \tag{4.153}$$

$$\mathbf{p} = \mathbf{U}^{-1} \, \mathbf{x} \tag{4.154}$$

The coefficient matrix is the modal matrix of the system. Because the columns of the modal matrix are linearly independent, the modal matrix is nonsingular, and the transformations have a one-to-one correspondence.

4.1.2   ★ Damped Systems

The best way to solve free vibrations of multi-DOF systems with general damping:

$$\mathbf{M} \, \ddot{\mathbf{x}} + \mathbf{C} \, \dot{\mathbf{x}} + \mathbf{K} \, \mathbf{x} = \mathbf{0} \tag{4.155}$$

is to convert the n second-order differential equation (4.222) of motion to $2n$ first-order differential equations and develop the exponential solutions:

$$^+\mathbf{M} \dot{\mathbf{y}} + \, ^+\mathbf{K} \, \mathbf{y} = \mathbf{0} \tag{4.156}$$

$$\mathbf{y} = \begin{bmatrix} \dot{\mathbf{x}} \\ \mathbf{x} \end{bmatrix} \tag{4.157}$$

$$^{+}\mathbf{M} = \begin{bmatrix} \mathbf{0} & \mathbf{M} \\ \mathbf{M} & \mathbf{C} \end{bmatrix} \tag{4.158}$$

$$^{+}\mathbf{K} = \begin{bmatrix} -\mathbf{M} & \mathbf{0} \\ \mathbf{0} & \mathbf{K} \end{bmatrix} \tag{4.159}$$

A solution to Eq. (4.156) is exponential function of time:

$$\mathbf{y} = A\,\mathbf{u}\,e^{-\lambda t} \tag{4.160}$$

where λ is the eigenvalues and $\mathbf{u}$ is eigenvectors of $^{+}\mathbf{M}^{-1}\,^{+}\mathbf{K}$. Therefore the complete solution will be a linear combination of all individual solutions. The coefficients A_i will be calculated from initial conditions:

$$\mathbf{y} = \sum_{i=1}^{n} A_i\,\mathbf{u}_i\,e^{-\lambda_i t} \tag{4.161}$$

Proof The general equations of motion of free damped multi-degree of freedom (DOF) vibrating system is:

$$\mathbf{M}\,\ddot{\mathbf{x}} + \mathbf{C}\,\dot{\mathbf{x}} + \mathbf{K}\,\mathbf{x} = \mathbf{0} \tag{4.162}$$

A general method of solution of free vibrations of damped system is to convert the equations to a set of first-order equations. The solution of the resultant set of first-order linear equations will be exponential. To convert the set of second-order equations of motion (4.162) is to define $2n$ new coordinates $\mathbf{y}$ for $\mathbf{x}$ and $\dot{\mathbf{x}}$:

$$\mathbf{y} = \begin{bmatrix} \mathbf{x} \\ \dot{\mathbf{x}} \end{bmatrix} \quad or \quad \mathbf{y} = \begin{bmatrix} \dot{\mathbf{x}} \\ \mathbf{x} \end{bmatrix} \tag{4.163}$$

Employing the new variables $\mathbf{y}$, we can rewrite Eq. (4.162) in the following form:

$$\begin{bmatrix} \mathbf{0} & \mathbf{M} \\ \mathbf{M} & \mathbf{C} \end{bmatrix} \begin{bmatrix} \ddot{\mathbf{x}} \\ \dot{\mathbf{x}} \end{bmatrix} + \begin{bmatrix} -\mathbf{M} & \mathbf{0} \\ \mathbf{0} & \mathbf{K} \end{bmatrix} \begin{bmatrix} \dot{\mathbf{x}} \\ \mathbf{x} \end{bmatrix} = \begin{bmatrix} \mathbf{0} \\ \mathbf{0} \end{bmatrix} \tag{4.164}$$

$$^{+}\mathbf{M}\,\dot{\mathbf{y}} + {}^{+}\mathbf{K}\,\mathbf{y} = \mathbf{0} \tag{4.165}$$

The solution of the resultant set of equations will be exponential function of time:

$$\mathbf{y} = \mathbf{u}\,e^{-\lambda t} \tag{4.166}$$

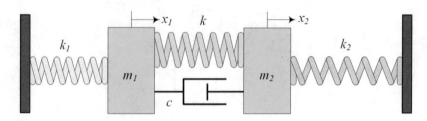

Fig. 4.6 A 2 DOF damped vibrating system

Substituting the solution into the equation yields:

$$\gamma \, ^+\mathbf{M}\,\mathbf{u} = \, ^+\mathbf{K}\,\mathbf{u} \tag{4.167}$$

$$\gamma \, \mathbf{u} = \, ^+\mathbf{M}^{-1}\, ^+\mathbf{K}\,\mathbf{u} \tag{4.168}$$

Hence, λ is the eigenvalues and $\mathbf{u}$ is eigenvectors of $^+\mathbf{M}^{-1}\, ^+\mathbf{K}$. The values of γ occur in complex conjugate pairs γ and $\bar{\gamma}$. Eigenvectors corresponding to complex conjugate eigenvalues are also complex conjugates of one another $\mathbf{u}$ and $\bar{\mathbf{u}}$. Eigenvectors corresponding to eigenvalues which are not complex conjugates satisfy the orthogonality relation:

$$\bar{\mathbf{u}}_i^T \, ^+\mathbf{M}\,\mathbf{u}_j = 0 \tag{4.169}$$

Therefore the complete solution is a linear combination of all individual solutions with coefficients A_i to be calculated from initial conditions:

$$\mathbf{y} = \sum_{i=1}^{n} A_i \, \mathbf{u}_i \, e^{-\lambda_i t} \tag{4.170}$$

∎

Example 141 A two DOF vibrating system with damping. If the damper is not connected to the fixed wall, it will damp the relative displacements between its two end, but a steady-state oscillation can remain in the system. Here is an example. Figure 4.6 illustrates a 2 DOF damped vibrating system:

$$K = \frac{1}{2}m_1\dot{x}_1^2 + \frac{1}{2}m_2\dot{x}_2^2 \tag{4.171}$$

$$D = \frac{1}{2}c\,(\dot{x}_1 - \dot{x}_2)^2 \tag{4.172}$$

$$P = \frac{1}{2}k_1 x_1^2 + \frac{1}{2}k_2 x_2^2 + \frac{1}{2}k\,(x_1 - x_2)^2 \tag{4.173}$$

The Lagrange method provides us with the following equations of motion:

$$\frac{d}{dt}\left(\frac{\partial K}{\partial \dot{x}_r}\right) - \frac{\partial K}{\partial x_r} + \frac{\partial D}{\partial \dot{x}_r} + \frac{\partial P}{\partial x_r} = \qquad r = 1, 2 \tag{4.174}$$

$$m_1\ddot{x}_1 + c\,(\dot{x}_1 - \dot{x}_2) + k_1 x_1 + k\,(x_1 - x_2) = 0 \tag{4.175}$$

$$m_2\ddot{x}_2 - c\,(\dot{x}_1 - \dot{x}_2) + k_2 x_2 - k\,(x_1 - x_2) = 0 \tag{4.176}$$

$$\mathbf{M}\,\ddot{\mathbf{x}} + \mathbf{K}\,\mathbf{x} = \mathbf{0} \tag{4.177}$$

$$\begin{bmatrix} m_1 & 0 \\ 0 & m_2 \end{bmatrix}\begin{bmatrix} \ddot{x}_1 \\ \ddot{x}_2 \end{bmatrix} + \begin{bmatrix} c & -c \\ -c & c \end{bmatrix}\begin{bmatrix} \dot{x}_1 \\ \dot{x}_2 \end{bmatrix}$$

$$+ \begin{bmatrix} k + k_1 & -k \\ -k & k + k_2 \end{bmatrix}\begin{bmatrix} x_1 \\ x_2 \end{bmatrix} = \begin{bmatrix} 0 \\ 0 \end{bmatrix} \tag{4.178}$$

The equivalent set of first-order equations is:

$$^{+}\mathbf{M}\,\dot{\mathbf{y}} + {}^{+}\mathbf{K}\,\mathbf{y} = \mathbf{0} \tag{4.179}$$

$$\mathbf{y} = \begin{bmatrix} \dot{x}_1 & \dot{x}_2 & x_1 & x_2 \end{bmatrix}^T \tag{4.180}$$

$$= \begin{bmatrix} y_1 & y_2 & y_3 & y_4 \end{bmatrix}^T \tag{4.181}$$

$$^{+}\mathbf{M} = \begin{bmatrix} \mathbf{0} & \mathbf{M} \\ \mathbf{M} & \mathbf{C} \end{bmatrix} = \begin{bmatrix} 0 & 0 & m_1 & 0 \\ 0 & 0 & 0 & m_2 \\ m_1 & 0 & c & -c \\ 0 & m_2 & -c & c \end{bmatrix} \tag{4.182}$$

$$^{+}\mathbf{K} = \begin{bmatrix} -\mathbf{M} & \mathbf{0} \\ \mathbf{0} & \mathbf{K} \end{bmatrix} = \begin{bmatrix} -m_1 & 0 & 0 & 0 \\ 0 & -m_2 & 0 & 0 \\ 0 & 0 & k + k_1 & -k \\ 0 & 0 & -k & k + k_2 \end{bmatrix} \tag{4.183}$$

$$^{+}\mathbf{M}\,\dot{\mathbf{y}} + {}^{+}\mathbf{K}\,\mathbf{y} = \begin{bmatrix} m_1\dot{y}_3 - m_1 y_1 \\ m_2\dot{y}_4 - m_2 y_2 \\ (k + k_1)\,y_3 + c\dot{y}_3 - c\dot{y}_4 - k y_4 + m_1\dot{y}_1 \\ (k + k_2)\,y_4 - c\dot{y}_3 + c\dot{y}_4 - k y_3 + m_2\dot{y}_2 \end{bmatrix} = 0 \tag{4.184}$$

The natural frequencies of the system are the square root of the eigenvalues of $^+\mathbf{M}^{-1}\,^+\mathbf{K}$:

$$^+\mathbf{M}^{-1}\,^+\mathbf{K} = \begin{bmatrix} \frac{c}{m_1} & -\frac{c}{m_1} & \frac{1}{m_1}(k+k_1) & -\frac{k}{m_1} \\ -\frac{c}{m_2} & \frac{c}{m_2} & -\frac{k}{m_2} & \frac{1}{m_2}(k+k_2) \\ -1 & 0 & 0 & 0 \\ 0 & -1 & 0 & 0 \end{bmatrix} \tag{4.185}$$

Assume a nonzero initial conditions and some numerical values for the parameters of the system:

$$\mathbf{x}(0) = \mathbf{x}_0 = \begin{bmatrix} 0.1 \\ 0 \end{bmatrix} \qquad \dot{\mathbf{x}}(0) = \dot{\mathbf{x}}_0 = \begin{bmatrix} 0 \\ 0 \end{bmatrix} \tag{4.186}$$

$$m_1 = 1\,\text{kg} \qquad m_2 = 1\,\text{kg} \qquad c = 100\,\text{N}/\text{m}/\text{s} \tag{4.187}$$

$$k_1 = 100\,\text{kg} \qquad k_2 = 100\,\text{kg} \qquad k = 100\,\text{kg} \tag{4.188}$$

$$^+\mathbf{M}^{-1}\,^+\mathbf{K} = \begin{bmatrix} 100 & -100 & 200 & -100 \\ -100 & 100 & -100 & 200 \\ -1 & 0 & 0 & 0 \\ 0 & -1 & 0 & 0 \end{bmatrix} \tag{4.189}$$

$$\Omega^2 = \begin{bmatrix} -10i & 0 & 0 & 0 \\ 0 & 10i & 0 & 0 \\ 0 & 0 & 100 - 10\sqrt{97} & 0 \\ 0 & 0 & 0 & 10\sqrt{97} + 100 \end{bmatrix} \tag{4.190}$$

$$\mathbf{U} = \begin{bmatrix} 10i & -10i & 100 - 10\sqrt{97} & 10\sqrt{97} + 100 \\ 10i & -10i & 10\sqrt{97} - 100 & -10\sqrt{97} - 100 \\ 1 & 1 & -1 & -1 \\ 1 & 1 & 1 & 1 \end{bmatrix} \tag{4.191}$$

$$\mathbf{y} = \sum_{i=1}^{4} A_i\,\mathbf{u}_i\,e^{-\lambda_i t} \tag{4.192}$$

$$\mathbf{y} = A_1 \begin{bmatrix} 10i \\ 10i \\ 1 \\ 1 \end{bmatrix} e^{-10i\,t} + A_3 \begin{bmatrix} 100 - 10\sqrt{97} \\ 10\sqrt{97} - 100 \\ -1 \\ 1 \end{bmatrix} e^{\left(100 - 10\sqrt{97}\right)t}$$

$$+ A_2 \begin{bmatrix} -10i \\ -10i \\ 1 \\ 1 \end{bmatrix} e^{10i\,t} + A_4 \begin{bmatrix} 10\sqrt{97} + 100 \\ -10\sqrt{97} - 100 \\ -1 \\ 1 \end{bmatrix} e^{\left(10\sqrt{97} + 100\right)t} \quad (4.193)$$

Employing Euler complex number identity and converting the solution to real functions of time yield:

$$y_1 = -\frac{3\sqrt{97}}{388} e^{\left(-100 + 10\sqrt{97}\right)t} + \frac{3\sqrt{97}}{388} e^{\left(-10\sqrt{97} - 100\right)t} - \frac{1}{2} \sin 10t \quad (4.194)$$

$$y_2 = \frac{3\sqrt{97}}{388} e^{\left(10\sqrt{97} - 100\right)t} - \frac{3\sqrt{97}}{388} e^{\left(-100 - 10\sqrt{97}\right)t} - \frac{1}{2} \sin 10t \quad (4.195)$$

$$y_3 = \left(\frac{1}{40} - \frac{\sqrt{97}}{388}\right) e^{\left(-100 - 10\sqrt{97}\right)t} + \left(\frac{1}{40} + \frac{\sqrt{97}}{388}\right) e^{\left(10\sqrt{97} - 100\right)t}$$

$$+ \frac{1}{20} \cos 10t \quad (4.196)$$

$$y_4 = \left(\frac{\sqrt{97}}{388} - \frac{1}{40}\right) e^{\left(-100 - 10\sqrt{97}\right)t} + \left(-\frac{\sqrt{97}}{388} - \frac{1}{40}\right) e^{\left(10\sqrt{97} - 100\right)t}$$

$$+ \frac{1}{20} \cos 10t \quad (4.197)$$

The time response of the system for the displacements x_1 and x_2 is illustrated in Fig. 4.7.

The damper kills the relative motion between m_1 and m_2, and eventually both masses move together to the right and left with same amplitude $X_1 = X_2 = 1/20$. The initial energy of the system was:

$$E_1 = \frac{1}{2} k_1 x_{01}^2 = \frac{1}{2} (100) (0.1)^2 = \frac{1}{2} \mathrm{N\,m} \quad (4.198)$$

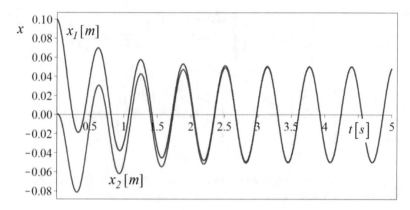

Fig. 4.7 Displacements of x_1 and x_2 versus time

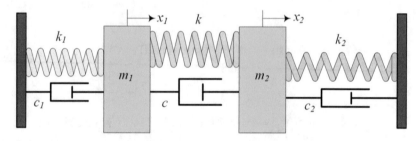

Fig. 4.8 A multi-DOF vibrating damped system with two oscillating masses

and the final level of energy of the system is:

$$E_2 = \frac{1}{2}k_1 X_1^2 + \frac{1}{2}k_2 X_2^2$$

$$= \frac{1}{2}(100)\left(\frac{1}{20}\right)^2 + \frac{1}{2}(100)\left(\frac{1}{20}\right)^2 = \frac{1}{4}\,\mathrm{N\,m} \qquad (4.199)$$

The difference between E_1 and E_2 is the amount of dissipated energy by the damper.

Example 142 A two DOF damped vibrating system. It is to compare free vibrations of a damped and undamped systems.

To investigate the effect of fixed end damping, let us consider free and damped case of the system in Fig. 4.1 as shown in Fig. 4.8:

$$\mathbf{M}\ddot{\mathbf{x}} + \mathbf{C}\dot{\mathbf{x}} + \mathbf{K}\mathbf{x} = \mathbf{0} \qquad (4.200)$$

$$\begin{bmatrix} m_1 & 0 \\ 0 & m_2 \end{bmatrix}\begin{bmatrix} \ddot{x}_1 \\ \ddot{x}_2 \end{bmatrix} + \begin{bmatrix} c + c_1 & -c \\ -c & c + c_2 \end{bmatrix}\begin{bmatrix} \dot{x}_1 \\ \dot{x}_2 \end{bmatrix} \qquad (4.201)$$

$$+ \begin{bmatrix} k + k_1 & -k \\ -k & k + k_2 \end{bmatrix} \begin{bmatrix} x_1 \\ x_2 \end{bmatrix} = \begin{bmatrix} 0 \\ 0 \end{bmatrix} \tag{4.202}$$

The equivalent set of first-order equations is:

$$^+\mathbf{M}\,\dot{\mathbf{y}} + \,^+\mathbf{K}\,\mathbf{y} = \mathbf{0} \tag{4.203}$$

$$\mathbf{y} = \begin{bmatrix} \dot{x}_1 & \dot{x}_2 & x_1 & x_2 \end{bmatrix}^T \tag{4.204}$$

$$= \begin{bmatrix} y_1 & y_2 & y_3 & y_4 \end{bmatrix}^T \tag{4.205}$$

$$^+\mathbf{M} = \begin{bmatrix} \mathbf{0} & \mathbf{M} \\ \mathbf{M} & \mathbf{C} \end{bmatrix} = \begin{bmatrix} 0 & 0 & m_1 & 0 \\ 0 & 0 & 0 & m_2 \\ m_1 & 0 & c + c_1 & -c \\ 0 & m_2 & -c & c + c_2 \end{bmatrix} \tag{4.206}$$

$$^+\mathbf{K} = \begin{bmatrix} -\mathbf{M} & \mathbf{0} \\ \mathbf{0} & \mathbf{K} \end{bmatrix} = \begin{bmatrix} -m_1 & 0 & 0 & 0 \\ 0 & -m_2 & 0 & 0 \\ 0 & 0 & k + k_1 & -k \\ 0 & 0 & -k & k + k_2 \end{bmatrix} \tag{4.207}$$

$$^+\mathbf{M}\,\dot{\mathbf{y}} + \,^+\mathbf{K}\,\mathbf{y} = \begin{bmatrix} m_1 \dot{y}_3 - m_1 y_1 \\ m_2 \dot{y}_4 - m_2 y_2 \\ (k + k_1)\, y_3 + (c + c_1)\, \dot{y}_3 - c\dot{y}_4 - k y_4 + m_1 \dot{y}_1 \\ (k + k_2)\, y_4 - c\dot{y}_3 + (c + c_2)\, \dot{y}_4 - k y_3 + m_2 \dot{y}_2 \end{bmatrix} = 0 \tag{4.208}$$

The natural frequencies of the system are the square root of the eigenvalues of $^+\mathbf{M}^{-1}\,^+\mathbf{K}$:

$$^+\mathbf{M}^{-1}\,^+\mathbf{K} = \begin{bmatrix} \frac{c+c_1}{m_1} & -\frac{c}{m_1} & \frac{k+k_1}{m_1} & -\frac{k}{m_1} \\ -\frac{c}{m_2} & \frac{c+c_2}{m_2} & -\frac{k}{m_2} & \frac{k+k_2}{m_2} \\ -1 & 0 & 0 & 0 \\ 0 & -1 & 0 & 0 \end{bmatrix} \tag{4.209}$$

A nonzero initial conditions and a set of given numerical values for the parameters of the system produce the following eigen and modal matrices:

$$\mathbf{x}\,(0) = \mathbf{x}_0 = \begin{bmatrix} 0.1 \\ 0 \end{bmatrix} \qquad \dot{\mathbf{x}}\,(0) = \dot{\mathbf{x}}_0 = \begin{bmatrix} 0 \\ 0 \end{bmatrix} \tag{4.210}$$

$$m_1 = 1\,\text{kg} \qquad m_2 = 1\,\text{kg} \tag{4.211}$$

$$k_1 = 100\,\text{kg} \qquad k_2 = 100\,\text{kg} \qquad k = 100\,\text{kg} \tag{4.212}$$

$$c_1 = 100\,\text{kg} \qquad c_2 = 2\,\text{kg} \qquad c = 2\,\text{kg} \tag{4.213}$$

$$^+\mathbf{M}^{-1}\,{}^+\mathbf{K} = \begin{bmatrix} 102 & -100 & 200 & -100 \\ -100 & 102 & -100 & 200 \\ -1 & 0 & 0 & 0 \\ 0 & -1 & 0 & 0 \end{bmatrix} \tag{4.214}$$

$$\mathbf{\Omega}^2 = \begin{bmatrix} \sqrt{9901}+101 & 0 & 0 & 0 \\ 0 & 101-\sqrt{9901} & 0 & 0 \\ 0 & 0 & 1-3i\sqrt{11} & 0 \\ 0 & 0 & 0 & 3i\sqrt{11}+1 \end{bmatrix} \tag{4.215}$$

$$\mathbf{U} = \begin{bmatrix} \sqrt{9901}+101 & 101-\sqrt{9901} & 3i\sqrt{11}-1 & -3i\sqrt{11}-1 \\ -\sqrt{9901}-101 & \sqrt{9901}-101 & 3i\sqrt{11}-1 & -3i\sqrt{11}-1 \\ -1 & -1 & 1 & 1 \\ 1 & 1 & 1 & 1 \end{bmatrix} \tag{4.216}$$

The solution of the equations will be:

$$\mathbf{y} = \sum_{i=1}^{4} A_i\,\mathbf{u}_i\,e^{-\lambda_i t} \tag{4.217}$$

$$y_1 = -0.075374\,e^{-1.4962\,t} + 0.075374\,e^{-200.5038\,t} - 0.502519\,e^{-t}\sin 9.95t \tag{4.218}$$

$$y_2 = 0.075374\,e^{-1.4962\,t} - 0.075374\,e^{-200.5038\,t} - 0.502519\,e^{-t}\sin 9.95t \tag{4.219}$$

$$y_3 = 0.050376\,e^{-1.4962\,t} - 0.00037592326\,e^{-200.5038\,t}$$
$$+ \frac{e^{-t}}{20}\cos 9.95t + 0.005025\,e^{-t}\sin 9.95t \tag{4.220}$$

$$y_4 = -0.050376\,e^{-1.4962\,t} + 0.00037592326\,e^{-200.5038\,t}$$
$$+ \frac{e^{-t}}{20}\cos 9.95t + 0.005025\,e^{-t}\sin 9.95t \tag{4.221}$$

The time response of x_1 and x_2 is illustrated in Fig. 4.9.

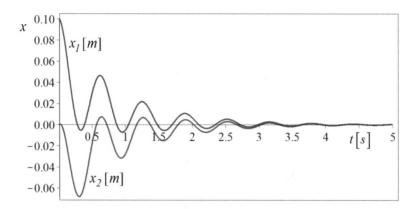

Fig. 4.9 Displacements of x_1 and x_2 versus time for the damped two DOF vibrating system

Example 143 Proportional damping. If damping matrix $\mathbf{C}$ is proportional to $\mathbf{M}$ or $\mathbf{K}$, the equations of motion can be decoupled.

The general equation of motion of free linear multi-degree of freedom (DOF) vibrating system is:

$$\mathbf{M}\,\ddot{\mathbf{x}} + \mathbf{C}\,\dot{\mathbf{x}} + \mathbf{K}\,\mathbf{x} = \mathbf{0} \tag{4.222}$$

Let us investigate the condition under which the damping matrix can be diagonalized by the principal coordinate transformation. To make the damping matrix diagonalized, it can be proportional to $\mathbf{M}$ or $\mathbf{K}$ or both:

$$\mathbf{C} = a\mathbf{M} \tag{4.223}$$

$$\mathbf{C} = b\mathbf{K} \tag{4.224}$$

$$\mathbf{C} = a\mathbf{M} + b\mathbf{K} \tag{4.225}$$

In such cases, the decoupling method by modal matrix $\mathbf{U}$ and principal coordinate $\mathbf{p}$:

$$\mathbf{x} = \mathbf{U}\,\mathbf{p} \tag{4.226}$$

Such damping is called proportional damping and often referred to as Rayleigh damping:

$$\mathbf{M}\,\ddot{\mathbf{x}} + (a\mathbf{M} + b\mathbf{K})\,\dot{\mathbf{x}} + \mathbf{K}\,\mathbf{x} = \mathbf{0} \tag{4.227}$$

Employing the modal matrix, $\mathbf{U}$, we can transform the equations of motion in terms of a set of generalized *principal coordinates* $\mathbf{p}$:

$$\mathbf{M}'\,\ddot{\mathbf{p}} + \left(a\mathbf{M}' + b\mathbf{K}'\right)\,\dot{\mathbf{p}} + \mathbf{K}'\,\mathbf{p} = \mathbf{0} \tag{4.228}$$

The new mass and stiffness matrices $\mathbf{M}'$ and $\mathbf{K}'$ are diagonal:

$$\mathbf{M}' = \mathbf{U}^T \mathbf{M} \mathbf{U} = \begin{bmatrix} m'_1 & 0 & \cdots & 0 \\ 0 & m'_2 & \cdots & 0 \\ \vdots & \vdots & \ddots & \vdots \\ 0 & 0 & \cdots & m'_n \end{bmatrix} \tag{4.229}$$

$$\mathbf{K}' = \mathbf{U}^T \mathbf{K} \mathbf{U} = \begin{bmatrix} k'_1 & 0 & \cdots & 0 \\ 0 & k'_2 & \cdots & 0 \\ \vdots & \vdots & \ddots & \vdots \\ 0 & 0 & \cdots & k'_n \end{bmatrix} \tag{4.230}$$

$$\mathbf{U} = \begin{bmatrix} \mathbf{u}_1 & \mathbf{u}_2 & \mathbf{u}_2 & \cdots & \mathbf{u}_n \end{bmatrix} \tag{4.231}$$

and, hence, the equations of motion are decoupled:

$$m'_i \, \ddot{p}_i + \left(am'_i + bk'_i\right) \, \dot{p}_i + k'_i \, p_i = 0 \qquad i = 1, 2, 3, \cdots, n \tag{4.232}$$

or

$$\ddot{p}_i + 2\xi_i \omega_i \, \dot{p}_i + \omega_i^2 \, p_i = 0 \qquad i = 1, 2, 3, \cdots, n \tag{4.233}$$

$$\omega_i^2 = \frac{k'_i}{m'_i} = \frac{\mathbf{u}_i^T \mathbf{K} \mathbf{u}_i}{\mathbf{u}_i^T \mathbf{M} \mathbf{u}_i} \tag{4.234}$$

$$\xi_i = \frac{1}{2} \left(\frac{a}{\omega_i} + b\omega_i \right) \tag{4.235}$$

$$\Omega^2 = \frac{\mathbf{K}'}{\mathbf{M}'} = \frac{\mathbf{U}^T \mathbf{K} \mathbf{U}}{\mathbf{U}^T \mathbf{M} \mathbf{U}} \tag{4.236}$$

The parameter ξ_i is called the modal damping ratio. The general solution of Eq. (4.233) for $\xi_i < 1$ will be:

$$p_i = e^{-\xi_i \omega_i t} \left(A_i \cos\left(\omega_i \sqrt{1 - \xi_i^2} \, t \right) + B_i \sin\left(\omega_i \sqrt{1 - \xi_i^2} \, t \right) \right) \tag{4.237}$$

where A_i and B_i are determined from initial conditions. The solution in the original generalized coordinates $\mathbf{x}$ will then be determined by substituting all p_i in Eq. (4.226). If the damping matrix $\mathbf{C}$ is not proportional to $\mathbf{M}$ and $\mathbf{K}$, the principal coordinates do not uncouple Eq. (4.222).

Example 144 Alternative method for converting damped system to first-order equations. We can use other methods to convert a set of n second-order damped equations to a set of $2n$ first-order equations to be solved by exponential matrix function.

Consider a multi-DOF damped linear vibrating system:

$$\mathbf{M}\ddot{\mathbf{x}} + \mathbf{C}\dot{\mathbf{x}} + \mathbf{K}\mathbf{x} = 0 \tag{4.238}$$

Let us define a new n component coordinate vector $\mathbf{z}$, to convert the set of equations to first-order equations:

$$\dot{\mathbf{x}} = \mathbf{z} \tag{4.239}$$

$$\mathbf{M}\dot{\mathbf{z}} + \mathbf{C}\mathbf{z} + \mathbf{K}\mathbf{x} = 0 \tag{4.240}$$

$$\dot{\mathbf{z}} = -\mathbf{M}^{-1}\mathbf{C}\mathbf{z} - \mathbf{M}^{-1}\mathbf{K}\mathbf{x} \tag{4.241}$$

Now we define a new $2n$ component coordinate vector $\mathbf{y}$:

$$\mathbf{y} = \begin{bmatrix} \mathbf{x} \\ \mathbf{z} \end{bmatrix} = \begin{bmatrix} \mathbf{x} \\ \dot{\mathbf{x}} \end{bmatrix} \tag{4.242}$$

and make $2n$ first-order equations:

$$\begin{bmatrix} \dot{\mathbf{x}} \\ \dot{\mathbf{z}} \end{bmatrix} = \begin{bmatrix} 0 & \mathbf{I} \\ -\mathbf{M}^{-1}\mathbf{K} & -\mathbf{M}^{-1}\mathbf{C} \end{bmatrix} \begin{bmatrix} \mathbf{x} \\ \mathbf{z} \end{bmatrix} \tag{4.243}$$

It can be written in a compact form:

$$\dot{\mathbf{y}} = \mathbf{A}\mathbf{y} \tag{4.244}$$

where:

$$\mathbf{A} = \begin{bmatrix} 0 & \mathbf{I} \\ -\mathbf{M}^{-1}\mathbf{K} & -\mathbf{M}^{-1}\mathbf{C} \end{bmatrix} \tag{4.245}$$

The solution of the set of coupled first-order linear homogeneous equations is:

$$\mathbf{y} = \sum_{j=1}^{2n} e^{-\lambda_j t} C_j \mathbf{u}_j \tag{4.246}$$

where $\mathbf{u}_j$ is the eigenvectors and λ_j, $j = 1, 2, \cdots, 2n$ are eigenvalues of the $2n \times 2n$ coefficient matrix $\mathbf{A}$. This solution can equivalently be expressed by:

$$\mathbf{y}(t) = e^{-\mathbf{A}t}\mathbf{C} = e^{-\mathbf{A}t}\mathbf{y}_0 \qquad (4.247)$$

where:

$$e^{-\mathbf{A}t} = \mathbf{I} - \mathbf{A}t + \frac{1}{2!}\mathbf{A}^2 t^2 - \frac{1}{3!}\mathbf{A}^3 t^3 + \frac{1}{4!}\mathbf{A}^4 t^4 - \cdots$$

$$= \sum_{j=0}^{\infty} (-1)^j \frac{(\mathbf{A}t)^j}{j!} \qquad (4.248)$$

We may also express the solution of the homogeneous equation (5.248) as:

$$\mathbf{x} = e^{-\mathbf{A}t}\mathbf{C} = \mathbf{U}\, e^{-\mathbf{\Lambda}t}\, \mathbf{U}^{-1}\, \mathbf{C} \qquad (4.249)$$

where $\mathbf{\Lambda}$ is the associated diagonalized matrix of $\mathbf{A}$ and $\mathbf{U}$ is the modal matrix of $\mathbf{A}$:

$$\mathbf{\Lambda} = \mathbf{U}^{-1}\,\mathbf{A}\,\mathbf{U} = \begin{bmatrix} \lambda_1 & 0 & \cdots & 0 \\ 0 & \lambda_2 & \cdots & 0 \\ \vdots & \vdots & \ddots & \vdots \\ 0 & 0 & \cdots & \lambda_{2n} \end{bmatrix} \qquad (4.250)$$

$$\mathbf{U} = \begin{bmatrix} \mathbf{u}_1 & \mathbf{u}_2 & \cdots & \mathbf{u}_{2n} \end{bmatrix} \qquad (4.251)$$

4.2   ★ Forced Vibrations

Assume that a vibrating system with n degrees of freedom is subjected to given force functions $\mathbf{f}(\mathbf{x}, \dot{\mathbf{x}}, t)$:

$$\mathbf{M}\,\ddot{\mathbf{x}} + \mathbf{C}\,\dot{\mathbf{x}} + \mathbf{K}\,\mathbf{x} = \mathbf{f}(\mathbf{x}, \dot{\mathbf{x}}, t) \qquad (4.252)$$

$$\mathbf{x} = \begin{bmatrix} x_1 & x_2 & \cdots & x_n \end{bmatrix}^T \qquad (4.253)$$

The complete solution of Eq. (4.252) has two parts: a homogenous solution $\mathbf{x}_h$, plus the a particular solution $\mathbf{x}_p$. The homogenous solution $\mathbf{x}_h$ is the solution of free vibrations for $\mathbf{f} = \mathbf{0}$, and the particular solution is the special solutions of the equations associated to the forcing function $\mathbf{f}(\mathbf{x}, \dot{\mathbf{x}}, t)$. The homogeneous solution depends on system parameters $\mathbf{M}$, $\mathbf{C}$, and $\mathbf{K}$, while the particular solution is the response due to the particular form of the excitation $\mathbf{f}(\mathbf{x}, \dot{\mathbf{x}}, t)$.

The particular solution $\mathbf{x}_p$ of a forced vibrating system is not possible to be found for general force functions $\mathbf{f}(\mathbf{x}, \dot{\mathbf{x}}, t)$. However, if the force functions are continuous functions of time $\mathbf{f}(t)$ and are combinations of following functions:

1. Constant, such as $\mathbf{f} = \mathbf{F}$
2. Polynomials in t, such as $\mathbf{f} = \mathbf{a}_0 + \mathbf{a}_1 t + \mathbf{a}_2 t^2 + \cdots + \mathbf{a}_n t^n$
3. Exponential functions, such as $\mathbf{f} = \mathbf{F} e^{at}$
4. Harmonic functions, such as $\mathbf{f} = \mathbf{F}_1 \sin \omega t + \mathbf{F}_2 \cos \omega t$

then the particular solution $\mathbf{x}_p(t)$ of the linear equation of motion will have the same form as the forcing terms:

1. $\mathbf{x}_p(t) = $ constant, such as $\mathbf{x}_p(t) = \mathbf{X}$
2. $\mathbf{x}_p(t) = $ polynomial of the same degree, such as $\mathbf{x}_p(t) = \mathbf{C}_0 + \mathbf{C}_1 t + \mathbf{C}_2 t^2 + \cdots + \mathbf{C}_n t^n$
3. $\mathbf{x}_p(t) = $ exponential functions, such as $\mathbf{x}_p(t) = \mathbf{C} e^{at}$
4. $\mathbf{x}_p(t) = $ harmonic functions of the same frequency, such as $\mathbf{x}_p(t) = \mathbf{A} \sin \omega t + \mathbf{B} \cos \omega t$

The coefficients of particular solutions are calculated by substituting the solution in the differential equation and solving the resultant algebraic equations to satisfy the differential equations. Every element of the force function $\mathbf{f}$ can be different than the other. The linearity of the equations allows us to use the superposition principle and determine the whole particular solutions $\mathbf{x}_p$ by adding the particular solutions to every individual component of $\mathbf{f}$. The initial conditions will appear in the coefficients of the homogenous solution of the differential equation of motion. The initial conditions must be applied after the particular solution $\mathbf{x}_p$ and the general homogenous solution $\mathbf{x}_h$ are found and added together to have the general solution $\mathbf{x} = \mathbf{x}_h + \mathbf{x}_p$.

The best method to solve forced vibrations of multi-DOF systems with general damping and general forcing functions is to convert the n second- order differential equation (4.252) of motion to $2n$ first-order differential equations and develop the exponential solutions:

$$\dot{\mathbf{y}} + \mathbf{A}\mathbf{y} = \mathbf{F} \tag{4.254}$$

$$\mathbf{A} = {}^+\mathbf{M}^{-1}\,{}^+\mathbf{K} \tag{4.255}$$

$$\mathbf{F} = {}^+\mathbf{M}^{-1}\,{}^+\mathbf{f} \tag{4.256}$$

$$\mathbf{y} = \begin{bmatrix} \dot{\mathbf{x}} \\ \mathbf{x} \end{bmatrix} \tag{4.257}$$

$$ {}^+\mathbf{M} = \begin{bmatrix} \mathbf{0} & \mathbf{M} \\ \mathbf{M} & \mathbf{C} \end{bmatrix} \tag{4.258}$$

$$^{+}\mathbf{K} = \begin{bmatrix} -\mathbf{M} & \mathbf{0} \\ \mathbf{0} & \mathbf{K} \end{bmatrix} \qquad (4.259)$$

$$^{+}\mathbf{f} = \begin{bmatrix} \mathbf{0} \\ \mathbf{f} \end{bmatrix} \qquad (4.260)$$

The homogenous solution of Eq. (4.254) has been developed in the previous section on free vibrations of damped systems. The solution of any linear forced vibrating system of any order can also be expressed by a set of $2n$ coupled first-order linear system (4.254):

$$\mathbf{y} = e^{-\mathbf{A}t} \int_0^t e^{\mathbf{A}s} \, \mathbf{F}(s) \, ds + e^{-\mathbf{A}t} \, \mathbf{y}_0 \qquad (4.261)$$

$$\mathbf{y} = \begin{bmatrix} y_1 & y_2 & y_3 & \cdots & y_{2n} \end{bmatrix}^T \qquad (4.262)$$

$$\mathbf{F} = \begin{bmatrix} F_1 & F_2 & F_3 & \cdots & F_{2n} \end{bmatrix}^T \qquad (4.263)$$

Proof Introducing a set of $2n$ new variables (4.257) will convert the n second-order differential equation (4.252) to $2n$ first-order Eq. (4.254). The set of equations can always be expressed in matrix form:

$$\dot{\mathbf{y}} + \mathbf{A}\,\mathbf{y} = \mathbf{F} \qquad (4.264)$$

$$\mathbf{y} = \begin{bmatrix} y_1 & y_2 & y_3 & \cdots & y_{2n} \end{bmatrix}^T \qquad (4.265)$$

$$\dot{\mathbf{y}} = \begin{bmatrix} \dot{y}_1 & \dot{y}_2 & \dot{y}_3 & \cdots & \dot{y}_{2n} \end{bmatrix}^T \qquad (4.266)$$

The solution of Eq. (4.264) is a combination of homogenous $\mathbf{x}_h$ and particular $\mathbf{x}_p$ solutions. Multiplying $e^{\mathbf{A}t}$ by both sides of Eq. (4.264) yields:

$$e^{\mathbf{A}t}\,(\dot{\mathbf{y}} + \mathbf{A}\,\mathbf{y}) = \frac{d}{dt}\left(e^{\mathbf{A}t}\,\mathbf{y}\right) = e^{\mathbf{A}t}\,\mathbf{F} \qquad (4.267)$$

Integrating this equation yields:

$$e^{\mathbf{A}t}\,\mathbf{y} = \int_0^t e^{\mathbf{A}s}\,\mathbf{F}\,ds + \mathbf{y}_0 \qquad (4.268)$$

or:

$$\mathbf{x} = e^{-\mathbf{A}t} \int_0^t e^{\mathbf{A}s} \, \mathbf{F} \, ds + e^{-\mathbf{A}t} \, \mathbf{y}_0 \qquad (4.269)$$

If the forced system is an undamped linear n degree of freedom (DOF) vibrating system

$$\mathbf{M}\ddot{\mathbf{x}} + \mathbf{K}\mathbf{x} = \mathbf{f} \qquad (4.270)$$

$$\mathbf{x} = \begin{bmatrix} x_1 \ x_2 \ \cdots \ x_n \end{bmatrix}^T \qquad (4.271)$$

then, employing the *principal coordinates* $\mathbf{p}$:

$$\mathbf{x} = \mathbf{U}\mathbf{p} \qquad (4.272)$$

the equations of motion can be decoupled:

$$\mathbf{M}'\ddot{\mathbf{p}} + \mathbf{K}'\mathbf{p} = \mathbf{F} \qquad (4.273)$$

The new mass and stiffness matrices $\mathbf{M}'$, $\mathbf{K}'$ and forcing term $\mathbf{F}$ are diagonal:

$$\mathbf{M}' = \mathbf{U}^T \mathbf{M} \mathbf{U} = \begin{bmatrix} m_1' & 0 & \cdots & 0 \\ 0 & m_2' & \cdots & 0 \\ \vdots & \vdots & \ddots & \vdots \\ 0 & 0 & \cdots & m_n' \end{bmatrix} \qquad (4.274)$$

$$\mathbf{K}' = \mathbf{U}^T \mathbf{K} \mathbf{U} = \begin{bmatrix} k_1' & 0 & \cdots & 0 \\ 0 & k_2' & \cdots & 0 \\ \vdots & \vdots & \ddots & \vdots \\ 0 & 0 & \cdots & k_n' \end{bmatrix} \qquad (4.275)$$

$$\mathbf{F} = \mathbf{U}^T \mathbf{f} = [F_1 \ F_2 \ \cdots \ F_n]^T \qquad (4.276)$$

$$\mathbf{U} = [\mathbf{u}_1 \ \mathbf{u}_2 \ \mathbf{u}_2 \ \cdots \ \mathbf{u}_n] \qquad (4.277)$$

Therefore, the equations of motion are decoupled:

$$m_i' \ddot{p}_i + k_i' p_i = F_i \qquad i = 1, 2, 3, \cdots, n \qquad (4.278)$$

The solution of the decoupled system of equations is addition of homogenous solution p_h and particular solution p_p:

$$p_i = p_h + p_p \qquad (4.279)$$

$$p_h = A_i \sin \omega_i t + B_i \sin \omega_i t \qquad (4.280)$$

The particular solution p_p depends on the force function F_i and will be calculated by the techniques of one DOF forced excitation in previous chapters.

The transformation matrix $\mathbf{U}$ is the *modal matrix* of the system. The columns of the $n \times n$ modal matrix $\mathbf{U}$ are the mode shapes of the system. The principal coordinates $\mathbf{p} = \mathbf{U}^T \mathbf{x}$ are also called the *natural coordinates* of the system. Undamped systems are the base to discover the principal characteristics of a system such as natural frequencies, mode shapes, modal matrix, and principal coordinates.

Having the solutions of the decoupled equations, $p_i(t)$, we can reconstruct the solution in the original coordinates x_i by inverse modal transformation:

$$\mathbf{x} = \mathbf{U}\,\mathbf{p} \tag{4.281}$$

$$\begin{bmatrix} x_1(t) \\ x_2(t) \\ \dots \\ x_n(t) \end{bmatrix} = \begin{bmatrix} \mathbf{u}_1 \ \mathbf{u}_2 \ \mathbf{u}_2 \ \cdots \ \mathbf{u}_n \end{bmatrix} \begin{bmatrix} p_1(t) \\ p_2(t) \\ \dots \\ p_n(t) \end{bmatrix} \tag{4.282}$$

∎

Example 145 ★ Proportional damping. The decoupling method is also applicable for forced vibrating systems with proportional damping $\mathbf{C}$.

Consider a multi-DOF linear vibrating system:

$$\mathbf{M}\,\ddot{\mathbf{x}} + \mathbf{C}\,\dot{\mathbf{x}} + \mathbf{K}\,\mathbf{x} = \mathbf{f}(\mathbf{x}, \dot{\mathbf{x}}, t) \tag{4.283}$$

with damping $\mathbf{C}$ to be proportional to mass and stiffness matrices:

$$\mathbf{C} = a\mathbf{M} + b\mathbf{K}$$

$$\mathbf{M}\,\ddot{\mathbf{x}} + (a\mathbf{M} + b\mathbf{K})\,\dot{\mathbf{x}} + \mathbf{K}\,\mathbf{x} = \mathbf{f} \tag{4.284}$$

In such cases, the decoupling method by modal matrix $\mathbf{U}$ and principal coordinate $\mathbf{p}$:

$$\mathbf{x} = \mathbf{U}\,\mathbf{p} \tag{4.285}$$

The modal matrix, $\mathbf{U}$, transforms the equations of motion to *principal coordinates* $\mathbf{p}$:

$$\mathbf{M}'\,\ddot{\mathbf{p}} + (a\mathbf{M}' + b\mathbf{K}')\,\dot{\mathbf{p}} + \mathbf{K}'\,\mathbf{p} = \mathbf{f}' \tag{4.286}$$

The new mass and stiffness matrices $\mathbf{M}'$ and $\mathbf{K}'$ are diagonal:

$$\mathbf{M}' = \mathbf{U}^T \mathbf{M} \mathbf{U} = \begin{bmatrix} m_1' & 0 & \cdots & 0 \\ 0 & m_2' & \cdots & 0 \\ \vdots & \vdots & \ddots & \vdots \\ 0 & 0 & \cdots & m_n' \end{bmatrix} \tag{4.287}$$

$$\mathbf{K}' = \mathbf{U}^T \mathbf{K} \mathbf{U} = \begin{bmatrix} k_1' & 0 & \cdots & 0 \\ 0 & k_2' & \cdots & 0 \\ \vdots & \vdots & \ddots & \vdots \\ 0 & 0 & \cdots & k_n' \end{bmatrix} \tag{4.288}$$

$$\mathbf{f}' = \mathbf{U}^T \mathbf{f} \tag{4.289}$$

$$\mathbf{U} = \begin{bmatrix} \mathbf{u}_1 & \mathbf{u}_2 & \mathbf{u}_2 & \cdots & \mathbf{u}_n \end{bmatrix} \tag{4.290}$$

and, hence, the equations of motion are decoupled:

$$m_i' \ddot{p}_i + \left(am_i' + bk_i' \right) \dot{p}_i + k_i' p_i = f_i' \qquad i = 1, 2, 3, \cdots, n \tag{4.291}$$

or:

$$\ddot{p}_i + 2\xi_i \omega_i \dot{p}_i + \omega_i^2 p_i = F_i \qquad i = 1, 2, 3, \cdots, n \tag{4.292}$$

$$F_i = \frac{f_i'}{m_i'} \tag{4.293}$$

$$\omega_i^2 = \frac{k_i'}{m_i'} = \frac{\mathbf{u}_i^T \mathbf{K} \mathbf{u}_i}{\mathbf{u}_i^T \mathbf{M} \mathbf{u}_i} \tag{4.294}$$

$$\xi_i = \frac{1}{2} \left(\frac{a}{\omega_i} + b\omega_i \right) \tag{4.295}$$

$$\Omega^2 = \frac{\mathbf{K}'}{\mathbf{M}'} = \frac{\mathbf{U}^T \mathbf{K} \mathbf{U}}{\mathbf{U}^T \mathbf{M} \mathbf{U}} \tag{4.296}$$

The homogenous solution of Eq. (4.292) for $\xi_i < 1$ will be:

$$p_{hi} = e^{-\xi_i \omega_i t} \left(A_i \cos \left(\omega_i \sqrt{1 - \xi_i^2}\, t \right) + B_i \sin \left(\omega_i \sqrt{1 - \xi_i^2}\, t \right) \right) \tag{4.297}$$

where A_i and B_i are determined from initial conditions:

$$p_{hi} = e^{-\xi_i \omega_i t} \left(x_{i0} \cos\left(\omega_i \sqrt{1 - \xi_i^2}\, t\right) + \frac{\dot{x}_{i0} + \xi_i \omega_i x_{i0}}{\omega_i \sqrt{1 - \xi_i^2}} \sin\left(\omega_i \sqrt{1 - \xi_i^2}\, t\right) \right)$$

(4.298)

The particular solution of Eq. (4.292) is:

$$p_{pi} = \frac{1}{m_i \omega_i \sqrt{1 - \xi_i^2}} \int_0^t e^{-\xi_i \omega_i (t-\tau)} F_i(\tau) \sin\left(\omega_i \sqrt{1 - \xi_i^2}\,(t-\tau)\right) d\tau$$

(4.299)

The solution in the original generalized coordinates $\mathbf{x}$ will then be determined by substituting all p_i in Eq. (4.292). If the damping matrix $\mathbf{C}$ cannot be converted to be proportional to $\mathbf{M}$ and $\mathbf{K}$, the principal coordinates do not uncouple Eq. (4.292).

Example 146 Converting forced system to first-order equations. There are other methods to convert a set of n second-order equations to another set of $2n$ first- order equations. Here is one of them.

Consider a multi-*DOF* forced linear vibrating system:

$$\mathbf{M}\,\ddot{\mathbf{x}} + \mathbf{C}\,\dot{\mathbf{x}} + \mathbf{K}\,\mathbf{x} = \mathbf{f}(\mathbf{x}, \dot{\mathbf{x}}, t)$$

(4.300)

Let us define a new n component coordinate vector $\mathbf{z}$:

$$\dot{\mathbf{x}} = \mathbf{z}$$

(4.301)

and convert the set of second-order equations to a set of first-order equations:

$$\mathbf{M}\,\dot{\mathbf{z}} + \mathbf{C}\,\mathbf{z} + \mathbf{K}\,\mathbf{x} = \mathbf{f}(\mathbf{x}, \mathbf{z}, t)$$

(4.302)

$$\dot{\mathbf{z}} = \mathbf{M}^{-1}\,\mathbf{f} - \mathbf{M}^{-1}\,\mathbf{C}\,\mathbf{z} - \mathbf{M}^{-1}\,\mathbf{K}\,\mathbf{x}$$

(4.303)

Now we define a new $2n$ component coordinate vector $\mathbf{y}$:

$$\mathbf{y} = \begin{bmatrix} \mathbf{x} \\ \mathbf{z} \end{bmatrix} = \begin{bmatrix} \mathbf{x} \\ \dot{\mathbf{x}} \end{bmatrix}$$

(4.304)

and make $2n$ first-order equations for the original n second-order equations:

$$\begin{bmatrix} \dot{\mathbf{x}} \\ \dot{\mathbf{z}} \end{bmatrix} = \begin{bmatrix} \mathbf{0} & \mathbf{I} \\ -\mathbf{M}^{-1}\,\mathbf{K} & -\mathbf{M}^{-1}\,\mathbf{C} \end{bmatrix} \begin{bmatrix} \mathbf{x} \\ \mathbf{z} \end{bmatrix} + \begin{bmatrix} \mathbf{0} \\ \mathbf{M}^{-1}\,\mathbf{f} \end{bmatrix}$$

(4.305)

or in a compact form:

$$\dot{\mathbf{y}} = \mathbf{A}\,\mathbf{y} + \mathbf{F}$$

(4.306)

where:

$$A = \begin{bmatrix} 0 & I \\ -M^{-1}K & -M^{-1}C \end{bmatrix} \qquad F = \begin{bmatrix} 0 \\ M^{-1}f \end{bmatrix} \qquad (4.307)$$

Example 147 A general two degree of freedom system with a harmonic force excitation. It is a practice on determining particular solution of a multi-DOF system when one component of the forcing functions is a harmonic.

Consider a two DOF vibrating system that its motion is expressed by two coupled differential equations. A harmonic forcing function is applied on only one of the coordinates:

$$m_1\ddot{x}_1 + c_{11}\dot{x}_1 + c_{12}\dot{x}_2 + k_{11}x_1 + k_{12}x_2 = F \sin \omega t \qquad (4.308)$$

$$m_2\ddot{x}_2 + c_{21}\dot{x}_1 + c_{22}\dot{x}_2 + k_{21}x_1 + k_{22}x_2 = 0 \qquad (4.309)$$

Let us only determine the particular solution here. Because the forcing function is harmonic, then the particular responses will be harmonic functions of the same frequency:

$$\mathbf{x}_p(t) = \mathbf{A} \sin \omega t + \mathbf{B} \cos \omega t \qquad (4.310)$$

$$x_1 = A_1 \sin \omega t + B_1 \cos \omega t \qquad (4.311)$$

$$x_2 = A_2 \sin \omega t + B_2 \cos \omega t \qquad (4.312)$$

$$\dot{x}_1 = A_1\omega \cos \omega t - B_1\omega \sin \omega t \qquad (4.313)$$

$$\dot{x}_2 = A_2\omega \cos \omega t - B_2\omega \sin \omega t \qquad (4.314)$$

$$\ddot{x}_1 = -A_1\omega^2 \sin \omega t - B_1\omega^2 \cos \omega t \qquad (4.315)$$

$$\ddot{x}_2 = -A_2\omega^2 \sin \omega t - B_2\omega^2 \cos \omega t \qquad (4.316)$$

Substituting the solution function into the equations of motion will provide us with two algebraic equations:

$$\left(A_1k_{11} + A_2 \left(k_{12} - m_1\omega^2 \right) - \omega \left(B_1c_{11} + B_2c_{12} \right) - F \right) \sin t\omega$$
$$+ \left(B_1k_{11} + B_2 \left(k_{12} - m_1\omega^2 \right) + \omega \left(A_1c_{11} + A_2c_{12} \right) \right) \cos t\omega = 0 \quad (4.317)$$

$$\left(A_1k_{21} + A_2 \left(k_{22} - m_2\omega^2 \right) - \omega \left(B_1c_{21} + B_2c_{22} \right) \right) \sin t\omega$$
$$+ \left(B_1k_{21} + B_2 (k_{22} - m_2) \omega^2 + \omega \left(A_1c_{21} + \omega c_{22} \right) \right) \cos t\omega = 0 \quad (4.318)$$

Because $\sin \omega t$ and $\cos \omega t$ are orthogonal functions, the only possible solution for these equations is having all coefficients of $\sin \omega t$ and $\cos \omega t$ to be zero. This will give us a set of algebraic equations to determine the coefficients $\mathbf{A}$ and $\mathbf{B}$:

$$
\begin{bmatrix}
k_{11} & k_{12} - m_1\omega^2 & -\omega c_{11} & -c_{12}\omega \\
\omega c_{11} & \omega c_{12} & k_{11} & k_{12} - m_1\omega^2 \\
k_{21} & k_{22} - m_2\omega^2 & -\omega c_{21} & -\omega c_{22} \\
\omega c_{21} & \omega c_{22} & k_{21} & k_{22} - m_2\omega^2
\end{bmatrix}
\begin{bmatrix}
A_1 \\ A_2 \\ B_1 \\ B_2
\end{bmatrix}
=
\begin{bmatrix}
F \\ 0 \\ 0 \\ 0
\end{bmatrix}
\tag{4.319}
$$

Determining A_1, A_2, B_1, and B_2 and substituting them in (4.311) and (4.311) provide the particular solutions.

Example 148 Impulse response. Impulse will only change the initial conditions, and, hence, the response of a multi-DOF system will be the transient response to new set of initial conditions. Here is an example of a 2 DOF undamped system.

Consider a statically coupled two degree of freedom system that an impulse I acting at m_1 when the system is at rest:

$$
m_1\ddot{x}_1 + k_{11}x_1 + k_{12}x_2 = f_1(t) \tag{4.320}
$$

$$
m_2\ddot{x}_2 + k_{21}x_1 + k_{22}x_2 = 0 \tag{4.321}
$$

$$
f_1(t) = I \tag{4.322}
$$

The impulse will change the velocity of m_1 according to impulse-momentom principle:

$$
I = m_1\dot{x}_1 \tag{4.323}
$$

The impulse will change the rest initial conditions:

$$
x_1(0) = 0 \qquad x_2(0) = 0 \qquad \dot{x}_1(0) = 0 \qquad \dot{x}_2(0) = 0 \tag{4.324}
$$

to:

$$
x_1(0) = 0 \qquad x_2(0) = 0 \qquad \dot{x}_1(0) = \frac{I}{m_1} \qquad \dot{x}_2(0) = 0 \tag{4.325}
$$

The solution of free vibration of this system is found as Eqs. (4.126)–(4.127):

$$
x_1 = A_1 \sin(\omega_1 t + \varphi_1) + A_2 \sin(\omega_2 t + \varphi_2) \tag{4.326}
$$

$$
x_2 = u_1 A_1 \sin(\omega_1 t + \varphi_1) + u_2 A_2 \sin(\omega_2 t + \varphi_2) \tag{4.327}
$$

Here we need to determine the unknown amplitudes A_1 and A_2 and phase angles φ_1 and φ_2 from the initial conditions (4.325):

$$A_1 \sin \varphi_1 + A_2 \sin \varphi_2 = 0 \tag{4.328}$$

$$u_1 A_1 \sin \varphi_1 + u_2 A_2 \sin \varphi_2 = 0 \tag{4.329}$$

$$A_1 \omega_1 \cos \varphi_1 - A_2 \omega_2 \cos \varphi_2 = \frac{I}{m_1} \tag{4.330}$$

$$u_1 A_1 \omega_1 \cos \varphi_1 - u_2 A_2 \omega_2 \cos \varphi_2 = 0 \tag{4.331}$$

Calculating A_1, A_2, φ_1, and φ_2 competes the solutions:

$$x_1 = \frac{I \sin \omega_1 t}{m_1^2 \omega_1 (1 - u_1/u_2)} + \frac{I \sin \omega_2 t}{m_1^2 \omega_2 (1 - u_2/u_1)} \tag{4.332}$$

$$x_2 = \frac{u_1 I \sin \omega_1 t}{m_1^2 \omega_1 (1 - u_1/u_2)} + \frac{u_2 I \sin \omega_2 t}{m_1^2 \omega_2 (1 - u_2/u_1)} \tag{4.333}$$

where ω_1, ω_2, u_1, and u_2 are calculated in Eqs. (4.118) and (4.122).

Considering the force $f_1(t)$ to be divided into a large number of impulses, the response of the masses m_1 and m_2 will be:

$$x_1 = \frac{1}{m_1 (u_2 - u_1)} \int_0^t \left(\frac{u_2}{\omega_1} \sin \omega_1 (t - \tau) - \frac{u_1}{\omega_2} \sin \omega_2 (t - \tau) \right) f_1 \, d\tau \tag{4.334}$$

$$x_2 = \frac{u_1 u_2}{m_1 (u_2 - u_1)} \int_0^t \left(\frac{1}{\omega_1} \sin \omega_1 (t - \tau) - \frac{1}{\omega_2} \sin \omega_2 (t - \tau) \right) f_1 \, d\tau \tag{4.335}$$

The response of the system to forces $f_1(t)$ applied to mass m_1 and $f_2(t)$ applied to mass m_2 can be formed from Eqs. (4.334) and (4.335) by manipulation of subscripts and superposition.

The response of the system to a force $f_1(t)$ applied to mass m_1 can also be determined by using expression of the equations in principal coordinates. Introducing the change of coordinates of Eqs. (4.131)–(4.132):

$$x_1 = A_1 p_1 + A_2 p_2 \tag{4.336}$$

$$x_2 = u_1 A_1 p_1 + u_2 A_2 p_2 \tag{4.337}$$

into Eqs. (4.320)–(4.321) and following the analysis of Eqs. (4.320)–(4.141) yield:

$$\ddot{p}_1 + \omega_1^2 \, p_1 = A_1 f_1 \tag{4.338}$$

$$\ddot{p}_2 + \omega_2^2 \, p_2 = A_2 f_1 \tag{4.339}$$

where A_1 and A_2 satisfy the normalizing conditions (4.129)–(4.130). The solutions of equations (4.338)–(4.339) are given in terms of Duhamel's integral as Equation (3.506) for no force and $\xi = 0$:

$$x_1 = \frac{A_1}{\omega_1} \int_0^t f_1(\tau) \sin \omega_1 (t - \tau) \, d\tau \tag{4.340}$$

$$x_2 = \frac{A_2}{\omega_2} \int_0^t f_1(\tau) \sin \omega_2 (t - \tau) \, d\tau \tag{4.341}$$

Returning to the original coordinates, x_1 and x_2, the response of masses m_1 and m_2 will be:

$$x_1 = \frac{A_1^2}{\omega_1} \int_0^t \sin \omega_1 (t - \tau) f_1 \, d\tau - \frac{A_2^2}{\omega_2} \int_0^t \sin \omega_2 (t - \tau) f_1 \, d\tau \tag{4.342}$$

$$x_2 = \frac{u_1 A_1^2}{\omega_1} \int_0^t \sin \omega_1 (t - \tau) f_1 \, d\tau - \frac{u_2 A_2^2}{\omega_2} \int_0^t \sin \omega_2 (t - \tau) f_1 \, d\tau \tag{4.343}$$

From (4.129) and (4.130), we have:

$$A_1^2 = \frac{1}{m_1 + m_2 u_1^2} \qquad A_2^2 = \frac{1}{m_1 + m_2 u_2^2} \tag{4.344}$$

and from the relation (4.123):

$$u_1 u_2 = -m_1 m_2 \tag{4.345}$$

we have:

$$A_1^2 = \frac{1}{m_1 (1 - u_1/u_2)} \qquad A_2^2 = \frac{1}{m_1 (1 - u_2/u_1)} \tag{4.346}$$

and, hence, the expressions (4.342) and (4.343) are equivalent to (4.334) and (4.335).

Example 149 General two *DOF* vibrating system. Study a general low *DOF* vibrating system will show all steps and features of analysis of forced multi-*DOF* systems.

Figure 4.10 illustrates a two *DOF* vibrating system. Its equations of motion are:

$$\mathbf{M}\ddot{x} + \mathbf{C}\dot{x} + \mathbf{K}x = \mathbf{f} \tag{4.347}$$

$$\begin{bmatrix} m_1 & 0 \\ 0 & m_2 \end{bmatrix} \begin{bmatrix} \ddot{x}_1 \\ \ddot{x}_2 \end{bmatrix} + \begin{bmatrix} c + c_1 & -c \\ -c & c + c_2 \end{bmatrix} \begin{bmatrix} \dot{x}_1 \\ \dot{x}_2 \end{bmatrix}$$

$$+ \begin{bmatrix} k + k_1 & -k \\ -k & k + k_2 \end{bmatrix} \begin{bmatrix} x_1 \\ x_2 \end{bmatrix} = \begin{bmatrix} f_1 \\ f_2 \end{bmatrix} \tag{4.348}$$

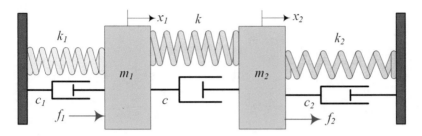

Fig. 4.10 A two degree of freedom forced system

Let us define a new set of variables **y**:

$$\mathbf{y} = \begin{bmatrix} \dot{\mathbf{x}} \\ \mathbf{x} \end{bmatrix} \tag{4.349}$$

to convert the equations of motion to a set of first-order equations:

$$\dot{\mathbf{y}} + \mathbf{A}\,\mathbf{y} = \mathbf{F} \tag{4.350}$$

$$\mathbf{A} = {}^{+}\mathbf{M}^{-1}\,{}^{+}\mathbf{K} = \begin{bmatrix} \frac{c+c_1}{m_1} & -\frac{c}{m_1} & \frac{k+k_1}{m_1} & -\frac{k}{m_1} \\ -\frac{c}{m_2} & \frac{c+c_2}{m_2} & -\frac{k}{m_2} & \frac{k+k_2}{m_2} \\ -1 & 0 & 0 & 0 \\ 0 & -1 & 0 & 0 \end{bmatrix} \tag{4.351}$$

$$\mathbf{F} = {}^{+}\mathbf{M}^{-1}\,{}^{+}\mathbf{f} = \begin{bmatrix} \frac{f_1}{m_1} & \frac{f_2}{m_2} & 0 & 0 \end{bmatrix}^{T} \tag{4.352}$$

$${}^{+}\mathbf{M} = \begin{bmatrix} \mathbf{0} & \mathbf{M} \\ \mathbf{M} & \mathbf{C} \end{bmatrix} = \begin{bmatrix} 0 & 0 & m_1 & 0 \\ 0 & 0 & 0 & m_2 \\ m_1 & 0 & c+c_1 & -c \\ 0 & m_2 & -c & c+c_2 \end{bmatrix} \tag{4.353}$$

$${}^{+}\mathbf{K} = \begin{bmatrix} -\mathbf{M} & \mathbf{0} \\ \mathbf{0} & \mathbf{K} \end{bmatrix} = \begin{bmatrix} -m_1 & 0 & 0 & 0 \\ 0 & -m_2 & 0 & 0 \\ 0 & 0 & k+k_1 & -k \\ 0 & 0 & -k & k+k_2 \end{bmatrix} \tag{4.354}$$

$${}^{+}\mathbf{f} = \begin{bmatrix} \mathbf{0} \\ \mathbf{f} \end{bmatrix} \tag{4.355}$$

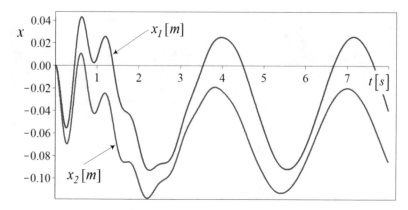

Fig. 4.11 Forced vibration response of a two DOF system

A zero initial condition and a set of given numerical values for the parameters of the system and a set of force functions produce response shown in Fig. 4.11:

$$\mathbf{x}\,(0) = \mathbf{x}_0 = \begin{bmatrix} 0 \\ 0 \end{bmatrix} \qquad \dot{\mathbf{x}}\,(0) = \dot{\mathbf{x}}_0 = \begin{bmatrix} 0 \\ 0 \end{bmatrix} \tag{4.356}$$

$$m_1 = 1\,\text{kg} \qquad m_2 = 1\,\text{kg} \tag{4.357}$$

$$k_1 = 100\,\text{kg} \qquad k_2 = 100\,\text{kg} \qquad k = 100\,\text{kg} \tag{4.358}$$

$$c_1 = 100\,\text{kg} \qquad c_2 = 2\,\text{kg} \qquad c = 2\,\text{kg} \tag{4.359}$$

$$f_1 = 10\sin 2t\,\text{N} \qquad f_2 = -10\,\text{N} \tag{4.360}$$

Example 150 Transient response of bicycle vehicle model to step input. Analysis and optimization of transient vibration response of vehicles is a step input which is an important stage in modeling and design of suspensions. In this example we review bicycle models of vehicle.

The analysis and optimization of the response of a vehicle to transient inputs are less important than frequency response optimization. However, there are applications where the transient response of vehicles becomes important. An optimized suspension due to frequency response provides us with ride comfort in steady-state driving conditions. The transient response of a vehicle has to be considered with sudden change of road conditions, the shock and jerk level of response, and stability of a vehicle.

The vertical, pitch, and roll vibrations are the three possible vibrations of the body of a vehicle. To analyze these modes in the simplest possible models, we use a quarter car for vertical vibrations, a bicycle car for pitch vibrations, and a half car for roll vibrations. The bicycle and half car models also contain the body bounce. Using these three models, we may be able to optimize the three different

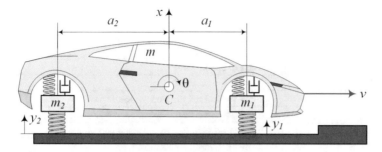

Fig. 4.12 A bicycle vehicle model going over a step

modes of oscillations of a vehicle. Then, using a full-car model, we may analyze the interaction of the three modes. Generally speaking, the rotational vibrations of roll and pitch of the body are more uncomfortable than the vertical vibration. The roll vibration of a car is the most unpleasant and the most uncomfortable motion for passengers.

Figure 4.12 illustrates a bicycle car model going over a step. The step input is a simple and multipurpose transient excitation, by which we discover many important transient behaviors of the systems. Although there would be a time lag between the front and rear wheels hitting the step, we first assume the step applies to front and rear wheels at the same time, and then we consider the time difference. The response of the car would be the solution of the following equations of motion:

$$[m]\ddot{\mathbf{x}} + [c]\dot{\mathbf{x}} + [k]\mathbf{x} = \mathbf{F} \tag{4.361}$$

$$[m] = \begin{bmatrix} m & 0 & 0 & 0 \\ 0 & I_y & 0 & 0 \\ 0 & 0 & m_1 & 0 \\ 0 & 0 & 0 & m_2 \end{bmatrix} \tag{4.362}$$

$$[c] = \begin{bmatrix} c_1 + c_2 & a_2c_2 - a_1c_1 & -c_1 & -c_2 \\ a_2c_2 - a_1c_1 & c_1a_1^2 + c_2a_2^2 & a_1c_1 & -a_2c_2 \\ -c_1 & a_1c_1 & c_1 & 0 \\ -c_2 & -a_2c_2 & 0 & c_2 \end{bmatrix} \tag{4.363}$$

$$[k] = \begin{bmatrix} k_1 + k_2 & a_2k_2 - a_1k_1 & -k_1 & -k_2 \\ a_2k_2 - a_1k_1 & k_1a_1^2 + k_2a_2^2 & a_1k_1 & -a_2k_2 \\ -k_1 & a_1k_1 & k_1 + k_{t_1} & 0 \\ -k_2 & -a_2k_2 & 0 & k_2 + k_{t_2} \end{bmatrix} \tag{4.364}$$

Table 4.1 Parameters of a bicycle vibrating vehicle

Parameter	Meaning
m	Half of body mass
m_1	Mass of a front wheel
m_2	Mass of a rear wheel
x	Body vertical displacement coordinate
x_1	Front wheel vertical displacement coordinate
x_2	Rear wheel vertical displacement coordinate
θ	Body pitch motion coordinate
y_1	Road excitation at the front wheel
y_2	Road excitation at the rear wheel
I_y	Half of body lateral mass moment
a_1	Absolute distance of C from front axle
a_2	Absolute distance of C from rear axle

$$
\mathbf{F} = \begin{bmatrix} 0 \\ 0 \\ k_{t_1} y_1 \\ k_{t_2} y_2 \end{bmatrix} \qquad \mathbf{x} = \begin{bmatrix} x \\ \theta \\ x_1 \\ x_2 \end{bmatrix} \tag{4.365}
$$

with respect to the symmetric road step input:

$$
y_1 = \begin{cases} Y & t > 0 \\ 0 & t \le 0 \end{cases} \qquad y_2 = \begin{cases} Y & t > 0 \\ 0 & t \le 0 \end{cases} \tag{4.366}
$$

The definitions of the parameters employed are in Table 4.1. Using a set of nominal numerical values:

$$
m = 420\,\text{kg} \qquad m_1 = 53\,\text{kg} \qquad m_2 = 76\,\text{kg} \qquad I_y = 1100\,\text{kg m}^2
$$

$$
a_1 = 1.4\,\text{m} \qquad a_2 = 1.47\,\text{m} \qquad l = a_1 + a_2 = 2.87\,\text{m}
$$

$$
k_1 = 10000\,\text{N}/\text{m} \qquad k_2 = 13000\,\text{N}/\text{m} \qquad k_{t_1} = k_{t_2} = 200000\,\text{N}/\text{m}
$$

$$
c_1 = 1000\,\text{N}\,\text{s}/\text{m} \qquad c_2 = 1000\,\text{N}\,\text{s}/\text{m} \tag{4.367}
$$

$$
Y = 0.1\,\text{m} \tag{4.368}
$$

the response of the car would be as shown in Fig. 4.13.

Examining the time response graphs indicates that the vibration frequencies of the wheels are higher than the vibration frequencies of the body. The wheels also stop vibrating much faster than the body. For the given data, the unpleasant pitch vibration θ would last longer than the other coordinates; however, pitch vibration has

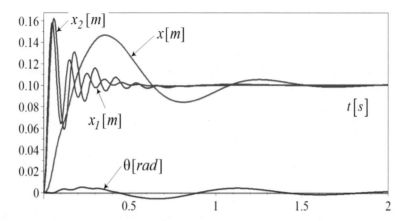

Fig. 4.13 Body bounce, x, body pitch, θ, front wheel, x_1, rear wheel, x_2, responses of a bicycle vehicle to a symmetric step excitation

only a small amplitude due to symmetry of input and not much difference between front and rear suspension parameters and geometry. The pitch vibration will have higher amplitude when the time lag between front and rear wheels is considered.

The real situation of a vehicle going over a step is to consider the time lag between the front and rear wheels excitations. Consider the car of Fig. 4.12 is moving with velocity v and hitting a step of height Y. Starting the time from the moment that the front wheel hits the step, we can calculate the time t_0 at which the rear wheel reaches the step:

$$t_0 = \frac{l}{v} \tag{4.369}$$

Therefore, the excitation of the rear wheel, y_2, would be a lagged step input:

$$y_1 = \begin{cases} Y & t > 0 \\ 0 & t \leq 0 \end{cases} \qquad y_2 = \begin{cases} Y\,H\left(t - \dfrac{l}{v}\right) & t > 0 \\ 0 & t \leq 0 \end{cases} \tag{4.370}$$

where $H(t - t_0)$ is the Heaviside function:

$$H\left(t - \frac{l}{v}\right) = \begin{cases} 1 & t > t_0 \\ 0 & t \leq t_0 \end{cases} \tag{4.371}$$

Let us assume a speed of v:

$$v = 10\,\mathrm{m/s} \tag{4.372}$$

The responses of the car are shown in Fig. 4.14.

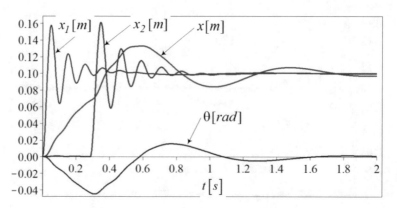

Fig. 4.14 Body bounce, x, body pitch, θ, front wheel, x_1, rear wheel, x_2, responses of a bicycle vehicle to a lagged step excitation

4.3 Summary

Time response analysis of an n degree of freedom (DOF) system is dependent on n natural frequencies ω_i and n mode shapes $\mathbf{u}_i$ of the system. An n DOF system is expressed by a set of n ordinary differential equations of the second order.

Free Vibrations: Undamped Systems

$$\mathbf{M}\ddot{\mathbf{x}} + \mathbf{K}\mathbf{x} = \mathbf{0} \tag{4.373}$$

$$\mathbf{x} = \begin{bmatrix} x_1 & x_2 & \cdots & x_n \end{bmatrix}^T \tag{4.374}$$

$$\mathbf{x}(0) = \mathbf{x}_0 \qquad \dot{\mathbf{x}}(0) = \dot{\mathbf{x}}_0 \tag{4.375}$$

A system shows its natural behavior when it is force-free and undamped. The time response of the free system is a linear combination of harmonic functions for each coordinate x_i, with ω_i as the *natural frequencies* and $\mathbf{u}_i$ as the *mode shapes* of the system:

$$\mathbf{x} = \sum_{i=1}^{n} \mathbf{u}_i \left(A_i \sin \omega_i t + B_i \cos \omega_i t \right) \quad i = 1, 2, 3, \cdots, n \tag{4.376}$$

$$= \sum_{i=1}^{n} X_i \mathbf{u}_i \sin \left(\omega_i t + \varphi_i \right) \quad i = 1, 2, 3, \cdots, n \tag{4.377}$$

$$X_i = \sqrt{A_i^2 + B_i^2} \qquad \tan \varphi_i = \frac{B_i}{A_i} \tag{4.378}$$

The coefficients A_i and B_i, or X_i and φ_i, will be determined from the initial conditions. The natural frequencies ω_i are solutions of the characteristic equation of the system:

$$\det\left[\mathbf{K} - \omega^2\,\mathbf{M}\right] = 0 \tag{4.379}$$

and the mode shape $\mathbf{u}_i$, corresponding to ω_i, is the solution of the eigenvector equation:

$$\left[\mathbf{K} - \omega_i^2\,\mathbf{M}\right]\mathbf{u}_i = 0 \tag{4.380}$$

Mode shapes are orthogonal with respect to the mass $\mathbf{M}$ and stiffness $\mathbf{K}$ matrices:

$$\mathbf{u}_j^T\,\mathbf{M}\,\mathbf{u}_i = 0 \qquad \mathbf{u}_j^T\,\mathbf{K}\,\mathbf{u}_i = 0 \qquad i \neq j \tag{4.381}$$

$$\omega_i^2 = \frac{\mathbf{u}_i^T\,\mathbf{K}\,\mathbf{u}_i}{\mathbf{u}_i^T\,\mathbf{M}\,\mathbf{u}_i} \tag{4.382}$$

Employing the modal matrix $\mathbf{U}$:

$$\mathbf{U} = \begin{bmatrix} \mathbf{u}_1 & \mathbf{u}_2 & \mathbf{u}_2 & \cdots & \mathbf{u}_n \end{bmatrix} \tag{4.383}$$

we can use a linear transformation to change the variables from the generalized coordinates $\mathbf{x}$ to principal coordinates $\mathbf{p}$ and make the equations decoupled:

$$\mathbf{x} = \mathbf{U}\,\mathbf{p} \tag{4.384}$$

$$\mathbf{M}'\ddot{\mathbf{p}} + \mathbf{K}'\mathbf{p} = 0 \tag{4.385}$$

$$m_i'\ddot{p}_i + k_i' p_i = 0 \tag{4.386}$$

$$\mathbf{M}' = \begin{bmatrix} m_1' & 0 & 0 & 0 \\ 0 & m_2' & 0 & 0 \\ \vdots & \vdots & \ddots & \vdots \\ 0 & 0 & \cdots & m_n' \end{bmatrix} \qquad \mathbf{K}' = \begin{bmatrix} k_1' & 0 & 0 & 0 \\ 0 & k_2' & 0 & 0 \\ \vdots & \vdots & \ddots & \vdots \\ 0 & 0 & \cdots & k_n' \end{bmatrix} \tag{4.387}$$

The uncoupled equations can be solved individually. The original solution will be found by combining the principal solutions and the transformation (4.15).

Free Vibrations: Damped Systems

To solve free vibrations of multi-DOF systems with general damping $\mathbf{C}$:

$$\mathbf{M}\,\ddot{\mathbf{x}} + \mathbf{C}\,\dot{\mathbf{x}} + \mathbf{K}\,\mathbf{x} = \mathbf{0} \tag{4.388}$$

we convert the n second-order differential equations of motion to $2n$ first-order differential equations and develop the exponential solutions:

$$^+\mathbf{M}\,\dot{\mathbf{y}} + {}^+\mathbf{K}\,\mathbf{y} = \mathbf{0} \tag{4.389}$$

$$\mathbf{y} = \begin{bmatrix} \dot{\mathbf{x}} \\ \mathbf{x} \end{bmatrix} \tag{4.390}$$

$$^+\mathbf{M} = \begin{bmatrix} \mathbf{0} & \mathbf{M} \\ \mathbf{M} & \mathbf{C} \end{bmatrix} \qquad {}^+\mathbf{K} = \begin{bmatrix} -\mathbf{M} & \mathbf{0} \\ \mathbf{0} & \mathbf{K} \end{bmatrix} \tag{4.391}$$

A solution of the new first-order equations of motion is exponential function of time:

$$\mathbf{y} = A\,\mathbf{u}\,e^{-\lambda t} \tag{4.392}$$

where λ is the eigenvalues and $\mathbf{u}$ is eigenvectors of $^+\mathbf{M}^{-1}\,{}^+\mathbf{K}$. Therefore, the complete solution will be a linear combination of all individual solutions. The coefficients A_i will be calculated from initial conditions:

$$\mathbf{y} = \sum_{i=1}^{n} A_i\,\mathbf{u}_i\,e^{-\lambda_i t} \tag{4.393}$$

Forced Vibrations

The complete solution of equations of a vibrating system with n degrees of freedom, subjected to given force functions $\mathbf{f}(\mathbf{x} : \dot{\mathbf{x}}, t)$

$$\mathbf{M}\,\ddot{\mathbf{x}} + \mathbf{C}\,\dot{\mathbf{x}} + \mathbf{K}\,\mathbf{x} = \mathbf{f}(\mathbf{x}, \dot{\mathbf{x}}, t) \tag{4.394}$$

$$\mathbf{x} = \begin{bmatrix} x_1 & x_2 & \cdots & x_n \end{bmatrix}^T \tag{4.395}$$

has two parts: a homogenous solution $\mathbf{x}_h$, plus the a particular solution $\mathbf{x}_p$. The homogenous solution $\mathbf{x}_h$ is the solution of free vibrations for $\mathbf{f} = \mathbf{0}$, and the particular solution is the special solutions of the equations associated to the forcing function $\mathbf{f}(\mathbf{x}, \dot{\mathbf{x}}, t)$.

To solve the forced vibrations, we convert the n second-order differential equations of motion to $2n$ first-order differential equations and develop the exponential solutions:

$$\dot{\mathbf{y}} + \mathbf{A}\,\mathbf{y} = \mathbf{F} \tag{4.396}$$

$$\mathbf{A} = {}^{+}\mathbf{M}^{-1}\,{}^{+}\mathbf{K} \qquad \mathbf{F} = {}^{+}\mathbf{M}^{-1}\,{}^{+}\mathbf{f} \tag{4.397}$$

$$\mathbf{y} = \begin{bmatrix} \dot{\mathbf{x}} \\ \mathbf{x} \end{bmatrix} \qquad {}^{+}\mathbf{f} = \begin{bmatrix} \mathbf{0} \\ \mathbf{f} \end{bmatrix} \tag{4.398}$$

$${}^{+}\mathbf{M} = \begin{bmatrix} \mathbf{0} & \mathbf{M} \\ \mathbf{M} & \mathbf{C} \end{bmatrix} \qquad {}^{+}\mathbf{K} = \begin{bmatrix} -\mathbf{M} & \mathbf{0} \\ \mathbf{0} & \mathbf{K} \end{bmatrix} \tag{4.399}$$

The solution of any linear forced vibrating system of any order can also be expressed by a set of $2n$ coupled first-order linear system:

$$\mathbf{y} = e^{-\mathbf{A}t} \int_{0}^{t} e^{\mathbf{A}s}\,\mathbf{F}\,(s)\,\,ds + e^{-\mathbf{A}t}\,\mathbf{y}_0 \tag{4.400}$$

$$\mathbf{y} = \begin{bmatrix} y_1 & y_2 & y_3 & \cdots & y_{2n} \end{bmatrix}^{T} \tag{4.401}$$

$$\mathbf{F} = \begin{bmatrix} F_1 & F_2 & F_3 & \cdots & F_{2n} \end{bmatrix}^{T} \tag{4.402}$$

4.4   Key Symbols

0	Zero vector, zero matrix
a, b	Coefficients of proportional damping
a, b, c, d	Constant parameters
a_i	Distance from mass center of vehicle, constant coefficients
b	Coefficient of proportional damping matrix
a	Constant shift coordinates
a, **b**	Real and imaginary parts of complex mode shapes
A, B	Coefficients of time response
A, **B**	Coefficients of time response
$\mathbf{A} = \mathbf{M}^{-1}\,\mathbf{K} = \mathbf{M}^{-1}\,\mathbf{K}$	Characteristic matrix
A	Coefficient matrix of set of first-order equation
$\mathbf{B} = \mathbf{A}^{-1} = \mathbf{K}^{-1}\,\mathbf{M}$	Inverse characteristic matrix, part of $[T]$
B	Transformation coordinate matrix
c	Damping

C	Constant coefficient, constant of integration, amplitude
$\mathbf{C}$	Damping matrix
C	Coefficient of polynomial solution
C	Coefficient of exponential solution
D	Dissipation function
e	Exponential function
E	Mechanical energy
f	Force, function, number of rigid modes
f	Characteristic equation
$\mathbf{f}$	Force
F	Force amplitude, force magnitude, constant force
$\mathbf{F}$	Forces function, force amplitude vector
g	Functions of displacement, gravitational acceleration
$\mathbf{g}$	Gravitational acceleration vector
i	Counting index, unit of imaginary numbers
I	Mass moment
$\mathbf{I}, [\mathbf{I}]$	Identity matrix
k	Stiffness
k_{ij}	Stiffness matrix elements
k_s	Sprung stiffness
k_u	Unsprung stiffness
$\mathbf{K} \equiv [k]$	Stiffness matrix
$\mathbf{K}'$	Diagonal stiffness matrix
$^{+}\mathbf{K}$	Stiffness matrix of equivalent set of first-order equations
l	Length, wheelbase
$\mathcal{L}$	Lagrangean
m	Mass
K	Kinetic energy
$\mathbf{M} \equiv [m]$	Mass matrix
$\mathbf{M}'$	Diagonal mass matrix
$^{+}\mathbf{M}$	Mass matrix of equivalent set of first-order equations
n	Number of DOF, number of equations, order of a matrix
n'	Rank of a matrix
p	Principal coordinate
$\mathbf{p}$	Principal coordinates
P	Potential energy
$\mathbf{P}$	Principal force
q	Generalized coordinate, temporal function of x
$\mathbf{q}$	Generalized coordinates
$\mathbf{Q}$	Generalized force
r	Frequency ratio, radial coordinate
t	Time
u	Components of $\mathbf{u}$
$\mathbf{u}$	eigenvector
$\mathbf{U}$	Modal matrix

$v \equiv \dot{x}$	Velocity
$v_0 = \dot{x}_0$	Initial velocity
W	Work
x	Displacement coordinate, displacement variable
$\mathbf{x}$	Displacement coordinates
x_0	Initial displacement
$\mathbf{x}_h$	Homogenous solution
$\mathbf{x}_p$	Particular solution
X	Displacement amplitude
$\mathbf{X}$	Displacement amplitudes
y	Displacement excitation, displacement, relative displacement
$\mathbf{y} = [\,\dot{\mathbf{x}}\; \mathbf{x}\,]$	State space vector
Y	Amplitude of displacement excitation
z	Variable, relative displacement
$\mathbf{z} = \dot{\mathbf{x}}$	Velocity variable vector
Z	Short name of expressions

Greek

α	Stiffness parameter, characteristic frequency, stiffness ratio
β	Stiffness parameter, characteristic frequency
γ	Complex eigenvalue
δ	Virtual notation
δ	Variation
ε	Mass ratio, coupling mass element
ε	Small parameter
λ	Eigenvalue
$\boldsymbol{\Lambda} \equiv \boldsymbol{\Omega}^2$	Eigen matrix
θ	Angular coordinate, angular variable, pitch motion
τ	Time constant
φ	Phase angle
μ	Stiffness coupling element
ω	Excitation frequency, angular frequency
ω_0	Critical excitation frequency
ω_i, ω_n	Natural frequency
$\boldsymbol{\Omega}$	Natural frequency matrix

Symbol

$H(t - t_0)$	Heaviside function
$\mathcal{L}$	Lagrangean
DOF	Degree of freedom
tr	Trace
det	Determinant

Exercises

1. Natural coordinates

   Consider a general 3 DOF system:

$$m_{11}\ddot{x}_1 + m_{12}\ddot{x}_2 + m_{13}\ddot{x}_3$$
$$+k_{11}x_1 + k_{12}x_2 + k_{13}x_3 = 0 \tag{4.403}$$
$$m_{21}\ddot{x}_1 + m_{22}\ddot{x}_2 + m_{23}\ddot{x}_3$$
$$+k_{21}x_1 + k_{22}x_2 + k_{23}x_3 = 0 \tag{4.404}$$
$$m_{31}\ddot{x}_1 + m_{32}\ddot{x}_2 + m_{33}\ddot{x}_3$$
$$+k_{31}x_1 + k_{32}x_2 + k_{33}x_3 = 0 \tag{4.405}$$

   Determine the natural frequencies, mode shapes, and the principal coordinates to decouple the equations. Then, derive the time response of the system for a general set of initial conditions.

2. Free vibrations of undamped multi-DOF system

   Consider a free and undamped multi-DOF system with the following mass and stiffness matrices:

$$\mathbf{M} = \begin{bmatrix} 1 & 0 & 0 \\ 0 & 1.5 & 0 \\ 0 & 0 & 0 \end{bmatrix} \tag{4.406}$$

$$K = \begin{bmatrix} 100 & -100 & 0 \\ -100 & 300 & -200 \\ 0 & -200 & 500 \end{bmatrix} \tag{4.407}$$

   Determine characteristic equation, natural frequencies, mode shapes, modal matrix, principal coordinates, and time response of the system.

3. Forced vibrations of undamped multi-DOF system

   Consider a forced and undamped multi-DOF system with the following matrices:

$$\mathbf{f} = \begin{bmatrix} 10\sin 10t \\ t \\ 0 \end{bmatrix} \qquad \mathbf{M} = \begin{bmatrix} 1 & 0 & 0 \\ 0 & 1.5 & 0 \\ 0 & 0 & 0 \end{bmatrix} \tag{4.408}$$

$$K = \begin{bmatrix} 100 & -100 & 0 \\ -100 & 300 & -200 \\ 0 & -200 & 500 \end{bmatrix} \tag{4.409}$$

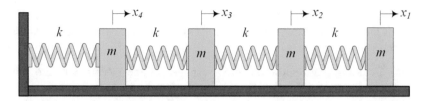

Fig. 4.15 A four DOF undamped vibrating system

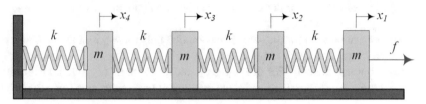

Fig. 4.16 A four DOF undamped forced vibrating system

Determine and plot the time response of the system using principal coordinates for zero initial conditions..

4. Free vibrations of a four DOF conservative system

Figure 4.15 illustrates a four DOF undamped system with $m = 1$ kg, $k = 10$ N / m:

(a) Determine the natural frequencies and mode shapes of the system.
(b) Determine and show the time response of the system for $x_1 = 0.1$, $x_2 = 0$, $x_3 = 0$, and $x_4 = 0$.
(c) Determine and show the time response of the system for $x_1 = 0$, $x_2 = 0.1$, $x_3 = 0$, and $x_4 = 0$.
(d) Determine and show the time response of the system for $x_1 = 0.1$, $x_2 = 0.1$, $x_3 = 0.1$, and $x_4 = 0.1$.
(d) Determine and show the time response of the system for $x_1 = 0.1$, $x_2 = 0.1$, $x_3 = -0.1$, and $x_4 = -0.1$.

5. Forced vibrations of a four DOF conservative system

Figure 4.16 illustrates a four DOF undamped system with $m = 1$ kg, $k = 100$ N / m:

(a) Determine the natural frequencies and mode shapes of the system.
(b) Determine and show the time response of the system for zero initial conditions and $f = 100$ N.
(c) Determine and show the time response of the system for zero initial conditions and $f = 100t$ N.
(d) Determine and show the time response of the system for zero initial conditions and $f = 100e^{-t}$ N.
(e) Determine and show the time response of the system for zero initial conditions and $f = 100 \sin 10t$ N.

6. Position of step force on vibrations of a four DOF conservative system
   Figure 4.16 illustrates a four DOF undamped system with $m = 1$ kg, $k = 100\,\text{N}/\text{m}$:

   (a) Determine and show the time response of the system for zero initial conditions and $f = 100\,\text{N}$. Plot the response of x_1, x_2, x_3, and x_4.
   (b) Repeat part (a) for the same force applied only on m_2. Plot the response of x_1, x_2, x_3, and x_4.
   (c) Repeat part (a) for the same force applied only on m_3. Plot the response of x_1, x_2, x_3, and x_4.
   (d) Repeat part (a) for the same force applied only on m_4. Plot the response of x_1, x_2, x_3, and x_4.
   (e) Compare the behavior of x_1, x_2, x_3, and x_4 for parts (a) to (d). Determine engineering comments about the position of the applied step force function.

7. Position of harmonic force on vibrations of a four DOF conservative system
   Figure 4.16 illustrates a four DOF undamped system with $m = 1$ kg, $k = 100\,\text{N}/\text{m}$:

   (a) Determine and show the time response of the system for zero initial conditions and $f = 100 \sin 10t$ N. Plot the response of x_1, x_2, x_3, and x_4.
   (b) Repeat part (a) for the same force applied only on m_2. Plot the response of x_1, x_2, x_3, and x_4.
   (c) Repeat part (a) for the same force applied only on m_3. Plot the response of x_1, x_2, x_3, and x_4.
   (d) Repeat part (a) for the same force applied only on m_4. Plot the response of x_1, x_2, x_3, and x_4.
   (e) Compare the behavior of x_1, x_2, x_3, and x_4 for parts (a) to (d). Determine engineering comments about the position of the applied harmonic force function.

8. Position of harmonic force on vibrations of a three DOF conservative system
   Figure 4.17 illustrates a three DOF undamped system with $m = 1$ kg, $k = 100\,\text{N}/\text{m}$:

   (a) Determine and show the time response of the system for zero initial conditions and $f = 100 \sin 10t$ N. Plot the response of x_1, x_2, x_3, and x_4.
   (b) Repeat part (a) for the same force applied only on m_2. Plot the response of x_1, x_2, x_3, and x_4.
   (c) Repeat part (a) for the same force applied only on m_3. Plot the response of x_1, x_2, x_3, and x_4.
   (d) Repeat part (a) for the same force applied only on m_4. Plot the response of x_1, x_2, x_3, and x_4.
   (e) Compare the behavior of x_1, x_2, x_3, and x_4 for parts (a) to (d). Determine engineering comments about the position of the applied harmonic force function.

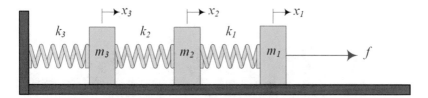

Fig. 4.17 A three DOF undamped forced vibrating system

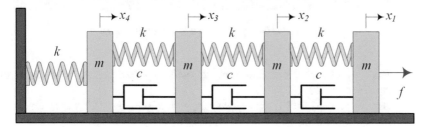

Fig. 4.18 A four DOF forced damped system

9. ★ Position of decaying harmonic force on vibrations of a three DOF conservative system

Figure 4.17 illustrates a three DOF undamped system with $m = 1$ kg, $k = 100 \, \text{N} / \text{m}$:

1. (a) Determine and show the time response of the system for zero initial conditions and $f = 100e^{-t} \sin 10t$ N. Plot the response of x_1, x_2, x_3, and x_4.
   (b) Repeat part (a) for the same force applied only on m_2. Plot the response of x_1, x_2, x_3, and x_4.
   (c) Repeat part (a) for the same force applied only on m_3. Plot the response of x_1, x_2, x_3, and x_4.
   (d) Repeat part (a) for the same force applied only on m_4. Plot the response of x_1, x_2, x_3, and x_4.
   (e) Compare the behavior of x_1, x_2, x_3, and x_4 for parts (a) to (d). Determine engineering comments about the position of the applied harmonic force function.
10. Forced vibrations of a four DOF damped system

Figure 4.18 illustrates a four DOF forced and damped system with $m = 1$ kg, $k = 100 \, \text{N} / \text{m}$, $c = 10 \, \text{N} \, \text{s} / \text{m}$:

1. (a) Determine and show the time response of the system for zero initial conditions and $f = 100$ N. Plot the response of x_1, x_2, x_3, and x_4.
   (b) Repeat part (a) for the same force applied only on m_2. Plot the response of x_1, x_2, x_3, and x_4.

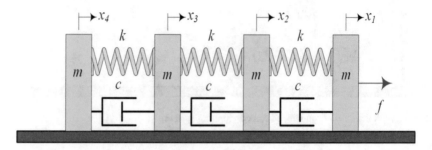

Fig. 4.19 A forced damped four DOF system with a zero natural frequency

(c) Repeat part (a) for the same force applied only on m_3. Plot the response of x_1, x_2, x_3, and x_4.

(d) Repeat part (a) for the same force applied only on m_4. Plot the response of x_1, x_2, x_3, and x_4.

(e) Compare the behavior of x_1, x_2, x_3, and x_4 for parts (a) to (d). Determine engineering comments about the position of the applied harmonic force function.

11. ★ Vibrations of a forced damped four DOF system with a zero natural frequency

   Figure 4.19 illustrates a four DOF forced damped system with a zero natural frequency and $m = 1$ kg, $k = 100$ N / m, $c = 10$ N s / m:

1. (a) Determine and show the time response of the system for zero initial conditions and $f = 100 \sin 10t$ N. Plot the response of x_1, x_2, x_3, and x_4.

   (b) Repeat part (a) for the same force applied only on m_2. Plot the response of x_1, x_2, x_3, and x_4.

   (c) Repeat part (a) for the same force applied only on m_3. Plot the response of x_1, x_2, x_3, and x_4.

   (d) Repeat part (a) for the same force applied only on m_4. Plot the response of x_1, x_2, x_3, and x_4.

   (e) Compare the behavior of x_1, x_2, x_3, and x_4 for parts (a) to (d). Determine engineering comments about the position of the applied harmonic force function.

12. Vibrations of a forced damped four DOF system with a zero natural frequency
   Figure 4.19 illustrates a four DOF forced damped system with a zero natural frequency and $m = 1$ kg, $k = 100$ N / m, $c = 10$ N s / m:

   (a) Determine and show the time response of the system for zero initial conditions and $f = 100 \cos 10t$ N. Plot the response of x_1, x_2, x_3, and x_4.

   (b) Repeat part (a) for the same force applied only on m_2. Plot the response of x_1, x_2, x_3, and x_4.

   (c) Repeat part (a) for the same force applied only on m_3. Plot the response of x_1, x_2, x_3, and x_4.

(d) Repeat part (*a*) for the same force applied only on m_4. Plot the response of x_1, x_2, x_3, and x_4.

(e) Compare the behavior of x_1, x_2, x_3, and x_4 for parts (*a*) to (*d*). Determine engineering comments about the position of the applied harmonic force function.

Chapter 5
First-Order Systems

There are cases where the behavior of a dynamic system can be modeled by first-order or reducible to first-order differential equations. For example, eliminating the mass of a vibrating system simplifies the equation of motion of the system to a first-order forced differential equation.

$$c\dot{x} + kx = f \tag{5.1}$$

Also eliminating the external force and the spring, as is shown in Fig. 5.1, reduces the equation of motion to a homogeneous first-order differential equation.

$$m\dot{v} + cv = 0 \tag{5.2}$$

In this chapter, we study the dynamics of systems whose equation of motion can be expressed or reduced to first-order equations.

5.1 Natural Motion

The motion of a system with no external force is called the natural motion. When there is no external force, the equation of motion of a first-order system can always be simplified to

$$\dot{x} + \alpha x = 0 \tag{5.3}$$

of which the solution would be

$$x = x(0) \, e^{-\alpha t} = x_0 e^{-t/\tau} \qquad \tau = \frac{1}{\alpha} \tag{5.4}$$

© The Author(s), under exclusive license to Springer Nature Switzerland AG 2022
R. N. Jazar, *Advanced Vibrations*, https://doi.org/10.1007/978-3-031-16356-2_5

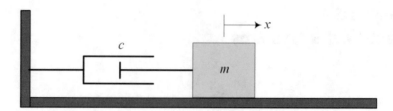

Fig. 5.1 A mass m attached to a damper c on a frictionless surface is a first-order system

The most important characteristic of first-order systems is their time constant τ. The response of a system after a period of one time constant $t = \tau$ reaches e^{-1} of its initial value. A time constant is passed when x drops by about %64 of its initial value x_0.

$$x = x(t + \tau) = \frac{x(t)}{e} \tag{5.5}$$

The natural motion of first-order systems is either exponentially decreasing or increasing function of time, and they do not show vibrations.

Proof Every vibrating system remains in equilibrium position if it is already in equilibrium and there is no excitation. If the system is under no excitation, then the nonzero initial condition is the only reason of motion. Any motion of a dynamic system that occurs only because of nonzero initial conditions is called *natural motion*. Such a motion will be characterized by an exponential function of time.

Equation (5.3) is the simplest differential equation in system dynamics. Let us rewrite the equation in a new form.

$$\dot{x}(t) = -\alpha x(t) \tag{5.6}$$

Exponential function is the solution of all first-order equations (5.6) because if $x(t)$ an exponential function $x(t) = Ce^{\alpha t}$, then $\dot{x}(t) = C\alpha e^{\alpha t} = \alpha x(t)$, and hence it satisfies the equation. Therefore,

$$x(t) = Ce^{-\alpha t} \tag{5.7}$$

is the solution of (5.6). If the initial condition of the equation is given as

$$x(0) = x_0 \tag{5.8}$$

then the solution would be

$$x(t) = x(0)\,e^{-\alpha t} = x_0 e^{-t/\tau} \qquad \tau = \frac{1}{\alpha} \tag{5.9}$$

The solution (5.9) is the unique function to satisfy the equation. To show this fact, let us assume $u(t)$ to also be a solution and $\dot{u} = -\alpha u$. The derivative of $u(t) e^{\alpha t}$ is

$$\frac{d}{dt}\left(u(t) e^{\alpha t}\right) = \dot{u} e^{\alpha t} + \alpha u e^{\alpha t}$$

$$= -\alpha u e^{\alpha t} + \alpha u e^{\alpha t} = 0 \qquad (5.10)$$

and therefore, $u(t) e^{\alpha t}$ is a constant C. So $u(t) = Ce^{-\alpha t}$, which indicates the uniqueness of the solution (5.9).

When $\alpha > 0$, and hence the exponent is negative, the solution indicates an exponentially decreasing function, starting from x_0 and approaching $x = 0$. In case $\alpha < 0$, the solution indicates an exponentially increasing function starting from x_0 and approaching $x = \infty$. The case of $\alpha = 0$ is a constant solution at $C = x(0)$.

The most important characteristic of first-order system is time constant τ. The following facts are applied for all first-order systems.

1. The value of $x(t)$ drops from $x = x(t_1)$ at $t = t_1$ to $x = x(t_1 + \tau) = x(t_1)/e = 0.368x(t_1)$ at $t = t_1 + \tau$. Therefore, starting from $x = x_0$, the solution drops to $x = x_0/e$ at $t = \tau$, drops to $x = x_0/e^2$ at $t = 2\tau$, and so on.
2. Beginning at $t = 0$ from $x = x_0$, a tangent line to the curve will intersect the t-axis at $t = \tau$. Then a tangent to the curve at $t = \tau$ will intersect the t-axis at $t = 2\tau$ and so on.
3. The value of $x(t)$ drops from $x = x(t_1)$ at $t = t_1$ to $x = \tau x(t_1)$ at $t = t_1 + 1$. Therefore, starting from $x = x_0$, the curve of x drops to $x = \tau x_0$ at $t = 1$, and then drops to $x = \tau^2 x_0$ at $t = 2$, and so on.
4. The slope of $x(t)$ at $t = 0$ hits the t-axis at $t = \tau$. The slope of $x(t)$ at $t = \tau$ hits the x-axis at $x = 2x_0/e$. The slope at $t = 2\tau$ hits the x-axis at $x = 3x_0/e^2$. Similarly, the slope at $t = n\tau$ hits the x-axis at $x = (n+1)x_0/e^n$.

These facts about the behavior of first-order systems (5.9) help a designer to predict the response of the system. Figure 5.2 illustrates the time response of a system assuming $\tau > 1$, and Fig. 5.3 shows the relation between the slopes to the $x(t)$-curve at $t = n\tau, n = 1, 2, 3, \cdots$. ∎

Example 151 A mass on a damper. If the spring of a vibrating system (m, c, k) is missing or broken or very low rate, the mass may be assumed to be attached to a base only by a damper. Such system will be modeled by a first-order equation. Here are possible natural motions of the system.

The mass m in Fig. 5.1 is attached to the wall by a damper with damping c. Assuming the initial velocity of m is $\dot{x}_0$, we may transform its equation of motion

$$m\ddot{x} + c\dot{x} = 0 \qquad (5.11)$$

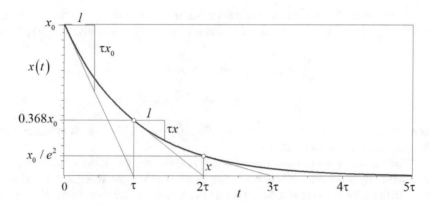

Fig. 5.2 The time response of the first-order and unexcited system

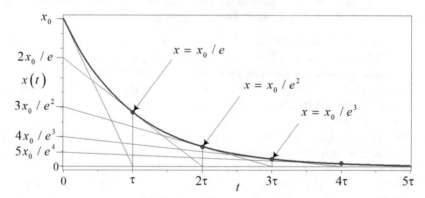

Fig. 5.3 The relationship of the slopes to the curve of time response of a first-order and unexcited system

to a first-order equation, by introducing a new variable v.

$$m\dot{v} + cv = 0 \qquad v = \dot{x} \tag{5.12}$$

Employing the initial velocity $\dot{x}_0 = v_0$, its response is

$$v = v_0 e^{-ct/m} = v_0 e^{-t/\tau} \qquad \tau = \frac{m}{c} \tag{5.13}$$

Setting

$$m = 1\,\text{kg} \qquad v_0 = 1\,\text{m}/\text{s} \tag{5.14}$$

we can draw the response for different c as shown in Fig. 5.4.

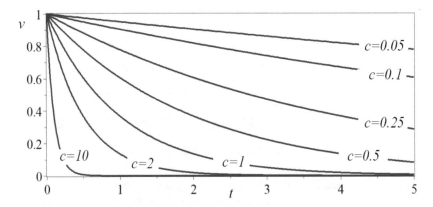

Fig. 5.4 Velocity of m as a function of time, for different c

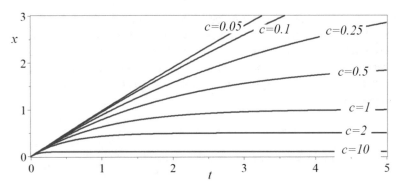

Fig. 5.5 Position of m as a function of time, for different c

The displacement of m is found by integration.

$$x = \int v\, dt = x_0 + \frac{m}{c}\dot{x}_0\left(1 - e^{-\frac{c}{m}t}\right) \tag{5.15}$$

Let us assume that $x_0 = 0$ and determine how far m will go because of the initial velocity. Figure 5.5 illustrates x for different c as a function of time. Theoretically, it takes an infinite time for the mass m to stop and get zero speed, $v = 0$. The distance x_{Max} that m will move before stopping can be calculable by limit.

$$x_{Max} = \lim_{t\to\infty} x(t) = \lim_{t\to\infty}\left(\frac{m}{c}\dot{x}_0\left(1 - e^{-\frac{c}{m}t}\right)\right) = \frac{m}{c}\dot{x}_0 \tag{5.16}$$

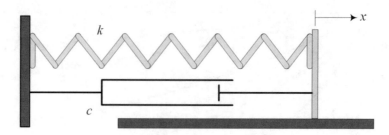

Fig. 5.6 A first-order system of a parallel damper and spring

Example 152 A practical first-order system. Assuming the mass of a vibrating system is too low compared to the rate of spring and damper, the equation of motion of the system may be modeled by a static balance of forces which will be a first-order equation.

A practical example of a first-order system is shown in Fig. 5.6. It is made by a parallel spring and damper that are connected to a massless plate whose motion is measured by x. The systems have an element to collect potential energy and an element to dissipate energy, but there is no element to collect kinetic energy. The governing equation of the system is a balance of the total applied forces on the massless plate.

$$c\dot{x} = -kx \tag{5.17}$$

Therefore, the time response of the system will be

$$x(t) = x_0 \, e^{-kt/c} = x_0 \, e^{-t/\tau} \qquad \tau = \frac{c}{k} \tag{5.18}$$

where $\tau = c/k$ is the time constant of the system.

$$x(\tau) = \frac{x_0}{e} = 0.36788 \, x_0 \tag{5.19}$$

At any instant, there are only two forces in balance: the spring force kx and the damping force $c\dot{x}$. The spring always attempts to bring the system to the equilibrium position at which the elastic force is zero, and the damping attempts to stop the system and make velocity zero. So, the natural motion of the system occurs only when there is a nonzero initial condition. Figure 5.2 illustrates the time response of the system assuming $\tau > 1$.

Example 153 Separable first-order equations. A solvable case of first-order systems with variable coefficient is when the variables are separable. Here is the theory.

A solvable first-order equation is the one in which the variables are separable. Such equations are in the following form.

$$\frac{dx}{dt} = g(x) \, h(t) \tag{5.20}$$

The solution of these equations is found by integration.

$$\int \frac{dx}{g\,(x)} = \int h\,(t)\,dt + C \tag{5.21}$$

The case of a parallel combination of a spring and a damper of Fig. 5.6 has a separable equation.

$$c\dot{x} = -kx \tag{5.22}$$

Separation of the variables and integration

$$\int \frac{dx}{x} = -\frac{k}{c} \int dt + C_1 \tag{5.23}$$

provides us with the solution.

$$\ln x = -\frac{k}{c}t + C_1 \tag{5.24}$$

Rearrangement of the solution makes it look familiar:

$$x = C_2 e^{-t/\tau} \qquad C_2 = x\,(0) = x_0 \qquad \tau = \frac{c}{k} \tag{5.25}$$

Example 154 ★ Homogeneous first-order equations. The general form of a first-order system with variable coefficients and with no external excitation is studied in this example. This is the most general natural motion of first-order systems.

A function $f\,(x,t)$ is called homogeneous of degree n if we have

$$f\,(hx, ht) = h^n f\,(x,t) \tag{5.26}$$

So, replacing the variables x and t by hx and ht, we can factor h^n out of the original function.

The first-order equation

$$N\,(x,t)\,dx + M\,(x,t)\,dt = 0 \tag{5.27}$$

is called homogeneous if the variable coefficients M and N are homogeneous functions of the same degree. The homogeneous equation can then be written in a new form:

$$\frac{dx}{dt} = -\frac{M\,(x,t)}{N\,(x,t)} = f\,(x,t) \tag{5.28}$$

A homonymous equation can always be transformed to a separable equation by introducing a new variable y.

$$y = \frac{x}{t} \tag{5.29}$$

Knowing that

$$f(hx, ht) = h^0 f(x, t) = f(x, t) \tag{5.30}$$

we may set $h = 1/x$ and then

$$f(x, t) = f(x/t, 1) = f(y, 1) \tag{5.31}$$

Because $x = yt$, we have

$$\frac{dx}{dt} = y + t\frac{dy}{dt} \tag{5.32}$$

and Eq. (5.28) becomes

$$y + t\frac{dy}{dt} = f(y, 1) \tag{5.33}$$

which is separable and solvable.

$$\frac{dy}{f(y, 1) - y} = \frac{dt}{t} \tag{5.34}$$

Example 155 ★ Total differential. If a general first-order partial differential equation is a total differential, it will be solvable. Here is the theory.

Consider a general linear partial differential equation.

$$A(x, t)\, dx + B(x, t)\, dy + C(x, t)\, dz + D(x, t)\, dt = 0 \tag{5.35}$$

If there happens to exist a function $f(x, y, z, t)$ such that

$$\frac{\partial f}{\partial x} = A(x, t) \qquad \frac{\partial f}{\partial y} = B(x, t)$$

$$\frac{\partial f}{\partial z} = C(x, t) \qquad \frac{\partial f}{\partial t} = D(x, t) \tag{5.36}$$

then we can write (5.35) in the following form.

$$df = \frac{\partial f}{\partial x}dx + \frac{\partial f}{\partial y}dy + \frac{\partial f}{\partial z}dz + \frac{\partial f}{\partial t}dt = 0 \tag{5.37}$$

The general solution of such a differential equation is

$$f(x, y, z, t) = C_1 \tag{5.38}$$

As an example, consider a differential equation

$$\sin y \, dx + (\sin z + x \cos y) \, dy + (\sin t + y \cos z) \, dz + z \cos t \, dt = 0 \tag{5.39}$$

that is a total differential because there is a function f such that the equation is df.

$$f(x, y, z, t) = x \sin y + y \sin z + z \sin t = C_1 \tag{5.40}$$

Example 156 ★ Bernoulli equation to a linear equation. There is a rich theory behind first-order differential equations. The general first-order forced system is modeled by Bernoulli differential equation, and here is the solution.

An ordinary differential equation (ODE) of the following form is an inhomogeneous linear first-order ODE.

$$\frac{dy}{dx} + p(x) \, y = q(x) \tag{5.41}$$

This equation is always solvable.

$$y = e^{-\int p(x) dx} \left(C_1 + \int q(x) \, e^{\int p(x) dx} dx \right) \tag{5.42}$$

There are many methods to change a variable and make a differential equation integrable. Transforming a Bernoulli equation to a linear differential form is one that is often applied. Consider the equation

$$dx + \left(a(t) \, x + b(t) \, x^C \right) dt = 0 \tag{5.43}$$

where C is a constant and $a(t)$ and $b(t)$ are arbitrary functions of t. When $C = 0$, the equation is a linear ODE, and when $C = 1$, the equation is separable. For all other values of C, we can introduce a new variable y

$$y = x^{1-C} \tag{5.44}$$

and transform the equation to a new form which is linear and integrable.

$$\dot{y} + ((1 - C) \, a(t) \, y + (1 - C) \, b(t)) = 0 \tag{5.45}$$

Equations of the form (5.43) are Bernoulli equations.

As another example, consider an equation of the following form which is not a Bernoulli equation in x.

$$\dot{x} - \frac{t}{t^2 x^2 + x^5} = 0 \tag{5.46}$$

However, we can exchange the dependent and independent variables

$$\frac{dt}{dx} - x^2 t - \frac{x^5}{t} = 0 \tag{5.47}$$

and make a Bernoulli equation in t for $C = -1$, similar to (5.43).

Example 157 ★ Integrable differential equation.
  Consider the differential equation

$$\left(te^x + 2x\right) dx + e^x dt = 0 \tag{5.48}$$

To check if the equation is total, we examine the required conditions:

$$df = \frac{\partial f}{\partial x} dx + \frac{\partial f}{\partial t} dt = 0 \tag{5.49}$$

$$\frac{\partial^2 f}{\partial t\, \partial x} = e^x \qquad \frac{\partial^2 f}{\partial x\, \partial t} = e^x \tag{5.50}$$

Therefore, the equation is integrable. To solve the equation, we integrate $\partial f / \partial t$.

$$f = \int e^x \, dt + g\,(x) = te^x + g\,(x) \tag{5.51}$$

$$\frac{\partial f}{\partial x} = te^x + \frac{dg\,(x)}{dx} \tag{5.52}$$

Because (5.52) must be equal to $te^x + 2x$, we have $dg\,(x)/dx = 2x$, which provides us with $g(x) = x^2$, and therefore, the solution is found.

$$te^x + x^2 = C \tag{5.53}$$

If the dynamic system is at $x = x_0$ when $t = 0$, then $C = 0$.
  If a first-order dynamic system is modeled by a differential equation in the form

$$A\,dx + B\,dy = 0 \tag{5.54}$$

where

$$\frac{\partial B}{\partial x} = \frac{\partial A}{\partial y} \tag{5.55}$$

then the solution is

$$A + \int \left(B - \frac{\partial M}{\partial y} \right) dy = C \qquad (5.56)$$

Example 158 ★ Integrating factor. The equation of motion of some first-order systems are integrable if there is an integrating factor for them. Here are the theory and example.

The conditions (5.55) guarantee that a given first-order differential equation (5.54) is integrable. However, there exist differential equations that are integrable, while the condition (5.55) is not fulfilled. This is because the given equation is divided by a common factor. The divisor is called integrating factor and indicated by μ.

We show the existence of μ for a first-order non-total equation:

$$A\,dx + B\,dy = 0 \qquad \frac{\partial A}{\partial y} \neq \frac{\partial B}{\partial x} \qquad (5.57)$$

Assume that there exists a function f for this differential such that

$$f(x, y) = C \qquad (5.58)$$

The total differential of such function f is

$$df = \frac{\partial f}{\partial x}dx + \frac{\partial f}{\partial y}dy \qquad (5.59)$$

Comparing (5.59) and (5.57) yields

$$\frac{\partial f / \partial x}{A} = \frac{\partial f / \partial y}{B} = \mu(x, y) \qquad (5.60)$$

where we denote the common ratio by $\mu(x, y)$. Therefore, if (5.57) is integrable, then it has at least one integrating factor:

$$\frac{\partial f}{\partial x} = \mu A \qquad \frac{\partial f}{\partial y} = \mu B \qquad (5.61)$$

$$\frac{\partial (\mu A)}{\partial y} = \frac{\partial (\mu B)}{\partial x} \qquad (5.62)$$

We can show that, if $f_1(f)$ is any function of f, then

$$\mu f_1(A\,dx + B\,dy) = f_1\,df = d\left(\int f_1(f)\,df \right) \qquad (5.63)$$

so $\mu f_1(f)$ is also an integrating factor.

To develop the required conditions to get a total differential equation, we begin by expanding (5.62).

$$\frac{1}{\mu}\left(B\frac{\partial\mu}{\partial x} - A\frac{\partial\mu}{\partial y}\right) = \frac{\partial A}{\partial y} - \frac{\partial B}{\partial x} \tag{5.64}$$

It is a partial differential equation for $\mu = \mu(x, y)$. However, any particular solution for $\mu(x, y)$ is enough to make the differential equation total. Let us examine an integrating factor as a function of x alone. In this case, Eq. (5.64) simplifies.

$$\frac{1}{\mu}\frac{d\mu}{dx} = \frac{1}{B}\left(\frac{\partial A}{\partial y} - \frac{\partial B}{\partial x}\right) \tag{5.65}$$

Because the left-hand side of this equation is only a function of x, the right-hand side must also be a function of x:

$$\frac{1}{B}\left(\frac{\partial A}{\partial y} - \frac{\partial B}{\partial x}\right) = g(x) \tag{5.66}$$

Therefore, we have

$$\frac{1}{\mu}\frac{d\mu}{dx} = g(x) \tag{5.67}$$

which can be integrated.

$$\ln\mu = \int g(x)\,dx \tag{5.68}$$

$$\mu = e^{\int g(x)dx} \tag{5.69}$$

We may also search for an integrating factor as a function of y and find

$$\mu = e^{\int h(y)dy} \tag{5.70}$$

where

$$h(y) = -\frac{1}{A}\left(\frac{\partial A}{\partial y} - \frac{\partial B}{\partial x}\right) \tag{5.71}$$

As an example, let us consider the differential equation

$$A\,dx + B\,dy = y\,dx + \left(x^2 y - x\right)dy = 0 \tag{5.72}$$

which shows that

$$\frac{\partial A}{\partial y} = 1 \qquad \frac{\partial B}{\partial x} = 2xy - 1 \qquad (5.73)$$

However, if we multiply the equation by $1/x^2$, it becomes a total differential:

$$\frac{y}{x^2} dx + \left(y - \frac{1}{x} \right) dy = d \left(\frac{y^2}{2} - \frac{y}{x} \right) = 0 \qquad (5.74)$$

Let us use Eq. (5.69) to find an integrating factor that is only a function of x:

$$g(x) = \frac{1}{B} \left(\frac{\partial A}{\partial y} - \frac{\partial B}{\partial x} \right) = -\frac{2}{x} \qquad (5.75)$$

$$\mu = e^{\int g(x) dx} = x^{-2} \qquad (5.76)$$

Example 159 Linear equation integrating factor. Linear first-order equations can always be integrated by integrating factor.

Consider a first-order dynamic system with an equation of the following form where $p(t)$ and $q(t)$ are arbitrary functions of t.

$$\dot{x} + p(t) x = q(t) \qquad (5.77)$$

This is an inhomogeneous linear first-order ODE. Linear equations have a general integrating factor μ:

$$\mu(t) = e^{\int p(t) dt} \qquad (5.78)$$

Multiplying (5.77) by μ yields

$$\frac{d}{dt} (x \mu) = q(t) \mu \qquad (5.79)$$

Therefore, Eq. (5.77) is always integrable.

$$x = \frac{1}{\mu} C + \frac{1}{\mu} \int q(t) \mu \, dx$$

$$= e^{-\int p(t) dx} \left(C + \int q(t) e^{\int p(t) dx} \, dt \right) \qquad (5.80)$$

Example 160 ★ Series solution. Series solution of first-order equations is another way to prove their solutions are exponential.

Consider a first-order system, and let us assume a solution exists in the form of a time series.

$$\dot{x} + \alpha x = 0 \qquad x(0) = x_0 \tag{5.81}$$

$$x = b_0 + b_1 t + b_2 t^2 + b_3 t^3 + \cdots \tag{5.82}$$

Substituting this solution in the equation yields

$$b_1 + 2b_2 t + 3b_3 t^2 + \cdots = -\alpha \left(b_0 + b_1 t + b_2 t^2 + b_3 t^3 + \cdots \right) \tag{5.83}$$

For this equation to hold at any time, we must have

$$b_1 = -\alpha b_0 \tag{5.84}$$

$$b_2 = -\frac{1}{2}\alpha b_1 = \frac{1}{2}\alpha^2 b_0 \tag{5.85}$$

$$b_3 = -\alpha \frac{1}{3} b_2 = -\frac{1}{3 \times 2}\alpha^3 b_0 = -\frac{1}{3!}\alpha^3 b_0 \tag{5.86}$$

$$\vdots$$

The value of b_0 is x at $t = 0$.

$$b_0 = x(0) = x_0 \tag{5.87}$$

Therefore, we can write the solution as a time series of exponential function.

$$x(t) = x_0 \left(1 - \alpha t + \frac{1}{2!}\alpha^2 t^2 - \frac{1}{3!}\alpha^3 t^3 + \frac{1}{4!}\alpha^4 t^4 - \cdots \right)$$

$$= x_0 e^{-\alpha t} \tag{5.88}$$

Searching a solution of Taylor series will end up to the same solution (5.88). Assume there exists an analytical solution in the form of a Taylor series for Eq. (5.81).

$$x = x(0) + \dot{x}(0)\, t + \frac{\ddot{x}(0)}{2!} t^2 + \cdots + \frac{x^{(n)}(0)}{n!} t^n + \cdots \tag{5.89}$$

To evaluate the coefficients of the series, we differentiate both sides of the differential equation of (5.81) infinitely many times with respect to time, which yields

$$\dot{x} = -\alpha x$$

$$\ddot{x} = -\alpha \dot{x} = \alpha^2 x$$

$$\dddot{x} = a^2 \dot{x} = -a^3 x$$

$$\cdots$$

$$x^{(n)} = (-1)^n a^n x \tag{5.90}$$

At $t = 0$, we have

$$\dot{x}(0) = -a x_0$$
$$\ddot{x}(0) = a^2 x_0$$
$$\dddot{x}(0) = -a^3 x_0$$

$$\cdots$$

$$x^{(n)}(0) = (-1)^n a^n x_0 \tag{5.91}$$

Substituting these results back in Eq. (5.89) yields

$$x(t) = \left(1 - \alpha t + \frac{1}{2!} \alpha^2 t^2 - \frac{1}{3!} \alpha^3 t^3 + \frac{1}{4!} \alpha^4 t^4 - \cdots \right) x_0$$

$$= x_0 e^{-\alpha t} \tag{5.92}$$

5.2 General Motion

The excited case of every linear first-order system can be expressed by

$$\dot{x} + \alpha x = f(t) \tag{5.93}$$

The solution $x = x(t)$, $t > 0$, is

$$x = e^{-\alpha t} \int e^{\alpha t} f(t) \, dt + C e^{-\alpha t} \tag{5.94}$$

Proof Let us assume that x is a solution of (5.93), then

$$\frac{d}{dt} \left(e^{\alpha t} x \right) = e^{\alpha t} f(t) \tag{5.95}$$

and

$$e^{\alpha t} x = \int e^{\alpha t} f(t) \, dt + C \tag{5.96}$$

which provides us with

$$x = e^{-\alpha t} \int e^{\alpha t} f(t) \, dt + C e^{-\alpha t} \tag{5.97}$$

The parameter $1/\alpha$ is the time constant of the system τ.

$$\tau = \frac{1}{\alpha} \tag{5.98}$$

Now, the solution of (5.93) can also be written as

$$x = e^{-t/\tau} \int e^{t/\tau} f(t) \, dt + C e^{-t/\tau} \tag{5.99}$$

The solution has two parts with different characteristics.

$$x = x_p + x_h \tag{5.100}$$

$$x_p = e^{-t/\tau} \int e^{t/\tau} f(t) \, dt \tag{5.101}$$

$$x_h = C e^{-t/\tau} \tag{5.102}$$

The first part, x_h, is the natural homogenous solution of the system and is the solution of the system without $f(t)$. The second part is the particular solution, x_p, and is the solution associated with the forcing term $f(t)$. The natural solution x_h is related to the characteristics of the system. In practical cases where $\alpha > 0$, the natural solution x_h will vanish by time, and the particular solution x_p will be the steady-state solution of the system. The constant of integration C is found by the initial condition $x(0) = x_0$ in the general solution (5.97). ■

Example 161 A constant force on a mass-damper. A mass-damper system is governed by a first-order equation. A forced mass-damper system is studied here. A constant force will move m at a steady-state velocity depending on the damping of the system.

The mass m in Fig. 5.7 is attached to the wall by a damper with damping c. Assuming a constant force $f = F$ is applied on m, we have a forced first-order system in terms of velocity.

$$m\dot{v} + cv = F \qquad v = \dot{x} \tag{5.103}$$

$$v(0) = 0 \qquad x(0) = 0 \tag{5.104}$$

Let us rewrite (5.103) as

$$\dot{v} + \frac{c}{m}v = \dot{v} + \alpha v = \frac{F}{m} \qquad \alpha = \frac{c}{m} \tag{5.105}$$

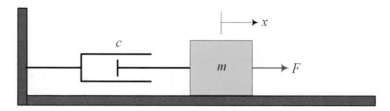

Fig. 5.7 A forced excited mass-damper system

and write the solution based on (5.97):

$$v = e^{-\alpha t} \int e^{\alpha t} \frac{F}{m} dt + C_1 e^{-\alpha t} = \frac{F}{m\alpha} + C_1 e^{-\alpha t} \tag{5.106}$$

Employing the initial velocity $v(0) = 0$, the response would be

$$v = \frac{F}{m\alpha} \left(1 - e^{-\alpha t}\right) \tag{5.107}$$

The position x is found by integration.

$$x = \int v \, dt = \int \frac{F}{m\alpha} \left(1 - e^{-\alpha t}\right) dt = \frac{F}{m\alpha^2} \left(e^{-\alpha t} + \alpha t\right) + C_2$$

$$= \frac{F}{m\alpha^2} \left(e^{-\alpha t} + \alpha t - 1\right) \tag{5.108}$$

$$C_2 = -\frac{F}{m\alpha^2} \tag{5.109}$$

Let us assume

$$m = 1\,\text{kg} \qquad c = 1\,\text{N m}/\text{s} \qquad F = 1\,\text{N} \tag{5.110}$$

and plot the solutions x and v in Fig. 5.8.

When $t \to \infty$, the transient part of the solution disappears, and they approach their steady-state values x_s and v_s.

$$x_s = \lim_{t \to \infty} x = \frac{F}{m\alpha} t \qquad v_s = \lim_{t \to \infty} v = \frac{F}{m\alpha} \tag{5.111}$$

Therefore, m will move with a constant velocity $v_s = \dot{x}_s = F/(m\alpha)$ forever. While m is not at its steady-state condition, its motion is in a transient mode in which the kinematics of m are as (5.107) and (5.108). How fast the acceleration of m vanishes depends on the time constant of the system $\tau = 1/\alpha$:

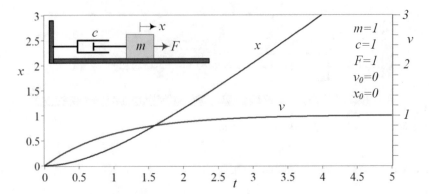

Fig. 5.8 Plot of the solutions x and v for $m = 1, c = 1, F = 1$, and zero initial conditions

$$\ddot{x} = \frac{dv}{dt} = \frac{F}{m}e^{-t\alpha} \tag{5.112}$$

To examine the effect of initial conditions, let us assume

$$v(0) = \dot{x}_0 \qquad x(0) = 0 \tag{5.113}$$

and find the solution.

$$v = \frac{F}{m\alpha} + e^{-\alpha t}\left(\dot{x}_0 - \frac{F}{m\alpha}\right) \tag{5.114}$$

$$x = \frac{F}{m\alpha^2}\left(e^{-\alpha t} + \alpha t - 1\right) + \frac{\dot{x}_0}{\alpha}\left(1 - e^{-\alpha t}\right) \tag{5.115}$$

$$\ddot{x} = \frac{F - m\alpha x_0}{m}e^{-t\alpha} \tag{5.116}$$

Figure 5.9 illustrates the velocity $v(t) = \dot{x}(t)$ for different initial values $v(0) = \dot{x}_0$. The force F will accelerate m when $\dot{x}_0 < \dot{x}_s$, and decelerate m when $\dot{x}_0 > \dot{x}_s$, until the speed of m takes the steady-state value of (5.111).

Example 162 An exponential force on a mass-damper system. The forcing function on a mass-damper system may be modeled as exponentially decreasing or increasing function of time. Here is the response of the system.

Let us examine the response of the mass m in Fig. 5.7 to an exponential force function.

$$m\dot{v} + cv = Fe^{-\alpha_1 t} \qquad v = \dot{x} \tag{5.117}$$

$$v(0) = 0 \qquad x(0) = 0 \tag{5.118}$$

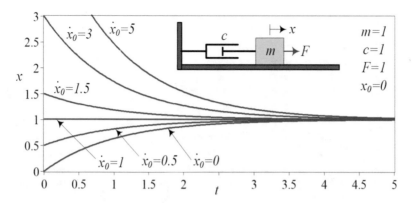

Fig. 5.9 The velocity $v(t) = \dot{x}(t)$ for different initial velocities $v(0) = \dot{x}_0$ of a mass-damper system

We may rewrite the equation of motion,

$$\dot{v} + \alpha v = \frac{F}{m} e^{-\alpha_1 t} \qquad \alpha = \frac{c}{m} \tag{5.119}$$

to solve the equation.

$$
\begin{aligned}
v &= e^{-\alpha t} \int e^{\alpha t} \frac{F}{m} e^{-\alpha_1 t}\, dt + C_1 e^{-\alpha t} \\
&= \left(\frac{F}{m} \frac{e^{(\alpha - \alpha_1)t}}{\alpha - \alpha_1} + C_1 \right) e^{-\alpha t}
\end{aligned} \tag{5.120}
$$

Employing a zero initial velocity $v(0) = 0$, the response would be

$$v = \frac{F}{m} \frac{e^{-\alpha_1 t} - e^{-\alpha t}}{\alpha - \alpha_1} \qquad C_1 = -\frac{F}{m} \frac{1}{\alpha - \alpha_1} \tag{5.121}$$

The position x is found by integration:

$$
\begin{aligned}
x &= \int v\, dt = \int \frac{F}{m(\alpha - \alpha_1)} \left(e^{-\alpha_1 t} - e^{-\alpha t} \right) dt \\
&= \frac{F}{m} \frac{\alpha e^{-\alpha_1 t} - \alpha_1 e^{-\alpha t}}{\alpha \alpha_1 (\alpha - \alpha_1)} + C_2 \\
&= \frac{F}{m} \frac{\alpha \left(1 - e^{-\alpha_1 t} \right) - \alpha_1 \left(1 - e^{-\alpha t} \right)}{\alpha \alpha_1 (\alpha - \alpha_1)}
\end{aligned} \tag{5.122}
$$

$$C_2 = -\frac{F}{m} \frac{1}{\alpha \alpha_1} \tag{5.123}$$

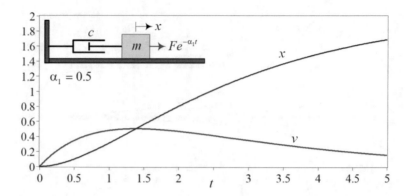

Fig. 5.10 The position and velocity response of a mass-damper system to an exponential force

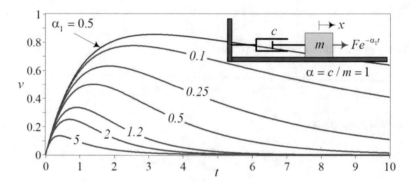

Fig. 5.11 The velocity response of a mass-damper system to different exponential forces

The response of the system for $t \to \infty$ is

$$x_s = \lim_{t \to \infty} x = \frac{F}{m} \frac{1}{\alpha \alpha_1} \qquad v_s = \lim_{t \to \infty} v = 0 \tag{5.124}$$

Let us assume

$$m = 1\,\text{kg} \qquad c = 1\,\text{N}\,\text{m}/\text{s} \qquad F = 1\,\text{N} \tag{5.125}$$

and draw the solutions x and v in Fig. 5.10. Figure 5.11 illustrates the velocity response of the system for different α_1. A good study would be the resonance case when $\alpha_1 = \alpha$.

Example 163 ★ Resonance excitation of a first-order system. When the forcing function on a first-order system is $f = \exp(-\alpha_1 t)$ and the time constant α of the system is equal to α_1 the system is at resonance. Here is the detail analysis of resonance case.

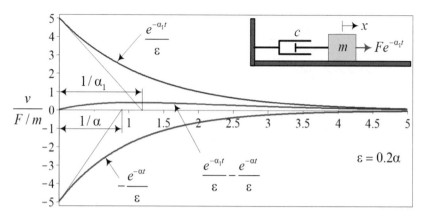

Fig. 5.12 The response of a first-order system to an exponential force function for $\varepsilon = 0.2\alpha$ near resonance

The velocity equation of the mass-damper system of Fig. 5.7 when an exponential force is applied on m is

$$\dot{v} + \alpha v = \frac{F}{m} e^{-\alpha_1 t} \qquad \alpha = \frac{c}{m} \qquad v = \dot{x} \qquad (5.126)$$

$$v(0) = 0 \qquad x(0) = 0 \qquad (5.127)$$

$$v = \frac{F}{m(\alpha - \alpha_1)} \left(e^{-\alpha_1 t} - e^{-\alpha t} \right) \qquad (5.128)$$

This solution does not explain exactly what will happen if $\alpha_1 = \alpha$. That is, what happens when the system is forced with a time function which is proportional to its own natural solution. Such a special case is called resonance.

Although the solution (5.128) shows $v \to 0/0$ when $\alpha_1 = \alpha$, the behavior of the system for $\alpha_1 = \alpha$ is not unusual and shows a response with similar characteristics as any other number for α_1. To resolve this indeterminate situation, we assume α_1 differs from α by a small amount ε, and we investigate the limiting case $\varepsilon \to 0$.

$$\alpha_1 = \alpha - \varepsilon \qquad \varepsilon \ll 1 \qquad (5.129)$$

Substituting (5.129) in the solution (5.128) yields

$$v = \frac{F}{m\varepsilon} \left(e^{-\alpha_1 t} - e^{-\alpha t} \right) = \frac{F e^{-\alpha_1 t}}{m\varepsilon} - \frac{F e^{-\alpha t}}{m\varepsilon} \qquad (5.130)$$

The solution v is the difference of two large functions $F e^{-\alpha_1 t} / (m\varepsilon)$ and $F e^{-\alpha t} / (m\varepsilon)$. Figure 5.12 illustrates the large functions and their difference for $\varepsilon = 0.2\alpha$.

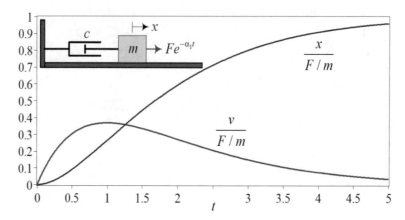

Fig. 5.13 The velocity v and position x of m at resonance for $\alpha = 1$

Let us rewrite the solution (5.128) as a series

$$v = \frac{F}{m\varepsilon}\left(e^{-\alpha_1 t} - e^{-\alpha t}\right) = \frac{F}{m}\frac{1}{\varepsilon e^{\alpha_1 t}}\left(1 - e^{-\varepsilon t}\right)$$

$$= \frac{F}{m}\frac{t}{e^{\alpha_1 t}}\left(1 - \frac{\varepsilon}{2!}t + \frac{\varepsilon^2}{3!}t^2 + \cdots\right) \tag{5.131}$$

and substitute $\varepsilon = 0$ to get the resonance solution for v.

$$v = \frac{F}{m}\frac{t}{e^{\alpha_1 t}} = \frac{F}{m}\frac{t}{e^{\alpha t}} \tag{5.132}$$

The position of m at resonance would be determined by integration.

$$x = \int v dt = \frac{F}{m\alpha^2}\left(1 - e^{-t\alpha}\left(1 + t\alpha\right)\right) \tag{5.133}$$

Figure 5.13 illustrates v and x at resonance for $\alpha = 1$.

Example 164 Linear first-order equations. There is a closed-form solution for the general linear first-order equation but no general solution for nonlinear case. Here is the solution for linear first-order equations.

There is no solution for the general first-order equation:

$$\frac{dx}{dt} = f(x, t) \tag{5.134}$$

However, the most important first-order equation is the linear equation in which the derivative is a linear function of the variable.

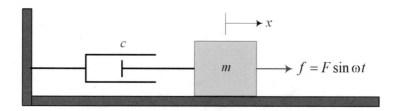

Fig. 5.14 A mass-damper under a sinusoidal harmonic force $f = F \sin \omega t$

$$\frac{dx}{dt} + p(t)x = q(t) \tag{5.135}$$

We may examine that

$$\frac{d}{dt}\left(x \, e^{\int pdx}\right) = x \, p \, e^{\int pdx} + e^{\int pdx}\frac{dx}{dt} = \left(\frac{dx}{dt} + px\right)e^{\int pdx} \tag{5.136}$$

so, if we multiply Eq. (5.135) by $e^{\int pdx}$, we have

$$\frac{d}{dt}\left(x \, e^{\int pdx}\right) = q(t) \, e^{\int pdx} \tag{5.137}$$

and therefore, the general solution of (5.135) will be found.

$$x = e^{-\int pdx}\left(\int q(t) \, e^{\int pdx} dt\right) + C e^{-\int pdx} \tag{5.138}$$

Example 165 A sinusoidal harmonic force on a mass-damper system. Harmonic force function will make a dynamic system to oscillate harmonically. The amplitude of the response as a function of excitation frequency is usually the most important response of the dynamic system to determine. Here is the time response analysis of first-order systems to sinusoidal excitations.

A sinusoidal force $f = F \sin \omega t$ is applied on the mass m of Fig. 5.14. The equation of motion of the system is

$$m\dot{v} + cv = F \sin \omega t \qquad v = \dot{x} \tag{5.139}$$

$$v(0) = \dot{x}_0 \qquad x(0) = x_0 \tag{5.140}$$

Let us rewrite the equation in the form of Eq. (5.135)

$$\dot{v} + \alpha v = \frac{F}{m}\sin \omega t \qquad \alpha = \frac{c}{m} \tag{5.141}$$

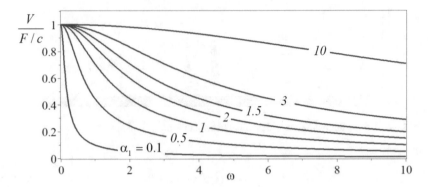

Fig. 5.15 Frequency response of a first-order system for different time constant α

and derive the solution (5.97) or (5.138).

$$v = e^{-\int \alpha dt} \left(\frac{F}{m} \int \sin \omega t \; e^{\int \alpha dt} dt + C_1 e^{-\alpha t} \right)$$

$$= \frac{F}{m} \frac{1}{\alpha^2 + \omega^2} (\alpha \sin \omega t - \omega \cos \omega t) + C_1 e^{-\alpha t} \qquad (5.142)$$

$$C_1 = \dot{x}_0 + \frac{F}{m} \frac{\omega}{\alpha^2 + \omega^2} \qquad (5.143)$$

The natural solution $v_h = C_1 e^{-\alpha t}$ will vanish after a while, and the particular solution remains. The particular solution is a harmonic oscillation of amplitude V.

$$V = \frac{F}{m\sqrt{\alpha^2 + \omega^2}} \qquad (5.144)$$

Figure 5.15 illustrates the frequency response V of the system for different time constant α. The maximum amplitude of V happens at zero frequency $\omega = 0$. The amplitudes drops to zero by increasing the excitation frequency. The lower the time constant α, the higher the rate of reduction V. In other words, the damper acts rigid at high frequencies.

The position of m will be found by integration:

$$x = -\frac{F}{m\omega} \frac{1}{\alpha^2 + \omega^2} (\alpha \cos t\omega + \omega \sin t\omega) - \frac{1}{\alpha} C_1 e^{-\alpha t} + C_2 \qquad (5.145)$$

$$C_2 = x_0 + \frac{F}{m} \frac{\alpha}{\omega} \frac{1}{\alpha^2 + \omega^2} + \frac{1}{\alpha} C_1 = x_0 + \frac{1}{\alpha} \dot{x}_0 + \frac{F}{m\alpha\omega} \qquad (5.146)$$

Assuming zero initial conditions

$$x_0 = 0 \qquad \dot{x}_0 = 0 \qquad (5.147)$$

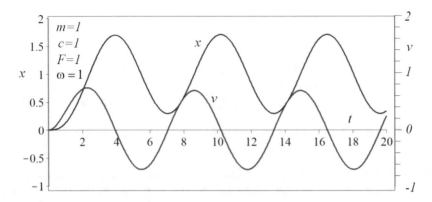

Fig. 5.16 The plot of x and v for a sinusoidal harmonic force excited first-order system

and

$$m = 1\,\text{kg} \qquad c = 1\,\text{N}\,\text{m}/\text{s} \qquad F = 1\,\text{N} \qquad \omega = 1\,\text{rad}/\text{s} \qquad (5.148)$$

we plot the solutions x and v in Fig. 5.16. The transient response $\frac{1}{\alpha}C_1 e^{-\alpha t}$ fades away very fast, and a harmonic steady-state response remains. However, the harmonic displacement oscillation of the system has an interesting characteristic. The harmonic displacement occurs about a center point at $x_e = C_2 \neq 0$.

$$x_e = C_2 = x_0 + \frac{1}{\alpha}\dot{x}_0 + \frac{F}{m\alpha\omega} \qquad (5.149)$$

The amplitude X of the oscillation is

$$X = x_e \pm \frac{F}{m\omega\sqrt{\alpha^2 + \omega^2}} \qquad (5.150)$$

Therefore, the position of m would be

$$-\frac{F}{m\omega\sqrt{\alpha^2 + \omega^2}} \leq (X - x_e) \leq \frac{F}{m\omega\sqrt{\alpha^2 + \omega^2}} \qquad (5.151)$$

Although every point of x can be an equilibrium point of the system, we may call x_e the equilibrium point of the system about which a steady-state oscillation occurs. The appearance of an off-center vibration is a characteristic of the sinusoidal force and the first-order mass-damper system. This is an interesting phenomenon that the steady-state response of a first-order system is different to sinusoidal and cosine harmonic excitation. This fact will be shown in the next example.

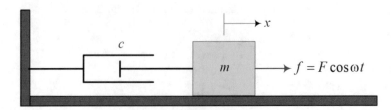

Fig. 5.17 A mass-damper under a cosine harmonic force $f = F \cos \omega t$

Example 166 A cosine force on a mass-damper. Harmonic force function will make a dynamic system to oscillate harmonically. The amplitude of the response as a function of excitation frequency is the most important response of the dynamic system. Here is the time response analysis of first-order systems to co-sinusoidal excitations. The importance of this example is to examine the difference of the response of a first-order system to sine and cosine harmonic excitations.

To analyze the effect of a sin or cos excitation of the equilibrium position, we examine a cosine forcing term on the system of Fig. 5.17. The equation of motion of the system is

$$\dot{v} + \alpha v = \frac{F}{m} \cos \omega t \qquad \alpha = \frac{c}{m} \qquad v = \dot{x} \tag{5.152}$$

$$v(0) = \dot{x}_0 \qquad\qquad x(0) = x_0 \tag{5.153}$$

The solution of the equation is

$$v = e^{-\int \alpha dt} \left(\frac{F}{m} \int \cos \omega t \, e^{\int \alpha dt} dt + C_1 e^{-\alpha t} \right)$$

$$= \frac{F}{m} \frac{1}{\alpha^2 + \omega^2} (\alpha \cos \omega t + \omega \sin \omega t) + C_1 e^{-\alpha t} \tag{5.154}$$

$$C_1 = \dot{x}_0 - \frac{F}{m} \frac{\alpha}{\alpha^2 + \omega^2} \tag{5.155}$$

After vanishing of the natural solution $v_h = C_1 e^{-\alpha t}$, the particular solution remains. The amplitude of the steady-state oscillation V is the same as (5.144) of a sinusoidal force.

$$V = \frac{F}{m\sqrt{\alpha^2 + \omega^2}} \tag{5.156}$$

Interpreting (5.154) provides us with the position of m.

$$x = \frac{F}{m} \frac{1}{\alpha^2 + \omega^2} \left(\frac{\alpha}{\omega} \sin \omega t - \cos \omega t \right) - \frac{1}{\alpha} C_1 e^{-\alpha t} + C_2 \tag{5.157}$$

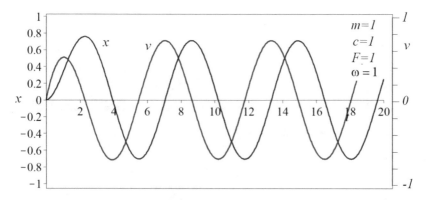

Fig. 5.18 The plot of x and v for a cosine force excited first-order system

$$C_2 = x_0 + \frac{F}{m}\frac{1}{\alpha^2 + \omega^2} + \frac{1}{\alpha}C_1 = x_0 + \frac{1}{\alpha}v_0 \qquad (5.158)$$

Assuming zero initial conditions

$$x_0 = 0 \qquad \dot{x}_0 = 0 \qquad (5.159)$$

and

$$m = 1\,\text{kg} \qquad c = 1\,\text{N}\,\text{m}\,/\,\text{s} \qquad F = 1\,\text{N} \qquad \omega = 1\,\text{rad}\,/\,\text{s} \qquad (5.160)$$

we plot the solutions x and v in Fig. 5.18. The transient response $\frac{1}{\alpha}C_1 e^{-\alpha t}$ fades out very fast, and a harmonic steady-state response remains. A harmonic oscillation occurs about $x = 0$, which shows no equilibrium offset appears for cosine harmonic force. The amplitude X of the oscillation is the same as (5.150) of the sinusoidal excitation.

$$X = \frac{F}{m\omega\sqrt{\alpha^2 + \omega^2}} \qquad (5.161)$$

Example 167 A mass under sinusoidal or cosine force. The previous examples about the frequency response of a first-order system under sine or cosine force function suggest to investigate the behavior of a mass under such forces. Here, we will see what would be the difference of behavior of a single mass under cosine or sine harmonic forces.

Consider the extreme case of a sinusoidal forced excited mass. The system is a single mass m under a force such as shown in Fig. 5.19. The equation of motion of m when the applied force is sinusoidal is reducible to a first-order equation.

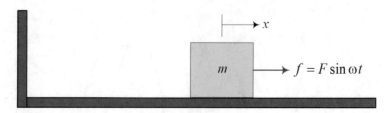

Fig. 5.19 A mass under sinusoidal force

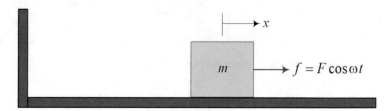

Fig. 5.20 A mass under cosine force

$$\dot{v} = \frac{F}{m} \sin \omega t \qquad v = \dot{x} \qquad\qquad (5.162)$$

$$v(0) = \dot{x}_0 \qquad\qquad x(0) = x_0 \qquad\qquad (5.163)$$

The solution of the equation of motion is

$$v = \frac{F}{m} \int \sin \omega t \, dt + C_1 = -\frac{F}{m\omega} \cos \omega t + C_1$$

$$= \dot{x}_0 + \frac{F}{m\omega} (1 - \cos \omega t) \qquad\qquad (5.164)$$

$$C_1 = \dot{x}_0 + \frac{F}{m\omega} \qquad\qquad (5.165)$$

$$x = \int v \, dt + C_2 = \dot{x}_0 t + \frac{F}{m\omega} t - \frac{F}{m\omega^2} \sin \omega t + C_2$$

$$= x_0 + \left(\dot{x}_0 + \frac{F}{m\omega} \right) t - \frac{F}{m\omega^2} \sin \omega t \qquad\qquad (5.166)$$

$$C_2 = x_0 \qquad\qquad (5.167)$$

Let us change the sinusoidal force for a cosine force as is shown in Fig. 5.20. The equation of motion of m when the applied force is cosine function of time is

$$\dot{v} = \frac{F}{m} \cos \omega t \qquad v = \dot{x} \tag{5.168}$$

$$v(0) = \dot{x}_0 \qquad x(0) = x_0 \tag{5.169}$$

The solution of the equation of motion is

$$v = \frac{F}{m} \int \cos \omega t \, dt + C_1 = \frac{F}{m\omega} \sin \omega t + C_1$$

$$= \dot{x}_0 + \frac{F}{m\omega} \sin \omega t \tag{5.170}$$

$$C_1 = \dot{x}_0 \tag{5.171}$$

$$x = \int v \, dt + C_2 = \dot{x}_0 t - \frac{F}{m\omega^2} \cos \omega t + C_2$$

$$= x_0 + \dot{x}_0 t + \frac{F}{m\omega^2} (1 - \cos \omega t) \tag{5.172}$$

$$C_2 = x_0 + \frac{F}{m\omega^2} \tag{5.173}$$

The responses of m to sinusoidal and cosine forces are very different. Assuming zero initial conditions, the mass m shows an oscillatory motion about $x = 0$ when the applied force is a cosine function, $f = F \cos \omega t$. However, the mass m will have an oscillatory motion, while it is moving with a constant speed $v = F/(m\omega)$, when the applied force is a sinusoidal function, $f = F \sin \omega t$.

Example 168 Frequency response of a first-order system. Frequency response of vibrating system usually refers to second-order systems; however, frequency response of first-order system may also be determined. Here is to calculate the frequency response of first-order systems.

In case the transient response of a first-order system is not as important as the steady-state oscillatory response, we may use the harmonic balance method to determine the frequency response of the system. Consider the system of Fig. 5.14 with the following equation of motion.

$$m\dot{v} + cv = F \sin \omega t \qquad v = \dot{x} \tag{5.174}$$

Substituting a harmonic solution for v

$$v = A_v \sin \omega t + B_v \cos \omega t \tag{5.175}$$

and balancing sin and cos coefficients provide us with two equations to determine A_v and B_v.

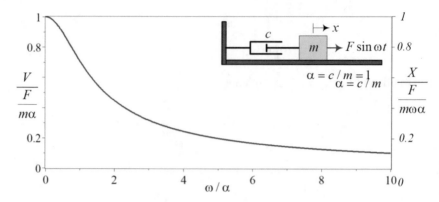

Fig. 5.21 The frequency response of a first-order system

$$cA_v - m\omega B_v = F \tag{5.176}$$

$$m\omega A_v + cB_v = 0 \tag{5.177}$$

$$\begin{bmatrix} c & -m\omega \\ m\omega & c \end{bmatrix} \begin{bmatrix} A_v \\ B_v \end{bmatrix} = \begin{bmatrix} F \\ 0 \end{bmatrix} \tag{5.178}$$

The unknown coefficients and the steady-state amplitude of v are compatible with (5.144)

$$\begin{bmatrix} A_v \\ B_v \end{bmatrix} = \begin{bmatrix} F\dfrac{c}{c^2 + m^2\omega^2} \\ -Fm\dfrac{\omega}{c^2 + m^2\omega^2} \end{bmatrix} \tag{5.179}$$

where

$$V = \sqrt{A_v^2 + B_v^2} = \frac{F}{\sqrt{c^2 + m^2\omega^2}} = \frac{F/m}{\sqrt{\alpha^2 + \omega^2}} \qquad \alpha = \frac{c}{m} \tag{5.180}$$

We may rewrite the frequency response in a dimensionless form for generality:

$$\frac{V}{F/(m\alpha)} = \frac{1}{\sqrt{1 + (\omega/\alpha)^2}} \tag{5.181}$$

As is shown in Fig. 5.21, the frequency response of a first-order system does not have any resonance zone. The amplitude of velocity oscillation reduces by increasing ω/α. The maximum velocity amplitude is at $\omega = 0$ which indicates the maximum speed of m.

Let us try the harmonic balance method again and determine the frequency response of motion of m. The equation of motion is

$$m\ddot{x} + c\dot{x} = F \sin \omega t \tag{5.182}$$

Substituting

$$x = A_x \sin \omega t + B_x \cos \omega t \tag{5.183}$$

provides us with another two algebraic equations to determine A_x and B_x.

$$- m\omega^2 A_x - c\omega B_x = F \tag{5.184}$$

$$c\omega A_x - m\omega^2 B_x = 0 \tag{5.185}$$

$$\begin{bmatrix} -m\omega^2 & -c\omega \\ c\omega & -m\omega^2 \end{bmatrix} \begin{bmatrix} A_x \\ B_x \end{bmatrix} = \begin{bmatrix} F \\ 0 \end{bmatrix} \tag{5.186}$$

The unknown coefficients and the steady-state amplitude of x are compatible with (5.150):

$$\begin{bmatrix} A_x \\ B_x \end{bmatrix} = \begin{bmatrix} -F \dfrac{m}{c^2 + m^2\omega^2} \\ -F \dfrac{c}{c^2\omega + m^2\omega^3} \end{bmatrix} \tag{5.187}$$

$$X = \sqrt{A_x^2 + B_x^2} = \frac{F}{\omega\sqrt{c^2 + m^2\omega^2}} = \frac{F}{m\omega\sqrt{\alpha^2 + \omega^2}} \tag{5.188}$$

We may rewrite the frequency response in a dimensionless form for generality:

$$\frac{X}{F/(m\omega\alpha)} = \frac{1}{\sqrt{1 + (\omega/\alpha)^2}} \tag{5.189}$$

The graph of (5.189) is similar to Fig. 5.21.

Example 169 ★ A third-order system. Attachment of a mass to a series attachment of a spring and a damper makes a complicated system whose equation of motion can be expressed by a third-order differential equation. Here is the analysis of such system.

Figure 5.22 illustrates a mass m that is connected to a fixed wall by a series of spring k and damper c. The system is governed by two equations, a first-order equation for the y-coordinate and a second-order equation for the x-coordinate:

$$m\ddot{x} = -k(x - y) \tag{5.190}$$

$$c\dot{y} = k(x - y) \tag{5.191}$$

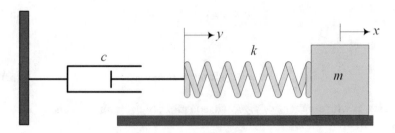

Fig. 5.22 A mass m that is connected to a fixed support by a series of spring k and damper c

To analyze the dynamics of the system, it is better to eliminate y between the equations to make a single equation. Such an elimination will end up with a third-order equation. To derive the equation, let us take a derivative of (5.190)

$$m\ddot{x} + k\dot{x} = k\dot{y} \tag{5.192}$$

and substitute for $\dot{y}$ from (5.191):

$$\frac{c}{k}\left(m\dddot{x} + k\ddot{x}\right) = k\left(x - y\right) \tag{5.193}$$

and substitute for $k\left(x - y\right)$ from (5.190):

$$mc\dddot{x} + mk\ddot{x} + kc\dot{x} = 0 \tag{5.194}$$

Introducing a new variable

$$v = \dot{x} \tag{5.195}$$

transforms the equation to a second-order one:

$$mc\ddot{v} + mk\dot{v} + kcv = 0 \tag{5.196}$$

The solution of the equation is

$$v = C_1 e^{s_1 t} + C_2 e^{s_2 t} \tag{5.197}$$

$$s_{1,2} = \frac{-km \pm \sqrt{km\left(km - 4c^2\right)}}{2cm} \tag{5.198}$$

Assuming the initial conditions

$$x\left(0\right) = 0 \qquad \dot{x}\left(0\right) = v\left(0\right) = \dot{x}_0 \qquad y\left(0\right) = 0 \tag{5.199}$$

yields

$$\ddot{x}(0) = \dot{v}(0) = 0 \tag{5.200}$$

and, therefore,

$$C_1 = \frac{s_2}{s_2 - s_1}\dot{x}_0 = \frac{\sqrt{k^2 m^2 - 4c^2 km} - km}{2\sqrt{k^2 m^2 - 4c^2 km}}\dot{x}_0 \tag{5.201}$$

$$C_2 = \frac{s_1}{s_2 - s_1}\dot{x}_0 = -\frac{\sqrt{k^2 m^2 - 4c^2 km} + km}{2\sqrt{k^2 m^2 - 4c^2 km}}\dot{x}_0 \tag{5.202}$$

The x-coordinate is calculated by integration.

$$x = \int v\, dt = \int \left(C_1 e^{s_1 t} + C_2 e^{s_2 t}\right) dt$$

$$= \frac{C_1}{s_1}\left(e^{s_1 t} - 1\right) + \frac{C_2}{s_2}\left(e^{s_2 t} - 1\right) \tag{5.203}$$

Now Eq. (5.191) for y is a forced first-order equation.

$$c\dot{y} + ky = kx = k\frac{C_1}{s_1}\left(e^{s_1 t} - 1\right) + k\frac{C_2}{s_2}\left(e^{s_2 t} - 1\right) \tag{5.204}$$

The time constant of this equation is

$$\tau = \frac{c}{k} \tag{5.205}$$

and the solution is

$$y = \left(\frac{k}{s_1\,(k + cs_1)}e^{s_1 t} - \frac{1}{s_1} + \frac{c}{k + cs_1}e^{-t/\tau}\right)C_1$$

$$+ \left(\frac{k}{s_2\,(k + cs_2)}e^{s_2 t} - \frac{1}{s_2} + \frac{c}{k + cs_2}e^{-t/\tau}\right)C_2 \tag{5.206}$$

Figure 5.23 illustrates the time response of the coordinates x and y for a set of nominal values:

$$m = 1 \qquad c = 1 \qquad k = 1 \qquad v_0 = 1 \tag{5.207}$$

To examine the behavior of the system, let us plot $x(t)$ for different c and k in Figs. 5.24 and 5.25, respectively. As Fig. 5.24 indicates, the motion of m approaches a first-order response when c is decreasing and approaches an undamped mass-spring vibration when c increases. There is a steady-state value of x for every $0 < c < \infty$.

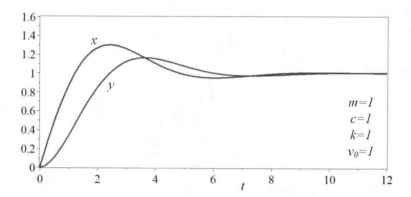

Fig. 5.23 The time response of the coordinates x and y for a set of nominal values

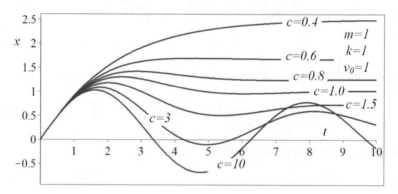

Fig. 5.24 The plot $x(t)$ for different c

$$x_{ss} = -\frac{C_1}{s_1} - \frac{C_2}{s_2} \tag{5.208}$$

Figure 5.26 also indicates that y has an almost similar behavior as x. The coordinate y approaches a first-order response when c is decreasing and shows vibration when c increases. The lag in y at $t = 0$ is not comparable with a first-order system, and it is because of interaction between the first- and second-order systems. There is a steady-state value of y for every $0 < c < \infty$. Because there is no friction, there would be no residual displacement in the spring when $t \to \infty$, and therefore, the steady-state values of x and y are equal:

$$\lim_{t \to \infty} x = \lim_{t \to \infty} y \tag{5.209}$$

Figure 5.25 illustrates the motion of m for different k. Increasing k makes the system approach a first-order response, and decreasing k makes it approach a second-order system. Figure 5.27 also illustrates y that shows almost similar behavior as x,

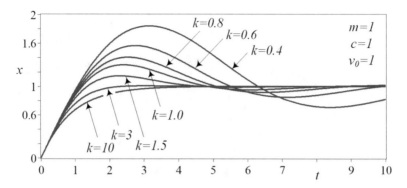

Fig. 5.25 The plot $x(t)$ for different k

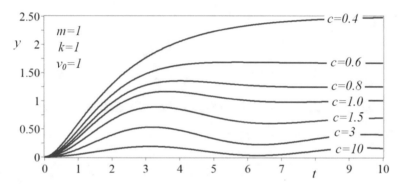

Fig. 5.26 The plot $y(t)$ for different c

although the response of y at $t = 0$ is not similar to a first-order system regardless of the value of k or c. The steady-state value of x and y for the given c and for every $0 < k < \infty$ is

$$\lim_{t \to \infty} x = \lim_{t \to \infty} y = 1 \tag{5.210}$$

Example 170 A base speed excited first-order system. Suspension is the isolators between a mass and a moving base. A moving base will make the mass to move. Here is the analysis of the motion of the mass for two different types of base motion, a constant velocity, and sinusoidal oscillation.

Figure 5.28 illustrates a base excited first-order system. Its equation of motion is

$$m\dot{v} + cv = c\dot{y} \qquad v = \dot{x} \qquad v(0) = 0 \qquad x(0) = 0 \tag{5.211}$$

Let us examine the reaction of m to a constant velocity base excitations.

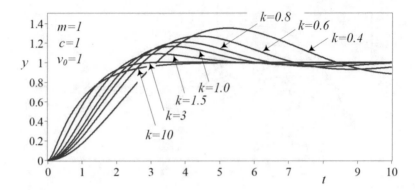

Fig. 5.27 The plot $y(t)$ for different k

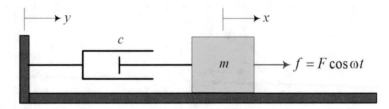

Fig. 5.28 A base excited first-order system

$$\dot{y} = \dot{y}_0 \tag{5.212}$$

The equation of motion becomes a first-order with a constant forcing term.

$$m\dot{v} + cv = c\dot{y}_0 \tag{5.213}$$

This case is similar to Example 161 in which the mass m is under a constant force $f = F$. Let us rewrite (5.213) as

$$\dot{v} + \alpha v = \alpha \dot{y}_0 \qquad \alpha = \frac{c}{m} \tag{5.214}$$

and derive the solution $v = v(t)$.

$$v = e^{-\alpha t} \int e^{\alpha t}\, \alpha \dot{y}_0\, dt + C_1 e^{-\alpha t} = \dot{y}_0 + C_1 e^{-t\alpha}$$

$$= \dot{y}_0 \left(1 - e^{-\alpha t}\right) \tag{5.215}$$

$$C_1 = -\dot{y}_0 \tag{5.216}$$

The mass m starts moving with an acceleration $\ddot{x} = \alpha \dot{y}_0$ which exponentially reduces to zero:

$$\ddot{x} = \alpha \dot{y}_0 e^{-\alpha t} \tag{5.217}$$

Therefore, the speed of m approaches $\dot{y}_0$ when $t \to \infty$. The position x would be

$$x = \int v\, dt = \frac{\dot{y}_0}{\alpha}\left(e^{-\alpha t} + \alpha t\right) + C_2 = \frac{\dot{y}_0}{\alpha}\left(e^{-\alpha t} + \alpha t - 1\right) \tag{5.218}$$

$$C_2 = -\frac{\dot{y}_0}{\alpha} \tag{5.219}$$

If we are interested in the relative displacement of m with respect to the base, we define $z = x - y$.

$$z = x - y = \frac{\dot{y}_0}{\alpha}\left(e^{-\alpha t} + \alpha t - 1\right) - \dot{y}_0 t = \frac{\dot{y}_0}{\alpha}\left(e^{-\alpha t} - 1\right) \tag{5.220}$$

If the initial distance between y and x is l, then their distance decreases from l to $l - \dot{y}_0/\alpha$ at $t = \infty$. Therefore, m will not necessarily hit the base, and it remains at a constant distance from base while moving with the same speed as the base. This result is correct as long as there is no resistance force such as friction. The existence of any resistance force on m causes a continuing reduction of the relative distance.

As another example, consider the system of Fig. 5.28 where the base motion y is a sinusoidal displacement.

$$y = Y \sin \omega t \tag{5.221}$$

The equation of motion of the system is

$$\dot{v} + \alpha v = \alpha \dot{y} \qquad v = \dot{x} \qquad \alpha = \frac{c}{m} \tag{5.222}$$

Substituting for y, we have

$$\dot{v} + \alpha v = \alpha \omega Y \cos \omega t \tag{5.223}$$

which yields

$$v = e^{-\int \alpha dt}\left(\alpha \omega Y \int \cos \omega t\, e^{\int \alpha dt}\, dt + C_1 e^{-\alpha t}\right)$$

$$= \frac{\alpha \omega Y}{\alpha^2 + \omega^2}(\alpha \cos \omega t + \omega \sin \omega t) + C_1 e^{-\alpha t} \tag{5.224}$$

$$C_1 = \dot{x}_0 - \frac{\alpha^2 \omega}{\alpha^2 + \omega^2}Y \tag{5.225}$$

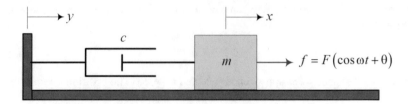

Fig. 5.29   A mass-damper under a cosine harmonic force $f = F \cos (\omega t + \theta)$

The amplitude of the steady-state oscillation of the velocity is

$$V = \frac{\alpha \omega}{\sqrt{\alpha^2 + \omega^2}} Y \tag{5.226}$$

Integrating v, we find the position of the mass.

$$x = \int v \, dt = \frac{\alpha \omega Y}{\alpha^2 + \omega^2} \left(\frac{\alpha}{\omega} \sin \omega t - \cos \omega t \right) - \frac{1}{\alpha} C_1 e^{-\alpha t} + C_2 \tag{5.227}$$

$$C_2 = x_0 + \frac{1}{\alpha} C_1 + \frac{\alpha \omega Y}{\alpha^2 + \omega^2} = x_0 + \frac{\dot{x}_0}{\alpha} \tag{5.228}$$

The amplitude of the oscillation of m is

$$X = Y \frac{\alpha}{\sqrt{\alpha^2 + \omega^2}} \tag{5.229}$$

The amplitude of m is a monotonically decreasing function of ω similar to Fig. 5.21, starting from Y at $\omega = 0$.

Example 171 ★ A phased harmonic force on a mass-damper. To analyze the effect of phase lag on the response of a first-order system to a harmonic excitation, we study a first-order system under the force $f = F \cos (\omega t + \theta)$.

The mass-damper system of Fig. 5.29 is under a harmonic force with a phase θ to control the type of the harmonic function. The equation of motion of the system is

$$\dot{v} + \alpha v = \frac{F}{m} \cos (\omega t + \theta) \qquad \alpha = \frac{c}{m} \qquad v = \dot{x} \tag{5.230}$$

$$v(0) = \dot{x}_0 \qquad\qquad x(0) = x_0 \tag{5.231}$$

The solution of the equation is

$$v = e^{-\int \alpha dt} \left(\frac{F}{m} \int \cos (\omega t + \theta) \, e^{\int \alpha dt} dt + C_1 e^{-\alpha t} \right)$$

$$= \frac{F}{m} \frac{1}{\alpha^2 + \omega^2} (\alpha \cos (\omega t + \theta) + \omega \sin (\omega t + \theta)) + C_1 e^{-\alpha t} \tag{5.232}$$

$$C_1 = \dot{x}_0 - \frac{F}{m} \frac{1}{\alpha^2 + \omega^2} (\alpha \cos \theta + \omega \sin \theta) \tag{5.233}$$

The natural solution $v_h = C_1 e^{-\alpha t}$ will vanish in time, and the particular solution remains. The particular solution is a harmonically oscillating motion with amplitude V:

$$V = \frac{F}{m\sqrt{\alpha^2 + \omega^2}} \tag{5.234}$$

The position of m is calculated by integration.

$$x = \int v \, dt$$

$$= \frac{F/m}{\alpha^2 + \omega^2} \left(\frac{\alpha}{\omega} \sin(\omega t + \theta) - \cos(\omega t + \theta) \right) - \frac{C_1}{\alpha} e^{-\alpha t} + C_2 \tag{5.235}$$

$$C_2 = x_0 + \frac{1}{\alpha} v_0 - \frac{F}{m\alpha\omega} \sin\theta \tag{5.236}$$

Assuming a steady-state response and zero initial conditions

$$x_0 = 0 \qquad \dot{x}_0 = 0 \tag{5.237}$$

yields

$$v = \frac{F}{m} \frac{1}{\alpha^2 + \omega^2} (\alpha \cos(\omega t + \theta) + \omega \sin(\omega t + \theta)) \tag{5.238}$$

$$x = \frac{F}{m} \frac{1}{\alpha^2 + \omega^2} \left(\frac{\alpha}{\omega} \sin(\omega t + \theta) - \cos(\omega t + \theta) \right) - \frac{F}{m\alpha\omega} \sin\theta \tag{5.239}$$

The position of m is a harmonic oscillation with amplitude X about the equilibrium point x_e, where θ is the phase angle of the force $f = F \cos(\omega t + \theta)$.

$$X = \frac{F}{m\omega\sqrt{\alpha^2 + \omega^2}} \tag{5.240}$$

$$x_e = -\frac{F}{m\alpha\omega} \sin\theta \tag{5.241}$$

When $\theta = 0$, the force is cosine and $x_e = 0$. This is compatible with example 166. When $\theta = -\pi/2$, the force is sinusoidal and $x_e = \frac{F}{m\alpha\omega}$, which is compatible with example 165. Depending on the phase θ, the position of m would be

$$-X \le \left(x + \frac{F}{m\alpha\omega} \sin\theta \right) \le X \tag{5.242}$$

Example 172 Transient vibrations. This is to show how the convolution integral method can be applied on first-order equations to determine the solution.

When the forcing function $f(t)$ disappears after a period of time, the system is under transient excitation.

$$\dot{x} + ax = f(t) \qquad x(0) = x_0 \tag{5.243}$$

The convolution integral method can also be used on first-order system to determine the transient response easier.

$$x(t) = x_0 e^{-at} + \int_0^t f(t-\tau) e^{-a\tau} d\tau \tag{5.244}$$

$$= x_0 e^{-at} + \int_0^t e^{-a(t-\tau)} f(\tau) d\tau \tag{5.245}$$

As an example of application, let us determine the response of a first-order system to a step function.

$$f(t) = \begin{cases} F & 0 < t \\ 0 & t < 0 \end{cases} \tag{5.246}$$

$$x(t) = x_0 e^{-at} + \int_0^t F e^{-a\tau} d\tau = x_0 e^{-at} + \frac{F}{a}\left(1 - e^{-at}\right)$$

$$= \frac{F}{a} + \left(x_0 - \frac{F}{a}\right) e^{-at} \tag{5.247}$$

5.3   ★ Coupled Systems

Any linear free vibrating system of any order can be expressed by a set of coupled first-order linear differential equations.

$$\dot{\mathbf{x}} + [A]\mathbf{x} = \mathbf{0} \tag{5.248}$$

$$[A] = \begin{bmatrix} a_{11} & a_{12} & a_{13} & \cdots & a_{1n} \\ a_{21} & a_{22} & a_{23} & \cdots & a_{2n} \\ \vdots & \vdots & \vdots & \vdots & \vdots \\ a_{n1} & a_{n2} & a_{n3} & \cdots & a_{nn} \end{bmatrix} \tag{5.249}$$

The solution of a set of coupled first-order linear homogeneous equations is

$$\mathbf{x} = \sum_{j=1}^{n} e^{-\lambda_j t} C_j \mathbf{u}_j \tag{5.250}$$

where $\mathbf{u}_j$ are the eigenvectors and λ_j, $j = 1, 2, \cdots, n$ are the eigenvalues of the $n \times n$ coefficient matrix $[A]$. This solution can also be expressed by

$$\mathbf{x}(t) = e^{-[A]t} \mathbf{C} = e^{-[A]t} \mathbf{x}_0 \tag{5.251}$$

where

$$e^{-[A]t} = [I] - [A]t + \frac{1}{2!}[A]^2 t^2 - \frac{1}{3!}[A]^3 t^3 + \frac{1}{4!}[A]^4 t^4 - \cdots$$

$$= \sum_{n=0}^{\infty} (-1)^n \frac{(At)^n}{n!} \tag{5.252}$$

We may also express the solution of the homogeneous equations (5.248) as

$$\mathbf{x} = e^{-[A]t} \mathbf{C} = [U] e^{-[\Lambda]t} [U]^{-1} \mathbf{C} \tag{5.253}$$

where $[\Lambda]$ is the associated diagonalized matrix of $[A]$ and $[U]$ is the modal matrix of $[A]$.

$$[\Lambda] = [U]^{-1} [A][U] = \begin{bmatrix} \lambda_1 & 0 & \cdots & 0 \\ 0 & \lambda_2 & \cdots & 0 \\ \vdots & \vdots & \ddots & \vdots \\ 0 & 0 & \cdots & \lambda_n \end{bmatrix} \tag{5.254}$$

$$[U] = \begin{bmatrix} \mathbf{u}_1 & \mathbf{u}_2 & \cdots & \mathbf{u}_n \end{bmatrix} \tag{5.255}$$

Any linear forced vibrating system of any order can also be expressed by a set of n coupled first-order linear system,

$$\dot{\mathbf{x}} + [A]\mathbf{x} = \mathbf{f}(t) \tag{5.256}$$

where its solution $\mathbf{x} = \mathbf{x}(t)$ is

$$\mathbf{x} = e^{-[A]t} \int e^{[A]t} \mathbf{f}(t) \, dt + e^{-[A]t} \mathbf{C} \tag{5.257}$$

$$\mathbf{x} = \begin{bmatrix} x_1 & x_2 & x_3 & \cdots & x_n \end{bmatrix}^T \tag{5.258}$$

$$\mathbf{f} = \begin{bmatrix} f_1 & f_2 & f_3 & \cdots & f_n \end{bmatrix}^T \tag{5.259}$$

$$\mathbf{C} = \begin{bmatrix} C_1 & C_2 & C_3 & \cdots & C_n \end{bmatrix}^T \tag{5.260}$$

Using the modal matrix $[U]$ and its diagonalized matrix $[\Lambda]$, the solution of the forced equations (5.256) can also be expressed as below based on a modal transformation and principal coordinates $\mathbf{p}$,

$$\mathbf{x} = [U]\mathbf{p} = [U]\left[e^{-[\Lambda]t} \int e^{[\Lambda]t} \mathbf{P} \, dt + e^{-[\Lambda]t} \mathbf{C} \right] \tag{5.261}$$

where

$$\mathbf{p} = [U]^{-1}\mathbf{x} \tag{5.262}$$

$$\mathbf{P} = [U]^{-1}\mathbf{f}(t) \tag{5.263}$$

$$\mathbf{p} = e^{-[\Lambda]t} \int e^{[\Lambda]t} \mathbf{P} \, dt + e^{-[\Lambda]t} \mathbf{C} \tag{5.264}$$

Proof Assume the equation of free motion of a vibrating system is expressed by an n-order linear differential equation.

$$a_n(t) \frac{d^n x}{dt^n} + a_{n-1}(t) \frac{d^{n-1}x}{dt^{n-1}} + \cdots + a_0(t) x = 0 \tag{5.265}$$

To convert the equation into a set of coupled first-order equations, we introduce n new variables as below.

$$x_1 = x \qquad x_2 = \dot{x} \qquad \cdots \qquad x_n = \frac{d^n x}{dt^n} \tag{5.266}$$

Employing the new set of variables $x_1, x_2, \cdots, x_n$, we make the following set of first-order equations.

$$\dot{x}_1 = x_2 \qquad \dot{x}_2 = x_3 \qquad \cdots \qquad \dot{x}_{n-1} = x_n \tag{5.267}$$

$$\dot{x}_n = -\frac{a_{n-1}(t) x_n + a_{n-2}(t) x_{n-1} + \cdots + a_0(t) x_1}{a_n(t)} \tag{5.268}$$

The set of equations can always be expressed in matrix form.

$$\dot{\mathbf{x}} + [A]\mathbf{x} = \mathbf{0} \tag{5.269}$$

$$\mathbf{x} = \begin{bmatrix} x_1 & x_2 & x_3 & \cdots & x_n \end{bmatrix}^T \tag{5.270}$$

$$\dot{\mathbf{x}} = \begin{bmatrix} \dot{x}_1 & \dot{x}_2 & \dot{x}_3 & \cdots & \dot{x}_n \end{bmatrix}^T \tag{5.271}$$

$$[A] = \begin{bmatrix} 0 & -1 & 0 & \cdots & 0 \\ 0 & 0 & -1 & \cdots & 0 \\ \vdots & \vdots & \vdots & \vdots & \vdots \\ 0 & 0 & 0 & 0 & -1 \\ \dfrac{a_0}{a_n} & \dfrac{a_1}{a_n} & \dfrac{a_2}{a_n} & \cdots & \dfrac{a_{n-1}}{a_n} \end{bmatrix} \tag{5.272}$$

To solve the set of homogeneous equations (5.248), let us check a guessed solution,

$$\mathbf{x} = e^{-\lambda t}\mathbf{u} \tag{5.273}$$

where $\mathbf{u}$ is a constant $n \times 1$ vector and λ is a constant value.

$$-\lambda \mathbf{u} e^{-\lambda t} + [A]\mathbf{u} e^{-\lambda t} = \mathbf{0} \tag{5.274}$$

Eliminating $e^{-\lambda t}$,

$$[A]\mathbf{u} - \lambda \mathbf{u} = 0 \tag{5.275}$$

indicates that the function (5.273) satisfies Eq. (5.248) if $\mathbf{u}$ is an eigenvector of $[A]$, and λ is an eigenvalue of $[A]$, associated with $\mathbf{u}$. Because the differential equations are linear, the function $\mathbf{x}$ is also a solution.

$$\mathbf{x} = e^{-\lambda t}C\mathbf{u} \tag{5.276}$$

Suppose the $n \times n$ coefficient matrix $[A]$ has eigenvalues λ_j and linearly independent eigenvectors $\mathbf{u}_j$, $j = 1, 2, \cdots, n$. Then each function $\mathbf{x}_j = e^{-\lambda_j t}\mathbf{u}_j$ is a solution of Eq. (5.248), and the general solution is a linear combination of the n solutions.

$$\mathbf{x} = \sum_{j=1}^{n} e^{-\lambda_j t}C_j\mathbf{u}_j \tag{5.277}$$

To prove Eq. (5.251) is the right solution, assume that there is a solution for Eq. (5.248) in the form of a time series.

$$\mathbf{x} = \mathbf{b}_0 + \mathbf{b}_1 t + \mathbf{b}_2 t^2 + \mathbf{b}_3 t^3 + \cdots \tag{5.278}$$

Substituting this solution in the equation yields

$$\mathbf{b}_1 + 2\mathbf{b}_2 t + 3\mathbf{b}_3 t^2 + \cdots = -[A]\left(\mathbf{b}_0 + \mathbf{b}_1 t + \mathbf{b}_2 t^2 + \mathbf{b}_3 t^3 + \cdots\right) \tag{5.279}$$

For this equation to hold at any time, we must have

$$\mathbf{b}_1 = -[A]\mathbf{b}_0 \tag{5.280}$$

$$\mathbf{b}_2 = -\frac{1}{2}[A]\mathbf{b}_1 = \frac{1}{2}[A]^2\mathbf{b}_0 \tag{5.281}$$

$$\mathbf{b}_3 = -\frac{1}{3}[A]\mathbf{b}_2 = -\frac{1}{3\times 2}[A]^3\mathbf{b}_0 = -\frac{1}{3!}[A]^3\mathbf{b}_0 \tag{5.282}$$

$$\vdots$$

The value of $\mathbf{b}_0$ is $\mathbf{x}$ at $t = 0$:

$$\mathbf{b}_0 = \mathbf{x}(0) = \mathbf{x}_0 \tag{5.283}$$

Therefore, we can write the solution as

$$\mathbf{x}(t) = \left([I] - [A]t + \frac{1}{2!}[A]^2 t^2 - \frac{1}{3!}[A]^3 t^3 + \frac{1}{4!}[A]^4 t^4 - \cdots\right)\mathbf{x}_0$$

$$= e^{-[A]t}\mathbf{x}_0 \tag{5.284}$$

The matrix $[A]$ turns into a diagonal matrix $[\Lambda]$ when we use the eigenvectors properly. Suppose the n by n matrix $[A]$ has n linearly independent eigenvectors $\mathbf{u}_1$, $\mathbf{u}_2, \cdots, \mathbf{u}_n$. Put them into columns of an eigenvector matrix $[U]$; then $[U]^{-1}[A][U]$ will be a diagonal matrix $[\Lambda]$ including the eigenvalues of $[A]$.

$$[\Lambda] = [U]^{-1}[A][U] = \begin{bmatrix} \lambda_1 & 0 & \cdots & 0 \\ 0 & \lambda_2 & \cdots & 0 \\ \vdots & \vdots & \ddots & \vdots \\ 0 & 0 & \cdots & \lambda_n \end{bmatrix} \tag{5.285}$$

$$[A] = [U][\Lambda][U]^{-1} \tag{5.286}$$

To find $[\Lambda]$, we note that when $\mathbf{u}_j$ is an eigenvector, multiplication $[A]$ by $\mathbf{u}_j$ is equal to multiplication $\mathbf{u}_j$ by a number λ_j.

$$[A]\mathbf{u}_j = \lambda_j\mathbf{u}_j \tag{5.287}$$

Let us multiply $[A]$ by its eigenvectors, which are the columns of $[U]$. Then the column number j of $[A][U]$ will be $[A]\mathbf{u}_j$, which is equal to $\lambda_j\mathbf{u}_j$. Therefore,

$$[A][U] = [A]\begin{bmatrix}\mathbf{u}_1 & \mathbf{u}_2 & \cdots & \mathbf{u}_n\end{bmatrix} = \begin{bmatrix}\lambda_1\mathbf{u}_1 & \lambda_2\mathbf{u}_2 & \cdots & \lambda_n\mathbf{u}_n\end{bmatrix} \tag{5.288}$$

The trick is to split this result into multiplication of $[U]$ and $[\Lambda]$.

$$\left[\lambda_1 \mathbf{u}_1 \ \lambda_2 \mathbf{u}_2 \ \cdots \ \lambda_n \mathbf{u}_n\right] = \left[\mathbf{u}_1 \ \mathbf{u}_2 \ \cdots \ \mathbf{u}_n\right] \begin{bmatrix} \lambda_1 & 0 & \cdots & 0 \\ 0 & \lambda_2 & \cdots & 0 \\ \vdots & \vdots & \ddots & \vdots \\ 0 & 0 & \cdots & \lambda_n \end{bmatrix} \tag{5.289}$$

$$= [U][\Lambda] \tag{5.290}$$

Hence, we have

$$[A][U] = [U][\Lambda] \tag{5.291}$$

$$[A] = [U][\Lambda][U]^{-1} \tag{5.292}$$

$$[\Lambda] = [U]^{-1}[A][U] \tag{5.293}$$

Knowing that

$$[A]^2 = [U][\Lambda][U]^{-1}[U][\Lambda][U]^{-1} = [U][\Lambda]^2[U]^{-1} \tag{5.294}$$

and

$$\begin{aligned} e^{[A]} &= [I] + [A]t + \frac{1}{2!}[A]^2 t^2 + \frac{1}{3!}[A]^3 t^3 + \frac{1}{4!}[A]^4 t^4 + \cdots \\ &= [I] + [U][\Lambda][U]^{-1} t + \frac{1}{2!}[U][\Lambda][U]^{-1}[U][\Lambda][U]^{-1} t^2 + \cdots \\ &= [U]\left[[I] + [\Lambda]t + \frac{1}{2!}[\Lambda]^2 t^2 + \frac{1}{3!}[\Lambda]^3 t^3 + \cdots\right][U]^{-1} \\ &= [U] e^{[\Lambda]} [U]^{-1} \end{aligned} \tag{5.295}$$

we have

$$[A]^k = [U][\Lambda]^k [U]^{-1} \qquad k \in \mathbb{N} \tag{5.296}$$

$$e^{[A]} = [U] e^{[\Lambda]} [U]^{-1} = e^{[\Lambda]} = \begin{bmatrix} e^{\lambda_1} & 0 & 0 & 0 \\ 0 & e^{\lambda_2} & 0 & 0 \\ 0 & 0 & \ddots & \vdots \\ 0 & 0 & \cdots & e^{\lambda_n} \end{bmatrix} \tag{5.297}$$

Therefore, the solution of Eq. (5.248) is

$$\mathbf{x}\,(t) = e^{-[A]t}\mathbf{x}_0 = [U]\,e^{-[\Lambda]t}\,[U]^{-1}\,\mathbf{x}_0$$
$$= e^{-\lambda_1 t}\mathbf{u}_1 + e^{-\lambda_2 t}\mathbf{u}_2 + \cdots + e^{-\lambda_n t}\mathbf{u}_n \tag{5.298}$$

When the system of equations is forced as Eq. (5.256),

$$\dot{\mathbf{x}} + [A]\,\mathbf{x} = \mathbf{f}\,(t) \tag{5.299}$$

then the solution of the equation will be a combination of homogenous $\mathbf{x}_h$ and particular $\mathbf{x}_p$ solutions. Multiplying $e^{-[A]t}$ by both sides of Eq. (5.256) yields

$$e^{[A]t}\,(\dot{\mathbf{x}} + [A]\,\mathbf{x}) = \frac{d}{dt}\left(e^{[A]t}\mathbf{x}\right) = e^{[A]t}\,\mathbf{f}\,(t) \tag{5.300}$$

Integrating this equation yields

$$e^{[A]t}\mathbf{x} = \int e^{[A]t}\,\mathbf{f}\,(t)\,dt + \mathbf{C} \tag{5.301}$$

or

$$\mathbf{x} = e^{-[A]t}\int e^{[A]t}\,\mathbf{f}\,(t)\,dt + e^{-[A]t}\mathbf{C} \tag{5.302}$$

∎

Example 173 Solution of a set of two equations. It is to show how a second-order equation will be equal to two first-order equations and checking a set of solutions.
   Consider a second-order equation.

$$\ddot{x} - 2 = 0 \tag{5.303}$$

An option is to select two axillary variables x_1 and x_2 to break the equation into two first-order equations and solve them individually.

$$\dot{x}_1 = 1 \qquad \dot{x}_2 = 2x_1 \tag{5.304}$$

$$x_1 = t + x_1\,(0) \qquad x_2 = t^2 + x_2\,(0) \tag{5.305}$$

As another method to break down the equation into a set of first-order equations, we may use different x_1 and x_2 and solve them.

$$\dot{x}_1 = x_2 \qquad \dot{x}_2 = 2 \tag{5.306}$$

$$x_2 = 2t + x_2\,(0) \tag{5.307}$$

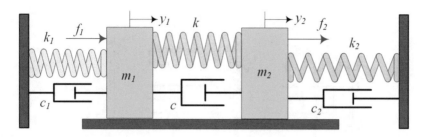

Fig. 5.30 A two degrees of freedom linear system, expressed by four first-order equations

$$x_1 = \int x_2 dt = \int (2t + x_2(0))\, dt = t^2 + t\, x_2(0) + x_1(0)$$
$$= t^2 + t\, \dot{x}_1(0) + x_1(0) \tag{5.308}$$

A set of second-order systems is expressed as

$$\ddot{x} + B\,\dot{x} + C\,x = 0 \tag{5.309}$$

This set of equations can be transformed to a set of first-order equations as below.

$$\mathbf{y} = \begin{bmatrix} \mathbf{x} \\ \dot{\mathbf{x}} \end{bmatrix} \qquad \dot{\mathbf{y}} = \begin{bmatrix} \dot{\mathbf{x}} \\ \ddot{\mathbf{x}} \end{bmatrix} \tag{5.310}$$

$$\begin{bmatrix} \dot{\mathbf{x}} \\ \ddot{\mathbf{x}} \end{bmatrix} = \begin{bmatrix} \mathbf{0} & \mathbf{1} \\ -\mathbf{C} & -\mathbf{B} \end{bmatrix} \begin{bmatrix} \mathbf{x} \\ \dot{\mathbf{x}} \end{bmatrix} \tag{5.311}$$

$$\dot{\mathbf{y}} = \mathbf{A}\,\mathbf{y} \tag{5.312}$$

As a practical example, let us examine the vibrating system of Fig. 5.30. It shows a two degrees of freedom linear vibrating system, that its motion can bed expressed by four first-order equations. Newton's second law yields the following coupled second-order ordinary differential equations that describe the motions of the masses.

$$\begin{bmatrix} m_1 & 0 \\ 0 & m_2 \end{bmatrix} \begin{bmatrix} \ddot{y}_1 \\ \ddot{y}_2 \end{bmatrix} + \begin{bmatrix} c_1 + c & -c \\ -c & c_2 + c \end{bmatrix} \begin{bmatrix} \dot{y}_1 \\ \dot{y}_2 \end{bmatrix}$$
$$+ \begin{bmatrix} k_1 + k & -k \\ -k & k_2 + k \end{bmatrix} \begin{bmatrix} y_1 \\ y_2 \end{bmatrix} = \begin{bmatrix} f_1 \\ f_2 \end{bmatrix} \tag{5.313}$$

$$y_1(0) = 0 \qquad y_2(0) = 0 \qquad \dot{y}_1(0) = 0 \qquad \dot{y}_2(0) = 0 \tag{5.314}$$

Introducing $x_1 = y_1$, $x_2 = \dot{y}_1$, $x_3 = y_2$, and $x_4 = \dot{y}_2$, we can express the equations of motion equivalently by a system of first-order ordinary differential equation.

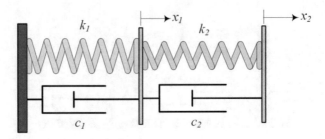

Fig. 5.31 A series of two first-order systems

$$\begin{bmatrix} \dot{x}_1 \\ \dot{x}_2 \\ \dot{x}_3 \\ \dot{x}_4 \end{bmatrix} + \begin{bmatrix} 0 & 1 & 0 & 0 \\ \dfrac{k_1+k}{m_1} & \dfrac{c_1+c}{m_1} & -\dfrac{k}{m_1} & -\dfrac{c}{m_1} \\ 0 & 0 & 0 & 1 \\ -\dfrac{k}{m_2} & -\dfrac{c}{m_2} & \dfrac{k_2+k}{m_2} & \dfrac{c_2+c}{m_2} \end{bmatrix} \begin{bmatrix} x_1 \\ x_2 \\ x_3 \\ x_4 \end{bmatrix} = \begin{bmatrix} 0 \\ \dfrac{f_1}{m_1} \\ 0 \\ \dfrac{f_2}{m_2} \end{bmatrix} \qquad (5.315)$$

$$x_1\,(0) = 0 \qquad x_2\,(0) = 0 \qquad x_3\,(0) = 0 \qquad x_4\,(0) = 0 \qquad (5.316)$$

Example 174 ★ A series of two first-order systems. Free response of connected first-order systems is still similar to a first-order system, without vibrations. As long as there are no mass and no periodic excitations, first-order systems will move from initial conditions to rest position exponentially.

Figure 5.31 illustrates a series of two first-order systems. The equations of motion for the coordinates x_1 and x_2 are

$$\begin{bmatrix} c_1+c_2 & -c_2 \\ -c_2 & c_2 \end{bmatrix} \begin{bmatrix} \dot{x}_1 \\ \dot{x}_2 \end{bmatrix} + \begin{bmatrix} k_1+k_2 & -k_2 \\ -k_2 & k_2 \end{bmatrix} \begin{bmatrix} x_1 \\ x_2 \end{bmatrix} = 0 \qquad (5.317)$$

Multiplying $[c]^{-1}$, we can transform the equations to the standard form of (5.248).

$$\begin{bmatrix} \dot{x}_1 \\ \dot{x}_2 \end{bmatrix} + \begin{bmatrix} \dfrac{k_1}{c_1} & 0 \\ \dfrac{k_1}{c_1} - \dfrac{k_2}{c_2} & \dfrac{k_2}{c_2} \end{bmatrix} \begin{bmatrix} x_1 \\ x_2 \end{bmatrix} = 0 \qquad (5.318)$$

The eigenvalues and eigenvectors of the coefficient matrix are

$$\lambda_1 = \dfrac{k_1}{c_1} \qquad\qquad \lambda_2 = \dfrac{k_2}{c_2} \qquad (5.319)$$

$$\mathbf{u}_1 = \begin{bmatrix} 1 \\ 1 \end{bmatrix} \qquad\qquad \mathbf{u}_2 = \begin{bmatrix} 0 \\ 1 \end{bmatrix} \qquad (5.320)$$

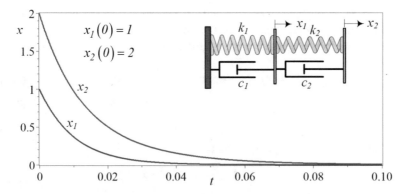

Fig. 5.32 Time response of the series of two first-order of Figure 5.31

Therefore, the time response of the system is

$$\begin{bmatrix} x_1 \\ x_2 \end{bmatrix} = C_1 e^{-k_1 t/c_1} \begin{bmatrix} 1 \\ 1 \end{bmatrix} + C_2 e^{-k_2 t/c_2} \begin{bmatrix} 0 \\ 1 \end{bmatrix} \tag{5.321}$$

The system has two time constants τ_1 and τ_2.

$$\tau_1 = \frac{1}{\lambda_1} = \frac{c_1}{k_1} \qquad \tau_2 = \frac{1}{\lambda_2} = \frac{c_2}{k_2} \tag{5.322}$$

If the initial conditions are

$$x_1\,(0) = x_{10} \qquad x_2\,(0) = x_{20} \tag{5.323}$$

then

$$C_1 = x_{10} \qquad C_2 = x_{20} - x_{10} \tag{5.324}$$

Let us assume

$$c_1 = c_2 = 1 \qquad k_1 = 100 \qquad k_2 = 50 \tag{5.325}$$
$$x_1\,(0) = 1 \qquad x_2\,(0) = 2 \tag{5.326}$$

and find the response

$$\begin{bmatrix} x_1 \\ x_2 \end{bmatrix} = e^{-100t} \begin{bmatrix} 1 \\ 1 \end{bmatrix} + e^{-50t} \begin{bmatrix} 0 \\ 1 \end{bmatrix} \tag{5.327}$$

Figure 5.32 illustrates the solutions for $t \le 0.1$.

Example 175 Homogeneous systems with constant coefficients. It is investigating two parametric equations to show the steps toward solution.

Consider the following system of two differential equations with two unknowns.

$$\dot{x}_1 = a_{11}x_1 + a_{12}x_2 \qquad x_1(0) = x_{10} \tag{5.328}$$

$$\dot{x}_2 = a_{21}x_1 + a_{22}x_2 \qquad x_2(0) = x_{20} \tag{5.329}$$

$$\dot{x} + [A]x = 0 \tag{5.330}$$

$$\begin{bmatrix} \dot{x}_1 \\ \dot{x}_2 \end{bmatrix} = \begin{bmatrix} a_{11} & a_{12} \\ a_{21} & a_{22} \end{bmatrix} \begin{bmatrix} x_1 \\ x_2 \end{bmatrix} \tag{5.331}$$

Assuming a set of solution of exponential functions yields

$$x_1 = C_1 e^{st} \qquad x_2 = C_2 e^{st} \tag{5.332}$$

$$C_1 s = a_{11}C_1 + a_{12}C_2 \tag{5.333}$$

$$C_2 s = a_{21}C_1 + a_{22}C_2 \tag{5.334}$$

$$(a_{11} - s)C_1 + a_{12}C_2 = 0 \tag{5.335}$$

$$a_{21}C_1 + (a_{22} - s)C_2 = 0 \tag{5.336}$$

These equations can be set in matrix form.

$$([A] - s[I])C = 0 \tag{5.337}$$

$$\left(\begin{bmatrix} a_{11} - s & a_{12} \\ a_{21} & a_{22} - s \end{bmatrix} - s \begin{bmatrix} 1 & 0 \\ 0 & 1 \end{bmatrix} \right) \begin{bmatrix} C_1 \\ C_2 \end{bmatrix} = \begin{bmatrix} 0 \\ 0 \end{bmatrix} \tag{5.338}$$

Equation (5.337) has nontrivial solutions only if the determinant of the coefficient matrix is zero.

$$||[A] - s[I]|| = 0 \tag{5.339}$$

$$(a_{11} - s)(a_{22} - s) - a_{12}a_{21} = 0 \tag{5.340}$$

This is the characteristic equation of Eqs. (5.335) and (5.336), and its roots provide us with the two eigenvalues of $[A]$.

$$s_1, s_2 = \frac{(a_{11} + a_{22}) \pm \sqrt{(a_{11} + a_{22})^2 - 4(a_{11}a_{22} - a_{12}a_{21})}}{2} \tag{5.341}$$

The general solution of the differential equations is a linear combination of exponential functions.

$$x_1 = C_{11}e^{s_1 t} + C_{12}e^{s_2 t} \tag{5.342}$$

$$x_2 = C_{21}e^{s_1 t} + C_{22}e^{s_2 t} \tag{5.343}$$

The constant coefficients C_{11}, C_{12}, C_{21}, and C_{22} are related to the initial conditions.

$$x_{10} = C_{11} + C_{12} \qquad x_{20} = C_{21} + C_{22} \tag{5.344}$$

Equations (5.335) and (5.336) yield

$$\frac{C_1}{C_2} = \frac{-a_{12}}{a_{11} - s} = \frac{a_{22} - s}{-a_{21}} \tag{5.345}$$

$$(a_{11} - s)\, C_1 + a_{12}C_2 = 0 \tag{5.346}$$

$$a_{21}C_1 + (a_{22} - s)\, C_2 = 0 \tag{5.347}$$

Example 176 ★Power series of matrix functions. Exponential function of a matrix can be determined by series. Here are the analytic calculations for $e^{\mathbf{A}t}$.

Similar to a power series expansion of a function of a scalar variable, $f(x)$,

$$f(x) = a_0 + a_1 x + a_2 x^2 + a_3 x^3 + \cdots \tag{5.348}$$

we may define power series expansion of a function of a square matrix, $f(\mathbf{A})$,

$$f(\mathbf{A}) = a_0 \mathbf{I} + a_1 \mathbf{A} + a_2 \mathbf{A}^2 + a_3 \mathbf{A}^3 + \cdots \tag{5.349}$$

where $\mathbf{I}$ is the unit matrix of the same order as $\mathbf{A}$. As an example, exponential function of a coefficient matrix $\mathbf{A}$ is

$$e^{\mathbf{A}t} = \mathbf{I} + \mathbf{A}t + \frac{1}{2}\mathbf{A}^2 t^2 + \frac{1}{3!}\mathbf{A}^3 t^3 + \frac{1}{4!}\mathbf{A}^4 t^4 + \cdots \tag{5.350}$$

Assume $\mathbf{A}$ is a square $n \times n$ matrix whose eigenvalues $\lambda_1, \lambda_2, \lambda_3, \cdots, \lambda_n$ are associated with its eigenvectors $\mathbf{u}_1, \mathbf{u}_2, \mathbf{u}_3, \cdots, \mathbf{u}_n$.

$$\mathbf{A}\mathbf{u}_i = \lambda_i \mathbf{u}_i \qquad i = 1, 2, 3, \cdots, n \tag{5.351}$$

$$\mathbf{u}_i = \begin{bmatrix} u_{1i} & u_{2i} & u_{3i} & \dots & u_{ni} \end{bmatrix}^T \tag{5.352}$$

We may combine all equations of (5.351) and write them in form of an equation,

$$\mathbf{A}\mathbf{U} = \mathbf{\Lambda}\mathbf{U} \tag{5.353}$$

where $\mathbf{\Lambda}$ is a diagonal eigenvalue matrix and $\mathbf{U}$ is the square eigenvector matrix, assuming all eigenvalues are real and distinct,

$$\mathbf{\Lambda} = \begin{bmatrix} \lambda_1 & 0 & 0 & \cdots \\ 0 & \lambda_2 & 0 & \cdots \\ 0 & 0 & \lambda_3 & \cdots \\ \vdots & \vdots & \vdots & \ddots \end{bmatrix} = diag\,(\lambda_1, \lambda_2, \lambda_3, \cdots) \tag{5.354}$$

$$\mathbf{U} = \begin{bmatrix} \mathbf{u}_1 \\ \mathbf{u}_2 \\ \mathbf{u}_3 \\ \vdots \end{bmatrix}^T = \begin{bmatrix} u_{11} & u_{21} & u_{31} & \cdots \\ u_{12} & u_{22} & u_{32} & \cdots \\ u_{13} & u_{23} & u_{33} & \cdots \\ \vdots & \vdots & \vdots & \ddots \end{bmatrix} \tag{5.355}$$

and therefore,

$$\mathbf{U}^{-1}\mathbf{A}\mathbf{U} = \mathbf{\Lambda} \tag{5.356}$$

$$\mathbf{A} = \mathbf{U}\mathbf{\Lambda}\mathbf{U}^{-1} \tag{5.357}$$

Now we may have the exponents of $\mathbf{A}$,

$$\mathbf{A}^2 = \left[\mathbf{U}\mathbf{\Lambda}\mathbf{U}^{-1}\right] \cdot \left[\mathbf{U}\mathbf{\Lambda}\mathbf{U}^{-1}\right] = \mathbf{U}\mathbf{\Lambda}^2\mathbf{U}^{-1} \tag{5.358}$$

$$\mathbf{A}^3 = \mathbf{A}^2\mathbf{A} = \left[\mathbf{U}\mathbf{\Lambda}^2\mathbf{U}^{-1}\right] \cdot \left[\mathbf{U}\mathbf{\Lambda}\mathbf{U}^{-1}\right] = \mathbf{U}\mathbf{\Lambda}^3\mathbf{U}^{-1} \tag{5.359}$$

$$\cdots = \cdots$$

$$\mathbf{A}^n = \mathbf{A}^{n-1}\mathbf{A} = \left[\mathbf{U}\mathbf{\Lambda}^{n-1}\mathbf{U}^{-1}\right] \cdot \left[\mathbf{U}\mathbf{\Lambda}\mathbf{U}^{-1}\right] = \mathbf{U}\mathbf{\Lambda}^n\mathbf{U}^{-1} \tag{5.360}$$

where

$$\mathbf{A}^n = \begin{bmatrix} \lambda_1^n & 0 & 0 & \cdots \\ 0 & \lambda_2^n & 0 & \cdots \\ 0 & 0 & \lambda_3^n & \cdots \\ \vdots & \vdots & \vdots & \ddots \end{bmatrix} \tag{5.361}$$

Hence, the power series of a matrix function (5.349) will be

$$\begin{aligned} f\,(\mathbf{A}) &= a_0\mathbf{I} + a_1\mathbf{A} + a_2\mathbf{A}^2 + a_3\mathbf{A}^3 + \cdots \\ &= a_0\mathbf{U}\mathbf{I}\mathbf{U}^{-1} + a_1\mathbf{U}\mathbf{\Lambda}\mathbf{U}^{-1} + a_2\mathbf{U}\mathbf{\Lambda}^2\mathbf{U}^{-1} + a_3\mathbf{U}\mathbf{\Lambda}^3\mathbf{U}^{-1} + \cdots \\ &= \mathbf{U}\left[a_0\mathbf{I} + a_1\mathbf{\Lambda} + a_2\mathbf{\Lambda}^2 + a_3\mathbf{\Lambda}^3 + \cdots\right]\mathbf{U}^{-1} \end{aligned} \tag{5.362}$$

where

$$\left[a_0 \mathbf{I} + a_1 \mathbf{\Lambda} + a_2 \mathbf{\Lambda}^2 + \cdots \right] = diag(a_0 + a_1\lambda_1 + a_2\lambda_1^2 \cdots ,$$
$$a_0 + a_1\lambda_2 + a_2\lambda_2^2 \cdots , \ldots$$
$$a_0 + a_1\lambda_n + a_2\lambda_n^2 \cdots) \tag{5.363}$$

As an example, let us consider

$$\mathbf{A} = \begin{bmatrix} 1 & 1 \\ 9 & 1 \end{bmatrix} \tag{5.364}$$

with eigenvalues and eigenvectors of

$$\lambda_1 = -2 \qquad \mathbf{u}_1 = \begin{bmatrix} -1/3 \\ 1 \end{bmatrix} \tag{5.365}$$

$$\lambda_2 = 4 \qquad \mathbf{u}_2 = \begin{bmatrix} 1/3 \\ 1 \end{bmatrix} \tag{5.366}$$

and therefore,

$$\mathbf{\Lambda} = \begin{bmatrix} -2 & 0 \\ 0 & 4 \end{bmatrix} \qquad \mathbf{U} = [\mathbf{u}_1 \ \mathbf{u}_2] = \begin{bmatrix} -1/3 & 1/3 \\ 1 & 1 \end{bmatrix} \tag{5.367}$$

and we may check that $\mathbf{A} = \mathbf{U}\mathbf{\Lambda}\mathbf{U}^{-1}$.

$$\mathbf{A} = \mathbf{U}\mathbf{\Lambda}\mathbf{U}^{-1} = \begin{bmatrix} -1/3 & 1/3 \\ 1 & 1 \end{bmatrix} \begin{bmatrix} -2 & 0 \\ 0 & 4 \end{bmatrix} \begin{bmatrix} -1/3 & 1/3 \\ 1 & 1 \end{bmatrix}^{-1}$$
$$= \begin{bmatrix} 1 & 1 \\ 9 & 1 \end{bmatrix} \tag{5.368}$$

Now we may calculate the exponents of $\mathbf{A}$,

$$\mathbf{A}^2 = \mathbf{U}\mathbf{\Lambda}^2\mathbf{U}^{-1} = \begin{bmatrix} 10 & 2 \\ 18 & 10 \end{bmatrix} \tag{5.369}$$

$$\mathbf{A}^3 = \mathbf{U}\mathbf{\Lambda}^3\mathbf{U}^{-1} = \begin{bmatrix} 28 & 12 \\ 108 & 28 \end{bmatrix} \tag{5.370}$$

$$\cdots$$

to have $e^{\mathbf{A}t}$.

$$e^{\mathbf{A}t} = \mathbf{I} + \mathbf{A}t + \frac{1}{2}\mathbf{A}^2 t^2 + \frac{1}{3!}\mathbf{A}^3 t^3 + \cdots$$

$$= \begin{bmatrix} 1 & 0 \\ 0 & 1 \end{bmatrix} + \begin{bmatrix} 1 & 1 \\ 9 & 1 \end{bmatrix} t + \frac{1}{2}\begin{bmatrix} 10 & 2 \\ 18 & 10 \end{bmatrix} t^2 + \cdots$$

$$\simeq \begin{bmatrix} \frac{14}{3}t^3 + 5t^2 + t + 1 & 2t^3 + t^2 + t \\ 18t^3 + 9t^2 + 9t & \frac{14}{3}t^3 + 5t^2 + t + 1 \end{bmatrix} \tag{5.371}$$

The exact value of $e^{\mathbf{A}t}$ is

$$e^{\mathbf{A}t} = \frac{1}{6}\begin{bmatrix} 3e^{-2t} + 3e^{4t} & e^{4t} - e^{-2t} \\ 9e^{4t} - 9e^{-2t} & 3e^{-2t} + 3e^{4t} \end{bmatrix} \tag{5.372}$$

Knowing that if we have a diagonal matrix, then the exponential function of the matrix would be

$$\exp\begin{bmatrix} a & 0 \\ 0 & b \end{bmatrix} = \begin{bmatrix} e^a & 0 \\ 0 & e^b \end{bmatrix} \tag{5.373}$$

We may also transform a given matrix $\mathbf{A}$ to its Jordan form

$$\begin{bmatrix} 1 & 1 \\ 9 & 1 \end{bmatrix} = \begin{bmatrix} 1 & 1 \\ -3 & 3 \end{bmatrix}\begin{bmatrix} -2 & 0 \\ 0 & 4 \end{bmatrix}\begin{bmatrix} 1 & 1 \\ -3 & 3 \end{bmatrix}^{-1} \tag{5.374}$$

to calculate the exact $e^{\mathbf{A}t}$.

$$\exp\begin{bmatrix} 1 & 1 \\ 9 & 1 \end{bmatrix} t = \begin{bmatrix} 1 & 1 \\ -3 & 3 \end{bmatrix}\exp\begin{bmatrix} -2t & 0 \\ 0 & 4t \end{bmatrix}\begin{bmatrix} 1 & 1 \\ -3 & 3 \end{bmatrix}^{-1}$$

$$= \begin{bmatrix} 1 & 1 \\ -3 & 3 \end{bmatrix}\begin{bmatrix} e^{-2t} & 0 \\ 0 & e^{4t} \end{bmatrix}\begin{bmatrix} 1 & 1 \\ -3 & 3 \end{bmatrix}^{-1}$$

$$= \frac{1}{6}\begin{bmatrix} 3e^{-2t} + 3e^{4t} & e^{4t} - e^{-2t} \\ 9e^{4t} - 9e^{-2t} & 3e^{-2t} + 3e^{4t} \end{bmatrix} \tag{5.375}$$

Example 177 Solving free vibrations of a one DOF second-order system as a coupled first-order system.

Consider a second-order vibrating system.

$$\ddot{x} + 2\xi\omega_n\dot{x} + \omega_n^2 x = 0 \tag{5.376}$$

$$x(0) = x_0 \qquad \dot{x}(0) = \dot{x}_0 \tag{5.377}$$

Let us rewrite the equation by two first-order equations.

$$x_1 = x \qquad x_2 = \dot{x} \tag{5.378}$$

$$\dot{x}_1 = x_2 \tag{5.379}$$

$$\dot{x}_2 = -2\xi\omega_n x_2 - \omega_n^2 x_1 \tag{5.380}$$

This system can be written in matrix form.

$$\begin{bmatrix} \dot{x}_1 \\ \dot{x}_2 \end{bmatrix} = A \begin{bmatrix} x_1 \\ x_2 \end{bmatrix} \tag{5.381}$$

$$A = \begin{bmatrix} 0 & 1 \\ -\omega_n^2 & -2\xi\omega_n \end{bmatrix} \tag{5.382}$$

The coefficient matrix of the system has the following eigenvalues and eigenvectors.

$$\lambda_1 = -\xi\omega_n + i\sqrt{1-\xi^2}\omega_n \qquad \mathbf{u}_1 = \begin{bmatrix} -\dfrac{1}{\omega_n}\left(\xi + i\sqrt{1-\xi^2}\right) \\ 1 \end{bmatrix} \tag{5.383}$$

$$\lambda_2 = -\xi\omega_n - i\sqrt{1-\xi^2}\omega_n \qquad \mathbf{u}_2 = \begin{bmatrix} -\dfrac{1}{\omega_n}\left(\xi - i\sqrt{1-\xi^2}\right) \\ 1 \end{bmatrix} \tag{5.384}$$

Therefore, the diagonal eigenvalue matrix, Λ, and the square eigenvector matrix, $\mathbf{U}$ are

$$\Lambda = \begin{bmatrix} -\xi\omega_n + i\sqrt{1-\xi^2}\omega_n & 0 \\ 0 & -\xi\omega_n - i\sqrt{1-\xi^2}\omega_n \end{bmatrix} \tag{5.385}$$

$$\mathbf{U} = \begin{bmatrix} -\dfrac{1}{\omega_n}\left(\xi + i\sqrt{1-\xi^2}\right) & -\dfrac{1}{\omega_n}\left(\xi - i\sqrt{1-\xi^2}\right) \\ 1 & 1 \end{bmatrix} \tag{5.386}$$

The first four exponents of A will make a good approximate solution for e^{At}.

$$A = \mathbf{U}\Lambda\mathbf{U}^{-1} = \begin{bmatrix} 0 & 1 \\ -\omega_n^2 & -2\xi\omega_n \end{bmatrix} \tag{5.387}$$

$$A^2 = \mathbf{U}\Lambda^2\mathbf{U}^{-1} = \begin{bmatrix} -\omega_n^2 & -2\xi\omega_n \\ 2\xi\omega_n^3 & \omega_n^2\left(4\xi^2-1\right) \end{bmatrix} \tag{5.388}$$

$$A^3 = \mathbf{U}\Lambda^3\mathbf{U}^{-1} = \begin{bmatrix} 2\xi\omega_n^3 & \omega_n^2\left(4\xi^2-1\right) \\ -\omega_n^4\left(4\xi^2-1\right) & -4\xi\omega_n^3\left(2\xi^2-1\right) \end{bmatrix} \tag{5.389}$$

$$\mathbf{A}^4 = \mathbf{U}\mathbf{\Lambda}^4\mathbf{U}^{-1} = \begin{bmatrix} -\omega_n^4 \left(4\xi^2 - 1\right) & -4\xi\omega_n^3 \left(2\xi^2 - 1\right) \\ 4\xi\omega_n^5 \left(2\xi^2 - 1\right) & \omega_n^4 \left(16\xi^4 - 12\xi^2 + 1\right) \end{bmatrix} \tag{5.390}$$

$$\cdots$$

$$e^{\mathbf{A}t} = \mathbf{I} + \mathbf{A}t + \frac{\mathbf{A}^2 t^2}{2} + \frac{\mathbf{A}^3 t^3}{3!} + \frac{\mathbf{A}^4 t^4}{4!} + \cdots \simeq \begin{bmatrix} r_{11} & r_{12} \\ r_{21} & r_{22} \end{bmatrix} \tag{5.391}$$

$$r_{11} = \left(\frac{1}{24} - \frac{1}{6}\xi^2\right)\omega_n^4 t^4 + \frac{1}{3}\xi\omega_n^3 t^3 - \frac{1}{2}\omega_n^2 t^2 + 1 \tag{5.392}$$

$$r_{21} = \frac{1}{6}\left(\left(2\xi^2 - 1\right)\xi\omega_n^3 t^3 + \left(1 - 4\xi^2\right)\omega_n^2 t^2 + 6\xi\omega_n t - 6\right)\omega_n^2 t \tag{5.393}$$

$$r_{12} = -\frac{1}{6}\left(\left(2\xi^2 - 1\right)\xi\omega_n^3 t^3 + \left(1 - 4\xi^2\right)\omega_n^2 t^2 + 6\xi\omega_n t - 6\right)t \tag{5.394}$$

$$r_{22} = \left(\frac{2}{3}\xi^4 - \frac{1}{2}\xi^2 + \frac{1}{24}\right)\omega_n^4 t^4 + \left(\frac{2}{3} - \frac{4}{3}\xi^2\right)\xi\omega_n^3 t^3$$
$$+ \left(2\xi^2 - \frac{1}{2}\right)\omega_n^2 t^2 - 2\xi\omega_n t + 1 \tag{5.395}$$

Alternatively, we can find the Jordan form of $\mathbf{A}$ to its Jordan form to calculate the exact $e^{\mathbf{A}t}$.

$$\mathbf{A} = \begin{bmatrix} 1 & 1 \\ \lambda_{2n} & \lambda_1 \end{bmatrix} \begin{bmatrix} \lambda_2 & 0 \\ 0 & \lambda_1 \end{bmatrix} \begin{bmatrix} 1 & 1 \\ \lambda_{2n} & \lambda_1 \end{bmatrix}^{-1} \tag{5.396}$$

$$\lambda_1 = -\xi\omega_n + i\sqrt{1 - \xi^2}\omega_n \qquad \lambda_2 = -\xi\omega_n - i\sqrt{1 - \xi^2}\omega_n \tag{5.397}$$

$$e^{\mathbf{A}t} = \begin{bmatrix} 1 & 1 \\ \lambda_2 & \lambda_1 \end{bmatrix} \exp\begin{bmatrix} \lambda_2 t & 0 \\ 0 & \lambda_1 t \end{bmatrix} \begin{bmatrix} 1 & 1 \\ \lambda_2 & \lambda_1 \end{bmatrix}^{-1}$$
$$= \begin{bmatrix} 1 & 1 \\ \lambda_2 & \lambda_1 \end{bmatrix} \begin{bmatrix} e^{\lambda_2 t} & 0 \\ 0 & e^{\lambda_1 t} \end{bmatrix} \begin{bmatrix} 1 & 1 \\ \lambda_2 & \lambda_1 \end{bmatrix}^{-1}$$
$$= \begin{bmatrix} \dfrac{\lambda_1 e^{\lambda_2 t} - \lambda_2 e^{\lambda_1 t}}{\lambda_1 - \lambda_2} & \dfrac{e^{\lambda_1 t} - e^{\lambda_2 t}}{\lambda_1 - \lambda_2} \\ -\dfrac{e^{\lambda_1 t} - e^{\lambda_2 t}}{\lambda_1 - \lambda_2}\lambda_1\lambda_2 & \dfrac{\lambda_1 e^{\lambda_1 t} - \lambda_2 e^{\lambda_2 t}}{\lambda_1 - \lambda_2} \end{bmatrix} \tag{5.398}$$

Example 178 ★ One *DOF* second-order system. Solving vibrating systems in state space provides us with position and velocity together, and it can be used to check the results.

Consider a second-order vibrating system.

$$\ddot{x} + 2\xi\omega_n\dot{x} + \omega_n^2 x = 0 \tag{5.399}$$

$$x(0) = x_0 \qquad \dot{x}(0) = \dot{x}_0 \tag{5.400}$$

This system can be rewritten in matrix form.

$$\dot{\mathbf{x}} = \mathbf{A}\mathbf{x} \qquad x_1 = x \qquad x_2 = \dot{x} \tag{5.401}$$

$$\begin{bmatrix} \dot{x}_1 \\ \dot{x}_2 \end{bmatrix} = \begin{bmatrix} 0 & 1 \\ -\omega_n^2 & -2\xi\omega_n \end{bmatrix} \begin{bmatrix} x_1 \\ x_2 \end{bmatrix} \tag{5.402}$$

The response of the system can be written as

$$\mathbf{x} = \mathbf{x}_0 \, e^{\mathbf{A}t} \tag{5.403}$$

Having the eigenvalues of $\mathbf{A}$ in (5.383) and (5.384), we have

$$\begin{bmatrix} 1 & \lambda_1 t \\ 1 & \lambda_2 t \end{bmatrix} \begin{bmatrix} \dfrac{\lambda_1 e^{\lambda_2 t} - \lambda_2 e^{\lambda_1 t}}{\lambda_1 - \lambda_2} \\ \dfrac{e^{\lambda_1 t} - e^{\lambda_2 t}}{(\lambda_1 - \lambda_2) t} \end{bmatrix} = \begin{bmatrix} e^{\lambda_1 t} \\ e^{\lambda_2 t} \end{bmatrix} \tag{5.404}$$

$$e^{\mathbf{A}t} = \frac{\lambda_1 e^{\lambda_2 t} - \lambda_2 e^{\lambda_1 t}}{\lambda_1 - \lambda_2} [\mathbf{I}] + \frac{e^{\lambda_1 t} - e^{\lambda_2 t}}{(\lambda_1 - \lambda_2)} \mathbf{A} \tag{5.405}$$

$$\mathbf{x} = \frac{\lambda_1 e^{\lambda_2 t} - \lambda_2 e^{\lambda_1 t}}{\lambda_1 - \lambda_2} \mathbf{x}_0 + \frac{e^{\lambda_1 t} - e^{\lambda_2 t}}{(\lambda_1 - \lambda_2)} \mathbf{A} \, \mathbf{x}_0 \tag{5.406}$$

In the case of underdamped system, $\xi < 1$, this solution will be simplified to the following:

$$x_1 = x_0 e^{-\xi\omega_n t}\left(\cos \omega_d t + \frac{\xi}{\sqrt{1-\xi^2}}\sin \omega_d t\right) + \frac{\dot{x}_0 e^{-\xi\omega_n t}}{\omega_d}\sin \omega_d t \tag{5.407}$$

$$x_2 = \frac{-x_0\omega_d e^{-\xi\omega_n t}}{1-\xi^2}\sin \omega_d t$$

$$+ \frac{\dot{x}_0 e^{-\xi\omega_n t}}{\sqrt{1-\xi^2}}\left(\sqrt{1-\xi^2}\cos \omega_d t - \xi \sin \omega_d t\right) \tag{5.408}$$

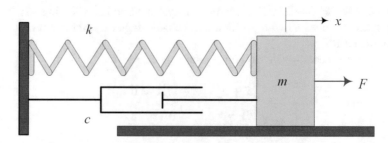

Fig. 5.33 A constant force on a mass-spring-damper system

Example 179 A step force on a first-order system. This is an example of analyzing the forced vibrations of a second-order system as a couple of first-order systems.

Consider the system of Fig. 5.33.

$$m\ddot{x} + c\dot{x} + kx = F \qquad (5.409)$$

$$x_1(0) = 0 \qquad x_2(0) = 0 \qquad (5.410)$$

Introducing two new coordinators x_1 and x_2, and utilizing the following numerical values,

$$x_1 = x \qquad x_2 = \dot{x} \qquad (5.411)$$

$$m = 1 \qquad c = 3 \qquad k = 2 \qquad F = 1 \qquad (5.412)$$

we rewrite the equations of motion as a set of first-order equations in terms of $x_1(t)$ and $x_2(t)$.

$$\dot{x}_1 - x_2 = 0 \qquad \dot{x}_2 + 2x_1 + 3x_2 = 1 \qquad (5.413)$$

$$\dot{\mathbf{x}} + [A]\mathbf{x} = \mathbf{f}(t) \qquad (5.414)$$

$$\begin{bmatrix} \dot{x}_1 \\ \dot{x}_2 \end{bmatrix} + \begin{bmatrix} 0 & -1 \\ 2 & 3 \end{bmatrix} \begin{bmatrix} x_1 \\ x_2 \end{bmatrix} = \begin{bmatrix} 0 \\ 1 \end{bmatrix} \qquad (5.415)$$

The eigenvalues and modal matrix of the coefficient matrix $[A]$ are

$$\lambda_1 = 1 \qquad \lambda_2 = 2 \qquad [U] = \begin{bmatrix} -1 & -1/2 \\ 1 & 1 \end{bmatrix} \qquad (5.416)$$

Therefore,

$$e^{[\Lambda]t} = \begin{bmatrix} e^t & 0 \\ 0 & e^{2t} \end{bmatrix} \tag{5.417}$$

and

$$e^{[A]t} = [U] e^{-[\Lambda]t} [U]^{-1} = \begin{bmatrix} 2e^t - e^{2t} & e^t - e^{2t} \\ 2e^{2t} - 2e^t & 2e^{2t} - e^t \end{bmatrix} \tag{5.418}$$

$$e^{-[A]t} = \begin{bmatrix} 2e^{-t} - e^{-2t} & e^{-t} - e^{-2t} \\ 2e^{-2t} - 2e^{-t} & 2e^{-2t} - e^{-t} \end{bmatrix} \tag{5.419}$$

The solutions of the equations are

$$\mathbf{x} = e^{-[A]t} \int e^{[A]t} \mathbf{f}(t)\, dt + e^{-[A]t} \mathbf{C}$$

$$= e^{-[A]t} \int \begin{bmatrix} 2e^t - e^{2t} & e^t - e^{2t} \\ 2e^{2t} - 2e^t & 2e^{2t} - e^t \end{bmatrix} \begin{bmatrix} 0 \\ 1 \end{bmatrix} dt + e^{-[A]t} \mathbf{C}$$

$$= \begin{bmatrix} \frac{1}{2} \\ 0 \end{bmatrix} + \begin{bmatrix} 2e^{-t} - e^{-2t} & e^{-t} - e^{-2t} \\ 2e^{-2t} - 2e^{-t} & 2e^{-2t} - e^{-t} \end{bmatrix} \begin{bmatrix} C_1 \\ C_2 \end{bmatrix} \tag{5.420}$$

Using the initial conditions

$$\begin{bmatrix} 0 \\ 0 \end{bmatrix} = \begin{bmatrix} -\frac{1}{2} \\ 0 \end{bmatrix} + \begin{bmatrix} 1 & 0 \\ 0 & 1 \end{bmatrix} \begin{bmatrix} C_1 \\ C_2 \end{bmatrix} = \begin{bmatrix} C_1 - \frac{1}{2} \\ C_2 \end{bmatrix} \tag{5.421}$$

we calculate the integral constants,

$$C_1 = \frac{1}{2} \qquad C_2 = 0 \tag{5.422}$$

and we determine the time response of the system:

$$\mathbf{x} = \begin{bmatrix} e^{-t} - \frac{1}{2} e^{-2t} + \frac{1}{2} \\ e^{-2t} - e^{-t} \end{bmatrix} \tag{5.423}$$

Example 180 ★ Derivative of exponential matrix. Convergent series act similar to regular functions, and we are able to take derivative and integral of them term by term. Here is the derivative of $e^{[A]t}$.

The exponential function of a matrix argument

$$e^{[A]t} = \sum_{i=0}^{\infty} \frac{1}{i!} [A]^i t^i = [\mathbf{I}] + [A]t + \frac{1}{2!} [A]^2 t^2 + \frac{1}{3!} [A]^3 t^3 + \cdots \qquad (5.424)$$

is convergent for all t. The convergence allows us to differentiate the series term by term:

$$\frac{d}{dt} e^{[A]t} = [A] + [A]^2 t + \frac{1}{2!} [A]^3 t^2 + \cdots + \frac{[A]^i t^{i-1}}{(i-1)!} + \cdots$$

$$= [A] \left([\mathbf{I}] + [A]t + \frac{1}{2!} [A]^2 t^2 + \cdots + \frac{[A]^{i-1} t^{i-1}}{(i-1)!} + \cdots \right)$$

$$= \left([\mathbf{I}] + [A]t + \frac{1}{2!} [A]^2 t^2 + \cdots + \frac{[A]^{i-1} t^{i-1}}{(i-1)!} + \cdots \right) [A]$$

$$= e^{[A]t} [A] \qquad (5.425)$$

We may also use the definition of the derivative and find the same result.

$$\frac{d}{dt} e^{t[A]} = \lim_{h \to 0} \frac{e^{(t+h)[A]} - e^{t[A]}}{h} = \lim_{h \to 0} \frac{e^{t[A]} e^{h[A]} - e^{t[A]}}{h}$$

$$= e^{t[A]} \lim_{h \to 0} \frac{e^{h[A]} - [\mathbf{I}]}{h} = [A] e^{t[A]} \qquad (5.426)$$

Example 181 ★ Exponential function of a diagonal matrix. Diagonal matrices are the simplest matrix to work with. Their inverse is calculated by the inversion of diagonal terms, and any other mathematical operation is only applied on diagonal terms and will end up a diagonal matrix. Here is to show how we calculate the exponential function of a diagonal matrix. The matrix is assumed to be generated by the diagonalization of a coefficient matrix of a set of first-order differential equations.

To find the exponential function of a diagonal matrix such as

$$[\Lambda] = \begin{bmatrix} 1 & 0 \\ 0 & -2 \end{bmatrix} \qquad (5.427)$$

we note that

$$[\Lambda]^n = \begin{bmatrix} 1^n & 0 \\ 0 & (-2)^n \end{bmatrix} \qquad (5.428)$$

Substituting $[\Lambda]^n$ in $e^{[\Lambda]t}$

$$e^{[\Lambda]t} = \sum_{i=0}^{\infty} \frac{1}{i!} [\Lambda]^i t^i = [\mathbf{I}] + [\Lambda]t + \frac{1}{2!}[\Lambda]^2 t^2 + \frac{1}{3!}[\Lambda]^3 t^3 + \cdots \quad (5.429)$$

yields

$$e^{[\Lambda]t} = \begin{bmatrix} 1 & 0 \\ 0 & 1 \end{bmatrix} + \begin{bmatrix} 1 & 0 \\ 0 & -2 \end{bmatrix} t + \frac{1}{2!}\begin{bmatrix} 1^n & 0 \\ 0 & (-2)^n \end{bmatrix} t^2 + \cdots$$

$$= \begin{bmatrix} \sum_{i=0}^{\infty} \frac{t^i}{i!} & 0 \\ 0 & \sum_{i=0}^{\infty} \frac{(-2)^i t^i}{i!} \end{bmatrix} = \begin{bmatrix} e^t & 0 \\ 0 & e^{-2t} \end{bmatrix} \quad (5.430)$$

In general, the exponential function $e^{[\Lambda]t}$ of a diagonal matrix $[\Lambda]$

$$[\Lambda] = \begin{bmatrix} \lambda_1 & 0 & 0 & 0 \\ 0 & \lambda_2 & 0 & 0 \\ 0 & 0 & \ddots & \vdots \\ 0 & 0 & \cdots & \lambda_n \end{bmatrix} \quad (5.431)$$

is

$$e^{[\Lambda]t} = \begin{bmatrix} e^{\lambda_1 t} & 0 & 0 & 0 \\ 0 & e^{\lambda_2 t} & 0 & 0 \\ 0 & 0 & \ddots & \vdots \\ 0 & 0 & \cdots & e^{\lambda_n t} \end{bmatrix} \quad (5.432)$$

When $[A]$ is not diagonal, but it is diagonalizable, to determine $[A]^n$, it is necessary to find the modal matrix $[U]$ to diagonalize $[A]$.

$$[\Lambda] = [U]^{-1} [A] [U] \quad (5.433)$$

The diagonal elements of matrix $[\Lambda]$ are the eigenvalues of $[A]$.

$$[\Lambda] = \begin{bmatrix} \lambda_1 & 0 & 0 & 0 \\ 0 & \lambda_2 & 0 & 0 \\ 0 & 0 & \ddots & \vdots \\ 0 & 0 & \cdots & \lambda_n \end{bmatrix} \quad (5.434)$$

Then, if $[\Lambda]$ is the diagonalized form of $[A]$, we have

$$[A] = [U] \ [\Lambda] \ [U]^{-1} \tag{5.435}$$

Then $[A]^n$ would be

$$[A]^n = \left[U \left[\Lambda \right] U^{-1} \right] \left[U \left[\Lambda \right]^{n-1} U^{-1} \right] = U \left[\Lambda \right]^n U^{-1} \tag{5.436}$$

and, therefore,

$$\begin{aligned}
e^{[A]t} &= [\mathbf{I}] + \left[U \left[\Lambda \right] U^{-1} \right] t + \left[U \left[\Lambda \right]^2 U^{-1} \right] \frac{t^2}{2!} + \left[U \left[\Lambda \right]^3 U^{-1} \right] \frac{t^3}{3!} + \cdots \\
&= [U] \left[[\mathbf{I}] + [\Lambda] t + \frac{1}{2!} [\Lambda]^2 t^2 + \frac{1}{3!} [\Lambda]^3 t^3 + \cdots \right] [U]^{-1} \tag{5.437}
\end{aligned}$$

The exponential matrix, $e^{[A]}$, appears by setting $t = 1$:

$$e^{[A]} = [U] \left[[\mathbf{I}] + [\Lambda] + \frac{1}{2!} [\Lambda]^2 + \frac{1}{3!} [\Lambda]^3 + \cdots \right] [U]^{-1} \tag{5.438}$$

$$= [U] \begin{bmatrix} e^{\lambda_1} & 0 & 0 & 0 \\ 0 & e^{\lambda_2} & 0 & 0 \\ 0 & 0 & \ddots & \vdots \\ 0 & 0 & \cdots & e^{\lambda_n} \end{bmatrix} [U]^{-1} \tag{5.439}$$

Applying similar method, we can show that any function $f(x)$ that is defined on the eigenvalues λ_i of a matrix $[A]$ is:

$$f([A]) = [U] f([\Lambda]) [U]^{-1} \tag{5.440}$$

Example 182 ★ Exponential of a non-diagonal matrix. Here is a numerical example of calculating exponential function $e^{[A]}$ of a non-diagonal matrix $[A]$ by utilizing diagonalization method.

To find the exponential function of a matrix such as

$$[A] = \begin{bmatrix} 1 & 3 \\ 4 & -2 \end{bmatrix} \tag{5.441}$$

we first determine its modal matrix $[U]$:

$$[U] = \begin{bmatrix} \dfrac{3}{8} - \dfrac{1}{8}\sqrt{57} & \dfrac{1}{8}\sqrt{57} + \dfrac{3}{8} \\ 1 & 1 \end{bmatrix} \tag{5.442}$$

Then the diagonal matrix $[\Lambda]$, associated with $[A]$, is

$$[\Lambda] = [U]^{-1} \begin{bmatrix} 1 & 3 \\ 4 & -2 \end{bmatrix} [U] \approx \begin{bmatrix} -4.275 & 0 \\ 0 & 3.275 \end{bmatrix} \tag{5.443}$$

The exponential function of $[\Lambda]$ is

$$e^{[\Lambda]} = \begin{bmatrix} e^{-4.275} & 0 \\ 0 & e^{3.275} \end{bmatrix} \tag{5.444}$$

and hence,

$$e^{[A]} = [U] \, e^{[\Lambda]} \, [U]^{-1} = \begin{bmatrix} 18.478 & 10.501 \\ 14.001 & 7.9769 \end{bmatrix} \tag{5.445}$$

If we are looking for $e^{[\Lambda]t}$ and $e^{[A]t}$, then

$$e^{[\Lambda]t} = \begin{bmatrix} e^{-4.275t} & 0 \\ 0 & e^{3.275t} \end{bmatrix} \tag{5.446}$$

$$e^{[A]t} = [U] \, e^{[\Lambda]t} \, [U]^{-1}$$
$$= \begin{bmatrix} 0.3e^{-4.275t} + 0.699e^{3.275t} & 0.397e^{3.275t} - 0.397e^{-4.275t} \\ 0.53e^{3.275t} - 0.53e^{-4.275t} & 0.699e^{-4.275t} + 0.3e^{3.275t} \end{bmatrix} \tag{5.447}$$

Because at least one of the exponents has a positive real part, the associated dynamic system will be unstable.

Assume the matrix $[A]$ belongs to a set of first-order systems.

$$\dot{\mathbf{x}} + [A]\mathbf{x} = \mathbf{0} \tag{5.448}$$

$$\begin{bmatrix} \dot{x}_1 \\ \dot{x}_2 \end{bmatrix} + \begin{bmatrix} 1 & 3 \\ 4 & -2 \end{bmatrix} \begin{bmatrix} x_1 \\ x_2 \end{bmatrix} = \begin{bmatrix} 0 \\ 0 \end{bmatrix} \tag{5.449}$$

Employing the eigenvalues and modal matrix of $[A]$,

$$\lambda_1 = -4.2749 \qquad \lambda_2 = 3.2749 \tag{5.450}$$

$$[U] = \begin{bmatrix} -0.5687 & 1.3187 \\ 1.0 & 1.0 \end{bmatrix} \tag{5.451}$$

the natural solution of the system is

$$\mathbf{x} = [U] e^{-[\Lambda]t} [U]^{-1} \mathbf{x}_0 = [U] \begin{bmatrix} e^{-4.2749t} & 0 \\ 0 & e^{3.2749t} \end{bmatrix} [U]^{-1} \mathbf{x}_0 \qquad (5.452)$$

or

$$x_1 = \left(0.30131 e^{-4.2749t} + 0.69869 e^{3.2749t} \right) x_{10}$$

$$+ \left(0.39734 e^{3.2749t} - 0.39734 e^{-4.2749t} \right) x_{20} \qquad (5.453)$$

$$x_2 = \left(0.52983 e^{3.2749t} - 0.52983 e^{-4.2749t} \right) x_{10}$$

$$+ \left(0.69869 e^{-4.2749t} + 0.30131 e^{3.2749t} \right) x_{20} \qquad (5.454)$$

Example 183 ★ Inverse of exponential matrix. The exponential function of matrices is often being used in linear vibrations of coupled first-order systems. The function will be a matrix, and hence, their inversion might be needed in further steps. Here is to show how to calculate the inversion of an exponential function of a matrix. It is an easy operation.

The distribution property of exponential functions

$$e^{[A]t} e^{[A]\tau} = e^{[A](t+\tau)} \qquad (5.455)$$

may be shown by a series expansion of the functions

$$e^{[A]t} e^{[A]\tau} = \left(\sum_{i=0}^{\infty} \frac{1}{i!} [A]^i \, t^i \right) \left(\sum_{i=0}^{\infty} \frac{1}{j!} [A]^j \, \tau^j \right) = \sum_{i=0}^{\infty} \sum_{i=0}^{\infty} [A]^{i+j} \frac{t^i \tau^j}{i! j!} \qquad (5.456)$$

Let us assume that

$$i + j = k \qquad (5.457)$$

to find

$$e^{[A]t} e^{[A]\tau} = \sum_{i=0}^{\infty} \sum_{k=i}^{\infty} [A]^k \frac{t^i \tau^{k-i}}{i! \, (k-i)!} = \sum_{k=0}^{\infty} \frac{1}{k!} [A]^k \sum_{i=0}^{\infty} \frac{k! \, t^i \tau^{k-i}}{i! \, (k-i)!}$$

$$= \sum_{k=0}^{\infty} \frac{1}{k!} [A]^k \, (t+\tau)^k = e^{[A](t+\tau)} \qquad (5.458)$$

If $\tau = -t$, then

$$e^{[A]t} e^{-[A]t} = e^{-[A]t} e^{[A]t} = e^{[A](t-t)} = [\mathbf{I}] \qquad (5.459)$$

Therefore, $e^{-[A]t}$ is the inverse of $e^{[A]t}$, and vice versa:

$$\left[e^{[A]t}\right]^{-1} = e^{-[A]t} \tag{5.460}$$

Example 184 ★ Principal coordinates. Converting a set of coupled first-order equations to an equivalent set of decoupled first-order equations is an effective practical method of analysis.

A practical method of solution for a set of first-order equations

$$\dot{\mathbf{x}} + [A]\mathbf{x} = \mathbf{f}(t) \tag{5.461}$$

is to transform the equations to the principal coordinates $\mathbf{p}$,

$$\mathbf{p} = [U]^{-1}\mathbf{x} \tag{5.462}$$

where $[U]$ is the modal matrix of the coefficient matrix $[A]$. Substituting $[U]\mathbf{p}$ for $\mathbf{x}$ in Eq. (5.461) yields

$$[U]\dot{\mathbf{p}} + [A][U]\mathbf{p} = \mathbf{f}(t) \tag{5.463}$$

which provides us with

$$\dot{\mathbf{p}} + [U]^{-1}[A][U]\mathbf{p} = [U]^{-1}\mathbf{f}(t) \tag{5.464}$$

The modal transformation (5.462) decouples the equations to

$$\dot{\mathbf{p}} + [\Lambda]\mathbf{p} = \mathbf{P} \tag{5.465}$$
$$[\Lambda] = [U]^{-1}[A][U] \tag{5.466}$$
$$\mathbf{P} = [U]^{-1}\mathbf{f}(t) \tag{5.467}$$

or

$$\dot{p}_1 + \lambda_1 p_1 = P_1$$
$$\dot{p}_2 + \lambda_2 p_2 = P_2$$
$$\vdots$$
$$\dot{p}_n + \lambda_n p_n = P_n \tag{5.468}$$

The solutions of the independent equations (5.468) are

$$p_1 = e^{-\lambda_1 t} \int e^{\lambda_1 t} \, P_1 \, dt + C_1 e^{-\lambda_1 t}$$

$$p_2 = e^{-\lambda_2 t} \int e^{\lambda_2 t} \, P_2 \, dt + C_2 e^{-\lambda_2 t}$$

$$\vdots$$

$$p_n = e^{-\lambda_n t} \int e^{\lambda_n t} \, P_n \, dt + C_n e^{-\lambda_n t} \tag{5.469}$$

or

$$\mathbf{p} = e^{-[\Lambda]t} \int e^{[\Lambda]t} \, \mathbf{P} \, dt + e^{-[\Lambda]t} \mathbf{C} \tag{5.470}$$

The solution in terms of the original coordinates appears after a reverse transformation:

$$\mathbf{x} = [U]\mathbf{p} = [U]\left[e^{-[\Lambda]t} \int e^{[\Lambda]t} \, \mathbf{P} \, dt + e^{-[\Lambda]t} \mathbf{C} \right] \tag{5.471}$$

Expansion of the result will show its equivalence to Eq. (5.257):

$$\mathbf{x} = [U]\left[[U]^{-1} e^{-[A]t} [U] \int [U]^{-1} e^{[A]t} [U] \mathbf{P} dt + [U]^{-1} e^{-[A]t} [U] \mathbf{C} \right]$$

$$= e^{-[A]t} \int e^{[A]t} \mathbf{f} dt + e^{-[A]t} \mathbf{C}' \tag{5.472}$$

Example 185 ★ Complex eigenvalues. The coefficient matrix of vibrating systems must have complex eigenvalues. The real parts will indicate growing or shrinking of the amplitude and, hence, show the stability of the system. The imaginary part will be converted to sin and cos functions to indicate oscillations.

Consider a set of homogeneous equations:

$$\dot{\mathbf{x}} + [A]\mathbf{x} = \mathbf{0} \tag{5.473}$$

where the eigenvalues and eigenvectors of $[A]$ are complex. The solution of the equations should be arranged in a real and imaginary part,

$$\mathbf{x}(t) = \mathbf{x}_R(t) + i\mathbf{x}_I(t) \tag{5.474}$$

where $\mathbf{x}_R(t)$ and $\mathbf{x}_I(t)$ are real-valued functions of t. Then both $\mathbf{x}_R(t)$ and $\mathbf{x}_I(t)$ are solutions of the set of Eqs. (5.473). There is always a complex conjugate eigenvalue and eigenvector for each complex eigenvalue and eigenvector

$$\lambda_i \equiv \bar{\lambda}_i \qquad \mathbf{u}_i \equiv \bar{\mathbf{u}}_i \tag{5.475}$$

where

$$([A] - s_i [I]) \mathbf{u}_i = 0 \qquad ([A] - s_i [I]) \bar{\mathbf{u}}_i = 0 \tag{5.476}$$

and, therefore,

$$\mathbf{x}_R (t) = \frac{1}{2} (\mathbf{x} (t) + \bar{\mathbf{x}} (t)) \tag{5.477}$$

$$\mathbf{x}_I (t) = -\frac{i}{2} (\mathbf{x} (t) - \bar{\mathbf{x}} (t)) \tag{5.478}$$

Let us write a pair of eigenvalues and eigenvectors of the coefficient matrix $[A]$ as

$$\lambda = a + ib \qquad \bar{\lambda} = a - ib \tag{5.479}$$

$$\mathbf{u} = \mathbf{a} + i\mathbf{b} \qquad \bar{\mathbf{u}} = \mathbf{a} - i\mathbf{b} \tag{5.480}$$

Then one solution would be

$$\mathbf{x} (t) = (\mathbf{a} + i\mathbf{b}) e^{a+bi} = (\mathbf{a} + i\mathbf{b}) e^{at} (\cos bt + i \sin bt)$$
$$= e^{at} (\mathbf{a} \cos bt - \mathbf{b} \sin bt) + i e^{at} (\mathbf{a} \sin bt + \mathbf{b} \cos bt) \tag{5.481}$$

The other solution would be the complex conjugate of this solution.

$$\mathbf{x} (t) = (\mathbf{a} - i\mathbf{b}) e^{a-bi} = (\mathbf{a} - i\mathbf{b}) e^{at} (\cos bt - i \sin bt)$$
$$= e^{at} (\mathbf{a} \cos bt - \mathbf{b} \sin bt) - i e^{at} (\mathbf{a} \sin bt + \mathbf{b} \cos bt) \tag{5.482}$$

Both the real and imaginary parts are solutions. Therefore, we have found two real solutions:

$$\mathbf{x}_1 (t) = \mathbf{x}_R (t) = e^{at} (\mathbf{a} \cos bt - \mathbf{b} \sin bt) \tag{5.483}$$

$$\mathbf{x}_2 (t) = \mathbf{x}_I (t) = e^{at} (\mathbf{a} \sin bt + \mathbf{b} \cos bt) \tag{5.484}$$

As an example, consider

$$\dot{\mathbf{x}} + \begin{bmatrix} 2 & 3 \\ -1 & 4 \end{bmatrix} \mathbf{x} = \mathbf{0} \tag{5.485}$$

The eigenvalues and eigenvectors of the coefficient matrix are

$$s_1 = 3 - i\sqrt{2} \qquad s_2 = 3 + i\sqrt{2} \tag{5.486}$$

$$\mathbf{u}_1 = \begin{bmatrix} 1 + i\sqrt{2} \\ 1 \end{bmatrix} \qquad \mathbf{u}_2 = \begin{bmatrix} 1 - i\sqrt{2} \\ 1 \end{bmatrix} \tag{5.487}$$

The solution is

$$\mathbf{x} = \sum_{i=1}^{n} e^{-s_i t} \mathbf{u}_i$$

$$= e^{-(3-i\sqrt{2})t} \begin{bmatrix} 1 + i\sqrt{2} \\ 1 \end{bmatrix} + e^{-(3+i\sqrt{2})t} \begin{bmatrix} 1 - i\sqrt{2} \\ 1 \end{bmatrix} \tag{5.488}$$

Employing the Euler equation,

$$e^{i\theta} = \cos\theta + i\sin\theta \tag{5.489}$$

we can transform the first solution to

$$\mathbf{x} = \begin{bmatrix} \left(1 + i\sqrt{2}\right) e^{-3t} \left(\cos\sqrt{2}t - i\sin\sqrt{2}t\right) \\ e^{-3t} \left(\cos\sqrt{2}t - i\sin\sqrt{2}t\right) \end{bmatrix} \tag{5.490}$$

$$= \begin{bmatrix} e^{-3t} \left(\left(\cos\sqrt{2}t + \sqrt{2}\sin\sqrt{2}t\right) + i\left(\sqrt{2}\cos\sqrt{2}t - \sin\sqrt{2}t\right)\right) \\ e^{-3t}\left(\cos\sqrt{2}t - i\sin\sqrt{2}t\right) \end{bmatrix}$$

$$\begin{bmatrix} e^{-3t}\left(\cos\sqrt{2}t + \sqrt{2}\sin\sqrt{2}t\right) \\ e^{-3t}\cos\sqrt{2}t \end{bmatrix} + i\begin{bmatrix} e^{-3t}\left(\sqrt{2}\cos\sqrt{2}t - \sin\sqrt{2}t\right) \\ -e^{-3t}\sin\sqrt{2}t \end{bmatrix}$$

where

$$\mathbf{x}_R(t) = \begin{bmatrix} \cos\sqrt{2}t + \sqrt{2}\sin\sqrt{2}t \\ \cos\sqrt{2}t \end{bmatrix} e^{-3t} \tag{5.491}$$

$$\mathbf{x}_I(t) = \begin{bmatrix} \sqrt{2}\cos\sqrt{2}t - \sin\sqrt{2}t \\ -\sin\sqrt{2}t \end{bmatrix} e^{-3t} \tag{5.492}$$

As another example, let us examine two coupled first-order systems obey the following equation.

$$\dot{\mathbf{x}} - \begin{bmatrix} 0 & 1 \\ -3 & -2 \end{bmatrix} \mathbf{x} = \mathbf{0} \tag{5.493}$$

The eigenvalues and eigenvectors of the coefficient matrix are

$$s_1 = -1 - i\sqrt{2} \qquad\qquad s_2 = -1 + i\sqrt{2} \tag{5.494}$$

$$\mathbf{u}_1 = \begin{bmatrix} \frac{1}{3}i\sqrt{2} - \frac{1}{3} \\ 1 \end{bmatrix} \qquad \mathbf{u}_2 = \begin{bmatrix} -\frac{1}{3}i\sqrt{2} - \frac{1}{3} \\ 1 \end{bmatrix} \tag{5.495}$$

The real solutions of the system are

$$\mathbf{x}_1(t) = \mathbf{x}_R(t) = e^{at}(\mathbf{a}\cos bt - \mathbf{b}\sin bt)$$

$$= e^{-t}\left(\begin{bmatrix} -1/3 \\ 1 \end{bmatrix}\cos\sqrt{2}t + \begin{bmatrix} \sqrt{2}/3 \\ 1 \end{bmatrix}\sin\sqrt{2}t\right) \qquad (5.496)$$

$$\mathbf{x}_2(t) = \mathbf{x}_I(t) = e^{at}(\mathbf{a}\sin bt + \mathbf{b}\cos bt)$$

$$= e^{-t}\left(\begin{bmatrix} -1/3 \\ 1 \end{bmatrix}\cos\sqrt{2}t - \begin{bmatrix} \sqrt{2}/3 \\ 1 \end{bmatrix}\sin\sqrt{2}t\right) \qquad (5.497)$$

If we define

$$x_1 = x \qquad x_2 = \dot{x} \qquad (5.498)$$

then the system of equations will be equivalent to

$$\ddot{x} + 2\dot{x} + 3x = 0 \qquad (5.499)$$

which is the equation of motion of a mass-damper-spring system with $m = 1, c = 2$, and $k = 3$.

5.4 Summary

There are cases where the behavior of a dynamic system can be modeled by first-order or reducible to first-order differential equations.

$$a\dot{x} + bx = f \qquad (5.500)$$

The motion of a first-order system with no external force is the natural motion.

$$\dot{x} + \alpha x = 0 \qquad (5.501)$$

The solution of natural motion is

$$x = x(0)\, e^{-\alpha t} = x_0 e^{-t/\tau} \qquad \tau = \frac{1}{\alpha} \qquad (5.502)$$

The most important characteristic of first-order systems is their time constant τ. The response of a system after a period of one time constant $t = \tau$ reaches e^{-1} of its initial value. A time constant is passed when x drops by about %64 of its initial value x_0.

$$x = x(t + \tau) = \frac{x(t)}{e} \tag{5.503}$$

The natural motion of first-order systems is either exponentially decreasing or increasing function of time, and they do not show vibrations.

The excited case of every linear first-order system is expressed by a full first-order equation.

$$\dot{x} + \alpha x = f(t) \tag{5.504}$$

The solution $x = x(t)$, for $t > 0$, is

$$x = e^{-\alpha t} \int e^{\alpha t} f(t) \, dt + C e^{-\alpha t} \tag{5.505}$$

Any linear free vibrating system of any order can be expressed by a set of coupled first-order linear differential equations.

$$\dot{x} + [A] x = 0 \tag{5.506}$$

$$[A] = \begin{bmatrix} a_{11} & a_{12} & a_{13} & \cdots & a_{1n} \\ a_{21} & a_{22} & a_{23} & \cdots & a_{2n} \\ \vdots & \vdots & \vdots & \vdots & \vdots \\ a_{n1} & a_{n2} & a_{n3} & \cdots & a_{nn} \end{bmatrix} \tag{5.507}$$

The solution of the set of coupled first-order linear homogeneous equations is

$$x = \sum_{j=1}^{n} e^{-\lambda_j t} C_j u_j \tag{5.508}$$

where u_j are the eigenvectors and λ_j, $j = 1, 2, \cdots, n$ are the eigenvalues of the $n \times n$ coefficient matrix $[A]$. This solution can also be expressed by

$$x(t) = e^{-[A]t} C = e^{-[A]t} x_0 \tag{5.509}$$

where

$$e^{-[A]t} = [I] - [A]t + \frac{1}{2!}[A]^2 t^2 - \frac{1}{3!}[A]^3 t^3 + \frac{1}{4!}[A]^4 t^4 - \cdots$$

$$= \sum_{n=0}^{\infty} (-1)^n \frac{(\mathbf{A}t)^n}{n!} \tag{5.510}$$

The forced vibration of any number of first-order equations can be expressed by a set of n coupled equations,

$$\dot{\mathbf{x}} + [A]\mathbf{x} = \mathbf{f}(t) \tag{5.511}$$

where its solution $\mathbf{x} = \mathbf{x}(t)$ is

$$\mathbf{x} = e^{-[A]t} \int e^{[A]t} \mathbf{f}(t) \, dt + e^{-[A]t} \mathbf{C} \tag{5.512}$$

$$\mathbf{x} = \begin{bmatrix} x_1 \; x_2 \; x_3 \; \cdots \; x_n \end{bmatrix}^T \tag{5.513}$$

$$\mathbf{f} = \begin{bmatrix} f_1 \; f_2 \; f_3 \; \cdots \; f_n \end{bmatrix}^T \tag{5.514}$$

$$\mathbf{C} = \begin{bmatrix} C_1 \; C_2 \; C_3 \; \cdots \; C_n \end{bmatrix}^T \tag{5.515}$$

5.5 Key Symbols

a, b	Coefficient
$\mathbf{b}$	$n \times 1$ coefficient vector
A, B	Coefficients of frequency response
A, B, C, D	Functions of x and t
$\mathbf{A}, \mathbf{B}$	Coefficient matrix
$[A]$	Coefficient matrix
c	Damping
C	Constant, constant of integration
$\mathbf{C}$	Vector of constants of integration
e	Exponential function
$E = K + P$	Mechanical energy
f	Force, function
F	Force amplitude, force magnitude, constant force
$\mathbf{f}, \mathbf{F}$	$n \times 1$ force set
g	Functions of displacement, gravitational acceleration
$g, \mathbf{g}$	Gravitational acceleration vector
h	Functions of time, homogeneity factor
i, j	Dummy index

$[\mathbf{I}]$	Identity matrix
k	Stuffiness
K	Kinetic energy
l	Distance, length
m	Mass
M, N	Homogeneous functions of x and t
n	Number of equations, number of coordinates, DOF
p, q	Functions of x
$\mathbf{p}$	Momentum, principal coordinate set
$\mathbf{P}$	$n \times 1$ principal force set
$\mathbf{P}$	$n \times 1$ principal coordinate
P	Potential energy
s	Characteristic value, eigenvalue
t	Time
u	Solution of equation, displacement
$\mathbf{u}$	Eigenvector
$[U]$	Modal matrix
$v \equiv \dot{x}$	Velocity
$v_0 = \dot{x}_0$	Initial velocity
V	Velocity amplitude
$x, \mathbf{x}$	Displacement, $n \times 1$ displacement set
x_0	Initial displacement
x_e	Equilibrium point
X	Displacement amplitude
y	Displacement excitation, variable, displacement
z	Variable

Greek

α	Constant, coefficient, inverse of time constant
$[\alpha], [A]$	Coefficient matrix
λ	Eigenvalue
$[\Lambda]$	Diagonalized coefficient matrix of $[\alpha]$
s	Eigenvalue
ω	Frequency
τ	Time constant
μ	Integrating factor

Subscript

I	Imaginary
h	Natural solution
p	Particular solution
R	Real
s	Steady state
v	Velocity coefficient
x	Position coefficient

Exercises

1. Sudden stop of the base of a mass-damper system.
   Figure 5.34 illustrates a mass-damper system. Assume the base and mass are moving at a constant speed until the base suddenly stops. Use $m = 1$ and $c = 1$, and determine

   (a) the velocity $\dot{x}$ of m as a function of time;
   (b) the position x of m as a function of time;
   (c) the final values of $\dot{x}_s$ and x_s at $t = \infty$;
   (d) the transmitted force to the base.

2. Step input from the base.
   Assume we apply the base of the system of Fig. 5.34 a step input from $y = 0$ to $y = Y$ at $t = 0$. Determine

   (a) the velocity $\dot{x}$ of m as a function of time;
   (b) the position x of m as a function of time;
   (c) the final values of $\dot{x}_s$ and x_s at $t = \infty$;
   (d) the magnitude of x and $\dot{x}$ after $t = \tau, 2\tau, 3\tau, \cdots$, where τ is the time constant of the system, using $m = 1$, $c = 1$, and $Y = 1$.

3. Cosine harmonic base excited first-order system.
   The base of the system of Fig. 5.34 is moving by a cosine displacement excitation, $y = Y \cos \omega t$. Determine

   (a) the velocity-time response of m;
   (b) the steady-state amplitude of velocity response for $Y = 1$ and different time constants;
   (c) the displacement-time response of m;
   (d) the steady-state amplitude of displacement response for $Y = 1$ and different time constants.

4. Sine harmonic base excited first-order system.
   The base of the system of Fig. 5.34 is moving by a cosine displacement excitation, $y = Y \sin \omega t$. Determine

   (a) the velocity-time response of m;

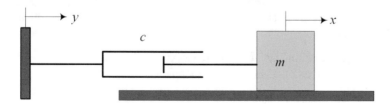

Fig. 5.34 A mass-damper system

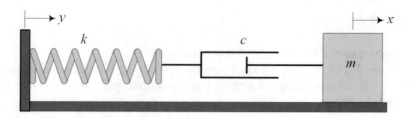

Fig. 5.35 A third-order system by a series of spring and damper

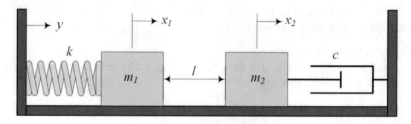

Fig. 5.36 A second-order and first-order system in interaction

    (b) the steady-state amplitude of velocity response for $Y = 1$ and different time constants;

    (c) the displacement-time response of m;

    (d) the steady-state amplitude of displacement response for $Y = 1$ and different time constants.

5. A third-order system.

  Figure 5.35 illustrates a mass m that is attached to a series of damper c and spring k. Determine

    (a) the equations of motion of the system;

    (b) a unique differential equation to express the x-coordinate;

    (c) the solution of the system coordinate x and y as functions of time;

    (d) the plot of $x\,(t)$ for various c and k around $m = 1$, $c = 1$, and $k = 1$, and assume $x\,(0) = 0$, $y\,(0) = 0$, and $\dot{x}\,(0) = 1$.

6. A second-order and first-order system in interaction.

  The spring behind the mass m_1 in Fig. 5.36 is compressed by $l_1 < l$ when m_1 is released from rest. The spring is not attached to m_1, and the masses can slide on the frictionless floor. The mass m_1 elastically hits m_2 after sliding the distance l. Determine

    (a) the speed and time at which m_1 reaches m_2;

    (b) the distance that m_2 slides before stopping.

  Now assume that m_1 plastically hits m_2 and sticks to it. Determine

    (c) ★ the position of equilibrium of the system;

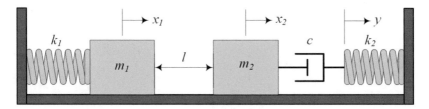

Fig. 5.37 A third-order and first-order system in interaction

(d) ★ x_2 as a function of time.

7. ★ Base excitation of a combined system.
   Consider the system in Fig. 5.36 which is at rest at the moment. Assume there is a base harmonic excitation as

$$y = Y \cos \omega t \tag{5.516}$$

and that the system can get into a steady-state oscillation as expected. Determine

(a) the value of Y as a function of ω where m_1 and m_2 barely touch each other if $\omega < \left(\omega_n = \sqrt{k/m_1}\right)$.
(b) the value of ω as a function of Y where m_1 and m_2 barely touch each other if $\omega < \left(\omega_n = \sqrt{k/m_1}\right)$.
(c) the value of Y as a function of ω where m_1 and m_2 barely touch each other if $\omega > \left(\omega_n = \sqrt{k/m_1}\right)$.
(d) the value of ω as a function of Y where m_1 and m_2 barely touch each other if $\omega > \left(\omega_n = \sqrt{k/m_1}\right)$.

Now assume that y is a step displacement $y = Y$ and determine

(e) the time response of m_1 and m_2.
(f) the time t_0 and condition at which m_1 and m_2 touch each other.
(g) the condition for Y that m_1 and m_2 touch each other before $\dot{x}_1$ gets negative.
   Then assume that y is a negative step displacement, $y = -Y$, and determine
(h) the time response of m_1 and m_2;
(i) the time t_0 and condition at which m_1 and m_2 touch each other.

8. ★ A third-order and first-order system in interaction.
   The spring behind the mass m_1 in Fig. 5.37 is compressed by $l_1 < l$. The mass m_1 is released from rest. The spring k_1 is not attached to m_1, and the masses can slide on the frictionless floor. If $m_1 = m_2$ and m_1 elastically hits m_2 after sliding the distance l, determine

(a) the speed and time at which m_1 reaches m_2;
(b) the distance that m_2 slides before final stopping.

Now assume that m_1 plastically hits m_2 and sticks to it. Determine

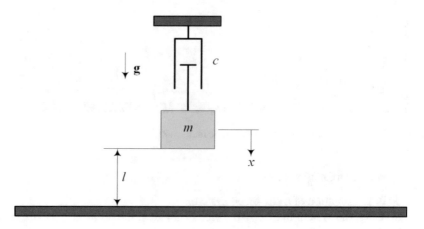

Fig. 5.38 A damped falling mass

(c) ★ the position of equilibrium of the system;

(d) ★ x_2 as a function of time.

9. A damped falling mass.

Figure 5.38 illustrates a mass m that is attached to the ceiling via a linear damper. The mass m is released from rest at a distance of l from the floor. If $m = 1$, determine

(a) the speed at which m reaches the floor;

(b) the time at which m reaches the floor;

(c) the plot the acceleration curve $\ddot{x}/g$ versus time.

10. A damped falling mass and a first-order support.

Figure 5.39 illustrates a mass m that is attached to the ceiling via a linear damper c_1. The mass m is released from rest at a distance of l from a plate on another damper c_2. If $m = 1$, $c_1 = c_2 = 1$, and $l_1 = l_2 = 1$, determine

(a) the speed at which m reaches the plate;

(b) the time at which m reaches the floor;

(c) the equation and plot of the acceleration curve $\ddot{x}/g$ and $\ddot{y}/g$ versus time;

(d) the equation and plot of the velocity curve $\dot{x}$ and $\dot{y}$ versus time;

(e) the equation and plot of the position curve x and y versus time.

11. Frequency response of a combined system.

Figure 5.40 illustrates a second- and a third-order system. Assume there is a base harmonic excitation

$$y = Y \cos \omega t \tag{5.517}$$

and the system can get into a steady-state oscillation. Determine

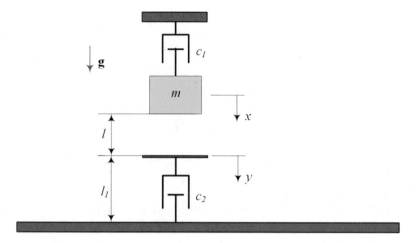

Fig. 5.39 A damped falling mass and a first-order support

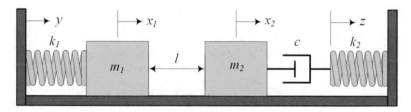

Fig. 5.40 Base excitation of a combined third- and second-order system

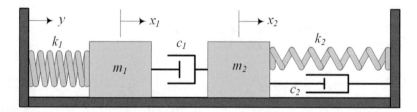

Fig. 5.41 A base excited mass-spring-damper system

(a) the frequency response of m_2.
(b) the amplitude Y of the steady-state variation of z.
(c) the condition for m_1 and m_2 to barely touch each other if $\omega < (\omega_n = \sqrt{k/m_1})$.
(d) the condition for m_1 and m_2 to barely touch each other if $\omega > (\omega_n = \sqrt{k/m_1})$.

12. ★ A base excited fourth-order system.
    A harmonically base excited two second-order system is illustrated in Fig. 5.41.
    Introducing the following coordinates and values

$$m_1 = 1 \qquad m_2 = 2 \qquad c_1 = 1 \qquad c_2 = 1 \qquad y = Y \sin \omega t$$

$$k_1 = 100 \qquad k_2 = 500 \tag{5.518}$$

$$Y = 0.1 \qquad x_1 = x_1 \qquad x_3 = \dot{x}_1 \qquad x_2 = x_2 \qquad x_4 = \dot{x}_2$$

write the equations of motion by a set of first-order equations and determine
$x_1(t)$ and $x_2(t)$.

13. A practice on first-order systems.
    Find the solutions of the following equations and define a mechanical system
    for each of them.

    (a) $\dot{x} + 2x = 1$.
    (b) $\dot{x} + x = e^x$.
    (c) $\dot{x} + 3x = t^2 + t$.

14. A fourth-order equation.
    Show that every solution of $x^{(4)} + x = 0$ has the form

$$x = e^{x/\sqrt{2}} \left(C_1 \cos \frac{x}{\sqrt{2}} + C_2 \sin \frac{x}{\sqrt{2}} \right)$$

$$+ e^{-x/\sqrt{2}} \left(C_3 \cos \frac{x}{\sqrt{2}} + C_4 \sin \frac{x}{\sqrt{2}} \right) \tag{5.519}$$

15. Resonance excitation of a first-order system.
    Consider the response of a mass m

$$\dot{v} + \alpha v = \frac{F}{m} e^{-\alpha_1 t} \qquad \alpha = \frac{c}{m} \qquad v = \dot{x} \tag{5.520}$$

$$v(0) = 0 \qquad x(0) = 0 \tag{5.521}$$

with the solution of

$$v = \frac{F}{m(\alpha - \alpha_1)} \left(e^{-\alpha_1 t} - e^{-\alpha t} \right) \tag{5.522}$$

(a) Determine the maximum of the response, v_M, and its associated time t_M.
(b) Determine the slope of v at $t = 0$.

16. ★ A bump excitation to a mass-damper system.
    The mass-damper system in Fig. 5.42 is moving with a constant speed v to pass
    the bump.

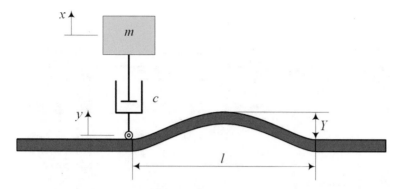

Fig. 5.42 A mass-damper system is moving with a constant speed v to pass the bump

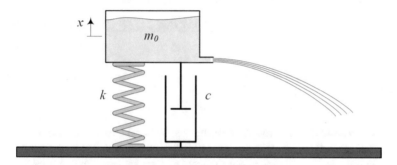

Fig. 5.43 A mass-spring-damper system with mass reduction

$$y = Y \sin^2 \omega t \qquad \omega = \frac{\pi l}{2v} \qquad (5.523)$$

Assume that there is no gravity and determine

(a) the vertical displacement of m during the excitation.
(b) the vertical velocity of m during the excitation.
(c) the vertical acceleration of m during the excitation.
(d) the vertical displacement of m after the excitation.
(e) the vertical velocity of m after the excitation.
(f) the vertical acceleration of m after the excitation.
(g) the maximum vertical speed of m.
(h) Is there any limit for the vertical displacement of m?

17. ★ Reducing mass.
    The box of the system of Fig. 5.43 is full of water. The water is leaking out of
    the box resulting in a mass reduction. Assume that the initial mass of the box
    is $m_0 = 1$ kg and its initial condition is $x(0) = 1$ and $\dot{x}(0) = 0$. The rate of
    losing mass is $m' = 0.01$ kg / s.

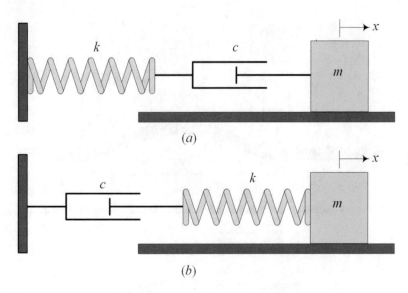

Fig. 5.44 Two third-order systems. (**a**) series of mass, damper, spring, (**b**) series of mass, spring, damper

    (a) Determine the time response of the system for the extreme cases of $m = m_0$ and $m \to 0$.

    (b) Is it possible to decide about the intermediate stages of losing mass?

18. Converting a set of third-order equations to a set of first-order equations.
Convert the third-order systems of Fig. 5.44 to a set of first-order equations in terms of new variable **z**.

$$\dddot{\mathbf{x}} + \mathbf{B}\ddot{\mathbf{x}} + \mathbf{C}\dot{\mathbf{x}} + \mathbf{D}\mathbf{x} = \mathbf{0} \tag{5.524}$$

$$\dot{\mathbf{z}} = \mathbf{A}\,\mathbf{z} \qquad \mathbf{z} = \begin{bmatrix} \mathbf{x} \\ \dot{\mathbf{x}} \\ \ddot{\mathbf{x}} \end{bmatrix} \tag{5.525}$$

19. A time-dependent stiffness system.
Assume there is a system that its governing model is expressed by the following differential equation.

$$\dot{x} + \frac{1}{1-t}x = 1 - t \qquad x\,(0) = 1 \tag{5.526}$$

The coefficient of x is time dependent, and if we assume it is a model for stiffness, then we have a time-dependent stiffness system. Also there is a time-dependent forcing term.

(a) make a mechanical system for the given equations.

(b) Using integration factor method, show that the time response of the system is

$$x = 1 - t^2 \tag{5.527}$$

20. Three first-order equations.

(a) Make a mechanical system for the following equations.

$$\dot{x}_1 = 9x_1 - 3x_2 \tag{5.528}$$

$$\dot{x}_2 = -3x_1 + 12x_2 - 3x_3 \tag{5.529}$$

$$\dot{x}_3 = -3x_2 + 9x_3 \tag{5.530}$$

(b) Determine the eigenvalues and eigenvectors of the coefficient matrix and find the time response of the system.

Part III
Frequency Response

The harmonic excitation is the most common continuous excitation in industry. Any combination of sin and cos functions is called harmonic functions. The response of linear discrete vibrating systems to harmonic excitation would be harmonic with proportional magnitude. The magnitude of the harmonic responses is functions of the excitation frequencies and depends on the characteristics of the system. The magnitude of the harmonic response versus excitation frequency is called the frequency response that indicates the behavior of the system at different excitation frequencies.

In this part we review the harmonic forced vibrations of vibrating systems. It is common that the response of mechanical systems to harmonic excitation is called the *forced vibration* and the response of electric system is called *frequency response*, although the forced vibration may not necessarily be a harmonic excitation.

Chapter 6
One Degree of Freedom Systems

The frequency response is the *steady-state* solution of the equation of motion at each frequency, when the system is *harmonically excited*. Steady-state response refers to the constant amplitude oscillations, after the effect of the initial conditions has disappeared. A harmonic excitation is any combination of *sinusoidal functions* that is applied on a vibrating system. If the system is linear, then a harmonic excitation generates a harmonic response with a frequency-dependent amplitude. In frequency response analysis, we are looking for the steady-state amplitude of the oscillation as a function of the excitation frequency.

A vast amount of vibrating systems can be modeled by a one DOF mass-spring-damper system. There are only four types of harmonically excited one DOF systems, as are shown in Fig. 6.1 symbolically:

1. Base excitation
2. Eccentric excitation
3. Eccentric base excitation
4. Forced excitation

Base excitation is the most practical model for vertical vibration of mechanical systems, such as vehicles, and structures. *Eccentric excitation* is a model for every type of rotary motor on a suspension, such as engine on engine mounts. *Eccentric base excitation* is a model for the vibration of any equipment mounted on an engine. *Forced excitation* has almost no practical application; however, it is the simplest model of forced vibrations with good pedagogical use.

Because of its simplicity, we first examine the frequency response of a harmonically forced vibrating system.

R. N. Jazar, *Advanced Vibrations*, https://doi.org/10.1007/978-3-031-16356-2_6

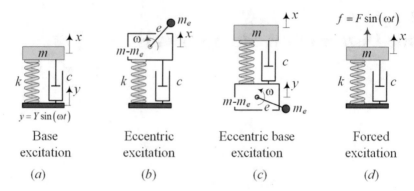

Fig. 6.1 The four practical types of one DOF harmonically excited systems: (**a**) base excitation, (**b**) eccentric excitation, (**c**) eccentric base excitation, and (**d**) forced excitation

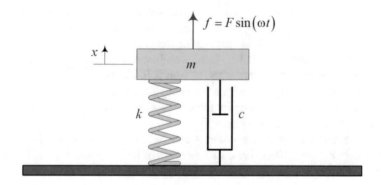

Fig. 6.2 A harmonically forced excited, single DOF system

6.1 Forced Excitation

Figure 6.2 illustrates a one DOF vibrating mass m supported by a spring k and a damper c. The absolute motion of m with respect to its equilibrium position is measured by the coordinate x. A sinusoidal excitation force is applied on m, which makes the system vibrate.

$$f = F \sin \omega t \tag{6.1}$$

The equation of motion of the system is

$$m\ddot{x} + c\dot{x} + kx = F \sin \omega t \tag{6.2}$$

which generates a frequency response expressed by either of the following functions:

$$x = A_1 \sin \omega t + B_1 \cos \omega t \tag{6.3}$$

$$= X \sin (\omega t - \varphi_x) \tag{6.4}$$

The steady-state response has an *amplitude X* and a *phase* φ_x.

$$X = \frac{1}{\sqrt{(k - m\omega^2)^2 + c^2\omega^2}} F \tag{6.5}$$

$$\tan \varphi_x = \frac{c\omega}{k - m\omega^2} \tag{6.6}$$

We usually express the frequency response X and *phase* φ_x in dimensionless form

$$\frac{X}{F/k} = \frac{1}{\sqrt{(1 - r^2)^2 + (2\xi r)^2}} \tag{6.7}$$

$$\varphi_x = \tan^{-1} \frac{2\xi r}{1 - r^2} \tag{6.8}$$

where we use the *frequency ratio r*, *natural frequency* ω_n, and *damping ratio* ξ.

$$r = \frac{\omega}{\omega_n} \tag{6.9}$$

$$\xi = \frac{c}{2\sqrt{km}} \tag{6.10}$$

$$\omega_n = \sqrt{\frac{k}{m}} \tag{6.11}$$

The phase φ_x indicates the *angular lag* of the response x with respect to the excitation f.

The frequency responses for X and φ_x as a function of r and ξ are plotted in Figs. 6.3 and 6.4.

Proof Applying Newton's method and using the free body diagram of the system, as shown in Fig. 6.5, generate the equation of motion.

$$m\ddot{x} + c\dot{x} + kx = F \sin \omega t \tag{6.12}$$

The steady-state solution of the linear equation (6.12) will be the same function as the excitation with an unknown amplitude and phase. Therefore, the solution can be (6.3) or (6.4). We substitute the solution in the equation of motion to find the amplitude and phase of the response. Let us examine the solution (6.3) and find the following equation.

$$-m\omega^2 (A_1 \sin \omega t + B_1 \cos \omega t) + c\omega (A_1 \cos \omega t - B_1 \sin \omega t)$$

$$+k (A_1 \sin \omega t + B_1 \cos \omega t) = F \sin \omega t \tag{6.13}$$

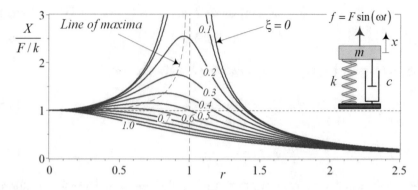

Fig. 6.3 The position frequency response for $\dfrac{X}{F/k}$

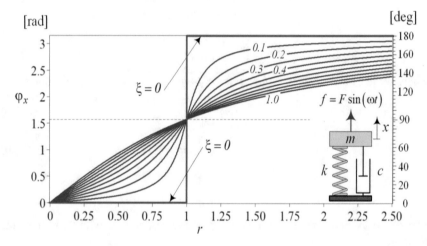

Fig. 6.4 The frequency response for φ_x

The harmonic functions $\sin \omega t$ and $\cos \omega t$ are orthogonal; therefore, their coefficients must be balanced on both sides of the equal sign. Balancing the coefficients of $\sin \omega t$ and $\cos \omega t$ provides us with a set of two algebraic equations for A_1 and B_1.

$$\begin{bmatrix} k - m\omega^2 & -c\omega \\ c\omega & k - m\omega^2 \end{bmatrix} \begin{bmatrix} A_1 \\ B_1 \end{bmatrix} = \begin{bmatrix} F \\ 0 \end{bmatrix} \tag{6.14}$$

Solving for coefficients A_1 and B_1

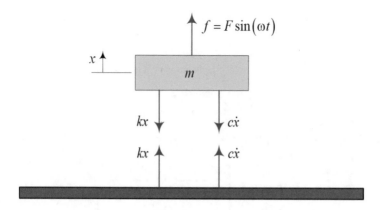

Fig. 6.5 Free body diagram of the harmonically forced excited, single DOF system shown in Fig. 6.2

$$\begin{bmatrix} A_1 \\ B_1 \end{bmatrix} = \begin{bmatrix} k - m\omega^2 & -c\omega \\ c\omega & k - m\omega^2 \end{bmatrix}^{-1} \begin{bmatrix} F \\ 0 \end{bmatrix}$$

$$= \begin{bmatrix} \dfrac{k - m\omega^2}{\left(k - m\omega^2\right)^2 + c^2\omega^2} F \\ \dfrac{-c\omega}{\left(k - m\omega^2\right)^2 + c^2\omega^2} F \end{bmatrix} \tag{6.15}$$

provides the steady-state solution (6.3).

$$x = F\frac{\left(k - m\omega^2\right)}{\left(k - m\omega^2\right)^2 + c^2\omega^2} \sin \omega t - F\frac{c\omega}{\left(k - m\omega^2\right)^2 + c^2\omega^2} \cos \omega t \tag{6.16}$$

The amplitude X and phase φ_x can be found by equating Equations (6.3) and (6.4):

$$A_1 \sin \omega t + B_1 \cos \omega t = X \sin \left(\omega t - \varphi_x\right)$$
$$= X \cos \varphi_x \sin \omega t - X \sin \varphi_x \cos \omega t \tag{6.17}$$

which shows that

$$A_1 = X \cos \varphi_x \qquad B_1 = -X \sin \varphi_x \tag{6.18}$$

and therefore,

$$X = \sqrt{A_1^2 + B_1^2} \qquad \tan \varphi_x = \frac{-B_1}{A_1} \tag{6.19}$$

Substituting A_1 and B_1 from (6.15) yields

$$X = \frac{1}{\sqrt{\left(k - m\omega^2\right)^2 + c^2\omega^2}} F \tag{6.20}$$

$$\tan \varphi_x = \frac{c\omega}{k - m\omega^2} \tag{6.21}$$

The coefficients A_1 and B_1 based on r and ξ are

$$\begin{bmatrix} A_1 \\ B_1 \end{bmatrix} = \begin{bmatrix} \dfrac{F}{k} \dfrac{1 - r^2}{\left(1 - r^2\right)^2 + (2\xi r)^2} \\ \dfrac{F}{k} \dfrac{-2\xi r}{\left(1 - r^2\right)^2 + (2\xi r)^2} \end{bmatrix} \tag{6.22}$$

The frequency responses of the amplitude X and phase φ_x are more practical if we express them by dimensionless parameters r and ξ as expressed in Eqs. (6.7) and (6.8)

When we apply a constant force $f = F$ on m, a constant steady-state displacement, δ_s, appears.

$$\delta_s = \frac{F}{k} \tag{6.23}$$

If we call δ_s "*static amplitude*" and X "*dynamic amplitude*," then X/δ_s is the ratio of dynamic to static amplitudes. So, we can plot $X/(F/k)$ instead of X to make the graph of frequency response dimensionless. We have $X/(F/k) = 1$ at $r = 0$, and $X \to 0$, when $r \to \infty$. However, X gets a high value when $r \to 1$ or $\omega \to \omega_n$. Theoretically, $X \to \infty$ if $\xi = 0$ and $r \to 1$. The frequency domain around the natural frequency is called the *resonance zone*. We reduce the high amplitude of vibration in the resonance zone by introducing damping.

The displacement frequency response function $X = X(\omega)$ is the most important frequency-dependent behavior of vibrating systems. Furthermore, we may use the frequency response of every characteristic of the system that is a function of the excitation frequency, such as velocity frequency response $\dot{X} = \dot{X}(\omega)$ and transmitted force frequency response $f_T = f_T(\omega)$.

In mechanics, there is no practical method to apply a periodic force on an object without attaching a mechanical device and applying a displacement. Hence, the forced vibrating system shown in Fig. 6.2 has no practical application in mechanics. However, it is possible to apply a periodic force on a mass m using a ferromagnetic material and applying an alternating or periodic magnetic force. ∎

Example 186 A forced vibrating system. A sample numerical example of frequency response of a forced excited system to visualize the phase lag between response and excitation.

Consider a mass-spring-damper system with the following numerical values.

$$m = 200 \, \text{kg} \qquad k = 10,000 \, \text{N} / \text{m} \qquad c = 100 \, \text{N} \, \text{s} / \text{m} \tag{6.24}$$

The natural frequency and damping ratio of the system are

$$\omega_n = \sqrt{\frac{k}{m}} = \sqrt{\frac{10,000}{200}} = 7.0711 \, \text{rad} / \text{s} \approx 1.125 \, \text{Hz} \tag{6.25}$$

$$\xi = \frac{c}{2\sqrt{km}} = \frac{100}{2\sqrt{10,000 \times 200}} = 0.0353 \tag{6.26}$$

If a harmonic force f is applied on m,

$$f = 100 \sin 10t \tag{6.27}$$

$$r = \frac{\omega}{\omega_n} = \frac{10}{7.0711} = 0.707 \tag{6.28}$$

then the steady-state amplitude of vibrations of the mass would be

$$X = \frac{F/k}{\sqrt{\left(1 - r^2\right)^2 + (2\xi r)^2}} = 0.0199 \, \text{m} \tag{6.29}$$

The phase φ_x of the vibration is

$$\varphi_x = \tan^{-1} \frac{2\xi r}{1 - r^2} = 0.1 \, \text{rad} \approx 5.73 \, \text{deg} \tag{6.30}$$

Therefore, the steady-state vibrations of m can be expressed by

$$x = 0.0199 \sin (10t - 0.1) \, \text{mm} \tag{6.31}$$

The values of X and φ_x may also be found from Figs. 6.3 and 6.4 approximately.

Figure 6.6 illustrates the effect of the phase lag φ_x on steady-state behavior of x. The excitation (6.27) starts from $f/k = 0$ at $t = 0$ and returns to its state after the period $T = \frac{2\pi}{\omega} = \frac{2\pi}{10} = 0.6283 \, \text{s}$ elapsed. The response of the system in (6.31) is at $x = -1.986 \, 7 \times 10^{-3}$ when $t = 0$, which is $\varphi_x = 0.1$ rad or $t_\varphi = 0.01$ s behind f.

Example 187 ★ Power consumed by steady-state vibrations. In all dynamic systems, the required power indicates the size of motor to run the system. Assuming a mass-spring-damper system is excited by a harmonic force, the required power is calculated in this example.

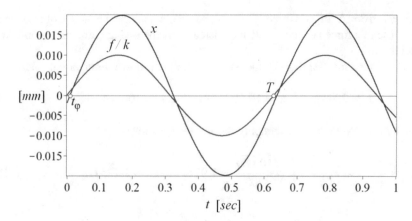

Fig. 6.6 The effect of the phase lag φ_x on displacement x

The power P transmitted to a forced excited system is equal to the applied force f on m times the velocity of m:

$$P(t) = f(t)\, \dot{x}(t) = F \sin \omega t \cdot X \omega \cos\left(\omega t - \varphi_x\right)$$

$$= F X \omega \sin \varphi_x \sin^2 \omega t + F X \omega \cos \varphi_x \cos \omega t \sin \omega t \qquad (6.32)$$

The second term is the active power which corresponds to a zero energy loss per period.

$$\int_0^{2\pi/\omega} F X \omega \cos \varphi_x \,(\cos \omega t \sin \omega t)\, dt = 0 \qquad (6.33)$$

The first term is the active power which is consumed by the damper and leads to a loss of energy per period.

$$E = \int_0^{2\pi/\omega} F X \omega \left(\sin \varphi_x \sin^2 \omega t\right) dt = \pi F X \sin \varphi_x \qquad (6.34)$$

Let us substitute for X and φ_x in the dissipated energy:

$$E = \frac{\pi c \omega F^2}{\left(k - m\omega^2\right)^2 + c^2 \omega^2} \qquad (6.35)$$

and determine the mean consumed power

$$P = \frac{E}{T} = \frac{1}{2} \frac{c \omega^2 F^2}{\left(k - m\omega^2\right)^2 + c^2 \omega^2} \qquad (6.36)$$

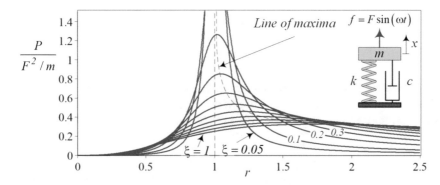

Fig. 6.7 Mean power consumed $P/\left(F^2/m\right)$ as a function of frequency ratio $r = \omega/\omega_n$ at different damping ratio ξ

which can be set in a dimensionless form.

$$\frac{P}{F^2/m} = \frac{\xi r^3}{\left(1 - r^2\right)^2 + (2\xi r)^2} \tag{6.37}$$

Figure 6.7 illustrates $P/\left(F^2/m\right)$ as a function of frequency ratio $r = \omega/\omega_n$ at different damping ratios ξ.

The maximum power occurs at $d\left(P/\left(F^2/m\right)\right)/dr = 0$

$$r^4 + \left(2 - 4\xi^2\right)r^2 - 3 = 0 \tag{6.38}$$

which provides us with the following frequency-damping relationship.

$$r^2_{Max} = 2\xi^2 \pm 2\sqrt{\xi^4 - \xi^2 + 1} - 1 \tag{6.39}$$

Example 188 Frequency response of a third-order system. The method of calculating frequency response of the second-order system is applicable for systems with any other order. Here, we apply the method on a third-order system.

Consider a harmonically excited third-order system.

$$a\dddot{x} + b\ddot{x} + c\dot{x} + dx = F\sin\omega t \tag{6.40}$$

The steady-state response of the system will be a full harmonic function with frequency ω.

$$x = A\sin\omega t + B\cos\omega t = X\sin\left(\omega t - \varphi_x\right) \tag{6.41}$$

Substituting the solution in the equation and balancing the functions $\sin\omega t$ and $\cos\omega t$ provide us with a set of two algebraic equations for A and B.

$$\left(Aa\omega^3 - Bb\omega^2 - Ac\omega + Bd\right) \sin \omega t$$

$$+ \left(-Ba\omega^3 - Ab\omega^2 + Bc\omega + Ad\right) \cos \omega t = F \sin \omega t \qquad (6.42)$$

The harmonic functions $\sin \omega t$ and $\cos \omega t$ are orthogonal; therefore, their coefficients must be balanced on both sides of the equal sign. Balancing the coefficients of $\sin \omega t$ and $\cos \omega t$ provides us with a set of two algebraic equations for A_1 and B_1.

$$\begin{bmatrix} a\omega^3 - c\omega & d - b\omega^2 \\ d - b\omega^2 & c\omega - a\omega^3 \end{bmatrix} \begin{bmatrix} A \\ B \end{bmatrix} = \begin{bmatrix} F \\ 0 \end{bmatrix} \qquad (6.43)$$

Solving for coefficients A and B

$$\begin{bmatrix} A \\ B \end{bmatrix} = \begin{bmatrix} a\omega^3 - c\omega & d - b\omega^2 \\ d - b\omega^2 & c\omega - a\omega^3 \end{bmatrix}^{-1} \begin{bmatrix} F \\ 0 \end{bmatrix}$$

$$= \begin{bmatrix} -F \dfrac{c\omega - a\omega^3}{a^2\omega^6 - 2ac\omega^4 + b^2\omega^4 - 2bd\omega^2 + c^2\omega^2 + d^2} \\ F \dfrac{d - b\omega^2}{a^2\omega^6 - 2ac\omega^4 + b^2\omega^4 - 2bd\omega^2 + c^2\omega^2 + d^2} \end{bmatrix} \qquad (6.44)$$

provides the steady-state solution (6.41). The amplitude X and phase φ_x will be

$$A \sin \omega t + B \cos \omega t = X \sin \left(\omega t - \varphi_x\right)$$

$$= X \cos \varphi_x \sin \omega t - X \sin \varphi_x \cos \omega t \qquad (6.45)$$

$$A = X \cos \varphi_x \qquad B = -X \sin \varphi_x \qquad (6.46)$$

$$X = \sqrt{A^2 + B^2} \qquad \tan \varphi_x = \frac{-B}{A} \qquad (6.47)$$

$$X = \frac{F}{\sqrt{a^2\omega^6 - 2ac\omega^4 + b^2\omega^4 - 2bd\omega^2 + c^2\omega^2 + d^2}} \qquad (6.48)$$

$$\tan \varphi_x = \frac{d - b\omega^2}{c\omega - a\omega^3} \qquad (6.49)$$

Graphical plots will show better how the frequency response of a third-order system may differ from a second-order system. Figure 6.8 illustrates a sample of frequency response of the third-order system (6.48).

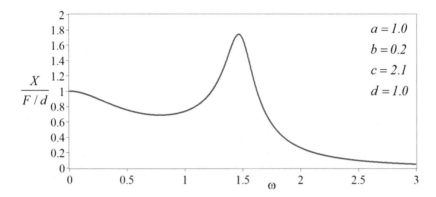

Fig. 6.8 A sample of frequency response of the third-order system (6.48)

Example 189 Line of maxima, quality, bandwidth, and half-power points. To design a vibrating system, the engineer must work based on critical, maximum, and minimum conditions. Here are the equations and plots of maximum amplitude and associated frequency ratio as functions of damping ratio.

When $\xi \neq 0$, the dynamic amplitude X increases by $r \rightarrow 1$. However, the maximum amplitude X_{Max} occurs at the frequency $r_{Max} < 1$ which is a function of ξ. To determine the associated frequency to X_{Max}, we take the derivative of X:

$$\frac{d}{dr}X = -\frac{2r^3 + 4r\xi^2 - 2r}{\sqrt{\left(r^4 + 4r^2\xi^2 - 2r^2 + 1\right)^3}} = 0 \tag{6.50}$$

and solve the equation.

$$r_{Max} = \sqrt{1 - 2\xi^2} \tag{6.51}$$

The value of X_{Max} would then be only a function of ξ:

$$\frac{X_{Max}}{F/k} = \frac{1}{2\xi\sqrt{1 - \xi^2}} \tag{6.52}$$

Figures 6.9 and 6.10 illustrate r_{Max} and X_{Max} as functions of ξ.

The value of X_{Max} as a function of ξ occurs at the frequency $\omega_{Max} < \omega_n$.

$$\omega_{Max} = \omega_n\sqrt{1 - 2\xi^2} \tag{6.53}$$

For light damping of $\xi \leq 0.05$, the curve of $X_{Max}/(F/k)$ is approximately symmetric around $r = 1$, and we have

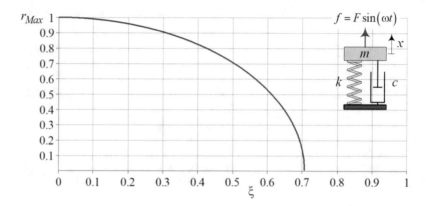

Fig. 6.9 The associated frequency ratio to the maximum amplitude is a function of damping ratio only

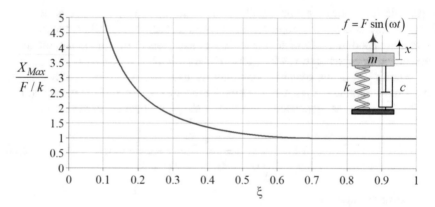

Fig. 6.10 The value of maximum amplitude is a function of damping ratio only

$$Q = \frac{X_{Max}}{F/k} \simeq \frac{1}{2\xi} \tag{6.54}$$

where Q is called the **quality factor**. Whenever vibration is desired in a system, to minimize the amount of required energy, we adjust $r = 1$ and make Q as high as possible. Figure 6.11 illustrates the frequency response of a system for $\xi = 0.04$.

Points P_1 and P_2 at which the maximum amplitude of oscillation reduces to $Q/\sqrt{2}$ are called **half-power points**. The absorbed power by the damper will be half of P_{Max} at these points. The associated excitation frequency difference between the half-power points P_1 and P_2 is called the **bandwidth** of the system $\Delta\omega$, which for small damping is

$$\Delta\omega = \omega_2 - \omega_1 = 2\xi\omega_n = \frac{\omega_n}{Q} \tag{6.55}$$

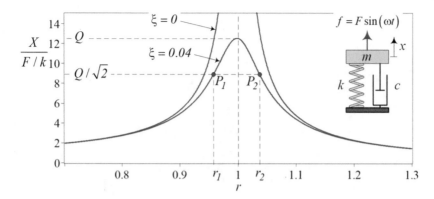

Fig. 6.11 Quality, bandwidth, and half-power points of a system for $\xi = 0.04$

Example 190 ★ Rotating vector representation. A harmonic motion can be represented by rotating vector in a plane with angular velocity equal to its harmonic frequency. A harmonic forced excitation makes a linear system to oscillate with the same frequency; hence, their vector's expression will remain relatively motionless and will rotate together. Here is to show steady-state oscillation of a forced excited system by vectors.

Representing excitation force, spring force, damper force, and displacement of a forced excited system by rotating vectors is a good tool to illustrate the meaning of relative amplitudes and phase φ_x between f and x. Consider the equation of motion of a forced excited m, k, c system.

$$m\ddot{x} + c\dot{x} + kx - F \sin \omega t = 0 \qquad (6.56)$$

This equation states that the sum of the dynamic forces acting on m is balanced with the inertia force $m\ddot{x}$. At the steady-state condition, we have

$$x = X \sin \left(\omega t - \varphi_x\right) \qquad (6.57)$$

and, therefore,

$$\dot{x} = X\omega \cos \left(\omega t - \varphi_x\right) = X\omega \sin \left(\omega t - \varphi_x + \frac{\pi}{2}\right) \qquad (6.58)$$

$$\ddot{x} = -X\omega^2 \sin \left(\omega t - \varphi_x\right) = X\omega^2 \sin \left(\omega t - \varphi_x + \pi\right) \qquad (6.59)$$

Now, we may interpret Equation (6.56) as the sum of the following four forces.

1. $f = F \sin \omega t$ is the active force that is ahead of the displacement $x = X \sin \left(\omega t - \varphi_x\right)$ by phase φ_x.
2. $f_m = m\ddot{x} = mX\omega^2 \sin \left(\omega t - \varphi_x + \pi\right)$ is the inertia force that is ahead of the displacement $x = X \sin \left(\omega t - \varphi_x\right)$ by phase π.

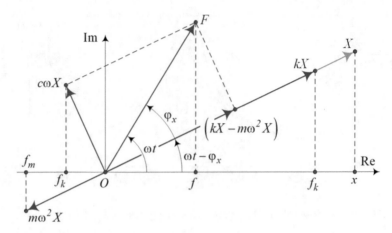

Fig. 6.12 Vectorial representation of steady-state behavior of forced excited m, k, c system

3. $f_c = c\dot{x} = cX\omega \sin\left(\omega t - \varphi_x + \frac{\pi}{2}\right)$ is the damping force that is ahead of the displacement $x = X \sin\left(\omega t - \varphi_x\right)$ by phase $\frac{\pi}{2}$.
4. $f_k = kx = kX \sin\left(\omega t - \varphi_x\right)$ is the elastic force that is in phase with the displacement $x = X \sin\left(\omega t - \varphi_x\right)$.

These forces are expressed by a set of vectors rotating at the same angular velocity ω, as shown in Fig. 6.12. The graphical representation of the forces can be used to calculate X and φ_x geometrically.

$$F^2 = \left(kX - m\omega^2 X\right)^2 + (c\omega X)^2 \tag{6.60}$$

Therefore, we have

$$X = \frac{F}{\sqrt{\left(k - m\omega^2\right)^2 + c^2\omega^2}} \tag{6.61}$$

$$\tan\varphi_x = \frac{c\omega}{k - m\omega^2} \tag{6.62}$$

Example 191 ★ Exponential solution. An alternative method to determine the frequency response of the forced excited (m, c, k) system is to consider the harmonic excitation force f to be the real part of an exponential function and determine the displacement as real part of an exponential function to determine its magnitude.

Consider an (m, c, k) system with the forced excitation f as the real part of

$$f = Fe^{i\omega t} \tag{6.63}$$

where F is the complex amplitude with unit of force.

$$\ddot{x} + 2\xi\omega_n\dot{x} + \omega_n^2 x = \frac{F}{m}e^{i\omega t} \tag{6.64}$$

Considering a steady-state solution as

$$x = Xe^{i\omega t} \tag{6.65}$$

yields

$$\left(\left(\omega_n^2 - \omega^2\right) + i2\xi\omega\omega_n\right)Xe^{i\omega t} = \frac{F}{m}e^{i\omega t} \tag{6.66}$$

which provides us with

$$X = \frac{F/m}{\left(\omega_n^2 - \omega^2\right) + i2\xi\omega\omega_n} \tag{6.67}$$

$$x = \frac{F/m}{\left(\omega_n^2 - \omega^2\right) + i2\xi\omega\omega_n}e^{i\omega t} \tag{6.68}$$

Let us call the ratio of $X/(F/k)$ the complex frequency response, denoting it by $H(\omega)$.

$$H(\omega) = \frac{X}{F/k} = \frac{x}{f/k} = \frac{1}{\left(1 - r^2\right) + i2\xi r} \tag{6.69}$$

Using the complex conjugate $H^*(\omega)$

$$H^*(\omega) = \frac{1}{\left(1 - r^2\right) - i2\xi r} \tag{6.70}$$

we find the amplitude of $H(\omega)$ by

$$|H(\omega)|^2 = H(\omega)H^*(\omega) = \frac{1}{\left(1 - r^2\right)^2 + (2\xi r)^2} \tag{6.71}$$

and therefore, the amplitude of the vibration X is

$$|X| = |H(\omega)|\left|\frac{F}{k}\right| = \frac{F/k}{\sqrt{\left(1 - r^2\right)^2 + (2\xi r)^2}} \tag{6.72}$$

Example 192 Inverse engineering for m, k, c. By nondimensionalization, we reduced the number of parameters of a harmonic oscillator from three (m, c, k) to two (ξ, r) for a forced excited system or to (ξ, ω_n) for a free vibrating system. Assuming we are able to quantify the value of ξ and ω_n, here we show how to evaluate m, k, c.

Consider a vibrating system with known ω_n, ξ. To determine the value of m, k, c, we need to have at least one of m, k, c.

Given m, we find k and c as

$$k = m\omega_n^2 \qquad c = 2\xi \sqrt{km} = \frac{2\xi k}{\omega_n} \tag{6.73}$$

Given k, we find m and c as

$$m = \frac{k}{\omega_n^2} \qquad c = 2\xi \sqrt{km} = 2\xi m \omega_n \tag{6.74}$$

Given c, we find k and m as

$$k = \frac{c\omega_n}{2\xi} \qquad m = \frac{c}{2\xi \omega_n} \tag{6.75}$$

Example 193 Relation between amplitude and damping ratio. This example shows an engineering application of harmonized adjustment of damping ratio and excitation frequency to keep a vibrating system at its maximum amplitude of oscillation.

Consider an m, k, c system with $\xi = 0.2$. The pick value of displacement amplitude of $X_{Max}/(F/k)$ happens at

$$\frac{X_{Max}}{F/k} = \frac{1}{2\xi \sqrt{1 - \xi^2}} = 2.5516 \tag{6.76}$$

$$r_{Max} = \sqrt{1 - 2\xi^2} = 0.9592 \tag{6.77}$$

If we keep r at $r_{Max} = 0.9592$ and increase ξ to $\xi = 0.3$, the amplitude of vibration reduces to

$$\frac{X}{F/k} = \frac{1}{\sqrt{\left(1 - r^2\right)^2 + (2\xi r)^2}} = 1.6971 \tag{6.78}$$

which is less than $X_{Max}/(F/k) = 1.7471$ for $\xi = 0.3$. The value of ξ for which $X_{Max}/(F/k) = 1.6971$ is

$$\xi = 0.30987 \tag{6.79}$$

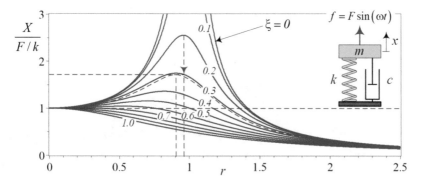

Fig. 6.13 Changing ξ will change the position and value of X_{Max} and r_{Max}

at

$$r_{Max} = 0.89887 \tag{6.80}$$

This analysis is illustrated in Fig. 6.13. Therefore, to keep a vibrating system oscillating at its maximum amplitude, any change in damping must be followed by an adjustment in excitation frequency:

$$r_{Max} = \sqrt{1 - 2\xi^2} \tag{6.81}$$

Example 194 Velocity and acceleration frequency responses. Although displacement amplitude is the most important frequency response that engineers work with, there are applications in which we would need velocity, acceleration, jerk, transmitted force, or other dynamic frequency responses. This example shows how to derive and visualize the velocity and acceleration frequency responses of the forced excited systems. It also shows the method to derive the frequency response of other variables.

Having the steady-state position frequency response, x, we are able to calculate the velocity and acceleration steady-state responses by the derivative.

$$x = A_1 \sin \omega t + B_1 \cos \omega t = X \sin \left(\omega t - \varphi_x\right) \tag{6.82}$$

$$\begin{aligned} \dot{x} &= A_1 \omega \cos \omega t - B_1 \omega \sin \omega t = X\omega \cos \left(\omega t - \varphi_x\right) \\ &= \dot{X} \cos \left(\omega t - \varphi_x\right) \end{aligned} \tag{6.83}$$

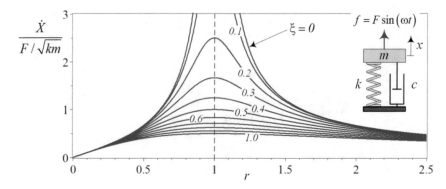

Fig. 6.14 The velocity frequency response for $\dfrac{\dot{X}}{F/\sqrt{km}}$

$$\ddot{x} = -A_1\omega^2 \sin \omega t - B_1\omega^2 \cos \omega t = -X\omega^2 \sin (\omega t - \varphi_x)$$

$$= \ddot{X} \sin (\omega t - \varphi_x) \tag{6.84}$$

The amplitude of velocity and acceleration frequency responses are shown by $\dot{X}$, $\ddot{X}$

$$\dot{X} = \frac{\omega}{\sqrt{(k - m\omega^2)^2 + c^2\omega^2}} F \tag{6.85}$$

$$\ddot{X} = \frac{\omega^2}{\sqrt{(k - m\omega^2)^2 + c^2\omega^2}} F \tag{6.86}$$

which can also be written in dimensionless forms.

$$\frac{\dot{X}}{F/\sqrt{km}} = \frac{r}{\sqrt{(1 - r^2)^2 + (2\xi r)^2}} \tag{6.87}$$

$$\frac{\ddot{X}}{F/m} = \frac{r^2}{\sqrt{(1 - r^2)^2 + (2\xi r)^2}} \tag{6.88}$$

The velocity and acceleration frequency responses (6.87) and (6.88) are plotted in Figs. 6.14 and 6.15.

Example 195 Force transmitted to the base. Transmitted force to the base is an engineering measured date to design the foundations. In this example, we calculate the transmitted force frequency response of a forced excited m, c, k system.

A forced excited system transmits a force, f_T, to the ground which is equal to the sum of forces in spring and damper:

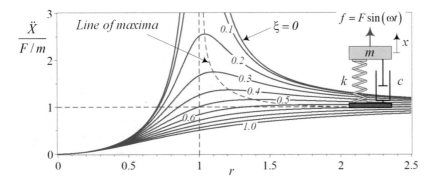

Fig. 6.15 The acceleration frequency response for $\dfrac{\ddot{X}}{F/m}$

$$f_T = f_k + f_c = kx + c\dot{x} \tag{6.89}$$

Substituting x from (6.3), and A_1, B_1 from (6.15), shows that the steady-state response of the transmitted force is

$$
\begin{aligned}
f_T &= k\,(A_1 \sin \omega t + B_1 \cos \omega t) + c\,\omega\,(A_1 \cos \omega t - B_1 \sin \omega t) \\
&= (kA_1 - c\omega B_1)\sin t\omega + (kB_1 + c\omega A_1)\cos t\omega \\
&= F_T \sin\left(\omega t - \varphi_{F_T}\right) \tag{6.90}
\end{aligned}
$$

The amplitude F_T and phase φ_{F_T} of f_T are

$$\frac{F_T}{F} = \frac{\sqrt{k + c^2\omega^2}}{\sqrt{\left(k - m\omega^2\right)^2 + c^2\omega^2}} = \frac{\sqrt{1 + (2\xi r)^2}}{\sqrt{\left(1 - r^2\right)^2 + (2\xi r)^2}} \tag{6.91}$$

$$\tan \varphi_{F_T} = \frac{c\omega}{k - m\omega^2} = \frac{2\xi r}{1 - r^2} \tag{6.92}$$

because

$$F_T = \sqrt{(kA_1 - c\omega B_1)^2 + (kB_1 + c\omega A_1)^2} \tag{6.93}$$

$$\tan \varphi_{F_T} = \frac{-(kB_1 + c\omega A_1)}{kA_1 - c\omega B_1} \tag{6.94}$$

It is also possible to use the equation of motion and substitute x from (6.3) to find the transmitted force frequency response of f_T.

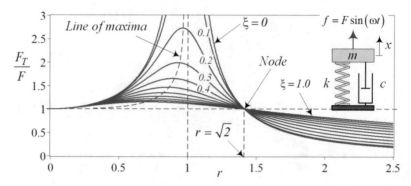

Fig. 6.16 The frequency response for $\dfrac{F_T}{F}$

$$f_T = F \sin \omega t - m\ddot{x}$$
$$= F \sin \omega t + m\omega^2 \left(A_1 \sin \omega t + B_1 \cos \omega t\right)$$
$$= \left(mA_1\omega^2 + F\right) \sin t\omega + m\omega^2 B_1 \cos t\omega$$
$$= F_T \sin \left(\omega t - \varphi_x\right) \tag{6.95}$$

The amplitude F_T and phase φ_{F_T} would be the same as (6.91) and (6.92), because

$$F_T = \sqrt{\left(mA_1\omega^2 + F\right)^2 + \left(m\omega^2 B_1\right)^2} = \frac{\sqrt{k + c^2\omega^2}}{\sqrt{\left(k - m\omega^2\right)^2 + c^2\omega^2}} \tag{6.96}$$

$$\tan \varphi_{F_T} = \frac{-m\omega^2 B_1}{m\omega^2 A_1 + F} = \frac{c\omega}{k - m\omega^2} \tag{6.97}$$

The dimensionless transmitted force frequency response F_T/F is plotted in Fig. 6.16, and because φ_{F_T} is the same as Eq. (6.8), the graph for φ_{F_T} is the same as Fig. 6.4.

Example 196 Turning point: low and high frequencies. Definition of low and high frequency is based on the frequency ration at $r = \sqrt{2}$ associated with the node of frequency responses.

The **node** in Fig. 6.16 is a turning point at which the relation between F_T/F and ξ reverses. Let us call the frequencies before the node **low frequency** and after the node **high frequency**. At low frequency, the transmitted force F_T increases by decreasing ξ, and at high frequency, F_T increases by increasing ξ. The transmitted force F_T is always greater than the applied force F at low frequency, and F_T is always less than the applied force F at high frequency. From a design engineer viewpoint, the suspension of m amplifies the transmitted force F_T with respect to the applied force F at low frequency, then the operating frequency of the system should be at high frequency. The high frequency zone is the safe zone to operate the system.

As a numerical example, consider a mass-spring-damper system with

$$m = 1\,kg \qquad k = 1000\,N\,/\,m \qquad c = 10\,N\,s\,/\,m \tag{6.98}$$

which is under a harmonic forced excitation.

$$f = 10 \sin 2\pi t \tag{6.99}$$

If the suspension of m is supposed to act as an isolator, then the transmitted force F_T to the ground must be less than the applied force $F = 10\,N$. To have $F_T \leq F$, we need

$$\frac{F_T}{F} = \frac{\sqrt{1 + (2\xi r)^2}}{\sqrt{(1 - r^2)^2 + (2\xi r)^2}} \leq 1 \tag{6.100}$$

or

$$\sqrt{1 + (2\xi r)^2} \leq \sqrt{(1 - r^2)^2 + (2\xi r)^2} \tag{6.101}$$

After simplification, we get the following condition:

$$r^2 \geq 2 \tag{6.102}$$

which is consistent with Fig. 6.16.

Example 197 ★ Frequency response is attractive. Every constant excitation frequency ω produces a steady-state oscillation at amplitude $X(\omega)$. That amplitude is attractive, which means any disturbance δX to move X to $X + \delta X$ will disappear after a while and the system will get back to the same oscillating amplitude $X(\omega)$.

Consider a mass-spring-damper under a harmonic force excitation at a given constant frequency ω_0.

$$m\ddot{x} + c\dot{x} + kx = F \sin \omega_0 t \tag{6.103}$$

The steady-state response of vibration would be

$$x_o = X_0 \sin(\omega_0 t - \varphi_x) \tag{6.104}$$

$$X_0 = \frac{F/k}{\sqrt{\left(1 - r^2\right)^2 + (2\xi r)^2}} \qquad r = \frac{\omega_0}{\omega_n} \tag{6.105}$$

Now assume that because of a sudden disturbance the amplitude changes to $X_0 + \delta X$ at the same excitation frequency ω_0 and the response of the system changes to $x = x_0 + \delta x$

$$x = x_0 + \delta x = X \sin\left(\omega t - \varphi_x\right) + \delta x \tag{6.106}$$

Substituting x in Eq. (6.103)

$$m\ddot{x}_0 + c\dot{x}_0 + kx_0 + m\,\delta\ddot{x} + c\,\delta\dot{x} + k\,\delta x = F \sin \omega_0 t \tag{6.107}$$

and considering that x_0 satisfies the equation, we find the equation to determine δx.

$$m\,\delta\ddot{x} + c\,\delta\dot{x} + k\,\delta x = 0 \tag{6.108}$$

The solution of this equation with initial conditions

$$\delta x\,(0) = \delta X \qquad \delta\dot{x}\,(0) = 0 \tag{6.109}$$

is

$$\delta x = \frac{1}{2}\frac{-c^2 - c\sqrt{c^2 - 4km} + 4km}{-c^2 + 4km}\delta X \exp\left(\frac{c - \sqrt{c^2 - 4km}}{-2m}t\right)$$

$$+ \frac{1}{2}\frac{-c^2 + c\sqrt{c^2 - 4km} + 4km}{-c^2 + 4km}\delta X \exp\left(\frac{c + \sqrt{c^2 - 4km}}{-2m}t\right) \tag{6.110}$$

As long as $c^2 > 4km$, we have

$$\lim_{t\to\infty} \delta x = 0 \tag{6.111}$$

and hence,

$$\lim_{t\to\infty} x = x_0 = X \sin\left(\omega t - \varphi_x\right) \tag{6.112}$$

Therefore, any disturbance to a steady-state vibrating system will disappear after a while, and the system will be back to its original steady-state vibration. This is the stability characteristic of frequency response and indicates that the frequency response curves are attractive for small amplitude disturbances as is shown in Fig. 6.17 for $\xi = 0.4$.

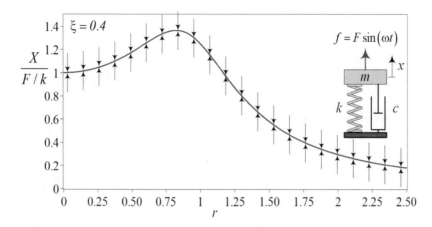

Fig. 6.17 The stability and attraction characteristic of frequency response

For instance, let us examine a system with

$$m = 1\,\text{kg} \qquad k = 100\,\text{N}/\text{m} \qquad c = 5\,\text{N}\,\text{s}/\text{m} \qquad (6.113)$$

that is under a force f

$$f = 10 \sin(2\pi t)\,\text{N} \qquad (6.114)$$

The steady-state amplitude of this system will be

$$X = \frac{F}{\sqrt{(k - m\omega^2)^2 + c^2\omega^2}} = 0.1466\,\text{m} \qquad (6.115)$$

Now suppose that, because of a disturbance, x jumps $\pm 30\%$ to an amplitude of $1.3X$ or $0.7X$. Figure 6.18 illustrates the time history of the motions and shows how the system settles down to X again.

Example 198 ★ Why do we consider $X/(F/k)$ and $r = \omega/\omega_n$? It is a dimensional analysis of harmonic oscillators. A great advantage of making equations nondimensional is to make it work with minimum number of parameters and variables. Second advantage is to make the solution to be applied on all similar systems regardless of their size.

Consider a forced excited undamped single DOF system.

$$m\,\ddot{x} + kx = F \sin \omega t \qquad (6.116)$$

The involved input, output, and system parameters are m, k, F, X, and $t \equiv 1/\omega$. Based on the *Buckingham π-theorem*, any complete physical relation of a system can be expressed in terms of a set of independent dimensionless products of its π

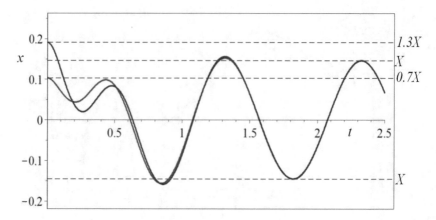

Fig. 6.18 The attraction process of the amplitude to steady-state value after a sudden jump to $1.3X$ or $0.7X$

terms, which is the products of relevant physical parameters of the system. Because each π term is dimensionless, we must have a balance for the dimensions involved:

$$X^a k^b m^c F^d \omega^e = L^0 M^0 T^0 \tag{6.117}$$

Let us substitute the parameters X, k, m, F, ω by their dimensions of mass M, length L, and time T:

$$L^a \left(\frac{M}{T^2}\right)^b M^c \left(\frac{ML}{T^2}\right)^d \left(\frac{1}{T}\right)^e = L^0 M^0 T^0 \tag{6.118}$$

so we have

$$L^{a+d} M^{b+c+d} T^{-2b-2d-e} = L^0 M^0 T^0 \tag{6.119}$$

and hence,

$$a + d = 0 \tag{6.120}$$

$$b + c + d = 0 \tag{6.121}$$

$$-2b - 2d - e = 0 \tag{6.122}$$

These three equations may be solved for any three out of five exponents a, b, c, d, e. The solution by assuming a and e known is

$$d = -a \qquad b = a - \frac{1}{2}e \qquad c = \frac{1}{2}e \tag{6.123}$$

Substituting them back, we have

$$X^a k^{a-e/2} m^{e/2} F^{-a} \omega^e = L^0 M^0 T^0 \tag{6.124}$$

which indicates

$$\left(\frac{X}{F/k}\right)^a \left(\frac{\omega}{\sqrt{k/m}}\right)^e = L^0 M^0 T^0 \tag{6.125}$$

The dimensionless π terms are

$$\pi_1 = \frac{X}{F/k} \qquad \pi_2 = \frac{\omega}{\sqrt{k/m}} \tag{6.126}$$

The π-theorem states that there is a function f of the π terms such that

$$f\left(\frac{X}{F/k}, \frac{\omega}{\sqrt{k/m}}\right) = 0 \tag{6.127}$$

So, $X/(F/k)$ is a function of $\omega/\sqrt{k/m}$. If the solution of the Eq. (6.116) is expressed based on π_1 and π_2, then it will be the solution of any vibrating system that could be modeled by Eq. (6.116).

Example 199 ★ Response to a periodic force. Employing Fourier series, any periodic force can be decomposed to a series of harmonic forces. Superposition principle of linear systems allows us to derive the solution for the harmonic forces one by one and add them together to find the solution to the periodic force. Here is how to do this method.

Assume that a periodic force $f(t) = f(t+T)$, $T = 2\pi/\omega$ is applied on a mass-spring-damper system.

$$m\ddot{x} + c\dot{x} + kx = f(t) \tag{6.128}$$

To determine the steady-state response of the system, we expand the force in its Fourier series

$$f(t) = \frac{1}{2}a_0 + \sum_{j=1}^{\infty} a_j \cos(j\omega t) + \sum_{j=1}^{\infty} b_j \sin(j\omega t) \tag{6.129}$$

$$a_0 = \frac{2}{T} \int_0^T f(t)\, dt \tag{6.130}$$

$$a_j = \frac{2}{T} \int_0^T f(t) \cos(j\omega t)\, dt \tag{6.131}$$

$$b_j = \frac{2}{T} \int_0^T f(t) \sin(j\omega t) \, dt \tag{6.132}$$

and substitute the forcing term.

$$m\ddot{x} + c\dot{x} + kx = \frac{a_0}{2} + \sum_{j=1}^{\infty} a_j \cos(j\omega t) + \sum_{j=1}^{\infty} b_j \sin(j\omega t) \tag{6.133}$$

Using the principle of superposition, we have three equations to solve.

$$m\ddot{x}_1 + c\dot{x}_1 + kx_1 = \frac{a_0}{2} \tag{6.134}$$

$$m\ddot{x}_2 + c\dot{x}_2 + kx_2 = a_j \cos(j\omega t) \tag{6.135}$$

$$m\ddot{x}_3 + c\dot{x}_3 + kx_3 = b_j \sin(j\omega t) \tag{6.136}$$

The steady-state solution of (6.134) is

$$X_1 = \frac{a_0}{2k} \tag{6.137}$$

and the steady-state solutions of (6.135) and (6.136) are

$$X_2 = \frac{a_j/k}{\sqrt{\left(1 - j^2 r^2\right)^2 + (2\xi jr)^2}} \tag{6.138}$$

$$X_3 = \frac{b_j/k}{\sqrt{\left(1 - j^2 r^2\right)^2 + (2\xi jr)^2}} \tag{6.139}$$

The phase of these two motions is

$$\varphi_j = \tan^{-1} \frac{2\xi jr}{1 - j^2 r^2} \tag{6.140}$$

Therefore, the complete steady-state solutions of (6.128) is

$$x = \frac{a_0}{2k} + \sum_{j=1}^{\infty} \frac{a_j/k}{\sqrt{\left(1 - j^2 r^2\right)^2 + (2\xi jr)^2}} \cos\left(j\omega t - \varphi_j\right)$$

$$+ \sum_{j=1}^{\infty} \frac{b_j/k}{\sqrt{\left(1 - j^2 r^2\right)^2 + (2\xi jr)^2}} \sin\left(j\omega t - \varphi_j\right) \tag{6.141}$$

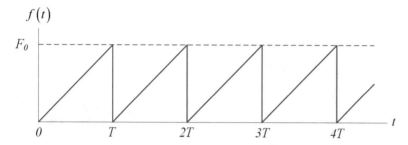

Fig. 6.19 A periodic saw-tooth force excitation

In mechanical vibrations, the average of $f(t)$ is usually zero, so $a_0 = 0$ and the steady-state solution (6.141) simplifies to

$$x = \sum_{j=1}^{\infty} \frac{a_j/k}{\sqrt{\left(1 - j^2 r^2\right)^2 + (2\xi j r)^2}} \cos\left(j\omega t - \varphi_j\right)$$

$$+ \sum_{j=1}^{\infty} \frac{b_j/k}{\sqrt{\left(1 - j^2 r^2\right)^2 + (2\xi j r)^2}} \sin\left(j\omega t - \varphi_j\right) \qquad (6.142)$$

with a steady-state amplitude of

$$X = \sum_{j=1}^{\infty} \frac{\sqrt{a_j^2 + b_j^2}/k}{\sqrt{\left(1 - j^2 r^2\right)^2 + (2\xi j r)^2}} \qquad (6.143)$$

As an example, consider a mass-spring-damper system under a periodic forcing function $f(t)$ such as the sawtooth excitation of Fig. 6.19.

$$m\ddot{x} + c\dot{x} + kx = f(t) \qquad (6.144)$$

Let us expand the force function by a Fourier series.

$$f(t) = \frac{1}{2}a_0 + \sum_{j=1}^{\infty} a_j \cos(j\omega t) + \sum_{j=1}^{\infty} b_j \sin(j\omega t)$$

$$= F_0 \left(\frac{1}{2} - \frac{1}{\pi} \sum_{j=1}^{\infty} \frac{1}{j} \sin(j\omega t)\right) \qquad (6.145)$$

$$a_0 = \frac{2}{T} \int_0^T f(t)\, dt = \frac{2F_0}{T^2} \int_0^T t\, dt = F_0 \tag{6.146}$$

$$a_j = \frac{2}{T} \int_0^T f(t) \cos(j\omega t)\, dt = \frac{2F_0}{T^2} \int_0^T t \cos(j\omega t)\, dt = 0 \tag{6.147}$$

$$b_j = \frac{2}{T} \int_0^T f(t) \sin(j\omega t)\, dt = \frac{2F_0}{T^2} \int_0^T t \sin(j\omega t)\, dt = -\frac{F_0}{j\pi} \tag{6.148}$$

The equation of motion would be

$$m\ddot{x} + c\dot{x} + kx = F_0 \left(\frac{1}{2} - \frac{1}{\pi} \sum_{j=1}^{\infty} \frac{1}{j} \sin(j\omega t) \right) \tag{6.149}$$

with the solution of

$$x = \frac{F_0}{2k} - \frac{F_0}{\pi} \sum_{j=1}^{\infty} \frac{1}{j} \frac{\sin(j\omega t - \varphi_j)}{\sqrt{(k - j^2 m\omega^2) + (m^2 \omega^2)}} \tag{6.150}$$

Example 200 ★ Complex form of response to a periodic force. Fourier series can be expressed by exponential functions as well. Therefore, the response of oscillating system with periodic excitation may also be treated by exponential expansion of periodic forcing functions.

Consider a force $f(t) = f(t + T)$, of period $T = 2\pi/\omega$, that is applied on a mass-spring-damper system.

$$m\ddot{x} + c\dot{x} + kx = f(t) \tag{6.151}$$

We express the force as the real part of its complex Fourier series

$$f(t) = \sum_{j=-\infty}^{\infty} C_j e^{ij\omega t} \qquad i^2 = -1 \tag{6.152}$$

$$C_j = \frac{1}{T} \int_{-T/2}^{T/2} f(t)\, e^{-ij\omega t}\, dt \qquad \omega = \frac{2\pi}{T} \tag{6.153}$$

where the C_j are the complex amplitudes in units of force and carrying information about the phase angle. If the average of the force $f(t)$ is zero, then it can also be denoted by

$$f(t) = \sum_{j=1}^{\infty} F_j e^{ij\omega t} \tag{6.154}$$

Now the equation of motion becomes

$$\ddot{x} + 2\xi\omega_n\dot{x} + \omega_n^2 x = \frac{1}{m}\sum_{j=1}^{\infty}F_j e^{ij\omega t} \tag{6.155}$$

Considering a steady-state solution as

$$x = \text{Re}\, X_j e^{ij\omega t} \tag{6.156}$$

yields

$$x = \text{Re}\sum_{j=1}^{\infty}\frac{F_j/k}{\left(1 - j^2 r^2\right) + i2\xi\, jr}e^{ij\omega t} = \text{Re}\sum_{j=1}^{\infty}H_j\frac{F_j}{k}e^{ij\omega t} \tag{6.157}$$

$$H_j = \frac{1}{\left(1 - j^2 r^2\right) + i2\xi\, jr} \tag{6.158}$$

Example 201 ★ Response to two harmonic forces. Multiple harmonic force function is an example of how Fourier series expansion of forcing function and the response of the vibrating system will be in expanded form. Here is an example of two harmonic force functions, the method of harmonic balance and determining the steady-state solution.

Let us determine the steady-state response of a system to two harmonic excitations with two different frequencies.

$$m\ddot{x} + c\dot{x} + kx = F_1\cos\omega_1 t + (F_2\cos\omega_2 t + F_3\sin\omega_2 t) \tag{6.159}$$

The first force is a cosine function, $F_1\cos\omega_1 t$, and the second force is a full harmonic function with both sine and cosine terms, $F_2\cos\omega_2 t + F_3\sin\omega_2 t$. The equation of motion can be rewritten as

$$\ddot{x} + 2\xi\omega_n\dot{x} + \omega_n^2 x = \frac{F_1}{m}\cos\omega_1 t + \left(\frac{F_2}{m}\cos\omega_2 t + \frac{F_3}{m}\sin\omega_2 t\right) \tag{6.160}$$

The steady-state solution must have two full harmonic functions of the frequencies.

$$x = A_1\sin\omega_1 t + B_1\cos\omega_1 t + A_2\sin\omega_2 t + B_2\cos\omega_2 t \tag{6.161}$$

Substituting the solution in the equation of motion yields

$$- \omega_1^2\left(A_1\sin t\omega_1 + B_1\cos t\omega_1\right) - \omega_2^2\left(A_2\sin t\omega_2 + B_2\cos t\omega_2\right)$$
$$+2\xi\omega_n\left(\omega_1 A_1\cos t\omega_1 + \omega_2 A_2\cos t\omega_2 - \omega_1 B_1\sin t\omega_1 - \omega_2 B_2\sin t\omega_2\right)$$
$$+\omega_n^2\left(A_1\sin\omega_1 t + B_1\cos\omega_1 t + A_2\sin\omega_2 t + B_2\cos\omega_2 t\right)$$
$$= \frac{F_1}{m}\cos\omega_1 t + \frac{F_2}{m}\cos\omega_2 t + \frac{F_3}{m}\sin\omega_2 t \tag{6.162}$$

Balancing the harmonic functions $\cos t\omega_1$, $\cos t\omega_2$, $\sin t\omega_1$, and $\sin t\omega_2$ on both sides provides us with four algebraic equations to determine A_1, B_1, A_2, and B_2.

$$A_1\omega_n^2 - 2\xi B_1\omega_1\omega_n - A_1\omega_1^2 = 0 \tag{6.163}$$

$$\omega_n^2 B_1 - \omega_1^2 B_1 - \frac{1}{m}F_1 + 2\xi\omega_1\omega_n A_1 = 0 \tag{6.164}$$

$$\frac{1}{m}F_3 - \omega_2^2 A_2 + \omega_n^2 A_2 - 2\xi\omega_2\omega_n B_2 = 0 \tag{6.165}$$

$$\frac{1}{m}F_2 - \omega_2^2 B_2 + \omega_n^2 B_2 + 2\xi\omega_2\omega_n A_2 = 0 \tag{6.166}$$

$$\begin{bmatrix} \omega_n^2 - \omega_1^2 & -2\xi\omega_1\omega_n & 0 & 0 \\ 2\xi\omega_1\omega_n & \omega_n^2 - \omega_1^2 & 0 & 0 \\ 0 & 0 & -\omega_2^2 + \omega_n^2 & -2\xi\omega_2\omega_n \\ 0 & 0 & 2\xi\omega_2\omega_n & -\omega_2^2 + \omega_n^2 \end{bmatrix} \begin{bmatrix} A_1 \\ B_1 \\ A_2 \\ B_2 \end{bmatrix} = \begin{bmatrix} 0 \\ \dfrac{1}{m}F_1 \\ -\dfrac{1}{m}F_3 \\ -\dfrac{1}{m}F_2 \end{bmatrix} \tag{6.167}$$

Therefore, the coefficients A_1, B_1, A_2, and B_2 are

$$A_1 = \frac{2}{m}\xi\omega_1\omega_n \frac{F_1}{4\xi^2\omega_1^2\omega_n^2 + \omega_1^4 - 2\omega_1^2\omega_n^2 + \omega_n^4} \tag{6.168}$$

$$B_1 = -\frac{1}{m}F_1 \frac{\omega_1^2 - \omega_n^2}{4\xi^2\omega_1^2\omega_n^2 + \omega_1^4 - 2\omega_1^2\omega_n^2 + \omega_n^4} \tag{6.169}$$

$$A_2 = \frac{1}{m}F_3 \frac{\omega_2^2 - \omega_n^2}{4\xi^2\omega_2^2\omega_n^2 + \omega_2^4 - 2\omega_2^2\omega_n^2 + \omega_n^4}$$
$$\quad - \frac{2}{m}\xi\omega_2\omega_n \frac{F_2}{4\xi^2\omega_2^2\omega_n^2 + \omega_2^4 - 2\omega_2^2\omega_n^2 + \omega_n^4} \tag{6.170}$$

$$B_2 = \frac{1}{m}F_2 \frac{\omega_2^2 - \omega_n^2}{4\xi^2\omega_2^2\omega_n^2 + \omega_2^4 - 2\omega_2^2\omega_n^2 + \omega_n^4}$$
$$\quad + \frac{2}{m}\xi\omega_2\omega_n \frac{F_3}{4\xi^2\omega_2^2\omega_n^2 + \omega_2^4 - 2\omega_2^2\omega_n^2 + \omega_n^4} \tag{6.171}$$

The steady-state solution of Eq. (6.159) is determined.

The steady-state solution is the sum of two harmonic functions $A_1 \sin \omega_1 t + B_1 \cos \omega_1 t$ and $A_2 \sin \omega_2 t + B_2 \cos \omega_2 t$ with frequencies ω_1 and ω_2. The actual value of the excitation frequencies ω_1 and ω_2 relative to the natural frequency ω_n

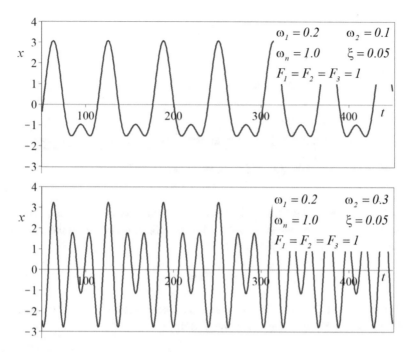

Fig. 6.20 Two samples of steady-state responses of two harmonic excitations of a linear oscillator

determines the pattern and the amplitude of the steady-state response. There is no frequency response plot for multiple frequency excitation because of the complexity of the patterns of solutions. When there are only two excitation frequencies, ω_1 and ω_2, then for any fixed value of ω_1, we may vary ω_2 and expect an steady-state amplitude for each value of ω_2. However, even for a fixed ω_1, and depending on the frequency ratios ω_2/ω_n and ω_2/ω_1, we may have a beating or two waves with amplitudes X_1 and X_2 interacting.

$$X_1 = \sqrt{A_1^2 + B_1^2} \qquad X_2 = \sqrt{A_2^2 + B_2^2} \tag{6.172}$$

Figure 6.20 illustrates two samples of steady-state response of the system when both excitation frequencies are much lower than the natural frequency.

6.2 Base Excitation

Figure 6.21 illustrates a one DOF base excited vibrating system made of a mass m supported by a spring k and a damper c. The base excited system is a good model for a vehicle suspension system or any equipment that is mounted on a vibrating

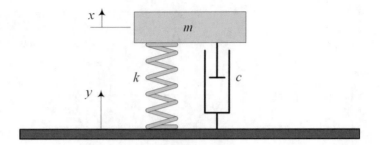

Fig. 6.21 A harmonically base excited single DOF system

base. The absolute motion of m with respect to its equilibrium position is measured by the coordinate x. A sinusoidal excitation motion, y, is applied to the base of the suspension that makes the system vibrate.

$$y = Y \sin \omega t \tag{6.173}$$

The equation of motion for the system can be expressed by either one of the following equations for the absolute displacement x.

$$m \ddot{x} + c \dot{x} + kx = cY\omega \cos \omega t + kY \sin \omega t \tag{6.174}$$

$$\ddot{x} + 2\xi \omega_n \dot{x} + \omega_n^2 x = 2\xi \omega_n \omega Y \cos \omega t + \omega_n^2 Y \sin \omega t \tag{6.175}$$

$$\omega_n = \sqrt{\frac{k}{m}} \qquad r = \frac{\omega}{\omega_n} \qquad \xi = \frac{c}{2\sqrt{km}} \tag{6.176}$$

The system may also be expressed by either one of the following equations for the relative displacement z.

$$m \ddot{z} + c \dot{z} + kz = m\omega^2 Y \sin \omega t \tag{6.177}$$

$$\ddot{z} + 2\xi \omega_n \dot{z} + \omega_n^2 z = \omega^2 Y \sin \omega t \tag{6.178}$$

$$z = x - y \tag{6.179}$$

The equations of motion generate the absolute and relative steady-state responses as

$$x = A_2 \sin \omega t + B_2 \cos \omega t \tag{6.180}$$

$$= X \sin \left(\omega t - \varphi_x\right) \tag{6.181}$$

$$z = A_3 \sin \omega t + B_3 \cos \omega t \tag{6.182}$$

$$= Z \sin \left(\omega t - \varphi_z\right) \tag{6.183}$$

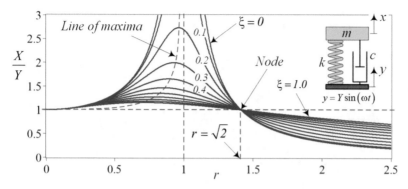

Fig. 6.22 The position frequency response for $\dfrac{X}{Y}$

The steady-state response of x has an amplitude X and phase lag φ_x, while the steady-state response of z has an amplitude Z and phase lag φ_z:

$$\frac{X}{Y} = \frac{\sqrt{1 + (2\xi r)^2}}{\sqrt{\left(1 - r^2\right)^2 + (2\xi r)^2}} \tag{6.184}$$

$$\frac{Z}{Y} = \frac{r^2}{\sqrt{\left(1 - r^2\right)^2 + (2\xi r)^2}} \tag{6.185}$$

$$\varphi_x = \tan^{-1} \frac{2\xi r^3}{1 - r^2 + (2\xi r)^2} \tag{6.186}$$

$$\varphi_z = \tan^{-1} \frac{2\xi r}{1 - r^2} \tag{6.187}$$

The phase φ_x indicates the *angular lag* of the response x with respect to the excitation y, and the phase φ_z is the *angular lag* of the relative motion response z with respect to the excitation y. The frequency responses for X, Z, and φ_x as a function of r and ξ are plotted in Figs. 6.22, 6.23, and 6.24.

Proof The free body diagram of the system, as shown in Fig. 6.25, generates the equation of motion

$$m\ddot{x} = -c\,(\dot{x} - \dot{y}) - k\,(x - y) \tag{6.188}$$

which, after substituting the excitation function (6.173), leads to the equation of motion (6.174). Equation (6.174) can be transformed to (6.175) by dividing over m and using the definitions (6.176) for natural frequency ω_n, damping ratio ξ, and frequency ratio r.

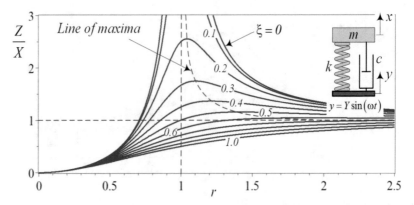

Fig. 6.23 The frequency response for $\dfrac{Z}{Y}$

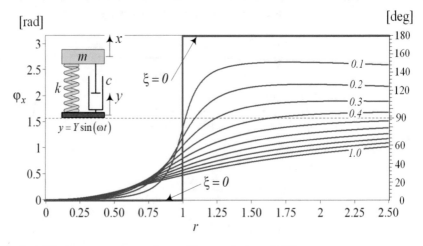

Fig. 6.24 The frequency response for φ_x of base excited one DOF system

A practical response for a base excited system is the relative displacement:

$$z = x - y \tag{6.189}$$

For every mechanical device that is mounted on a suspension, we usually need to determine and control the maximum or minimum distance between the base and the device. Therefore, the relative displacement z is the physical variable to be measured. Taking derivatives from (6.189)

$$\ddot{z} = \ddot{x} - \ddot{y} \tag{6.190}$$

and substituting in (6.188)

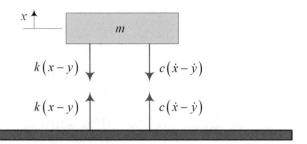

Fig. 6.25 A harmonically based excited single DOF system

$$m\,(\ddot{z} + \ddot{y}) = -c\,\dot{z} - kz \tag{6.191}$$

generate Eqs. (6.177) and (6.178).

The steady-state solution of Eq. (6.174) can be (6.180) or (6.181). To calculate the amplitude and phase of the response, we substitute the solution (6.180) in the equation of motion (6.174).

$$-m\omega^2\,(A_2 \sin \omega t + B_2 \cos \omega t) + c\omega\,(A_2 \cos \omega t - B_2 \sin \omega t)$$

$$+k\,(A_2 \sin \omega t + B_2 \cos \omega t) = cY\omega \cos \omega t + kY \sin \omega t \tag{6.192}$$

The coefficients of the functions $\sin \omega t$ and $\cos \omega t$ must balance on both sides of the equation.

$$kA_2 - mA_2\omega^2 - cB_2\omega = Yk \tag{6.193}$$

$$kB_2 - m\omega^2 B_2 + c\omega A_2 = Yc\omega \tag{6.194}$$

Therefore, we get two algebraic equations to calculate A_2 and B_2.

$$\begin{bmatrix} k - m\omega^2 & -c\omega \\ c\omega & k - m\omega^2 \end{bmatrix} \begin{bmatrix} A_2 \\ B_2 \end{bmatrix} = \begin{bmatrix} Yk \\ Yc\omega \end{bmatrix} \tag{6.195}$$

Solving for the coefficients A_2 and B_2 provides us with the steady-state solution (6.180).

$$\begin{bmatrix} A_2 \\ B_2 \end{bmatrix} = \begin{bmatrix} k - m\omega^2 & -c\omega \\ c\omega & k - m\omega^2 \end{bmatrix}^{-1} \begin{bmatrix} Yk \\ Yc\omega \end{bmatrix}$$

$$= \begin{bmatrix} \dfrac{k\,(k - m\omega^2) + c^2\omega^2}{(k - m\omega^2)^2 + c^2\omega^2}\,Y \\[4mm] \dfrac{c\omega\,(k - m\omega^2) - ck\omega}{(k - m\omega^2)^2 + c^2\omega^2}\,Y \end{bmatrix} \tag{6.196}$$

The amplitude X and phase φ_x can be found by

$$X = \sqrt{A_2^2 + B_2^2} \tag{6.197}$$

$$\tan \varphi_x = \frac{-B_2}{A_2} \tag{6.198}$$

which, after substituting A_2 and B_2 from (6.196), will be

$$X = \frac{\sqrt{k^2 + c^2\omega^2}}{\sqrt{(k - m\omega^2)^2 + c^2\omega^2}} Y \tag{6.199}$$

$$\tan \varphi_x = \frac{-cm\omega^3}{k(k - m\omega^2) + c^2\omega^2} \tag{6.200}$$

More practical expressions for X and φ_x are Eqs. (6.184) and (6.186), which can be found by employing r and ξ.

To find the relative displacement frequency response (6.185), we substitute Equation (6.182) in (6.177).

$$-m\omega^2 (A_3 \sin \omega t + B_3 \cos \omega t) + c\omega (A_3 \cos \omega t - B_3 \sin \omega t)$$

$$+k (A_3 \sin \omega t + B_3 \cos \omega t) = m\omega^2 Y \sin \omega t \tag{6.201}$$

Balancing the coefficients of $\sin \omega t$ and $\cos \omega t$

$$kA_3 - mA_3\omega^2 - cB_3\omega = m\omega^2 Y \tag{6.202}$$

$$kB_3 - m\omega^2 B_3 + c\omega A_3 = 0 \tag{6.203}$$

provides us with two algebraic equations to find A_3 and B_3:

$$\begin{bmatrix} k - m\omega^2 & -c\omega \\ c\omega & k - m\omega^2 \end{bmatrix} \begin{bmatrix} A_3 \\ B_3 \end{bmatrix} = \begin{bmatrix} m\omega^2 Y \\ 0 \end{bmatrix} \tag{6.204}$$

Solving for the coefficients A_3 and B_3, we have

$$\begin{bmatrix} A_3 \\ B_3 \end{bmatrix} = \begin{bmatrix} k - m\omega^2 & -c\omega \\ c\omega & k - m\omega^2 \end{bmatrix}^{-1} \begin{bmatrix} m\omega^2 Y \\ 0 \end{bmatrix}$$

$$= \begin{bmatrix} \dfrac{m\omega^2 (k - m\omega^2)}{(k - m\omega^2)^2 + c^2\omega^2} Y \\ -\dfrac{mc\omega^3}{(k - m\omega^2)^2 + c^2\omega^2} Y \end{bmatrix} \tag{6.205}$$

which provides us with the steady-state solution (6.182). The amplitude Z and phase φ_z can be found by

$$Z = \sqrt{A_3^2 + B_3^2} \tag{6.206}$$

$$\tan \varphi_z = \frac{-B_3}{A_3} \tag{6.207}$$

which, after substituting A_3 and B_3 from (6.205), leads to the following solutions.

$$Z = \frac{m\omega^2}{\sqrt{\left(k - m\omega^2\right)^2 + c^2\omega^2}} Y \tag{6.208}$$

$$\tan \varphi_z = \frac{c\omega}{k - m\omega^2} \tag{6.209}$$

More practical expressions for Z and φ_z are Eqs. (6.185) and (6.187) employing r and ξ.   ∎

Example 202 A base excited system. It is a numerical example to work with equations to understand and compare the order of magnitudes.

Consider a mass-spring-damper system with

$$m = 2\,\text{kg} \qquad k = 10,000\,\text{N}/\text{m} \qquad c = 100\,\text{N}\,\text{s}/\text{m} \tag{6.210}$$

If a harmonic base excitation y is applied to the system,

$$y = 0.02 \sin 140t \tag{6.211}$$

then the absolute and relative steady-state amplitude of vibrations of the mass, X and Z, would be

$$X = \frac{Y\sqrt{1 + (2\xi r)^2}}{\sqrt{\left(1 - r^2\right)^2 + (2\xi r)^2}} = 0.0106\,\text{m} \tag{6.212}$$

$$Z = \frac{Yr^2}{\sqrt{\left(1 - r^2\right)^2 + (2\xi r)^2}} = 0.0242\,\text{m} \tag{6.213}$$

because

$$\omega_n = \sqrt{\frac{k}{m}} = 70.71\,\text{rad}/\text{s} \approx 11.254\,\text{Hz} \tag{6.214}$$

$$\xi = \frac{c}{2\sqrt{km}} = 0.353 \tag{6.215}$$

$$r = \frac{\omega}{\omega_n} = 1.98 \tag{6.216}$$

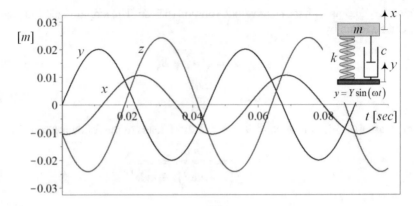

Fig. 6.26 The excitation y and the steady-state responses x and z at $r = 1.98$

The phases φ_x and φ_z for x and z are

$$\varphi_x = \tan^{-1} \frac{2\xi r^3}{1 - r^2 + (2\xi r)^2} = 1.744 \, \text{rad} \approx 49.96 \, \text{deg} \qquad (6.217)$$

$$\varphi_z = \tan^{-1} \frac{2\xi r}{1 - r^2} = 2.694 \, \text{rad} \approx 77.19 \, \text{deg} \qquad (6.218)$$

Therefore, the steady-state vibrations of the mass m can be expressed by

$$x = 0.0106 \sin (140t - 1.744) \qquad (6.219)$$

$$z = 0.0242 \sin (140t - 2.694) \qquad (6.220)$$

The excitation y and the steady-state responses x and z are plotted in Fig. 6.26 to illustrate their relative phase and amplitude at $r = 1.98$.

Example 203 Comparison between frequency responses. Several equations and graphs of different dynamic characteristics of one *DOF* harmonically excited systems in dimensionless form are similar. Here are a few examples.

A comparison shows that Eq. (6.185) is equal to Eq. (6.87), and therefore the relative frequency response $\frac{Z}{Y}$ for a base excited system is the same as the acceleration frequency response $\frac{\ddot{X}}{F/m}$ for a forced excited system. Also the graph for φ_z would be the same as Fig. 6.4.

Comparing Eqs. (6.184) and (6.91) indicates that the amplitude frequency response of a base excited system $\frac{X}{Y}$ is the same as the transmitted force frequency response of a harmonically forced excited system $\frac{F_T}{F}$. However, the phases of these two responses are different.

Example 204 Absolute velocity and acceleration of a base excited system. This example shows how to derive and visualize the velocity and acceleration frequency responses of the base excited systems.

Having the position frequency response of a base excited system,

$$x = A_2 \sin \omega t + B_2 \cos \omega t = X \sin \left(\omega t - \varphi_x \right) \tag{6.221}$$

we are able to calculate the velocity and acceleration frequency responses:

$$\dot{x} = A_2 \omega \cos \omega t - B_2 \omega \sin \omega t$$
$$= X \omega \cos \left(\omega t - \varphi_x \right) = \dot{X} \cos \left(\omega t - \varphi_x \right) \tag{6.222}$$

$$\ddot{x} = -A_2 \omega^2 \sin \omega t - B_2 \omega^2 \cos \omega t$$
$$= -X \omega^2 \sin \left(\omega t - \varphi_x \right) = \ddot{X} \sin \left(\omega t - \varphi_x \right) \tag{6.223}$$

The amplitude of velocity and acceleration frequency responses, $\dot{X}$, $\ddot{X}$, are

$$\dot{X} = \frac{\omega \sqrt{k^2 + c^2 \omega^2}}{\sqrt{\left(k - m \omega^2 \right)^2 + c^2 \omega^2}} Y \tag{6.224}$$

$$\ddot{X} = \frac{\omega^2 \sqrt{k^2 + c^2 \omega^2}}{\sqrt{\left(k - m \omega^2 \right)^2 + c^2 \omega^2}} Y \tag{6.225}$$

which can also be written in dimensionless forms.

$$\frac{\dot{X}}{\omega_n Y} = \frac{r \sqrt{1 + (2 \xi r)^2}}{\sqrt{\left(1 - r^2 \right)^2 + (2 \xi r)^2}} \tag{6.226}$$

$$\frac{\ddot{X}}{\omega_n^2 Y} = \frac{r^2 \sqrt{1 + (2 \xi r)^2}}{\sqrt{\left(1 - r^2 \right)^2 + (2 \xi r)^2}} \tag{6.227}$$

The velocity and acceleration frequency responses (6.226) and (6.227) are plotted in Figs. 6.27 and 6.28.

There is a **switching point** or **node** in both figures, at which the behavior of $\dot{X}$ and $\ddot{X}$ switches. Before the node, $\dot{X}$ and $\ddot{X}$ increase by increasing ξ, while they decrease after the node. To find the node, we can find the intersection between frequency response curves for $\xi = 0$ and $\xi = \infty$ because the amplitude at the node is independent of the damping. Let us apply this method to the acceleration frequency response:

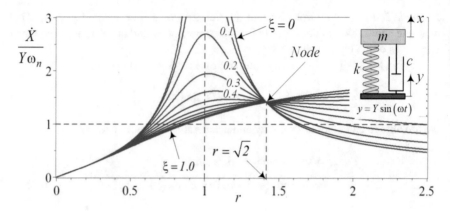

Fig. 6.27 The velocity frequency response

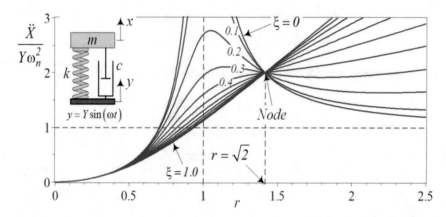

Fig. 6.28 The acceleration frequency response

$$\lim_{\xi \to 0} \frac{\ddot{X}}{\omega_n^2 Y} = \pm \frac{r^2}{\left(1 - r^2\right)} \tag{6.228}$$

$$\lim_{\xi \to \infty} \frac{\ddot{X}}{\omega_n^2 Y} = \pm r^2 \tag{6.229}$$

Therefore, the frequency ratio r at the intersection of these two limits is the solution of

$$r^2 \left(r^2 - 2\right) = 0 \tag{6.230}$$

The nodal frequency response is then equal to

$$r = \sqrt{2} \tag{6.231}$$

The value of acceleration frequency response at the node is only a function of ξ.

$$\lim_{r \to \sqrt{2}} \frac{\ddot{X}}{\omega_n^2 Y} = \frac{\sqrt{2\xi^2 + 1}}{\sqrt{8\xi^2 + 1}} \qquad (6.232)$$

Applying the same method for the velocity frequency response results in the same nodal frequency ratio, $r = \sqrt{2}$, with a different amplitude:

$$\lim_{r \to \sqrt{2}} \frac{\dot{X}}{\omega_n Y} = \sqrt{2}\frac{\sqrt{2\xi^2 + 1}}{\sqrt{8\xi^2 + 1}} \qquad (6.233)$$

Example 205 Relative velocity and acceleration of a base excited system. As a practice of determining frequency response of relative variables, here is to show how relative velocity and relative acceleration frequency responses of base excited one *DOF* system will be calculated.

We may use the relative displacement frequency response of a base excited system

$$z = A_3 \sin \omega t + B_3 \cos \omega t = Z \sin\left(\omega t - \varphi_z\right) \qquad (6.234)$$

and calculate the relative velocity and acceleration frequency responses:

$$\dot{z} = A_3 \omega \cos \omega t - B_3 \omega \sin \omega t$$

$$= Z\omega \cos\left(\omega t - \varphi_z\right) = \dot{Z} \cos\left(\omega t - \varphi_z\right) \qquad (6.235)$$

$$\ddot{z} = -A_3 \omega^2 \sin \omega t - B_3 \omega^2 \cos \omega t$$

$$= -Z\omega^2 \sin\left(\omega t - \varphi_z\right) = \ddot{Z} \sin\left(\omega t - \varphi_z\right) \qquad (6.236)$$

The amplitude of velocity and acceleration frequency responses, $\dot{Z}$, $\ddot{Z}$,

$$\dot{Z} = \frac{m\omega^3}{\sqrt{\left(k - m\omega^2\right)^2 + c^2\omega^2}}Y \qquad (6.237)$$

$$\ddot{Z} = \frac{m\omega^4}{\sqrt{\left(k - m\omega^2\right)^2 + c^2\omega^2}}Y \qquad (6.238)$$

can also be written in dimensionless forms.

$$\frac{\dot{Z}}{\omega_n Y} = \frac{r^3}{\sqrt{\left(1 - r^2\right)^2 + (2\xi r)^2}} \tag{6.239}$$

$$\frac{\ddot{Z}}{\omega_n^2 Y} = \frac{r^4}{\sqrt{\left(1 - r^2\right)^2 + (2\xi r)^2}} \tag{6.240}$$

Example 206 Force transmitted to the base of a base excited system. Transmitted force to the base is always an important design factor. Here is the frequency response of the transmitted force to the base of a base excited system.

The force transmitted f_T to the ground by a base excited system is equal to the sum of forces in the spring and damper:

$$f_T = f_k + f_c = k\,(x - y) + c\,(\dot{x} - \dot{y}) \tag{6.241}$$

which, based on the equation of motion (6.188), is also equal to

$$f_T = -m\ddot{x} \tag{6.242}$$

Substituting $\ddot{x}$ from (6.223) and (6.227) shows that the frequency response of the transmitted force is

$$\frac{F_T}{kY} = \frac{\omega^2 \sqrt{k^2 + c^2 \omega^2}}{\sqrt{\left(k - m\omega^2\right)^2 + c^2 \omega^2}} \tag{6.243}$$

$$= \frac{r^2 \sqrt{1 + (2\xi r)^2}}{\sqrt{\left(1 - r^2\right)^2 + (2\xi r)^2}} \tag{6.244}$$

The frequency response of $\frac{F_T}{kY}$ is the same as is shown in Fig. 6.28.

Example 207 ★ Line of maxima in X/Y. The curves of X/Y for different ξ have maximums. The value of X/Y at maximum and the associated frequency ratio are important to design engineers. Here is to derive them analytically and graphically.

The peak value of the absolute displacement frequency response X/Y occurs at different r depending on ξ. To find this relationship, we take the derivative of X/Y, given in Eq. (6.184), with respect to r and solve the equation.

$$\frac{d}{dr}\frac{X}{Y} = \frac{2r\left(1 - r^2 - 2r^4 \xi^2\right)}{\sqrt{1 + 4r^2 \xi^2}\left(\left(1 - r^2\right)^2 + (2\xi r)^2\right)^{\frac{3}{2}}} = 0 \tag{6.245}$$

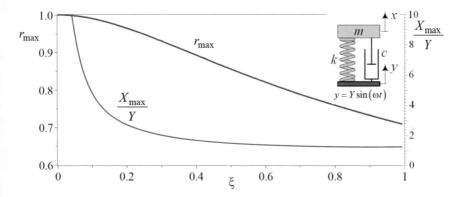

Fig. 6.29 The peak amplitude $X_{\max}$ and the associated frequency $r_{\max}$, as a function of ξ

Let us denote the peak amplitude by $X_{\max}$ and the associated frequency by $r_{\max}$. The value of $r_{\max}^2$ is only a function of ξ.

$$r_{\max}^2 = \frac{1}{4\xi^2}\left(-1 \pm \sqrt{1 + 8\xi^2}\right) \tag{6.246}$$

Substituting the positive sign of (6.246) in (6.184) determines the peak amplitude $X_{\max}$:

$$\frac{X_{\max}}{Y} = \frac{2\sqrt{2}\xi^2 \sqrt[4]{8\xi^2 + 1}}{\sqrt{8\xi^2 + \left(8\xi^4 - 4\xi^2 - 1\right)\sqrt{8\xi^2 + 1} + 1}} \tag{6.247}$$

Figure 6.29 shows $X_{\max}$ and $r_{\max}$ as functions of ξ.

Example 208 ★ Line of maxima in Z/Y. The value of Z/Y at maximum and the associated frequency ratio are important to design engineers. Here is to derive them analytically and graphically.

The peak value of the relative displacement frequency response Z/Y occurs at $r > 1$, depending on ξ. To find this relationship, we take the derivative of Z/Y, given in Eq. (6.185), with respect to r and solve the equation.

$$\frac{d}{dr}\frac{Z}{Y} = \frac{2r\left(1 - r^2 - 2r^4\xi^2\right)}{\left(\left(1 - r^2\right)^2 + (2\xi r)^2\right)^{\frac{3}{2}}} = 0 \tag{6.248}$$

Let us indicate the peak amplitude by $Z_{\max}$ and the associated frequency by $r_{\max}$. The value of $r_{\max}^2$ is

$$r_{\max} = \frac{1}{\sqrt{1 - 2\xi^2}} \tag{6.249}$$

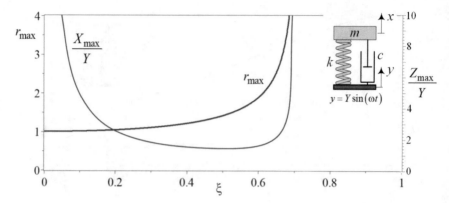

Fig. 6.30 The peak amplitude Z_{max} and the associated frequency r_{max}, as a function of ξ

which has a real value for

$$\xi < \frac{\sqrt{2}}{2} \tag{6.250}$$

Substituting (6.249) in (6.185) determines the peak amplitude Z_{max}.

$$\frac{Z_{max}}{Y} = \frac{1}{2\xi\sqrt{1 - 2\xi^2}} \tag{6.251}$$

Figure 6.30 shows Z_{max} and r_{max} as a function of ξ.

As an example, the maximum amplitude of the system with

$$m = 2\,\text{kg} \qquad k = 10,000\,\text{N}/\text{m} \qquad c = 100\,\text{N}\,\text{s}/\text{m}$$
$$\omega = 140\,\text{rad}/\text{s} \qquad r = 1.97 \qquad \xi = 0.353 \qquad Y = 0.02\,\text{m} \tag{6.252}$$

is

$$Z_{max} = \frac{Y}{2\xi\sqrt{1 - 2\xi^2}} = 1.633\,\text{m} \tag{6.253}$$

which occurs at

$$r_{max} = \frac{1}{\sqrt[4]{1 - 2\xi^2}} = 1.155 \tag{6.254}$$

Example 209 ★ Response to a periodic base excitation. A periodic base excitation can be expanded into Fourier series to be broken to its harmonics. Solving the system for individual harmonics and superposing them provide us with the solution to the periodic excitation.

Assume a periodic base excitation $y(t) = y(t + T)$, $T = 2\pi/\omega$, is applied on the base of a mass-spring-damper system.

$$m\ddot{z} + c\dot{z} + kz = -m\ddot{y} \qquad z = x - y \tag{6.255}$$

To determine the steady-state response of the system, we expand $y(t)$ to its Fourier series and take the double derivative and substitute in Eq. (6.178).

$$y(t) = \frac{1}{2}a_0 + \sum_{j=1}^{\infty} a_j \cos(j\omega t) + \sum_{j=1}^{\infty} b_j \sin(j\omega t) \tag{6.256}$$

$$\ddot{y}(t) = -\sum_{j=1}^{\infty} a_j j^2 \omega^2 \cos(j\omega t) - \sum_{j=1}^{\infty} b_j j^2 \omega^2 \sin(j\omega t) \tag{6.257}$$

$$a_0 = \frac{2}{T} \int_0^T y(t)\, dt \qquad a_j = \frac{2}{T} \int_0^T y(t) \cos(j\omega t)\, dt \tag{6.258}$$

$$b_j = \frac{2}{T} \int_0^T y(t) \sin(j\omega t)\, dt \tag{6.259}$$

$$\ddot{z} + 2\xi\omega_n \dot{z} + \omega_n^2 z = \sum_{j=1}^{\infty} a_j j^2 \omega^2 \cos(j\omega t) \sum_{j=1}^{\infty} b_j j^2 \omega^2 \sin(j\omega t) \tag{6.260}$$

Using the principle of superposition, we have two equations of motion:

$$\ddot{z}_1 + 2\xi\omega_n \dot{z}_1 + \omega_n^2 z_1 = a_j j^2 \omega^2 \cos(j\omega t) \tag{6.261}$$

$$\ddot{z}_2 + 2\xi\omega_n \dot{z}_2 + \omega_n^2 z_2 = b_j j^2 \omega^2 \sin(j\omega t) \tag{6.262}$$

The steady-state solutions of (6.261) and (6.262) are

$$Z_1 = \frac{a_j j^2 r^2}{\sqrt{\left(1 - j^2 r^2\right)^2 + (2\xi jr)^2}} \tag{6.263}$$

$$Z_2 = \frac{b_j j^2 r^2}{\sqrt{\left(1 - j^2 r^2\right)^2 + (2\xi jr)^2}} \tag{6.264}$$

The phase of these two motions is

$$\varphi_j = \tan^{-1} \frac{2\xi jr}{1 - j^2 r^2} \tag{6.265}$$

Therefore, the complete steady-state solution of (6.255) is

$$
x = \sum_{j=1}^{\infty} \frac{a_j j^2 r^2}{\sqrt{\left(1 - j^2 r^2\right)^2 + (2\xi jr)^2}} \cos\left(j\omega t - \varphi_j\right)
$$

$$
+ \sum_{j=1}^{\infty} \frac{b_j j^2 r^2}{\sqrt{\left(1 - j^2 r^2\right)^2 + (2\xi jr)^2}} \sin\left(j\omega t - \varphi_j\right) \tag{6.266}
$$

whose steady-state amplitude is

$$
X = \sum_{j=1}^{\infty} \frac{j^2 r^2 \sqrt{a_j^2 + b_j^2}}{\sqrt{\left(1 - j^2 r^2\right)^2 + (2\xi jr)^2}} \tag{6.267}
$$

6.3   ★ Vibration Isolation

Vibration isolator refers to a *suspension* to isolate a primary mass from a source of vibration. In this section, we examine a linear, one DOF, base excited vibration isolator system as the simplest model for a suspension to separate a mass m from the vibrating base. Based on a root mean square (RMS) optimization method, we develop a design chart to determine the optimal damper and spring for the best vibration isolation.

6.3.1   *Mathematical Model*

Figure 6.31 illustrates a one DOF base excited linear vibrating system. It can represent a model of vertical vibrations of a vehicle or the suspension of any equipment mounted on a vibrating base. As a model of a vehicle, a one-fourth (1/4) of the mass of the body is modeled as a solid mass m denoted as *sprung mass*. A spring of stiffness k and a shock absorber with viscous damping c support the sprung mass and represent the main suspension of the vehicle. The suspension parameters k and c are the equivalent stiffness and damping for one wheel, measured at the center of the wheel.

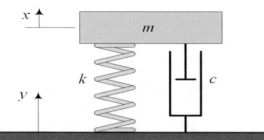

Fig. 6.31 A base excited linear suspension

The equations of motion of the system for the absolute displacement x and the relative displacement z are

$$m\ddot{x} + c\dot{x} + kx = c\dot{y} + ky \tag{6.268}$$

$$m\ddot{z} + c\dot{z} + kz = -m\ddot{y} \tag{6.269}$$

$$z = x - y \tag{6.270}$$

The variable x is the absolute displacement of the body, and y is the absolute displacement of the ground. The equations of motion (6.268) and (6.269), which are dependent on the three parameters (m, c, k), can be transformed to the following two parameter-dependent equations:

$$\ddot{x} + 2\xi\omega_n \dot{x} + \omega_n^2 x = 2\xi\omega_n \dot{y} + \omega_n^2 y \tag{6.271}$$
$$\ddot{z} + 2\xi\omega_n \dot{z} + \omega_n^2 z = -\ddot{y} \tag{6.272}$$

by introducing the *natural frequency* ω_n and the *damping ratio* ξ:

$$\xi = \frac{c}{2\sqrt{km}} \tag{6.273}$$

$$\omega_n = \sqrt{\frac{k}{m}} = 2\pi f_n \tag{6.274}$$

Proof The kinetic energy, potential energy, and dissipation function of the system are

$$K = \frac{1}{2}m\dot{x}^2 \tag{6.275}$$

$$P = \frac{1}{2}k(x - y)^2 \tag{6.276}$$

$$D = \frac{1}{2}c\,(\dot{x} - \dot{y})^2 \tag{6.277}$$

Employing the Lagrange equation

$$\frac{d}{dt}\left(\frac{\partial K}{\partial \dot{x}}\right) - \frac{\partial K}{\partial x} + \frac{\partial D}{\partial \dot{x}} + \frac{\partial P}{\partial x} = 0 \tag{6.278}$$

we find the equation of motion

$$\frac{d}{dt}(m\,\dot{x}) + c\,(\dot{x} - \dot{y}) + k\,(x - y) = 0 \tag{6.279}$$

which can be simplified to Eq. (6.268). Introducing a relative position variable, $z = x - y$, we have

$$\dot{z} = \dot{x} - \dot{y} \tag{6.280}$$

$$\ddot{z} = \ddot{x} - \ddot{y} \tag{6.281}$$

so we can rewrite Equation (6.279) as

$$m\frac{d}{dt}(\ddot{z} + \ddot{y}) + c\dot{z} + kz = 0 \tag{6.282}$$

which is equivalent to (6.269).

Dividing Equations (6.268) and (6.269) by m and using (6.273) and (6.274) generate the equivalent Equations (6.271) and (6.272), respectively. ∎

Example 210 Different model for front and rear parts of a vehicle. Vibration isolator system for model of a vehicle will be different for front and rear parts of the vehicle because the mass center is not in the middle and, hence, the loads on front and rear wheels are unequal.

Consider a vehicle with the following information:

$$car\ mass = 1500\,\text{kg} \qquad wheel\ mass = 50\,\text{kg}$$

$$F_{z_1} = 3941.78\,\text{N} \qquad\qquad F_{z_2} = 3415.6\,\text{N} \tag{6.283}$$

where F_{z_1} and F_{z_2} are the front and rear tire loads, respectively. The mass m of the vibrating model for the front of the car will be

$$m = \frac{F_{z_1}}{F_{z_1} + F_{z_2}} \times (1500 - 4 \times 50) = 696.49\,\text{kg} \tag{6.284}$$

and the rear of the car will be

$$m = \frac{F_{z_2}}{F_{z_1} + F_{z_2}} \times (1500 - 4 \times 50) = 603.51\,\text{kg} \tag{6.285}$$

Example 211 Function of an isolator and rubber mount. The function of a vibration isolator and the assumption in analysis are reviewed here.

The function of an isolator is to reduce the magnitude of motion transmitted from a vibrating foundation to the equipment, or to reduce the magnitude of force transmitted from the equipment to its foundation, in both time and frequency domains.

In the simplest approach to suspension analysis, the parameters m, k, and c are considered constant and independent of the excitation frequency or behavior of the foundation. This assumption is equivalent to considering an infinitely stiff and massive foundation. For rubber mounts, the damping coefficient usually decreases, and the stiffness coefficient increases with excitation frequency. Moreover, neither the engine nor the body can be assumed to be an infinitely stiff rigid body at high frequencies.

6.3.2 Frequency Response

The most important frequency responses of base excited system are absolute displacement G_0, relative displacement S_2, and absolute acceleration G_2 and force transmitted to m:

$$G_0 = \left| \frac{X}{Y} \right| = \frac{\sqrt{1 + (2\xi r)^2}}{\sqrt{(1 - r^2)^2 + (2\xi r)^2}} \tag{6.286}$$

$$S_2 = \left| \frac{Z}{Y} \right| = \frac{r^2}{\sqrt{(1 - r^2)^2 + (2\xi r)^2}} \tag{6.287}$$

$$G_2 = \left| \frac{\ddot{X}}{Y \omega_n^2} \right| = \frac{r^2 \sqrt{1 + (2\xi r)^2}}{\sqrt{(1 - r^2)^2 + (2\xi r)^2}} \tag{6.288}$$

where

$$r = \frac{\omega}{\omega_n} \qquad \xi = \frac{c}{2\sqrt{km}} \qquad \omega_n = \sqrt{\frac{k}{m}} \tag{6.289}$$

Proof Applying a harmonic excitation,

$$y = Y \sin \omega t \tag{6.290}$$

the equation of motion (6.272) reduces to

$$\ddot{z} + 2\xi \omega_n \dot{z} + \omega_n^2 z = \omega^2 Y \sin \omega t \tag{6.291}$$

Considering a harmonic solution

$$z = A_3 \sin \omega t + B_3 \cos \omega t \tag{6.292}$$

and substituting it in the equation of motion

$$-A_3 \omega^2 \sin \omega t - B_3 \omega^2 \cos \omega t + 2\xi \omega_n \left(A_3 \omega \cos \omega t - B_3 \omega \sin \omega t \right)$$
$$+\omega_n^2 \left(A_3 \sin \omega t + B_3 \cos \omega t \right) = \omega^2 Y \sin \omega t \tag{6.293}$$

we find a set of equations to calculate A_3 and B_3.

$$\begin{bmatrix} \omega_n^2 - \omega^2 & -2\xi \omega \omega_n \\ 2\xi \omega \omega_n & \omega_n^2 - \omega^2 \end{bmatrix} \begin{bmatrix} A_3 \\ B_3 \end{bmatrix} = \begin{bmatrix} Y\omega^2 \\ 0 \end{bmatrix} \tag{6.294}$$

The first row of the set (6.294) is a balance of the coefficients of $\sin \omega t$ in Eq. (6.293), and the second row is a balance of the coefficients of $\cos \omega t$. Therefore, the coefficients A_3 and B_3 can be found as

$$\begin{bmatrix} A_3 \\ B_3 \end{bmatrix} = \begin{bmatrix} \omega_n^2 - \omega^2 & -2\xi \omega \omega_n \\ 2\xi \omega \omega_n & \omega_n^2 - \omega^2 \end{bmatrix}^{-1} \begin{bmatrix} Y\omega^2 \\ 0 \end{bmatrix}$$

$$= \begin{bmatrix} -\dfrac{\omega^2 - \omega_n^2}{4\xi^2 \omega^2 \omega_n^2 + \omega^4 - 2\omega^2 \omega_n^2 + \omega_n^4} Y\omega^2 \\ -\dfrac{2\xi \omega \omega_n}{4\xi^2 \omega^2 \omega_n^2 + \omega^4 - 2\omega^2 \omega_n^2 + \omega_n^4} Y\omega^2 \end{bmatrix} \tag{6.295}$$

These equations may be transformed to a simpler form, by using r and ξ.

$$\begin{bmatrix} A_3 \\ B_3 \end{bmatrix} = \begin{bmatrix} \dfrac{1 - r^2}{\left(1 - r^2\right)^2 + (2\xi r)^2} r^2 Y \\ \dfrac{-2\xi r}{\left(1 - r^2\right)^2 + (2\xi r)^2} r^2 Y \end{bmatrix} \tag{6.296}$$

The relative displacement amplitude Z provides us with $S_2 = |Z/Y|$ in Eq. (6.287).

$$Z = \sqrt{A_3^2 + B_3^2} = \frac{r^2}{\sqrt{\left(1 - r^2\right)^2 + (2\xi r)^2}} Y \tag{6.297}$$

To find the absolute frequency response G_0, we assume

$$x = A_2 \sin \omega t + B_2 \cos \omega t = X \sin \left(\omega t - \varphi_x\right) \tag{6.298}$$

and write

$$z = x - y \tag{6.299}$$

$$A_3 \sin \omega t + B_3 \cos \omega t = A_2 \sin \omega t + B_2 \cos \omega t - Y \sin \omega t \tag{6.300}$$

which shows

$$A_2 = A_3 + Y \tag{6.301}$$

$$B_2 = B_3 \tag{6.302}$$

The absolute displacement amplitude X then provides us with $G_0 = |X/Y|$ of Eq. (6.286).

$$X = \sqrt{A_2^2 + B_2^2} = \sqrt{(A_3 + Y)^2 + B_3^2}$$

$$= \frac{\sqrt{1 + (2\xi r)^2}}{\sqrt{\left(1 - r^2\right)^2 + (2\xi r)^2}} Y \tag{6.303}$$

The absolute acceleration frequency response, $\ddot{X}$, can be found by twice differentiating the displacement frequency response (6.298).

$$\ddot{x} = -X\omega^2 \sin \left(\omega t - \varphi_x\right) = -\ddot{X} \sin \left(\omega t - \varphi_x\right) \tag{6.304}$$

If we denote the amplitude of the absolute acceleration by $\ddot{X}$, then we can define $\ddot{X}$ by

$$\left| \frac{\ddot{X}}{Y\omega_n^2} \right| = \frac{r^2\sqrt{1 + (2\xi r)^2}}{\sqrt{\left(1 - r^2\right)^2 + (2\xi r)^2}} \tag{6.305}$$

which provides us with $G_2 = |\ddot{X}/ (\omega_n^2 Y)|$ as in Eq. (6.288).   ∎

Example 212 Principal method for absolute motion X. The frequency response of the absolute displacement is calculated here.

To find the absolute frequency response G_0, we substitute the base excitation function

$$y = Y \sin \omega t \tag{6.306}$$

and a harmonic solution for x

$$x = A_2 \sin \omega t + B_2 \cos \omega t \tag{6.307}$$

in Eq. (6.271)

$$\ddot{x} + 2\xi \omega_n \dot{x} + \omega_n^2 x = 2\xi \omega_n \dot{y} + \omega_n^2 y \tag{6.308}$$

and solve for $X = \sqrt{A_2^2 + B_2^2}$.

$$-\omega^2 A_2 \sin \omega t - \omega^2 B_2 \cos \omega t + 2\xi \omega_n \omega \left(A_2 \cos \omega t - B_2 \sin \omega t \right)$$
$$+\omega_n^2 \left(A_2 \sin \omega t + B_2 \cos \omega t \right) = 2\xi \omega_n \omega Y \cos \omega t + \omega_n^2 Y \sin \omega t \tag{6.309}$$

The set of equations for A_2 and B_2 from the coefficients of sin and cos results in the following solution:

$$\begin{bmatrix} \omega_n^2 - \omega^2 & -2\xi \omega \omega_n \\ 2\xi \omega \omega_n & \omega_n^2 - \omega^2 \end{bmatrix} \begin{bmatrix} A_2 \\ B_2 \end{bmatrix} = \begin{bmatrix} Y \omega_n^2 \\ 2Y \xi \omega \omega_n \end{bmatrix} \tag{6.310}$$

$$\begin{bmatrix} A_2 \\ B_2 \end{bmatrix} = \begin{bmatrix} \omega_n^2 - \omega^2 & -2\xi \omega \omega_n \\ 2\xi \omega \omega_n & \omega_n^2 - \omega^2 \end{bmatrix}^{-1} \begin{bmatrix} Y \omega_n^2 \\ 2Y \xi \omega \omega_n \end{bmatrix}$$

$$= \begin{bmatrix} \dfrac{-\left(\omega^2 - \omega_n^2 \right) \omega_n^2 + 4\xi^2 \omega^2 \omega_n^2}{4\xi^2 \omega^2 \omega_n^2 + \omega^4 - 2\omega^2 \omega_n^2 + \omega_n^4} Y \\ \dfrac{-2\xi \omega \omega_n^3}{4\xi^2 \omega^2 \omega_n^2 + \omega^4 - 2\omega^2 \omega_n^2 + \omega_n^4} Y \end{bmatrix}$$

$$= \begin{bmatrix} \dfrac{(2\xi r)^2 - \left(1 - r^2 \right)}{\left(1 - r^2 \right)^2 + (2\xi r)^2} Y \\ \dfrac{-2\xi r^3}{\left(1 - r^2 \right)^2 + (2\xi r)^2} Y \end{bmatrix} \tag{6.311}$$

Therefore, the amplitude of the absolute displacement X would be the same as (6.303).

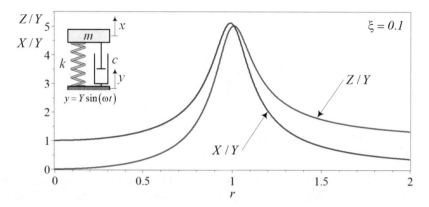

Fig. 6.32 Absolute and relative displacement frequency responses for a vehicle

Example 213 $G_0 \neq S_2 + 1$. The relative displacement frequency response is not equal to the difference of frequency responses of the relative displacements. Here is to show this fact.

We can find the absolute frequency response $G_0 = |X/Y|$, from the result for S_2. The frequency response S_2 is

$$S_2 = \frac{Z}{Y} \tag{6.312}$$

However,

$$S_2 \neq \frac{X}{Y} - 1$$
$$\neq G_0 - 1 \tag{6.313}$$

because the amplitude of the relative displacement Z is not equal to the amplitude of the absolute displacement X minus the amplitude of the excitation Y.

$$Z \neq X - Y \tag{6.314}$$

Example 214 Absolute and relative frequency response curve comparison. Absolute and relative frequency responses are two main characteristics of vibration isolation systems. Here is a graphical illustration.

Consider an isolator system with damping ratio $\xi = 0.08$. The absolute and relative displacements frequency responses of the system are shown in Fig. 6.32. The relative displacement starts at *zero* and ends at *one*, while the absolute displacement starts at *one* and ends up at *zero*.

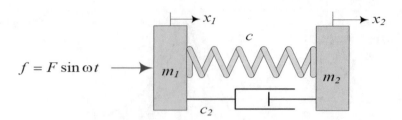

Fig. 6.33 A base forced excited system

Example 215 A base forced excited system. Vibration isolator is any suspension system separating a mass from a source of vibrations. Here is another example of isolation system between two masses where there is a harmonic force of one of them.

Figure 6.33 illustrates a vibrating system with a massive body m_1 under a forced excitation along with a secondary system (m_2, k, c). Employing the energies and dissipation function of the system,

$$K = \frac{1}{2}m_1\dot{x}_1^2 + \frac{1}{2}m_2\dot{x}_2^2 \tag{6.315}$$

$$P = \frac{1}{2}k\,(x_1 - x_2)^2 \tag{6.316}$$

$$D = \frac{1}{2}c\,(\dot{x}_1 - \dot{x}_2)^2 \tag{6.317}$$

the equations of motion of the system will be derived.

$$m_1\ddot{x}_1 + c\,(\dot{x}_1 - \dot{x}_2) + k\,(x_1 - x_2) = f \tag{6.318}$$

$$m_2\ddot{x}_2 - c\,(\dot{x}_1 - \dot{x}_2) - k\,(x_1 - x_2) = 0 \tag{6.319}$$

$$f = F\sin\omega t \tag{6.320}$$

Subtracting the equations in relative displacement z yields

$$\frac{m_1 m_2}{m_1 + m_2}\ddot{z} + c\dot{z} + kz = \frac{m_2}{m_1 + m_2}f \tag{6.321}$$

$$z = x_1 - x_2 \tag{6.322}$$

Defining the damping ratio ξ and natural frequency ω_n,

$$\xi = \frac{c}{2\sqrt{k\dfrac{m_1 m_2}{m_1 + m_2}}} \tag{6.323}$$

$$\omega_n = \sqrt{\frac{k(m_1 + m_2)}{m_1 m_2}} \qquad (6.324)$$

the equation of relative motion transforms to

$$\ddot{z} + \xi \dot{z} + \omega_n z = \frac{1}{m_1} f = \frac{F}{m_1} \sin \omega t \qquad (6.325)$$

The equation of relative motion (6.325) is the same as the absolute displacement of the one DOF force excitation (6.2). Therefore, the amplitude of the relative motion is S_0 as in (6.7)

$$S_0 = \frac{Z}{F/k} = \frac{1}{\sqrt{\left(1 - r^2\right)^2 + (2\xi r)^2}} \qquad (6.326)$$

and the force transmitted to m_2 or the force of suspension on m_1 is G_0 as in (6.91).

$$F_T = kz + c\dot{z} \qquad (6.327)$$

$$G_0 = \frac{F_T}{F} = \frac{\sqrt{1 + (2\xi r)^2}}{\sqrt{\left(1 - r^2\right)^2 + (2\xi r)^2}} \qquad (6.328)$$

To determine the frequency response of m_1 and m_2, we use Eqs. (6.319) and (6.322).

$$\begin{bmatrix} m_1 & 0 \\ 0 & m_2 \end{bmatrix} \begin{bmatrix} \ddot{x}_1 \\ \ddot{x}_2 \end{bmatrix} + \begin{bmatrix} c & -c \\ -c & c \end{bmatrix} \begin{bmatrix} \dot{x}_1 \\ \dot{x}_2 \end{bmatrix}$$
$$+ \begin{bmatrix} k & -k \\ -k & k \end{bmatrix} \begin{bmatrix} x_1 \\ x_2 \end{bmatrix} = \begin{bmatrix} f \\ 0 \end{bmatrix} \qquad (6.329)$$

Substituting harmonic solutions for x_1 and x_2,

$$\begin{bmatrix} x_1 \\ x_2 \end{bmatrix} = \begin{bmatrix} A_1 \sin \omega t + B_2 \cos \omega t \\ A_2 \sin \omega t + B_2 \cos \omega t \end{bmatrix} \qquad (6.330)$$

we find the coefficients A_1, A_2, B_1, and B_2 by using (7.632).

$$\begin{bmatrix} \mathbf{A} \\ \mathbf{B} \end{bmatrix} = \begin{bmatrix} A_1 \\ A_2 \\ B_1 \\ B_2 \end{bmatrix} = \begin{bmatrix} [k] - \omega^2 [m] & -\omega [c] \\ \omega [c] & [k] - \omega^2 [m] \end{bmatrix}^{-1} \begin{bmatrix} F \\ 0 \\ 0 \\ 0 \end{bmatrix}$$

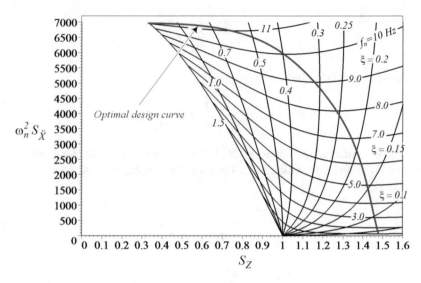

Fig. 6.34 Design chart for optimal suspension parameters of equipments

$$= \begin{bmatrix} k - \omega^2 m_1 & -k & -c\omega & c\omega \\ -k & k - \omega^2 m_2 & c\omega & -c\omega \\ c\omega & -c\omega & k - \omega^2 m_1 & -k \\ -c\omega & c\omega & -k & k - \omega^2 m_2 \end{bmatrix} \quad (6.331)$$

The frequency response of m_1 and m_2 are

$$X_1 = \sqrt{A_1^2 + B_1^2} \qquad X_2 = \sqrt{A_2^2 + B_2^2} \qquad (6.332)$$

6.3.3 ★ *Root Mean Square (RMS) Optimization*

Figure 6.34 is a design chart for optimal suspension parameters of base excited systems. The horizontal axis is the root mean square of the relative displacement, $S_Z = RMS(S_2)$, and the vertical axis is the root mean square of the absolute acceleration, $S_{\ddot{x}} = RMS(G_2)$. There are two sets of curves that make the mesh. The first set, which is almost parallel horizontal at the right end, is the constant natural frequency f_n, and the second set, which spreads from $S_Z = 1$, is the constant damping ratio ξ. There is a curve, called the *optimal design curve*, which indicates the optimal suspension parameters.

Most pieces of equipment that are mounted on vehicles have natural frequencies around $f_n = 10\,\text{Hz}$, while the main natural frequencies of the vehicle are around

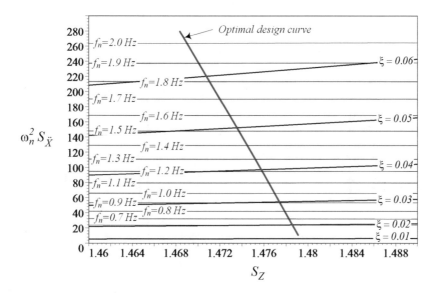

Fig. 6.35 Design chart for optimal suspension parameters of vehicles

$f_n = 1$ Hz. So, we use Fig. 6.34 to design the suspension of base excited equipment and use the magnified chart shown in the zoomed-in Fig. 6.35 to design vehicle suspensions.

The optimal design curve is the result of the following optimization strategy:

$$Minimize \ S_{\ddot{X}} \ with \ respect \ to \ S_Z \qquad (6.333)$$

which states that the minimum RMS of the absolute acceleration with respect to the RMS of the relative displacement makes a suspension optimal. Mathematically, it is equivalent to the following minimization problem:

$$\frac{\partial S_{\ddot{X}}}{\partial S_Z} = 0 \qquad (6.334)$$

$$\frac{\partial^2 S_{\ddot{X}}}{\partial S_Z^2} > 0 \qquad (6.335)$$

To use the optimization chart and determine the optimal stiffness k and damping c, we begin from an estimated value of S_X on the horizontal axis and draw a vertical line to hit the optimal curve. The intersection point indicates the optimal f_n and ξ for the S_X to have the best vibration isolation.

Figure 6.36 illustrates a sample application for $S_X = 1$, which indicates $\xi \approx 0.4$ and $f_n \approx 10$ Hz as the optimal suspension parameters. The f_n, ξ, and the mass of the equipment determine the optimal value of k and c.

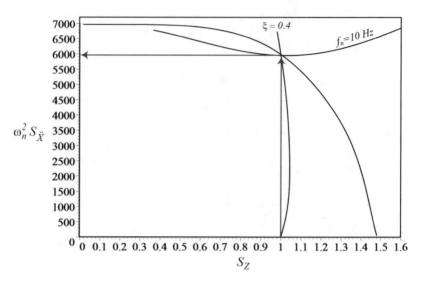

Fig. 6.36 Application of the design chart for $S_X = 1$ that indicates the optimal values $\xi \approx 0.4$ and $f_n \approx 10\,\text{Hz}$

Proof Let us define a working frequency range $0 < f < 20\,\text{Hz}$ to include almost all ground vehicles, especially road vehicles, and denote the RMS of S_2 and G_2 by

$$S_Z = RMS(S_2) \tag{6.336}$$

$$S_{\ddot{X}} = RMS(G_2) \tag{6.337}$$

In applied vehicle dynamics, we usually measure frequencies in [Hz] instead of [rad / s]; so we develop design calculations based on cyclic frequencies f and f_n in [Hz] and analytic calculations based on angular frequencies ω and ω_n in [rad / s].

To calculate S_Z and $S_{\ddot{X}}$ over the working frequency range,

$$S_Z = \sqrt{\frac{1}{40\pi} \int_0^{40\pi} S_2^2 d\omega} \tag{6.338}$$

$$S_{\ddot{X}} = \sqrt{\frac{1}{40\pi} \int_0^{40\pi} G_2^2 d\omega} \tag{6.339}$$

we first find the integrals of S_2^2 and G_2.

$$\int S_2^2 d\omega = Z_1\omega - \frac{Z_2}{Z_3\sqrt{Z_4}} \tan^{-1} \frac{\omega}{\sqrt{Z_4}} + \frac{Z_5}{Z_6\sqrt{Z_7}} \tan^{-1} \frac{\omega}{\sqrt{Z_7}} \tag{6.340}$$

$$\omega_n^4 \int G_2^2 d\omega = Z_8\omega + \frac{1}{3}Z_9\omega^3 + \frac{Z_{10}}{Z_{11}\sqrt{Z_{12}}} \tan^{-1} \frac{\omega}{\sqrt{Z_{12}}}$$

$$+ \frac{Z_{13}}{Z_{14}\sqrt{Z_{15}}} \tan^{-1} \frac{\omega}{\sqrt{Z_{15}}} \tag{6.341}$$

The parameters Z_1 through Z_{15} are as follows:

$$Z_1 = 1 \tag{6.342}$$

$$Z_2 = \omega_n^2 \left(8\xi^6 - 12\xi^4 + 4\xi^2 - \left(-8\xi^4 + 8\xi^2 - 1 \right) \xi \sqrt{1 - \xi^2} \right) \tag{6.343}$$

$$Z_3 = -4\xi^2 \left(1 - \xi^2 \right) \tag{6.344}$$

$$Z_4 = \omega_n^2 \left(-1 + 2\xi^2 + 2\xi \sqrt{1 - \xi^2} \right) \tag{6.345}$$

$$Z_5 = \omega_n^2 \left(8\xi^6 - 12\xi^4 + 4\xi^2 - \left(8\xi^4 - 8\xi^2 + 1 \right) \xi \sqrt{1 - \xi^2} \right) \tag{6.346}$$

$$Z_6 = -4\xi^2 \left(1 - \xi^2 \right) \tag{6.347}$$

$$Z_7 = \omega_n^2 \left(-1 + 2\xi^2 - 2\xi \sqrt{1 - \xi^2} \right) \tag{6.348}$$

$$Z_8 = \omega_n^4 \left(-16\xi^4 + 8\xi^2 + 1 \right) \tag{6.349}$$

$$Z_9 = 4\omega_n^2 \xi^2 \tag{6.350}$$

$$Z_{10} = \omega_n^6 \left(128\xi^{10} - 256\xi^8 + 144\xi^6 - 12\xi^4 - 4\xi^2 \right)$$

$$- \omega_n^6 \left(-128\xi^8 + 192\xi^6 - 64\xi^4 - 4\xi^2 + 1 \right) \xi \sqrt{1 - \xi^2} \tag{6.351}$$

$$Z_{11} = -4\xi^2 \left(1 - \xi^2 \right) \tag{6.352}$$

$$Z_{12} = \omega_n^2 \left(-1 + 2\xi^2 + 2\xi \sqrt{1 - \xi^2} \right) \tag{6.353}$$

$$Z_{13} = \omega_n^6 \left(128\xi^{10} - 256\xi^8 + 144\xi^6 - 12\xi^4 - 4\xi^2 \right)$$
$$- \omega_n^6 \left(128\xi^8 - 192\xi^6 + 64\xi^4 + 4\xi^2 - 1 \right) \xi \sqrt{1 - \xi^2} \qquad (6.354)$$

$$Z_{14} = -4\xi^2 \left(1 - \xi^2 \right) \qquad (6.355)$$

$$Z_{15} = \omega_n^2 \left(-1 + 2\xi^2 - 2\xi \sqrt{1 - \xi^2} \right) \qquad (6.356)$$

Therefore, S_Z and $S_{\ddot{X}}$ over the frequency range $0 < f < 20\,\mathrm{Hz}$ can be calculated analytically from Eqs. (6.338) and (6.339).

Integrating over the excitation frequency ω eliminates ω and shows both $S_{\ddot{X}}$ and S_Z are functions of only two parameters ω_n and ξ.

$$S_{\ddot{X}} = S_{\ddot{X}} (\omega_n, \xi) \qquad (6.357)$$

$$S_Z = S_Z (\omega_n, \xi) \qquad (6.358)$$

Therefore, any pair of design parameters (ω_n, ξ) determines $S_{\ddot{X}}$ and S_Z uniquely. It is also theoretically possible to define ω_n and ξ as two functions of the variables $S_{\ddot{X}}$ and S_Z.

$$\omega_n = \omega_n \left(S_{\ddot{X}}, S_Z \right) \qquad (6.359)$$

$$\xi = \xi \left(S_{\ddot{X}}, S_Z \right) \qquad (6.360)$$

Hence, we would be able to determine the required ω_n and ξ for a specific value of $S_{\ddot{X}}$ and S_Z.

Using Eqs. (6.357) and (6.358), we may draw Fig. 6.37 to illustrate how $S_{\ddot{X}}$ behaves with respect to S_Z when f_n and ξ vary. Keeping f_n constant and varying ξ, it is possible to minimize $S_{\ddot{X}}$ with respect to S_Z. The minimum points make the optimal curve and determine the best f_n and ξ. The key to use the optimal design curve is to adjust, determine, or estimate a value for S_Z or $S_{\ddot{X}}$ and find the associated point on the design curve.

To justify the optimization principle (6.333), we plot $\omega_n^2 S_{\ddot{X}}/S_Z$ versus f_n for different values of ξ in Fig. 6.38. It shows that increasing either one of ξ or f_n increases the value of $\omega_n^2 S_{\ddot{X}}/S_Z$. It is equivalent to making the suspension more rigid, which causes an increase in acceleration or decrease in relative displacement. On the contrary, decreasing ξ or f_n decreases the value of $\omega_n^2 S_{\ddot{X}}/S_Z$, which is equivalent to making the suspension softer.

Softening of a suspension decreases the body acceleration; however, it requires a larger room for relative displacement. Due to physical constraints, the relative displacement or *wheel travel* is limited, and hence, we must design the suspension to use the available wheel travel as much as possible and decrease the body acceleration as low as possible. Mathematically, it is equivalent to (6.334) and (6.335). ∎

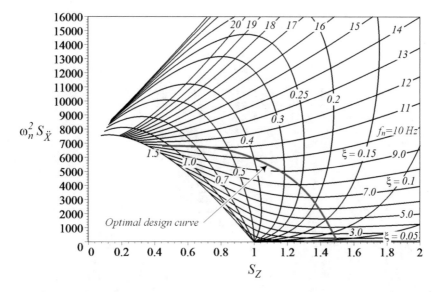

Fig. 6.37 Behavior of $S_{\ddot{X}}$ with respect to S_Z when f_n and ξ are varied

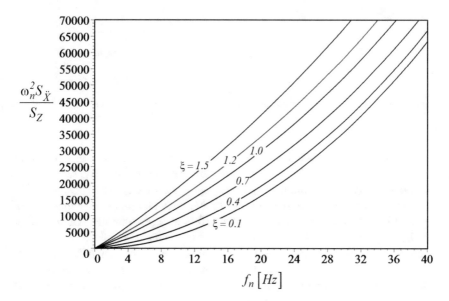

Fig. 6.38 A plot of ratio $\omega_n^2 S_{\ddot{X}}/S_Z$ versus f_n for different values of ξ

Example 216 Wheel travel calculation. The *RMS* optimization design can be well understood by investigating a vehicle suspension. In the theoretical study, we assume the wheel travel to be symmetric, although in practice, it may not be symmetric.

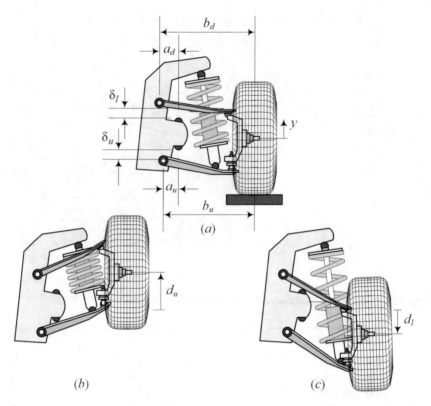

Fig. 6.39 A double A-arm suspension mechanism at (**a**) equilibrium, (**b**) upper limit, and (**c**) lower limit

Figure 6.39a illustrates a double A-arm suspension mechanism at its equilibrium position. To limit the motion of the wheel with respect to the body, two stoppers must be employed. There are many possible options for the type and position of the stoppers. Most stoppers are made of stiff rubber balls and mounted somewhere on the body or suspension mechanism or both. It is also possible that the bottom up damper acts as a stopper. Figure 6.39a shows an example of stoppers.

The gap sizes δ_u and δ_l indicate the upper and lower distances that the suspension mechanism can move. However, the maximum motion of the wheel must be calculated at the center of the wheel. So, we transfer δ_u and δ_l to the center of the wheel and show them by d_u and d_l.

$$d_u \approx \frac{b_u}{a_u}\delta_u \qquad d_l \approx \frac{b_l}{a_l}\delta_l \qquad (6.361)$$

Figure 6.39b and c show the mechanism at the upper and lower limits, respectively. The distance d_u is called the upper wheel travel, and d_l is called the lower wheel travel. The upper wheel travel is important in ride comfort, and the lower wheel

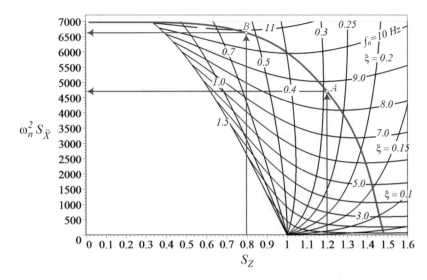

Fig. 6.40 Comparing two suspensions A and B with $S_{Z_A} = 1.2$ and $S_{Z_B} = 0.8$

travel is important for safety. To have better ride comfort, the upper wheel travel should be as high as possible to make the suspension as soft as possible.

Although the upper and lower wheel travels may be different, for practical purposes, we may assume $d_l = d_u$ and design the suspension based on a unique wheel travel. Wheel travel is also called suspension travel, suspension room, and suspension clearance.

Example 217 Soft and hard suspensions. A practical example to learn how to use the design chat for optimal suspension.

Consider two pieces of equipment, A and B, under a base excitation with an average amplitude $Y = 1$ cm ≈ 0.5 in. Equipment A has a suspension travel $d_A = 1.2$ cm ≈ 0.6 in, and equipment B has $d_B = 0.8$ cm ≈ 0.4 in. Let us assume $S_Z = d_u/Y$.

$$S_{Z_A} \approx 1.2 \qquad S_{Z_B} \approx 0.8 \qquad (6.362)$$

Using the design chart in Fig. 6.40, the optimal suspensions for A and B are

$$f_{n_A} \approx 8.53\,\text{Hz} \qquad \xi_A \approx 0.29$$
$$f_{n_B} \approx 10.8\,\text{Hz} \qquad \xi_B \approx 0.56 \qquad (6.363)$$

Assuming a mass m

$$m = 300\,\text{kg} \approx 660\,\text{lb} \qquad (6.364)$$

we calculate the optimal springs and dampers as

$$k_A = (2\pi f_{n_A})^2 m = 8.6175 \times 10^5 \, \text{N} / \text{m} \tag{6.365}$$

$$k_B = (2\pi f_{n_B})^2 m = 13.814 \times 10^5 \, \text{N} / \text{m} \tag{6.366}$$

$$c_A = 2\xi_A \sqrt{k_A m} = 9325.7 \, \text{N} \, \text{s} / \text{m} \tag{6.367}$$

$$c_B = 2\xi_B \sqrt{k_B m} = 22{,}800 \, \text{N} \, \text{s} / \text{m} \tag{6.368}$$

Equipment B has a harder suspension compared to equipment A. This is because equipment B has less suspension travel, and, hence, it has higher acceleration level $\omega_n^2 S_{\ddot{X}}$. Figure 6.40 shows that

$$\omega_n^2 S_{\ddot{X}_A} \approx 4700 \, 1/\text{s}^2 \tag{6.369}$$

$$\omega_n^2 S_{\ddot{X}_B} \approx 6650 \, 1/\text{s}^2 \tag{6.370}$$

Example 218 Soft and hard vehicle suspensions. Applying the suspension optimization to vehicle suspension is based on a small part of the design chart due to physical restrictions and limit range of suspension spring. Here is an example for a soft and hard suspensions.

Consider two vehicles A and B that are moving on a bumpy road with an average amplitude $Y = 10 \, \text{cm} \approx 3.937 \, \text{in}$. Vehicle A has a suspension travel $d_A = 14.772 \, \text{cm} \approx 5.816 \, \text{in}$, and vehicle B has $d_B = 14.714 \, \text{cm} \approx 5.793 \, \text{in}$. Let us assume that $S_Z = d_u / Y$.

$$S_{Z_A} = 1.4772 \qquad S_{Z_B} = 1.4714 \tag{6.371}$$

Using design chart Fig. 6.41, the optimal suspensions for vehicles A and B are

$$f_{n_A} \approx 0.7 \, \text{Hz} \qquad \xi_A \approx 0.023 \tag{6.372}$$

$$f_{n_B} \approx 1.85 \, \text{Hz} \qquad \xi_B \approx 0.06 \tag{6.373}$$

Assuming a mass m

$$m = 300 \, \text{kg} \approx 660 \, \text{lb} \tag{6.374}$$

we calculate the optimal springs and dampers as

$$k_A = (2\pi f_{n_A})^2 m \approx 5803 \, \text{N} / \text{m} \tag{6.375}$$

$$k_B = (2\pi f_{n_B})^2 m \approx 40{,}534 \, \text{N} / \text{m} \tag{6.376}$$

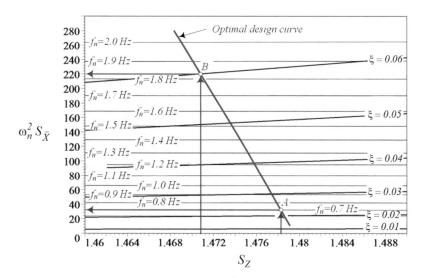

Fig. 6.41 Comparing two suspensions, A and B, with $S_{Z_A} = 1.4772$ and $S_{Z_B} = 1.4714$

$$c_A = 2\xi_A \sqrt{k_A m} \approx 60.7 \, \text{N s} / \text{m} \tag{6.377}$$

$$c_B = 2\xi_B \sqrt{k_B m} \approx 418.5 \, \text{N s} / \text{m} \tag{6.378}$$

These are equivalent dampers and springs at the center of the wheel. The actual values of the suspension parameters depend on the geometry of the suspension mechanism and installment of the spring and damper. Because $k_B > k_A$ and $c_B > c_A$, the suspension of B is harder than A. This is because vehicle B has less wheel travel and hence, it has higher acceleration level, $\omega_n^2 S_{\ddot{X}}$. Figure 6.41 shows

$$\omega_n^2 S_{\ddot{X}_B} \approx 220 \, 1/\text{s}^2 \qquad \omega_n^2 S_{\ddot{X}_A} \approx 28 \, 1/\text{s}^2 \tag{6.379}$$

Example 219 Sensitivity of $S_{\ddot{X}}$ with respect to S_Z on the optimal curve. The optimal points of the optimal design chart are at the minimum of $S_{\ddot{X}}$. Minimum or maximum point of a curve has minimum sensitivity to change of the independent variable. Here is the discussion on suspension optimization.

Because $S_{\ddot{X}}$ is minimum on the optimal curve, the sensitivity of acceleration RMS with respect to relative displacement RMS is minimum at any point on the optimal curve. Therefore, an optimal suspension has the least sensitivity to the available wheel travel variation. If a suspension is optimized for a certain wheel travel, it is still near optimal when the wheel travel changes a little over time or because of weight changes.

Example 220 Suspension trade-off and trivial optimization. The trade-off between suspension travel and acceleration is discussed here.

Reduction of the absolute acceleration is the main goal in the optimization of suspensions, because it represents the transmitted force to the body. A vibration isolator reduces the absolute acceleration by increasing deflection of the isolator. The relative deflection is a measure of the clearance known as the working space of the isolator. The clearance should be minimized due to safety and the physical constraints in the mechanical design.

There is a trade-off between the acceleration and relative motion. The ratio of $\omega_n^4 S_{\ddot{x}}$ to S_Z is a monotonically increasing function of ω_n and ξ. Keeping S_Z constant increases $\omega_n^4 S_{\ddot{x}}$ by increasing both ω_n and ξ. However, keeping $\omega_n^4 S_{\ddot{x}}$ constant, we decrease S_Z by increasing ω_n and ξ. Hence, $\omega_n^4 S_{\ddot{x}}$ and S_Z have opposite behaviors. These behaviors show that the suspension of $\omega_n = 0$ and $\xi = 0$ is the trivial and non-practical solution for the best isolation.

Example 221 ★ $RMS(G_0) \equiv RMS(X/Y)$. Although the absolute displacement is not involved in RMS optimization method, the RMS of absolute displacement involves in objective functions of many optimization algorithms. Here is the analytic calculation of the function.

For the RMS of the absolute displacement, S_X, one needs the integral of $G_0^2 \equiv (X/Y)^2$, which is determined as follows:

$$\int G_0^2 d\omega = \frac{Z_{16}}{Z_{17}\sqrt{Z_{18}}} \tan^{-1} \frac{\omega}{\sqrt{Z_{18}}} + \frac{Z_{19}}{Z_{20}\sqrt{Z_{21}}} \tan^{-1} \frac{\omega}{\sqrt{Z_{21}}} \qquad (6.380)$$

$$Z_{16} = \omega_n^2 \left(-8\xi^6 + 8\xi^4 - \left(8\xi^4 - 4\xi^2 - 1 \right) \xi \sqrt{1 - \xi^2} \right) \qquad (6.381)$$

$$Z_{17} = -4\xi^2 \left(1 - \xi^2 \right) \qquad (6.382)$$

$$Z_{18} = \omega_n^2 \left(1 - 2\xi^2 - 2\xi \sqrt{1 - \xi^2} \right) \qquad (6.383)$$

$$Z_{19} = \omega_n^2 \left(8\xi^6 - 8\xi^4 - \left(8\xi^4 - 4\xi^2 - 1 \right) \xi \sqrt{1 - \xi^2} \right) \qquad (6.384)$$

$$Z_{20} = Z_{17} = -4\xi^2 \left(1 - \xi^2 \right) \qquad (6.385)$$

$$Z_{21} = -\omega_n^2 \left(1 - 2\xi^2 - 2\xi \sqrt{1 - \xi^2} \right) \qquad (6.386)$$

Example 222 ★ Alternative optimization methods. Optimization of vibration isolator has a rich background with many different methods. Here is introducing a couple of other methods than the *RMS* method.

There are various approaches and suggested methods for vibration isolator optimization, depending on the application and mathematical modeling. However, there is not a universally accepted method applicable to every application. Every optimization strategy can be transformed to a minimization of a function called the cost function. Considerable attention has been given to minimization of the absolute displacement, known as the main transmissibility. However, for a vibration isolator, the cost function may include any state variable such as absolute and relative displacements, velocities, accelerations, and jerks.

Constraints may determine the domain of acceptable design parameters by dictating an upper and lower limits for ω_n and ξ. For vehicle suspensions, it is generally desired to select ω_n and ξ such that the absolute acceleration of the system is minimized and the relative displacement does not exceed a prescribed level. The most common optimization strategies are the following.

Minimax the absolute acceleration $S_{\ddot{X}}$ for specified relative displacement S_{Z_0}. Specify the allowable relative displacement, and then find the minimax of absolute acceleration:

$$\frac{\partial S_{\ddot{X}}}{\partial \omega_n} = 0 \qquad \frac{\partial S_{\ddot{X}}}{\partial \xi} = 0 \qquad S_Z = S_{Z_0} \qquad (6.387)$$

Minimax the relative displacement S_Z for specified absolute acceleration $S_{\ddot{X}_0}$. Specify the allowable absolute acceleration, and then find the minimax relative displacement:

$$\frac{\partial S_Z}{\partial \omega_n} = 0 \qquad \frac{\partial S_Z}{\partial \xi} = 0 \qquad S_{\ddot{X}} = S_{\ddot{X}_0} \qquad (6.388)$$

Example 223 ★ Further application of the design chart. The *RMS* optimization method can be applied on any two characteristics that have a trade-off property. Here is more explanation.

The *RMS* optimization criterion

$$\frac{\partial S_{\ddot{X}}}{\partial S_Z} = 0 \qquad \frac{\partial^2 S_{\ddot{X}}}{\partial S_Z^2} > 0 \qquad (6.389)$$

is based on the root mean square of S_2 and G_2 over a working frequency range.

$$S_Z = \sqrt{\frac{1}{40\pi} \int_0^{40\pi} S_2^2 d\omega} \qquad (6.390)$$

$$S_{\ddot{X}} = \sqrt{\frac{1}{40\pi} \int_0^{40\pi} G_2 d\omega} \qquad (6.391)$$

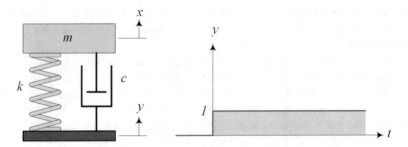

Fig. 6.42 A 1/8 car model and a unit step displacement base excitation

The design curve is the optimal condition for suspension of a base excited system using the following functions:

$$S_2 = \frac{Z_B}{Y} \qquad G_2 = \frac{\ddot{X}_B}{\omega_n^2 Y} \tag{6.392}$$

However, because

$$S_2 = \frac{\ddot{X}_F}{F/m} = \frac{Z_B}{Y} = \frac{X_E}{e\varepsilon_E} = \frac{Z_R}{e\varepsilon_R} \tag{6.393}$$

$$G_2 = \frac{\ddot{X}_B}{\omega_n^2 Y} = \frac{F_{TB}}{kY} = \frac{F_{TE}}{e\omega^2 m_e} = \frac{F_{TR}}{e\omega^2 m_e}\left(1 + \frac{m_a}{m}\right) \tag{6.394}$$

the optimal design curve can also be expressed as a minimization condition for any other G_2-function with respect to any other S_2-function, such as the force transmitted to the base $F_{TE}/\left(e\omega^2 m_e\right)$ for an eccentric excited system $X_E/(e\varepsilon_E)$. This minimization is equivalent to the optimization of an engine mount.

6.3.4 ★ *Time Response Optimization*

The transient response optimization depends on the type of transient excitation, as well as the definition of the cost function. Figure 6.42 illustrates a 1/8 car model and a unit step displacement.

$$y = \begin{cases} 1 & t > 0 \\ 0 & t \leq 0 \end{cases} \tag{6.395}$$

If the transient excitation is a step function, and the optimization criteria is minimization of the peak value of the acceleration versus peak value of the relative

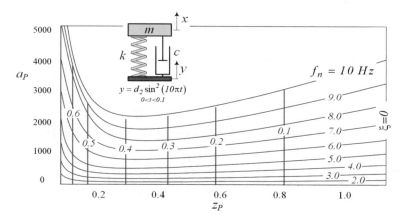

Fig. 6.43 Peak value of acceleration versus peak value of relative displacement for different ξ and f_n

displacement, then there is optimal $\xi^\star$ for any f_n that provides us with the best transient behavior of a $1/8$ car model. This behavior is shown in the design chart of Fig. 6.43.

$$\xi^\star = 0.4 \tag{6.396}$$

Proof The equation of motion for the base excited one DOF system shown in Fig. 6.42 is

$$\ddot{x} + 2\xi\omega_n\,\dot{x} + \omega_n^2\,x = 2\xi\omega_n\,\dot{y} + \omega_n^2\,y \tag{6.397}$$

Substituting $y = 1$ in Eq. (6.397) provides us with the following initial value problem to determine the absolute displacement of the mass m.

$$\ddot{x} + 2\xi\omega_n\,\dot{x} + \omega_n^2\,x = \omega_n^2 \tag{6.398}$$

$$y(0) = 0 \qquad \dot{y}(0) = 0 \tag{6.399}$$

The solution of the differential equation with zero initial conditions is

$$x = 1 - \frac{1}{2}\frac{A}{ib}e^{-A\omega_n t} + \frac{1}{2}\frac{\overline{A}}{ib}e^{-\overline{A}\omega_n t} \tag{6.400}$$

where A and $\overline{A}$ are two complex conjugate numbers.

$$A = \xi + i\sqrt{1 - \xi^2} \tag{6.401}$$

$$\overline{A} = \xi - i\sqrt{1 - \xi^2} \tag{6.402}$$

Having x and $y = 1$ is enough to calculate the relative displacement $z = x - y$.

$$z = x - y = -\frac{1}{2}\frac{A}{ib}e^{-A\omega_n t} + \frac{1}{2}\frac{\overline{A}}{ib}e^{-\overline{A}\omega_n t} \tag{6.403}$$

The absolute velocity and acceleration of the mass m can be obtained from Eq. (6.400).

$$\dot{x} = \frac{1}{2}\frac{A^2\omega_n}{ib}e^{-A\omega_n t} - \frac{1}{2}\frac{\overline{A}^2\omega_n}{ib}e^{-\overline{A}\omega_n t} \tag{6.404}$$

$$\ddot{x} = -\frac{1}{2}\frac{A^3\omega_n^2}{ib}e^{-A\omega_n t} + \frac{1}{2}\frac{\overline{A}^3\omega_n^2}{ib}e^{-\overline{A}\omega_n t} \tag{6.405}$$

The peak value of the relative displacement is z_P which occurs when $\dot{z} = 0$ at time t_1.

$$z_P = \exp\left(\frac{\cos^{-1}\left(2\xi^2 - 1\right)}{\omega_n\sqrt{1 - \xi^2}}\right) \tag{6.406}$$

$$t_1 = \frac{-\xi\cos^{-1}\left(2\xi^2 - 1\right)}{\sqrt{1 - \xi^2}} \tag{6.407}$$

The peak value of the absolute acceleration is a_P which occurs at the beginning of the excitation, $t = 0$, or at the time instant when $\dddot{x} = 0$ at time t_2.

$$a_P = \omega_n^2\exp\left(-\xi\frac{2\cos^{-1}\left(2\xi^2 - 1\right) - \pi}{\sqrt{1 - \xi^2}}\right) \tag{6.408}$$

$$t_2 = \frac{2\cos^{-1}\left(2\xi^2 - 1\right) - \pi}{\omega_n\sqrt{1 - \xi^2}} \tag{6.409}$$

Figure 6.43 is a plot for a_P versus z_P for different ξ and f_n. The minimum of the curves occur at $\xi = 0.4$ for every f_n. The optimal ξ can be found analytically by finding the minimum point of a_P versus z_P. The optimal ξ is the solution of the following transcendental equation which is $\xi = 0.4$.

$$2\xi\cos^{-1}\left(2\xi^2 - 1\right) - \pi - 4\xi\sqrt{1 - \xi^2} = 0 \tag{6.410}$$

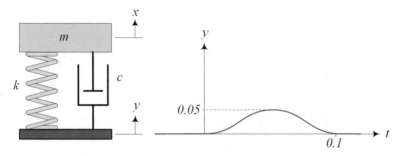

Fig. 6.44 A base excited *one* DOF system and a sine square bump input

The minimum peak value of the absolute acceleration with respect to the relative displacement is independent of the value of natural frequency f_n. ∎

Example 224 ★ Optimal design curve and time response. To examine if the optimal value of ξ based on step input, as well as the optimal suspension based on the optimal curve of *RMS* methods, is also working for other inputs, here we study a base excited system on a bump transient input.

To examine transient response of suspensions on the optimal design curve, we compare the response of three suspensions to a base excited equipment. The characteristics of the three suspensions may be shown by three points on Fig. 6.40.

$$Point\ P_1 \qquad f_n \approx 10\,\text{Hz} \qquad \xi \approx 0.15 \qquad (6.411)$$

$$Point\ P_2 \qquad f_n \approx 10\,\text{Hz} \qquad \xi \approx 0.4 \qquad (6.412)$$

$$Point\ P_3 \qquad f_n \approx 5\,\text{Hz} \qquad \xi \approx 0.15 \qquad (6.413)$$

Point P_1 indicates an off-optimal suspension, and points P_2 and P_3 are both on the optimal curve of Fig. 6.40. Points P_2 and P_3 are two alternative optimizations for point P_1. Point P_2 has $\xi \approx 0.4$ with the same natural frequency as P_1, and point P_3 has $f_n \approx 5\,\text{Hz}$ with the same damping as point P_1.

Figure 6.44 illustrates a base excited one *DOF* system and a sine square bump input.

$$y = \begin{cases} d_2 \sin^2 \dfrac{2\pi v}{d_1} t & 0 < t < 0.1 \\ 0 & t \le 0, t \ge 0.1 \end{cases} \qquad (6.414)$$

$$d_2 = 0.05\,\text{m} \qquad v = 10\,\text{m}\,/\,\text{s} \qquad d_1 = 1\,\text{m} \qquad (6.415)$$

The absolute and relative displacement time responses of the system at points 1, 2, and 3 are shown in Figs. 6.45, and 6.46, respectively. The absolute acceleration of m is shown in Fig. 6.47. System 3 has a lower relative displacement peak value and a lower absolute acceleration peak value, but it takes more time to settle down.

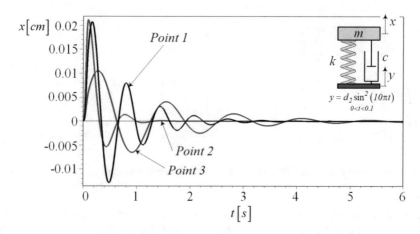

Fig. 6.45 Absolute displacement time response of the system for three different suspensions

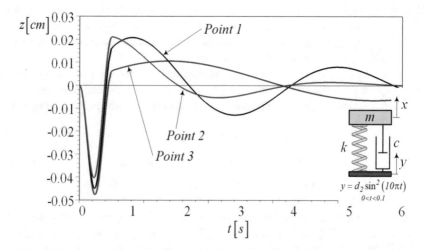

Fig. 6.46 Relative displacement time response of the system for three different suspensions

6.4   Eccentric Excitation

Figure 6.48 illustrates a one DOF eccentric excited vibrating system with a mass m supported by a suspension made of a spring k and a damper c. There is an unbalance mass m_e at a distance e that is rotating with an angular velocity ω. An eccentric excited vibrating system is the proper model for vibration analysis of an engine of a vehicle or any rotary motor that is mounted on a stationary base with flexible suspension.

The absolute motion of m with respect to its equilibrium position is measured by the coordinate x. When the lateral motion of m is protected, a harmonic excitation force is applied on m in the x-direction and makes the system vibrate.

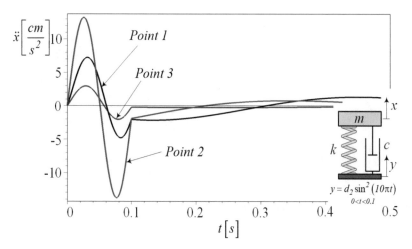

Fig. 6.47 Absolute acceleration time response of the system for three different suspensions

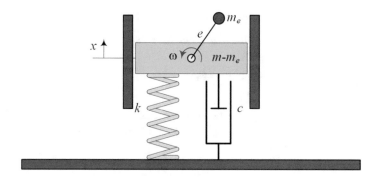

Fig. 6.48 An eccentric excited single DOF system

$$f_x = m_e e \omega^2 \sin \omega t \qquad (6.416)$$

The distance e is called the *eccentricity* and m_e is called the *eccentric mass*.
   The equation of motion of the system is

$$m \ddot{x} + c \dot{x} + kx = m_e e \omega^2 \sin \omega t \qquad (6.417)$$

or equivalently is

$$\ddot{x} + 2\xi \omega_n \dot{x} + \omega_n^2 x = \varepsilon e \omega^2 \sin \omega t \qquad (6.418)$$

$$\omega_n = \sqrt{\frac{k}{m}} \qquad r = \frac{\omega}{\omega_n} \qquad \xi = \frac{c}{2\sqrt{km}} \qquad \varepsilon = \frac{m_e}{m} \qquad (6.419)$$

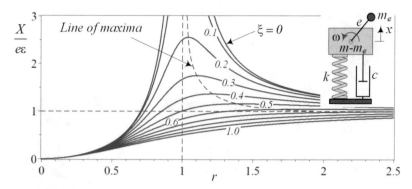

Fig. 6.49 The position frequency response for $\dfrac{X}{e\varepsilon}$

The absolute displacement response of the system is

$$x = A_4 \sin \omega t + B_4 \cos \omega t \tag{6.420}$$

$$= X \sin (\omega t - \varphi_e) \tag{6.421}$$

which has an amplitude X and phases φ_e.

$$\frac{X}{e\varepsilon} = \frac{r^2}{\sqrt{\left(1 - r^2\right)^2 + (2\xi r)^2}} \tag{6.422}$$

$$\varphi_e = \tan^{-1} \frac{2\xi r}{1 - r^2} \tag{6.423}$$

The phase φ_e indicates the angular lag of the response x with respect to the excitation $m_e e \omega^2 \sin \omega t$. The frequency responses for X and φ_e as a function of r and ξ are plotted in Figs. 6.49 and 6.50.

Proof Employing the free body diagram of the system, as shown in Fig. 6.51, and applying Newton's method in the x-direction generate the equation of motion

$$m \ddot{x} = -c \dot{x} - kx + m_e e \omega^2 \sin \omega t \tag{6.424}$$

Equation (6.417) can be transformed to (6.418) by dividing over m and using the definitions in (6.418) for natural frequency, damping ratio, and frequency ratio. The parameter $\varepsilon = m_e/m$ is the *mass ratio*, the ratio between the eccentric mass m_e and the total mass m.

The steady-state solution of Eqs. (6.417) can be (6.420) or (6.421). To find the amplitude and phase of the response, we substitute the solution (6.420) in the equation of motion.

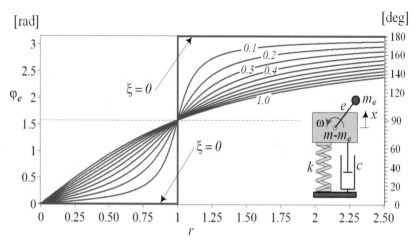

Fig. 6.50 The frequency response for φ_e

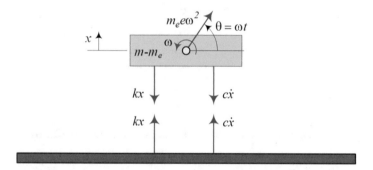

Fig. 6.51 Free body diagram of an eccentric excited single DOF system

$$-m\omega^2 (A_4 \sin \omega t + B_4 \cos \omega t) + c\omega (A_4 \cos \omega t - B_4 \sin \omega t)$$
$$+k (A_4 \sin \omega t + B_4 \cos \omega t) = m_e e \omega^2 \sin \omega t \qquad (6.425)$$

The coefficients of the functions $\sin \omega t$ and $\cos \omega t$ must be balanced on both sides of the equation.

$$kA_4 - m A_4 \omega^2 - c B_4 \omega = m_e e \omega^2 \qquad (6.426)$$
$$kB_4 - m\omega^2 B_4 + c\omega A_4 = 0 \qquad (6.427)$$

Therefore, we find two algebraic equations to calculate A_4 and B_4.

$$\begin{bmatrix} k - \omega^2 m & -c\omega \\ c\omega & k - \omega^2 m \end{bmatrix} \begin{bmatrix} A_4 \\ B_4 \end{bmatrix} = \begin{bmatrix} e\omega^2 m_e \\ 0 \end{bmatrix} \qquad (6.428)$$

$$\begin{bmatrix} A_4 \\ B_4 \end{bmatrix} = \begin{bmatrix} k - \omega^2 m & -c\omega \\ c\omega & k - \omega^2 m \end{bmatrix}^{-1} \begin{bmatrix} e\omega^2 m_e \\ 0 \end{bmatrix}$$

$$= \begin{bmatrix} \dfrac{k - m\omega^2 - \omega^2 m_e}{\left(k - \omega^2 m\right)^2 + c^2\omega^2} e\omega^2 m_e \\ \dfrac{-c\omega}{\left(k - \omega^2 m\right)^2 + c^2\omega^2} e\omega^2 m_e \end{bmatrix} \tag{6.429}$$

The solution provides us with the steady-state solution (6.420). The amplitude X and phase φ_e can be found by

$$X = \sqrt{A_4^2 + B_4^2} \tag{6.430}$$

$$\tan \varphi_e = \frac{-B_4}{A_4} \tag{6.431}$$

which, after substituting A_4 and B_4 from (6.429), yield:

$$X = \frac{\omega^2 e m_e}{\sqrt{\left(k - m\omega^2\right)^2 + c^2\omega^2}} \tag{6.432}$$

$$\tan \varphi_e = \frac{c\omega}{k - m\omega^2} \tag{6.433}$$

A more practical expression for X and φ_e are Eqs. (6.422) and (6.423) by employing r and ξ.

All rotating machines such as engines, turbines, generators, and turning machines have imperfections in their rotating components or have irregular mass distribution, which creates dynamic imbalances. When the unbalanced components rotate, an eccentric load applies to the structure. The load can be decomposed into two perpendicular harmonic forces in the plane of rotation in the lateral and normal directions. If the lateral force component is balanced by a reaction of a support, the normal component provides us with a harmonically variable force with an amplitude depending on the eccentricity $m_e e$. Unbalanced rotating machines are a common source of vibration excitation in electromechanical systems. ■

Example 225 An eccentric excited system. It is a numerical study of an eccentric excited system to work with equations and compare the order of magnitudes.

Consider an engine with a mass m

$$m = 110 \, \text{kg} \tag{6.434}$$

that is supported by four engine mounts, each with the following stiffness and damping.

$$k = 100,000 \, \text{N} / \text{m} \qquad c = 1000 \, \text{N} \, \text{s} / \text{m} \qquad (6.435)$$

The engine is running at

$$\omega = 5000 \, rpm \approx 523.6 \, \text{rad} / \text{s} \approx 83.333 \, \text{Hz} \qquad (6.436)$$

with the following eccentric parameters.

$$m_e = 0.001 \, \text{kg} \qquad e = 0.12 \, \text{m} \qquad (6.437)$$

The natural frequency ω_n, damping ratio ξ, and mass ratio ε of the system and frequency ratio r are

$$\omega_n = \sqrt{\frac{k}{m}} = \sqrt{\frac{100,000}{110}} = 30.15 \, \text{rad} / \text{s} \approx 4.8 \, \text{Hz} \qquad (6.438)$$

$$\xi = \frac{c}{2\sqrt{km}} = 0.015 \qquad (6.439)$$

$$\varepsilon = \frac{m_e}{m} = \frac{0.001}{110} = 9.0909 \times 10^{-6} \qquad (6.440)$$

$$r = \frac{\omega}{\omega_n} = \frac{523.60}{30.15} = 17.366 \qquad (6.441)$$

The engine's amplitude of vibration is

$$X = \frac{r^2 e \varepsilon}{\sqrt{\left(1 - r^2\right)^2 + (2\xi r)^2}} = 1.094 \times 10^{-6} \, \text{m} \qquad (6.442)$$

However, if the speed of the engine is at the natural frequency of the system,

$$\omega = 30.15 \, \text{rad} / \text{s} \approx 4.8 \, \text{Hz} \qquad r = 1 \qquad (6.443)$$

then the amplitude of the engine's vibration increases to

$$X = \frac{r^2 e \varepsilon}{\sqrt{\left(1 - r^2\right)^2 + (2\xi r)^2}} = 3.61 \times 10^{-6} \, \text{m} \qquad (6.444)$$

Example 226 Absolute velocity and acceleration of an eccentric excited system. The velocity and acceleration frequency responses of the eccentric excited systems are calculated and visualized in this example.

Using the position frequency response of an eccentric excited system, we can find the velocity and acceleration frequency responses.

$$x = A_4 \sin \omega t + B_4 \cos \omega t = X \sin \left(\omega t - \varphi_e \right) \tag{6.445}$$

$$\dot{x} = A_4 \omega \cos \omega t - B_4 \omega \sin \omega t$$

$$= X \omega \cos \left(\omega t - \varphi_e \right) = \dot{X} \cos \left(\omega t - \varphi_e \right) \tag{6.446}$$

$$\ddot{x} = -A_4 \omega^2 \sin \omega t - B_4 \omega^2 \cos \omega t$$

$$= -X \omega^2 \sin \left(\omega t - \varphi_e \right) = \ddot{X} \sin \left(\omega t - \varphi_e \right) \tag{6.447}$$

The amplitude of velocity and acceleration frequency responses, $\dot{X}$, $\ddot{X}$, are

$$\dot{X} = \frac{\omega^3 e m_e}{\sqrt{\left(k - m \omega^2 \right)^2 + c^2 \omega^2}} \tag{6.448}$$

$$\ddot{X} = \frac{\omega^4 e m_e}{\sqrt{\left(k - m \omega^2 \right)^2 + c^2 \omega^2}} \tag{6.449}$$

which can be rewritten in dimensionless forms.

$$\frac{\dot{X}}{e \varepsilon \omega_n} = \frac{r^3}{\sqrt{\left(1 - r^2 \right)^2 + (2 \xi r)^2}} \tag{6.450}$$

$$\frac{\ddot{X}}{e \varepsilon \omega_n^2} = \frac{r^4}{\sqrt{\left(1 - r^2 \right)^2 + (2 \xi r)^2}} \tag{6.451}$$

Figures 6.52 and 6.53 illustrate the velocity and acceleration frequency responses $\frac{\dot{X}}{e \varepsilon \omega_n}$ and $\frac{\ddot{X}}{e \varepsilon \omega_n^2}$, respectively.

Example 227 Force transmitted to the base of an eccentric excited system. For eccentric excited system, the transmitted force to the base foundation is a key factor of designing a foundation or managing installation of the machine in higher levels of a building. Here is deriving frequency response of transmitted force to the base.

The transmitted force f_T to the ground by an eccentric excited system is equal to the sum of forces in the spring and damper:

$$f_T = F_T \sin \left(\omega t - \varphi_T \right) \tag{6.452}$$

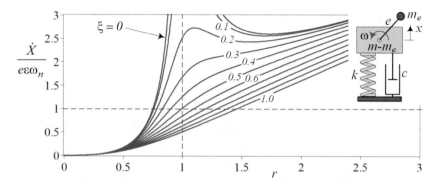

Fig. 6.52 The velocity frequency response for $\dfrac{\dot{X}}{e\varepsilon\omega_n}$

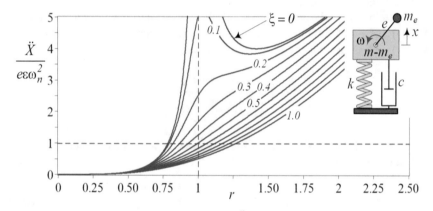

Fig. 6.53 The acceleration frequency response for $\dfrac{\ddot{X}}{e\varepsilon\omega_n^2}$

$$f_T = f_k + f_c = kx + c\dot{x} \qquad (6.453)$$

Substituting x and $\dot{x}$ from (6.420) shows that

$$f_T = (kA_4 - c\omega B_4)\sin t\omega + (kB_4 + c\omega A_4)\cos t\omega \qquad (6.454)$$

therefore, the amplitude of the transmitted force is

$$
\begin{aligned}
F_T &= \sqrt{(kA_4 - c\omega B_4)^2 + (kB_4 + c\omega A_4)^2} \\[2mm]
&= e\omega^2 m_e \sqrt{\dfrac{c^2\omega^2 + k^2}{\left(k - m\omega^2\right)^2 + c^2\omega^2}}
\end{aligned}
\qquad (6.455)
$$

The frequency response of the transmitted force can be simplified to the following applied dimensionless equation.

$$\frac{F_T}{e\omega^2 m_e} = \frac{\sqrt{1 + (2\xi r)^2}}{\sqrt{(1 - r^2)^2 + (2\xi r)^2}} \tag{6.456}$$

The graph of $F_T / (e\omega^2 m_e)$ is the same as X/Y in Fig. 6.22 for the absolute displacement frequency response of a base excited system.

Example 228 Line of maxima in $X/(e\varepsilon)$. Any equation with maximum or minimum or a negative sign between two terms in denominator is interesting to engineers because they can be used to determine the optimal or critical value of a variable. Here is determination of maximums in $X/(e\varepsilon)$ for different r.

Depending on ξ, the peak value of the absolute displacement frequency response $X/(e\varepsilon)$ occurs at different r. To find this relationship, we take a derivative of $X/(e\varepsilon)$ with respect to r and solve the equation.

$$\frac{d}{dr}\frac{X}{e\varepsilon} = \frac{2r\left(1 - r^2 + 2r^4\xi^2\right)}{\left((1 - r^2)^2 + (2\xi r)^2\right)^{\frac{3}{2}}} = 0 \tag{6.457}$$

Let us indicate the peak amplitude by X_{max} and the associated frequency by r_{max}. The value of r_{max}^2 is

$$r_{max}^2 = \frac{1}{4\xi^2}\left(\pm 1 - \sqrt{1 - 8\xi^2}\right) \tag{6.458}$$

Substituting the positive sign of (6.458) in (6.422) determines the peak amplitude X_{max}:

$$\frac{X_{max}}{e\varepsilon} = \frac{\sqrt{2}\left(1 - \sqrt{1 - 8\xi^2}\right)}{2\sqrt{-8\xi^2 + 16\xi^4 - \left(1 - 4\xi^2 + 8\xi^4\right)\sqrt{1 - 8\xi^2} + 1}} \tag{6.459}$$

Figure 6.49 shows the line of maxima.

6.5   ★ Eccentric Base Excitation

Figure 6.54 illustrates a one DOF eccentric base excited vibrating system with a mass m suspended by a spring k and a damper c on a base with mass m_b. The base has an unbalance mass m_e at a eccentricity distance e that is rotating with angular

Fig. 6.54 An eccentric base excited single DOF system

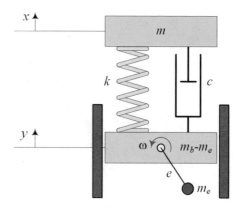

velocity ω. The eccentric base excited system is the proper model for vibration analysis of different equipment that are attached to the engine of a vehicle or any equipment mounted on a rotary motor.

The absolute motion of the mass m and the base are measured by variables x and y, respectively. Using the relative motion of m with respect to the base,

$$z = x - y \tag{6.460}$$

we develop the equation of motion as

$$\frac{mm_b}{m_b + m}\ddot{z} + c\dot{z} + kz = \frac{mm_e}{m_b + m}e\omega^2 \sin \omega t \tag{6.461}$$

or, equivalently, as

$$\ddot{z} + 2\xi\omega_n\dot{z} + \omega_n^2 z = \varepsilon e\omega^2 \sin \omega t \tag{6.462}$$

$$\omega_n = \sqrt{\frac{k}{m}} \qquad r = \frac{\omega}{\omega_n} \qquad \xi = \frac{c}{2\sqrt{km}} \qquad \varepsilon = \frac{m_e}{m_b} \tag{6.463}$$

The relative displacement response of the system is

$$z = A_5 \sin \omega t + B_5 \cos \omega t \tag{6.464}$$

$$= Z \sin (\omega t - \varphi_b) \tag{6.465}$$

which has an amplitude Z and phase φ_b.

$$\frac{Z}{e\varepsilon} = \frac{r^2}{\sqrt{\left(1 - r^2\right)^2 + (2\xi r)^2}} \tag{6.466}$$

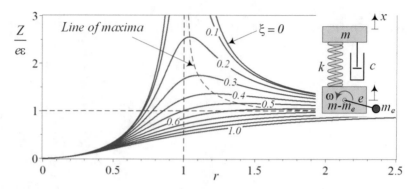

Fig. 6.55 The position frequency response for $\dfrac{Z}{e\varepsilon}$

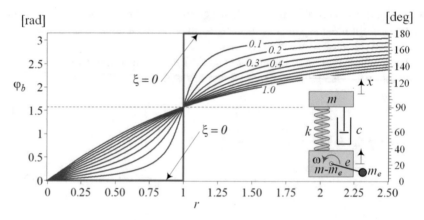

Fig. 6.56 The frequency response for φ_b

$$\varphi_b = \tan^{-1} \frac{2\xi r}{1 - r^2} \tag{6.467}$$

The frequency responses for Z and φ_b as functions of r and ξ are plotted in Figs. 6.55 and 6.56.

Proof The free body diagram in Fig. 6.57, along with Newton's equation in the x-direction, will be used to find the equations of the absolute motions.

$$m\ddot{x} = -c\,(\dot{x} - \dot{y}) - k\,(x - y) \tag{6.468}$$

$$m_b\ddot{y} = c\,(\dot{x} - \dot{y}) + k\,(x - y) - m_e e\omega^2 \sin \omega t \tag{6.469}$$

Using $z = x - y$, and its derivatives,

$$\dot{z} = \dot{x} - \dot{y} \tag{6.470}$$

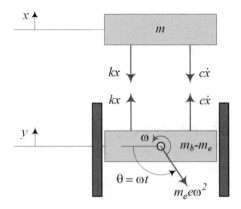

Fig. 6.57 Free body diagram of an eccentric base excited single DOF system

$$\ddot{z} = \ddot{x} - \ddot{y} \qquad (6.471)$$

we combine Eqs. (6.468) and (6.469) to find the equation of relative motion.

$$\frac{mm_b}{m_b + m}\ddot{z} + c\dot{z} + kz = \frac{mm_e}{m_b + m}e\omega^2 \sin\omega t \qquad (6.472)$$

Equation (6.472) can be transformed to (6.462) if we divide it by $\frac{mm_b}{m_b+m}$ and use the following definitions:

$$\xi = \frac{c}{2\sqrt{k\dfrac{mm_b}{m_b + m}}} \qquad (6.473)$$

$$\omega_n = \sqrt{k\Big/\frac{mm_b}{m_b + m}} = \sqrt{k\frac{m_b + m}{mm_b}} \qquad (6.474)$$

The parameter $\varepsilon = m_e/m_b$ is the *mass ratio* and indicates the ratio between the eccentric mass m_e and the total base mass m_b.

The steady-state solutions of Eq. (6.462) are (6.464) and (6.465). To find the amplitude and phase of the response, we substitute the solution (6.464) into the equation of motion.

$$-\omega^2 (A_5 \sin\omega t + B_5 \cos\omega t) + 2\xi\omega_n\omega (A_5 \cos\omega t - B_5 \sin\omega t)$$
$$+\omega_n^2 (A_5 \sin\omega t + B_5 \cos\omega t) = \varepsilon e\omega^2 \sin\omega t \qquad (6.475)$$

The coefficients of the functions $\sin\omega t$ and $\cos\omega t$ must be balance on both sides of the equations.

$$\omega_n^2 A_5 - \omega^2 A_5 - 2\xi\omega\omega_n B_5 = \varepsilon\omega^2 e \qquad (6.476)$$

$$2\xi A_5 \omega \omega_n - B_5 \omega^2 + B_5 \omega_n^2 = 0 \tag{6.477}$$

Therefore, we find two algebraic equations to calculate A_5 and B_5.

$$\begin{bmatrix} \omega_n^2 - \omega^2 & -2\xi \omega \omega_n \\ 2\xi \omega \omega_n & \omega_n^2 - \omega^2 \end{bmatrix} \begin{bmatrix} A_5 \\ B_5 \end{bmatrix} = \begin{bmatrix} \varepsilon \omega^2 e \\ 0 \end{bmatrix} \tag{6.478}$$

Solving them for the coefficients A_5 and B_5 provides us with the steady-state solution (6.464).

$$\begin{bmatrix} A_5 \\ B_5 \end{bmatrix} = \begin{bmatrix} \omega_n^2 - \omega^2 & -2\xi \omega \omega_n \\ 2\xi \omega \omega_n & \omega_n^2 - \omega^2 \end{bmatrix}^{-1} \begin{bmatrix} \varepsilon \omega^2 e \\ 0 \end{bmatrix}$$
$$= \begin{bmatrix} \dfrac{\omega_n^2 - \omega^2}{\left(\omega_n^2 - \omega^2\right)^2 + (2\xi \omega \omega_n)^2} \varepsilon \omega^2 e \\ \dfrac{-2\xi \omega \omega_n}{\left(\omega_n^2 - \omega^2\right)^2 + (2\xi \omega \omega_n)^2} \varepsilon \omega^2 e \end{bmatrix} \tag{6.479}$$

The amplitude Z and phase φ_b will be

$$X = \sqrt{A_5^2 + B_5^2} \tag{6.480}$$

$$\tan \varphi_b = \frac{-B_5}{A_5} \tag{6.481}$$

which, after substituting A_5 and B_5 from (6.479), yields

$$Z = \frac{\omega^2 e \varepsilon}{\sqrt{\left(\omega_n^2 - \omega^2\right)^2 + (2\xi \omega \omega_n)^2}} \tag{6.482}$$

$$\tan \varphi_b = \frac{2\xi \omega \omega_n}{\omega_n^2 - \omega^2} \tag{6.483}$$

Equations (6.482) and (6.483) can be simplified to the more practical expressions (6.466) and (6.467) by employing $r = \omega/\omega_n$. ∎

Example 229 ★ A base eccentric excited system. A numerical substitution of an eccentric base excited system will show how equations work and what the order of magnitudes are.

Consider an engine with a mass m_b

$$m_b = 110 \, \text{kg} \tag{6.484}$$

and an air intake device with a mass m that is mounted on the engine using an elastic mount with the following equivalent stiffness and damping.

$$m = 2\,\text{kg} \qquad k = 10{,}000\,\text{N}/\text{m} \qquad c = 100\,\text{N}\,\text{s}/\text{m} \tag{6.485}$$

The engine is running at

$$\omega = 576.0\,rpm \approx 60.302\,\text{rad}/\text{s} \approx 9.6\,\text{Hz} \tag{6.486}$$

with the eccentric parameters m_e and e.

$$m_e = 0.001\,\text{kg} \qquad e = 0.12\,\text{m} \tag{6.487}$$

The natural frequency ω_n, damping ratio ξ, mass ratio ε of the system, and frequency ratio r are

$$\omega_n = \sqrt{k\frac{m_b + m}{mm_b}} = 71.35\,\text{rad}/\text{s} \approx 11.356\,\text{Hz} \tag{6.488}$$

$$\xi = \frac{c}{2\sqrt{k\dfrac{mm_b}{m_b + m}}} = 0.357 \tag{6.489}$$

$$\varepsilon = \frac{m_e}{m_b} = 0.5 \times 10^{-3} \tag{6.490}$$

$$r = \frac{\omega}{\omega_n} = 0.845 \tag{6.491}$$

The relative amplitude of the device's vibration is

$$Z = \frac{e\varepsilon r^2}{\sqrt{\left(1 - r^2\right)^2 + (2\xi r)^2}} = 64.225 \times 10^{-6}\,\text{m} \tag{6.492}$$

Example 230 ★ Absolute displacement of the upper mass of an eccentric base excited system. The absolute displacement of the upper mass and relative displacement between the upper and lower masses are the two important measurements for practical machineries. Here is to derive the equation to determine the absolute displacement of the upper mass.

Equation (6.468), along with the solution (6.464), can be used to calculate the displacement frequency response of the upper mass m for an eccentric base excited system such as shown in Fig. 6.54.

$$\ddot{x} = -\frac{c}{m}(\dot{x} - \dot{y}) - \frac{k}{m}(x - y) = -\frac{c}{m}\dot{z} - \frac{k}{m}z \tag{6.493}$$

Assuming a steady-state displacement

$$x = A_6 \sin \omega t + B_6 \cos \omega t = X \sin \left(\omega t - \varphi_{bx} \right) \tag{6.494}$$

we have

$$- \omega^2 \left(A_6 \sin \omega t + B_6 \cos \omega t \right) = -\frac{c}{m} \dot{z} - \frac{k}{m} z$$

$$= -\frac{c}{m} \omega \left(A_5 \cos \omega t - B_5 \sin \omega t \right) - \frac{k}{m} \left(A_5 \sin \omega t + B_5 \cos \omega t \right)$$

$$= \left(\frac{c}{m} \omega B_5 - \frac{k}{m} A_5 \right) \sin t\omega + \left(-\frac{k}{m} B_5 - \frac{c}{m} \omega A_5 \right) \cos t\omega \tag{6.495}$$

and, therefore,

$$- \omega^2 A_6 = \frac{c}{m} \omega B_5 - \frac{k}{m} A_5 \tag{6.496}$$

$$- \omega^2 B_6 = -\frac{k}{m} B_5 - \frac{c}{m} \omega A_5 \tag{6.497}$$

Substituting A_5 and B_5 from (6.479) and using

$$X = \sqrt{A_6^2 + B_6^2} \tag{6.498}$$

$$\tan \varphi_{bx} = \frac{-B_6}{A_6} \tag{6.499}$$

yield

$$A_6 = -\frac{2c\xi\omega^2\omega_n + k \left(\omega_n^2 - \omega^2 \right)}{\left(\omega_n^2 - \omega^2 \right)^2 + \left(2\xi\omega\omega_n \right)^2} \frac{1}{m} \varepsilon e \tag{6.500}$$

$$B_6 = \frac{-c \left(\omega_n^2 - \omega^2 \right) + 2k\xi\omega_n}{\left(\omega_n^2 - \omega^2 \right)^2 + \left(2\xi\omega\omega_n \right)^2} \frac{1}{m} \varepsilon \omega e \tag{6.501}$$

and therefore, the amplitude X of steady-state vibration of the upper mass in an eccentric base excited system is

$$X = \frac{\sqrt{c^2\omega^2 + k^2}}{\sqrt{\left(\omega_n^2 - \omega^2 \right)^2 + \left(2\xi\omega\omega_n \right)^2}} \frac{\varepsilon}{m} e \tag{6.502}$$

$$\tan \varphi_{bx} = -\frac{-c \left(\omega_n^2 - \omega^2 \right) + 2k\xi\omega_n}{2c\xi\omega^2\omega_n + k \left(\omega_n^2 - \omega^2 \right)} \omega \tag{6.503}$$

6.6 Classification for the Frequency Responses of One *DOF* Forced Vibration Systems

A harmonically excited one *DOF* system can be one of the four systems shown in Fig. 6.58. Any steady-state response of these systems is equal to one of the following Eqs. (6.504)–(6.511), and the phase of the motion is equal to one of the Eqs. (6.512)–(6.515).

$$S_0 = \frac{1}{\sqrt{\left(1 - r^2\right)^2 + (2\xi r)^2}} \tag{6.504}$$

$$S_1 = \frac{r}{\sqrt{\left(1 - r^2\right)^2 + (2\xi r)^2}} = r S_0 \tag{6.505}$$

$$S_2 = \frac{r^2}{\sqrt{\left(1 - r^2\right)^2 + (2\xi r)^2}} = r^2 S_0 \tag{6.506}$$

$$S_3 = \frac{r^3}{\sqrt{\left(1 - r^2\right)^2 + (2\xi r)^2}} = r^3 S_0 \tag{6.507}$$

$$S_4 = \frac{r^4}{\sqrt{\left(1 - r^2\right)^2 + (2\xi r)^2}} = r^4 S_0 \tag{6.508}$$

$$G_0 = \frac{\sqrt{1 + (2\xi r)^2}}{\sqrt{\left(1 - r^2\right)^2 + (2\xi r)^2}} \tag{6.509}$$

$$G_1 = \frac{r\sqrt{1 + (2\xi r)^2}}{\sqrt{\left(1 - r^2\right)^2 + (2\xi r)^2}} = r G_0 \tag{6.510}$$

$$G_2 = \frac{r^2\sqrt{1 + (2\xi r)^2}}{\sqrt{\left(1 - r^2\right)^2 + (2\xi r)^2}} = r^2 G_0 \tag{6.511}$$

$$\Phi_0 = \tan^{-1} \frac{2\xi r}{1 - r^2} \tag{6.512}$$

$$\Phi_1 = \tan^{-1} \frac{1 - r^2}{-2\xi r} \tag{6.513}$$

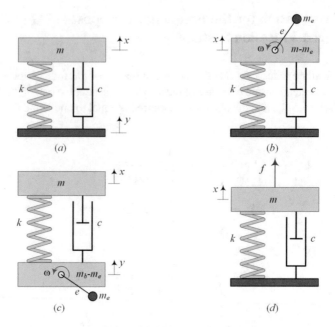

Fig. 6.58 The four practical types of *one* DOF harmonically excited systems: (**a**) base excitation, (**b**) eccentric excitation, (**c**) eccentric base excitation, and (**d**) forced excitation

$$\Phi_2 = \tan^{-1} \frac{-2\xi r}{1 - r^2} \tag{6.514}$$

$$\Phi_3 = \tan^{-1} \frac{2\xi r^3}{\left(1 - r^2\right)^2 + (2\xi r)^2} \tag{6.515}$$

The functions S_0 and G_0 are the main parts of all kinds of frequency response of single DOF harmonically excited systems of Fig. 6.58. To have a visual intuition of the behavior of the different responses, we usually plot them as a function of r and use ξ as a parameter. The mass m, stiffness k, and damping c of the system are fixed, and hence, the excitation frequency ω is the only quasi-static variable. We combine m, k, c, and ω, and we define the two parameters r and ξ to express the frequency responses by only two variable functions.

To classify the frequency responses of one DOF forced vibration systems, let us show the absolute displacement of the main mass by x, displacement excitation of the base by y, and relative displacement by $z = x - y$. We indicate the frequency responses related to the systems of Fig. 6.58 by adding a subscript and express their different responses as follows:

1. For a *base excited system*, we use the frequency responses of the relative and absolute motions denoted by Z_B, $\dot{Z}_B$, and $\ddot{Z}_B$, for relative displacement, velocity, and acceleration, and X_B, $\dot{X}_B$, and $\ddot{X}_B$, for absolute displacement, velocity, and acceleration of m, and the transmitted force to the base by F_{T_B}.
2. For an *eccentric excited system*, we use the frequency responses of the absolute motions X_E, $\dot{X}_E$, and $\ddot{X}_E$ and the transmitted force to the base by F_{T_E}.
3. For an *eccentric base excited system*, we use the frequency responses of the relative and absolute motions denoted by Z_R, $\dot{Z}_R$, and $\ddot{Z}_R$, for relative displacement, velocity, and acceleration, and X_R, $\dot{X}_R$, and $\ddot{X}_R$, for absolute displacement, velocity, and acceleration of m, and the transmitted force to the base by F_{T_R}.
4. For a *forced excited system*, we use the frequency responses of the absolute motions X_F, $\dot{X}_F$, and $\ddot{X}_F$ and the transmitted force to the base by F_{T_F}.

We summarized and label the frequency response of different characteristics of the four systems of Fig. 6.58 by the following nondimensionalized equations:

$$S_0 = \frac{X_F}{F/k} \tag{6.516}$$

$$S_1 = \frac{\dot{X}_F}{F/\sqrt{km}} \tag{6.517}$$

$$S_2 = \frac{\ddot{X}_F}{F/m} = \frac{Z_B}{Y} = \frac{X_E}{e\varepsilon_E} = \frac{Z_R}{e\varepsilon_R} \tag{6.518}$$

$$S_3 = \frac{\dot{Z}_B}{\omega_n Y} = \frac{\dot{X}_E}{e\varepsilon_E \omega_n} = \frac{\dot{Z}_R}{e\varepsilon_R \omega_n} \tag{6.519}$$

$$S_4 = \frac{\ddot{Z}_B}{\omega_n^2 Y} = \frac{\ddot{X}_E}{e\varepsilon_E \omega_n^2} = \frac{\ddot{Z}_R}{e\varepsilon_R \omega_n^2} \tag{6.520}$$

$$G_0 = \frac{F_{T_F}}{F} = \frac{X_B}{Y} \tag{6.521}$$

$$G_1 = \frac{\dot{X}_B}{\omega_n Y} \tag{6.522}$$

$$G_2 = \frac{\ddot{X}_B}{\omega_n^2 Y} = \frac{F_{T_B}}{kY} = \frac{F_{T_E}}{e\omega_n^2 m_e} = \frac{F_{T_R}}{e\omega_n^2 m_e}\left(1 + \frac{m_b}{m}\right) \tag{6.523}$$

Figures A.1, A.2, A.3, A.4, A.5, A.6, A.7 and A.8 in Appendix A visualize these frequency responses to guide design engineers to evaluate the behavior of the systems. The exact value of the responses should be found from the associated equations.

Proof The equations of motion of a harmonically forced vibrating one DOF system can always be modeled by a second-order differential equation, where the variable q is a general coordinate to show the absolute displacement x or relative displacement $z = x - y$.

$$m\ddot{q} + c\dot{q} + kq = f(q, \dot{q}, t) \tag{6.524}$$

The forcing term $f(x, \dot{x}, t)$ is a harmonic function, which in the general case can be a combination of $\sin \omega t$ and $\cos \omega t$, where ω is the excitation frequency.

$$f(q, \dot{q}, t) = a \sin \omega t + b \cos \omega t \tag{6.525}$$

Depending on the system and the frequency response that we are investigating, the coefficients a and b can be zero, constant, or proportional to ω, ω^2, ω^3, ω^4, ..., ω^n. To cover every practical harmonically forced vibrating system and derive Equations (6.504)–(6.511), let us assume a and b to be polynomial functions of ω.

$$a = a_0 + a_1\omega + a_2\omega^2 \tag{6.526}$$

$$b = b_0 + b_1\omega + b_2\omega^2 \tag{6.527}$$

We divide the equation of motion (6.524) by m to express it with ξ and ω_n.

$$\ddot{q} + 2\xi\omega_n\dot{q} + \omega_n^2 q = \left(A_0 + A_1\omega + A_2\omega^2\right)\sin \omega t$$

$$+ \left(B_0 + B_1\omega + B_2\omega^2\right)\cos \omega t \tag{6.528}$$

$$A_0 + A_1\omega + A_2\omega^2 = \frac{1}{m}\left(a_0 + a_1\omega + a_2\omega^2\right) \tag{6.529}$$

$$B_0 + B_1\omega + B_2\omega^2 = \frac{1}{m}\left(b_0 + b_1\omega + b_2\omega^2\right) \tag{6.530}$$

The solution of the equation of motion would be a harmonic response with unknown coefficients:

$$q = A \sin \omega t + B \cos \omega t = Q \sin (\omega t - \varphi) \tag{6.531}$$

To calculate the steady-state amplitude of the response Q and phase φ, we should substitute the solution (6.531) in the equation of motion.

$$-\omega^2 \left(A \sin \omega t + B \cos \omega t\right) + 2\xi \omega_n \omega \left(A \cos \omega t - B \sin \omega t\right)$$
$$+\omega_n^2 \left(A \sin \omega t + B \cos \omega t\right)$$
$$= \left(A_0 + A_1\omega + A_2\omega^2\right) \sin \omega t + \left(B_0 + B_1\omega + B_2\omega^2\right) \cos \omega t \qquad (6.532)$$

The coefficients of the functions $\sin \omega t$ and $\cos \omega t$ must balance on both sides of the equation.

$$\omega_n^2 A - \omega^2 A - 2\xi \omega \omega_n B = A_0 + A_1\omega + A_2\omega^2 \qquad (6.533)$$

$$2\xi A\omega\omega_n - B\omega^2 + B\omega_n^2 = B_0 + B_1\omega + B_2\omega^2 \qquad (6.534)$$

Therefore, we find two algebraic equations to calculate A and B:

$$\begin{bmatrix} \omega_n^2 - \omega^2 & -2\xi \omega \omega_n \\ 2\xi \omega \omega_n & \omega_n^2 - \omega^2 \end{bmatrix} \begin{bmatrix} A \\ B \end{bmatrix} = \begin{bmatrix} A_0 + A_1\omega + A_2\omega^2 \\ B_0 + B_1\omega + B_2\omega^2 \end{bmatrix} \qquad (6.535)$$

Solving for the coefficients A and B provides us with the steady-state solution amplitude Q and phase φ.

$$\begin{bmatrix} A \\ B \end{bmatrix} = \begin{bmatrix} \omega_n^2 - \omega^2 & -2\xi \omega \omega_n \\ 2\xi \omega \omega_n & \omega_n^2 - \omega^2 \end{bmatrix}^{-1} \begin{bmatrix} A_0 + A_1\omega + A_2\omega^2 \\ B_0 + B_1\omega + B_2\omega^2 \end{bmatrix}$$
$$= \begin{bmatrix} \dfrac{Z_1}{\left(1 - r^2\right)^2 + (2\xi r)^2} \\ \dfrac{Z_2}{\left(1 - r^2\right)^2 + (2\xi r)^2} \end{bmatrix} \qquad (6.536)$$

$$Z_1 = 2\xi r \frac{1}{\omega_n^2} \left(B_2\omega^2 + B_1\omega + B_0\right) + \frac{1}{\omega_n^2} \left(1 - r^2\right) \left(A_2\omega^2 + A_1\omega + A_0\right) \qquad (6.537)$$

$$Z_2 = \frac{1}{\omega_n^2} \left(1 - r^2\right) \left(B_2\omega^2 + B_1\omega + B_0\right) - 2\xi r \frac{1}{\omega_n^2} \left(A_2\omega^2 + A_1\omega + A_0\right) \qquad (6.538)$$

$$Q = \sqrt{A^2 + B^2} \qquad (6.539)$$

$$\tan \varphi = \frac{-B}{A} \qquad (6.540)$$

We are able to reproduce any of the steady-state responses S_i and G_i by setting the coefficients A_0, A_1, A_2, B_0, B_1, and B_2 properly. ∎

Example 231 Base excited frequency responses. This is an example to show that the steady-state frequency responses S_i and G_i can be recovered by substituting A_0, A_1, A_2, B_0, B_1, and B_2 from the equation of motion of the system.

A one DOF base excited vibrating system is shown in Fig. 6.58a. The equation of relative motion $z = x - y$ with a harmonic base excitation $y = Y \sin \omega t$ is

$$\ddot{z} + 2\xi \omega_n \dot{z} + \omega_n^2 z = \omega^2 Y \sin \omega t \tag{6.541}$$

This equation can be found from Eq. (6.528) if $q = z$ and

$$A_0 = 0 \qquad A_1 = 0 \qquad A_2 = Y$$
$$B_0 = 0 \qquad B_1 = 0 \qquad B_2 = 0 \tag{6.542}$$

Therefore, the frequency response of the system would be

$$Z = Q = \sqrt{A^2 + B^2}$$

$$= \sqrt{\left(\frac{1}{\omega_n^2} \frac{(1 - r^2) Y \omega^2}{(1 - r^2)^2 + (2\xi r)^2} \right)^2 + \left(\frac{1}{\omega_n^2} \frac{-2\xi r Y \omega^2}{(1 - r^2)^2 + (2\xi r)^2} \right)^2}$$

$$= \frac{r^2}{\sqrt{(1 - r^2)^2 + (2\xi r)^2}} Y \tag{6.543}$$

Example 232 Two-term forcing. Let us use the superposition principle and the general solutions S_i and G_i to determine frequency response of a system to multi-harmonic forcing function.

Consider a forced vibration system as shown in Fig. 6.58d

$$m\ddot{x} + c\dot{x} + kx = f \tag{6.544}$$

which the following harmonic forcing load is applied to.

$$f = F_1 \sin \omega t + F_2 \cos \omega t \tag{6.545}$$

Employing the parameters

$$A_0 = \frac{F_1}{m} \qquad A_1 = 0 \qquad A_2 = 0$$

$$B_0 = \frac{F_2}{m} \qquad B_1 - 0 \qquad B_2 = 0 \tag{6.546}$$

provides us with the following frequency response:

$$x = A \sin \omega t + B \cos \omega t = Q \sin (\omega t - \varphi) \tag{6.547}$$

$$A = \frac{1}{m\omega_n^2} \frac{2\xi r F_2 + \left(1 - r^2\right) F_1}{\left(1 - r^2\right)^2 + (2\xi r)^2} \tag{6.548}$$

$$B = \frac{1}{m\omega_n^2} \frac{\left(1 - r^2\right) F_2 - 2\xi r F_1}{\left(1 - r^2\right)^2 + (2\xi r)^2} \tag{6.549}$$

$$Q = \frac{1}{m\omega_n^2} \frac{\sqrt{F_1^2 + F_2^2}}{\sqrt{\left(1 - r^2\right)^2 + (2\xi r)^2}} \tag{6.550}$$

$$\varphi = \tan^{-1} \frac{2\xi r F_1 - \left(1 - r^2\right) F_2}{2\xi r F_2 + \left(1 - r^2\right) F_1} \tag{6.551}$$

Let us also solve the problem using the superposition principle. Two harmonic forces are applied on the system.

$$f_1 = F_1 \sin \omega t \qquad f_2 = F_2 \cos \omega t \tag{6.552}$$

Employing the solution (6.7), we have

$$\frac{X_1}{F_1/k} = \frac{1}{\sqrt{\left(1 - r^2\right)^2 + (2\xi r)^2}} \tag{6.553}$$

$$\frac{X_2}{F_2/k} = \frac{1}{\sqrt{\left(1 - r^2\right)^2 + (2\xi r)^2}} \tag{6.554}$$

$$Q = \sqrt{X_1^2 + X_2^2} = \frac{1}{k} \frac{\sqrt{F_1^2 + F_2^2}}{\sqrt{\left(1 - r^2\right)^2 + (2\xi r)^2}} \tag{6.555}$$

which is consistent with (6.550).

6.7   Summary

Frequency response analysis of one DOF systems is the most important analysis of vibrating systems in industry, because a vast amount of vibrating systems can be modeled by a one DOF mass-spring-damper system. The harmonic excitation is the most common excitation in industry. The response of a systems to harmonic excitation will be harmonic. The frequency response is the *steady-state* solution of the equation of motion at each frequency, when the system is *harmonically excited*. In frequency response analysis, we are interested in the steady-state amplitude of the oscillation as a function of the excitation frequency.

There are only four types of harmonically excited one DOF systems:

1. Base excitation
2. Eccentric excitation
3. Eccentric base excitation
4. Forced excitation

Base excitation is the most practical model for vertical vibration of mechanical systems. *Eccentric excitation* is a model for every type of rotary motor on a suspension. *Eccentric base excitation* is a model for the vibration of any equipment mounted on an engine. *Forced excitation* has almost no practical application; however, it is the simplest model to study forced vibration analysis methods.

Forced Excitation
The equation of motion of a harmonically forced system is

$$m\ddot{x} + c\dot{x} + kx = F \sin \omega t \tag{6.556}$$

which generates a frequency response expressed by either of the following functions:

$$x = A_1 \sin \omega t + B_1 \cos \omega t \tag{6.557}$$

$$= X \sin \left(\omega t - \varphi_x \right) \tag{6.558}$$

$$\begin{bmatrix} A_1 \\ B_1 \end{bmatrix} = \begin{bmatrix} \dfrac{F}{k} \dfrac{1 - r^2}{\left(1 - r^2\right)^2 + (2\xi r)^2} \\ \dfrac{F}{k} \dfrac{-2\xi r}{\left(1 - r^2\right)^2 + (2\xi r)^2} \end{bmatrix} \tag{6.559}$$

The steady-state response has an *amplitude* $X = \sqrt{A^2 + B^2}$ and a *phase* φ_x.

$$\frac{X}{F/k} = \frac{1}{\sqrt{\left(1 - r^2\right)^2 + (2\xi r)^2}} \tag{6.560}$$

$$\varphi_x = \tan^{-1} \frac{2\xi r}{1 - r^2} \tag{6.561}$$

where we use the *frequency ratio* r, *natural frequency* ω_n, and *damping ratio* ξ.

$$r = \frac{\omega}{\omega_n} \qquad \xi = \frac{c}{2\sqrt{km}} \qquad \omega_n = \sqrt{\frac{k}{m}} \tag{6.562}$$

Base Excitation

The base excited system is a good model for a vehicle suspension system or any equipment that is mounted on a vibrating base. The equation of motion of the system can be expressed by either one of the following equations for the absolute displacement x.

$$m\ddot{x} + c\dot{x} + kx = cY\omega \cos \omega t + kY \sin \omega t \tag{6.563}$$

$$\ddot{x} + 2\xi\omega_n \dot{x} + \omega_n^2 x = 2\xi\omega_n \omega Y \cos \omega t + \omega_n^2 Y \sin \omega t \tag{6.564}$$

$$\omega_n = \sqrt{\frac{k}{m}} \qquad r = \frac{\omega}{\omega_n} \qquad \xi = \frac{c}{2\sqrt{km}} \tag{6.565}$$

The system can also be expressed by either one of the following equations for the relative displacement z.

$$m\ddot{z} + c\dot{z} + kz = m\omega^2 Y \sin \omega t \tag{6.566}$$

$$\ddot{z} + 2\xi\omega_n \dot{z} + \omega_n^2 z = \omega^2 Y \sin \omega t \tag{6.567}$$

$$z = x - y \tag{6.568}$$

The equations of motion generate the absolute and relative steady-state responses as

$$x = A_2 \sin \omega t + B_2 \cos \omega t = X \sin \left(\omega t - \varphi_x\right) \tag{6.569}$$

$$z = A_3 \sin \omega t + B_3 \cos \omega t = Z \sin \left(\omega t - \varphi_z\right) \tag{6.570}$$

$$\frac{X}{Y} = \frac{\sqrt{1 + (2\xi r)^2}}{\sqrt{\left(1 - r^2\right)^2 + (2\xi r)^2}} \tag{6.571}$$

$$\frac{Z}{Y} = \frac{r^2}{\sqrt{\left(1 - r^2\right)^2 + (2\xi r)^2}} \tag{6.572}$$

$$\varphi_x = \tan^{-1} \frac{2\xi r^3}{1 - r^2 + (2\xi r)^2} \tag{6.573}$$

$$\varphi_z = \tan^{-1} \frac{2\xi r}{1 - r^2} \tag{6.574}$$

The phase φ_x indicates the *angular lag* of the response x with respect to the excitation y, and the phase φ_z is the *angular lag* of the relative motion response z with respect to the excitation y.

Eccentric Excitation
There is an unbalance mass m_e at an *eccentricity* distance e that is rotating with an angular velocity ω. An eccentric excited vibrating system is the proper model for vibration analysis of an engine of a vehicle or any rotary motor that is mounted on a stationary base with flexible suspension. The equation of motion of the system is

$$m\ddot{x} + c\dot{x} + kx = m_e e\omega^2 \sin \omega t \tag{6.575}$$

or equivalently is

$$\ddot{x} + 2\xi \omega_n \dot{x} + \omega_n^2 x = \varepsilon e\omega^2 \sin \omega t \tag{6.576}$$

$$\omega_n = \sqrt{\frac{k}{m}} \qquad r = \frac{\omega}{\omega_n} \qquad \xi = \frac{c}{2\sqrt{km}} \qquad \varepsilon = \frac{m_e}{m} \tag{6.577}$$

The absolute displacement response of the system is

$$x = A_4 \sin \omega t + B_4 \cos \omega t = X \sin (\omega t - \varphi_e) \tag{6.578}$$

$$\frac{X}{e\varepsilon} = \frac{r^2}{\sqrt{(1 - r^2)^2 + (2\xi r)^2}} \tag{6.579}$$

$$\varphi_e = \tan^{-1} \frac{2\xi r}{1 - r^2} \tag{6.580}$$

The phase φ_e indicates the angular lag of the response x with respect to the excitation $m_e e\omega^2 \sin \omega t$.

Eccentric Base Excitation
The base of a one DOF eccentric base excited vibrating system has an unbalance mass m_e at a eccentricity distance e that is rotating with angular velocity ω. The eccentric base excited system is the proper model for vibration analysis of different equipment that are attached to the engine of a vehicle or any equipment mounted on a rotary motor.

Using the relative motion z of m with respect to the base, the equation of motion of the system is

$$\frac{mm_b}{m_b + m}\ddot{z} + c\dot{z} + kz = \frac{mm_e}{m_b + m}e\omega^2 \sin\omega t \qquad (6.581)$$

$$z = x - y \qquad (6.582)$$

or, equivalently, is

$$\ddot{z} + 2\xi\omega_n\dot{z} + \omega_n^2 z = \varepsilon e\omega^2 \sin\omega t \qquad (6.583)$$

$$\omega_n = \sqrt{\frac{k}{m}} \qquad r = \frac{\omega}{\omega_n} \qquad \xi = \frac{c}{2\sqrt{km}} \qquad \varepsilon = \frac{m_e}{m_b} \qquad (6.584)$$

The relative displacement response of the system is

$$z = A_5 \sin\omega t + B_5 \cos\omega t = Z \sin\left(\omega t - \varphi_b\right) \qquad (6.585)$$

which has an amplitude Z and phase φ_b.

$$\frac{Z}{e\varepsilon} = \frac{r^2}{\sqrt{\left(1 - r^2\right)^2 + (2\xi r)^2}} \qquad (6.586)$$

$$\varphi_b = \tan^{-1}\frac{2\xi r}{1 - r^2} \qquad (6.587)$$

Any harmonically excited one DOF system is one of the above four systems. Any steady-state response of these systems is equal to one of the following equations.

$$S_0 = \frac{1}{\sqrt{\left(1 - r^2\right)^2 + (2\xi r)^2}} \qquad (6.588)$$

$$S_1 = \frac{r}{\sqrt{\left(1 - r^2\right)^2 + (2\xi r)^2}} = r S_0 \qquad (6.589)$$

$$S_2 = \frac{r^2}{\sqrt{\left(1 - r^2\right)^2 + (2\xi r)^2}} = r^2 S_0$$

$$S_3 = \frac{r^3}{\sqrt{\left(1 - r^2\right)^2 + (2\xi r)^2}} = r^3 S_0 \tag{6.590}$$

$$S_4 = \frac{r^4}{\sqrt{\left(1 - r^2\right)^2 + (2\xi r)^2}} = r^4 S_0 \tag{6.591}$$

$$G_0 = \frac{\sqrt{1 + (2\xi r)^2}}{\sqrt{\left(1 - r^2\right)^2 + (2\xi r)^2}} \tag{6.592}$$

$$G_1 = \frac{r\sqrt{1 + (2\xi r)^2}}{\sqrt{\left(1 - r^2\right)^2 + (2\xi r)^2}} = r G_0 \tag{6.593}$$

$$G_2 = \frac{r^2\sqrt{1 + (2\xi r)^2}}{\sqrt{\left(1 - r^2\right)^2 + (2\xi r)^2}} = r^2 G_0 \tag{6.594}$$

$$\Phi_0 = \tan^{-1} \frac{2\xi r}{1 - r^2} \tag{6.595}$$

$$\Phi_1 = \tan^{-1} \frac{1 - r^2}{-2\xi r} \tag{6.596}$$

$$\Phi_2 = \tan^{-1} \frac{-2\xi r}{1 - r^2} \tag{6.597}$$

$$\Phi_3 = \tan^{-1} \frac{2\xi r^3}{\left(1 - r^2\right)^2 + (2\xi r)^2} \tag{6.598}$$

The functions S_0 and G_0 are the main parts of all kinds of frequency response of single DOF harmonically excited systems. To classify the frequency responses of one DOF forced vibration systems, let us show the absolute displacement of the main mass by x, displacement excitation of the base by y, and relative displacement by $z = x - y$. We indicate the frequency responses related to the systems by adding a subscript and express their different responses as follows:

1. For a *base excited system,* we use the frequency responses of the relative and absolute motions denoted by Z_B, $\dot{Z}_B$, and $\ddot{Z}_B$, for relative displacement, velocity, and acceleration, and X_B, $\dot{X}_B$, and $\ddot{X}_B$, for absolute displacement, velocity, and acceleration of m, and the transmitted force to the base by F_{T_B}.

2. For an *eccentric excited system*, we use the frequency responses of the absolute motions X_E, $\dot{X}_E$, and $\ddot{X}_E$ and the transmitted force to the base by F_{T_E}.
3. For an *eccentric base excited system*, we use the frequency responses of the relative and absolute motions denoted by Z_R, $\dot{Z}_R$, and $\ddot{Z}_R$, for relative displacement, velocity, and acceleration, and X_R, $\dot{X}_R$, and $\ddot{X}_R$, for absolute displacement, velocity, and acceleration of m, and the transmitted force to the base by F_{T_R}.
4. For a *forced excited system*, we use the frequency responses of the absolute motions X_F, $\dot{X}_F$, and $\ddot{X}_F$ and the transmitted force to the base by F_{T_F}.

We summarized and label the frequency response of different characteristics of the four systems by the following nondimensionalized equations.

$$S_0 = \frac{X_F}{F/k} \tag{6.599}$$

$$S_1 = \frac{\dot{X}_F}{F/\sqrt{km}} \tag{6.600}$$

$$S_2 = \frac{\ddot{X}_F}{F/m} = \frac{Z_B}{Y} = \frac{X_E}{e\varepsilon_E} = \frac{Z_R}{e\varepsilon_R} \tag{6.601}$$

$$S_3 = \frac{\dot{Z}_B}{\omega_n Y} = \frac{\dot{X}_E}{e\varepsilon_E \omega_n} = \frac{\dot{Z}_R}{e\varepsilon_R \omega_n} \tag{6.602}$$

$$S_4 = \frac{\ddot{Z}_B}{\omega_n^2 Y} = \frac{\ddot{X}_E}{e\varepsilon_E \omega_n^2} = \frac{\ddot{Z}_R}{e\varepsilon_R \omega_n^2} \tag{6.603}$$

$$G_0 = \frac{F_{T_F}}{F} = \frac{X_B}{Y} \tag{6.604}$$

$$G_1 = \frac{\dot{X}_B}{\omega_n Y} \tag{6.605}$$

$$G_2 = \frac{\ddot{X}_B}{\omega_n^2 Y} = \frac{F_{T_B}}{kY} = \frac{F_{T_E}}{e\omega_n^2 m_e} = \frac{F_{T_R}}{e\omega_n^2 m_e}\left(1 + \frac{m_b}{m}\right) \tag{6.606}$$

Figures A.1, A.1, A.2, A.3, A.4, A.5, A.6, A.7 and A.8 in Appendix A visualize these frequency responses to guide design engineers to evaluate the behavior of the systems.

6.8   Key Symbols

$a \equiv \ddot{x}$	Acceleration
a, b, C	Fourier coefficient
a, b, c, d, e	Dimension exponent
A, B	Weight factor, coefficients for frequency responses
$A, \bar{A}$	Complex conjugate coefficients
c	Damping
C	Coefficient
d_1	Road wave length
d_2	Road wave amplitude
D	Dissipation function
e	Eccentricity arm, exponential function
E	Mechanical energy,
E_c	Consumed energy of a damper
$f = 1/T$	Cyclic frequency [Hz]
$f, F, \mathbf{F}$	Force
f, g	Function, periodic function
f_c	Damper force
f_e	Equivalent force
f_k	Spring force
f_m	Required force to move a mass m
f_T	Transmitted force
F	Amplitude of the harmonic force $f = F \sin \omega t$
F_T	Amplitude of transmitted force
F_0	Constant force
g	Gravitational acceleration, function
$g\left(r^2\right)$	Characteristic equation
G	General frequency response with node
H	Complex amplitude
H^*	Complex conjugate of H
j	Dummy index
k	Stiffness
k_e	Equivalent stiffness

K	Kinetic energy
l	Length
L	Length dimension
$\mathcal{L}$	Lagrangian
m	Mass, number of octave
m_b	Device mass
m_e	Eccentric mass, equivalent mass
M	Mass dimension
p	Pitch of a coil
P	Power, point
P	Potential energy
q	General coordinate
Q	Quality factor, general amplitude
r	Frequency ratio
$r_i,\ i \in N$	Nodal frequency ratio
$r_n = \omega_n/\omega_s$	Natural frequency ratio
$S_u = RMS(u)$	RMS of u
$S_\eta = RMS(\eta)$	RMS of η
S	General frequency response without node
t	Time
T	Period, time dimension
$v \equiv \dot{x}$	Velocity
$x, y, z,\ \mathbf{x}$	Displacement
x_0	Initial displacement
x_s	Sprung mass displacement
x_u	Unsprung mass displacement
X	Steady-state amplitude of x
X_s	Steady-state amplitude of x_s
X_u	Steady-state amplitude of x_u
$\dot{x}, \dot{y}, \dot{z}$	Velocity, time derivative of x, y, z
$\dot{x}_0$	Initial velocity
X, Y, Z	Amplitude

Y	Amplitude of base displacement
$\dot{y}$	Velocity
z	Relative displacement
Z	Amplitude of relative displacement
Z_1, Z_2	Short notation

Greek

α	Angle
δ	Difference
δ_s	Static deflection
Δ	Difference
γ	Angle
λ	Eigenvalue
Λ	Eigenvalue matrix
ε	Mass ratio
ε	Small coefficient

θ	Angular motion
Π	Buckingham dimensionless term
ρ	Length mass density
ξ	Damping ratio
ω	Angular frequency
ω_n	Natural frequency
φ, Φ	Phase angle
$\alpha = \omega_s/\omega_u$	Sprung mass ratio
$\varepsilon = m_s/m_u$	Sprung mass ratio
$\eta = |Z/Y|$	Sprung mass relative frequency response
$\mu = |X_s/Y|$	Sprung mass frequency response
$\xi = c_s/\left(2\sqrt{k_s m_s}\right)$	Damping ratio
$\xi^\star$	Optimal damping ratio
$\tau = |X_u/Y|$	Unsprung mass frequency response

$\omega = 2\pi f$ Angular frequency [rad / s]

$\omega_s = \sqrt{k_s/m_s}$ Sprung mass frequency

$\omega_u = \sqrt{k_u/m_u}$ Unsprung mass frequency

Subscript

$i \in N$ Node number

n Natural

s Sprung

u Unsprung

Symbols

DOF Degree of freedom

FBD Free body diagram

ODF One degree of freedom

Re   Real part

RMS Root mean square

Exercises

1. Forced excitation and spring stiffness.
   A forced excited mass-spring-damper system has $m = 200\,kg$ and $c = 100\,Ns/m$. Determine the stiffness of the spring, k, such that the natural frequency of the system is 1 Hz. What would be the amplitude of displacement, velocity, and acceleration of m if this force F is applied to the mass m:

$$F = 100 \sin 10t \tag{6.607}$$

2. Forced excitation and system parameters.
   A forced excited m-k-c system is under a force F.

$$F = 100 \sin 10t \tag{6.608}$$

   If the mass $m = 120\,kg$ should not have a dimensionless steady-state amplitude higher than 2 when it is excited at the natural frequency, determine c, k, X, φ_x, and F_T.

3. Base excited system and spring stiffness.
   A base excited m-k-c system has $m = 200\,kg$ and $c = 100\,Ns/m$. Determine the stiffness of the spring, k, such that the steady-state amplitude of m is less than 0.07 m when the base is excited by y at the natural frequency of the system.

$$y = 0.05 \sin 2\pi t \tag{6.609}$$

4. Base excited system and absolute acceleration.
   Assume a base excited m-k-c system is vibrating at the node of its absolute acceleration frequency response. If the base is excited according to

$$y = 0.06 \sin 2\pi t \tag{6.610}$$

   determine ω_n, $\ddot{X}$, and X.

5. Eccentric excitation and transmitted force.
   An engine with mass $m = 170\,kg$ and eccentricity $m_e e = 0.4 \times 0.1\,kg\,m$ is turning at $\omega_e = 4000\,rpm$.

   (a) Determine the steady-state amplitude of its vibration, if there are four engine mounts, each with $k = 9800\,N/m$ and $c = 100\,Ns/m$.
   (b) Determine the transmitted force to the base.

6. ★ Eccentric base excitation and absolute displacement.
   An eccentric base excited system has $m = 3\,kg$, $m_b = 175\,kg$, $m_e e = 0.4 \times 0.1\,kg\,m$, and $\omega = 4100\,rpm$. If $Z/(e\varepsilon) = 2$ at $r = 1$, calculate X and Y.

7. Characteristic values and free vibrations.
   An m-k-c system has $m = 250\,\text{kg}$, $k = 8500\,\text{N}/\text{m}$, and $c = 1000\,\text{N}\,\text{s}/\text{m}$. Determine the characteristic values of the system and its free vibration response for zero initial conditions.

8. The frequency r_{Max} of forced vibrations.
   Figure 6.9 illustrates the frequency r_{Max} associated with the maximum amplitude $X_{Max}/(F/k)$ as a function of ξ.

   (a) What is the maximum ξ at which r_{Max} reduces to zero? Let us denote this value by ξ_c and call it the critical damping ratio.
   (b) The r_{Max} is very sensitive to ξ when ξ is close to ξ_c, and it is not sensitive when ξ is close to zero. Explain, justify, and compare the systems.

9. Maximum amplitude of forced vibrations.
   Figure 6.10 illustrates the maximum amplitude $X_{Max}/(F/k)$ as a function of ξ.

   (a) What is the value of ξ at which $X_{Max}/(F/k)$ is minimum? Let us denote this value by ξ_{min}.
   (b) Is ξ_{min} equal to ξ_c? The ξ_c is the damping ratio at which r_{Max} reduces to zero.
   (c) Is Fig. 6.10 consistent with Fig. 6.11?
   (d) What is the meaning of having $X_{Max}/(F/k) > 1$ for $\xi > \xi_{min}$?

10. Forced excitation and spring stiffness.
    A forced excited mass-spring-damper system has $m = 200\,\text{kg}$ and $c = 1000\,\text{N}\,\text{s}/\text{m}$. Determine the stiffness of the spring, k, such that the natural frequency of the system is 1 Hz. What would be the amplitude of displacement, velocity, and acceleration of m if a force F is applied to the mass m.

$$F = 100\sin 10t \tag{6.611}$$

11. Forced excitation and system parameters.
    A forced excited m-k-c system is under a force F. If the mass $m = 200\,\text{kg}$ should not have a dimensionless steady-state amplitude higher than 2 when it is excited at the natural frequency, determine m, c, k, X, φ_x, and F_T.

$$F = 100\sin 10t \tag{6.612}$$

12. Base excited system and spring stiffness.
    A base excited m-k-c system has $m = 200\,\text{kg}$ and $c = 1000\,\text{N}\,\text{s}/\text{m}$. Determine the stiffness of the spring, k, such that the steady-state amplitude of m is less than 0.07 m when the base is excited by y at the natural frequency of the system.

$$y = 0.05\sin 2\pi t \tag{6.613}$$

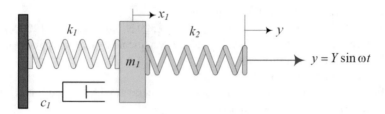

Fig. 6.59 A mass-spring-damper system that is excited by a harmonic displacement through a flexible connection

13. Base excited system and absolute acceleration.
    Assume that a base excited m-k-c system is vibrating at the node of its absolute acceleration frequency response. If the base is excited by y, determine ω_n, $\ddot{X}$, and X.

$$y = 0.05 \sin 2\pi t \qquad\qquad (6.614)$$

14. Frequency response of a flexible displacement excitation.
    Figure 6.59 illustrates a mass-spring-damper system that is excited by a harmonic displacement through a flexible connection. Determine the equation of motion and frequency response, and plot the frequency response for different damping.
15. Eccentric excitation and transmitted force.
    An engine with mass $m = 175\,\text{kg}$ and eccentricity $m_e e = 0.4 \times 0.1\,\text{kg m}$ is turning at $\omega_e = 4000\,rpm$.

    (a) Determine the steady-state amplitude of its vibration, if there are four engine mounts, each with $k = 10{,}000\,\text{N}/\text{m}$ and $c = 100\,\text{N s}/\text{m}$.
    (b) Determine the force transmitted to the base.

16. ★ Eccentric base excitation and absolute displacement.
    An eccentric base excited system has $m = 3\,\text{kg}$, $m_b = 175\,\text{kg}$, $m_e e = 0.4 \times 0.1\,\text{kg m}$, and $\omega = 4000\,rpm$. If $Z/(e\varepsilon) = 2$ at $r = 1$, calculate X and Y.
17. ★ Horizontal distance of X_{Max} curves.
    In forced excited one DOF systems, the maximum amplitude X_{Max} occurs at different frequencies. Determine the horizontal distance of X_{Max} from 1, as a function of the damping ratio ξ.
18. ★ The maximum ξ and X_{Max}.
    In forced excited one DOF systems, there is maximum value of damping ratio ξ_X for which a maximum amplitude X_{Max} exists.

    (a) Determine the maximum damping ratio ξ_X such that there is no X_{Max} for $\xi > \xi_X$.
    (b) Prove that $X/(F/k) < 1$ for $\xi > \xi_X$.

19. ★ Frequency ratio of $\dot{X}_{Max}$ curves.
    Prove that in forced excited one DOF systems, the maximum amplitude of $\dot{X}_{Max}$ always occurs at $r = 1$.

20. ★ Frequency ratio of $\ddot{X}_{Max}$ curves.
    In forced excited one DOF systems, determine the frequency at which $\ddot{X}/(F/m)$ is maximum, and plot r_{Max} as a function of ξ.

21. ★ Horizontal distance of $\ddot{X}_{Max}$ curves.
    In forced excited one DOF systems, the maximum amplitude of $\ddot{X}_{Max}$ occurs at different frequencies. Determine the horizontal distance of $\ddot{X}_{Max}$ from 1, as a function of the damping ratio ξ.

22. ★ The maximum ξ and $\ddot{X}_{Max}$.
    In forced excited one DOF systems, there is a maximum value of damping ratio $\xi_{\ddot{X}}$ for which a maximum amplitude $\ddot{X}_{Max}$ exists.

    (a) Determine the maximum damping ratio ξ_M such that there is no $\ddot{X}_{Max}$ for $\xi > \xi_M$.
    (b) Prove that $\ddot{X}/(F/m) < 1$ for $\xi > \xi_{\ddot{X}}$.

23. ★ Transmitted force and amplitude of vibrations.
    In forced excited one DOF systems, plot the transmitted F_T/F versus vibration amplitude $X/(F/k)$, and explain the accuracy of this sentence: decreasing amplitude of vibration increases the transmitted force.

24. ★ Frequency ratio of $F_{T_{Max}}$ curves.
    In forced excited one DOF systems, determine the frequency at which F_T/F is maximum, and plot r_{Max} as a function of ξ.

25. ★ Horizontal distance of $F_{T_{Max}}$ curves.
    In forced excited one DOF systems, the maximum amplitude of F_T/F occurs at different frequencies. Determine the horizontal distance of $F_{T_{Max}}$ from 1, as a function of the damping ratio ξ.

26. ★ Frequency ratio of Z/Y and X/Y curves.
    In base excited one DOF systems,

    (a) determine the frequency at which Z/Y is maximum and plot r_{Max} as a function of ξ;
    (b) determine the frequency at which X/Y is maximum and plot r_{Max} as a function of ξ.

27. ★ Horizontal distance of Z/Y and X/Y curves.
    In base excited one DOF systems, the maximum amplitudes of Z/Y and X/Y occur at different frequencies.

    (a) Determine the horizontal distance of Z_{Max}/Y from 1, as a function of the damping ratio ξ.
    (b) Determine the horizontal distance of X_{Max}/Y from 1, as a function of the damping ratio ξ.

28. ★ The maximum ξ and Z_{Max}/Y.
In base excited one DOF systems, there is maximum value of damping ratio ξ_M for which a maximum amplitude Z_{Max}/Y exists.

(a) Determine the maximum damping ratio ξ_M such that there is no Z_{Max}/Y for $\xi > \xi_M$.
(b) Prove that $Z/Y < 1$ for $\xi > \xi_Z$.

29. ★ Frequency ratio of $\dot{X}/Y$ and $\ddot{X}/Y$ curves.
In base excited one DOF systems,

(a) determine the frequency at with $\dot{X}/Y$ is maximum and plot r_{Max} as a function of ξ;
(b) determine the frequency at with $\ddot{X}/Y$ is maximum and plot r_{Max} as a function of ξ.

30. ★ Horizontal distance of $\dot{X}/Y$ and $\ddot{X}/Y$ curves.
In base excited one DOF systems, the maximum amplitudes of $\dot{X}/Y$ and $\ddot{X}/Y$ occur at different frequencies.

(a) Determine the horizontal distance of $\dot{X}_{Max}/Y$ from 1, as a function of damping ratio ξ.
(b) Determine the horizontal distance of $\ddot{X}_{Max}/Y$ from 1, as a function of damping ratio ξ.

31. ★ Variable amplitude of base excitation.
Consider a base excited one DOF system. If the base is moving by y, what would be the frequency response of the absolute and relative displacements X and Z?

$$y = Y \sin \omega t \qquad\qquad (6.615)$$

$$Y = \frac{\sqrt{1 + (2\xi r)^2}}{\sqrt{\left(1 - r^2\right)^2 + (2\xi r)^2}} \qquad\qquad (6.616)$$

32. 1/8 car model.
Consider a one-eight car model as a base excited one DOF system. Determine its natural ω_n and damped natural frequencies ω_d if

$$m = 1245\,\text{kg} \qquad k = 60{,}000\,\text{N}/\text{m} \qquad c = 2400\,\text{N}\,\text{s}/\text{m}$$

33. Quarter car model.
Consider a quarter car model. Determine its natural frequencies and mode shapes if

$$m_s = 1085/4\,\text{kg} \qquad m_u = 40\,\text{kg}$$
$$k_s = 10{,}000\,\text{N}/\text{m} \qquad k_u = 150{,}000\,\text{N}/\text{m} \qquad c_s = 800\,\text{N}\,\text{s}/\text{m}$$

Chapter 7
Multi Degrees of Freedom Systems

Systems with more than one degree of freedom (DOF), such as the one in Fig. 7.1, introduce multi natural frequencies, mode shapes, and mode interaction. These characteristics make the multi DOF systems different from one DOF systems. This chapter is to study the frequency response of multi DOF systems.

7.1 Natural Frequency and Mode Shape

Free and undamped vibrations of a system are the basic response of the system which express its natural behavior. We call a system with no external excitation a *free system* and with no damping an *undamped system*. Using the set of generalized coordinates,

$$\mathbf{x} = \left[x_1 \ x_2 \ \cdots \ x_n \right]^T \tag{7.1}$$

a linear undamped-free system is governed by the following set of differential equations.

$$[m]\,\ddot{\mathbf{x}} + [k]\,\mathbf{x} = \mathbf{0} \tag{7.2}$$

As long as the mass matrix $[m]$ and stiffness matrix $[k]$ are constant, the time response of the free system is harmonic:

$$\mathbf{x} = \sum_{i=1}^{n} \mathbf{u}_i \left(A_i \sin \omega_i t + B_i \cos \omega_i t \right) \quad i = 1, 2, 3, \cdots, n$$

$$= \sum_{i=1}^{n} C_i \mathbf{u}_i \sin \left(\omega_i t + \varphi_i \right) \quad i = 1, 2, 3, \cdots, n \tag{7.3}$$

© The Author(s), under exclusive license to Springer Nature Switzerland AG 2022
R. N. Jazar, *Advanced Vibrations*, https://doi.org/10.1007/978-3-031-16356-2_7

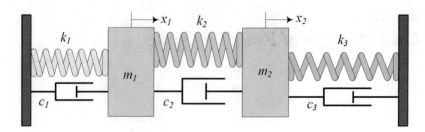

Fig. 7.1 A two-*DOF* vibrating system

$$C_i = \sqrt{A_i^2 + B_i^2} \qquad (7.4)$$

$$\tan \varphi_i = \frac{B_i}{A_i} \qquad (7.5)$$

where ω_i are the *natural frequencies* and $\mathbf{u}_i$ are the *mode shapes* of the system. The unknown coefficients A_i and B_i, or C_i and φ_i, must be determined from the initial conditions.

The natural frequencies ω_i are solutions of the characteristic equation of the system

$$\det \left[[k] - \omega^2 [m] \right] = 0, \qquad (7.6)$$

and the mode shape $\mathbf{u}_i$, corresponding to ω_i, is the solution of the eigenvector equation.

$$\left[[k] - \omega_i^2 [m] \right] \mathbf{u}_i = 0 \qquad (7.7)$$

Mode shapes are orthogonal with respect to the mass $[m]$ and stiffness $[k]$ matrices.

$$\mathbf{u}_j^T [m] \mathbf{u}_i = 0 \qquad i \neq j \qquad (7.8)$$

$$\mathbf{u}_j^T [k] \mathbf{u}_i = 0 \qquad i \neq j \qquad (7.9)$$

Employing the orthogonality property of the mode shapes, we can determine the generalized mass m_i and stiffness k_i, associated to $\mathbf{u}_i$.

$$\mathbf{u}_i^T [m] \mathbf{u}_i = m_i \qquad (7.10)$$

$$\mathbf{u}_i^T [k] \mathbf{u}_i = k_i \qquad (7.11)$$

Proof The general equations of motion of any linear multi degree-of-freedom (DOF) vibrating system is:

$$[m]\ddot{\mathbf{x}} + [c]\dot{\mathbf{x}} + [k]\mathbf{x} = \mathbf{F} \tag{7.12}$$

Eliminating the force and damping terms from the general equations of motion provides us with the equations of the undamped-free system.

$$[m]\ddot{\mathbf{x}} + [k]\mathbf{x} = \mathbf{0} \tag{7.13}$$

Let us search for a possible solution of the following form.

$$\mathbf{x} = \mathbf{u}\,q(t) \tag{7.14}$$

$$x_i = u_i\,q(t) \qquad i = 1, 2, 3, \cdots, n \tag{7.15}$$

$$\mathbf{u} = \begin{bmatrix} \mathbf{u}_1 & \mathbf{u}_2 & \cdots & \mathbf{u}_n \end{bmatrix}^T \tag{7.16}$$

This solution implies that the amplitude ratio of any two coordinates during motion does not depend on time. Substituting (7.14) into Eq. (7.13)

$$[m]\mathbf{u}\,\ddot{q}(t) + [k]\mathbf{u}\,q(t) = \mathbf{0} \tag{7.17}$$

and separating the time dependent terms yield:

$$-\frac{\ddot{q}(t)}{q(t)} = [[m]\mathbf{u}]^{-1}\,[[k]\mathbf{u}] = \frac{\sum_{j=1}^{n} k_{ij}u_j}{\sum_{j=1}^{n} m_{ij}u_j} = \omega^2 \qquad i = 1, 2, \cdots, n \tag{7.18}$$

Because the right-hand side of this equation is time independent and the left-hand side is independent of the index i, both sides must be equal to a constant. Let us assume the constant be a positive number ω^2. Hence, Eq. (7.18) can be separated into two equations,

$$\ddot{q}(t) + \omega^2 q(t) = 0 \tag{7.19}$$

and

$$\left[[k] - \omega^2\,[m] \right]\mathbf{u} = 0 \tag{7.20}$$

$$\sum_{j=1}^{n} \left(k_{ij} - \omega^2 m_{ij} \right) u_j = 0 \qquad i = 1, 2, 3, \cdots n. \tag{7.21}$$

The solution of (7.19) is:

$$q(t) = \sin(\omega t + \varphi) = \cos\varphi \sin\omega t + \sin\varphi \cos\omega t \tag{7.22}$$

which shows that all the coordinates of the system, x_i, have a harmonic motion with identical frequency ω and identical phase angle φ. The frequency ω will be determined from Eq. (7.20), which is a set of homogeneous equations for the unknown $\mathbf{u}$.

Both left and right sides of Eq. (7.18) must be equal to a constant. The sign of the constant ω^2 is dictated by physical considerations. A free and undamped vibrating system is conservative and has a constant mechanical energy, so the amplitude of vibration must remain finite when $t \to \infty$. If the constant is positive, then the response will be harmonic with a constant amplitude; however, if the constant is negative, the response will be hyperbolic with an exponentially increasing amplitude. Therefore, the sign of the constant ω^2 must be positive.

The set of Eqs. (7.20) always has a *trivial solution* $\mathbf{u} = \mathbf{0}$, which is the *rest position* of the system and shows no motion. To have a *nontrivial solution*, the determinant of the coefficient matrix must be zero:

$$\det\left[[k] - \omega^2[m]\right] = 0 \tag{7.23}$$

Determining the constant ω, such that the set of Eqs. (7.20) provide us with a nontrivial solution, is called the *eigenvalue problem*. Expanding the determinant (7.23) provides us with the algebraic *characteristic equation*. The characteristic equation is an nth order equation in terms of ω^2 and provides us with n natural frequencies ω_i. We usually set the natural frequencies ω_i in numerical order:

$$\omega_1 \leq \omega_2 \leq \omega_3 \leq \cdots \leq \omega_n \tag{7.24}$$

Having n values for ω indicates that the solution (7.22) is possible for n different frequencies $\omega_i, i = 1, 2, 3, \cdots, n$.

We may multiply Eq. (7.13) by $[m]^{-1}$:

$$\ddot{\mathbf{x}} + [m]^{-1}[k]\mathbf{x} = \ddot{\mathbf{x}} + [A]\mathbf{x} = \mathbf{0} \tag{7.25}$$

and find the characteristic equation (7.23) as

$$\det[[A] - \lambda\mathbf{I}] = 0 \tag{7.26}$$

where

$$[A] = [m]^{-1}[k] \tag{7.27}$$

Therefore, determination of the natural frequencies ω_i would be equivalent to determining the eigenvalues λ_i of the *characteristic matrix* $[A] = [m]^{-1}[k]$:

$$\lambda_i = \omega_i^2 \tag{7.28}$$

Determination of the vectors $\mathbf{u}_i$ to satisfy Eq. (7.20) is called the *eigenvector problem*. To determine $\mathbf{u}_i$, we substitute ω_i into Eq. (7.20) and solve (7.29) to find n different $\mathbf{u}_i$.

$$\left[[k] - \omega_i^2 [m] \right] \mathbf{u}_i = 0 \tag{7.29}$$

In mechanical vibrations, the eigenvector $\mathbf{u}_i$ corresponding to the eigenvalue ω_i is called the *mode shape*. Alternatively, we may find the eigenvectors of the matrix $[A] = [m]^{-1} [k]$ instead of finding the mode shapes from (7.29).

$$[[A] - \lambda_i \mathbf{I}] \mathbf{u}_i = 0 \tag{7.30}$$

Equations (7.29) are homogeneous, so if $\mathbf{u}_i$ is a solution, then $a\mathbf{u}_i$, $a \in \mathbb{R}$, is also a solution. Hence, the eigenvectors are not unique and may be expressed with any length. However, the ratio of any two elements of an eigenvector is unique and, therefore, $\mathbf{u}_i$ has a unique shape. If one of the elements of $\mathbf{u}_i$ is assigned, the remaining $n - 1$ elements are uniquely determined. The shape of an eigenvector $\mathbf{u}_i$ indicates the relative amplitudes of the coordinates of the system in vibration. Mode shapes are the principal elements of all possible free vibrations of a multi *DOF* system.

Because the length of an eigenvector is not uniquely defined, there are many options to express $\mathbf{u}_i$. The most common expressions are:

1. Normalization
2. Normal form
3. High-unit
4. First-unit
5. Last-unit

In the *normalization* expression, we adjust the length of $\mathbf{u}_i$ such that

$$\mathbf{u}_i^T [m] \mathbf{u}_i = 1 \tag{7.31}$$

or

$$\mathbf{u}_i^T [k] \mathbf{u}_i = 1 \tag{7.32}$$

and call $\mathbf{u}_i$ a *normal mode* with respect to $[m]$ or $[k]$, respectively.

In the *normal form* expression, we adjust $\mathbf{u}_i$ such that its length has a unity value.

In the *high-unit* expression, we adjust the length of $\mathbf{u}_i$ such that the largest element has a unity value.

In the *first-unit* expression, we adjust the length of $\mathbf{u}_i$ such that the first element has a unity value.

In the *last-unit* expression, we adjust the length of $\mathbf{u}_i$ such that the last element has a unity value.

Let us write Eq. (7.7) for two different natural frequencies $\omega_i, \omega_j, i \neq j$, as

$$[k]\,\mathbf{u}_i = \omega_i^2\,[m]\,\mathbf{u}_i \tag{7.33}$$

$$[k]\,\mathbf{u}_j = \omega_j^2\,[m]\,\mathbf{u}_j \tag{7.34}$$

Multiplying $\mathbf{u}_j^T$ by Eq. (7.33) and multiplying the transpose of (7.34) by $\mathbf{u}_i$, respectively, yield:

$$\mathbf{u}_j^T\,[k]\,\mathbf{u}_i = \omega_i^2\,\mathbf{u}_j^T\,[m]\,\mathbf{u}_i \tag{7.35}$$

$$\mathbf{u}_j^T\,[k]^T\,\mathbf{u}_i = \omega_j^2\,\mathbf{u}_j^T\,[m]^T\,\mathbf{u}_i \tag{7.36}$$

Assuming $[k]$ and $[m]$ are symmetric matrices, $[k] = [k]^T$, and we can write Eq. (7.36) as

$$\mathbf{u}_j^T\,[k]\,\mathbf{u}_i = \omega_j^2\,\mathbf{u}_j^T\,[m]\,\mathbf{u}_i \tag{7.37}$$

Subtracting (7.37) from (7.35)

$$\left(\omega_i^2 - \omega_j^2\right)\,\mathbf{u}_j^T\,[m]\,\mathbf{u}_i = 0 \tag{7.38}$$

we conclude that

$$\begin{aligned}\mathbf{u}_j^T\,[m]\,\mathbf{u}_i &= 0 \quad i \neq j \\ \mathbf{u}_j^T\,[m]\,\mathbf{u}_i &\neq 0 \quad i = j\end{aligned} \tag{7.39}$$

provided that $\omega_i \neq \omega_j$ and $[m]$ is a positive definite matrix. The mass matrix must be positive definite to guarantee that $\mathbf{u}_i^T\,[m]\,\mathbf{u}_i$ is not equal to zero for the nonzero mode shape $\mathbf{u}_i$. Therefore, we may write Eq. (7.39) as

$$\begin{aligned}\mathbf{u}_j^T\,[m]\,\mathbf{u}_i &= 0 \quad i \neq j \\ \mathbf{u}_j^T\,[m]\,\mathbf{u}_i &= m_i \quad i = j\end{aligned} \tag{7.40}$$

where m_i is called the *generalized mass* of the system. We may also write Eq. (7.7) for two different natural frequencies as:

$$\frac{1}{\omega_i^2}\,[k]\,\mathbf{u}_i = [m]\,\mathbf{u}_i \tag{7.41}$$

$$\frac{1}{\omega_j^2}\,[k]\,\mathbf{u}_j = [m]\,\mathbf{u}_j \tag{7.42}$$

Multiplying $\mathbf{u}_j^T$ by Eq. (7.41) and multiplying the transpose of (7.42) by $\mathbf{u}_i$ and taking advantage of symmetry of $[k]$ and $[m]$ lead to:

$$\frac{1}{\omega_i^2} \mathbf{u}_j^T [k] \mathbf{u}_i = \mathbf{u}_j^T [m] \mathbf{u}_i \tag{7.43}$$

$$\frac{1}{\omega_j^2} \mathbf{u}_j^T [k] \mathbf{u}_i = \mathbf{u}_j^T [m] \mathbf{u}_i \tag{7.44}$$

Subtracting (7.43) from (7.44)

$$\left(\frac{1}{\omega_i^2} - \frac{1}{\omega_j^2} \right) \mathbf{u}_j^T [k] \mathbf{u}_i = 0 \tag{7.45}$$

yields

$$\begin{aligned} \mathbf{u}_j^T [k] \mathbf{u}_i &= 0 \quad i \neq j \\ \mathbf{u}_j^T [k] \mathbf{u}_i &= k_i \quad i = j \end{aligned} \tag{7.46}$$

where k_i is the *generalized stiffness* of the system.

Any type of free, transient, or excited response of a linear vibrating system is dominated by its natural frequencies, mode shapes, and interaction of excitation frequencies. Determination of the natural frequencies ω_i and their associated mode shapes $\mathbf{u}_i$ is the first step in analysis of a multi DOF vibrating system. There exists at least one mode shape corresponding to each natural frequency. If an $n \times n$ matrix $[A]$ has n distinct eigenvalues, then there exist exactly n linearly independent eigenvectors, one associated with each eigenvalue.   ■

Example 233 Eigenvalues and eigenvectors of a 2×2 matrix. It is a numerical example on a small matrix to practice determination of characteristic equation, eigenvalue, and eigenvector.

Consider a 2×2 characteristic matrix $[A]$.

$$[A] = \begin{bmatrix} 5 & 3 \\ 3 & 6 \end{bmatrix} \tag{7.47}$$

To find the eigenvalues λ_i of $[A]$, we find the characteristic equation of the matrix by subtracting an unknown λ from the main diagonal elements and taking the determinant.

$$\det\left[[A] - \lambda \mathbf{I}\right] = \det\left[\begin{bmatrix} 5 & 3 \\ 3 & 6 \end{bmatrix} - \lambda \begin{bmatrix} 1 & 0 \\ 0 & 1 \end{bmatrix}\right]$$

$$= \det\begin{bmatrix} 5 - \lambda & 3 \\ 3 & 6 - \lambda \end{bmatrix} = \lambda^2 - 11\lambda + 21 \qquad (7.48)$$

The solutions of the characteristic equation (7.48) are:

$$\lambda_1 = 8.5414 \qquad \lambda_2 = 2.4586 \qquad (7.49)$$

To find the corresponding eigenvectors $\mathbf{u}_1$ and $\mathbf{u}_2$, we solve the eigenvector equations.

$$[[A] - \lambda_1 \mathbf{I}]\, \mathbf{u}_1 = 0 \qquad [[A] - \lambda_2 \mathbf{I}]\, \mathbf{u}_2 = 0 \qquad (7.50)$$

Let us denote the eigenvectors by

$$\mathbf{u}_1 = \begin{bmatrix} u_{11} \\ u_{12} \end{bmatrix} \qquad \mathbf{u}_2 = \begin{bmatrix} u_{21} \\ u_{22} \end{bmatrix} \qquad (7.51)$$

therefore,

$$[[A] - \lambda_1 \mathbf{I}]\, \mathbf{u}_1 = \left[\begin{bmatrix} 5 & 3 \\ 3 & 6 \end{bmatrix} - 8.5414 \begin{bmatrix} 1 & 0 \\ 0 & 1 \end{bmatrix}\right] \begin{bmatrix} u_{11} \\ u_{12} \end{bmatrix}$$

$$= \begin{bmatrix} 3u_{12} - 3.5414u_{11} \\ 3u_{11} - 2.5414u_{12} \end{bmatrix} = 0 \qquad (7.52)$$

$$[[A] - \lambda_2 \mathbf{I}]\, \mathbf{u}_2 = \left[\begin{bmatrix} 5 & 3 \\ 3 & 6 \end{bmatrix} - 2.4586 \begin{bmatrix} 1 & 0 \\ 0 & 1 \end{bmatrix}\right] \begin{bmatrix} u_{21} \\ u_{22} \end{bmatrix}$$

$$= \begin{bmatrix} 2.5414u_{21} + 3u_{22} \\ 3u_{21} + 3.5414u_{22} \end{bmatrix} = 0 \qquad (7.53)$$

To have last-unit eigenvectors, we assign

$$u_{12} = 1 \qquad u_{22} = 1 \qquad (7.54)$$

which leads to:

$$\mathbf{u}_1 = \begin{bmatrix} -1.1805 \\ 1.0 \end{bmatrix} \qquad \mathbf{u}_2 = \begin{bmatrix} 0.84713 \\ 1.0 \end{bmatrix} \qquad (7.55)$$

To show that the ratio of the elements of eigenvectors is unique for every eigenvalue, we examine the eigenvectors $\mathbf{u}_1$ and $\mathbf{u}_2$.

$$\mathbf{u}_1 = \begin{bmatrix} 3u_{12} - 3.5414u_{11} \\ 3u_{11} - 2.5414u_{12} \end{bmatrix} \tag{7.56}$$

$$\mathbf{u}_2 = \begin{bmatrix} 2.5414u_{21} + 3u_{22} \\ 3u_{21} + 3.5414u_{22} \end{bmatrix} \tag{7.57}$$

The ratio u_{11}/u_{12} from the first row of $\mathbf{u}_1$ in (7.56) is:

$$\frac{u_{11}}{u_{12}} = \frac{3}{3.5414} = 0.84712 \tag{7.58}$$

and from the second row is:

$$\frac{u_{11}}{u_{12}} = \frac{2.5414}{3} = 0.84713 \tag{7.59}$$

which shows their equality. The ratio u_{21}/u_{22} may also be found from the first or second row of $\mathbf{u}_2$ in (7.57) to check their equality.

$$\frac{u_{21}}{u_{22}} = -\frac{3}{2.5414} = -\frac{3.5414}{3} = -1.1805 \tag{7.60}$$

Example 234 Eigenvalues of a 2×2 matrix. This is a parametric example of determination of characteristic equation, eigenvalue, and eigenvector of a 2×2 matrix.

In most cases, determination of natural frequency of discrete vibrating systems reduces to eigenvalue problem of a symmetric matrix. Consider a 2×2 symmetric matrix:

$$[A] = \begin{bmatrix} a & b \\ b & c \end{bmatrix} \tag{7.61}$$

The eigenvalue problem of the matrix

$$|A - \lambda_i \mathbf{I}| = 0 \tag{7.62}$$

$$\left| \begin{bmatrix} a & b \\ b & c \end{bmatrix} - \lambda \begin{bmatrix} 1 & 0 \\ 0 & 1 \end{bmatrix} \right| = 0 \tag{7.63}$$

leads to a second-degree characteristic equation

$$\lambda^2 - (a + c)\lambda + \left(ac - b^2 \right) = 0 \tag{7.64}$$

which yields:

$$\lambda_1 = Z_1 - \sqrt{Z_3} \qquad \lambda_2 = Z_1 + \sqrt{Z_3} \tag{7.65}$$

$$Z_1 = \frac{a+c}{2} \qquad Z_3 = \left(\frac{a-c}{2}\right)^2 + b^2 \tag{7.66}$$

The eigenvectors of the matrix are:

$$\mathbf{u}_1 = \begin{bmatrix} Z_1 - \sqrt{Z_3} \\ b \end{bmatrix} \qquad \mathbf{u}_2 = \begin{bmatrix} Z_1 + \sqrt{Z_3} \\ b \end{bmatrix} \tag{7.67}$$

If the 2×2 matrix is asymmetric,

$$[A] = \begin{bmatrix} a & b \\ d & c \end{bmatrix} \tag{7.68}$$

then the eigenvalue problem of the matrix leads to

$$\left\| \begin{bmatrix} a & b \\ d & c \end{bmatrix} - \lambda \begin{bmatrix} 1 & 0 \\ 0 & 1 \end{bmatrix} \right\| = 0 \tag{7.69}$$

$$\lambda^2 - (a+c)\lambda + (ac - bd) = 0 \tag{7.70}$$

which yields:

$$\lambda_1 = Z_1 - \sqrt{Z_2} \qquad \lambda_2 = Z_1 + \sqrt{Z_2} \tag{7.71}$$

$$Z_1 = \frac{a+c}{2} \qquad Z_2 = \left(\frac{a-c}{2}\right)^2 + bd \tag{7.72}$$

The eigenvectors of the matrix are:

$$\mathbf{u}_1 = \begin{bmatrix} Z_1 - \sqrt{Z_2} \\ d \end{bmatrix} \qquad \mathbf{u}_2 = \begin{bmatrix} Z_1 + \sqrt{Z_2} \\ d \end{bmatrix} \tag{7.73}$$

These equations are helpful in the analysis of two DOF systems.

Example 235 ★ Characteristics of undamped-free vibrating systems. There are some engineering and mathematical points of linear vibrating system that are summarized in this example. All of them will be proven and shown in this chapter in the following sections.

The undamped-free vibrating systems have two characteristics:

1. Natural frequencies
2. Mode shapes

An $n\,DOF$ vibrating system will have n natural frequencies ω_i and n mode shapes $\mathbf{u}_i$. The natural frequencies ω_i are centers for the system's resonance zones, and the eigenvectors $\mathbf{u}_i$ show the relative vibrations of different coordinates of the system at the resonance ω_i. The largest element of each mode shape $\mathbf{u}_i$ indicates the coordinate or the component of the system which is most willing to vibrate at ω_i.

The response of undamped-free systems is the principal components for all other possible responses of the vibrating system. When there is damping, then the response of the system is bounded by the free undamped solution. When there is a forcing function, then the natural frequencies of the system indicate the resonance zones at which the amplitude of the response may go to infinity if an excitation frequency of the force function matches the natural frequency.

An $n\,DOF$ system needs n independent variable coordinates to be specified. For such a system, there would also be n second-order differential equations of the coordinates. Furthermore, the system has n natural frequencies ω_i and n mode shapes $\mathbf{u}_i$. Although the number of natural frequencies and the number of required coordinates are equal, the natural frequencies are fixed characteristics of the system and are not dependent on the set of coordinates of the system. The natural frequency ω_i is not associated with any particular coordinate of the system; for example, ω_2 is not related to the coordinate x_2 or any particular x_i. Selection of different coordinates changes the differential equations of motion of the system, but it will not change the natural frequencies. The mode shapes $\mathbf{u}_i$, associated to ω_i, are indicators of the relative motion of the coordinates, and they are coordinate dependent.

Example 236 ★ Quarter car natural frequencies and mode shapes. The quarter model is a great mathematical model for vertical vibration analysis of vehicles. Here is a mathematical model and a numerical example to calculate natural frequencies and mode shapes of the model.

Figure 7.2 illustrates a quarter car model which is made of two solid masses m_s and m_u denoted as *sprung* and *unsprung* masses, respectively. The sprung mass m_s represents $1/4$ of the body of the vehicle, and the unsprung mass m_u represents one wheel of the vehicle. A spring of stiffness k_s and a shock absorber with equivalent viscous damping coefficient c_s support the sprung mass. The unsprung mass m_u is in direct contact with the ground through a spring k_u and a damper c_u representing the tire stiffness and damping.

The governing differential equations of motion for the quarter car model are:

$$m_s\,\ddot{x}_s = -k_s\,(x_s - x_u) - c_s\,(\dot{x}_s - \dot{x}_u) \tag{7.74}$$

$$m_u\,\ddot{x}_u = k_s\,(x_s - x_u) + c_s\,(\dot{x}_s - \dot{x}_u)$$
$$-k_u\,(x_u - y) - c_u\,(\dot{x}_u - \dot{y}) \tag{7.75}$$

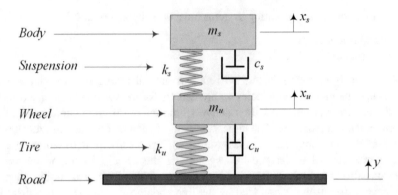

Fig. 7.2 A quarter car model

which can be expressed in a matrix form

$$[M]\ddot{\mathbf{x}} + [c]\dot{\mathbf{x}} + [k]\mathbf{x} = \mathbf{F} \qquad (7.76)$$

$$\begin{bmatrix} m_s & 0 \\ 0 & m_u \end{bmatrix}\begin{bmatrix} \ddot{x}_s \\ \ddot{x}_u \end{bmatrix} + \begin{bmatrix} c_s & -c_s \\ -c_s & c_s + c_u \end{bmatrix}\begin{bmatrix} \dot{x}_s \\ \dot{x}_u \end{bmatrix} +$$

$$\begin{bmatrix} k_s & -k_s \\ -k_s & k_s + k_u \end{bmatrix}\begin{bmatrix} x_s \\ x_u \end{bmatrix} = \begin{bmatrix} 0 \\ k_u y + c_u \dot{y} \end{bmatrix} \qquad (7.77)$$

To find the natural frequencies and mode shapes of the quarter car model, we have to drop the damping and forcing terms and analyze the following set of equations.

$$\begin{bmatrix} m_s & 0 \\ 0 & m_u \end{bmatrix}\begin{bmatrix} \ddot{x}_s \\ \ddot{x}_u \end{bmatrix} + \begin{bmatrix} k_s & -k_s \\ -k_s & k_s + k_u \end{bmatrix}\begin{bmatrix} x_s \\ x_u \end{bmatrix} = 0 \qquad (7.78)$$

Consider a vehicle with the following characteristics.

$$m_s = 375\,\text{kg} \qquad\qquad m_u = 75\,\text{kg}$$
$$k_u = 193000\,\text{N}/\text{m} \qquad k_s = 35000\,\text{N}/\text{m} \qquad (7.79)$$

The equations of vibration motion for this vehicle are:

$$\begin{bmatrix} 375 & 0 \\ 0 & 75 \end{bmatrix}\begin{bmatrix} \ddot{x}_s \\ \ddot{x}_u \end{bmatrix} + \begin{bmatrix} 35000 & -35000 \\ -35000 & 2.28 \times 10^5 \end{bmatrix}\begin{bmatrix} x_s \\ x_u \end{bmatrix} = 0 \qquad (7.80)$$

The natural frequencies of the vehicle can be found by solving its characteristic equation:

$$\det\left[[k] - \omega^2[m]\right] = \det\left[\begin{bmatrix} 35000 & -35000 \\ -35000 & 2.28 \times 10^5 \end{bmatrix} - \omega^2\begin{bmatrix} 375 & 0 \\ 0 & 75 \end{bmatrix}\right]$$

$$= \det\begin{bmatrix} 35000 - 375\omega^2 & -35000 \\ -35000 & 2.28 \times 10^5 - 75\omega^2 \end{bmatrix}$$

$$= 28125\omega^4 - 8.8125 \times 10^7\omega^2 + 6.755 \times 10^9 \qquad (7.81)$$

$$\omega_1 = 8.8671 \text{ rad}/\text{s} \approx 1.41 \text{ Hz} \qquad (7.82)$$

$$\omega_2 = 55.269 \text{ rad}/\text{s} \approx 8.79 \text{ Hz} \qquad (7.83)$$

To find the corresponding mode shapes, we use Eq. (7.29).

$$\left[[k] - \omega_1^2[m]\right]\mathbf{u}_1$$

$$= \left[\begin{bmatrix} 35000 & -35000 \\ -35000 & 2.28 \times 10^5 \end{bmatrix} - 3054.7\begin{bmatrix} 375 & 0 \\ 0 & 75 \end{bmatrix}\right]\begin{bmatrix} u_{11} \\ u_{12} \end{bmatrix}$$

$$= \begin{bmatrix} -1.1105 \times 10^6 u_{11} - 35000 u_{12} \\ -35000 u_{11} - 1102.5 u_{12} \end{bmatrix} = 0 \qquad (7.84)$$

$$\left[[k] - \omega_2^2[m]\right]\mathbf{u}_2$$

$$= \left[\begin{bmatrix} 35000 & -35000 \\ -35000 & 2.28 \times 10^5 \end{bmatrix} - 78.625\begin{bmatrix} 375 & 0 \\ 0 & 75 \end{bmatrix}\right]\begin{bmatrix} u_{21} \\ u_{22} \end{bmatrix}$$

$$= \begin{bmatrix} 5515.6 u_{21} - 35000 u_{22} \\ 2.221 \times 10^5 u_{22} - 35000 u_{21} \end{bmatrix} = 0 \qquad (7.85)$$

Searching for the first-unit expression of $\mathbf{u}_1$ and $\mathbf{u}_2$ provides us with the following mode shapes:

$$\mathbf{u}_1 = \begin{bmatrix} 1 \\ -3.1729 \times 10^{-3} \end{bmatrix} \qquad \mathbf{u}_2 = \begin{bmatrix} 1 \\ 0.157\,58 \end{bmatrix} \qquad (7.86)$$

Therefore, the free vibrations of the quarter car would be

$$\mathbf{x} = \sum_{i=1}^{n}\mathbf{u}_i\left(A_i \sin \omega_i t + B_i \cos \omega_i t\right) \qquad i = 1, 2 \qquad (7.87)$$

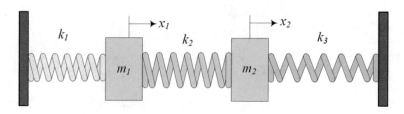

Fig. 7.3 A two DOF vibrating system

$$\begin{bmatrix} x_s \\ x_u \end{bmatrix} = \begin{bmatrix} 1 \\ -3.1729 \times 10^{-3} \end{bmatrix} (A_1 \sin 8.8671t + B_1 \cos 8.8671t)$$

$$+ \begin{bmatrix} 1 \\ 0.157\,58 \end{bmatrix} (A_2 \sin 55.269t + B_2 \cos 55.269t) \tag{7.88}$$

Example 237 ★ Two DOF vibrating systems, mode shape illustration, nodes, possible free vibrations of multi DOF systems. This example shows how to derive the equations of motion of multi DOF vibrating systems using Lagrange method and how to determine the natural frequencies and mode shapes analytically. It also geometrically illustrates the mode shapes to clarify their meaning. Any possible free vibration of a multi DOF system will be a linear combination of its mode shapes. Illustration of the mode shapes gives us a vision of how the system will vibrate at each mode.

Figure 7.3 illustrates a two DOF vibrating system. The kinetic and potential energies of the system are:

$$K = \frac{1}{2}m_1\dot{x}_1^2 + \frac{1}{2}m_2\dot{x}_2^2 \tag{7.89}$$

$$P = \frac{1}{2}k_1x_1^2 + \frac{1}{2}k_2(x_1 - x_2)^2 + \frac{1}{2}k_3x_2^2 \tag{7.90}$$

Using the Lagrangean of the system $\mathcal{L} = K - P$, and employing the Lagrange equation,

$$\frac{d}{dt}\left(\frac{\partial \mathcal{L}}{\partial \dot{x}_i}\right) - \frac{\partial \mathcal{L}}{\partial x_i} = 0 \qquad i = 1, 2 \tag{7.91}$$

the equations of motion are:

$$m_1\ddot{x}_1 + k_1x_1 + k_2(x_1 - x_2) = 0 \tag{7.92}$$

$$m_2\ddot{x}_2 - k_2(x_1 - x_2) + k_3x_2 = 0 \tag{7.93}$$

We may rearrange the equations in matrix form:

$$\begin{bmatrix} m_1 & 0 \\ 0 & m_2 \end{bmatrix} \begin{bmatrix} \ddot{x}_1 \\ \ddot{x}_2 \end{bmatrix} + \begin{bmatrix} k_1 + k_2 & -k_2 \\ -k_2 & k_2 + k_3 \end{bmatrix} \begin{bmatrix} x_1 \\ x_2 \end{bmatrix} = \begin{bmatrix} 0 \\ 0 \end{bmatrix} \tag{7.94}$$

The natural frequencies of the system can be found from $[A]$,

$$\begin{aligned} [A] &= \begin{bmatrix} m_1 & 0 \\ 0 & m_2 \end{bmatrix}^{-1} \begin{bmatrix} k_1 + k_2 & -k_2 \\ -k_2 & k_2 + k_3 \end{bmatrix} \\ &= \begin{bmatrix} \dfrac{1}{m_1}(k_1 + k_2) & -\dfrac{k_2}{m_1} \\ -\dfrac{k_2}{m_2} & \dfrac{1}{m_2}(k_2 + k_3) \end{bmatrix} \end{aligned} \tag{7.95}$$

which is a symmetric matrix and has the following eigenvalues.

$$\lambda_1 = Z_1 - \sqrt{Z_2} \qquad \lambda_2 = Z_1 + \sqrt{Z_2} \tag{7.96}$$

$$Z_1 = \frac{1}{2}\left(\frac{k_1 + k_2}{m_1} + \frac{k_2 + k_3}{m_2} \right) \tag{7.97}$$

$$Z_2 = \frac{1}{4}\left(\frac{k_1 + k_2}{m_1} - \frac{k_2 + k_3}{m_2} \right)^2 + \frac{k_2^2}{m_1 m_2} \tag{7.98}$$

In the symmetrical case of Fig. 7.4 where

$$m_1 = m_2 = m \qquad k_1 = k_2 = k_3 = k \tag{7.99}$$

the eigenvalues λ_i and natural frequencies ω_i simplify to:

$$\lambda_1 = \frac{k}{m} \qquad \lambda_2 = 3\frac{k}{m} \tag{7.100}$$

$$\omega_1 = \sqrt{\lambda_1} = \sqrt{\frac{k}{m}} \qquad \omega_2 = \sqrt{\lambda_2} = \sqrt{3}\sqrt{\frac{k}{m}} \tag{7.101}$$

of which the eigenvectors, respectively, are:

$$\mathbf{u}_1 = \begin{bmatrix} 1 \\ 1 \end{bmatrix} \qquad \mathbf{u}_2 = \begin{bmatrix} -1 \\ 1 \end{bmatrix} \tag{7.102}$$

Mode shapes are the principal elements for all possible vibrations of a free system and indicate the relative displacements at each natural frequency. Figure 7.5

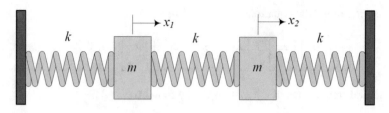

Fig. 7.4 A symmetrical two DOF vibrating system with $m_1 = m_2 = m$, $k_1 = k_2 = k_3 = k$

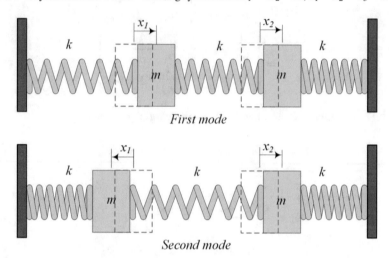

First mode

Second mode

Fig. 7.5 The first and second mode shapes of a symmetric two DOF vibrating system

illustrates the relative position of the masses in the first and second modes of the two DOF symmetric system of Fig. 7.4.

To distinguish between masses and springs, let us use the notation of Fig. 7.3. In the first mode, both masses move in the same direction with equal amplitudes. The middle spring k_2 has a constant length with no stretch and applies no force on m_1 and m_2. Therefore, m_1 is only under the force of k_1, and similarly m_2 is only under the force of k_2. In the second mode, the masses move in the opposite directions with equal amplitudes. The middle spring k_2 stretches twice more than k_1 and k_2 so that the midpoint of k_2 remains stationary. So, k_2 applies twice as much force as k_1 or k_3, and, therefore, m_1 and m_2 are under three times more force than in the first mode.

The motion of the system of the first mode can be expressed by

$$\mathbf{x}_1 = C_1 \mathbf{u}_1 \sin(\omega_1 t + \varphi_1) \tag{7.103}$$

and its motion in the second mode can be expressed by

$$\mathbf{x}_2 = C_2 \mathbf{u}_2 \sin(\omega_2 t + \varphi_2) \tag{7.104}$$

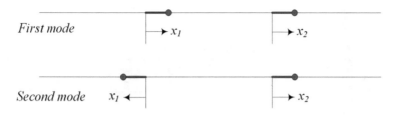

Fig. 7.6 Physical mode shape illustration

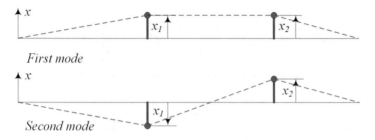

Fig. 7.7 Geometric mode shape illustration

Any possible free vibration of the system is a linear combination of its two modes shapes, where C_1 and C_2 are weight factors of the mode shapes.

$$\mathbf{x} = C_1\mathbf{u}_1 \sin(\omega_1 t + \varphi_1) + C_2\mathbf{u}_2 \sin(\omega_2 t + \varphi_2) \tag{7.105}$$

To visualize mode shapes, we illustrate each mode shape symbolically by the position of its elements with respect to the equilibrium position as shown physically in Figs. 7.5 and 7.6 or geometrically in Fig. 7.7. The node in the geometric mode shape illustration of Fig. 7.7 indicates that the midpoint of spring k_2 is motionless. The motionless points of mode shapes are called nodes. Furthermore, geometric illustration depicts the relative motion of any point of the springs.

Example 238 Free end of a two DOF system. This example shows how the natural frequencies of a two DOF system will get lower when only one solid wall is attached to the system. Generally speaking, more solid constraints will make higher natural frequencies.

Figure 7.8 illustrates the same system of Fig. 7.3 after removing the connection of m_2 and the wall. This is equivalent to substitute $k_3 = 0$ in system of Fig. 7.8 to derive the equations of motion.

$$m_1\ddot{x}_1 + k_1 x_1 + k_2 (x_1 - x_2) = 0 \tag{7.106}$$

$$m_2\ddot{x}_2 - k_2 (x_1 - x_2) = 0 \tag{7.107}$$

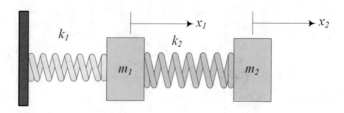

Fig. 7.8 An open end two DOF vibrating system

$$\begin{bmatrix} m_1 & 0 \\ 0 & m_2 \end{bmatrix} \begin{bmatrix} \ddot{x}_1 \\ \ddot{x}_2 \end{bmatrix} + \begin{bmatrix} k_1 + k_2 & -k_2 \\ -k_2 & k_2 \end{bmatrix} \begin{bmatrix} x_1 \\ x_2 \end{bmatrix} = \begin{bmatrix} 0 \\ 0 \end{bmatrix} \tag{7.108}$$

Employing the $[A]$-matrix, we find the eigenvalues of the system.

$$[A] = [m]^{-1}\,[k] = \begin{bmatrix} m_1 & 0 \\ 0 & m_2 \end{bmatrix}^{-1} \begin{bmatrix} k_1 + k_2 & -k_2 \\ -k_2 & k_2 \end{bmatrix}$$

$$= \begin{bmatrix} (k_1 + k_2)\,/m_1 & -k_2/m_1 \\ -k_2/m_2 & k_2/m_2 \end{bmatrix} \tag{7.109}$$

$$\lambda_1 = Z_1 - \sqrt{Z_2} \qquad \lambda_2 = Z_1 + \sqrt{Z_2} \tag{7.110}$$

$$Z_1 = \frac{1}{2}\left(\frac{k_1 + k_2}{m_1} + \frac{k_2}{m_2}\right) \tag{7.111}$$

$$Z_2 = \frac{1}{4}\left(\frac{k_1 + k_2}{m_1} - \frac{k_2}{m_2}\right)^2 + \frac{k_2^2}{m_1 m_2} \tag{7.112}$$

When $m_1 = m_2$ and $k_1 = k_2$, the natural frequencies ω_1, ω_2 and modes shapes $\mathbf{u}_1$, $\mathbf{u}_2$ simplify.

$$\omega_1 = \sqrt{\frac{k}{m}}\sqrt{\frac{3 - \sqrt{5}}{2}} \approx 0.618\sqrt{\frac{k}{m}} \tag{7.113}$$

$$\omega_2 = \sqrt{\frac{k}{m}}\sqrt{\frac{3 + \sqrt{5}}{2}} \approx 1.618\sqrt{\frac{k}{m}} \tag{7.114}$$

$$\mathbf{u}_1 = \begin{bmatrix} 0.618 \\ 1 \end{bmatrix} = \begin{bmatrix} 1 \\ 1.618 \end{bmatrix} \tag{7.115}$$

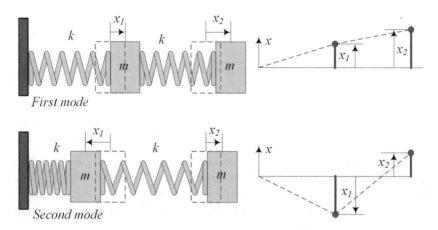

Fig. 7.9 The first and second mode shapes of the symmetric case of the open end two *DOF* vibrating system

$$\mathbf{u}_2 = \begin{bmatrix} -1.618 \\ 1 \end{bmatrix} = \begin{bmatrix} -1 \\ 0.618 \end{bmatrix} \tag{7.116}$$

Figure 7.9 illustrates the mode shapes and relative positions of the masses in the first and second modes.

Let us compare the free-end system of Fig. 7.8 with the constrained system of Fig. 7.3. Generally speaking, the masses of the constrained system of Fig. 7.3 are kept stiffer than the system of Fig. 7.8. Higher stiffness means higher natural frequencies. Hence, we expect the system (Fig. 7.5) to have higher natural frequencies than system (Fig. 7.9). By eliminating the second wall connection, the first natural frequency reduces from $\omega_1 = \sqrt{k/m}$ to $\omega_1 = 0.618\sqrt{k/m}$, which shows around 40% reduction, and the second natural frequency reduces from $\omega_1 = 1.732\sqrt{k/m}$ to $\omega_1 = 1.618\sqrt{k/m}$, which shows around 10% reduction.

Both masses of the constrained system of Fig. 7.5 move in the same direction with equal amplitudes in the first mode. However, although both masses of the free-end system of Fig. 7.9 move in the same direction in the first mode, the second mass has a higher amplitude. Having higher amplitude is another property of a more flexible system. The masses of the constrained system of Fig. 7.5 move in the opposite directions with equal amplitudes in the second mode. The masses of the free-end system of Fig. 7.9 also move in opposite directions in the second mode, and the amplitude of the first mass is larger compared to a unit displacement of the second mass. This behavior changes the position of the node from the middle point of k_2 to the point of $(1 - 1/1.618)\, l = 0.382l$ from m_1, where l is the length of k_2.

Example 239 Relative natural frequencies. Comparison of order of magnitude of natural frequencies relives many information about the nature of behavior of a system. We use the system of Fig. 7.3 to show natural frequency comparison.

Let us rewrite the mass and stiffness matrices of the two DOF systems of Fig. 7.3.

$$[m] = \begin{bmatrix} m_1 & 0 \\ 0 & m_2 \end{bmatrix} \tag{7.117}$$

$$[k] = \begin{bmatrix} k_{12} & -k_2 \\ -k_2 & k_{23} \end{bmatrix} = \begin{bmatrix} k_1 + k_2 & -k_2 \\ -k_2 & k_2 + k_3 \end{bmatrix} \tag{7.118}$$

The middle spring k_2 is the only connection between m_1 and m_2 and appears as the off-diagonal element of $[k]$. If there is no k_2, then the system breaks into two one DOF systems with natural frequencies $\sqrt{k_1/m_1}$ and $\sqrt{k_2/m_2}$. To examine the effect of k_2 on the overall natural frequencies of the system, we solve the characteristic equation of the system.

$$\det\Big[[k] - \omega^2\,[m]\Big] = 0 \tag{7.119}$$

$$\begin{vmatrix} -m_1\omega^2 + k_{12} & -k_2 \\ -k_2 & -m_2\omega^2 + k_{23} \end{vmatrix} = 0 \tag{7.120}$$

$$m_1 m_2 \omega^4 - (m_2 k_{12} + m_1 k_{23})\,\omega^2 + k_{12}k_{23} - k_2^2 = 0 \tag{7.121}$$

The roots of the equation are:

$$\omega^2 = \frac{m_2 k_{12} + m_1 k_{23}}{2 m_1 m_2} \pm \sqrt{\left(\frac{m_2 k_{12} + m_1 k_{23}}{2 m_1 m_2}\right)^2 - \frac{k_{12}k_{23} - k_2^2}{m_1 m_2}} \tag{7.122}$$

Let us rearrange the natural frequencies ω_1 and ω_2 of the system.

$$\omega_1^2 = \frac{\alpha\,(1+\varepsilon)}{2} + \frac{\Omega_1^2 + \Omega_2^2}{2}$$
$$+ \sqrt{\left(\frac{\alpha\,(1+\varepsilon)}{2} + \frac{\Omega_1^2 + \Omega_2^2}{2}\right)^2 - \alpha\left(\Omega_2^2 + \varepsilon\Omega_1^2\right) - \Omega_1^2\Omega_2^2} \tag{7.123}$$

$$\omega_2^2 = \frac{\alpha\,(1+\varepsilon)}{2} + \frac{\Omega_1^2 + \Omega_2^2}{2}$$
$$- \sqrt{\left(\frac{\alpha\,(1+\varepsilon)}{2} + \frac{\Omega_1^2 + \Omega_2^2}{2}\right)^2 - \alpha\left(\Omega_2^2 + \varepsilon\Omega_1^2\right) - \Omega_1^2\Omega_2^2} \tag{7.124}$$

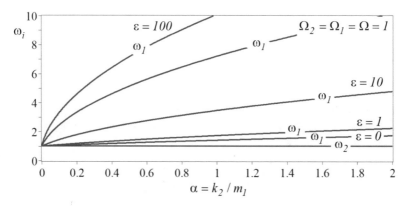

Fig. 7.10 Natural frequencies of the two DOF system of Fig. 7.3

$$\alpha = \frac{k_2}{m_1} \qquad \varepsilon = \frac{m_1}{m_2} \qquad \Omega_1^2 = \frac{k_1}{m_1} \qquad \Omega_2^2 = \frac{k_3}{m_2} \qquad (7.125)$$

If

$$\Omega_1^2 = \Omega_2^2 = \Omega^2 \qquad (7.126)$$

then

$$\omega_1 = \sqrt{\Omega^2 + \alpha^2 (\varepsilon + 1)^2} \qquad \omega_2 = \Omega \qquad (7.127)$$

When $k_2 = 0$, both the left and right systems have independent and equal natural frequencies. However, increasing k_2 increases ω_1 and separates the two natural frequencies. The difference also increases by $\varepsilon = m_1/m_2$. Figure 7.10 illustrates the natural frequencies ω_1 and ω_2 as functions of α. Increasing ε while keeping Ω constant means the ratio of k_1/k_2 will increase with the same order of ε. Therefore, increasing ε makes the first mass heavier and the first spring stiffer. That makes ω_1/ω_2 higher.

Example 240 ★ Energy of mode shapes. The free and undamped multi DOF systems are energy conserved. The initial conditions determine how much energy the system has. The initial conditions and the amount of energy determine the amplitude, mode, and mode interaction of oscillation. This example explains this phenomenon.

Each mode shape is associated to a specific natural frequency. If we order the natural frequencies from lowest to highest, then the system is vibrating slower in lower mode and faster in higher mode shapes. Consider the two DOF symmetric system of Fig. 7.4 with natural frequencies ω_1, ω_2 and modes shapes $\mathbf{u}_1$, $\mathbf{u}_2$.

$$\omega_1 = \sqrt{\frac{k}{m}} \qquad \omega_2 = \sqrt{3}\sqrt{\frac{k}{m}} \tag{7.128}$$

$$\mathbf{u}_1 = \begin{bmatrix} 1 \\ 1 \end{bmatrix} \qquad \mathbf{u}_2 = \begin{bmatrix} -1 \\ 1 \end{bmatrix} \tag{7.129}$$

Let us assume the system is oscillating only in the first mode or only in the second mode, both with amplitude X. The displacement vector $\mathbf{x}$ of m_i at first and second modes can be expressed as:

$$\mathbf{x}_1 = \begin{bmatrix} X \\ X \end{bmatrix} \sin \omega_1 t \qquad \mathbf{x}_2 = \begin{bmatrix} -X \\ X \end{bmatrix} \sin \omega_2 t \tag{7.130}$$

The maximum kinetic and maximum potential energies of the system in the first mode are:

$$K_{Max_1} = \frac{1}{2} m \omega_1^2 X^2 + \frac{1}{2} m \omega_1^2 X^2 = m \omega_1^2 X^2 \tag{7.131}$$

$$P_{Max_1} = \frac{1}{2} k X^2 + \frac{1}{2} k (X - X)^2 + \frac{1}{2} k X^2 = k X^2 \tag{7.132}$$

and K_{Max} and P_{Max} in the second mode are:

$$K_{Max_2} = \frac{1}{2} m \omega_2^2 X^2 + \frac{1}{2} m \omega_2^2 X^2 = m \omega_2^2 X^2 = 3 K_{Max_1} \tag{7.133}$$

$$P_{Max_2} = \frac{1}{2} k X^2 + \frac{1}{2} k (-X - X)^2 + \frac{1}{2} k X^2 = 3 k X^2 = 3 V_{Max_1} \tag{7.134}$$

The system vibrates in the second mode $\sqrt{3}$ times faster than the first mode. Therefore, it passes through the equilibrium position $\sqrt{3}$ times faster in the second mode than the first mode. The amount of required energy to vibrate in the second mode is three times of the required energy to vibrate in the first mode for the same amplitude. If the amount of energy is constant, then the amplitude of vibration in the second mode will be $1/\sqrt{3}$ smaller than the amplitude of vibrations in the first mode. To make the system to vibrate in the first mode, the initial conditions must match with the first mode, such as $\mathbf{x}(0) = \begin{bmatrix} 1 & 1 \end{bmatrix}$. Similarly, to make the system to vibrate in the second mode, the initial conditions must match with the second mode, such as $\mathbf{x}(0) = \begin{bmatrix} -1 & 1 \end{bmatrix}$. The amount of required energy to make the initial conditions in each mode can be calculated and be felt by the amount of extension and contraction of the springs in Fig. 7.5.

Any real system has some damping and needs an external source of energy to keep vibrating. The level of injection of the external energy depends on the mode and amplitude of vibrations. That is why when the source of energy is low, every system is willing to vibrate in its first mode.

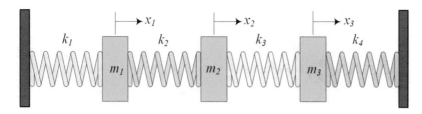

Fig. 7.11 A three DOF vibrating system

Example 241 Three DOF vibrating system. The mode shapes get more complicated by increasing the number of degrees of freedom. This is an example to determine and illustrate mode shapes of a three DOF vibrating system.

Figure 7.11 illustrates an undamped and free three DOF vibrating system. The kinetic and potential energies of the system are:

$$K = \frac{1}{2}m_1\dot{x}_1^2 + \frac{1}{2}m_2\dot{x}_2^2 + \frac{1}{2}m_3\dot{x}_3^2 \tag{7.135}$$

$$P = \frac{1}{2}k_1x_1^2 + \frac{1}{2}k_2(x_1 - x_2)^2 + \frac{1}{2}k_3(x_2 - x_3)^2 + \frac{1}{2}k_4x_3^2 \tag{7.136}$$

and its equations of motion are:

$$\begin{bmatrix} m_1 & 0 & 0 \\ 0 & m_2 & 0 \\ 0 & 0 & m_3 \end{bmatrix} \begin{bmatrix} \ddot{x}_1 \\ \ddot{x}_2 \\ \ddot{x}_3 \end{bmatrix} + \begin{bmatrix} k_1 + k_2 & -k_2 & 0 \\ -k_2 & k_2 + k_3 & -k_3 \\ 0 & -k_3 & k_3 + k_4 \end{bmatrix} \begin{bmatrix} x_1 \\ x_2 \\ x_3 \end{bmatrix} = 0 \tag{7.137}$$

The natural frequencies of the system can be found from matrix $[A]$.

$$[A] = \begin{bmatrix} m_1 & 0 & 0 \\ 0 & m_2 & 0 \\ 0 & 0 & m_3 \end{bmatrix}^{-1} \begin{bmatrix} k_1 + k_2 & -k_2 & 0 \\ -k_2 & k_2 + k_3 & -k_3 \\ 0 & -k_3 & k_3 + k_4 \end{bmatrix}$$

$$= \begin{bmatrix} (k_1 + k_2)/m_1 & -k_2/m_1 & 0 \\ -k_2/m_2 & (k_2 + k_3)/m_2 & -k_3/m_2 \\ 0 & -k_3/m_3 & (k_3 + k_4)/m_3 \end{bmatrix} \tag{7.138}$$

In the symmetrical case, where,

$$m_1 = m_2 = m_3 = m \tag{7.139}$$

$$k_1 = k_2 = k_3 = k_4 = k \tag{7.140}$$

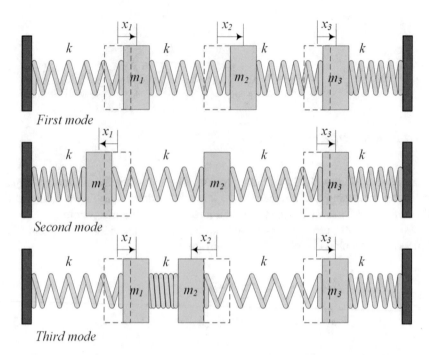

Fig. 7.12 The physical mode shapes of a symmetrical three DOF vibrating system with $m_1 = m_2 = m_3 = m$, $k_1 = k_2 = k_3 = k_4 = k$

the eigenvalues λ_i and natural frequencies ω_i and their eigenvectors $\mathbf{u}_i$ will be:

$$\lambda_1 = \frac{k}{m}\left(2 - \sqrt{2}\right) \qquad \lambda_2 = 2\frac{k}{m} \qquad \lambda_3 = \frac{k}{m}\left(2 + \sqrt{2}\right) \qquad (7.141)$$

$$\omega_1 = \sqrt{2 - \sqrt{2}}\sqrt{\frac{k}{m}} \qquad \omega_2 = \sqrt{2}\sqrt{\frac{k}{m}} \qquad \omega_3 = \sqrt{2 + \sqrt{2}}\sqrt{\frac{k}{m}} \qquad (7.142)$$

$$\mathbf{u}_1 = \begin{bmatrix} 1 \\ \sqrt{2} \\ 1 \end{bmatrix} \qquad \mathbf{u}_2 = \begin{bmatrix} -1 \\ 0 \\ 1 \end{bmatrix} \qquad \mathbf{u}_3 = \begin{bmatrix} 1 \\ -\sqrt{2} \\ 1 \end{bmatrix} \qquad (7.143)$$

The system in the three mode shapes is shown in Fig. 7.12. The geometric mode shape illustration in Fig. 7.13 indicates there is no node in the first mode, one node at m_2 in the second mode, and two nodes in the third mode.

Determination of the position of nodes is important in the design of coupling and connection of mechanical components. Any coupling, connecting joints, or clutch discs must be installed at the nodes of the most possible mode of vibrations.

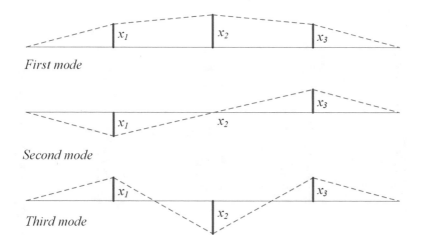

Fig. 7.13 The geometric mode shapes of a symmetrical three DOF vibrating system with $m_1 = m_2 = m_3 = m, k_1 = k_2 = k_3 = k_4 = k$

Example 242 Lower natural frequency for lower stiff system. Lowering stuffiness of a system will lower the value of natural frequencies of the system.

Let us examine the effect of removing the elastic connection of m_3 and the fixed wall. Figure 7.14a illustrates the vibrating system after removing the elastic connection of m_3 and the fixed wall. Removing the connection reduces the overall constraints and stiffness of the system, so we expect to have a system with relatively lower natural frequencies and higher steady state amplitudes. Eliminating k_4 provides us with the following equations of motion.

$$\begin{bmatrix} m_1 & 0 & 0 \\ 0 & m_2 & 0 \\ 0 & 0 & m_3 \end{bmatrix} \begin{bmatrix} \ddot{x}_1 \\ \ddot{x}_2 \\ \ddot{x}_3 \end{bmatrix} + \begin{bmatrix} k_1 + k_2 & -k_2 & 0 \\ -k_2 & k_2 + k_3 & -k_3 \\ 0 & -k_3 & k_3 \end{bmatrix} \begin{bmatrix} x_1 \\ x_2 \\ x_3 \end{bmatrix} = 0 \qquad (7.144)$$

Employing the $[A]$-matrix and assuming a symmetric system,

$$m_1 = m_2 = m_3 = m \qquad k_1 = k_2 = k_3 = k \qquad (7.145)$$

leads to:

$$[A] = [m]^{-1} [k] = \begin{bmatrix} m & 0 & 0 \\ 0 & m & 0 \\ 0 & 0 & m \end{bmatrix}^{-1} \begin{bmatrix} 2k & -k & 0 \\ -k & 2k & -k \\ 0 & -k & k \end{bmatrix}$$

$$= \begin{bmatrix} 2\dfrac{k}{m} & -\dfrac{k}{m} & 0 \\ -\dfrac{k}{m} & 2\dfrac{k}{m} & -\dfrac{k}{m} \\ 0 & -\dfrac{k}{m} & \dfrac{k}{m} \end{bmatrix} \qquad (7.146)$$

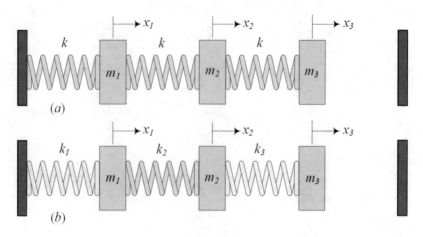

Fig. 7.14 A free-end three DOF vibrating system. (**a**) general, (**b**) equal stiffnesses and masses

$$\omega_1 \approx 0.445\sqrt{\frac{k}{m}} \qquad \omega_2 \approx 1.247\sqrt{\frac{k}{m}} \qquad \omega_3 \approx 1.802\sqrt{\frac{k}{m}} \qquad (7.147)$$

$$\mathbf{u}_1 \approx \begin{bmatrix} 0.445 \\ 0.802 \\ 1 \end{bmatrix} \qquad \mathbf{u}_2 \approx \begin{bmatrix} -1.247 \\ -0.555 \\ 1 \end{bmatrix} \qquad \mathbf{u}_3 \approx \begin{bmatrix} 1.8019 \\ -2.247 \\ 1 \end{bmatrix} \qquad (7.148)$$

Comparing these natural frequencies with (7.142) indicates that ω_1 is reduced from $0.765\sqrt{k/m}$ to $0.445\sqrt{k/m}$, ω_2 is reduced from $1.414\sqrt{k/m}$ to $1.247\sqrt{k/m}$, and ω_3 is reduced from $1.848\sqrt{k/m}$ to $1.802\sqrt{k/m}$. A comparison of the mode shapes also indicates a higher relative amplitude for the free-end system of Fig. 7.14b.

Example 243 More constraints on a free-end three DOF system. Here we add stiffness to the system of three serial mass-spring connection of system of Fig. 7.14a to examine the effect of such extra stiffness on natural frequencies and mode shapes of the system.

Consider the system of Fig. 7.15 which is similar to the system of Fig. 7.14a with an extra spring, connecting m_1 to m_3. The kinetic and potential energies of the system are:

$$K = \frac{1}{2}m_1\dot{x}_1^2 + \frac{1}{2}m_2\dot{x}_2^2 + \frac{1}{2}m_3\dot{x}_3^2 \qquad (7.149)$$

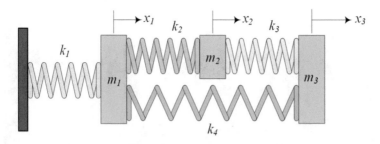

Fig. 7.15 A free-end 3 DOF vibrating system with connected masses

$$P = \frac{1}{2}k_1 x_1^2 + \frac{1}{2}k_2 (x_1 - x_2)^2$$

$$+ \frac{1}{2}k_3 (x_2 - x_3)^2 + \frac{1}{2}k_4 (x_1 - x_3)^2 \qquad (7.150)$$

which provide us with the equations of motion of the system:

$$\begin{bmatrix} m_1 & 0 & 0 \\ 0 & m_2 & 0 \\ 0 & 0 & m_3 \end{bmatrix} \begin{bmatrix} \ddot{x}_1 \\ \ddot{x}_2 \\ \ddot{x}_3 \end{bmatrix}$$

$$+ \begin{bmatrix} k_1 + k_2 + k_4 & -k_2 & -k_4 \\ -k_2 & k_2 + k_3 & -k_3 \\ -k_4 & -k_3 & k_3 + k_4 \end{bmatrix} \begin{bmatrix} x_1 \\ x_2 \\ x_3 \end{bmatrix} = 0 \qquad (7.151)$$

Employing the $[A]$-matrix for

$$m_1 = m_2 = m_3 = m \qquad k_1 = k_2 = k_3 = k_4 = k \qquad (7.152)$$

yields:

$$[A] = [m]^{-1} [k] = \begin{bmatrix} m & 0 & 0 \\ 0 & m & 0 \\ 0 & 0 & m \end{bmatrix}^{-1} \begin{bmatrix} 3k & -k & -k \\ -k & 2k & -k \\ -k & -k & 2k \end{bmatrix}$$

$$= \begin{bmatrix} 3k/m & -k/m & -k/m \\ -k/m & 2k/m & -k/m \\ -k/m & -k/m & 2k/m \end{bmatrix} \qquad (7.153)$$

$$\omega_1 = \sqrt{2 - \sqrt{3}}\sqrt{\frac{k}{m}} \approx 0.517\sqrt{\frac{k}{m}} \qquad (7.154)$$

$$\omega_2 = \sqrt{3}\sqrt{\frac{k}{m}} \approx 1.732\sqrt{\frac{k}{m}} \qquad (7.155)$$

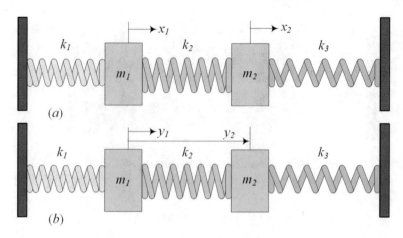

Fig. 7.16 An undamped two DOF system

$$\omega_3 = \sqrt{2+\sqrt{3}}\sqrt{\frac{k}{m}} \approx 1.932\sqrt{\frac{k}{m}} \tag{7.156}$$

$$\mathbf{u}_1 \approx \begin{bmatrix} 0.732 \\ 1 \\ 1 \end{bmatrix} \qquad \mathbf{u}_2 \approx \begin{bmatrix} 0 \\ -1 \\ 1 \end{bmatrix} \qquad \mathbf{u}_3 \approx \begin{bmatrix} -2.732 \\ 1 \\ 1 \end{bmatrix} \tag{7.157}$$

The spring k_4 makes the system stiffer, and hence, we expect the natural frequencies to be higher. However, predication of the changes in mode shapes is not as easy as prediction of changes in natural frequencies. Comparison of mode shapes (7.157) to (7.148) indicates that mode shapes have changed significantly.

Example 244 ★ Change of coordinates. The mode shapes illustrates the relative amplitude of coordinates at each natural frequency. Hence, the mode shapes are related to the coordinates we pick to express the motion of the system. This example shows how the modes shapes will change when we change the cordialness. It suggests if there is a superior coordinate system for a vibrating system. The answer is yes, and such superior coordinate system is called the principal or natural coordinate system.

The equations of motion and mode shapes of a vibrating system are coordinate dependent, but the natural frequencies are not. To examine this fact, let us consider the two DOF system of Fig. 7.16a and b. In Fig. 7.16a we are using the absolute coordinate x_1 and x_2 to express the position of m_1 and m_2. In Fig. 7.16b we replace the absolute coordinate x_2 with the relative coordinate y_2.

The kinetic and potential energies and the equations of motion of the system of Fig. 7.16a are:

$$K = \frac{1}{2}m_1\dot{x}_1^2 + \frac{1}{2}m_2\dot{x}_2^2 \tag{7.158}$$

$$P = \frac{1}{2}k_1x_1^2 + \frac{1}{2}k_2(x_2 - x_1)^2 + \frac{1}{2}k_3x_2^2 \tag{7.159}$$

$$\begin{bmatrix} m_1 & 0 \\ 0 & m_2 \end{bmatrix}\begin{bmatrix} \ddot{x}_1 \\ \ddot{x}_2 \end{bmatrix} + \begin{bmatrix} k_1 + k_2 & -k_2 \\ -k_2 & k_2 + k_3 \end{bmatrix}\begin{bmatrix} x_1 \\ x_2 \end{bmatrix} = \begin{bmatrix} 0 \\ 0 \end{bmatrix} \tag{7.160}$$

The characteristic equation, $\left|[m]^{-1}[k] - \lambda_i \mathbf{I}\right| = 0$, the eigenvalues λ_i, and the mode shapes $\mathbf{u}_i$ of the system are:

$$\lambda^2 - \left(\frac{k_1 + k_2}{m_1} + \frac{k_2 + k_3}{m_2}\right)\lambda + \left(\frac{k_1 + k_2}{m_1}\frac{k_2 + k_3}{m_2} - \frac{k_2^2}{m_1 m_2}\right) = 0 \tag{7.161}$$

$$\lambda_1 = Z_1 - \sqrt{Z_2} \qquad \lambda_2 = Z_1 + \sqrt{Z_2} \tag{7.162}$$

$$Z_1 = \frac{1}{2}\left(\frac{k_1 + k_2}{m_1} + \frac{k_2 + k_3}{m_2}\right) \tag{7.163}$$

$$Z_2 = \frac{1}{4}\left(\frac{k_1 + k_2}{m_1} - \frac{k_2 + k_3}{m_2}\right)^2 + \frac{k_2^2}{m_1 m_2} \tag{7.164}$$

$$\mathbf{u}_1 = \begin{bmatrix} Z_1 - \sqrt{Z_2} \\ k_2/m_2 \end{bmatrix} \qquad \mathbf{u}_2 = \begin{bmatrix} Z_1 + \sqrt{Z_2} \\ k_2/m_2 \end{bmatrix} \tag{7.165}$$

When the system is symmetric,

$$m_1 = m_2 = m \qquad k_1 = k_2 = k_3 = k \tag{7.166}$$

then we have

$$\lambda_1 = \frac{k}{m} \qquad \lambda_2 = 3\frac{k}{m} \tag{7.167}$$

$$\mathbf{u}_1 = \begin{bmatrix} 1 \\ 1 \end{bmatrix} \qquad \mathbf{u}_2 = \begin{bmatrix} -1 \\ 1 \end{bmatrix} \tag{7.168}$$

The kinetic and potential energies and the equations of motion of the system of Fig. 7.16b are:

$$K = \frac{1}{2}m_1\dot{y}_1^2 + \frac{1}{2}m_2(\dot{y}_1 + \dot{y}_2)^2 \tag{7.169}$$

$$P = \frac{1}{2}k_1 y_1^2 + \frac{1}{2}k_2 y_2^2 + \frac{1}{2}k_3 (y_1 + y_2)^2 \tag{7.170}$$

$$\begin{bmatrix} m_1 + m_2 & m_2 \\ m_2 & m_1 \end{bmatrix} \begin{bmatrix} \ddot{y}_1 \\ \ddot{y}_2 \end{bmatrix} + \begin{bmatrix} k_1 + k_3 & k_3 \\ k_3 & k_2 + k_3 \end{bmatrix} \begin{bmatrix} y_1 \\ y_2 \end{bmatrix} = \begin{bmatrix} 0 \\ 0 \end{bmatrix} \tag{7.171}$$

The characteristic equation, eigenvalues λ_i, and mode shapes $\mathbf{u}_i$ of the system are:

$$\lambda^2 - \frac{m_1 (k_1 + k_2 + 2k_3) + m_2 (k_2 - k_3)}{m_1^2 - m_2^2 + m_1 m_2} \lambda + \frac{k_1 k_2 + k_1 k_3 + k_2 k_3}{m_1^2 + m_1 m_2 - m_2^2} = 0 \tag{7.172}$$

$$\lambda_1 = Z_1 - \sqrt{Z_2} \qquad \lambda_2 = Z_1 + \sqrt{Z_2} \tag{7.173}$$

$$Z_1 = \frac{1}{2} \left(\frac{m_1 (k_1 + k_2 + 2k_3) + (k_2 - k_3) m_2}{m_1^2 - m_2^2 + m_1 m_2} \right) \tag{7.174}$$

$$Z_2 = \frac{1}{4} \left(\frac{k_1 m_1 - k_2 m_1 - k_2 m_2 - k_3 m_2}{m_1^2 - m_2^2 + m_1 m_2} \right)^2$$

$$+ (k_1 m_2 - k_3 m_1) \frac{k_2 m_2 - k_3 m_1 + k_3 m_2}{\left(m_1^2 - m_2^2 + m_1 m_2 \right)^2} \tag{7.175}$$

$$\mathbf{u}_1 = \begin{bmatrix} Z_1 - \sqrt{Z_2} \\ k_1 m_2 - k_3 m_1 \\ m_1^2 + m_1 m_2 - m_2^2 \end{bmatrix} \qquad \mathbf{u}_2 = \begin{bmatrix} Z_1 + \sqrt{Z_2} \\ k_1 m_2 - k_3 m_1 \\ m_1^2 + m_1 m_2 - m_2^2 \end{bmatrix} \tag{7.176}$$

When the system is symmetric,

$$m_1 = m_2 = m \qquad k_1 = k_2 = k_3 = k \tag{7.177}$$

then we have

$$\lambda_1 = \frac{k}{m} \qquad \lambda_2 = 3\frac{k}{m} \tag{7.178}$$

$$\mathbf{u}_1 = \begin{bmatrix} 1 \\ 0 \end{bmatrix} \qquad \mathbf{u}_2 = \begin{bmatrix} -1/2 \\ 1 \end{bmatrix} \tag{7.179}$$

Example 245 ★ Other forms of the characteristic equation. Besides the method of characteristic equation, we can determine the natural frequencies from eigenvalues of $[A]$ or $[B] = [A]^{-1}$. Here is the explanation.

Let us multiply the undamped equations of motion by $[k]^{-1}$.

$$[m]\ddot{\mathbf{x}} + [k]\mathbf{x} = \mathbf{0} \tag{7.180}$$

$$[k]^{-1}[m]\ddot{\mathbf{x}} + \mathbf{x} = [B]\ddot{\mathbf{x}} + \mathbf{x} = \mathbf{0} \tag{7.181}$$

The product $[k]^{-1}[m]$ is the inverse of the $[A]$-matrix in (7.27).

$$[B] = [k]^{-1}[m] = [A]^{-1} = \left[[m]^{-1}[k]\right]^{-1} \tag{7.182}$$

Assuming a set of harmonic solutions

$$\mathbf{x} = \mathbf{X}\sin(\omega t + \varphi) = \mathbf{X}\sin\left(\sqrt{\lambda}t + \varphi\right) \tag{7.183}$$

we find the characteristic equation in terms of $\eta = 1/\lambda$.

$$\det[[B] - \eta\mathbf{I}] = 0 \qquad \eta = \frac{1}{\lambda} \tag{7.184}$$

Therefore, depending on the simplicity, the characteristic equation of a vibrating system may be calculated from any of the following equations:

$$|[k] - \lambda[m]| = 0 \tag{7.185}$$

$$|[A] - \lambda\mathbf{I}| = \left|[m]^{-1}[k] - \lambda\mathbf{I}\right| = 0 \tag{7.186}$$

$$|[B] - \eta\mathbf{I}| = \left|[k]^{-1}[m] - \frac{1}{\lambda}\mathbf{I}\right| = 0 \tag{7.187}$$

However, using the characteristic equation is not the best method for determination of the eigenvalue.

Example 246 ★ General two *DOF* systems and eigenvalues. This is an analytic discussion on the natural frequencies of two *DOF* coupled system when the mass coupling term ε and stiffness coupling term μ varies.

By a proper scaling of the coordinates x_1 and x_2, the most general form of the equations of motion of an undamped two *DOF* system can be written in the following form.

$$\ddot{x}_1 + \alpha^2 x_1 + \varepsilon\ddot{x}_2 + \mu\alpha\beta x_2 = 0 \tag{7.188}$$

$$\ddot{x}_2 + \beta^2 x_2 + \varepsilon\ddot{x}_1 + \mu\alpha\beta x_1 = 0 \tag{7.189}$$

$$\begin{bmatrix} 1 & \varepsilon \\ \varepsilon & 1 \end{bmatrix} \begin{bmatrix} \ddot{x}_1 \\ \ddot{x}_2 \end{bmatrix} + \begin{bmatrix} \alpha^2 & \mu\alpha\beta \\ \mu\alpha\beta & \beta^2 \end{bmatrix} \begin{bmatrix} x_1 \\ x_2 \end{bmatrix} = \begin{bmatrix} 0 \\ 0 \end{bmatrix} \tag{7.190}$$

Substituting harmonic solutions (7.183) into the equations

$$\left(\alpha^2 - \lambda\right) x_1 + (\mu\alpha\beta - \varepsilon\lambda) x_2 = 0 \tag{7.191}$$

$$(\mu\alpha\beta - \varepsilon\lambda) x_1 + \left(\beta^2 - \lambda\right) x_2 = 0 \tag{7.192}$$

reduces the characteristic equation $\|[k] - \lambda [m]\| = 0$ to:

$$f(\lambda) = \left(\alpha^2 - \lambda\right)\left(\beta^2 - \lambda\right) - (\mu\alpha\beta - \varepsilon\lambda)^2 = 0 \tag{7.193}$$

We observe that

$$f\left(\alpha^2\right) = -\alpha^2 (\mu\beta - \varepsilon\alpha)^2 < 0 \tag{7.194}$$

$$f\left(\beta^2\right) = -\beta^2 (\mu\alpha - \varepsilon\beta)^2 < 0 \tag{7.195}$$

$$f(0) = \alpha^2\beta^2 \left(1 - \mu^2\right) > 0 \tag{7.196}$$

$$f(\infty) = \lambda^2 \left(1 - \varepsilon^2\right) > 0 \tag{7.197}$$

Therefore, the characteristic values of λ_1 and λ_2 are real and positive. Furthermore, one of them is smaller than α^2, and the other is greater than β^2. The interval $\left[\alpha \ \beta\right]$ separates the natural frequencies $\omega_1 = \sqrt{\lambda_1}$ and $\omega_2 = \sqrt{\lambda_2}$ and is in the interval $\left[\omega_1 \ \omega_2\right]$. Therefore, when two single DOF systems with natural frequencies α and β, $\alpha < \beta$ are attached to make a two DOF system, then the new system will have two natural frequencies ω_1 and ω_2, such that $\omega_1 < \alpha$ and $\omega_2 > \beta$.

To see the effect of mass coupling term ε, and stiffness coupling term μ, let us assume that the original systems have equal natural frequencies.

$$\alpha = \beta \tag{7.198}$$

The characteristic equation reduces to

$$\left(\alpha^2 - \lambda\right)^2 - \left(\mu\alpha^2 - \varepsilon\lambda\right)^2 = 0 \tag{7.199}$$

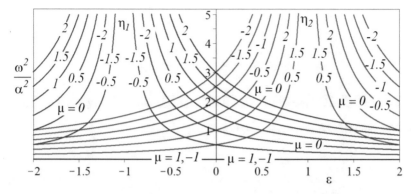

Fig. 7.17 Variation of $\eta_1 = \omega_1/\alpha$ and $\eta_2 = \omega_2/\alpha$ as functions of ε for various μ

or

$$(1 - \eta)^2 - (\mu - \varepsilon\eta)^2 = 0 \tag{7.200}$$

$$\eta = \frac{\lambda}{\alpha^2} = \frac{\omega^2}{\alpha^2} \tag{7.201}$$

which provides us with

$$\eta_1 = \frac{1 + \mu}{1 + \varepsilon} \qquad \eta_2 = \frac{-1 + \mu}{-1 + \varepsilon} \tag{7.202}$$

Figure 7.17 illustrates the variation of $\eta_1 = \omega_1^2/\alpha^2$ and $\eta_2 = \omega_2^2/\alpha^2$ as functions of ε for various μ.

When $\mu = 0$ and the system is only mass coupled, any nonzero value of ε provides us with two values for η such that always one is less than 1 and the other is greater than 1. The same situation appears when $\varepsilon = 0$ and the system is only stiffness coupled.

Example 247 Zero natural frequency and rigid mode. If a vibrating system is not attached to the ground, it can move respect to the ground. Such motion is indicated by a zero natural frequency and is called the rigid mode shape. Here is the introduction of how such zero natural frequency appears.

When a vibrating system can have any position with respect to a fixed coordinate frame, it has a zero natural frequency which indicates a rigid motion or rigid mode. As an example, consider the two DOF system in Fig. 7.18 with the following equations of motion:

$$\begin{bmatrix} m_1 & 0 \\ 0 & m_2 \end{bmatrix} \begin{bmatrix} \ddot{x}_1 \\ \ddot{x}_2 \end{bmatrix} + \begin{bmatrix} k & -k \\ -k & k \end{bmatrix} \begin{bmatrix} x_1 \\ x_2 \end{bmatrix} = \begin{bmatrix} 0 \\ 0 \end{bmatrix} \tag{7.203}$$

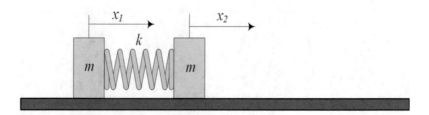

Fig. 7.18 A two *DOF* system with a rigid mode

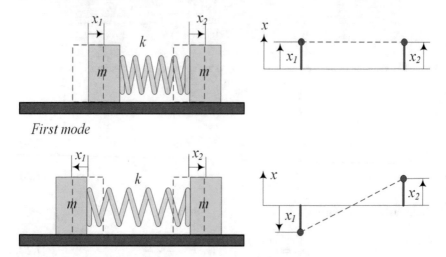

Fig. 7.19 The mode shapes of a two *DOF* system with a rigid mode

The eigenvalues and eigenvectors of the system are:

$$\omega_1 = 0 \qquad\qquad \omega_2 = \sqrt{k\dfrac{m_1 + m_2}{m_1 m_2}} \qquad (7.204)$$

$$u_1 = \begin{bmatrix} 1 \\ 1 \end{bmatrix} \qquad u_2 = \begin{bmatrix} -m_2/m_1 \\ 1 \end{bmatrix} \qquad (7.205)$$

The first mode shape shows a non-vibrating motion of the system, while the second mode indicates a vibrating motion. These modes are illustrated in Fig. 7.19 for $m_1 = m_2$, physically and geometrically. There is a node in the middle of the spring in the second mode of Fig. 7.19. This point can be moving with constant speed because of the first mode. The actual motion of the system is determined by the initial conditions of the masses.

The rank of a matrix $[A]$ is the order of the largest non-singular submatrix of $[A]$. A non-singular matrix has a nonzero determinant. For an n *DOF* vibrating system, $[A] = [m]^{-1}[k]$ would be an $n \times n$ matrix. If the rank of $[A]$ is n', then

$$f = n - n'$$ (7.206)

indicates the number of rigid modes of the system.

Example 248 ★ Rigid mode elimination. Mode shapes are coordinate dependent, and hence, we may use different coordinates to derive different mode shapes. Relative coordinates can simplify the equations of motions to hide the rigid body mode. Here is to show how.

It is possible to use the relative displacement and eliminate the zero natural frequency of a system. Consider the system of Fig. 7.18 whose equations of motion are:

$$m_1\ddot{x}_1 + kx_1 - kx_2 = 0$$ (7.207)

$$m_2\ddot{x}_2 + kx_2 - kx_1 = 0$$ (7.208)

If we define the relative displacement z,

$$z = x_1 - x_2$$ (7.209)

then we can subtract the equations to combine them as a single differential equation in terms of z.

$$m_1 m_2 \ddot{z} + k(m_1 + m_2)z = 0$$ (7.210)

This equation defines the behavior of relative displacement of the masses. The only natural frequency of this equation is equal to the nonzero natural frequency of the system.

$$\omega_n = \sqrt{k\frac{m_1 + m_2}{m_1 m_2}}$$ (7.211)

Example 249 A system with two rigid modes. Here is a more complicated vibrating system with more than one zero natural frequency.

Figure 7.20 illustrates four masses on a table that are connected to each other by similar springs. Let us measure the displacement of the masses by absolute coordinates x_i and y_i from the rest configurations. Each mass may move in X and Y-direction only. Let us for simplicity assume that all masses are equal.

$$m_1 = m_2 = m_3 = m_4 = m$$ (7.212)

The system is at rest when we have:

$$x_1 = x_2 = x_3 = x_4 = y_1 = y_2 = y_3 = y_4 = 0$$ (7.213)

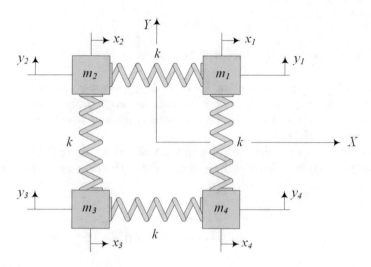

Fig. 7.20 Four masses on a table that are connected to each other by similar springs

The kinetic and potential energies of the system are:

$$\frac{2}{m}K = \dot{x}_1^2 + \dot{y}_1^2 + \dot{x}_2^2 + \dot{y}_2^2 + \dot{x}_3^2 + \dot{y}_3^2 + \dot{x}_4^2 + \dot{y}_4^2 \tag{7.214}$$

$$\frac{2}{k}P = (x_1 - x_2)^2 + (y_1 - y_2)^2 + (x_2 - x_3)^2 + (y_2 - y_3)^2$$
$$+ (x_3 - x_4)^2 + (y_3 - y_4)^2 + (x_4 - x_1)^2 + (y_4 - y_1)^2 \tag{7.215}$$

The Lagrange equation provides us with the equations of motion

$$m\ddot{x}_1 + k\,(2x_1 - x_2 - x_4) = 0 \tag{7.216}$$

$$m\ddot{y}_1 + k\,(2y_1 - y_2 - y_4) = 0 \tag{7.217}$$

$$m\ddot{x}_2 + k\,(2x_2 - x_3 - x_1) = 0 \tag{7.218}$$

$$m\ddot{y}_2 + k\,(2y_2 - y_3 - y_1) = 0 \tag{7.219}$$

$$m\ddot{x}_3 + k\,(2x_3 - x_4 - x_2) = 0 \tag{7.220}$$

$$m\ddot{y}_3 + k\,(2y_3 - y_4 - y_2) = 0 \tag{7.221}$$

$$m\ddot{x}_4 + k\,(2x_4 - x_1 - x_3) = 0 \tag{7.222}$$

$$m\ddot{y}_4 + k\,(2y_4 - y_1 - y_3) = 0 \tag{7.223}$$

or

$$[m]\ddot{\mathbf{x}} + [k]\mathbf{x} = 0 \tag{7.224}$$

where

$$\mathbf{x} = \begin{bmatrix} x_1 & y_1 & x_2 & y_2 & x_3 & y_3 & x_4 & y_4 \end{bmatrix}^T \tag{7.225}$$

$$[m] = m \begin{bmatrix} 1 & 0 & 0 & 0 & 0 & 0 & 0 & 0 \\ 0 & 1 & 0 & 0 & 0 & 0 & 0 & 0 \\ 0 & 0 & 1 & 0 & 0 & 0 & 0 & 0 \\ 0 & 0 & 0 & 1 & 0 & 0 & 0 & 0 \\ 0 & 0 & 0 & 0 & 1 & 0 & 0 & 0 \\ 0 & 0 & 0 & 0 & 0 & 1 & 0 & 0 \\ 0 & 0 & 0 & 0 & 0 & 0 & 1 & 0 \\ 0 & 0 & 0 & 0 & 0 & 0 & 0 & 1 \end{bmatrix} \tag{7.226}$$

$$[k] = k \begin{bmatrix} 2 & 0 & -1 & 0 & 0 & 0 & -1 & 0 \\ 0 & 2 & 0 & -1 & 0 & 0 & 0 & -1 \\ -1 & 0 & 2 & 0 & -1 & 0 & 0 & 0 \\ 0 & -1 & 0 & 2 & 0 & -1 & 0 & 0 \\ 0 & 0 & -1 & 0 & 2 & 0 & -1 & 0 \\ 0 & 0 & 0 & -1 & 0 & 2 & 0 & -1 \\ -1 & 0 & 0 & 0 & -1 & 0 & 2 & 0 \\ 0 & -1 & 0 & 0 & 0 & -1 & 0 & 2 \end{bmatrix} \tag{7.227}$$

The order of $[A] = [m]^{-1}[k]$ is $n = 8$, and the rank of $[A]$ is $n' = 6$. Therefore, this system has two rigid modes.

The natural frequencies and mode shapes of the system are:

$$\omega_{1,2} = 0 \qquad \omega_{3,4,5,6} = \sqrt{2}\sqrt{\frac{k}{m}} \qquad \omega_{7,8} = 2\sqrt{\frac{k}{m}} \tag{7.228}$$

$$\mathbf{u}_1 = \begin{bmatrix} 1 & 0 & 1 & 0 & 1 & 0 & 1 & 0 \end{bmatrix}^T \tag{7.229}$$

$$\mathbf{u}_2 = \begin{bmatrix} 0 & 1 & 0 & 1 & 0 & 1 & 0 & 1 \end{bmatrix}^T \tag{7.230}$$

$$\mathbf{u}_3 = \begin{bmatrix} -1 & 0 & 0 & 0 & 1 & 0 & 0 & 0 \end{bmatrix}^T \tag{7.231}$$

$$\mathbf{u}_4 = \begin{bmatrix} 0 & -1 & 0 & 0 & 0 & 1 & 0 & 0 \end{bmatrix}^T \tag{7.232}$$

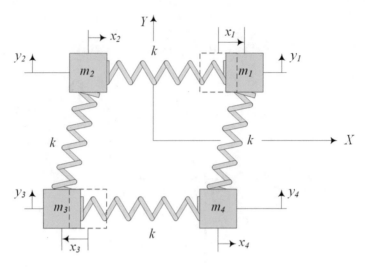

Fig. 7.21 The third mode shape $\mathbf{u}_3$

$$\mathbf{u}_5 = \begin{bmatrix} 0 & 0 & -1 & 0 & 0 & 0 & 1 & 0 \end{bmatrix}^T \tag{7.233}$$

$$\mathbf{u}_6 = \begin{bmatrix} 0 & 0 & 0 & -1 & 0 & 0 & 0 & 1 \end{bmatrix}^T \tag{7.234}$$

$$\mathbf{u}_7 = \begin{bmatrix} -1 & 0 & 1 & 0 & -1 & 0 & 1 & 0 \end{bmatrix}^T \tag{7.235}$$

$$\mathbf{u}_8 = \begin{bmatrix} 0 & -1 & 0 & 1 & 0 & -1 & 0 & 1 \end{bmatrix}^T \tag{7.236}$$

The first and second mode shapes indicate rigid motion of the whole system in X and Y-direction, respectively. Figure 7.21 illustrates the third mode, $\mathbf{u}_3$. The third, fourth, fifth, and sixth modes have similar relative displacements in different directions. Figure 7.22 depicts the seventh mode shape $\mathbf{u}_7$. The eighth mode shape, $\mathbf{u}_8$, is similar to $\mathbf{u}_7$ in the y-direction.

Example 250 Translational and rotational coordinates. A multi DOF vibrating system may have a combination of translational and rotational coordinates. The analysis of such system mathematically has no difference; however, the mode shapes are more complicated to be expressed. The smart may to show mode shapes is to convert all coordinates to be translational or rotational.

Consider the two DOF system in Fig. 7.23. A beam with mass m and mass moment I about the mass center C is sitting on two springs k_1 and k_2. This is good model to study half car model to determine and compare pitch and bounce vibrations of vehicles. The translational coordinate x of C and the rotational coordinate θ are the usual generalized coordinates that we use to measure the kinematics of the beam.

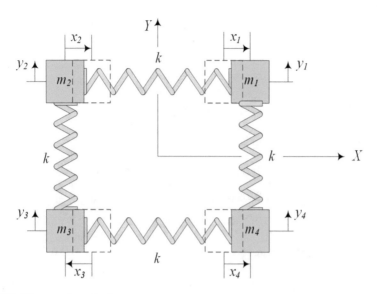

Fig. 7.22 The seventh mode shape $\mathbf{u}_7$

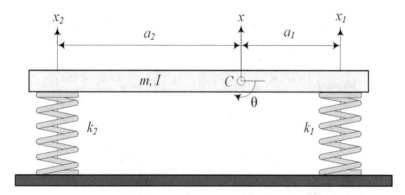

Fig. 7.23 A beam with mass m and mass moment I and siting on two springs k_1 and k_2

The equations of motion and the mode shapes of the system are functions of the chosen coordinates.

The kinetic and potential energies of the beam for small motions are:

$$K = \frac{1}{2}m\dot{x}^2 + \frac{1}{2}I\dot{\theta}^2 \tag{7.237}$$

$$P = \frac{1}{2}k_2\,(x + a_2\theta)^2 + \frac{1}{2}k_1\,(x - a_1\theta)^2 \tag{7.238}$$

The energies provide us with the equations of motion.

$$m\ddot{x} + k_2\,(x + a_2\theta) + k_1\,(x - a_1\theta) = 0 \tag{7.239}$$

$$I\ddot{\theta} + a_2k_2\,(x + a_2\theta) - a_1k_1\,(x - a_1\theta) = 0 \tag{7.240}$$

$$\begin{bmatrix} m & 0 \\ 0 & I \end{bmatrix} \begin{bmatrix} \ddot{x} \\ \ddot{\theta} \end{bmatrix} + \begin{bmatrix} k_1 + k_2 & a_2k_2 - a_1k_1 \\ a_2k_2 - a_1k_1 & a_2^2k_2 + a_1^2k_1 \end{bmatrix} \begin{bmatrix} x \\ \theta \end{bmatrix} = 0 \tag{7.241}$$

Let us use a set of numeric data:

$$
\begin{aligned}
m &= 1000\,\text{kg} & I &= 200\,\text{kg}\,\text{m}^2 \\
k_1 &= 8200\,\text{N}\,/\,\text{m} & k_2 &= 9000\,\text{N}\,/\,\text{m} \\
a_1 &= 0.75\,\text{m} & a_2 &= 1.1\,\text{m}
\end{aligned} \tag{7.242}
$$

to find the following equations:

$$\begin{bmatrix} 1000 & 0 \\ 0 & 200 \end{bmatrix} \begin{bmatrix} \ddot{x} \\ \ddot{\theta} \end{bmatrix} + \begin{bmatrix} 17200 & 3750 \\ 3750 & 15503 \end{bmatrix} \begin{bmatrix} x \\ \theta \end{bmatrix} = 0 \tag{7.243}$$

The natural frequencies and mode shapes of the system are calculated based on eigenvalues and eigenvectors of $[A] = \left[[m]^{-1}\,[k]\right]$:

$$[A] = \begin{bmatrix} 1000 & 0 \\ 0 & 200 \end{bmatrix}^{-1} \begin{bmatrix} 17200 & 3750 \\ 3750 & 15503 \end{bmatrix} = \begin{bmatrix} 17.2 & 3.75 \\ 18.75 & 77.51 \end{bmatrix} \tag{7.244}$$

The ω_i and $\mathbf{u}_i$ are:

$$\omega_1 = 4.007\,\text{rad}\,/\,\text{s} \qquad \omega_2 = 8.868\,\text{rad}\,/\,\text{s} \tag{7.245}$$

$$\mathbf{u}_1 = \begin{bmatrix} 0.95647 \\ -0.29182 \end{bmatrix} \qquad \mathbf{u}_2 = \begin{bmatrix} 0.06091 \\ 0.99814 \end{bmatrix} \tag{7.246}$$

Because of the chosen coordinate vector $\begin{bmatrix} x & \theta \end{bmatrix}^T$, the first component of the mode shapes refers to the displacement x, and the second components refer to the rotation θ. Assuming the components of the mode shapes are comparable, the mode shapes in Fig. 7.24 show that the displacement x is the most observable motion in the first mode and the rotation θ in the second mode. However, because the mode shapes are coordinate dependent, when the coordinates are measured in different dimensions, the rotational coordinate would not necessarily get the biggest number in any mode. These mode shapes do not show the configuration of the system.

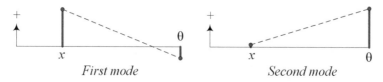

Fig. 7.24 Mode shapes of the two DOF beam with coordinates x and θ

To compare the mode shapes of a system practically, we recommend to choose the generalized coordinates with the same dimension. Let us employ the coordinates x_1 and x_2 instead of x and θ, as shown in Fig. 7.23. The energies of the system in this set of coordinates will be:

$$K = \frac{1}{2}m\left(\frac{a_2\dot{x}_1 + a_1\dot{x}_2}{a_1 + a_2}\right)^2 + \frac{1}{2}I\left(\frac{\dot{x}_1 - \dot{x}_2}{a_1 + a_2}\right)^2 \tag{7.247}$$

$$P = \frac{1}{2}k_2x_2^2 + \frac{1}{2}k_1x_1^2 \tag{7.248}$$

The energies provide us with the following equations of motion.

$$m\frac{a_2}{a_1 + a_2}\frac{a_2\ddot{x}_1 + a_1\ddot{x}_2}{a_1 + a_2} + I\frac{1}{a_1 + a_2}\frac{\ddot{x}_1 - \ddot{x}_2}{a_1 + a_2} + k_1x_1 = 0 \tag{7.249}$$

$$m\frac{a_1}{a_1 + a_2}\frac{a_2\ddot{x}_1 + a_1\ddot{x}_2}{a_1 + a_2} - I\frac{1}{a_1 + a_2}\frac{\ddot{x}_1 - \ddot{x}_2}{a_1 + a_2} + k_2x_2 = 0 \tag{7.250}$$

$$\begin{bmatrix} \dfrac{ma_2^2 + I}{(a_1 + a_2)^2} & \dfrac{ma_1a_2 - I}{(a_1 + a_2)^2} \\ \dfrac{ma_1a_2 - I}{(a_1 + a_2)^2} & \dfrac{ma_1^2 + I}{(a_1 + a_2)^2} \end{bmatrix}\begin{bmatrix} \ddot{x}_1 \\ \ddot{x}_2 \end{bmatrix} + \begin{bmatrix} k_1 & 0 \\ 0 & k_2 \end{bmatrix}\begin{bmatrix} x_1 \\ x_2 \end{bmatrix} = 0 \tag{7.251}$$

Using the data (7.242), the equations of motion simplify to:

$$\begin{bmatrix} 411.98 & 182.62 \\ 182.62 & 222.79 \end{bmatrix}\begin{bmatrix} \ddot{x}_1 \\ \ddot{x}_2 \end{bmatrix} + \begin{bmatrix} 8200 & 0 \\ 0 & 9000 \end{bmatrix}\begin{bmatrix} x_1 \\ x_2 \end{bmatrix} = 0 \tag{7.252}$$

The natural frequencies and mode shapes of the system in the new coordinates are calculated based on eigenvalues and eigenvectors of new $[A] = \left[[m]^{-1}[k]\right]$:

$$[A] = \begin{bmatrix} 411.98 & 182.62 \\ 182.62 & 222.79 \end{bmatrix}^{-1}\begin{bmatrix} 8200 & 0 \\ 0 & 9000 \end{bmatrix}$$

$$= \begin{bmatrix} 31.263 & -28.127 \\ -25.627 & 63.452 \end{bmatrix} \tag{7.253}$$

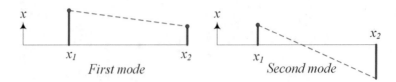

Fig. 7.25 Mode shapes of the two DOF beam with coordinates x_1 and x_2

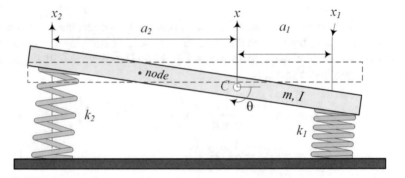

Fig. 7.26 The second mode shape of the beam with a node on the beam

The natural frequencies ω_i will be the same as (7.245), but $\mathbf{u}_i$ will be new.

$$\omega_1 = 4.007 \, \text{rad} \, / \, \text{s} \qquad \omega_2 = 8.868 \, \text{rad} \, / \, \text{s} \qquad (7.254)$$

$$\mathbf{u}_1 = \begin{bmatrix} 0.87965 \\ 0.47562 \end{bmatrix} \qquad \mathbf{u}_2 = \begin{bmatrix} 0.51034 \\ -0.85997 \end{bmatrix} \qquad (7.255)$$

These mode shapes are illustrated in Fig. 7.25. They well show the configuration of the system. Because the components of the mode shapes have the same dimension, more information can be extracted from them compared to Fig. 7.24. The second mode shape shows a node on the beam. The node is a motionless point of the beam such that the whole beam will rotate about the node as is shown in Fig. 7.26. Although in a real situation the springs will tilt when $\theta \neq 0$, we ignore this, and we assume that the springs remain vertical.

The first mode shape does not show a node on the beam, but because the two ends of the beam have different displacements, there is a node on the extension of the centerline of the beam, as is shown in Fig. 7.27.

Example 251 ★ Mode shape design. The natural frequencies and mode shapes of a given system are functions of the stiffness and inertia of the system. To adjust natural frequencies to a desired value, or designing the mode shapes to a desired shape, the stiffness and inertia of the system must be adjusted.

Although the components of the mode shapes of a given system are dependent on the chosen generalized coordinates to describe the configuration of the system,

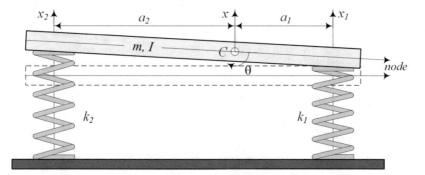

Fig. 7.27 The first mode shape of the beam with a node on extension of the beam

when the mode shapes are used to express the actual configuration of the system, every set of generalized coordinates ends up with the same configuration. Therefore, besides the natural frequencies, the actual configuration of a given system at each natural frequency is also coordinate independent. Consequently, we are unable to change the mode shapes while keeping the natural frequencies unchanged and vice versa. Both the natural frequencies and their associated mode shapes of a system are dependent on the stiffness and inertia characteristics of the system, as well as their geometric properties.

As an example, let us examine the effect of the value of k_1 on the natural frequencies of the beam of Fig. 7.23. Employing the coordinates x_1 and x_2, the equations of motion of the system are given in Eq. (7.251). Assuming an unknown k_1 and using the rest of the data (7.242), we have

$$\begin{bmatrix} 411.98 & 182.62 \\ 182.62 & 222.79 \end{bmatrix} \begin{bmatrix} \ddot{x}_1 \\ \ddot{x}_2 \end{bmatrix} + \begin{bmatrix} k_1 & 0 \\ 0 & 9000 \end{bmatrix} \begin{bmatrix} x_1 \\ x_2 \end{bmatrix} = 0 \tag{7.256}$$

which yields:

$$[A] = \left[[m]^{-1} [k] \right] = \begin{bmatrix} 3.8126 \times 10^{-3} k_1 & -28.127 \\ -3.1252 \times 10^{-3} k_1 & 63.452 \end{bmatrix} \tag{7.257}$$

The natural frequencies are:

$$\omega_1^2 = 1.9 \times 10^{-3} k_1$$
$$-0.5\sqrt{1.45 \times 10^{-5} k_1^2 - 0.132 k_1 + 4026.2} + 31.72 \tag{7.258}$$
$$\omega_2^2 = 1.9 \times 10^{-3} k_1$$
$$+0.5\sqrt{1.45 \times 10^{-5} k_1^2 - 0.132 k_1 + 4026.2} + 31.72 \tag{7.259}$$

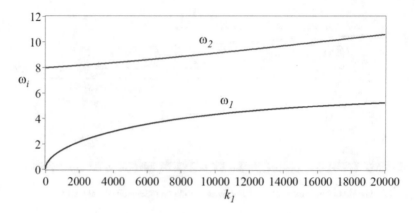

Fig. 7.28 The two natural frequencies of the beam as a function of k_1

A plot of the natural frequencies as a function of k_1 is shown in Fig. 7.28. The figure indicates that both frequencies are functions of k_1 and get higher values by increasing k_1.

$$k_1 = 1000 \qquad \mathbf{u}_1 = \begin{bmatrix} 0.99869 \\ 0.05109 \end{bmatrix} \qquad \mathbf{u}_2 = \begin{bmatrix} 0.45047 \\ -0.97826 \end{bmatrix} \qquad (7.260)$$

$$k_1 = 5000 \qquad \mathbf{u}_1 = \begin{bmatrix} .95878 \\ .28414 \end{bmatrix} \qquad \mathbf{u}_2 = \begin{bmatrix} 0.48043 \\ -0.90061 \end{bmatrix} \qquad (7.261)$$

$$k_1 = 8200 \qquad \mathbf{u}_1 = \begin{bmatrix} 0.87965 \\ 0.47561 \end{bmatrix} \qquad \mathbf{u}_2 = \begin{bmatrix} 0.51074 \\ -0.86066 \end{bmatrix} \qquad (7.262)$$

$$k_1 = 9000 \qquad \mathbf{u}_1 = \begin{bmatrix} 0.85438 \\ 0.51963 \end{bmatrix} \qquad \mathbf{u}_2 = \begin{bmatrix} 0.51963 \\ -0.85438 \end{bmatrix} \qquad (7.263)$$

$$k_1 = 20000 \qquad \mathbf{u}_1 = \begin{bmatrix} 0.53868 \\ 0.9349 \end{bmatrix} \qquad \mathbf{u}_2 = \begin{bmatrix} 0.61551 \\ -0.78812 \end{bmatrix} \qquad (7.264)$$

The effect of variation of k_1 on mode shapes, for some examples, is indicated in Eqs. (7.260)–(7.264). When the components of the mode shape have opposite signs, then there exists a node on the beam. Comparing $\mathbf{u}_2$ for different k_1 indicates that by increasing k_1, the node gets closer to the k_2-end of the beam. Similarly, when the components of the mode shape have different values with the same sign, there exists a node on the extension of the beam. Comparing $\mathbf{u}_1$ for different k_1 indicates that by increasing k_1, the node goes away to infinity from the k_1-end and finally comes close to the other side of the beam from infinity. Therefore, there must be a specific value of k_1 for which the components of $\mathbf{u}_1$ are equal and, hence, the beam is oscillating up and down without tilting. For such a mode shape, the springs act parallel, and therefore the associated natural frequency must be:

$$\omega_1 = \sqrt{\frac{k_1 + k_2}{m}} = \sqrt{\frac{k_1 + 9000}{1000}} \tag{7.265}$$

Employing (7.265) and (7.258) provides us with the design value of k_1:

$$k_1 = 13200 \tag{7.266}$$

At $k_1 = 13200$ we will have:

$$\omega_1 = 4.711 \, \text{rad} \, / \, \text{s} \qquad \omega_2 = 9.569 \, \text{rad} \, / \, \text{s} \tag{7.267}$$

$$\mathbf{u}_1 = \begin{bmatrix} 0.7071 \\ 0.7071 \end{bmatrix} \qquad \mathbf{u}_2 = \begin{bmatrix} 0.5733 \\ -0.8409 \end{bmatrix} \tag{7.268}$$

Example 252 Orthogonality of mode shapes. The mode shapes are orthogonal, and that is the reason we are able to express the time response of multi DOF systems by a linear combination of mode shapes. Here is an example of the meaning of mode shapes orthogonality.

To examine the orthogonality conditions (7.8)–(7.11), let us use the three DOF system of Fig. 7.12. The equations of motion of the system are:

$$\begin{bmatrix} m & 0 & 0 \\ 0 & m & 0 \\ 0 & 0 & m \end{bmatrix} \begin{bmatrix} \ddot{x}_1 \\ \ddot{x}_2 \\ \ddot{x}_3 \end{bmatrix} + \begin{bmatrix} 2k & -k & 0 \\ -k & 2k & -k \\ 0 & -k & 2k \end{bmatrix} \begin{bmatrix} x_1 \\ x_2 \\ x_3 \end{bmatrix} = 0 \tag{7.269}$$

and its natural frequencies and mode shapes are:

$$\lambda_1 = \frac{k}{m} \left(2 - \sqrt{2} \right) \qquad \lambda_2 = 2\frac{k}{m} \qquad \lambda_3 = \frac{k}{m} \left(2 + \sqrt{2} \right) \tag{7.270}$$

$$\omega_1 = \sqrt{2 - \sqrt{2}} \sqrt{\frac{k}{m}} \qquad \omega_2 = \sqrt{2} \sqrt{\frac{k}{m}} \qquad \omega_3 = \sqrt{2 + \sqrt{2}} \sqrt{\frac{k}{m}} \tag{7.271}$$

$$\mathbf{u}_1 = \begin{bmatrix} 1 \\ \sqrt{2} \\ 1 \end{bmatrix} \qquad \mathbf{u}_2 = \begin{bmatrix} -1 \\ 0 \\ 1 \end{bmatrix} \qquad \mathbf{u}_3 = \begin{bmatrix} 1 \\ -\sqrt{2} \\ 1 \end{bmatrix} \tag{7.272}$$

The orthogonality with respect to the mass matrix,

$$\mathbf{u}_j^T \, [m] \, \mathbf{u}_i = 0 \qquad i \neq j \tag{7.273}$$

is examined by

$$\mathbf{u}_1^T [m] \mathbf{u}_2 = \begin{bmatrix} 1 \\ \sqrt{2} \\ 1 \end{bmatrix}^T \begin{bmatrix} m & 0 & 0 \\ 0 & m & 0 \\ 0 & 0 & m \end{bmatrix} \begin{bmatrix} -1 \\ 0 \\ 1 \end{bmatrix} = 0 \qquad (7.274)$$

$$\mathbf{u}_2^T [m] \mathbf{u}_3 = \begin{bmatrix} -1 \\ 0 \\ 1 \end{bmatrix}^T \begin{bmatrix} m & 0 & 0 \\ 0 & m & 0 \\ 0 & 0 & m \end{bmatrix} \begin{bmatrix} 1 \\ -\sqrt{2} \\ 1 \end{bmatrix} = 0 \qquad (7.275)$$

$$\mathbf{u}_3^T [m] \mathbf{u}_1 = \begin{bmatrix} 1 \\ -\sqrt{2} \\ 1 \end{bmatrix}^T \begin{bmatrix} m & 0 & 0 \\ 0 & m & 0 \\ 0 & 0 & m \end{bmatrix} \begin{bmatrix} 1 \\ \sqrt{2} \\ 1 \end{bmatrix} = 0 \qquad (7.276)$$

The orthogonality with respect to the stiffness matrix,

$$\mathbf{u}_j^T [k] \mathbf{u}_i = 0 \qquad i \neq j \qquad (7.277)$$

is examined by:

$$\mathbf{u}_1^T [k] \mathbf{u}_2 = \begin{bmatrix} 1 \\ \sqrt{2} \\ 1 \end{bmatrix}^T \begin{bmatrix} 2k & -k & 0 \\ -k & 2k & -k \\ 0 & -k & 2k \end{bmatrix} \begin{bmatrix} -1 \\ 0 \\ 1 \end{bmatrix} = 0 \qquad (7.278)$$

$$\mathbf{u}_2^T [k] \mathbf{u}_3 = \begin{bmatrix} -1 \\ 0 \\ 1 \end{bmatrix}^T \begin{bmatrix} 2k & -k & 0 \\ -k & 2k & -k \\ 0 & -k & 2k \end{bmatrix} \begin{bmatrix} 1 \\ -\sqrt{2} \\ 1 \end{bmatrix} = 0 \qquad (7.279)$$

$$\mathbf{u}_3^T [k] \mathbf{u}_1 = \begin{bmatrix} 1 \\ -\sqrt{2} \\ 1 \end{bmatrix}^T \begin{bmatrix} 2k & -k & 0 \\ -k & 2k & -k \\ 0 & -k & 2k \end{bmatrix} \begin{bmatrix} 1 \\ \sqrt{2} \\ 1 \end{bmatrix} = 0 \qquad (7.280)$$

The generalized masses are:

$$m_1 = \mathbf{u}_1^T [m] \mathbf{u}_1 = \begin{bmatrix} 1 \\ \sqrt{2} \\ 1 \end{bmatrix}^T \begin{bmatrix} m & 0 & 0 \\ 0 & m & 0 \\ 0 & 0 & m \end{bmatrix} \begin{bmatrix} 1 \\ \sqrt{2} \\ 1 \end{bmatrix} = 4m \qquad (7.281)$$

$$m_2 = \mathbf{u}_2^T [m] \mathbf{u}_2 = \begin{bmatrix} -1 \\ 0 \\ 1 \end{bmatrix}^T \begin{bmatrix} m & 0 & 0 \\ 0 & m & 0 \\ 0 & 0 & m \end{bmatrix} \begin{bmatrix} -1 \\ 0 \\ 1 \end{bmatrix} = 2m \qquad (7.282)$$

$$m_3 = \mathbf{u}_3^T \,[m]\, \mathbf{u}_3 = \begin{bmatrix} 1 \\ -\sqrt{2} \\ 1 \end{bmatrix}^T \begin{bmatrix} m & 0 & 0 \\ 0 & m & 0 \\ 0 & 0 & m \end{bmatrix} \begin{bmatrix} 1 \\ -\sqrt{2} \\ 1 \end{bmatrix} = 4m \qquad (7.283)$$

and the generalized stiffness are:

$$k_1 = \mathbf{u}_1^T \,[k]\, \mathbf{u}_1 = \begin{bmatrix} 1 \\ \sqrt{2} \\ 1 \end{bmatrix}^T \begin{bmatrix} 2k & -k & 0 \\ -k & 2k & -k \\ 0 & -k & 2k \end{bmatrix} \begin{bmatrix} 1 \\ \sqrt{2} \\ 1 \end{bmatrix}$$

$$= -4k \left(\sqrt{2} - 2 \right) \qquad (7.284)$$

$$k_2 = \mathbf{u}_2^T \,[k]\, \mathbf{u}_2 = \begin{bmatrix} -1 \\ 0 \\ 1 \end{bmatrix}^T \begin{bmatrix} 2k & -k & 0 \\ -k & 2k & -k \\ 0 & -k & 2k \end{bmatrix} \begin{bmatrix} -1 \\ 0 \\ 1 \end{bmatrix}$$

$$= 4k \qquad (7.285)$$

$$k_3 = \mathbf{u}_3^T \,[k]\, \mathbf{u}_3 = \begin{bmatrix} 1 \\ -\sqrt{2} \\ 1 \end{bmatrix}^T \begin{bmatrix} 2k & -k & 0 \\ -k & 2k & -k \\ 0 & -k & 2k \end{bmatrix} \begin{bmatrix} 1 \\ -\sqrt{2} \\ 1 \end{bmatrix}$$

$$= 4k \left(\sqrt{2} + 2 \right) \qquad (7.286)$$

Example 253 ★ Mode shapes are linearly independent. Orthogonality is equivalent to be linearly independent and vice versa.

A set of vectors $\mathbf{u}_1, \mathbf{u}_2, \mathbf{u}_3, \cdots$ is linearly independent if their linear combination is only zero when the coefficients $a_1, a_2, a_3, \cdots$ are all zero.

$$a_1\mathbf{u}_1 + a_2\mathbf{u}_2 + a_3\mathbf{u}_3 + \cdots = 0 \qquad (7.287)$$

To show that mode shapes are linearly independent, we assume that (7.287) is correct. Multiplying (7.287) by $\mathbf{u}_i^T \,[m]$ and using the orthogonality condition (7.40) yield:

$$a_i\mathbf{u}_j^T \,[m]\, \mathbf{u}_i = a_i m_i = 0 \qquad (7.288)$$

Because $m_i \neq 0$, the coefficient a_i must be zero.

Example 254 ★ Complex mode shape. The mode shapes of a system may be expressed by complex vectors. In such case, the natural frequency will be dependent on the real and imaginary parts of the mode shape.

Let us recall Eqs. (7.19) and (7.20) where ω is the natural frequency and $\mathbf{u}$ is its associated mode shape.

$$\ddot{q}(t) + \omega^2 q(t) = 0 \tag{7.289}$$

$$\left[[k] - \omega^2 [m] \right] \mathbf{u} = 0 \tag{7.290}$$

Assuming a general complex mode shape $\mathbf{u} = \mathbf{a} + i\mathbf{b}$ yields:

$$[k](\mathbf{a} + i\mathbf{b}) = \omega^2 [m](\mathbf{a} + i\mathbf{b}) \qquad i^2 = -1 \tag{7.291}$$

Multiplying $(\mathbf{a} - i\mathbf{b})^T$ by (7.291) and using the symmetry of $[k]$ and $[m]$,

$$\mathbf{a}^T [k] \mathbf{b} = \left(\mathbf{a}^T [k] \mathbf{b} \right)^T = \mathbf{b}^T [k] \mathbf{a} \tag{7.292}$$

$$\mathbf{a}^T [m] \mathbf{b} = \left(\mathbf{a}^T [m] \mathbf{b} \right)^T = \mathbf{b}^T [m] \mathbf{a} \tag{7.293}$$

we have

$$\mathbf{a}^T [k] \mathbf{a} + \mathbf{b}^T [k] \mathbf{b} = \omega^2 \mathbf{a}^T [m] \mathbf{a} + \omega^2 \mathbf{b}^T [m] \mathbf{b} \tag{7.294}$$

and, therefore,

$$\omega^2 = \frac{\mathbf{a}^T [k] \mathbf{a} + \mathbf{b}^T [k] \mathbf{b}}{\mathbf{a}^T [m] \mathbf{a} + \mathbf{b}^T [m] \mathbf{b}} > 0 \tag{7.295}$$

Example 255 ★ The smallest natural frequency and Dunkerley formula. There are several approximation methods to estimate the largest and the smallest natural frequencies, ω_m, ω_M, of a multi DOF system. Considering all natural frequencies will be in the domain of $[\omega_m \ \omega_M]$; it will help the designer to know the range of resonance zone of frequency domain. Here is the method of evaluating the smallest natural frequency.

An acceptable approximation of the first natural frequency ω_1 of a vibrating system is:

$$\frac{1}{\omega_1^2} \approx \text{tr} \left[[k]^{-1} [m] \right] \tag{7.296}$$

To show this, consider the characteristic equation of a multi DOF system

$$\left| [k] - \omega^2 [m] \right| = 0 \tag{7.297}$$

which can also be written as

$$\left| -\frac{1}{\omega^2} [I] + [k]^{-1} [m] \right| = \left| -\frac{1}{\omega^2} [I] + [\alpha] [m] \right| = 0 \tag{7.298}$$

$$[\alpha] = [k]^{-1} \tag{7.299}$$

The equation reduces to

$$\begin{vmatrix} \alpha_{11}m_1 - \dfrac{1}{\omega^2} & \alpha_{12}m_2 & \cdots & \alpha_{1n}m_n \\[2mm] \alpha_{21}m_1 & \alpha_{22}m_2 - \dfrac{1}{\omega^2} & \cdots & \alpha_{2n}m_n \\[2mm] \vdots & \vdots & \ddots & \vdots \\[2mm] \alpha_{n1}m_1 & \alpha_{n2}m_2 & \cdots & \alpha_{nn}m_n - \dfrac{1}{\omega^2} \end{vmatrix} = 0 \tag{7.300}$$

The expanded form of this determinant will be

$$\left(\frac{1}{\omega^2}\right)^n - (\alpha_{11}m_1 + \alpha_{22}m_2 + \cdots + \alpha_{nn}m_n)\left(\frac{1}{\omega^2}\right)^{n-1}$$

$$+ \big(\alpha_{11}\alpha_{22}m_1m_2 + \alpha_{11}\alpha_{23}m_1m_3 + \cdots + \alpha_{n-1,n-1}\alpha_{nn}m_{n-1}m_n$$

$$-\alpha_{12}\alpha_{21}m_1m_2 - \cdots - \alpha_{n-1,n}\alpha_{n,n-1}m_{n-1}m_n\Big) \left(\frac{1}{\omega^2}\right)^{n-2}$$

$$- \cdots = 0 \tag{7.301}$$

This is a polynomial equation of nth degree for $(1/\omega^2)$ and may be written as:

$$\left(\frac{1}{\omega^2} - \frac{1}{\omega_1^2}\right)\left(\frac{1}{\omega^2} - \frac{1}{\omega_2^2}\right) \cdots \left(\frac{1}{\omega^2} - \frac{1}{\omega_n^2}\right)$$

$$= \left(\frac{1}{\omega^2}\right)^n - \left(\frac{1}{\omega_1^2} + \frac{1}{\omega_2^2} + \cdots + \frac{1}{\omega_n^2}\right)\left(\frac{1}{\omega^2}\right)^{n-1} - \cdots = 0 \tag{7.302}$$

Equating the coefficients of (7.301) and (7.302) yields:

$$\frac{1}{\omega_1^2} + \frac{1}{\omega_2^2} + \cdots + \frac{1}{\omega_n^2} = \alpha_{11}m_1 + \alpha_{22}m_2 + \cdots + \alpha_{nn}m_n \tag{7.303}$$

If we assume that ω_1 is much smaller than the other natural frequencies, then

$$\frac{1}{\omega_1^2} << \frac{1}{\omega_i^2} \tag{7.304}$$

and we may determine ω_1 approximately.

$$\frac{1}{\omega_1^2} \approx \alpha_{11} m_1 + \alpha_{22} m_2 + \cdots + \alpha_{nn} m_n \tag{7.305}$$

This equation is called the **Dunkerley formula**.

As an example, let us evaluate the first natural frequency of the system of Fig. 7.11 for symmetric case.

$$m_1 = m_2 = m_3 = m \qquad k_1 = k_2 = k_3 = k_4 = k \tag{7.306}$$

Multiplying the inverse of the stiffness matrix $[k]$ by $[m]$, we have:

$$[k]^{-1}[m] = \begin{bmatrix} m/k & m/k & m/k \\ m/k & 5m/(3k) & 4m/(3k) \\ m/k & 4m/(3k) & 5m/(3k) \end{bmatrix} \tag{7.307}$$

Therefore,

$$\frac{1}{\omega_1^2} \approx \text{tr}\left[[k]^{-1}[m]\right] = \frac{13}{3k}m \tag{7.308}$$

which yields

$$\omega_1 \approx \sqrt{\frac{3k}{13m}} = 0.48038\sqrt{\frac{k}{m}} \tag{7.309}$$

Comparing the approximate value with the exact one

$$\omega_1 = \sqrt{2 - \sqrt{3}}\sqrt{\frac{k}{m}} = 0.51764\sqrt{\frac{k}{m}} \tag{7.310}$$

indicates that the Dunkerley formula predicts a value smaller than the actual ω_1.

$$\text{tr}\left[[k]^{-1}[m]\right] = \alpha_{11} m_1 + \alpha_{22} m_2 + \cdots + \alpha_{nn} m_n < \omega_1 \tag{7.311}$$

7.2 Coupling and Decoupling

The equations of motion of any undamped linear n degree-of-freedom (DOF) vibrating system in terms of a set of generalized coordinates $\mathbf{q}$ are:

$$[m]\,\ddot{\mathbf{q}} + [k]\,\mathbf{q} = \mathbf{Q} \tag{7.312}$$

$$\mathbf{q} = \begin{bmatrix} q_1 & q_2 & \cdots & q_n \end{bmatrix}^T \tag{7.313}$$

If the mass matrix $[m]$ has any nonzero cross element, the system is called *mass* or *inertially* or *dynamically coupled*, and if the stiffness matrix $[k]$ has any nonzero cross element, the system is called *stiffness* or *elastically* or *statically coupled*.

We can always change the set of generalized coordinates $\mathbf{q}$ to a set of *principal coordinates* $\mathbf{p}$:

$$\mathbf{q} = [U]\,\mathbf{p} \tag{7.314}$$

and change the equations of motion to a set of decoupled equations.

$$\left[m'\right]\ddot{\mathbf{p}} + \left[k'\right]\mathbf{p} = \mathbf{P} \tag{7.315}$$

$$\left[m'\right] = [U]^T\,[m]\,[U] \tag{7.316}$$

$$\left[k'\right] = [U]^T\,[k]\,[U] \tag{7.317}$$

$$\mathbf{P} = [U]^T\,\mathbf{Q} \tag{7.318}$$

The new mass and stiffness matrices $\left[m'\right]$ and $\left[k'\right]$ are diagonal:

$$[m'] = \begin{bmatrix} m'_1 & 0 & \cdots & 0 \\ 0 & m'_2 & \cdots & 0 \\ \vdots & \vdots & \ddots & \vdots \\ 0 & 0 & \cdots & m'_n \end{bmatrix} \qquad m'_{ij} = \begin{cases} m'_i & i = j \\ 0 & i \neq j \end{cases} \tag{7.319}$$

$$[k'] = \begin{bmatrix} k'_1 & 0 & \cdots & 0 \\ 0 & k'_2 & \cdots & 0 \\ \vdots & \vdots & \ddots & \vdots \\ 0 & 0 & \cdots & k'_n \end{bmatrix} \qquad k'_{ij} = \begin{cases} k'_i & i = j \\ 0 & i \neq j \end{cases} \tag{7.320}$$

and the equations of motion are decoupled.

$$m'_i\,\ddot{p}_i + k'_i\,p_i = F_i \qquad i = 1, 2, 3, \cdots, n \tag{7.321}$$

The square transformation matrix $[U]$ is called the *modal matrix* of the system. The columns of the $n \times n$ modal matrix $[U]$ are the mode shapes of the system:

$$[U] = \begin{bmatrix} \mathbf{u}_1 & \mathbf{u}_2 & \mathbf{u}_2 & \cdots & \mathbf{u}_n \end{bmatrix} \tag{7.322}$$

Using the modal matrix $[U]$, the natural frequencies of the system can be found easier.

$$[m']^{-1}[k'] = [U]^T[A][U] = \begin{bmatrix} \omega_1^2 & 0 & \cdots & 0 \\ 0 & \omega_2^2 & \cdots & 0 \\ \vdots & \vdots & \ddots & \vdots \\ 0 & 0 & \cdots & \omega_n^2 \end{bmatrix} \tag{7.323}$$

The principal coordinates $\mathbf{p} = [U]^T\mathbf{q}$ are also called the *natural coordinates* of the system. Decoupling of the equations of motion is applicable only for undamped systems. There is not always possible to decouple vibrating systems with damping. However, undamped systems are the base to discover the principal characteristics of a system such as natural frequencies, mode shapes, modal matrix, and principal coordinates.

Proof Consider a linear n DOF vibrating system with equations of motion based on a set of n generalized coordinates x_i, $i = 1, 2, 3, \cdots, n$.

$$[m]\ddot{\mathbf{x}} + [c]\dot{\mathbf{x}} + [k]\mathbf{x} = \mathbf{F} \tag{7.324}$$

The kinetic and potential energies and dissipation function of the system are:

$$K = \frac{1}{2}\dot{\mathbf{x}}^T[m]\dot{\mathbf{x}} \qquad m_{ij} = m_{ji} \tag{7.325}$$

$$P = \frac{1}{2}\mathbf{x}^T[k]\mathbf{x} \qquad k_{ij} = k_{ji} \tag{7.326}$$

$$D = \frac{1}{2}\dot{\mathbf{x}}^T[c]\dot{\mathbf{x}} \qquad c_{ij} = c_{ji} \tag{7.327}$$

where the mass matrix $[m]$, the stiffness matrix $[k]$ and the damping matrix $[c]$ are symmetric. Any other set of generalized coordinates q_i, $i = 1, 2, 3, \cdots, n$ will be a linear combination of x_j, $j = 1, 2, 3, \cdots, n$ and vice versa,

$$\mathbf{x} = [B]\mathbf{q} \tag{7.328}$$

where $[B]$ must be a nonsingular square matrix to have an inverse coordinate transformation.

$$\mathbf{q} = [B]^{-1}\mathbf{x} \tag{7.329}$$

Using the linear combination of the generalized velocities $\dot{q}_i$ and $\dot{x}_j$

$$\dot{\mathbf{x}} = [B]\,\dot{\mathbf{q}} \tag{7.330}$$

the kinetic, potential, and dissipation functions in terms of the new coordinates $\mathbf{q}$ will be:

$$K = \frac{1}{2}\dot{\mathbf{x}}^T\,[m]\,\dot{\mathbf{x}} = \frac{1}{2}\dot{\mathbf{q}}^T\,[B]^T\,[m]\,[B]\,\dot{\mathbf{q}} = \frac{1}{2}\dot{\mathbf{q}}^T\,[m']\,\dot{\mathbf{q}} \tag{7.331}$$

$$[m'] = [B]^T\,[m]\,[B] \tag{7.332}$$

$$P = \frac{1}{2}\mathbf{x}^T\,[k]\,\mathbf{x} = \frac{1}{2}\mathbf{q}^T\,[B]^T\,[k]\,[B]\,\mathbf{q} = \frac{1}{2}\mathbf{q}^T\,[k']\,\mathbf{q} \tag{7.333}$$

$$[k'] = [B]^T\,[k]\,[B] \tag{7.334}$$

$$D = \frac{1}{2}\dot{\mathbf{x}}^T\,[c]\,\dot{\mathbf{x}} = \frac{1}{2}\dot{\mathbf{q}}^T\,[B]^T\,[c]\,[B]\,\dot{\mathbf{q}} = \frac{1}{2}\dot{\mathbf{q}}^T\,[c']\,\dot{\mathbf{q}} \tag{7.335}$$

$$[c'] = [B]^T\,[c]\,[B] \tag{7.336}$$

The new mass, stiffness, and damping matrices $[m']$, $[k']$, and $[c']$ are also symmetric:

$$m'_{ij} = m'_{ji} \tag{7.337}$$

$$k'_{ij} = k'_{ji} \tag{7.338}$$

$$c'_{ij} = c'_{ji} \tag{7.339}$$

The components of the force vector $\mathbf{F}$ are related to the generalized coordinates $\mathbf{x}$. Employing the Lagrange equation (2.547), the equations of motion in terms of the new coordinates $\mathbf{q}$ would be:

$$[m']\,\ddot{\mathbf{q}} + [c']\,\mathbf{q} + [k']\,\mathbf{q} = \mathbf{Q} \tag{7.340}$$

A change of coordinates will generate a new force vector $\mathbf{Q}$, which is a linear combination of the elements of $\mathbf{F}$ and vice versa. To determine $\mathbf{Q}$, we need to recalculate the virtual work δW.

$$\delta W = \mathbf{F}^T\,\delta\mathbf{x} = \mathbf{F}^T\,[B]\,\delta\mathbf{q} = \mathbf{Q}^T\,\delta\mathbf{q} \tag{7.341}$$

$$\mathbf{Q} = [B]^T\,\mathbf{F} \tag{7.342}$$

Equations (7.332)–(7.335) indicate that employing a new set of coordinates will change the coupling of the system. They also indicate that coupling is not an inherent property of the system but depends on the coordinate system.

Let us eliminate the damping terms and rename $[m']$ and $[k']$ of Eq. (7.340) as $[m]$ and $[k]$ to have the equations of motion of the undamped system.

$$[m]\ddot{\mathbf{q}} + [k]\mathbf{q} = \mathbf{Q} \tag{7.343}$$

Using the *modal matrix* $[U]$, we define a coordinate transformation from any set of the generalized coordinates $\mathbf{q}$ to another set of generalized coordinates $\mathbf{p}$, called the *principal coordinates*, which decouples the equations of motion.

$$\mathbf{q} = [U]\mathbf{p} \tag{7.344}$$

$$\mathbf{p} = [U]^{-1}\mathbf{q} \tag{7.345}$$

The kinetic and potential energies of the system in terms of the new coordinates are:

$$K = \frac{1}{2}\dot{\mathbf{q}}^T [m]\dot{\mathbf{q}} = \frac{1}{2}\dot{\mathbf{p}}^T [U]^T [m][U]\dot{\mathbf{p}} = \frac{1}{2}\dot{\mathbf{p}}^T [m']\dot{\mathbf{p}} \tag{7.346}$$

$$P = \frac{1}{2}\mathbf{q}^T [k]\mathbf{q} = \frac{1}{2}\mathbf{p}^T [u]^T [k][u]\mathbf{p} = \frac{1}{2}\mathbf{p}^T [k']\mathbf{p} \tag{7.347}$$

$$[m'] = [U]^T [m][U] \qquad m'_{ij} = m'_{ji} \tag{7.348}$$

$$[k'] = [U]^T [k][U] \qquad k'_{ij} = k'_{ji} \tag{7.349}$$

Employing the Lagrange equation (2.547), the equations of motion in terms of the new coordinates p_i would be:

$$[m']\ddot{\mathbf{p}} + [k']\mathbf{p} = \mathbf{P} \tag{7.350}$$

$$\mathbf{P} = [U]^T \mathbf{Q} \tag{7.351}$$

Because the mode shapes are orthogonal with respect to the mass $[m]$ and stiffness $[k]$,

$$\mathbf{u}_j^T [m]\mathbf{u}_i = 0 \qquad i \neq j \tag{7.352}$$

$$\mathbf{u}_j^T [k]\mathbf{u}_i = 0 \qquad i \neq j \tag{7.353}$$

we have:

$$\mathbf{u}_i^T [m] \mathbf{u}_i = m_i \tag{7.354}$$

$$\mathbf{u}_i^T [k] \mathbf{u}_i = k_i \tag{7.355}$$

where m_i and k_i are scalars with mass and stiffness dimensions, respectively. Combining Eqs. (7.352)–(7.355), we define a modal matrix $[U]$ which provides us with a transformation to make new diagonal mass and stiffness matrices:

$$[U] = \begin{bmatrix} \mathbf{u}_1 & \mathbf{u}_2 & \mathbf{u}_2 & \cdots & \mathbf{u}_1 \end{bmatrix} \tag{7.356}$$

$$[m'] = [U]^T [m] [U] = \begin{bmatrix} m_1 & 0 & 0 & 0 \\ 0 & m_2 & 0 & 0 \\ \vdots & \vdots & \ddots & \vdots \\ 0 & 0 & \cdots & m_n \end{bmatrix} \tag{7.357}$$

$$[k'] = [U]^T [k] [U] = \begin{bmatrix} k_1 & 0 & 0 & 0 \\ 0 & k_2 & 0 & 0 \\ \vdots & \vdots & \ddots & \vdots \\ 0 & 0 & \cdots & k_n \end{bmatrix} \tag{7.358}$$

The diagonal mass and stiffness matrices make the equations of motion independent in principal coordinates $\mathbf{p}$.

$$[m'] \ddot{\mathbf{p}} + [k'] \mathbf{p} = \mathbf{P} \tag{7.359}$$

$$\begin{bmatrix} m_1 & 0 & 0 & 0 \\ 0 & m_2 & 0 & 0 \\ \vdots & \vdots & \ddots & \vdots \\ 0 & 0 & \cdots & m_n \end{bmatrix} \begin{bmatrix} \ddot{p}_1 \\ \ddot{p}_2 \\ \vdots \\ \ddot{p}_n \end{bmatrix} + \begin{bmatrix} k_1 & 0 & 0 & 0 \\ 0 & k_2 & 0 & 0 \\ \vdots & \vdots & \ddots & \vdots \\ 0 & 0 & \cdots & k_n \end{bmatrix} \begin{bmatrix} p_1 \\ p_2 \\ \vdots \\ p_n \end{bmatrix} = \begin{bmatrix} P_1 \\ P_2 \\ \vdots \\ P_n \end{bmatrix} \tag{7.360}$$

Therefore,

$$m_i \ddot{p}_i + k_i p_i = P_i \tag{7.361}$$

and hence, the natural frequency of the ith equation is:

$$\omega_i^2 = \frac{k_i}{m_i} \tag{7.362}$$

and, therefore,

$$\begin{bmatrix} \omega_1^2 & 0 & 0 & 0 \\ 0 & \omega_2^2 & 0 & 0 \\ \vdots & \vdots & \ddots & \vdots \\ 0 & 0 & \cdots & \omega_n^2 \end{bmatrix} = [m']^{-1}[k']$$

$$= \left[[U]^T [m][U] \right]^{-1} [U]^T [k][U]$$

$$= [U]^T [m]^{-1} [U]^{-T} [U]^T [k][U]$$

$$= [U]^T [m]^{-1} \left[[U][U]^{-1} \right]^T [k][U]$$

$$= [U]^T [m]^{-1} [k][U] = [U]^T [A][U] \tag{7.363}$$

The set of coordinates p_i, $i = 1, 2, 3, \cdots, n$ that make the equations independent is the *principal* or *natural coordinates* of the system. The orthogonality of mode shapes is the property that we utilize to show the modal matrix $[U]$ can transform any set of generalized coordinates to the set of principal coordinates and render the equations uncoupled.

The mode shapes are characteristics of the undamped equations of motion of a system and are only orthogonal with respect to the mass and stiffness matrices. Therefore, the damping matrix $[c]$ will not necessarily be transformed to a diagonal matrix by a principal coordinate transformation.

Being *statically coupled* means that a static displacement in q_i will displace at least one of the coordinates q_j, $j \neq i$. Therefore, a static displacement in q_i will not displace any of the coordinates q_j, $j \neq i$, if the system is statically decoupled.

Being *dynamically coupled* means that a sudden displacement in q_i will move at least one of the coordinates q_j, $j \neq i$. Therefore, a sudden displacement in q_i will not move any of the coordinates q_j, $j \neq i$, if the system is dynamically decoupled.

∎

Example 256 Physical meaning of the stiffness matrix elements. The element k_{ij} of a stiffness matrix $[k]$ indicates the required force F_i at x_i to produce a unit displacement at x_i while $x_j = 0, i \neq j$.

Consider the two *DOF* system in Fig. 7.29 with absolute coordinates x_1 and x_2. The system would have a 2×2 stiffness matrix $[k]$,

$$[k] = \begin{bmatrix} k_{11} & k_{12} \\ k_{21} & k_{22} \end{bmatrix} \tag{7.364}$$

Let us apply a unit displacement to x_1 while keeping $x_2 = 0$ as is shown in Fig. 7.30a. Using this configuration, we can determine k_{11} and k_{21}. To determine k_{11} and k_{21}, we calculate the required forces to keep the system in the configuration of Fig. 7.30a as is shown in Fig. 7.31a. A free-body-diagram shows that:

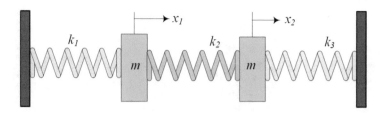

Fig. 7.29 A linear two DOF system at equilibrium position

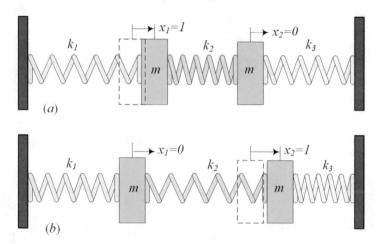

Fig. 7.30 The linear two DOF system at unit displacement positions. (**a**) $x_1 = 1$, $x_2 = 0$, and (**b**) $x_1 = 0$, $x_2 = 1$

$$k_{11} = k_1 + k_2 \qquad k_{21} = -k_2 \qquad (7.365)$$

Similarly, Fig. 7.30b depicts a unit displacement of x_2 while keeping $x_1 = 0$. Using this configuration, we can determine k_{22} and k_{12}. The required forces to keep the system in the configuration of Fig. 7.30b as is shown in Fig. 7.31b are:

$$k_{22} = k_2 + k_3 \qquad k_{12} = -k_2 \qquad (7.366)$$

Therefore,

$$[k] = \begin{bmatrix} k_1 + k_2 & -k_2 \\ -k_2 & k_2 + k_3 \end{bmatrix} \qquad (7.367)$$

This is the same stiffness matrix as in Example 237.

Static coupling means that it is impossible to have $x_i = 1$, without a set of required forces to keep $x_j = 0$, $i \neq j$. Using absolute coordinates guarantees that the mass matrix is diagonal and therefore the equations for motion of the system are:

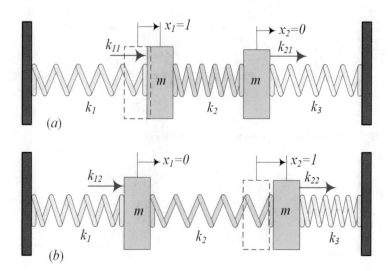

Fig. 7.31 The required force on the two DOF system to provide unit displacements. (**a**) $x_1 = 1$, $x_2 = 0$, and (**b**) $x_1 = 0$, $x_2 = 1$

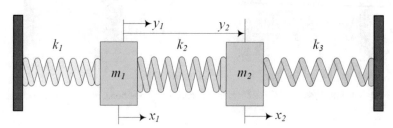

Fig. 7.32 A two DOF vibrating system

$$[m]\ddot{\mathbf{x}} + [k]\mathbf{x} = \mathbf{0} \tag{7.368}$$

$$\begin{bmatrix} m_1 & 0 \\ 0 & m_2 \end{bmatrix} \begin{bmatrix} \ddot{x}_1 \\ \ddot{x}_2 \end{bmatrix} + \begin{bmatrix} k_1 + k_2 & -k_2 \\ -k_2 & k_2 + k_3 \end{bmatrix} \begin{bmatrix} x_1 \\ x_2 \end{bmatrix} = \begin{bmatrix} 0 \\ 0 \end{bmatrix} \tag{7.369}$$

Example 257 Change of coordinates of a two DOF vibrating system. An example to derive and transform the equations of motion in absolute and relative coordinates.

Consider the two DOF system of Example 237 as is shown in Fig. 7.32. Its equations of motion are:

$$[m]\ddot{\mathbf{x}} + [k]\mathbf{x} = \mathbf{0} \tag{7.370}$$

$$\begin{bmatrix} m_1 & 0 \\ 0 & m_2 \end{bmatrix} \begin{bmatrix} \ddot{x}_1 \\ \ddot{x}_2 \end{bmatrix} + \begin{bmatrix} k_1 + k_2 & -k_2 \\ -k_2 & k_2 + k_3 \end{bmatrix} \begin{bmatrix} x_1 \\ x_2 \end{bmatrix} = \begin{bmatrix} 0 \\ 0 \end{bmatrix} \tag{7.371}$$

If we use the relative coordinates

$$y_1 = x_1 \qquad y_2 = x_2 - x_1 \qquad (7.372)$$

then the transformation matrix between **x** and **y** coordinates would be:

$$\mathbf{x} = [B]\,\mathbf{y} \qquad (7.373)$$

$$\begin{bmatrix} x_1 \\ x_2 \end{bmatrix} = \begin{bmatrix} 1 & 0 \\ -1 & 1 \end{bmatrix} \begin{bmatrix} y_1 \\ y_2 \end{bmatrix} \qquad (7.374)$$

The mass and stiffness matrices in terms of the new coordinates are:

$$[m'] = [B]^T\,[m]\,[B] = \begin{bmatrix} 1 & 0 \\ -1 & 1 \end{bmatrix}^T \begin{bmatrix} m_1 & 0 \\ 0 & m_2 \end{bmatrix} \begin{bmatrix} 1 & 0 \\ -1 & 1 \end{bmatrix}$$

$$= \begin{bmatrix} m_1 + m_2 & -m_2 \\ -m_2 & m_2 \end{bmatrix} \qquad (7.375)$$

$$[k'] = [B]^T\,[k]\,[B] = \begin{bmatrix} 1 & 0 \\ -1 & 1 \end{bmatrix}^T \begin{bmatrix} k_1 + k_2 & -k_2 \\ -k_2 & k_2 + k_3 \end{bmatrix} \begin{bmatrix} 1 & 0 \\ -1 & 1 \end{bmatrix}$$

$$\begin{bmatrix} k_1 + 4k_2 + k_3 & -2k_2 - k_3 \\ -2k_2 - k_3 & k_2 + k_3 \end{bmatrix} \qquad (7.376)$$

The equations of motion in terms of the new coordinates are:

$$[m']\,\ddot{\mathbf{y}} + [k']\,\mathbf{y} = \mathbf{0} \qquad (7.377)$$

$$\begin{bmatrix} m_1 + m_2 & -m_2 \\ -m_2 & m_2 \end{bmatrix} \begin{bmatrix} \ddot{y}_1 \\ \ddot{y}_2 \end{bmatrix}$$

$$+ \begin{bmatrix} k_1 + 4k_2 + k_3 & -2k_2 - k_3 \\ -2k_2 - k_3 & k_2 + k_3 \end{bmatrix} \begin{bmatrix} y_1 \\ y_2 \end{bmatrix} = \begin{bmatrix} 0 \\ 0 \end{bmatrix} \qquad (7.378)$$

In case that

$$m_1 = m_2 = m \qquad k_1 = k_2 = k_3 = k \qquad (7.379)$$

the original equations of motion in **x**-coordinates will be:

$$\begin{bmatrix} m & 0 \\ 0 & m \end{bmatrix} \begin{bmatrix} \ddot{x}_1 \\ \ddot{x}_2 \end{bmatrix} + \begin{bmatrix} 2k & -k \\ -k & 2k \end{bmatrix} \begin{bmatrix} x_1 \\ x_2 \end{bmatrix} = \begin{bmatrix} 0 \\ 0 \end{bmatrix} \qquad (7.380)$$

and the equations of motion in **y**-coordinates will be:

$$\begin{bmatrix} 2m & -m \\ -m & m \end{bmatrix} \begin{bmatrix} \ddot{y}_1 \\ \ddot{y}_2 \end{bmatrix} + \begin{bmatrix} 6k & -3 \\ -3k & 2k \end{bmatrix} \begin{bmatrix} y_1 \\ y_2 \end{bmatrix} = \begin{bmatrix} 0 \\ 0 \end{bmatrix}$$
(7.381)

The equations of motion in **x**-coordinates are statically coupled, and dynamically independent. However, the equations are both statically and dynamically coupled in the relative **y**-coordinates.

Example 258 Decoupling of a symmetric two DOF system. Decoupling the equations of motion is a great and easy method to solve multi DOF vibrating systems in terms of principal coordinates.

The symmetric case of the system of Fig. 7.32 with

$$m_1 = m_2 = m \qquad k_1 = k_2 = k_3 = k$$
(7.382)

and equations of motion (7.380) has the mode shapes of

$$\mathbf{u}_1 = \begin{bmatrix} 1 \\ 1 \end{bmatrix} \qquad \mathbf{u}_2 = \begin{bmatrix} -1 \\ 1 \end{bmatrix}$$
(7.383)

Therefore, the modal matrix of the system is:

$$[U] = \begin{bmatrix} \mathbf{u}_1 & \mathbf{u}_2 \end{bmatrix} = \begin{bmatrix} 1 & -1 \\ 1 & 1 \end{bmatrix}$$
(7.384)

Using $[U]$, the principal coordinates of the system are:

$$\mathbf{x} = [U] \, \mathbf{p} = \begin{bmatrix} p_1 - p_2 \\ p_1 + p_2 \end{bmatrix}$$
(7.385)

$$\mathbf{p} = [U]^{-1} \mathbf{x} = \begin{bmatrix} \frac{1}{2}(x_1 + x_2) \\ \frac{1}{2}(x_2 - x_1) \end{bmatrix}$$
(7.386)

Having $[U]$, we can find the diagonal mass and stiffness matrices and determine the natural frequencies of the system.

$$[m'] = [U]^T [m] [U] = \begin{bmatrix} 2m & 0 \\ 0 & 2m \end{bmatrix}$$
(7.387)

$$[k'] = [U]^T [k] [U] = \begin{bmatrix} 2k & 0 \\ 0 & 6k \end{bmatrix}$$
(7.388)

$$\begin{bmatrix} \omega_1^2 & 0 \\ 0 & \omega_2^2 \end{bmatrix} = [m']^{-1}[k'] = [U]^T[A][U] = \begin{bmatrix} \dfrac{k}{m} & 0 \\ 0 & 3\dfrac{k}{m} \end{bmatrix} \tag{7.389}$$

$$\omega_1 = \sqrt{\dfrac{k}{m}} \qquad \omega_2 = \sqrt{3}\sqrt{\dfrac{k}{m}} \tag{7.390}$$

The equations of motion of the system in terms of the principal coordinates $\mathbf{p}$ would be decoupled.

$$\begin{bmatrix} 2m & 0 \\ 0 & 2m \end{bmatrix}\begin{bmatrix} \ddot{p}_1 \\ \ddot{p}_2 \end{bmatrix} + \begin{bmatrix} 2k & 0 \\ 0 & 6k \end{bmatrix}\begin{bmatrix} p_1 \\ p_2 \end{bmatrix} = \begin{bmatrix} 0 \\ 0 \end{bmatrix} \tag{7.391}$$

The decoupled equations of motion are two harmonic equations.

$$2m\,\ddot{p}_1 + 2k\,p_1 = 0 \tag{7.392}$$

$$2m\,\ddot{p}_2 + 6k\,p_2 = 0 \tag{7.393}$$

The solution of the equations are simple harmonic equations of frequencies ω_1 and ω_2, respectively.

$$p_1 = A_1 \sin \omega_1 t + B_1 \cos \omega_1 t \tag{7.394}$$

$$p_2 = A_2 \sin \omega_2 t + B_2 \cos \omega_2 t \tag{7.395}$$

Substituting for the principal coordinates p_1 and p_2 in terms of the absolute coordinates x_1 and x_2 provides us with the solutions of the original equations (7.380).

$$\frac{1}{2}(x_1 + x_2) = A_1 \sin \omega_1 t + B_1 \cos \omega_1 t \tag{7.396}$$

$$\frac{1}{2}(x_2 - x_1) = A_2 \sin \omega_2 t + B_2 \cos \omega_2 t \tag{7.397}$$

$$x_1 = A_1 \sin \omega_1 t + B_1 \cos \omega_1 t - A_2 \sin \omega_2 t + B_2 \cos \omega_2 t \tag{7.398}$$

$$x_2 = A_2 \sin \omega_2 t + B_2 \cos \omega_2 t + A_2 \sin \omega_2 t + B_2 \cos \omega_2 t \tag{7.399}$$

As a practical example, let us use a set of sample numerical values,

$$k_1 = 100\,\text{N}/\text{m} \qquad k_2 = 80\,\text{N}/\text{m} \qquad k_1 = 120\,\text{N}/\text{m}$$

$$m_1 = 1\,\text{kg} \qquad\qquad m_2 = 1.2\,\text{kg} \tag{7.400}$$

and rewrite the equations of motion (7.380).

$$\begin{bmatrix} 1 & 0 \\ 0 & 1.2 \end{bmatrix} \begin{bmatrix} \ddot{x}_1 \\ \ddot{x}_2 \end{bmatrix} + \begin{bmatrix} 180 & -80 \\ -80 & 200 \end{bmatrix} \begin{bmatrix} x_1 \\ x_2 \end{bmatrix} = \begin{bmatrix} 0 \\ 0 \end{bmatrix} \qquad (7.401)$$

The matrix $[A] = [m]^{-1} [k]$ is:

$$[A] = [m]^{-1} [k] = \begin{bmatrix} 180 & -80 \\ -66.667 & 166.67 \end{bmatrix} \qquad (7.402)$$

and therefore, the natural frequencies and mode shapes of the system are:

$$\omega_1 = 10 \, \text{rad} \, / \, \text{s} \qquad \omega_2 = 15.706 \, \text{rad} \, / \, \text{s} \qquad (7.403)$$

$$\mathbf{u}_1 = \begin{bmatrix} 0.7071 \\ 0.7071 \end{bmatrix} \qquad \mathbf{u}_2 = \begin{bmatrix} 0.76821 \\ -0.64019 \end{bmatrix} \qquad (7.404)$$

Using the modal matrix

$$[U] = \begin{bmatrix} \mathbf{u}_1 & \mathbf{u}_2 \end{bmatrix} = \begin{bmatrix} 0.7071 & 0.76821 \\ 0.7071 & -0.64019 \end{bmatrix} \qquad (7.405)$$

we can determine the relationship between $\mathbf{x}$ and the principal coordinates $\mathbf{p}$:

$$\mathbf{x} = [U] \, \mathbf{p} \qquad (7.406)$$

$$\begin{bmatrix} x_1 \\ x_2 \end{bmatrix} = \begin{bmatrix} 0.70711 & 0.76821 \\ 0.70711 & -0.64019 \end{bmatrix} \begin{bmatrix} p_1 \\ p_2 \end{bmatrix}$$

$$= \begin{bmatrix} 0.707 \, p_1 + 0.768 \, p_2 \\ 0.707 \, p_1 - 0.64 \, p_2 \end{bmatrix} \qquad (7.407)$$

$$\mathbf{p} = [U]^{-1} \mathbf{x} \qquad (7.408)$$

$$\begin{bmatrix} p_1 \\ p_2 \end{bmatrix} = \begin{bmatrix} 0.70711 & 0.76821 \\ 0.70711 & -0.64019 \end{bmatrix}^{-1} \begin{bmatrix} x_1 \\ x_2 \end{bmatrix}$$

$$= \begin{bmatrix} 0.642x_1 + 0.771x_2 \\ 0.71x_1 - 0.71x_2 \end{bmatrix} \qquad (7.409)$$

to make the equations of motion decoupled.

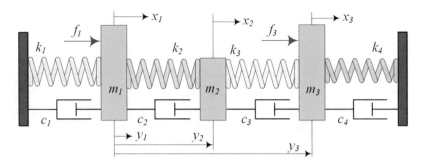

Fig. 7.33 A general 3 *DOF* system

$$[m'] = [U]^T [m][U] \approx \begin{bmatrix} 1.1 & -1.27 \times 10^{-5} \\ -1.27 \times 10^{-5} & 1.082 \end{bmatrix}$$

$$\approx \begin{bmatrix} 1.1 & 0 \\ 0 & 1.082 \end{bmatrix} \tag{7.410}$$

$$[k'] = [U]^T [k][U] \approx \begin{bmatrix} 110 & -1.27 \times 10^{-3} \\ -1.27 \times 10^{-3} & 266.88 \end{bmatrix}$$

$$\approx \begin{bmatrix} 110 & 0 \\ 0 & 266.88 \end{bmatrix} \tag{7.411}$$

$$\begin{bmatrix} 1.1 & 0 \\ 0 & 1.082 \end{bmatrix} \begin{bmatrix} \ddot{p}_1 \\ \ddot{p}_2 \end{bmatrix} + \begin{bmatrix} 110 & 0 \\ 0 & 266.88 \end{bmatrix} \begin{bmatrix} p_1 \\ p_2 \end{bmatrix} = \begin{bmatrix} 0 \\ 0 \end{bmatrix} \tag{7.412}$$

Example 259 Coordinate change for a three *DOF* system. Coordinate transformation is the best way to work with different coordinate systems. Here we derive the equations of motion in the absolute coordinates and then transform them to the relative coordinates by transformation method.

Figure 7.33 illustrates a three *DOF* forced and damped vibrating system. The required quadratures of the system to find the equations of motion are:

$$K = \frac{1}{2} m_1 \dot{x}_1^2 + \frac{1}{2} m_2 \dot{x}_2^2 + \frac{1}{2} m_3 \dot{x}_3^2 = \frac{1}{2} \dot{\mathbf{x}}^T [m] \mathbf{x} \tag{7.413}$$

$$[m] = \begin{bmatrix} m_1 & 0 & 0 \\ 0 & m_2 & 0 \\ 0 & 0 & m_3 \end{bmatrix} \tag{7.414}$$

$$P = \frac{1}{2}k_1 x_1^2 + \frac{1}{2}k_2 (x_1 - x_2)^2 + \frac{1}{2}k_3 (x_2 - x_3)^2 + \frac{1}{2}k_4 x_3^2$$

$$= \frac{1}{2}\mathbf{x}^T [k] \mathbf{x} \tag{7.415}$$

$$[k] = \begin{bmatrix} k_1 + k_2 & -k_2 & 0 \\ -k_2 & k_2 + k_3 & -k_3 \\ 0 & -k_3 & k_3 + k_4 \end{bmatrix} \tag{7.416}$$

$$D = \frac{1}{2}c_1 \dot{x}_1^2 + \frac{1}{2}c_2 (\dot{x}_1 - \dot{x}_2)^2 + \frac{1}{2}c_3 (\dot{x}_2 - \dot{x}_3)^2 + \frac{1}{2}c_4 \dot{x}_3^2$$

$$= \frac{1}{2}\dot{\mathbf{x}}^T [c] \dot{\mathbf{x}} \tag{7.417}$$

$$[c] = \begin{bmatrix} c_1 + c_2 & -c_2 & 0 \\ -c_2 & c_2 + c_3 & -c_3 \\ 0 & -c_3 & c_3 + c_4 \end{bmatrix} \tag{7.418}$$

Thus, the equations of motion are:

$$[m]\ddot{\mathbf{x}} + [c]\dot{\mathbf{x}} + [k]\mathbf{x} = \mathbf{F} = \begin{bmatrix} f_1 \\ 0 \\ f_3 \end{bmatrix} \tag{7.419}$$

To express the system in terms of the relative coordinates $\mathbf{y}$, we establish the relationship of $\mathbf{x}$ and $\mathbf{y}$.

$$y_1 = x_1 \qquad y_2 = x_2 - x_1 \qquad y_3 = x_3 - x_1 \tag{7.420}$$

Therefore, we have:

$$\mathbf{y} = [B]^{-1} \mathbf{x} \tag{7.421}$$

$$\begin{bmatrix} y_1 \\ y_2 \\ y_3 \end{bmatrix} = \begin{bmatrix} 1 & 0 & 0 \\ -1 & 1 & 0 \\ -1 & 0 & 1 \end{bmatrix} \begin{bmatrix} x_1 \\ x_2 \\ x_3 \end{bmatrix} \tag{7.422}$$

$$\mathbf{x} = [B]\mathbf{y} \tag{7.423}$$

$$\begin{bmatrix} x_1 \\ x_2 \\ x_3 \end{bmatrix} = \begin{bmatrix} 1 & 0 & 0 \\ 1 & 1 & 0 \\ 1 & 0 & 1 \end{bmatrix} \begin{bmatrix} y_1 \\ y_2 \\ y_3 \end{bmatrix} \tag{7.424}$$

Using the $[B]$-matrix, the new mass, stiffness, and damping matrices are:

$$[m'] = [B]^T [m][B] = \begin{bmatrix} m_1 + m_2 + m_3 & m_2 & m_3 \\ m_2 & m_2 & 0 \\ m_3 & 0 & m_3 \end{bmatrix} \quad (7.425)$$

$$[k'] = [B]^T [k][B] = \begin{bmatrix} k_1 + k_4 & 0 & k_4 \\ 0 & k_2 + k_3 & -k_3 \\ k_4 & -k_3 & k_3 + k_4 \end{bmatrix} \quad (7.426)$$

$$[c'] = [B]^T [c][B] = \begin{bmatrix} c_1 + c_4 & 0 & c_4 \\ 0 & c_2 + c_3 & -c_3 \\ c_4 & -c_3 & c_3 + c_4 \end{bmatrix} \quad (7.427)$$

In the last step, we find the forcing functions in the new coordinate system:

$$\mathbf{Q} = [B]^T \mathbf{F} = \begin{bmatrix} f_1 + f_3 \\ 0 \\ f_3 \end{bmatrix} \quad (7.428)$$

Therefore, the equations of motion in **y**-coordinate system would be:

$$[m'] \ddot{\mathbf{y}} + [c'] \dot{\mathbf{y}} + [k'] \mathbf{y} = \mathbf{Q} = \begin{bmatrix} f_1 + f_3 \\ 0 \\ f_3 \end{bmatrix} \quad (7.429)$$

Example 260 Principal coordinates of a three DOF system. The principal coordinate system cannot be found by trial and error. There is a systematic calculation method based on modal matrix. Here is another example for a three DOF system.

Let us assume

$$m_1 = m_2 = m_3 = m \quad (7.430)$$

$$k_1 = k_2 = k_3 = k_4 = k \quad (7.431)$$

and determine the principal coordinates of the three DOF undamped and unforced system of Fig. 7.33. The $[A]$-matrix of the system

$$[A] = [m]^{-1}[k] = \begin{bmatrix} m & 0 & 0 \\ 0 & m & 0 \\ 0 & 0 & m \end{bmatrix}^{-1} \begin{bmatrix} 2k & -k & 0 \\ -k & 2k & -k \\ 0 & -k & 2k \end{bmatrix}$$

$$= \frac{k}{m} \begin{bmatrix} 2 & -1 & 0 \\ -1 & 2 & -1 \\ 0 & -1 & 2 \end{bmatrix} \tag{7.432}$$

yields:

$$\omega_1 = \sqrt{2 - \sqrt{2}}\sqrt{\frac{k}{m}} \qquad \omega_2 = \sqrt{2}\sqrt{\frac{k}{m}} \qquad \omega_3 = \sqrt{2 + \sqrt{2}}\sqrt{\frac{k}{m}} \tag{7.433}$$

$$\mathbf{u}_1 = \begin{bmatrix} 1 \\ \sqrt{2} \\ 1 \end{bmatrix} \qquad \mathbf{u}_2 = \begin{bmatrix} -1 \\ 0 \\ 1 \end{bmatrix} \qquad \mathbf{u}_3 = \begin{bmatrix} 1 \\ -\sqrt{2} \\ 1 \end{bmatrix} \tag{7.434}$$

Using the modal matrix $[U]$

$$[U] = \begin{bmatrix} \mathbf{u}_1 & \mathbf{u}_2 & \mathbf{u}_3 \end{bmatrix} = \begin{bmatrix} 1 & -1 & 1 \\ \sqrt{2} & 0 & -\sqrt{2} \\ 1 & 1 & 1 \end{bmatrix} \tag{7.435}$$

we determine the relationship between $\mathbf{x}$ and the principal coordinates $\mathbf{p}$

$$\mathbf{x} = [U]\,\mathbf{p} \tag{7.436}$$

$$\begin{bmatrix} x_1 \\ x_2 \\ x_3 \end{bmatrix} = \begin{bmatrix} 1 & -1 & 1 \\ \sqrt{2} & 0 & -\sqrt{2} \\ 1 & 1 & 1 \end{bmatrix} \begin{bmatrix} p_1 \\ p_2 \\ p_3 \end{bmatrix} = \begin{bmatrix} p_1 - p_2 + p_3 \\ \sqrt{2}p_1 - \sqrt{2}p_3 \\ p_1 + p_2 + p_3 \end{bmatrix} \tag{7.437}$$

and therefore,

$$\mathbf{p} = [U]^{-1}\mathbf{x} \tag{7.438}$$

$$\begin{bmatrix} p_1 \\ p_2 \\ p_3 \end{bmatrix} = \begin{bmatrix} 1 & -1 & 1 \\ \sqrt{2} & 0 & -\sqrt{2} \\ 1 & 1 & 1 \end{bmatrix}^{-1} \begin{bmatrix} x_1 \\ x_2 \\ x_3 \end{bmatrix} = \frac{1}{4} \begin{bmatrix} x_1 + x_3 + \sqrt{2}x_2 \\ 2x_3 - 2x_1 \\ x_1 + x_3 - \sqrt{2}x_2 \end{bmatrix} \tag{7.439}$$

Let us also determine the principal coordinates of the three DOF system of Fig. 7.33 when we use the y-coordinates. This analysis will show that natural frequencies are not coordinate dependent while mode shapes are. Employing the assumptions (7.430) and (7.431) and the mass and stiffness matrices (7.425) and (7.426),

$$[m'] = [B]^T [m][B] = \begin{bmatrix} 3m & m & m \\ m & m & 0 \\ m & 0 & m \end{bmatrix} \tag{7.440}$$

$$[k'] = [B]^T [k][B] = \begin{bmatrix} 2k & 0 & k \\ 0 & 2k & -k \\ k & -k & 2k \end{bmatrix} \tag{7.441}$$

the $[A]$-matrix of the system will be:

$$[A] = [m]^{-1}[k] = \frac{k}{m} \begin{bmatrix} 1 & -1 & 0 \\ -1 & 3 & -1 \\ 0 & 0 & 2 \end{bmatrix} \tag{7.442}$$

The natural frequencies and mode shapes of the system are:

$$\omega_1 = \sqrt{2 - \sqrt{2}}\sqrt{\frac{k}{m}} \qquad \omega_2 = \sqrt{2}\sqrt{\frac{k}{m}} \qquad \omega_3 = \sqrt{2 + \sqrt{2}}\sqrt{\frac{k}{m}} \tag{7.443}$$

$$\mathbf{u}_1 = \begin{bmatrix} \sqrt{2}+1 \\ 1 \\ 0 \end{bmatrix} \qquad \mathbf{u}_2 = \begin{bmatrix} -1/2 \\ 1/2 \\ 1 \end{bmatrix} \qquad \mathbf{u}_3 = \begin{bmatrix} 1-\sqrt{2} \\ 1 \\ 0 \end{bmatrix} \tag{7.444}$$

Using the modal matrix

$$[U] = \begin{bmatrix} \mathbf{u}_1 & \mathbf{u}_2 & \mathbf{u}_3 \end{bmatrix} = \begin{bmatrix} \sqrt{2}+1 & -1/2 & 1-\sqrt{2} \\ 1 & 1/2 & 1 \\ 0 & 1 & 0 \end{bmatrix} \tag{7.445}$$

we determine the relationship between $\mathbf{y}$ and the new principal coordinates $\mathbf{p}$.

$$\mathbf{y} = [U]\,\mathbf{p} \tag{7.446}$$

$$\begin{bmatrix} y_1 \\ y_2 \\ y_3 \end{bmatrix} = \begin{bmatrix} \sqrt{2}+1 & -1/2 & 1-\sqrt{2} \\ 1 & 1/2 & 1 \\ 0 & 1 & 0 \end{bmatrix} \begin{bmatrix} p_1 \\ p_2 \\ p_3 \end{bmatrix}$$

$$= \begin{bmatrix} p_1\left(\sqrt{2}+1\right) - \frac{1}{2}p_2 - p_3\left(\sqrt{2}-1\right) \\ p_1 + \frac{1}{2}p_2 + p_3 \\ p_2 \end{bmatrix} \tag{7.447}$$

$$\mathbf{p} = [U]^{-1}\,\mathbf{y} \tag{7.448}$$

$$\begin{bmatrix} p_1 \\ p_2 \\ p_3 \end{bmatrix} = \begin{bmatrix} \sqrt{2}+1 & -1/2 & 1-\sqrt{2} \\ 1 & 1/2 & 1 \\ 0 & 1 & 0 \end{bmatrix}^{-1} \begin{bmatrix} y_1 \\ y_2 \\ y_3 \end{bmatrix}$$

$$= \frac{1}{4} \begin{bmatrix} y_3\left(\sqrt{2}-1\right) - y_2\left(\sqrt{2}-2\right) + \sqrt{2}y_1 \\ y_3 \\ \sqrt{2}y_2\left(\sqrt{2}+1\right) - \frac{1}{2}\sqrt{2}y_3\left(\sqrt{2}+2\right) - \sqrt{2}y_1 \end{bmatrix} \qquad (7.449)$$

Example 261 Arbitrariness of the length of mode shapes. Mode shapes are indicating the relative position of coordinates at their associated natural frequency. Hence the length of the mode shapes as vectors is not important.

Let us assume that the mode shapes of the system in Fig. 7.32

$$\begin{bmatrix} m & 0 \\ 0 & m \end{bmatrix} \begin{bmatrix} \ddot{x}_1 \\ \ddot{x}_2 \end{bmatrix} + \begin{bmatrix} 2k & -k \\ -k & 2k \end{bmatrix} \begin{bmatrix} x_1 \\ x_2 \end{bmatrix} = \begin{bmatrix} 0 \\ 0 \end{bmatrix} \qquad (7.450)$$

with

$$m_1 = m_2 = m \qquad k_1 = k_2 = k_3 = k \qquad (7.451)$$

are expressed in different lengths such as:

$$\mathbf{u}_1 = \begin{bmatrix} 2 \\ 2 \end{bmatrix} \qquad \mathbf{u}_2 = \begin{bmatrix} -1 \\ 1 \end{bmatrix} \qquad (7.452)$$

Therefore, the modal matrix of the system is:

$$[U] = \begin{bmatrix} \mathbf{u}_1 & \mathbf{u}_2 \end{bmatrix} = \begin{bmatrix} 2 & -1 \\ 2 & 1 \end{bmatrix} \qquad (7.453)$$

Using $[U]$, the diagonal mass and stiffness matrices are

$$[m'] = [U]^T [m] [U] = \begin{bmatrix} 8m & 0 \\ 0 & 2m \end{bmatrix} \qquad (7.454)$$

$$[k'] = [U]^T [k] [U] = \begin{bmatrix} 8k & 0 \\ 0 & 6k \end{bmatrix} \qquad (7.455)$$

They show that the equations of motion in principal coordinates

$$\begin{bmatrix} 8m & 0 \\ 0 & 2m \end{bmatrix} \begin{bmatrix} \ddot{p}_1 \\ \ddot{p}_2 \end{bmatrix} + \begin{bmatrix} 8k & 0 \\ 0 & 6k \end{bmatrix} \begin{bmatrix} p_1 \\ p_2 \end{bmatrix} = \begin{bmatrix} 0 \\ 0 \end{bmatrix} \qquad (7.456)$$

are equivalent to Eq. (7.391). The natural frequencies of the system are:

$$[m']^{-1}[k'] = \begin{bmatrix} k/m & 0 \\ 0 & 3k/m \end{bmatrix} = \begin{bmatrix} \omega_1^2 & 0 \\ 0 & \omega_2^2 \end{bmatrix} \qquad (7.457)$$

Therefore, expressing the mode shapes of a system in different length will not affect its natural characteristics and the principal equations of motion of the system.

Example 262 Inverse of the modal matrix. Inverse of modal matrix can also be calculated from $[m']$ or $[k']$.

The inverse of the modal matrix is needed to calculate the principal coordinates and forces in (7.345) and (7.351).

$$\mathbf{p} = [U]^T \mathbf{q} \qquad \mathbf{P} = [U]^T \mathbf{Q} \qquad (7.458)$$

We may use the orthogonality of the modal matrix with respect to the mass and stiffness matrices to determine $[U]^T$.

$$[m'] = [U]^T [m] [U] \qquad (7.459)$$

$$[k'] = [U]^T [k] [U] \qquad (7.460)$$

Multiplying $[m']^{-1}$ in (7.459) provides us with

$$[m']^{-1} [U]^T [m] [U] = [I] \qquad (7.461)$$

where $[I]$ is an $n \times n$ identity matrix. Now, multiplying (7.461) in $[U]^{-1}$ provides us with the required equation to calculate the inverse of the modal matrix.

$$[U]^{-1} = [m']^{-1} [U]^T [m] \qquad (7.462)$$

Similarly we can find $[U]^{-1}$ from $[k]$.

$$[U]^{-1} = [k']^{-1} [U]^T [k] \qquad (7.463)$$

Example 263 Expanded form of coupling proof. It is an analytic exercise of transformation from one set of generalized coordinate to another set of generalized coordinate system.

Consider an $n\,DOF$ undamped vibrating system. Employing a set of n generalized coordinates q_i, $i = 1, 2, 3, \cdots, n$, the kinetic and potential energies of the system would be defined based on q_i.

$$K = \frac{1}{2}\dot{\mathbf{q}}^T [m]\,\dot{\mathbf{q}} = \frac{1}{2}\sum_{i=1}^{n}\sum_{i=1}^{n} m_{ij}\dot{q}_i\dot{q}_j \qquad m_{ij} = m_{ji} \qquad (7.464)$$

$$P = \frac{1}{2}\mathbf{q}^T [k]\,\mathbf{q} = \frac{1}{2}\sum_{i=1}^{n}\sum_{i=1}^{n} k_{ij}q_i q_j \qquad k_{ij} = k_{ji} \qquad (7.465)$$

The mass matrix $[m]$ and the stiffness matrix $[k]$ are symmetric. Any other set of generalized coordinates p_i, $i = 1, 2, 3, \cdots, n$ will be a linear combination of q_j and vice versa.

$$q_i = \sum_{j=1}^{n} u_{ij} p_j \qquad\qquad (7.466)$$

Using the linear combination of the generalized velocities $\dot{p}_i$ and $\dot{q}_j$,

$$\dot{q}_i = \sum_{j=1}^{n} u_{ij} \dot{p}_j \qquad\qquad (7.467)$$

the kinetic and potential energies in terms of the new coordinates are:

$$\begin{aligned}
K &= \frac{1}{2}\dot{\mathbf{q}}^T [m]\,\dot{\mathbf{q}} = \frac{1}{2}\sum_{i=1}^{n}\sum_{i=1}^{n} m_{ij}\dot{q}_i\dot{q}_j \\
&= \frac{1}{2}\sum_{i=1}^{n}\sum_{i=1}^{n} m_{ij} \sum_{r=1}^{n} u_{ir}\dot{p}_r \sum_{s=1}^{n} u_{js}\dot{p}_s \\
&= \frac{1}{2}\sum_{r=1}^{n}\sum_{s=1}^{n} \dot{p}_r\dot{p}_s \sum_{i=1}^{n}\sum_{i=1}^{n} m_{ij} u_{ir} u_{js} \\
&= \frac{1}{2}\sum_{r=1}^{n}\sum_{s=1}^{n} m'_{rs}\dot{p}_r\dot{p}_s = \frac{1}{2}\dot{\mathbf{p}}^T [m']\,\dot{\mathbf{p}} \qquad (7.468)
\end{aligned}$$

$$m'_{rs} = \sum_{i=1}^{n}\sum_{i=1}^{n} m_{ij} u_{ir} u_{js} = m'_{sr} \qquad (7.469)$$

$$P = \frac{1}{2}\mathbf{q}^T\,[k]\,\mathbf{q} = \frac{1}{2}\sum_{i=1}^{n}\sum_{i=1}^{n}k_{ij}q_iq_j$$

$$= \frac{1}{2}\sum_{r=1}^{n}\sum_{s=1}^{n}k'_{rs}\,p_r\,p_s = \frac{1}{2}\mathbf{p}^T\,[k']\,\mathbf{p} \tag{7.470}$$

$$k'_{rs} = \sum_{i=1}^{n}\sum_{i=1}^{n}k_{ij}u_{ir}u_{js} = k'_{sr} \tag{7.471}$$

The new mass and stiffness matrices are also symmetric.

$$m'_{ij} = m'_{ji} \qquad k'_{ij} = k'_{ji} \tag{7.472}$$

Example 264 ★ Proper coordinate transformation. A general coordinate transformation will have a transformation matrix and might also have a constant shift. Here is the analytical treatment of such general transformation.

The general transformation from a set of generalized coordinates $\mathbf{q}$ to another set of generalized coordinates $\mathbf{x}$ may also have a constant shift vector $\mathbf{a}$. Then the mathematical relationship of the coordinates would be:

$$\mathbf{x} = [B]\,\mathbf{q} + \mathbf{a} \tag{7.473}$$

$$\mathbf{q} = [B]^{-1}\,(\mathbf{x} - \mathbf{a}) \tag{7.474}$$

The equations of motion of the system in terms of $\mathbf{x}$ are:

$$[m]\,\ddot{\mathbf{x}} + [c]\,\dot{\mathbf{x}} + [k]\,\mathbf{x} = \mathbf{F} \tag{7.475}$$

where $\mathbf{x} = \mathbf{0}$ indicates the equilibrium configuration of the system. The kinetic and potential energies and dissipation function of the system are:

$$K = \frac{1}{2}\dot{\mathbf{x}}^T\,[m]\,\dot{\mathbf{x}} \qquad m_{ij} = m_{ji} \tag{7.476}$$

$$P = \frac{1}{2}\mathbf{x}^T\,[k]\,\mathbf{x} \qquad k_{ij} = k_{ji} \tag{7.477}$$

$$D = \frac{1}{2}\dot{\mathbf{x}}^T\,[c]\,\dot{\mathbf{x}} \qquad c_{ij} = c_{ji} \tag{7.478}$$

The constant vector $\mathbf{a}$ disappears in the velocity relationship.

$$\dot{\mathbf{x}} = [B]\,\dot{\mathbf{q}} \tag{7.479}$$

Therefore, the kinetic, potential, and dissipation functions in terms of the new coordinates $\mathbf{q}$ are:

$$K = \frac{1}{2}\dot{\mathbf{x}}^T [m]\dot{\mathbf{x}} = \frac{1}{2}\dot{\mathbf{q}}^T [B]^T [m][B]\dot{\mathbf{q}} = \frac{1}{2}\dot{\mathbf{q}}^T [m']\dot{\mathbf{q}} \tag{7.480}$$

$$[m'] = [B]^T [m][B] \tag{7.481}$$

$$P = \frac{1}{2}\mathbf{x}^T [k]\mathbf{x} = \frac{1}{2}\left(\mathbf{q}^T [B]^T + \mathbf{a}^T\right)[k]\left([B]\mathbf{q} + \mathbf{a}\right)$$

$$= \frac{1}{2}\mathbf{q}^T [B]^T [k][B]\mathbf{q} + \frac{1}{2}\mathbf{q}^T [B]^T [k]\mathbf{a} + \frac{1}{2}\mathbf{a}^T [k][B]\mathbf{q} + \mathbf{a}^T [k]\mathbf{a}$$

$$= \frac{1}{2}\mathbf{q}^T [k']\mathbf{q} + \frac{1}{2}\mathbf{q}^T [k'][B]^T \mathbf{a} + \frac{1}{2}\mathbf{a}^T [B][k']\mathbf{q} + \mathbf{a}^T [k]\mathbf{a} \tag{7.482}$$

$$[k'] = [B]^T [k][B] \tag{7.483}$$

$$D = \frac{1}{2}\dot{\mathbf{x}}^T [c]\dot{\mathbf{x}} = \frac{1}{2}\dot{\mathbf{q}}^T [B]^T [c][B]\dot{\mathbf{q}} = \frac{1}{2}\dot{\mathbf{q}}^T [c']\dot{\mathbf{q}} \tag{7.484}$$

$$[c'] = [B]^T [c][B] \tag{7.485}$$

To use the Lagrange equation (2.547), we need to take the required quadrature derivatives:

$$\frac{\partial K}{\partial \dot{\mathbf{q}}} = [m']\dot{\mathbf{q}} \tag{7.486}$$

$$\frac{\partial P}{\partial \mathbf{q}} = [k']\mathbf{q} + \frac{1}{2}\left([k'][B]^T \mathbf{a} + \mathbf{a}^T [B][k']\right)$$

$$= [k']\mathbf{q} + \frac{1}{2}\left([B]^T [k]\mathbf{a} + \mathbf{a}^T [k][B]\right) \tag{7.487}$$

$$\frac{\partial D}{\partial \dot{\mathbf{q}}} = [c']\dot{\mathbf{q}} \tag{7.488}$$

Employing the Lagrange equation (2.547),

$$\frac{d}{dt}\frac{\partial K}{\partial \dot{\mathbf{x}}} + \frac{\partial K}{\partial \mathbf{x}} + \frac{\partial D}{\partial \dot{\mathbf{x}}} + \frac{\partial P}{\partial \mathbf{x}} = \mathbf{F} \tag{7.489}$$

we find the equation of motion,

$$[m']\ddot{\mathbf{q}} + [c']\dot{\mathbf{q}} + [k']\mathbf{q} = \frac{1}{2}\left([k'][B]^T \mathbf{a} + \mathbf{a}^T [B][k']\right) + \mathbf{Q} \tag{7.490}$$

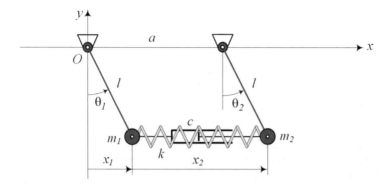

Fig. 7.34 A *two* pendulum system

where the new force vector $\mathbf{Q}$ is a linear combination of the elements of $\mathbf{F}$ and is calculated based on the virtual work δW.

$$\delta W = \mathbf{F}^T \, \delta \mathbf{x} = \mathbf{F}^T \, [B] \, \delta \mathbf{q} = \mathbf{Q}^T \, \delta \mathbf{q} \tag{7.491}$$

$$\mathbf{Q} = [B]^T \, \mathbf{F} \tag{7.492}$$

The constant term $\frac{1}{2} \left([B]^T \, [k] \, \mathbf{a} + \mathbf{a}^T \, [k] \, [B] \right)$ indicates the steady-state values $\mathbf{q}_s$ of $\mathbf{q}$ in the absence of the forcing terms.

$$\begin{aligned} \mathbf{q}_s &= \frac{1}{2} \left[k' \right]^{-1} \left(\left[k' \right] [B]^T \, \mathbf{a} + \mathbf{a}^T \, [B] \left[k' \right] \right) \\ &= \frac{1}{2} \left([B]^T \, \mathbf{a} + \left[k' \right]^{-1} \mathbf{a}^T \, [B] \left[k' \right] \right) \end{aligned} \tag{7.493}$$

However, having the constant term $\frac{1}{2} \left([B]^T \, [k] \, \mathbf{a} + \mathbf{a}^T \, [k] \, [B] \right)$ introduces a constant force on the system. Whenever there is no constant force on the original system, coordinate change should not generate such a force. A proper coordinate transformation must provide us with a zero value in the coordinate $\mathbf{q}$ at $\mathbf{x} = \mathbf{0}$ and vice versa. Therefore, the coordinate transformation (7.473) is not a proper transformation as long as $\mathbf{a} \neq \mathbf{0}$.

Example 265 ★ Improper coordinate change of a two pendulum system. Here is an applied example of improper coordinate transformation.

Figure 7.34 illustrates a two connected pendulum system with two different tip masses. The masses are connected by a spring k and a damper c. Assuming small angles, the energies and dissipation functions of the system are:

$$K = \frac{1}{2}m_1 l^2 \dot\theta_1^2 + \frac{1}{2}m_2 l^2 \dot\theta_2^2 \tag{7.494}$$

$$P = \frac{1}{2}kl^2 (\theta_1 - \theta_2)^2 - m_1 gl \left(1 - \frac{\theta_1^2}{2}\right) - m_2 gl \left(1 - \frac{\theta_2^2}{2}\right) \tag{7.495}$$

$$D = \frac{1}{2}cl^2 (\dot\theta_1 - \dot\theta_2)^2 \tag{7.496}$$

The definition of P indicates that the free length of the spring is a and when the pendulums are hanging down the potential energy in the spring is zero. The equations of motion of the system are:

$$\begin{bmatrix} m_1 l^2 & 0 \\ 0 & m_2 l^2 \end{bmatrix} \begin{bmatrix} \ddot\theta_1 \\ \ddot\theta_2 \end{bmatrix} + \begin{bmatrix} cl^2 & -cl^2 \\ -cl^2 & cl^2 \end{bmatrix} \begin{bmatrix} \dot\theta_1 \\ \dot\theta_2 \end{bmatrix}$$
$$+ \begin{bmatrix} kl^2 + m_1 gl & -kl^2 \\ -kl^2 & kl^2 + m_2 gl \end{bmatrix} \begin{bmatrix} \theta_1 \\ \theta_2 \end{bmatrix} = \begin{bmatrix} 0 \\ 0 \end{bmatrix} \tag{7.497}$$

Let us change the coordinates form θ to $\mathbf{x}$ using their relationship,

$$l\theta_1 = x_1 \qquad a + l\theta_2 = x_1 + x_2 \tag{7.498}$$

$$\begin{bmatrix} x_1 \\ x_2 \end{bmatrix} = \begin{bmatrix} l & 0 \\ -l & l \end{bmatrix} \begin{bmatrix} \theta_1 \\ \theta_2 \end{bmatrix} + \begin{bmatrix} 0 \\ a \end{bmatrix} \tag{7.499}$$

$$\begin{bmatrix} \theta_1 \\ \theta_2 \end{bmatrix} = \begin{bmatrix} l & 0 \\ -l & l \end{bmatrix}^{-1} \left(\begin{bmatrix} x_1 \\ x_2 \end{bmatrix} - \begin{bmatrix} 0 \\ a \end{bmatrix} \right) \tag{7.500}$$

and therefore,

$$\begin{bmatrix} \theta_1 \\ \theta_2 \end{bmatrix} = \begin{bmatrix} x_1/l \\ x_1/l - (a - x_2)/l \end{bmatrix} \tag{7.501}$$

Substituting (7.501) into Eqs. (7.494)–(7.496) yields:

$$K = \frac{1}{2}m_1 \dot x_1^2 + \frac{1}{2}m_2 (\dot x_1 + \dot x_2)^2 \tag{7.502}$$

$$P = \frac{1}{2}kl^2 \left(\frac{x_1}{l} - \left(\frac{x_1}{l} - \frac{a}{l} + \frac{x_2}{l}\right)\right)^2 - m_1 gl \left(1 - \frac{x_1^2}{2l^2}\right) - m_2 gl \left(1 - \frac{x_2^2}{2l^2}\right)$$
$$= \frac{1}{2}k (a - x_2)^2 + \frac{1}{2}m_1 \frac{g}{l} \left(x_1^2 - 2l^2\right) + \frac{1}{2}m_2 \frac{g}{l} \left(x_2^2 - 2l^2\right) \tag{7.503}$$

$$D = \frac{1}{2}cl^2 \left(\frac{\dot{x}_1}{l} - \left(\frac{\dot{x}_1}{l} + \frac{\dot{x}_2}{l} \right) \right)^2 = \frac{1}{2}c\dot{x}_2^2 \tag{7.504}$$

The equations of motion of the system in terms of the new coordinates are:

$$m_1\ddot{x}_1 + m_2(\ddot{x}_1 + \ddot{x}_2) + m_1\frac{g}{l}x_1 = 0 \tag{7.505}$$

$$m_2(\ddot{x}_1 + \ddot{x}_2) + c\dot{x}_2 - k(a - x_2) + m_2\frac{g}{l}x_2 = 0 \tag{7.506}$$

or

$$\begin{bmatrix} m_1 + m_2 & m_2 \\ m_2 & m_2 \end{bmatrix} \begin{bmatrix} \ddot{x}_1 \\ \ddot{x}_2 \end{bmatrix} + \begin{bmatrix} 0 & 0 \\ 0 & c \end{bmatrix} \begin{bmatrix} \dot{x}_1 \\ \dot{x}_2 \end{bmatrix}$$
$$+ \begin{bmatrix} m_1 g/l & 0 \\ 0 & k + m_2 g/l \end{bmatrix} \begin{bmatrix} x_1 \\ x_2 \end{bmatrix} = \begin{bmatrix} 0 \\ ka \end{bmatrix} \tag{7.507}$$

Having a constant term ka indicates that the transformation (7.498) is an improper transformation.

Example 266 Coordinate transformation. Here is an example of applying transformation of coordinates directly to the equations of motion.

Assume that we wish to change the generalized codominants of a system from a set $\mathbf{x}$ to a set $\mathbf{q}$.

$$[m]\ddot{\mathbf{q}} + [k]\mathbf{q} = \mathbf{Q} \tag{7.508}$$

In a coordinate change, instead of defining the energies and dissipation function based on a new set of generalized coordinates to determine the new set of equations of motion, we may use a direct coordinate transformation method. Let us begin with the equations of motion of the beam of Fig. 7.23 in x and θ (7.241) and transform to the coordinates x_1 and x_2 (7.251).

The coordinates x and θ are related to the coordinates x_1 and x_2 by a matrix transformation

$$\mathbf{x} = [B]\mathbf{q} \tag{7.509}$$

where

$$\mathbf{x} = \begin{bmatrix} x \\ \theta \end{bmatrix} \qquad \mathbf{q} = \begin{bmatrix} x_1 \\ x_2 \end{bmatrix} \qquad [B] = \begin{bmatrix} \dfrac{a_2}{a_1 + a_2} & \dfrac{a_1}{a_1 + a_2} \\ \dfrac{1}{a_1 + a_2} & \dfrac{-1}{a_1 + a_2} \end{bmatrix} \tag{7.510}$$

Substituting this transformation into Eq. (7.241) provides us with:

$$\begin{bmatrix} m & 0 \\ 0 & I \end{bmatrix} [B]\mathbf{q} + \begin{bmatrix} k_1 + k_2 & a_2 k_2 - a_1 k_1 \\ a_2 k_2 - a_1 k_1 & a_2^2 k_2 + a_1^2 k_1 \end{bmatrix} [B]\mathbf{q} = 0 \tag{7.511}$$

To keep the symmetry of the coefficient matrices, we also pre-multiply the equation by $[B]^T$, which yields:

$$[B]^T [m][B]\mathbf{q} + [B]^T [k][B]\mathbf{q} = [m']\mathbf{q} + [k']\mathbf{q} = 0 \tag{7.512}$$

where

$$[m'] = [B]^T [m][B] = \begin{bmatrix} \dfrac{ma_2^2 + I}{(a_1 + a_2)^2} & -\dfrac{I - ma_1 a_2}{(a_1 + a_2)^2} \\ -\dfrac{I - ma_1 a_2}{(a_1 + a_2)^2} & \dfrac{ma_1^2 + I}{(a_1 + a_2)^2} \end{bmatrix} \tag{7.513}$$

$$[k'] = [A]^T [k][A]$$

$$= \begin{bmatrix} \dfrac{k_1 (a_1 - a_2)^2 + 4a_2^2 k_2}{(a_1 + a_2)^2} & -2(a_1 - a_2)\dfrac{a_1 k_1 - a_2 k_2}{(a_1 + a_2)^2} \\ -2(a_1 - a_2)\dfrac{a_1 k_1 - a_2 k_2}{(a_1 + a_2)^2} & \dfrac{k_2 (a_1 - a_2)^2 + 4a_1^2 k_1}{(a_1 + a_2)^2} \end{bmatrix} \tag{7.514}$$

Example 267 Stiffness matrix elements in natural coordinates. The elements of stiffness matrix in principal coordinate have similar definition to the elements of the matrix in absolute coordinates. The element k_{ij} of a stiffness matrix $[k]$ indicates the required force F_i at x_i to produce a unit displacement at x_i while $x_j = 0, i \neq j$. Here is an analytic example to analyze this fact.

In case of

$$m_1 = m_2 = m \qquad k_1 = k_2 = k_3 = k \tag{7.515}$$

the equations of motion of the system of Fig. 7.29 are:

$$\begin{bmatrix} m & 0 \\ 0 & m \end{bmatrix} \begin{bmatrix} \ddot{x}_1 \\ \ddot{x}_2 \end{bmatrix} + \begin{bmatrix} 2k & -k \\ -k & 2k \end{bmatrix} \begin{bmatrix} x_1 \\ x_2 \end{bmatrix} = \begin{bmatrix} 0 \\ 0 \end{bmatrix} \tag{7.516}$$

The natural coordinates $\mathbf{p}$ of the system are:

$$\mathbf{p} = [U]^{-1}\mathbf{x} = \begin{bmatrix} \frac{1}{2}(x_1 + x_2) \\ \frac{1}{2}(x_2 - x_1) \end{bmatrix} \tag{7.517}$$

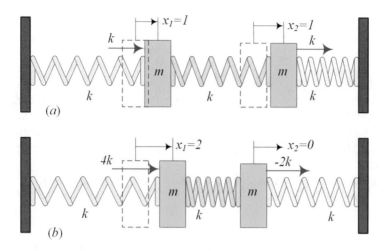

Fig. 7.35 The required natural force on a two DOF system to provide unit displacements in natural coordinates. (a) $p_1 = 1$, $p_2 = 0$, and (b) $p_1 = 0$, $p_2 = 1$

which makes the equations of motion decoupled.

$$\begin{bmatrix} m & 0 \\ 0 & m \end{bmatrix} \begin{bmatrix} \ddot{p}_1 \\ \ddot{p}_2 \end{bmatrix} + \begin{bmatrix} k & 0 \\ 0 & 3k \end{bmatrix} \begin{bmatrix} p_1 \\ p_2 \end{bmatrix} = \begin{bmatrix} 0 \\ 0 \end{bmatrix} \tag{7.518}$$

Because $k_{21} = 0$, a unit displacement in p_1 will not produce any displacement in p_2, and hence, no force is required to keep $p_2 = 0$. Similarly, because of $k_{12} = 0$, a unit displacement in p_2 will not produce any displacement in p_1, and hence, no force is required to keep $p_1 = 0$. Figures 7.35a and b illustrate the situations of $p_1 = 1$, $p_2 = 0$, and $p_2 = 1$, $p_1 = 0$ and the required forces, respectively.

If the force system of the original system is

$$\mathbf{F} = \begin{bmatrix} F_1 \\ F_2 \end{bmatrix} \tag{7.519}$$

then the force system in the natural coordinates would be

$$\mathbf{P} = [U]^T \mathbf{F} = \begin{bmatrix} \mathbf{u}_1 & \mathbf{u}_2 \end{bmatrix}^{-1} \mathbf{F} = \begin{bmatrix} 1 & -1 \\ 1 & 1 \end{bmatrix}^{-1} \mathbf{F}$$

$$= \begin{bmatrix} 1 & -1 \\ 1 & 1 \end{bmatrix}^{-1} \begin{bmatrix} F_1 \\ F_2 \end{bmatrix} = \frac{1}{2} \begin{bmatrix} F_1 + F_2 \\ F_2 - F_1 \end{bmatrix} = \begin{bmatrix} P_1 \\ P_2 \end{bmatrix} \tag{7.520}$$

Applying a force $P_1 = (F_1 + F_2)/2 = (k + k)/2 = k$ in Fig. 7.35a produces a unit displacement in the natural coordinate $p_1 = (x_1 + x_2)/2$. It does not need any nonzero force P_2 to keep $p_2 = (x_2 - x_1)/2 = 0$. Also a force $P_2 = (F_2 - F_1)/2 =$

$(-2k - 4k)/2 = 3k$ in Fig. 7.35b produces a unit displacement in the natural coordinate $p_2 = (x_1 - x_2)/2$ and does not need any nonzero force P_1 to keep $p_1 = (x_2 + x_1)/2 = 0$.

Example 268 Flexibility matrix. The flexibility matrix is the inverse of the stiffness matrix. Calculation of the elements of the flexibility matrix of a system is simpler to the calculation of the elements of stiffness matrix. The inverse of the calculated matrix will give us the stiffness matrix to derive the equations of motion and vibration characteristics of the system. Here is the analytic study of flexibility matrix.

The inverse of stiffness matrix $[k]$ is called the flexibility matrix $[a]$.

$$[k]^{-1} = [a] \qquad (7.521)$$

The element a_{ij} of a flexibility matrix $[a]$ is the displacement at x_j caused by a unit force F_i at x_i. Consider a system which is under a force F_i at the point $x = x_i$. Let us denote the displacement at $x = x_j$ due to F_i by x_{ij}. For a linear system, the deflection increases proportionally with the force.

$$x_{ij} = a_{ij} F_i \qquad (7.522)$$

The superposition principle of linear system implies that if the system is under the influence of a number of forces F_i, $i = 1, 2, \cdots, n$, then x_j will be a linear combination of the displacements caused by all forces.

$$x_j = \sum_{i=1}^{n} x_{ij} = \sum_{i=1}^{n} a_{ij} F_i \qquad i = 1, 2, \cdots, n \qquad (7.523)$$

Now we define the stiffness coefficient k_{ij} as the force at $x = x_i$ due to a unit displacement at $x = x_j$ while the points $x \neq x_j$ are fixed.

$$F_i = \sum_{j=1}^{n} k_{ij} x_j \qquad i = 1, 2, \cdots, n \qquad (7.524)$$

Equations (7.523) and (7.524) may also be represented in matrix forms

$$\mathbf{x} = [a]\,\mathbf{F} \qquad (7.525)$$

$$\mathbf{F} = [k]\,\mathbf{x} \qquad (7.526)$$

in which the matrices $[a]$ and $[k]$ are the flexibility and stiffness matrices, respectively.

$$[k] = [a]^{-1} \qquad (7.527)$$

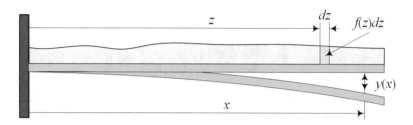

Fig. 7.36 A continuous beam and the flexibility and deflection in continuous system

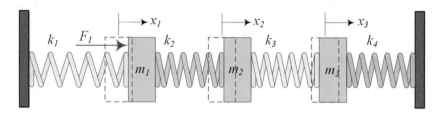

Fig. 7.37 A linear three DOF system

The same concept is applied when x_i and x_j are in a continuum media. The flexibility function $a(x, z)$ is defined as the deflection at point x due to a unit load at a point z as is illustrated in Fig. 7.36. Therefore, the increment of the deflection due to the load $f(z)\,dz$ is

$$dy(x, z) = a(x, z) f(z)\,dz \tag{7.528}$$

and hence, the total deflection at x is calculated by integrating over the length l.

$$y(x) = \int_0^l a(x, z) f(z)\,dz \tag{7.529}$$

Example 269 Flexibility and stiffness matrices of a 3 DOF system. This is an example to show how we can experimentally calculate the flexibility matrix.

To determine the flexibility matrix of the three DOF system of Fig. 7.37, we apply a unit force F_1 to the mass m_1 at x_1 and calculate the displacements $x_{11} = a_{11}$, $x_{12} = a_{12}$, and $x_{13} = a_{13}$. The mass m_1 is attached to a spring k_1 at the left and a spring $1/\sum_{i=2}^4 (1/k_i)$ on the right. The left and right springs of m_1 are parallel. Therefore, the displacement of x_1 would be:

$$
\begin{aligned}
x_{11} = a_{11} &= \frac{F_1 = 1}{k_1 + \dfrac{1}{1/k_2 + 1/k_3 + 1/k_4}} = \frac{\sum_{i=2}^4 (1/k_i)}{k_1 \sum_{i=1}^4 (1/k_i)} \\
&= \frac{k_2 k_3 + k_2 k_4 + k_3 k_4}{k_1 k_2 k_3 + k_1 k_2 k_4 + k_1 k_3 k_4 + k_2 k_3 k_4}
\end{aligned}
\tag{7.530}
$$

The displacement of x_3 would be the force on the right of m_1 divided by k_4.

$$x_{13} = a_{13} = \frac{a_{11}/\sum_{i=2}^{4}(1/k_i)}{k_4} = \frac{1}{k_1 k_4 \sum_{i=1}^{4}(1/k_i)} \tag{7.531}$$

The displacement of x_2 would be the force on the right of m_1 divided by the series of springs k_3 and k_4.

$$x_{12} = a_{12} = \frac{a_{11}/\sum_{i=2}^{4}(1/k_i)}{\dfrac{1}{1/k_3 + 1/k_4}} = \frac{k_3 + k_4}{k_1 k_3 k_4 \sum_{i=1}^{4}(1/k_i)} \tag{7.532}$$

Similarly, applying a unit force at x_2 yields:

$$
\begin{aligned}
a_{22} &= \frac{1}{\dfrac{1}{1/k_1 + 1/k_2} + \dfrac{1}{1/k_3 + 1/k_4}} \\
&= \frac{(k_1 + k_2)(k_3 + k_4)}{k_1 k_2 k_3 + k_1 k_2 k_4 + k_1 k_3 k_4 + k_2 k_3 k_4}
\end{aligned}
\tag{7.533}
$$

$$
\begin{aligned}
a_{21} &= \frac{a_{22}\left(\dfrac{1}{1/k_1 + 1/k_2}\right)}{k_1} = \frac{k_2(k_3 + k_4)}{k_1 k_2 k_3 + k_1 k_2 k_4 + k_1 k_3 k_4 + k_2 k_3 k_4} \\
&= \frac{k_3 + k_4}{k_1 k_3 k_4 \sum_{i=1}^{4}(1/k_i)} = a_{12}
\end{aligned}
\tag{7.534}
$$

$$
\begin{aligned}
a_{23} &= \frac{a_{22}\left(\dfrac{1}{1/k_3 + 1/k_4}\right)}{k_4} = \frac{k_3(k_1 + k_2)}{k_1 k_2 k_3 + k_1 k_2 k_4 + k_1 k_3 k_4 + k_2 k_3 k_4} \\
&= \frac{k_1 + k_2}{k_1 k_3 k_4 \sum_{i=1}^{4}(1/k_i)}
\end{aligned}
\tag{7.535}
$$

Applying a unit force at x_3 provides us with:

$$a_{33} = \frac{\sum_{i=1}^{3}(1/k_i)}{k_4 \sum_{i=1}^{4}(1/k_i)} \tag{7.536}$$

$$a_{31} = \frac{a_{33}/\sum_{i=2}^{4}(1/k_i)}{k_1} = \frac{1}{k_1 k_4 \sum_{i=1}^{4}(1/k_i)} = a_{13} \tag{7.537}$$

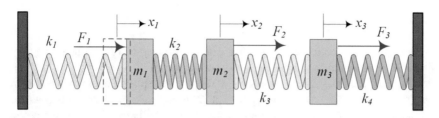

Fig. 7.38 A linear three DOF system and stiffness coefficients

$$a_{32} = \frac{a_{33}/\sum_{i=2}^{4}(1/k_i)}{\dfrac{1}{1/k_3 + 1/k_4}} = \frac{k_3 + k_4}{k_1 k_3 k_4 \sum_{i=1}^{4}(1/k_i)} = a_{23} \qquad (7.538)$$

Therefore, the flexibility matrix of the system is:

$$[a] = \frac{\begin{bmatrix} k_2 k_3 + k_2 k_4 + k_3 k_4 & k_2(k_3 + k_4) & k_2 k_3 \\ k_2(k_3 + k_4) & (k_1 + k_2)(k_3 + k_4) & k_3(k_1 + k_2) \\ k_2 k_3 & k_3(k_1 + k_2) & k_1 k_2 + k_1 k_3 + k_2 k_3 \end{bmatrix}}{k_1 k_2 k_3 + k_1 k_2 k_4 + k_1 k_3 k_4 + k_2 k_3 k_4} \qquad (7.539)$$

To determine the stiffness matrix $[k]$, we apply a force system such that $x_1 = 1$, and $x_2 = x_3 = 0$, as is shown in Fig. 7.38. The stiffness elements k_{11}, k_{21}, k_{31} can be found from this configuration.

$$k_{11} = k_1 + k_2 \qquad k_{21} = -k_2 \qquad k_{31} = 0 \qquad (7.540)$$

Similarly, imagine a force system that provides a unit displacement at x_2, while $x_1 = x_3 = 0$ to find k_{12}, k_{22}, and k_{32}.

$$k_{12} = -k_2 \qquad k_{22} = k_2 + k_3 \qquad k_{32} = -k_3 \qquad (7.541)$$

Another force system that provides a unit displacement at x_3, while $x_1 = x_3 = 0$, determines k_{13}, k_{23}, and k_{33}.

$$k_{13} = 0 \qquad k_{23} = -k_3 \qquad k_{33} = k_3 + k_4 \qquad (7.542)$$

Therefore, the stiffness matrix is

$$[k] = \begin{bmatrix} k_1 + k_2 & -k_2 & 0 \\ -k_2 & k_2 + k_3 & -k_3 \\ 0 & -k_3 & k_3 + k_4 \end{bmatrix} \qquad (7.543)$$

It can be examined that $[k]^{-1} = [a]$.

Example 270 ★ Reciprocity theorem. The reciprocity theorem states that for a linear elastic structure subject to two sets of forces F_i, $i = 1, ..., n$ and Q_j, $j = 1, ..., n$, the work done by the set F through the displacements produced by the set Q is equal to the work done by the set Q through the displacements produced by the set F. The reciprocity theorem makes the flexibility matrix to be symmetric.

The flexibility matrix $[a]$ is symmetric.

$$a_{ij} = a_{ji} \tag{7.544}$$

To prove it, let us calculate the work W_i done by a force f_i that is applied at x_i.

$$W_i = f_i x_i = \frac{1}{2} f_i^2 a_{ii} \tag{7.545}$$

Now applying a force f_j at x_j provides us with a work W_j plus an additional work W_{ij} done by f_i.

$$W_j = f_j x_j = \frac{1}{2} f_j^2 a_{jj} \tag{7.546}$$

$$W_{ij} = f_i x_{ij} = f_i f_j a_{ij} \tag{7.547}$$

Therefore, the total work of the force system is:

$$W = \frac{1}{2} f_i^2 a_{ii} + \frac{1}{2} f_j^2 a_{jj} + f_i f_j a_{ij} \tag{7.548}$$

If we reverse the order of loading, the total work will be:

$$W = \frac{1}{2} f_i^2 a_{ii} + \frac{1}{2} f_j^2 a_{jj} + f_j f_j a_{ji} \tag{7.549}$$

Because in a linear system the order of loading will not affect the total work done by the force system, we must have $a_{ij} = a_{ji}$.

The reciprocity theorem, also known as Betti's theorem, or Maxwell-Betti theorem, is discovered by an Italian mathematician, Enrico Betti (1823–1892) in 1872.

Example 271 ★ Double pendulum and dynamic coupling. Pendulums are nonlinear vibrating systems; however, assuming small oscillations, the equations of motion can be linearized. It is an example of a linearized system with dynamic coupled and statically decoupled.

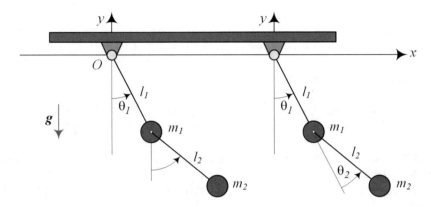

Fig. 7.39 A double pendulum

The double pendulum of Fig. 7.39 is made of two massless rods with lengths l_1 and l_2 and two point masses m_1 and m_2. The absolute variables θ_1 and θ_2 act as the generalized coordinates to express the configuration of the system. The Lagrangean of the system is:

$$
\begin{aligned}
\mathcal{L} &= K - P \\
&= \frac{1}{2}m_1 l_1^2 \dot{\theta}_1^2 + \frac{1}{2}m_2 \left(l_1^2 \dot{\theta}_1^2 + l_2^2 \dot{\theta}_2^2 + 2l_1 l_2 \dot{\theta}_1 \dot{\theta}_2 \cos(\theta_1 - \theta_2) \right) \\
&\quad + m_1 g l_1 \cos\theta_1 + m_2 g (l_1 \cos\theta_1 + l_2 \cos\theta_2)
\end{aligned}
\tag{7.550}
$$

where

$$
\begin{aligned}
K &= \frac{1}{2}m_1 v_1^2 + \frac{1}{2}m_2 v_2^2 \\
&= \frac{1}{2}m_1 l_1^2 \dot{\theta}_1^2 + \frac{1}{2}m_2 \left(l_1^2 \dot{\theta}_1^2 + l_2^2 \dot{\theta}_2^2 + 2l_1 l_2 \dot{\theta}_1 \dot{\theta}_2 \cos(\theta_1 - \theta_2) \right)
\end{aligned}
\tag{7.551}
$$

and

$$
\begin{aligned}
P &= m_1 g y_1 + m_2 g y_2 \\
&= -m_1 g l_1 \cos\theta_1 - m_2 g (l_1 \cos\theta_1 + l_2 \cos\theta_2)
\end{aligned}
\tag{7.552}
$$

Employing the Lagrange method (2.499), we find the equations of motion.

$$
\begin{aligned}
(m_1 + m_2) l_1^2 \ddot{\theta}_1 + m_2 l_1 l_2 \ddot{\theta}_2 \cos(\theta_1 - \theta_2) \\
- m_2 l_1 l_2 \dot{\theta}_2^2 \sin(\theta_1 - \theta_2) + (m_1 + m_2) l_1 g \sin\theta_1 = 0
\end{aligned}
\tag{7.553}
$$

$$m_2 l_2^2 \ddot{\theta}_2 + m_2 l_1 l_2 \ddot{\theta}_1 \cos{(\theta_1 - \theta_2)}$$

$$+ m_2 l_1 l_2 \dot{\theta}_1^2 \sin{(\theta_1 - \theta_2)} + m_2 l_2 g \sin{\theta_2} = 0 \qquad (7.554)$$

The linearized equations for small angles are:

$$\begin{bmatrix} (m_1 + m_2) l_1 & m_2 l_2 \\ l_1 & l_2 \end{bmatrix} \begin{bmatrix} \ddot{\theta}_1 \\ \ddot{\theta}_2 \end{bmatrix} + \begin{bmatrix} (m_1 + m_2) g & 0 \\ 0 & g \end{bmatrix} \begin{bmatrix} \theta_1 \\ \theta_2 \end{bmatrix} = \begin{bmatrix} 0 \\ 0 \end{bmatrix} \qquad (7.555)$$

which indicate the equation of the double pendulum is dynamically coupled and statically decoupled.

Employing the relative coordinates, the following equations of motion will be derived.

$$\left((m_1 + m_2) l_1^2 + m_2 l_2 (l_2 + 2 l_1 \cos{\theta_2}) \right) \ddot{\theta}_1$$

$$+ m_2 l_2 (l_2 + l_1 \cos{\theta_2}) \ddot{\theta}_2 - m_2 l_1 l_2 \dot{\theta}_2 (2 \dot{\theta}_1 + \dot{\theta}_2) \sin{\theta_2}$$

$$+ (m_1 + m_2) g l_1 \sin{\theta_1} + m_2 g l_2 \sin{(\theta_1 + \theta_2)} = 0 \qquad (7.556)$$

$$(m_2 l_2 (l_2 + l_1 \cos{\theta_2})) \ddot{\theta}_1 + m_2 l_2^2 \ddot{\theta}_2$$

$$+ m_2 l_1 l_2 \dot{\theta}_1^2 \sin{\theta_2} + m_2 g l_2 \sin{(\theta_1 + \theta_2)} = 0 \qquad (7.557)$$

The linearized form of these equations for small angles are both statically and dynamically coupled.

$$\begin{bmatrix} m_1 l_1^2 + m_2 (l_1 + l_2)^2 & m_2 l_2 (l_2 + l_1) \\ m_2 l_2 (l_2 + l_1) & m_2 l_2^2 \end{bmatrix} \begin{bmatrix} \ddot{\theta}_1 \\ \ddot{\theta}_2 \end{bmatrix}$$

$$+ \begin{bmatrix} (m_1 + m_2) g l_1 + m_2 g l_2 & m_2 g l_2 \\ m_2 g l_2 & m_2 g l_2 \end{bmatrix} \begin{bmatrix} \theta_1 \\ \theta_2 \end{bmatrix} = \begin{bmatrix} 0 \\ 0 \end{bmatrix} \qquad (7.558)$$

Example 272 ★ Normalized mode shapes. The modal matrix can be normalized according to mass or stiffness matrices. It simplifies calculation and determination of natural frequencies.

Let us divide each element of the mode shape $\mathbf{u}_i$ by $\sqrt{m_i}$ and call it $\mathbf{u}_{im}$ to indicate the normalized mode shape with respect to mass matrix

$$\mathbf{u}_{im} = \frac{\mathbf{u}_i}{\sqrt{m_i}} = \frac{\mathbf{u}_i}{\sqrt{\mathbf{u}_i^T [m] \mathbf{u}_i}} \qquad i = 1, 2, \cdots, n \qquad (7.559)$$

where

$$m_i = \mathbf{u}_i^T [m] \mathbf{u}_i \qquad (7.560)$$

Therefore,

$$\frac{\mathbf{u}_i^T}{\sqrt{m_i}} [m] \frac{\mathbf{u}_i}{\sqrt{m_i}} = \mathbf{u}_{im}^T [m] \mathbf{u}_{im} = [I] \qquad (7.561)$$

which shows that:

$$[U]_m^T [m] [U]_m = [I] \qquad (7.562)$$

The matrix $[U]_m$ is the modal matrix whose columns are the normalized mode shapes with respect to mass matrix. The normalized modal matrix $[U]_m$ leads to

$$[U]_m^T [k] [U]_m = \left[\omega_i^2 \right] \qquad (7.563)$$

where $\left[\omega_i^2 \right]$ is the diagonal matrix (7.323) because:

$$\mathbf{u}_{im}^T [k] \mathbf{u}_{im} = \frac{1}{m_i} \mathbf{u}_i^T [k] \mathbf{u}_i = \frac{k_i}{m_i} = \omega_i^2 \qquad (7.564)$$

Similarly, we may divide each element of the mode shape $\mathbf{u}_i$ by $\sqrt{k_i}$ and call it $\mathbf{u}_{ik}$ to indicate the normalized mode shape with respect to the stiffness matrix

$$\mathbf{u}_{im} = \frac{\mathbf{u}_i}{\sqrt{k_i}} = \frac{\mathbf{u}_i}{\sqrt{\mathbf{u}_i^T [k] \mathbf{u}_i}} \qquad i = 1, 2, \cdots, n \qquad (7.565)$$

where

$$k_i = \mathbf{u}_i^T [k] \mathbf{u}_i \qquad (7.566)$$

Therefore,

$$\frac{\mathbf{u}_i^T}{\sqrt{k_i}} [k] \frac{\mathbf{u}_i}{\sqrt{k_i}} = \mathbf{u}_{ik}^T [k] \mathbf{u}_{ik} = [I] \qquad (7.567)$$

which shows that:

$$[U]_m^T [k] [U]_m = [I] \qquad (7.568)$$

The matrix $[U]_k$ is the modal matrix whose columns are the normalized mode shapes with respect to stiffness matrix. The normalized modal matrix $[U]_k$ yields

$$[U]_k^T [m] [U]_k = \left[\frac{1}{\omega_i^2} \right] \tag{7.569}$$

because:

$$\mathbf{u}_{ik}^T [m] \mathbf{u}_{ik} = \frac{1}{k_i} \mathbf{u}_i^T [m] \mathbf{u}_i = \frac{m_i}{k_i} = \frac{1}{\omega_i^2} \tag{7.570}$$

The matrix $\left[1/\omega_i^2 \right]$ is the diagonal matrix whose elements are $1/\omega_i^2$.

Example 273 ★ Proportional damping and the general decoupling damping. If the damping matrix $[c]$ is a linear combination of $[m]$ and $[k]$, the equations of motion of a damped system can be decoupled.

Because the mode shapes are characteristics of the undamped system, they are only orthogonal with respect to mass and stiffness matrices, and hence, only undamped equations can be decoupled. Therefore, the damping matrix $[c]$ will not necessarily be transformed to a diagonal matrix by a modal or principal coordinate transformation. However, if the damping matrix can be written as a linear combination of the mass and stiffness matrices, we will be able to decouple the equations of motion of a damped system.

Let us assume that the damping matrix $[c]$ of a set of coupled equations of motion is proportional to $[m]$ and $[k]$,

$$[m] \ddot{\mathbf{x}} + [c] \dot{\mathbf{x}} + [k] \mathbf{x} = \mathbf{F} \tag{7.571}$$

$$[c] = a [m] + b [k] \tag{7.572}$$

where a and b are constant weight factors. Substituting (7.572) into (7.571) leads:

$$[m] \ddot{\mathbf{x}} + (a [m] + b [k]) \dot{\mathbf{x}} + [k] \mathbf{x} = \mathbf{F} \tag{7.573}$$

A modal transformation of the coordinates from $\mathbf{x}$ to the undamped principal coordinates $\mathbf{p}$,

$$\mathbf{x} = [U] \mathbf{p} \tag{7.574}$$

yields:

$$[m] [U] \ddot{\mathbf{p}} + (a [m] + b [k]) [U] \dot{\mathbf{p}} + [k] [U] \mathbf{p} = \mathbf{F} \tag{7.575}$$

Multiplying $[U]^T$ makes the equations decoupled

$$[U]^T [m] [U] \ddot{\mathbf{p}} + [U]^T (a [m] + b [k]) [U] \dot{\mathbf{p}} + [U]^T [k] [U] \mathbf{p}$$
$$= [m'] \ddot{\mathbf{p}} + (a [m'] + b [k']) \dot{\mathbf{p}} + [k'] \mathbf{p} = [U]^T \mathbf{F}$$
$$= [m'] \ddot{\mathbf{p}} + [c'] \dot{\mathbf{p}} + [k'] \mathbf{p} = \mathbf{Q} \tag{7.576}$$

where

$$[m'] = [U]^T [m] [U] \tag{7.577}$$

$$[k'] = [U]^T [k] [U] \tag{7.578}$$

and the new damping matrix is also diagonal.

$$[c'] = a [U]^T [m] [U] + b [U]^T [k] [U] = a [m'] + b [k'] \tag{7.579}$$

$$c_i' = am_i' + bk_i' \tag{7.580}$$

The equations of motion will be decoupled.

$$[m'] \ddot{\mathbf{p}} + [c'] \dot{\mathbf{p}} + [k'] \mathbf{p} = \mathbf{Q} \tag{7.581}$$

$$m_i \ddot{p} + c_i \dot{p} + k_i p = Q \tag{7.582}$$

The general expression for the viscous damping $[c]$ to provide decoupled equations of motion is:

$$[k] [m]^{-1} [c] = [c] [m]^{-1} [k] \tag{7.583}$$

This is because the matrix $[U]^T [c] [U]$ must be a diagonal matrix. Multiplying two diagonal matrices generates another diagonal matrix. Therefore, both sides of

$$[U]^T [c] [U] [m']^{-1} [k'] = [k'] [m']^{-1} [U]^T [c] [U] \tag{7.584}$$

must also be diagonal. Using (7.316) and (7.317) provides us with

$$[U]^T [c] [m]^{-1} [k] [U] = [U]^T [k] [m]^{-1} [c] [U] \tag{7.585}$$

which reduces to (7.584).

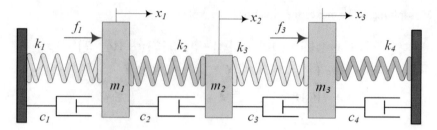

Fig. 7.40 A *three DOF* damped vibrating system

Example 274 Decoupling a damped system with proportional $[c]$. It is a practical example of a damped system with proportional damping to become decoupled.

Figure 7.40 illustrates a three DOF damped vibrating system with the following equations of motion

$$[m]\,\ddot{\mathbf{x}} + [c]\,\dot{\mathbf{x}} + [k]\,\mathbf{x} = \mathbf{F} = \begin{bmatrix} f_1 \\ 0 \\ f_3 \end{bmatrix} \tag{7.586}$$

where

$$[m] = \begin{bmatrix} m & 0 & 0 \\ 0 & m & 0 \\ 0 & 0 & m \end{bmatrix} \tag{7.587}$$

$$[k] = \begin{bmatrix} k_1 + k_2 & -k_2 & 0 \\ -k_2 & k_2 + k_3 & -k_3 \\ 0 & -k_3 & k_3 + k_4 \end{bmatrix} \tag{7.588}$$

$$[c] = \begin{bmatrix} c_1 + c_2 & -c_2 & 0 \\ -c_2 & c_2 + c_3 & -c_3 \\ 0 & -c_3 & c_3 + c_4 \end{bmatrix} \tag{7.589}$$

If the damping is supposed to be a linear combination of the mass and stiffness matrices, then we must have:

$$\begin{aligned} [c] &= a\,[m] + b\,[k] \\ &= \begin{bmatrix} am + bk_1 + bk_2 & -bk_2 & 0 \\ -bk_2 & am + bk_2 + bk_3 & -bk_3 \\ 0 & -bk_3 & am + bk_3 + bk_4 \end{bmatrix} \end{aligned} \tag{7.590}$$

Comparing (7.589) and (7.590) indicates that it is not possible to have a damping matrix that satisfies the proportionality condition (7.572) for $a \neq 0$. However, if we choose $a = 0$, then the proportionality condition will be satisfied if the damping is selected proportional to stiffness.

Let us assume that

$$m_1 = 2\,\mathrm{kg} \qquad m_2 = 1\,\mathrm{kg} \qquad m_3 = 2\,\mathrm{kg} \tag{7.591}$$

$$k_1 = k_2 = k_3 = k_4 = k = 1000\,\mathrm{N/m} \tag{7.592}$$

$$c_1 = c_2 = c_3 = c_4 = 5\,\mathrm{N\,s/m} \tag{7.593}$$

and therefore,

$$[m] = \begin{bmatrix} 2 & 0 & 0 \\ 0 & 1 & 0 \\ 0 & 0 & 2 \end{bmatrix} \tag{7.594}$$

$$[k] = \begin{bmatrix} k_1 + k_2 & -k_2 & 0 \\ -k_2 & k_2 + k_3 & -k_3 \\ 0 & -k_3 & k_3 + k_4 \end{bmatrix} = \frac{1}{1000} \begin{bmatrix} 2 & -1 & 0 \\ -1 & 2 & -1 \\ 0 & -1 & 2 \end{bmatrix} \tag{7.595}$$

$$[c] = \begin{bmatrix} c_1 + c_2 & -c_2 & 0 \\ -c_2 & c_2 + c_3 & -c_3 \\ 0 & -c_3 & c_3 + c_4 \end{bmatrix} = \begin{bmatrix} 10 & -5 & 0 \\ -5 & 10 & -5 \\ 0 & -5 & 10 \end{bmatrix} \tag{7.596}$$

The damping matrix is proportional to stiffness matrix.

$$[c] = 0.005\,[k] \tag{7.597}$$

The mode shapes and the modal matrix of the system are:

$$\mathbf{u}_1 = \begin{bmatrix} 1 \\ \sqrt{2} \\ 1 \end{bmatrix} \qquad \mathbf{u}_2 = \begin{bmatrix} -1 \\ 0 \\ 1 \end{bmatrix} \qquad \mathbf{u}_3 = \begin{bmatrix} 1 \\ -\sqrt{2} \\ 1 \end{bmatrix} \tag{7.598}$$

$$[U] = \begin{bmatrix} \mathbf{u}_1 & \mathbf{u}_2 & \mathbf{u}_3 \end{bmatrix} = \begin{bmatrix} 1 & -1 & 1 \\ \sqrt{2} & 0 & -\sqrt{2} \\ 1 & 1 & 1 \end{bmatrix} \tag{7.599}$$

Therefore, the principal coordinates $\mathbf{p} = [U]^{-1}\mathbf{x}$ make the equations of motion decoupled.

$$
\begin{bmatrix} p_1 \\ p_2 \\ p_3 \end{bmatrix} = \begin{bmatrix} 1 & -1 & 1 \\ \sqrt{2} & 0 & -\sqrt{2} \\ 1 & 1 & 1 \end{bmatrix}^{-1} \begin{bmatrix} x_1 \\ x_2 \\ x_3 \end{bmatrix} = \frac{1}{4} \begin{bmatrix} x_1 + x_3 + \sqrt{2}x_2 \\ 2x_3 - 2x_1 \\ x_1 + x_3 - \sqrt{2}x_2 \end{bmatrix} \tag{7.600}
$$

$$
[m'] = [U]^T [m] [U] = \begin{bmatrix} 4m & 0 & 0 \\ 0 & 2m & 0 \\ 0 & 0 & 4m \end{bmatrix} \tag{7.601}
$$

$$
[k'] = [U]^T [k] [U] = \begin{bmatrix} -4k\left(\sqrt{2}-2\right) & 0 & 0 \\ 0 & 4k & 0 \\ 0 & 0 & 4k\left(\sqrt{2}+2\right) \end{bmatrix} \tag{7.602}
$$

$$
[c'] = [U]^T [c] [U] = 0.005 \begin{bmatrix} -4k\left(\sqrt{2}-2\right) & 0 & 0 \\ 0 & 4k & 0 \\ 0 & 0 & 4k\left(\sqrt{2}+2\right) \end{bmatrix} \tag{7.603}
$$

The $[A]$-matrix of the system will be:

$$
[A] = [m']^{-1} [k'] = \frac{k}{m} \begin{bmatrix} -\left(\sqrt{2}-2\right) & 0 & 0 \\ 0 & 2 & 0 \\ 0 & 0 & \left(\sqrt{2}+2\right) \end{bmatrix} \tag{7.604}
$$

that provides us with the following natural frequencies and mode shapes in principal coordinates $\mathbf{p}$.

$$
\omega_1 = \sqrt{2 - \sqrt{2}}\sqrt{\frac{k}{m}} \qquad \omega_2 = \sqrt{2}\sqrt{\frac{k}{m}} \qquad \omega_3 = \sqrt{2 + \sqrt{2}}\sqrt{\frac{k}{m}} \tag{7.605}
$$

$$
\mathbf{u}_1 = \begin{bmatrix} 1 \\ 0 \\ 0 \end{bmatrix} \qquad \mathbf{u}_2 = \begin{bmatrix} 0 \\ 1 \\ 0 \end{bmatrix} \qquad \mathbf{u}_3 = \begin{bmatrix} 0 \\ 0 \\ 1 \end{bmatrix} \tag{7.606}
$$

The decoupled equations of motion of the system in principal coordinates are:

$$
[m']\ddot{\mathbf{p}} + [c']\dot{\mathbf{p}} + [k']\mathbf{p} = \mathbf{P} = \begin{bmatrix} \dfrac{1}{4}f_1 + \dfrac{1}{4}f_3 \\ \dfrac{1}{2}f_3 - \dfrac{1}{2}f_1 \\ \dfrac{1}{4}f_1 + \dfrac{1}{4}f_3 \end{bmatrix} \tag{7.607}
$$

where

$$\mathbf{P} = [U]^T \mathbf{F} \qquad (7.608)$$

Example 275 ★ Similarity of matrices and diagonalizability. Two matrices with the same eigenvalues and eigenvectors are called similar. The modal matrix is the tool to make smaller matrices.

Two matrices $[\alpha]$ and $[\beta]$ are called similar if there exist a nonsingular matrix $[U]$ such that:

$$[\beta] = [U]^T [\alpha][U] \qquad (7.609)$$

Similar matrices have the same eigenvalues and eigenvectors. A matrix is nonsingular if it is square and its determinant is not zero. The product $[U]^T [\alpha][U]$ is called a similarity transformation on $[\alpha]$.

A square matrix $[\alpha]$ is called diagonalizable when $[\alpha]$ is similar to a diagonal matrix. A complete set of eigenvectors of an $n \times n$ matrix $[\alpha]$ is any set of n linearly independent eigenvectors of $[\alpha]$. Matrices that do not have complete sets of eigenvectors may be called deficient or defective matrices. An $n \times n$ matrix $[\alpha]$ is diagonalizable if and only if it has a complete set of eigenvectors. The modal matrix $[U]$ of $[\alpha]$ diagonalizes $[\alpha]$ and provides us with a diagonal matrix $[\beta]$. The diagonal elements of $[\beta]$ are the eigenvalues λ_i of $[\alpha]$.

$$[\beta] = \begin{bmatrix} \lambda_1 & 0 & \cdots & 0 \\ 0 & \lambda_2 & \cdots & 0 \\ \vdots & \vdots & \ddots & \vdots \\ 0 & 0 & \cdots & \lambda_n \end{bmatrix} \qquad (7.610)$$

Example 276 Polar coordinate. The equations of motion of a planar or spatial vibrating system can be transformed to any orthogonal coordinate system for simpler forms of the equations. The polar coordinate system is the simplest alternative to try for planar vibrating system. Figure 7.41 illustrates a mass m that is supported by two similar and perpendicular springs with stiffness k. Let us express the dynamics of the system in polar coordinates (r, θ). The force system of m when it is out of equilibrium is:

$$F_r = -kr \qquad F_\theta = -kr\theta \qquad (7.611)$$

The equations of motion are:

$$F_r = ma_r = m\left(\ddot{r} - r\dot{\theta}^2\right) \qquad (7.612)$$

$$F_r = ma_\theta = m\left(r\ddot{\theta} + 2\dot{r}\dot{\theta}\right) \qquad (7.613)$$

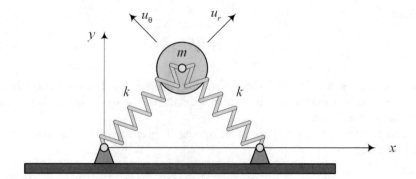

Fig. 7.41 A mass m supported by two similar and perpendicular springs

Substituting (7.611) into (7.612) and (7.613) yields

$$\ddot{r} - r\dot{\theta}^2 + \frac{k}{m}r = 0 \tag{7.614}$$

$$r\ddot{\theta} + 2\dot{r}\dot{\theta} + \frac{k}{m}r\theta = 0 \tag{7.615}$$

Linearizing the equations provides us with a set of two uncoupled similar equations,

$$\ddot{r} + \frac{k}{m}r = 0 \tag{7.616}$$

$$\ddot{\theta} + \frac{k}{m}\theta = 0 \tag{7.617}$$

with equal natural frequencies

$$\omega_n = \sqrt{\frac{k}{m}} \tag{7.618}$$

7.3   Harmonic Excitation

The general equations of motion of a linear n degree of freedom (DOF) vibrating system with a single frequency harmonic excitation is:

$$[m]\ddot{\mathbf{x}} + [c]\dot{\mathbf{x}} + [k]\mathbf{x} = \mathbf{F}_1 \sin \omega t + \mathbf{F}_2 \cos \omega t \tag{7.619}$$

where $\mathbf{F}_1$ and $\mathbf{F}_2$ are the external acting forces on the system. It can also be expressed in the expanded form.

$$\sum_{j=1}^{n} m_{ij}\ddot{x}_j + \sum_{j=1}^{n} c_{ij}\dot{x}_j + \sum_{j=1}^{n} k_{ij}x_j = F_{1i} \sin\omega t + F_{2i} \cos\omega t \qquad (7.620)$$

$$i = 1, 2, 3, \cdots, n$$

The steady-state response of the system is:

$$\mathbf{x} = \mathbf{A} \sin\omega t + \mathbf{B} \cos\omega t \qquad (7.621)$$

or

$$x_i = A_i \sin\omega t + B_i \cos\omega t \qquad (7.622)$$

$$= X_i \sin\left(\omega t - \varphi_i\right) \qquad (7.623)$$

where

$$\begin{bmatrix} \mathbf{A} \\ \mathbf{B} \end{bmatrix} = \begin{bmatrix} [k] - \omega^2 [m] & -\omega [c] \\ \omega [c] & [k] - \omega^2 [m] \end{bmatrix}^{-1} \begin{bmatrix} \mathbf{F}_1 \\ \mathbf{F}_2 \end{bmatrix} \qquad (7.624)$$

and the amplitude and phase of the frequency responses are:

$$X_i(\omega) = \sqrt{A_i^2 + B_i^2} \qquad i = 1, 2, \cdots, n \qquad (7.625)$$

$$\tan\varphi_i = \frac{-B_i}{A_i} \qquad (7.626)$$

Proof The steady-state solution of the n DOF system (7.619) is a set of n harmonic equations with the same frequency as the excitation

$$\mathbf{x} = \mathbf{A} \sin\omega t + \mathbf{B} \cos\omega t \qquad (7.627)$$

where $\mathbf{A}$ and $\mathbf{B}$ are $n \times 1$ coefficient matrices. We substitute the solution (7.627) into the equations of motion (7.619) and balance the coefficients of $\cos\omega t$ and $\sin\omega t$.

$$- \omega^2 [m] (\mathbf{A} \sin\omega t + \mathbf{B} \cos\omega t) + \omega [c] (\mathbf{A} \cos\omega t - \mathbf{B} \sin\omega t)$$

$$+ [k] (\mathbf{A} \sin\omega t + \mathbf{B} \cos\omega t) = \mathbf{F}_1 \sin\omega t + \mathbf{F}_2 \cos\omega t \qquad (7.628)$$

$$- \omega^2 [m] \mathbf{A} - \omega [c] \mathbf{B} + [k] \mathbf{A} = \mathbf{F}_1 \qquad (7.629)$$

$$-\omega^2 [m] \mathbf{B} + \omega [c] \mathbf{A} + [k] \mathbf{B} = \mathbf{F}_2 \qquad (7.630)$$

These equations provide us with a set of $2n$ algebraic equations to determine the coefficients $\mathbf{A}$ and $\mathbf{B}$:

$$\begin{bmatrix} [k] - \omega^2 [m] & -\omega [c] \\ \omega [c] & [k] - \omega^2 [m] \end{bmatrix} \begin{bmatrix} \mathbf{A} \\ \mathbf{B} \end{bmatrix} = \begin{bmatrix} \mathbf{F}_1 \\ \mathbf{F}_2 \end{bmatrix} \tag{7.631}$$

where

$$\begin{bmatrix} \mathbf{A} \\ \mathbf{B} \end{bmatrix} = \begin{bmatrix} [k] - \omega^2 [m] & -\omega [c] \\ \omega [c] & [k] - \omega^2 [m] \end{bmatrix}^{-1} \begin{bmatrix} \mathbf{F}_1 \\ \mathbf{F}_2 \end{bmatrix} \tag{7.632}$$

Having the coefficients matrices $\mathbf{A}$ and $\mathbf{B}$, the steady- state value of the coordinate x_i would be (7.622) or (7.623). Equating (7.622) and (7.623),

$$A_i \sin \omega t + B_i \cos \omega t = X_i \cos \varphi_i \sin \omega t - X_i \sin \varphi_i \cos \omega t \tag{7.633}$$

we can calculate the amplitude X_i and phase φ_i of every coordinate x_i.

$$X_i (\omega) = \sqrt{A_i^2 + B_i^2} \qquad i = 1, 2, \cdots, n \tag{7.634}$$

$$\tan \varphi_i = \frac{-B_i}{A_i} \qquad i = 1, 2, \cdots, n \tag{7.635}$$

The frequency response X_i is a function of the excitation frequency. The level of X_i at each frequency depends on the damping of the system. In case of no damping, all X_i would approach infinity at n natural frequencies of the system.

Depending on the type of excitation, each element of $\mathbf{F}_1$ and $\mathbf{F}_2$ may be a constant or proportional to ω, ω^2, ω^3, $\cdots$. ∎

Example 277 Forced vibration of a two DOF system and dimensionless frequency response. It is a detailed example to show how the frequency response of multi DOF will be determined and how they look like.

Figure 7.42 illustrates a forced excited two DOF vibrating system. The kinetic and potential energies and the dissipative function of the system are:

$$K = \frac{1}{2} m_1 \dot{x}_1^2 + \frac{1}{2} m_2 \dot{x}_2^2 \tag{7.636}$$

$$D = \frac{1}{2} c_1 \dot{x}_1^2 + \frac{1}{2} c_2 (\dot{x}_1 - \dot{x}_2)^2 + \frac{1}{2} c_3 \dot{x}_2^2 \tag{7.637}$$

$$P = \frac{1}{2} k_1 x_1^2 + \frac{1}{2} k_2 (x_1 - x_2)^2 + \frac{1}{2} k_3 x_2^2 \tag{7.638}$$

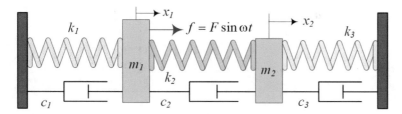

Fig. 7.42 A two DOF system under harmonic forced excitation

Employing the Lagrange equation

$$\frac{d}{dt}\left(\frac{\partial K}{\partial \dot{x}_i}\right) - \frac{\partial K}{\partial x_i} + \frac{\partial D}{\partial \dot{x}_i} + \frac{\partial P}{\partial x_i} = f_i \qquad i = 1, 2, \cdots n \tag{7.639}$$

the equations of motion will be calculated.

$$m_1\ddot{x}_1 + c_1\dot{x}_1 + c_2\,(\dot{x}_1 - \dot{x}_2) + k_1 x_1 + k_2\,(x_1 - x_2) = F \sin \omega t \tag{7.640}$$

$$m_2\ddot{x}_2 - c_2\,(\dot{x}_1 - \dot{x}_2) + c_3\dot{x}_2 - k_2\,(x_1 - x_2) + k_3 x_2 = 0 \tag{7.641}$$

To find the frequency response of the system, let us rearrange the equations in a matrix form:

$$\begin{bmatrix} m_1 & 0 \\ 0 & m_2 \end{bmatrix}\begin{bmatrix} \ddot{x}_1 \\ \ddot{x}_2 \end{bmatrix} + \begin{bmatrix} c_1 + c_2 & -c_2 \\ -c_2 & c_2 + c_3 \end{bmatrix}\begin{bmatrix} \dot{x}_1 \\ \dot{x}_2 \end{bmatrix}$$

$$+ \begin{bmatrix} k_1 + k_2 & -k_2 \\ -k_2 & k_2 + k_3 \end{bmatrix}\begin{bmatrix} x_1 \\ x_2 \end{bmatrix} = \begin{bmatrix} F \sin \omega t \\ 0 \end{bmatrix} \tag{7.642}$$

and substitute a harmonic solution

$$\begin{bmatrix} x_1 \\ x_2 \end{bmatrix} = \begin{bmatrix} A_1 \\ A_2 \end{bmatrix} \sin \omega t + \begin{bmatrix} B_1 \\ B_2 \end{bmatrix} \cos \omega t \tag{7.643}$$

to calculate the submatrices of the coefficient matrix of (7.632).

$$[k] - \omega^2\,[m] = \begin{bmatrix} -m_1\omega^2 + k_1 + k_2 & -k_2 \\ -k_2 & -m_2\omega^2 + k_2 + k_3 \end{bmatrix} \tag{7.644}$$

$$\omega\,[c] = \begin{bmatrix} \omega\,(c_1 + c_2) & -\omega c_2 \\ -\omega c_2 & \omega\,(c_2 + c_3) \end{bmatrix} \tag{7.645}$$

Therefore, the set of equations (7.632) to calculate the coefficients **A** and **B** are:

$$\begin{bmatrix} a_{11} & -k_2 & a_{13} & c_2\omega \\ -k_2 & a_{22} & c_2\omega & a_{24} \\ a_{31} & -c_2\omega & a_{33} & -k_2 \\ -c_2\omega & a_{42} & -k_2 & a_{44} \end{bmatrix} \begin{bmatrix} A_1 \\ A_2 \\ B_1 \\ B_2 \end{bmatrix} = \begin{bmatrix} F \\ 0 \\ 0 \\ 0 \end{bmatrix} \tag{7.646}$$

$$a_{11} = a_{33} = k_1 + k_2 - m_1\omega^2 \tag{7.647}$$

$$a_{22} = a_{44} = k_2 + k_3 - m_2\omega^2 \tag{7.648}$$

$$a_{31} = -a_{13} = \omega(c_1 + c_2) \tag{7.649}$$

$$a_{42} = -a_{24} = \omega(c_2 + c_3) \tag{7.650}$$

where

$$\begin{bmatrix} A_1 \\ A_2 \\ B_1 \\ B_2 \end{bmatrix} = \begin{bmatrix} a_{11} & -k_2 & a_{13} & c_2\omega \\ -k_2 & a_{22} & c_2\omega & a_{24} \\ a_{31} & -c_2\omega & a_{33} & -k_2 \\ -c_2\omega & a_{42} & -k_2 & a_{44} \end{bmatrix}^{-1} \begin{bmatrix} F \\ 0 \\ 0 \\ 0 \end{bmatrix} \tag{7.651}$$

$$X_1 = \sqrt{A_1^2 + B_1^2} \qquad X_2 = \sqrt{A_2^2 + B_2^2} \tag{7.652}$$

To calculate $\mathbf{A}$ and $\mathbf{B}$ and plot the frequency response curves, we need to examine a numerical example. Let us calculate the frequency response of the system for this set of sample data.

$$m_1 = 30\,\text{kg} \qquad m_2 = 20\,\text{kg}$$

$$c_1 = 10\,\text{N s}/\text{m} \qquad c_2 = 30\,\text{N s}/\text{m} \qquad c_3 = 50\,\text{N s}/\text{m}$$

$$k_1 = 6000\,\text{N}/\text{m} \qquad k_2 = 2000\,\text{N}/\text{m} \qquad k_3 = 1000\,\text{N}/\text{m}$$

$$F = 20\,\text{N} \tag{7.653}$$

Figure 7.43 illustrates the amplitudes X_1 and X_2 as functions of excitation frequency ω. Because the system has two DOF, it has two natural frequencies and, hence, two resonance zones in which the amplitudes are high. The natural frequencies of the system are:

$$\omega_1 = 10.39\,\text{rad}/\text{s} \qquad \omega_2 = 17.57\,\text{rad}/\text{s} \tag{7.654}$$

To have a better illustration, it is better to make the frequency response equations nondimensional. In case we apply constant forces on m_1 and m_2, then they will have static displacements X_{1s} and X_{2s}.

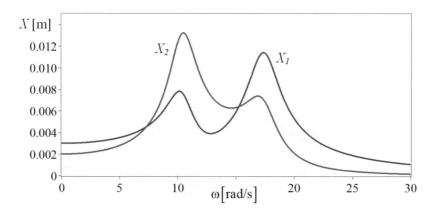

Fig. 7.43 Frequency responses X_1 and X_2

$$\begin{bmatrix} k_1 + k_2 & -k_2 \\ -k_2 & k_2 + k_3 \end{bmatrix} \begin{bmatrix} X_{1s} \\ X_{2s} \end{bmatrix} = \begin{bmatrix} F \\ 0 \end{bmatrix} \tag{7.655}$$

$$\begin{bmatrix} X_{1s} \\ X_{2s} \end{bmatrix} = \frac{F}{k_1 k_2 + k_1 k_3 + k_2 k_3} \begin{bmatrix} k_2 + k_3 \\ k_2 \end{bmatrix} \tag{7.656}$$

We can make the frequency responses (7.652) dimensionless by dividing frequency responses by static displacements.

$$\frac{X_1}{X_{1s}} = \frac{\sqrt{A_1^2 + B_1^2}}{X_{1s}} \qquad \frac{X_2}{X_{2s}} = \frac{\sqrt{A_2^2 + B_2^2}}{X_{2s}} \tag{7.657}$$

Figure 7.44 illustrates these frequency responses X_1/X_{1s} and X_2/X_{2s}. However, the best way to make frequency responses dimensionless is to divide all X_i by only the most important oscillating coordinate. Therefore, frequency response curves will be proportional to the actual displacements and, hence, then can be compared. Let us assume displacement of m_1 is more important to us and make the frequency responses (7.652) dimensionless by dividing frequency responses over X_{1s}.

$$\frac{X_1}{X_{1s}} = \frac{\sqrt{A_1^2 + B_1^2}}{X_{1s}} \qquad \frac{X_2}{X_{1s}} = \frac{\sqrt{A_2^2 + B_2^2}}{X_{1s}} \tag{7.658}$$

Figure 7.45 illustrates these frequency responses X_1/X_{1s} and X_2/X_{1s}.

Example 278 ★ Inverse of a large matrix. In analysis of multi *DOF* system with many degrees of freedom, we will need to calculate inverse of large matrices. The calculation effort of matrix inversion exponentially increases by the size of matrices.

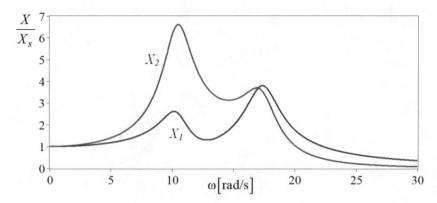

Fig. 7.44 Dimensionless frequency responses X_1/X_{1s} and X_2/X_{2s}

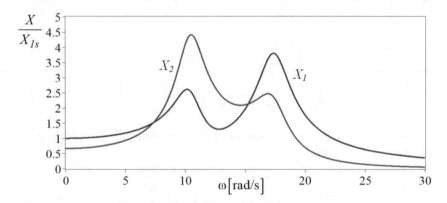

Fig. 7.45 Dimensionless frequency responses X_1/X_{1s} and X_2/X_{1s}

This example teaches a trick to break a large matrix into four smaller matrices and use them to calculate the inversion of the large matrix.

Let us assume $[T]$ to be a large matrix made of combination of four submatrices $[A]$, $[B]$, $[C]$, and $[D]$.

$$[T] = \begin{bmatrix} [A] & [B] \\ [C] & [D] \end{bmatrix} \tag{7.659}$$

Then, the inverse of T can be calculated by employing $[A]$, $[B]$, $[C]$, and $[D]$.

$$T^{-1} = \begin{bmatrix} A^{-1} + FCA^{-1} & -F \\ -E^{-1}CA^{-1} & E^{-1} \end{bmatrix} \tag{7.660}$$

$$[E] = D - CA^{-1}B \qquad [F] = A^{-1}BE^{-1} \tag{7.661}$$

As an example, let us examine the method on an arbitrary 2×2 matrix.

$$[T] = \begin{bmatrix} a & b \\ c & d \end{bmatrix} \tag{7.662}$$

Calculating $[E]$ and $[F]$

$$[E] = d - ca^{-1}b = \frac{1}{a}(ad - bc) \tag{7.663}$$

$$[F] = a^{-1}b\frac{a}{ad - bc} = \frac{b}{ad - bc} \tag{7.664}$$

we have

$$T^{-1} = \begin{bmatrix} \dfrac{d}{ad - bc} & -\dfrac{b}{ad - bc} \\ -\dfrac{c}{ad - bc} & \dfrac{a}{ad - bc} \end{bmatrix} \tag{7.665}$$

The frequency response of a multi DOF system is the outcome of equation (7.624).

$$\begin{bmatrix} \mathbf{A} \\ \mathbf{B} \end{bmatrix} = \begin{bmatrix} [k] - \omega^2 [m] & -\omega [c] \\ \omega [c] & [k] - \omega^2 [m] \end{bmatrix}^{-1} \begin{bmatrix} \mathbf{F}_1 \\ \mathbf{F}_2 \end{bmatrix} \tag{7.666}$$

Let us use the method of (7.660) to determine the inverse of the coefficient matrix

$$[T] = \begin{bmatrix} [A] & [B] \\ [C] & [D] \end{bmatrix} \tag{7.667}$$

where

$$[A] = [D] = [k] - \omega^2 [m] \tag{7.668}$$

$$[B] = -[C] = -\omega [c] \tag{7.669}$$

To determine T^{-1} we need

$$[E]^{-1} = \left[[k] - \omega^2 [m] + \omega^2 [c] \left[[k] - \omega^2 [m] \right]^{-1} [c] \right]^{-1} \tag{7.670}$$

$$[F] = A^{-1}BE^{-1} = -\omega \left[[k] - \omega^2 [m] \right]^{-1} [c] [E]^{-1} \tag{7.671}$$

and

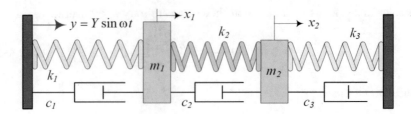

Fig. 7.46 A two *DOF* system with base displacement excitation

$$- E^{-1}CA^{-1} = -\omega E^{-1} [c] \left[[k] - \omega^2 [m] \right]^{-1} \tag{7.672}$$

$$A^{-1} + FCA^{-1} = \left[[k] - \omega^2 [m] \right]^{-1} \left([\mathbf{I}] - \omega [c] E^{-1}CA^{-1} \right) \tag{7.673}$$

Example 279 Base excited two *DOF* system. When the excitation frequency of a vibrating base is in a resonance zone of the system, the amplitude of oscillation of the other components of a multi *DOF* can reach to several times of the excitation amplitude. Here is the analysis of such system.

Figure 7.46 illustrates a two *DOF* system with base displacement excitation. Employing Lagrange method and the energy and dissipation functions of the system, we derive the equations of motion of the system.

$$K = \frac{1}{2} m_1 \dot{x}_1^2 + \frac{1}{2} m_2 \dot{x}_2^2 \tag{7.674}$$

$$P = \frac{1}{2} k_1 (y - x_1)^2 + \frac{1}{2} k_2 (x_1 - x_2)^2 + \frac{1}{2} k_3 x_2^2 \tag{7.675}$$

$$D = \frac{1}{2} c_1 (\dot{y} - \dot{x}_1)^2 + \frac{1}{2} c_2 (\dot{x}_1 - \dot{x}_2)^2 + \frac{1}{2} c_3 \dot{x}_2^2 \tag{7.676}$$

$$[m] \ddot{\mathbf{x}} + [c] \dot{\mathbf{x}} + [k] \mathbf{x} = \mathbf{F} \tag{7.677}$$

$$[m] = \begin{bmatrix} m_1 & 0 \\ 0 & m_2 \end{bmatrix} \qquad [k] = \begin{bmatrix} k_1 + k_2 & -k_2 \\ -k_2 & k_2 + k_3 \end{bmatrix}$$

$$\mathbf{F} = \begin{bmatrix} k_1 y + c_1 \dot{y} \\ 0 \end{bmatrix} \qquad [c] = \begin{bmatrix} c_1 + k_2 & -c_2 \\ -k_2 & c_2 + c_3 \end{bmatrix} \tag{7.678}$$

If *y* is a harmonically displacement excitation

$$y = Y \sin \omega t \tag{7.679}$$

then, we have:

$$\mathbf{F} = \begin{bmatrix} k_1 Y \sin \omega t + c_1 Y \omega \cos \omega t \\ 0 \end{bmatrix} \tag{7.680}$$

Substituting a harmonic solution

$$\begin{bmatrix} x_1 \\ x_2 \end{bmatrix} = \begin{bmatrix} A_1 \\ A_2 \end{bmatrix} \sin \omega t + \begin{bmatrix} B_1 \\ B_2 \end{bmatrix} \cos \omega t \tag{7.681}$$

provides us with a set of equations to calculate the coefficients $\mathbf{A}$ and $\mathbf{B}$.

$$\begin{bmatrix} A_1 \\ A_2 \\ B_1 \\ B_2 \end{bmatrix} = \begin{bmatrix} a_{11} & -k_2 & a_{13} & c_2\omega \\ -k_2 & a_{22} & c_2\omega & a_{24} \\ a_{31} & -c_2\omega & a_{33} & -k_2 \\ -c_2\omega & a_{42} & -k_2 & a_{44} \end{bmatrix}^{-1} \begin{bmatrix} k_1 Y \\ 0 \\ c_1\omega Y \\ 0 \end{bmatrix} \tag{7.682}$$

$$a_{11} = a_{33} = k_1 + k_2 - m_1\omega^2 \tag{7.683}$$

$$a_{22} = a_{44} = k_2 + k_3 - m_2\omega^2 \tag{7.684}$$

$$a_{31} = -a_{13} = \omega(c_1 + c_2) \tag{7.685}$$

$$a_{42} = -a_{24} = \omega(c_2 + c_3) \tag{7.686}$$

Let us examine the frequency response of the system for a set of numerical values.

$$m_2 = 1 \qquad m_1 = 1 \qquad c_1 = 1 \qquad c_2 = 1 \qquad c_3 = 1$$
$$k_1 = 100 \qquad k_2 = 100 \qquad k_3 = 100 \qquad Y = 1 \tag{7.687}$$

Figure 7.47 illustrates the frequency response of m_1 and m_2, and Fig. 7.48 illustrates the phase angle of X_1 and X_2.

When there is no damping, the set of equations of the system reduces to:

$$\begin{bmatrix} A_1 \\ A_2 \\ B_1 \\ B_2 \end{bmatrix} = \begin{bmatrix} a_{11} & -k_2 & 0 & 0 \\ -k_2 & a_{22} & 0 & 0 \\ 0 & 0 & a_{33} & -k_2 \\ 0 & 0 & -k_2 & a_{44} \end{bmatrix}^{-1} \begin{bmatrix} k_1 Y \\ 0 \\ 0 \\ 0 \end{bmatrix} \tag{7.688}$$

$$a_{11} = a_{33} = k_1 + k_2 - m_1\omega^2 \tag{7.689}$$

$$a_{22} = a_{44} = k_2 + k_3 - m_2\omega^2 \tag{7.690}$$

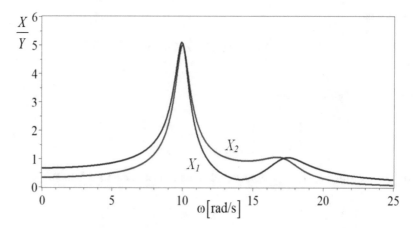

Fig. 7.47 The frequency response of m_1 and m_2

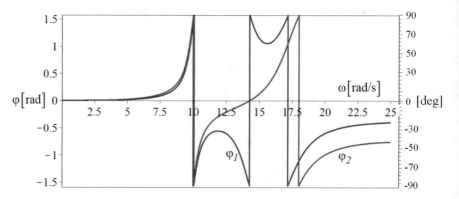

Fig. 7.48 The phase angle of X_1 and X_2

and the coefficients **A** and **B** will have simpler expressions.

$$
\begin{bmatrix} A_1 \\ A_2 \\ B_1 \\ B_2 \end{bmatrix} = \begin{bmatrix} -Yk_1 \dfrac{a_{22}}{k_2^2 - a_{11}a_{22}} \\ -Yk_1 \dfrac{k_2}{k_2^2 - a_{11}a_{22}} \\ 0 \\ 0 \end{bmatrix}
\tag{7.691}
$$

The frequency response and phase of the undamped system would be:

$$
X_1 = \frac{k_1 \left(k_2 + k_3 - m_2\omega^2\right)}{\left(k_1 + k_2 - m_1\omega^2\right)\left(k_2 + k_3 - m_2\omega^2\right) - k_2^2} Y
\tag{7.692}
$$

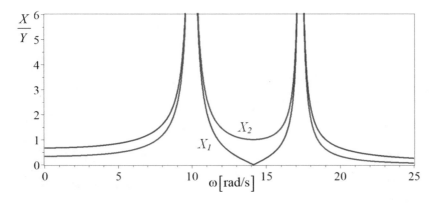

Fig. 7.49 The absolute value of frequency response of X_1 and X_2

$$X_2 = \frac{k_1 k_2}{\left(k_1 + k_2 - m_1\omega^2\right)\left(k_2 + k_3 - m_2\omega^2\right) - k_2^2}Y \qquad (7.693)$$

From a practical engineer, the value of amplitude is important, and hence, the frequency response curves are usually plotted for the absolute value of the equations. The dimensionless absolute value of the amplitudes is shown in Fig. 7.49. However, oscillation of the masses may be in phase or out of the phase with respect to the excitons. Figure 7.50 depicts the amplitudes and their sign. The phases of the motion are included in such illustration. The figures are all for the following data:

$$m_2 = m_1 = 1 \qquad k_1 = k_2 = k_3 = 100 \qquad Y = 1 \qquad (7.694)$$

$$X_1 = \frac{m\omega^2 - 2k}{\left(m\omega^2 - 2k\right)^2 - k^2}Y \qquad X_2 = \frac{k^2}{\left(m\omega^2 - 2k\right)^2 - k^2}Y \qquad (7.695)$$

The natural frequencies of the system for the given value of (7.694) are:

$$\omega_1 = \sqrt{Z_1 - \sqrt{Z_2}} \qquad \omega_2 = \sqrt{Z_1 + \sqrt{Z_2}} \qquad (7.696)$$

$$Z_1 = \frac{1}{2}\left(\frac{k_1 + k_2}{m_1} + \frac{k_2 + k_3}{m_2}\right) \qquad (7.697)$$

$$Z_2 = \frac{1}{4}\left(\frac{k_1 + k_2}{m_1} - \frac{k_2 + k_3}{m_2}\right)^2 + \frac{k_2^2}{m_1 m_2} \qquad (7.698)$$

$$\omega_1 = \sqrt{300} = 17.321 \qquad \omega_2 = \sqrt{100} = 10 \qquad (7.699)$$

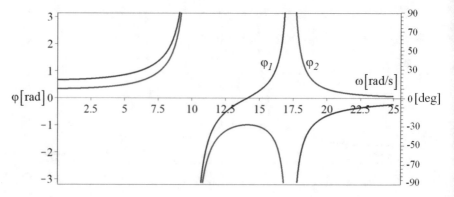

Fig. 7.50 The frequency response of X_1 and X_2

The frequency $\omega_0 = \sqrt{200} = 14.142$ is the one at which $X_1 = 0$.

$$\omega_0 = \sqrt{\frac{k_2 + k_3}{m_2}} \tag{7.700}$$

When $\omega = \omega_0$, regardless of the order of Y, the first mass m_1 will not move as long as there is no damping in the system.

$$\omega = \omega_0 \qquad X_1 = 0 \qquad X_2 = -\frac{k_1}{k_2} Y \tag{7.701}$$

At this frequency, the mass m_2 acts as vibration absorber of the system to make m_1 motionless.

Example 280 Superposition. When there are several excitations applied on a linear vibrating system, we can determine the response of the system to the excitations one by one and combine all responses to find the response of the system to all excitations combined.

Consider a two *DOF* vibrating system with the following equations of motion.

$$[m]\ddot{\mathbf{x}} + [c]\dot{\mathbf{x}} + [k]\mathbf{x} = \mathbf{F}_1 + \mathbf{F}_2 \tag{7.702}$$

$$[m] = \begin{bmatrix} m_1 & 0 \\ 0 & m_2 \end{bmatrix} \qquad [k] = \begin{bmatrix} k_1 + k_2 & -k_2 \\ -k_2 & k_2 + k_3 \end{bmatrix}$$

$$[c] - \begin{bmatrix} c_1 + k_2 & -c_2 \\ -k_2 & c_2 + c_3 \end{bmatrix} \tag{7.703}$$

$$\mathbf{F}_1 = \begin{bmatrix} k_1 y \\ 0 \end{bmatrix} \qquad \mathbf{F}_2 = \begin{bmatrix} c_1 \dot{y} \\ 0 \end{bmatrix} \tag{7.704}$$

If y is a harmonic function

$$y = Y \sin \omega t \tag{7.705}$$

then, we have:

$$\mathbf{F}_1 = \begin{bmatrix} k_1 Y \sin \omega t \\ 0 \end{bmatrix} \qquad \mathbf{F}_2 = \begin{bmatrix} c_1 Y \omega \cos \omega t \\ 0 \end{bmatrix} \tag{7.706}$$

Substituting a harmonic solution

$$\begin{bmatrix} x_{11} \\ x_{12} \end{bmatrix} = \begin{bmatrix} A_{11} \\ A_{12} \end{bmatrix} \sin \omega t + \begin{bmatrix} B_{11} \\ B_{12} \end{bmatrix} \cos \omega t \tag{7.707}$$

into the equations of motion with only one applied force $\mathbf{F}_1$

$$[m]\ddot{\mathbf{x}} + [c]\dot{\mathbf{x}} + [k]\mathbf{x} = \mathbf{F}_1 \tag{7.708}$$

provides us with a set of equations to calculate the coefficients $\mathbf{A}$ and $\mathbf{B}$

$$\begin{bmatrix} A_{11} \\ A_{12} \\ B_{11} \\ B_{12} \end{bmatrix} = \begin{bmatrix} a_{11} & -k_2 & a_{13} & c_2\omega \\ -k_2 & a_{22} & c_2\omega & a_{24} \\ a_{31} & -c_2\omega & a_{33} & -k_2 \\ -c_2\omega & a_{42} & -k_2 & a_{44} \end{bmatrix}^{-1} \begin{bmatrix} k_1 Y \\ 0 \\ 0 \\ 0 \end{bmatrix} \tag{7.709}$$

$$a_{11} = a_{33} = k_1 + k_2 - m_1\omega^2 \tag{7.710}$$
$$a_{22} = a_{44} = k_2 + k_3 - m_2\omega^2 \tag{7.711}$$
$$a_{31} = -a_{13} = \omega(c_1 + c_2) \tag{7.712}$$
$$a_{42} = -a_{24} = \omega(c_2 + c_3) \tag{7.713}$$

Let us show the frequency responses of the system to $\mathbf{F}_1$, by X_{11} and X_{12}.

$$X_{11} = \sqrt{A_{11} + B_{11}} \qquad X_{12} = \sqrt{A_{12} + B_{12}} \tag{7.714}$$

The plot of X_{11} and X_{12} are shown in Fig. 7.51 for the following date.

$$m_2 = 1 \qquad m_1 = 1 \qquad c_1 = 1 \qquad c_2 = 1 \qquad c_3 = 1$$
$$k_1 = 100 \qquad k_2 = 100 \qquad k_3 = 100 \qquad Y = 1 \tag{7.715}$$

In the next step, substituting a harmonic solution,

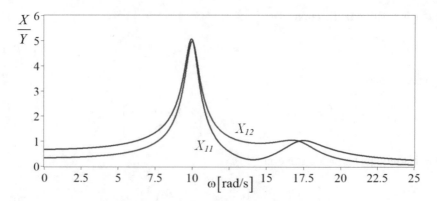

Fig. 7.51 The frequency responses X_{11} and X_{12}, of the system to $\mathbf{F}_1$

$$\begin{bmatrix} x_{21} \\ x_{22} \end{bmatrix} = \begin{bmatrix} A_{21} \\ A_{22} \end{bmatrix} \sin \omega t + \begin{bmatrix} B_{21} \\ B_{22} \end{bmatrix} \cos \omega t \qquad (7.716)$$

into the equation for $\mathbf{F}_2$,

$$[m]\,\ddot{\mathbf{x}} + [c]\,\dot{\mathbf{x}} + [k]\,\mathbf{x} = \mathbf{F}_2 \qquad (7.717)$$

provides us with a new set of equations to calculate the coefficients $\mathbf{A}$ and $\mathbf{B}$:

$$\begin{bmatrix} A_{21} \\ A_{22} \\ B_{21} \\ B_{22} \end{bmatrix} = \begin{bmatrix} a_{11} & -k_2 & a_{13} & c_2\omega \\ -k_2 & a_{22} & c_2\omega & a_{24} \\ a_{31} & -c_2\omega & a_{33} & -k_2 \\ -c_2\omega & a_{42} & -k_2 & a_{44} \end{bmatrix}^{-1} \begin{bmatrix} 0 \\ 0 \\ c_1\omega Y \\ 0 \end{bmatrix} \qquad (7.718)$$

where a_{ij} are the same as (7.710)–(7.713). Let us show the frequency responses of the system to $\mathbf{F}_1$, by X_{21} and X_{22}.

$$X_{21} = \sqrt{A_{21} + B_{21}} \qquad X_{22} = \sqrt{A_{22} + B_{22}} \qquad (7.719)$$

The plot of X_{21} and X_{22} are shown in Fig. 7.52 for the same data (7.715). The order of magnitude of X_{21} and X_{22} is roughly one tenth of X_{11} and X_{12}.

To determine the response of the system when both forces, $\mathbf{F}_1$ and $\mathbf{F}_2$, are applied, we can superpose the individual solutions

$$\begin{bmatrix} x_1 \\ x_2 \end{bmatrix} = \begin{bmatrix} x_{11} \\ x_{12} \end{bmatrix} + \begin{bmatrix} x_{21} \\ x_{22} \end{bmatrix}$$

$$= \begin{bmatrix} A_{11} + A_{21} \\ A_{12} + A_{22} \end{bmatrix} \sin \omega t + \begin{bmatrix} B_{11} + B_{21} \\ B_{12} + B_{22} \end{bmatrix} \cos \omega t$$

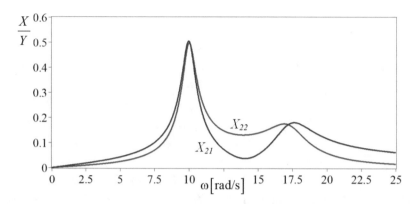

Fig. 7.52 The frequency responses X_{21} and X_{22}, of the system to $\mathbf{F}_2$

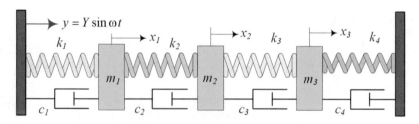

Fig. 7.53 A three DOF system with base displacement excitation

$$= \begin{bmatrix} A_1 \\ A_2 \end{bmatrix} + \sin \omega t + \begin{bmatrix} B_1 \\ B_2 \end{bmatrix} \cos \omega t \tag{7.720}$$

and determine the steady-state amplitude of the system.

$$X_1 = \sqrt{A_1 + B_1} \qquad X_2 = \sqrt{A_2 + B_2} \tag{7.721}$$

The steady-state amplitude of the superposed problem is shown in Fig. 7.47. The result is exactly the same as the solution of Example 279.

Example 281 Base excited three DOF system. To expand the application of the method to determine frequency response of vibrating systems under harmonic excitation, we apply the method on a three DOF system.

Figure 7.53 illustrates a three DOF system with a base displacement excitation. The equations of motion of the system are:

$$[m]\ddot{\mathbf{x}} + [c]\dot{\mathbf{x}} + [k]\mathbf{x} = \mathbf{F} \tag{7.722}$$

where

$$[k] = \begin{bmatrix} k_1 + k_2 & -k_2 & 0 \\ -k_2 & k_2 + k_3 & -k_3 \\ 0 & -k_3 & k_3 + k_4 \end{bmatrix} \tag{7.723}$$

$$[c] = \begin{bmatrix} c_1 + c_2 & -c_2 & 0 \\ -c_2 & c_2 + c_3 & -c_3 \\ 0 & -c_3 & c_3 + c_4 \end{bmatrix} \tag{7.724}$$

$$[m] = \begin{bmatrix} m_1 & 0 & 0 \\ 0 & m_2 & 0 \\ 0 & 0 & m_3 \end{bmatrix} \qquad F = \begin{bmatrix} k_1 y + c_1 \dot{y} \\ 0 \\ 0 \end{bmatrix} \tag{7.725}$$

Assume y is a harmonically displacement excitation

$$y = Y \sin \omega t \tag{7.726}$$

then,

$$F = \begin{bmatrix} k_1 Y \sin \omega t + c_1 Y \omega \cos \omega t \\ 0 \\ 0 \end{bmatrix} \tag{7.727}$$

Substituting a harmonic solution in the equations of motion,

$$\begin{bmatrix} x_1 \\ x_2 \\ x_3 \end{bmatrix} = \begin{bmatrix} A_1 \\ A_2 \\ A_3 \end{bmatrix} \sin \omega t + \begin{bmatrix} B_1 \\ B_2 \\ B_3 \end{bmatrix} \cos \omega t \tag{7.728}$$

provides us with a set of equations to calculate the coefficients $\mathbf{A}$ and $\mathbf{B}$ and frequency responses X_1, X_2, and X_3.

$$\begin{bmatrix} A_1 \\ A_2 \\ A_3 \\ B_1 \\ B_2 \\ B_3 \end{bmatrix} = \begin{bmatrix} [k] - \omega^2 [m] & -\omega [c] \\ \omega [c] & [k] - \omega^2 [m] \end{bmatrix}^{-1} \begin{bmatrix} k_1 Y \\ 0 \\ 0 \\ c_1 \omega Y \\ 0 \\ 0 \end{bmatrix} \tag{7.729}$$

$$X_1 = \sqrt{A_1 + B_1} \qquad X_2 = \sqrt{A_2 + B_2} \qquad X_2 = \sqrt{A_3 + B_3} \tag{7.730}$$

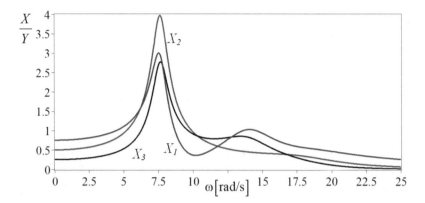

Fig. 7.54 The frequency response of three connected m_1, m_2, and m_3 with a harmonic base excitation

Let us examine the frequency response of the system for

$$c_1 = c_2 = c_3 = c_4 = 2 \qquad m_1 = m_2 = m_3 = 1$$
$$k_1 = k_2 = k_3 = k_4 = 100 \qquad Y = 1 \tag{7.731}$$

Figure 7.54 illustrates the frequency response of m_1, m_2, and m_3. The three resonance zones do not all show up for this set of data. To visualize the resonance zones, we must determine the natural frequencies and plot the frequency response of the undamped system. The natural frequencies of the system are:

$$\omega_1 = 7.6537 \qquad \omega_2 = 14.142 \qquad \omega_3 = 18.478 \tag{7.732}$$

and an illustration of the frequency response of the undamped system is shown in Fig. 7.55.

$$X_1 = -Yk \frac{3k^2 - 4km\omega^2 + m^2\omega^4}{-4k^3 + 10k^2m\omega^2 - 6km^2\omega^4 + m^3\omega^6} \tag{7.733}$$

$$X_2 = Y \frac{k^2}{2k^2 - 4km\omega^2 + m^2\omega^4} \tag{7.734}$$

$$X_3 = -Y \frac{k^3}{-4k^3 + 10k^2m\omega^2 - 6km^2\omega^4 + m^3\omega^6} \tag{7.735}$$

For the undamped system, there are two excitation frequencies at which m_1 stops vibrating. Setting the denominator of X_1 equal to zero, we determine these critical frequencies ω_{01} and ω_{02}.

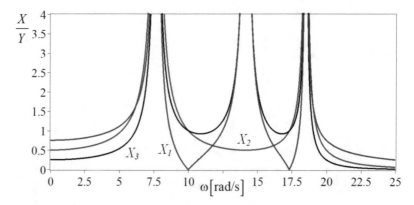

Fig. 7.55 The frequency response of the undamped system of three connected m_1, m_2 and m_3 with a harmonic base excitation

$$\omega_{01} = \sqrt{\frac{k}{m}} \qquad \omega_{02} = \sqrt{3\frac{k}{m}} \qquad\qquad (7.736)$$

Example 282 ★ Dynamic stiffness and receptance matrices. The control vision on vibration likes to treat systems with input and output relationship. In this view, the dynamics of the system will be collected as a coefficient of the input to produce output. In vibrations, the excitation is considered input, and the displacement is considered output such as $\mathbf{X} = [a\,(\omega)]\,\mathbf{F}$. The coefficient matrix $[a\,(\omega)]$ is called receptance matrix.

When there is no damping in a vibrating system, its equations of motion for the forced excited case can be written as follows.

$$[m]\,\ddot{\mathbf{x}} + [k]\,\mathbf{x} = \mathbf{F}\sin\omega t \qquad\qquad (7.737)$$

The steady-state response of the system would be a set of harmonic equations with the same frequency ω and amplitude $\mathbf{X}$ to be an $n \times 1$ matrix.

$$\mathbf{x} = \mathbf{X}\sin\omega t \qquad\qquad (7.738)$$

Substituting (7.738) into the equations of motion (7.737) provides us with:

$$\left([k] - \omega^2\,[m]\right)\mathbf{X} = \mathbf{F} \qquad\qquad (7.739)$$

Because this equation looks similar to a spring force balance equation, $F = kx$, the coefficient matrix of $\mathbf{X}$ is called the dynamic stiffness matrix and denoted by $[Z\,(\omega)]$.

$$[Z\,(\omega)] = [k] - \omega^2\,[m] \qquad\qquad (7.740)$$

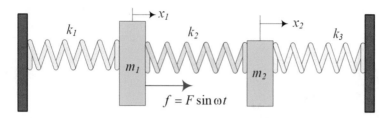

Fig. 7.56 A two DOF system under a harmonic force excitation

To determine the amplitude **X**, we need to find the inverse of the dynamic stiffness matrix. It is called the receptance matrix.

$$[a\,(\omega)] = \left[[k] - \omega^2\,[m]\right]^{-1} \tag{7.741}$$

$$\mathbf{X} = [a\,(\omega)]\,\mathbf{F} \tag{7.742}$$

The dynamic stiffness and receptance matrices are symmetric, which is an outcome of the reciprocity nature of linear systems. The reciprocity property states that the response at a coordinate x_i due to a force f applied at x_j is the same as the response at x_j due to the same force applied at x_i.

The determinant of dynamic stiffness matrix is the characteristic equation of the system.

$$f = \left|[k] - \omega^2\,[m]\right| \tag{7.743}$$

We may plot the characteristic function f versus ω to determine the natural frequencies of the system graphically. As an example, consider the system of Fig. 7.56. The characteristic function of the system for the following data

$$m_2 = 10\,\text{kg} \qquad m_1 = 20\,\text{kg} \qquad F = 2\,\text{N}$$
$$k_1 = 60\,\text{N}\,/\,\text{m} \qquad k_2 = 50\,\text{N}\,/\,\text{m} \qquad k_3 = 40\,\text{N}\,/\,\text{m} \tag{7.744}$$

is:

$$f = \left|[k] - \omega^2\,[m]\right| = m_1 m_2 \omega^4 + (-k_1 m_2 - k_2 m_1 - k_2 m_2 - k_3 m_1)\,\omega^2$$
$$+ (k_1 k_2 + k_1 k_3 + k_2 k_3) = 200\omega^4 - 2900\omega^2 + 7400 \tag{7.745}$$

The characteristic function f is shown in Fig. 7.57.
The receptance matrix of the system is:

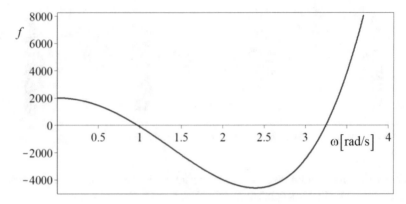

Fig. 7.57   Frequency function of a two *DOF* forced excited system

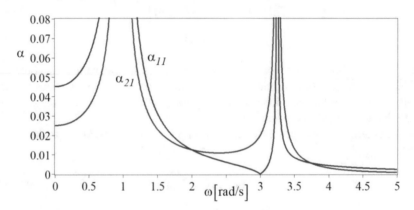

Fig. 7.58   Receptance function of a two *DOF* forced excited system

$$\left[[k] - \omega^2 [m]\right]^{-1} = \begin{bmatrix} -\dfrac{\omega^2 - 9}{20\omega^4 - 290\omega^2 + 740} & \dfrac{1}{4\omega^4 - 58\omega^2 + 148} \\ \dfrac{1}{4\omega^4 - 58\omega^2 + 148} & -\dfrac{2\omega^2/5 - 11/5}{4\omega^4 - 58\omega^2 + 148} \end{bmatrix}$$
(7.746)

When there is only one forcing term at point x_j, the element a_{ij} (ω) will be the frequency response at x_i due to the a unit force at x_j. In this example, there is only one force at x_1, and therefore a_{11} is the frequency response of m_1, and a_{21} is the frequency response of m_2 for a unit force at x_1. Figure 7.58 illustrates a_{11} (ω) and a_{21} (ω).

As a double check, let us calculate frequency response of the system by harmonic balance method. The equations of motion for $c_1 = c_2 = 0$ are as follows.

$$\begin{bmatrix} m_1 & 0 \\ 0 & m_2 \end{bmatrix} \begin{bmatrix} \ddot{x}_1 \\ \ddot{x}_2 \end{bmatrix} + \begin{bmatrix} k_1 + k_2 & -k_2 \\ -k_2 & k_2 + k_3 \end{bmatrix} \begin{bmatrix} x_1 \\ x_2 \end{bmatrix} = \begin{bmatrix} F \sin \omega t \\ 0 \end{bmatrix}$$
(7.747)

A harmonic solution for the system

$$\begin{bmatrix} x_1 \\ x_2 \end{bmatrix} = \begin{bmatrix} A_1 \\ A_2 \end{bmatrix} \sin \omega t + \begin{bmatrix} B_1 \\ B_2 \end{bmatrix} \cos \omega t \tag{7.748}$$

provides us with the following set of equations to calculate the coefficients **A** and **B**.

$$\begin{bmatrix} a_{11} & -k_2 & 0 & 0 \\ -k_2 & a_{22} & 0 & 0 \\ 0 & 0 & a_{33} & -k_2 \\ 0 & 0 & -k_2 & a_{44} \end{bmatrix} \begin{bmatrix} A_1 \\ A_2 \\ B_1 \\ B_2 \end{bmatrix} = \begin{bmatrix} F \\ 0 \\ 0 \\ 0 \end{bmatrix} \tag{7.749}$$

$$a_{11} = a_{33} = k_1 + k_2 - m_1 \omega^2 \qquad a_{22} = a_{44} = k_2 + k_3 - m_2 \omega^2 \tag{7.750}$$

The coefficients **A** and **B** for the data (7.744) are:

$$\begin{bmatrix} A_1 \\ A_2 \\ B_1 \\ B_2 \end{bmatrix} = \begin{bmatrix} -2\dfrac{\omega^2 - 9}{20\omega^4 - 290\omega^2 + 740} \\ \dfrac{1}{2\omega^4 - 29\omega^2 + 74} \\ 0 \\ 0 \end{bmatrix} \tag{7.751}$$

which are equal to $2a_{11}(\omega)$ and $2a_{21}(\omega)$ because of $F = 2\,\text{N}$.

Example 283 ★ Natural coordinate. Having a proportional damping matrix allows us to decouple equations of motion and solve them individually, each as a single *DOF* system.

The equation of motion and frequency response of the forced excitation of a single *DOF* system to a harmonic forcing term are as below.

$$m\ddot{x} + c\dot{x} + kx = F \sin \omega t \tag{7.752}$$

$$x = X \sin (\omega t - \varphi_x) \tag{7.753}$$

$$\frac{X}{F/k} = \frac{1}{\sqrt{(1 - r^2)^2 + (2\xi r)^2}} \tag{7.754}$$

$$r = \frac{\omega}{\omega_n} \qquad \xi = \frac{c}{2\sqrt{km}} \qquad \omega_n = \sqrt{\frac{k}{m}} \tag{7.755}$$

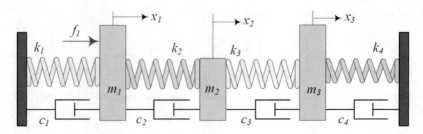

Fig. 7.59 A 3 *DOF* damped vibrating system

Figure 7.59 illustrates a three *DOF* system with proportional damped with the following equations of motion.

$$[m]\ddot{\mathbf{x}} + [c]\dot{\mathbf{x}} + [k]\mathbf{x} = \mathbf{F} = \begin{bmatrix} f_1 \\ 0 \\ 0 \end{bmatrix} = \begin{bmatrix} F\sin\omega t \\ 0 \\ 0 \end{bmatrix} \qquad (7.756)$$

$$[m] = \begin{bmatrix} m & 0 & 0 \\ 0 & m & 0 \\ 0 & 0 & m \end{bmatrix} \qquad (7.757)$$

$$[k] = \begin{bmatrix} k_1 + k_2 & -k_2 & 0 \\ -k_2 & k_2 + k_3 & -k_3 \\ 0 & -k_3 & k_3 + k_4 \end{bmatrix} \qquad (7.758)$$

$$[c] = b[k] = \begin{bmatrix} k_1 + k_2 & -k_2 & 0 \\ -k_2 & k_2 + k_3 & -k_3 \\ 0 & -k_3 & k_3 + k_4 \end{bmatrix} \qquad (7.759)$$

$$b = 0.005 \, \text{s} \qquad (7.760)$$

Using the modal matrix of the system,

$$[U] = \begin{bmatrix} \mathbf{u}_1 & \mathbf{u}_2 & \mathbf{u}_3 \end{bmatrix} = \begin{bmatrix} 1 & -1 & 1 \\ \sqrt{2} & 0 & -\sqrt{2} \\ 1 & 1 & 1 \end{bmatrix} \qquad (7.761)$$

we determine the principal coordinates of the system, $\mathbf{p} = [U]^{-1}\mathbf{x}$.

$$\begin{bmatrix} p_1 \\ p_2 \\ p_3 \end{bmatrix} = \begin{bmatrix} 1 & -1 & 1 \\ \sqrt{2} & 0 & -\sqrt{2} \\ 1 & 1 & 1 \end{bmatrix}^{-1} \begin{bmatrix} x_1 \\ x_2 \\ x_3 \end{bmatrix} = \frac{1}{4} \begin{bmatrix} x_1 + x_3 + \sqrt{2}x_2 \\ 2x_3 - 2x_1 \\ x_1 + x_3 - \sqrt{2}x_2 \end{bmatrix} \qquad (7.762)$$

The equations of motion are decoupled in the p coordinates:

$$[m']\ddot{\mathbf{p}} + [c']\dot{\mathbf{p}} + [k']\mathbf{p} = \mathbf{P} = \frac{1}{4}\begin{bmatrix} f_1 \\ -2f_1 \\ f_1 \end{bmatrix} = \frac{1}{4}\begin{bmatrix} F\sin\omega t \\ -2F\sin\omega t \\ F\sin\omega t \end{bmatrix} \qquad (7.763)$$

where

$$\mathbf{P} = [U]^T \mathbf{F} \qquad (7.764)$$

$$[m'] = [U]^T [m][U] = \begin{bmatrix} 4m & 0 & 0 \\ 0 & 2m & 0 \\ 0 & 0 & 4m \end{bmatrix} \qquad (7.765)$$

$$[k'] = [U]^T [k][U] = \begin{bmatrix} -4k\left(\sqrt{2}-2\right) & 0 & 0 \\ 0 & 4k & 0 \\ 0 & 0 & 4k\left(\sqrt{2}+2\right) \end{bmatrix} \qquad (7.766)$$

$$[c'] = [U]^T [c][U] = 0.005 \begin{bmatrix} -4k\left(\sqrt{2}-2\right) & 0 & 0 \\ 0 & 4k & 0 \\ 0 & 0 & 4k\left(\sqrt{2}+2\right) \end{bmatrix} \qquad (7.767)$$

and hence, the $[A]$-matrix of the system will be:

$$[A] = [m']^{-1}[k'] = \frac{k}{m}\begin{bmatrix} -\left(\sqrt{2}-2\right) & 0 & 0 \\ 0 & 2 & 0 \\ 0 & 0 & \left(\sqrt{2}+2\right) \end{bmatrix} \qquad (7.768)$$

It provides us with the following natural frequencies and mode shapes in the principal coordinates $\mathbf{p}$.

$$\omega_1 = \sqrt{2 - \sqrt{2}}\sqrt{\frac{k}{m}} \qquad \omega_2 = \sqrt{2}\sqrt{\frac{k}{m}} \qquad \omega_3 = \sqrt{2 + \sqrt{2}}\sqrt{\frac{k}{m}} \qquad (7.769)$$

$$\mathbf{u}_1 = \begin{bmatrix} 1 \\ 0 \\ 0 \end{bmatrix} \qquad \mathbf{u}_2 = \begin{bmatrix} 0 \\ 1 \\ 0 \end{bmatrix} \qquad \mathbf{u}_3 = \begin{bmatrix} 0 \\ 0 \\ 1 \end{bmatrix} \qquad (7.770)$$

The individual equations of motion of the system in principal coordinates are:

$$4m\ddot{p}_1 - 4bk\left(\sqrt{2} - 2\right)\dot{p}_1 - 4k\left(\sqrt{2} - 2\right)p_1 = \frac{1}{4}F\sin\omega t \qquad (7.771)$$

$$2m\ddot{p}_1 + 4bk\dot{p}_1 + 4kp_1 = -\frac{1}{2}F\sin\omega t \qquad (7.772)$$

$$4m\ddot{p}_1 + 4bk\left(\sqrt{2} + 2\right)\dot{p}_1 + 4k\left(\sqrt{2} + 2\right)p_1 = \frac{1}{4}F\sin\omega t \qquad (7.773)$$

The frequency responses of the equations are all similar to (7.753) and (7.754) with proper substitution of the parameters.

7.4 Summary

Systems with multi degree of freedom (DOF) introduce multi natural frequencies, mode shapes, and mode interaction. This chapter is to study the frequency response of multi DOF systems.

The general equations of motion of a linear n DOF vibrating system with a single frequency harmonic excitation is:

$$[m]\ddot{\mathbf{x}} + [c]\dot{\mathbf{x}} + [k]\mathbf{x} = \mathbf{F}_1\sin\omega t + \mathbf{F}_2\cos\omega t \qquad (7.774)$$

or in expanded form is:

$$\sum_{j=1}^{n} m_{ij}\ddot{x}_j + \sum_{j=1}^{n} c_{ij}\dot{x}_j + \sum_{j=1}^{n} k_{ij}x_j = F_{1i}\sin\omega t + F_{2i}\cos\omega t \qquad (7.775)$$

$$i = 1, 2, 3, \cdots, n$$

where $\mathbf{F}_1$ and $\mathbf{F}_2$ are the external acting forces on the system.

The steady-state response of the system is:

$$\mathbf{x} = \mathbf{A}\sin\omega t + \mathbf{B}\cos\omega t \qquad (7.776)$$

or

$$x_i = A_i\sin\omega t + B_i\cos\omega t = X_i\sin\left(\omega t - \varphi_i\right) \qquad (7.777)$$

where

$$\begin{bmatrix} \mathbf{A} \\ \mathbf{B} \end{bmatrix} = \begin{bmatrix} [k] - \omega^2[m] & -\omega[c] \\ \omega[c] & [k] - \omega^2[m] \end{bmatrix}^{-1} \begin{bmatrix} \mathbf{F}_1 \\ \mathbf{F}_2 \end{bmatrix} \qquad (7.778)$$

and the amplitude and phase of the frequency responses are:

$$X_i(\omega) = \sqrt{A_i^2 + B_i^2} \qquad i = 1, 2, \cdots, n \tag{7.779}$$

$$\tan \varphi_i = \frac{-B_i}{A_i} \tag{7.780}$$

Free and undamped vibrations of a system are the basic response of the system which express its natural behavior. Using a set of generalized coordinates, $\mathbf{x}$, a linear undamped-free system is governed by the following set of differential equations.

$$[m]\ddot{\mathbf{x}} + [k]\mathbf{x} = \mathbf{0} \tag{7.781}$$

$$\mathbf{x} = \begin{bmatrix} x_1 & x_2 & \cdots & x_n \end{bmatrix}^T \tag{7.782}$$

The natural frequencies of the system, ω_i, are solutions of the characteristic equation of the system

$$\det\left[[k] - \omega^2 [m] \right] = 0 \tag{7.783}$$

and the mode shape $\mathbf{u}_i$, corresponding to ω_i, is the solution of the eigenvector equation.

$$\left[[k] - \omega_i^2 [m] \right] \mathbf{u}_i = 0 \tag{7.784}$$

Mode shapes are orthogonal with respect to the mass $[m]$ and stiffness $[k]$ matrices.

$$\mathbf{u}_j^T [m]\mathbf{u}_i = 0 \qquad \mathbf{u}_j^T [k]\mathbf{u}_i = 0 \qquad i \neq j \tag{7.785}$$

Employing the orthogonality property of the mode shapes, we can determine the generalized mass m_i and stiffness k_i, associated to $\mathbf{u}_i$.

$$\mathbf{u}_i^T [m]\mathbf{u}_i = m_i \qquad \mathbf{u}_i^T [k]\mathbf{u}_i = k_i \tag{7.786}$$

The length of an eigenvector is not defined, and hence, there are many options to express $\mathbf{u}_i$. The most common expressions are: (1) normalization, (2) normal form, (3) high-unit, (4) first-unit, and (5) last-unit.

Any type of free, transient, or excited response of a linear vibrating system is dominated by its natural frequencies, mode shapes, and interaction of excitation frequencies. Determination of the natural frequencies ω_i and their associated mode shapes $\mathbf{u}_i$ is the first step in analysis of a multi DOF vibrating system. There exists at least one mode shape corresponding to each natural frequency. If an $n \times n$

matrix $[A]$ has n distinct eigenvalues, then there exist exactly n linearly independent eigenvectors, one associated with each eigenvalue.

The equations of motion of any undamped linear n DOF vibrating system in terms of a set of generalized coordinates $\mathbf{q}$ are:

$$[m]\ddot{\mathbf{q}} + [k]\mathbf{q} = \mathbf{Q} \tag{7.787}$$

$$\mathbf{q} = \begin{bmatrix} q_1 & q_2 & \cdots & q_n \end{bmatrix}^T \tag{7.788}$$

We can always change the set of generalized coordinates $\mathbf{q}$ to a set of *principal coordinates* $\mathbf{p}$:

$$\mathbf{q} = [U]\mathbf{p} \tag{7.789}$$

$$[U] = \begin{bmatrix} \mathbf{u}_1 & \mathbf{u}_2 & \mathbf{u}_2 & \cdots & \mathbf{u}_n \end{bmatrix} \tag{7.790}$$

and change the equations of motion to a set of decoupled equations.

$$[m']\ddot{\mathbf{p}} + [k']\mathbf{p} = \mathbf{P} \tag{7.791}$$

$$[m'] = [U]^T [m][U] \tag{7.792}$$

$$[k'] = [U]^T [k][U] \tag{7.793}$$

$$\mathbf{P} = [U]^T \mathbf{Q} \tag{7.794}$$

The new mass and stiffness matrices $[m']$ and $[k']$ are diagonal:

$$m'_{ij} = \begin{cases} m'_i & i = j \\ 0 & i \neq j \end{cases} \qquad k'_{ij} = \begin{cases} k'_i & i = j \\ 0 & i \neq j \end{cases} \tag{7.795}$$

and the equations of motion are decoupled.

$$m'_i \ddot{p}_i + k'_i p_i = F_i \qquad i = 1, 2, 3, \cdots, n \tag{7.796}$$

The square transformation matrix $[U]$ is the *modal matrix* of the system. Using the modal matrix $[U]$, the natural frequencies of the system can be found easier.

$$[m']^{-1}[k'] = [U]^T [A][U] = \begin{bmatrix} \omega_1^2 & 0 & \cdots & 0 \\ 0 & \omega_2^2 & \cdots & 0 \\ \vdots & \vdots & \ddots & \vdots \\ 0 & 0 & \cdots & \omega_n^2 \end{bmatrix} \tag{7.797}$$

7.5 Key Symbols

0	Zero vector, zero matrix
a, b, c, d	Constant parameters
a_i	Distance from mass center, constant coefficients
a_{ij}	Flexibility matrix element
$[a]$	$= [k]^{-1}$, flexibility matrix
b	Coefficient of proportional damping matrix
a	Constant shift coordinates
a, **b**	Real and imaginary parts of complex mode shapes
A, B	Coefficients of frequency response
$[A]$	$= [m]^{-1}[k]$, characteristic matrix, part of $[T]$
$[B]$	$= [A]^{-1} = [k]^{-1}[m]$, inverse characteristic matrix, part of $[T]$
$[B]$	Transformation coordinate matrix
c	Damping
$[c]$	Damping matrix
$[C]$	Part of $[T]$
C	Constant coefficient, constant of integration, amplitude
D	Dissipation function
$[D]$	Part of $[T]$
e	Exponential function
E	Mechanical energy
$[E]$	Part of $[T]$
f	Force, function, number of rigid modes, characteristic equation
F	Force amplitude, force magnitude, constant force
F	Forces
$[F]$	Part of $[T]$
g	Functions of displacement, gravitational acceleration
g	Gravitational acceleration vector
i	Counting index, unit of imaginary numbers
I	Mass moment
I, **[I]**	Identity matrix
k	Stiffness
k_{ij}	Stiffness matrix elements

k_s Sprung stiffness

k_u Unsprung stiffness

$[k]$ Stiffness matrix

$[k']$ Diagonal stiffness matrix

l Length

$\mathcal{L}$ Lagrangean

m Mass

K Kinetic energy

$[m]$ Mass matrix

$[m']$ Diagonal mass matrix

n Number of DOF, number of equations, order of a matrix

n' Rank of a matrix

p Principal coordinate

$\mathbf{p}$ Principal coordinates

P Potential energy

$\mathbf{P}$ Principal force

q Generalized coordinate, temporal function of x

$\mathbf{q}$ Generalized coordinates

$\mathbf{Q}$ Generalized force

r Frequency ratio, radial coordinate

t Time

$[T]$ A general large matrix

u Components of $\mathbf{u}$

$\mathbf{u}$ Eigenvector

$[U]$ Modal matrix

$v \equiv \dot{x}$ Velocity

$v_0 = \dot{x}_0$ Initial velocity

W Work

x Displacement coordinate, displacement variable

$\mathbf{x}$ Displacement coordinates

x_0 Initial displacement

X Displacement amplitude

$\mathbf{X}$ Displacement amplitudes

y Displacement excitation, displacement, relative displacement

y   Displacements
Y Amplitude of displacement excitation
z Variable, relative displacement
Z Short name of expressions

Greek

α Stiffness parameter, characteristic frequency, stiffness ratio
β Stiffness parameter, characteristic frequency
δ Virtual notation
δ Variation
ε Mass ratio, coupling mass element
ε Small parameter
λ Eigenvalue
η $= \dfrac{1}{\lambda}$ inverse eigenvalue, square frequency ratio
θ Angular coordinate, angular variable
τ Time constant

φ Phase angle
μ Stiffness coupling element
ω Excitation frequency, angular frequency
ω_0 Critical excitation frequency
ω_i, ω_n Natural frequency
Ω Characteristic frequency

Symbol

$\mathcal{L}$ Lagrangean
DOF Degree of freedom
tr   Trace
det   Determinant

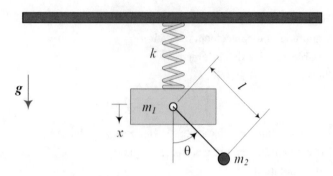

Fig. 7.60 A two DOF vibrating system

Exercises

1. Free vibrations.
   Use the figure and data of Problem 2.

   (a) Determine the free vibrations of the system if the initial conditions are:

$$x_1(0) = 0.05\,\text{m} \qquad x_2(0) = -0.05\,\text{m}$$

$$\dot{x}_1(0) = 0 \qquad\qquad \dot{x}_2(0) = -0$$

   (b) Adjust k_2 such that $\omega_2 = 2\omega_1$, if possible.
   (c) Adjust k_3 such that $\omega_2 = 2\omega_1$, if possible.
   (d) Adjust k_1 such that $\omega_2 = 2\omega_1$, if possible.
   (e) Adjust k_1, k_2, k_3 such that the given initial condition puts the system in one of the mode shapes, if possible.

2. Linearization of equations of motion.
   A two DOF vibrating system is shown in Fig. 7.60.

   (a) Determine the equations of motion.
   (b) Determine the linearized form of the equations of motion in matrix form.
   (c) Determine the natural frequencies and mode shapes.
   (d) Use the vertical distance y of m_2 with respect to m_1 and find the linearized equations of motion in generalized coordinates x, y.
   (e) Discuss the mass or stiffness couplings of the equations.
   (f) Determine the natural frequencies and mode shapes in the new generalized coordinates.
   (g) Determine the principal coordinates and decoupled equations of motion.

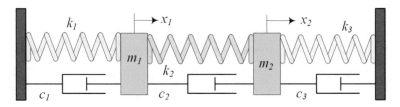

Fig. 7.61 A two DOF system

3. Different mode shape arrangements.
   A two DOF system is shown in Fig. 7.61. Assume

$$m_1 = 1\,\text{kg} \qquad m_2 = 2\,\text{kg}$$

$$k_1 = 100\,\text{N}\,/\,\text{m} \qquad k_2 = 110\,\text{N}\,/\,\text{m} \qquad k_3 = 120\,\text{N}\,/\,\text{m}$$

$$c_1 = 70\,\text{N}\,\text{s}\,/\,\text{m} \qquad c_2 = 60\,\text{N}\,\text{s}\,/\,\text{m} \qquad c_3 = 50\,\text{N}\,\text{s}\,/\,\text{m}$$

   and determine

   (a) The equations of motion.
   (b) Natural frequencies.
   (c) Mode shapes in mass normalized form.
   (d) Mode shapes in stiffness normalized form.
   (e) Mode shapes in high-unit form.
   (f) Mode shapes in first-unit form.
   (g) Mode shapes in last-unit form.
   (h) Number and position of vibration nodes.
   (i) Show the orthogonality of mode shapes and the mass matrix.
   (j) Show the orthogonality of mode shapes and the stiffness matrix.

4. ★ Number of nodes.
   Consider only translational vibrating systems. Is it true that a one DOF system has no node in absolute amplitude frequency response, a two DOF system may have one node, a three DOF system may have two nodes, and so on? Is it possible to prove the maximum number of nodes?

5. The equations of motion of three parallel pendulums.
   Figure 7.62 illustrates three coupled parallel pendulums.

   (a) Determine the equations of motion.
   (b) Linearize the equations of motion and show them in matrix form.
   (c) Discuss the static and dynamic couplings of the equations.
   (d) Determine the natural frequencies and mode shapes. Show the mode shapes using geometrical and physical illustrations.
   (e) Determine the natural coordinates of the undamped system and decouple the equations of motion.

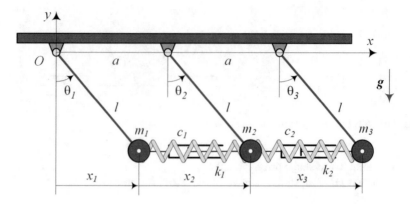

Fig. 7.62 Three coupled parallel pendulums

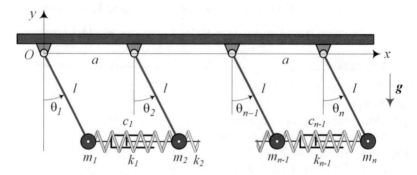

Fig. 7.63 *n* coupled parallel pendulums

6. ★ Many parallel pendulums.
   Figure 7.63 illustrates *n* coupled parallel pendulums.

   (a) Determine the equations of motion.
   (b) Linearize the equations of motion and show them in matrix form.
   (c) Discuss the static and dynamic couplings of the equations.
   (d) Determine the natural frequencies and mode shapes. Show the mode shapes
       using geometrical and physical illustrations.
   (e) Determine the natural coordinates of the undamped system and decouple the
       equations of motion.

7. Effect of attaching more vibration absorber on frequency response.
   Consider a base harmonic excitation

   $$y = Y \cos \omega t$$

   and the values of

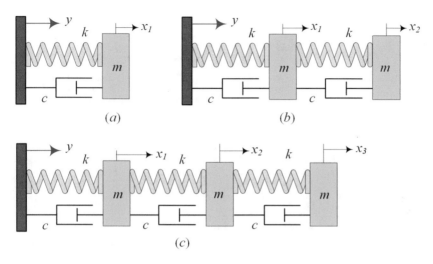

Fig. 7.64 Three base excited systems

$$Y = 1 \qquad m = 1 \qquad k = 1000 \qquad c = 10$$

(a) Determine the frequency response X_1 of the one DOF system of Fig. 7.64a.
(b) Determine the frequency response X_1 of the two DOF system of Fig. 7.64b.
(c) Determine the frequency response X_1 of the three DOF system of Fig. 7.64c.
(d) Plot the frequency response X_1 of the three systems of Fig. 7.64a,b,c on a graph and discuss the effect of increasing DOF.
(e) Describe your expectation of the frequency response X_1 by adding a fourth similar DOF.

8. Effect of more DOF on frequency response.
   Consider a base harmonic excitation

$$y = Y \cos \omega t$$

and the values of

$$Y = 1 \qquad m = 1 \qquad k = 1000 \qquad c = 10$$

(a) Determine the frequency response X_1 of the one DOF system of Fig. 7.65a.
(b) Determine the frequency response X_1 of the two DOF system of Fig. 7.65b.
(c) Determine the frequency response X_1 of the three DOF system of Fig. 7.65c.
(d) Plot the frequency response X_1 of the three systems of Fig. 7.65a,b,c on a graph and discuss the effect of increasing DOF.

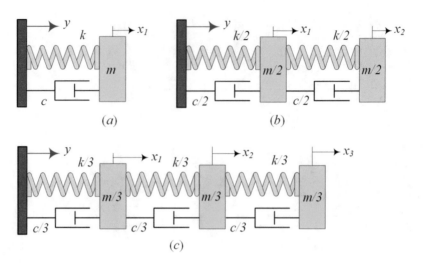

Fig. 7.65 Three base excited systems

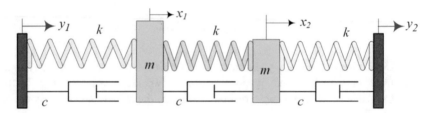

Fig. 7.66 A two DOF system with two base excitations y_1 and y_2

(e) Describe your expectation of the frequency response X_1 by adding a fourth similar DOF.

9. Frequency response and mode shapes.

Figure 7.66 illustrates a two DOF system with two base excitations y_1 and y_2. Assume $m = 1, k = 1$.

(a) Determine the natural frequencies of the system.
(b) Determine the mode shapes of the system and draw them.
(c) Determine and plot the frequency response of X_1 and X_2 for $c = 1$, $y_1 = 0.1 \cos \omega t$, $y_2 = 0$.
(d) At the first natural frequency, compare the ratio of X_1/X_2 and the first mode shape.
(e) At the second natural frequency, compare the ratio of X_1/X_2 and the first mode shape.
(f) At the first natural frequency, compare the ratio of X_1/X_2 and the second mode shape.

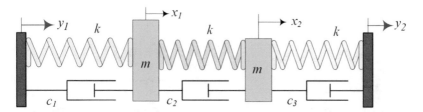

Fig. 7.67 A two DOF system with two base excitations y_1 and y_2 and unequal dampens

(g) At the second natural frequency, compare the ratio of X_1/X_2 and the second mode shape.

(h) Determine and plot the frequency response of X_1 and X_2 for $c = 1$, $y_1 = 0.1 \cos \omega t$, $y_2 = 0.1 \cos \omega t$.

(i) Determine and plot the frequency response of X_1 and X_2 for $c = 1$, $y_1 = 0.1 \cos \omega t$, $y_2 = -0.1 \cos \omega t$.

(j) Determine and plot the frequency response of X_1 and X_2 for $c = 1$, $y_1 = 0.1 \sin \omega t$, $y_2 = 0.1 \cos \omega t$.

(k) Determine and plot the frequency response of X_1 and X_2 for $c = 1$, $y_1 = 0.1 \cos \omega t$, $y_2 = 0.1 \sin \omega t$.

(l) Determine and plot the frequency response of X_1 and X_2 for $c = 1$, $y_1 = 0.1 \cos \omega t$, $y_2 = 0.2 \cos \omega t$.

(m) ★ Determine and plot the frequency response of X_1 and X_2 for $c = 1$, $y_1 = 0.1 \cos \omega t$, $y_2 = 0.1 \cos 2\omega t$.

10. Frequency response and mode shapes.

   Figure 7.67 illustrates a two DOF system with two base excitations y_1 and y_2 and unequal dampers. Assume $m = 1$, $k = 1$, $y_1 = 0.1 \cos \omega t$, $y_2 = 0.1 \cos \omega t$.

(a) Determine the frequency response of X_1 and X_2 for $c_1 = 1$, $c_2 = c_3 = 0$.

(b) Determine the frequency response of X_1 and X_2 for $c_2 = 1$, $c_1 = c_3 = 0$.

(c) Determine the frequency response of X_1 and X_2 for $c_3 = 1$, $c_2 = c_1 = 0$.

(d) Determine the frequency response of X_1 and X_2 for $c_1 = 1$, $c_2 = 1$, $c_3 = 0$.

(e) Determine the frequency response of X_1 and X_2 for $c_1 = 1$, $c_3 = 1$, $c_2 = 0$.

(f) Determine the frequency response of X_1 and X_2 for $c_3 = 1$, $c_2 = 1$, $c_1 = 0$.

(g) Determine the frequency response of X_1 and X_2 for $c_1 = 1$, $c_2 = 1$, $c_3 = 1$.

11. Zero natural frequency in mode of a train model.

   Figure 7.68 illustrates a simple model of a train.

(a) Determine the natural frequencies of the train.

(b) Determine and show the mode shapes of the train.

(c) Determine and show the frequency response of the train for $m = 1$, $k = 1$, $c = 0.1$, and $f = \sin \omega t$.

(d) Determine and show the frequency response of the train for $m = 1$, $k = 1$, $c = 0.1$, and $f = \cos \omega t$.

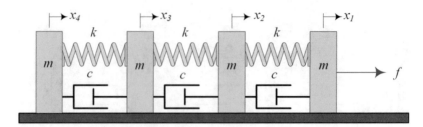

Fig. 7.68 A mathematical model of a train

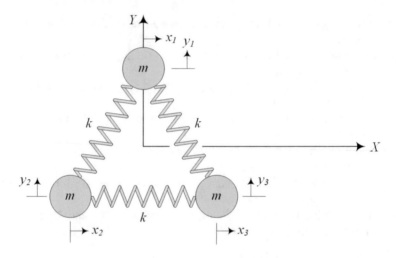

Fig. 7.69 Three masses at the corners of an equilateral triangle

(e) ★ Eliminate the fourth wagon to have a three wagon train. Determine the frequency response of the train for $m = 1$, $k = 1$, $c = 0.1$, and $f = \cos \omega t$. Compare the results with part d and discuss the effect of the fourth wagon.

12. ★ Zero natural frequency of a triangle.
   Figure 7.69 illustrates three masses at the corners of an equilateral triangle.

   (a) Determine the equations of motion of the system.
   (b) Determine the natural frequencies of the system and discuss the zero natural frequencies.
   (c) Determine and show the mode shapes of the system and discuss the rigid modes.

13. Rigid mode of a moveable pendulum.
   Figure 7.70 illustrates a pendulum hanging from a cart with heavy disc wheels.

   (a) Determine the equations of motion and set the linearized equations in matrix form using x_1 and θ as the generalized coordinates.
   (b) Determine the natural frequencies and mode shapes in x_1 and θ.

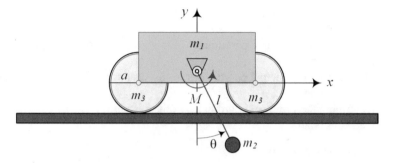

Fig. 7.70 A cart with a hanging pendulum

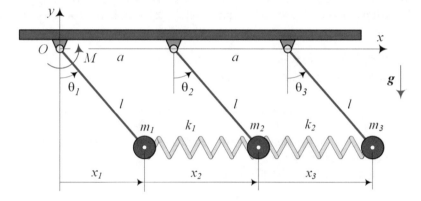

Fig. 7.71 Three connected pendulums

(c) Determine the equations of motion and set the linearized equations in matrix form using x_1 and x_2 as the generalized coordinates.

(d) Determine the natural frequencies and mode shapes in x_1 and x_2.

(e) ★ Determine the frequency response and dynamics of the system if $m_1 = 2$, $m_2 = 1$, $m_3 = 1$, $l = 1$, $M = \cos \omega t$.

(f) ★ Determine the frequency response and dynamics of the system if $m_1 = 2$, $m_2 = 1$, $m_3 = 1$, $l = 1$, $M = \sin \omega t$.

14. Principal coordinates of three pendulums.

   Three pendulums are connected as are shown in Fig. 7.71. Assume $m_1 = m_2 = m_3 = m$ and $k_1 = k_2 = k$.

   (a) Determine the linearized equations of motion using $\theta_1, \theta_2, \theta_3$ as generalized coordinates.

   (b) ★ Determine the frequency response of the system if $m = 1$, $l = 1$, $k = 1$, $M = \cos \omega t$.

   (c) Determine the linearized equations of motion using x_1, x_2, x_3 as generalized coordinates.

(d) Determine the linearized equations of motion using y_1, y_2, y_3 as generalized coordinates.

(e) Determine the principal coordinates of the system for θ_1, θ_2, θ_3 as generalized coordinates.

(f) Determine the principal coordinates of the system for x_1, x_2, x_3 as generalized coordinates.

(g) Determine the principal coordinates of the system for y_1, y_2, y_3 as generalized coordinates.

15. Principal coordinates of three systems.

(a) Determine the principal coordinates of the systems in Exercise 7.

(b) Determine the principal coordinates of the systems in Exercise 8.

(c) Discuss the difference of the principal coordinates in the two exercises, if any.

16. Coordinates change.

(a) In Exercise 11, use the coordinates

$$y_1 = x_1 - x_2 \qquad y_2 = x_1 - x_3 \qquad y_3 = x_1 - x_4$$

and transform the kinetic energy, potential energy, and equations of motion from coordinates **x** to **y**.

(b) In Exercise 11, use the coordinates

$$z_1 = x_1 - x_2 \qquad z_2 = x_2 - x_3 \qquad z_3 = x_3 - x_4$$

and transform the kinetic energy, potential energy, and equations of motion from coordinates **x** to **z**.

(c) Using the results of the previous parts, transform the kinetic energy, potential energy, and equations of motion from coordinates **y** to **z**.

17. Principal coordinates.

(a) Determine the principal coordinates of the system in Exercise 11; use the coordinates **x**.

(b) ★ Determine the principal coordinates of the system in Exercise 11; use the coordinates **y**.

$$y_1 = x_1 - x_2 \qquad y_2 = x_1 - x_3 \qquad y_3 = x_1 - x_4$$

(c) ★ Determine the principal coordinates of the system in Exercise 11; use the coordinates **z**.

$$z_1 = x_1 - x_2 \qquad z_2 = x_2 - x_3 \qquad z_3 = x_3 - x_4$$

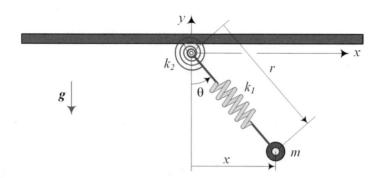

Fig. 7.72 An elastic pendulum

(d) ★ Determine the transformation between the principal coordinates based on **x**, **y**, and **z**.

18. Elements of the stiffness matrix.

(a) In Exercise 11, determine the stiffness matrix $[k]$ using the definition of k_{ij}.
(b) In Exercise 11, determine the stiffness matrix $[k]$ using the potential energy and compare the elements of k_{ij} with the result of the previous part.
(c) In Exercise 11, determine the flexibility matrix $[a]$ and show that $[a] = [k]^{-1}$.

19. ★ Polar coordinates.
Figure 7.72 depicts an elastic pendulum with a rotational spring. Assume $k_1 = 1, m = 1, k_2 = 1$ when needed.

(a) Derive the equation of the elastic pendulum in the coordinates r, and θ.
(b) Linearize the equations of motion.
(c) Determine the natural frequencies and mode shapes of the system in coordinates r and θ.
(d) Is it possible to change k_1 such that the two natural frequencies become equal?
(e) Is it possible to change k_2 such that the two natural frequencies become equal?
(f) Is there any combination of k_1 and k_2 such that the two natural frequencies become equal?
(g) Transform the linearized equations to x and y coordinates.
(h) Determine the natural frequencies and mode shapes of the system in coordinates x and y.
(i) Determine the principal coordinates in terms of r and θ.

20. ★ Two masses on a wire in the shape of $y = f(x)$.
Consider a wire in an arbitrary shape given by $y = f(x)$ as is shown in Fig. 7.73. Two masses m_1 and m_2, connected by a linear spring k, can freely slide on the wire. Determine the equation of motion of the masses. Is there any

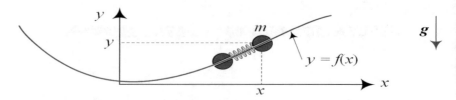

Fig. 7.73 Two connected masses on a wire in shape of $y = f(x)$

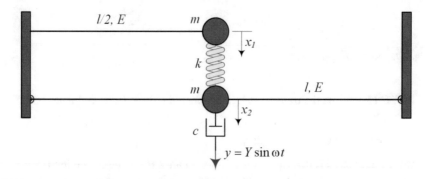

Fig. 7.74 A system of beams and masses with a harmonic displacement excitation

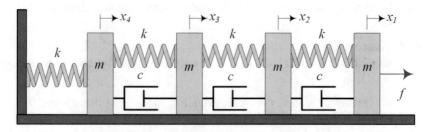

Fig. 7.75 A four DOF damped system under harmonic force function

linearized equation of motion? Can we find any natural frequency and mode shape?

21. Displacement excited two degrees of freedom system.

Figure 7.74 illustrates a system of beams and masses with a harmonic displacement excitation.

(a) Derive the equations of motion.
(b) Determine the frequency responses of the system.

22. ★ Frequency response of a four DOF damped system.

Figure 7.75 illustrates a four DOF damped system with $m = 1 \, \text{kg}$, $k = 100 \, \text{N} / \text{m}$.

1. (a) Determine and show the frequency responses of the system for $f = 100 \sin \omega t$ N. Plot the frequency response of x_1, x_2, x_3, x_4.
   (b) Repeat part (a) for the same force applied only on m_2. Plot the response of x_1, x_2, x_3, x_4.
   (c) Repeat part (a) for the same force applied only on m_3. Plot the response of x_1, x_2, x_3, x_4.
   (d) Repeat part (a) for the same force applied only on m_4. Plot the response of x_1, x_2, x_3, x_4.
   (e) Compare the behavior of x_1, x_2, x_3, x_4 for parts (a) to (d). Determine engineering comments about the position of the applied harmonic force function.

Chapter 8
★ Two Degrees of Freedom Systems

There are many practical engineering situations where we model a vibrating system by attaching a secondary vibrating system m_2, k_2, c_2 to a primary system m_1, k_1, c_1. All practical situations are shown in Fig. 8.1. The secondary system makes the whole system to have two degrees of freedom (DOF) and affects the primary system, making it to deviate from its basic frequency responses of Chap. 6.

The harmonic excitation of the primary system can only be one of the following types:

1. base excitation;
2. eccentric excitation;
3. eccentric base excitation;
4. forced excitation.

The *base excitation* is the most practical model for vertical vibration of several mechanical systems including vehicles and structures. The *eccentric excitation* is a model for equipment attached to a rotary motor on a suspension. The *eccentric base excitation* is a model for vibration of any double equipment mounted on an engine. While the *forced excitation* has almost no practical application, it is known the simplest two DOF model of forced vibrations to serve as a good pedagogical application. Because of its simplicity, we first examine the frequency response of a harmonically forced vibrating system.

8.1 Forced Excitation

Figure 8.2 illustrates a two DOF forced vibrating system such that a secondary system (m_2, k_2, c_2) is attached to a primary forced excited system (m_1, k_1, c_1). The equations of motion of the system are

$$m_1\ddot{x}_1 + c_1\dot{x}_1 + c_2(\dot{x}_1 - \dot{x}_2) + k_1x_1 + k_2(x_1 - x_2) = F\sin\omega t \qquad (8.1)$$

© The Author(s), under exclusive license to Springer Nature Switzerland AG 2022
R. N. Jazar, *Advanced Vibrations*, https://doi.org/10.1007/978-3-031-16356-2_8

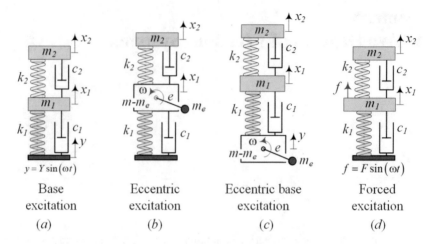

Fig. 8.1 The practical harmonically excited two DOF systems

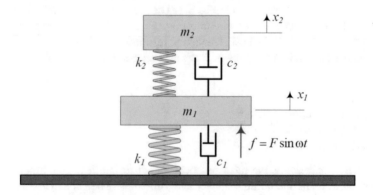

Fig. 8.2 The model of the two DOF forced excited system

$$m_2\ddot{x}_2 - c_2(\dot{x}_1 - \dot{x}_2) - k_2(x_1 - x_2) = 0 \tag{8.2}$$

and the steady-state amplitudes X_1 and X_2 for vibration of m_1 and m_2 are

$$\left(\frac{X_1}{F}\right)^2 = \frac{\left(k_2 - \omega^2 m_2\right)^2 + \omega^2 c_2^2}{Z_1^2 + \omega^2 c_2^2 Z_2^2 + c_1^2 \omega^2 Z_3^2 + \omega^4 c_1 c_2 \left(c_1 c_2 + 2\omega^2 m_2^2\right)} \tag{8.3}$$

$$\left(\frac{X_2}{F}\right)^2 = \frac{k_2^2 + \omega^2 c_2^2}{Z_1^2 + \omega^2 c_2^2 Z_2^2 + c_1^2 \omega^2 Z_3^2 + \omega^4 c_1 c_2 \left(c_1 c_2 + 2\omega^2 m_2^2\right)} \tag{8.4}$$

where

$$Z_1 = \left(k_1 - \omega^2 m_1\right)\left(k_2 - \omega^2 m_2\right) - \omega^2 m_2 k_2 \tag{8.5}$$

$$Z_2 = k_1 - \omega^2 m_1 - \omega^2 m_2 \tag{8.6}$$

$$Z_3 = \left(k_2 - \omega^2 m_2\right) \tag{8.7}$$

Introducing the following parameters,

$$\varepsilon = \frac{m_2}{m_1} \tag{8.8}$$

$$\xi_1 = \frac{c_1}{2m_1\omega_1} \tag{8.9}$$

$$\xi_2 = \frac{c_2}{2m_2\omega_2} \tag{8.10}$$

$$\omega_1 = \sqrt{\frac{k_1}{m_1}} \tag{8.11}$$

$$\omega_2 = \sqrt{\frac{k_2}{m_2}} \tag{8.12}$$

$$\alpha = \frac{\omega_2}{\omega_1} = \sqrt{\frac{k_2}{\varepsilon k_1}} \tag{8.13}$$

$$r = \frac{\omega}{\omega_1} \tag{8.14}$$

$$\mu = \frac{X_1}{F/k_1} \tag{8.15}$$

$$\tau = \frac{X_2}{F/k_1} \tag{8.16}$$

we can transform the frequency responses (8.3) and (8.4) to the following dimensionless forms.

$$\mu^2 = \left(\frac{X_1}{F/k_1}\right)^2 = \frac{4\alpha^2\xi_2^2 r^2 + \left(r^2 - \alpha^2\right)^2}{4\xi_2^2 r^2\alpha^2 Z_5^2 + Z_4^2 + 8\xi_1\xi_2\alpha r^4 Z_5 + 4r^6\xi_1^2} \tag{8.17}$$

$$\tau^2 = \left(\frac{X_2}{F/k_1}\right)^2 = \frac{\alpha^2\left(4\xi_2^2 r^2 + \alpha^2\right)}{4\xi_2^2 r^2\alpha^2 Z_5^2 + Z_4^2 + 8\xi_1\xi_2\alpha r^4 Z_5 + 4r^6\xi_1^2} \tag{8.18}$$

$$Z_4 = \varepsilon \alpha^2 r^2 - \left(r^2 - 1\right)\left(r^2 - \alpha^2\right) \tag{8.19}$$

$$Z_5 = r^2 \left(1 + \varepsilon\right) - 1 \tag{8.20}$$

Proof The kinetic and potential energies and dissipation function of the system are

$$K = \frac{1}{2} m_1 \dot{x}_1^2 + \frac{1}{2} m_2 \dot{x}_2^2 \tag{8.21}$$

$$P = \frac{1}{2} k_1 x_1^2 + \frac{1}{2} k_2 (x_1 - x_2)^2 \tag{8.22}$$

$$D = \frac{1}{2} c_1 \dot{x}_1^2 + \frac{1}{2} c_2 (\dot{x}_1 - \dot{x}_2)^2 \tag{8.23}$$

Employing the Lagrange equation (2.547), we find the equations of motion as (8.1) and (8.2) or equivalently in the following matrix form.

$$\begin{bmatrix} m_1 & 0 \\ 0 & m_2 \end{bmatrix} \begin{bmatrix} \ddot{x}_1 \\ \ddot{x}_2 \end{bmatrix} + \begin{bmatrix} c_1 + c_2 & -c_2 \\ -c_2 & c_2 \end{bmatrix} \begin{bmatrix} \dot{x}_1 \\ \dot{x}_2 \end{bmatrix}$$
$$+ \begin{bmatrix} k_1 + k_2 & -k_2 \\ -k_2 & k_2 \end{bmatrix} = \begin{bmatrix} F \sin \omega t \\ 0 \end{bmatrix} \tag{8.24}$$

Substituting a set of harmonic solutions with the same excitation frequency ω

$$x_1 = A_1 \sin \omega t + B_1 \cos \omega t \tag{8.25}$$

$$x_2 = A_2 \sin \omega t + B_2 \cos \omega t \tag{8.26}$$

provides us with a set of four algebraic equations to calculate the coefficients A_1, B_1, A_2, and B_2.

$$\begin{bmatrix} a_{11} & -k_2 & -\omega (c_1 + c_2) & \omega c_2 \\ -k_2 & a_{22} & \omega c_2 & -\omega c_2 \\ \omega (c_1 + c_2) & -\omega c_2 & a_{33} & -k_2 \\ -\omega c_2 & \omega c_2 & -k_2 & a_{44} \end{bmatrix} \begin{bmatrix} A_1 \\ A_2 \\ B_1 \\ B_2 \end{bmatrix} = \begin{bmatrix} F \\ 0 \\ 0 \\ 0 \end{bmatrix} \tag{8.27}$$

$$a_{11} = a_{33} = k_2 + k_1 - m_1 \omega^2 \tag{8.28}$$

$$a_{22} = a_{44} = k_2 - m_2 \omega^2 \tag{8.29}$$

The steady-state amplitudes of m_1 and m_2 are X_1 and X_2 which, after simplification, can be written as (8.3) and (8.4).

$$X_1 = \sqrt{A_1^2 + B_1^2} \tag{8.30}$$

$$X_2 = \sqrt{A_2^2 + B_2^2} \tag{8.31}$$

Converting the equations and solutions of multi-DOF systems in dimensionless form is not necessary and not unique, and this is not based on any standard method. However, we can make the solutions general with minimum parameters by nondimensionalization. Using the parameters (8.8)–(8.16), we transform the frequency responses (8.3) and (8.4)–(8.17) and (8.18). The original frequency responses (8.3) and (8.4) are expressed by the *ten* parameters and variables, $m_1, m_2,$ $k_1, k_2, c_1, c_2, F, \omega, X_1,$ and X_2, while the dimensionless forms of (8.17) and (8.18) are expressed by the *seven* parameters and variables, $\varepsilon, \alpha, \xi_1, \xi_2, r, \mu,$ and τ.

The parameters ξ_1 and ξ_2 are defined as the damping ratios of the individual single DOF subsystems and are proportional to c_1 and c_2. The parameter $\alpha = \omega_2/\omega_1 = \sqrt{k_2/(\varepsilon k_1)}$ indicates the stiffness ratio of the secondary and primary systems. The frequencies ω_1 and ω_2 are the natural frequencies of the separated primary and secondary subsystems and are not the natural frequencies of the combined system. Comparing them with the natural frequencies of the system indicates what would be the effect of attaching the two systems compared to the disassembled subsystems. The mass ratio ε denotes the mass of the added system compared to the primary one. The frequency responses μ and τ indicate the dynamic amplitude of m_1 and m_2 compared to the static deflection of m_1 under the applied force.

To examine the effectiveness of the secondary system in magnifying or damping the vibrations of the primary system, we determine the ratio of the amplitudes of m_1 and m_2:

$$\frac{X_1^2}{X_2^2} = \frac{\left(k_2 - \omega^2 m_2\right)^2 + \omega^2 c_2^2}{k_2^2 + \omega^2 c_2^2} \tag{8.32}$$

$$\frac{\mu^2}{\tau^2} = \frac{4\alpha^2 \xi_2^2 r^2 + \left(r^2 - \alpha^2\right)^2}{4\alpha^2 \xi_2^2 r^2 + \alpha^4} \tag{8.33}$$

Interestingly, the primary parameters $m_1, k_1,$ and c_1 do not affect the amplitude ratio. Therefore, although changing the primary parameters will change X_1 and X_2, the values of $m_1, k_1,$ and c_1 will not change the ratio of X_1/X_2. In dimensionless form, X_1/X_2 is a function of $\alpha, \xi_2,$ and r and not a function of ε and ξ_1.

In a forced excited system, we assume that the excitation comes from an ideal source such that none of the characteristics or reactions of the system affects it. The source can provide as much energy and power as needed. Although such assumptions are realistic in practice, the analysis provides us with a base to analyze and compare systems in particular cases. ∎

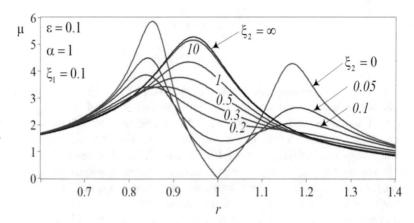

Fig. 8.3 The frequency response $\mu = \frac{X_1}{F/k_1}$ for various ξ_2 and for $\varepsilon = 0.1, \xi_1 = 0.1$, and $\alpha = 1$

Example 284 Frequency response of the forced two DOF system. Frequency response of linear vibrating system can always be calculated analytically. The equations for frequency responses of multi-DOF system are amplitude X_i as function of the excitation frequency ω, but they are usually dependent on more than one parameter to be able to show the frequency response curves in two dimensional graphs. The right way to visualize their frequency response is to make the results dimensionless to minimize the number of parameters and generalize the result to be valid for all similar systems. Then we may set all parameters to their nominal values except one. Then we plot the frequency response curves for different values of the parameter. This is engineering treatment of frequency response of multi-DOF systems. Here is the engineering treatment of forced excitation of two DOF system.

Because the ratio of X_1/X_2 is only a function of α, ξ_2, and r, it is reasonable to show the frequency responses (8.17) and (8.18) by choosing ε and ξ_1 and varying α and ξ_2 as functions of r. The frequency responses μ and τ are functions of too many parameters to be shown on a plane or in three dimensions. Having the analytic equations is enough to plot the frequency response for any particular set of parameters. Showing two examples, Figs. 8.3 and 8.4 depict the frequency responses μ and τ for $\varepsilon = 0.1, \xi_1 = 0.1$, and $\alpha = 1$, and Figs. 8.5 and 8.6 show μ and τ for $\varepsilon = 0.1, \xi_1 = 0.4$, and $\alpha = 1$.

Example 285 Relative displacement. For any two-mass vibrating system, the relative motion of the masses is important to engineers because there is always a limited space for their relative movement. The relative displacement frequency response is always a part of optimization objective function for two-mass vibrating system. This example shows how to calculate relative displacement frequency response and how to illustrate it.

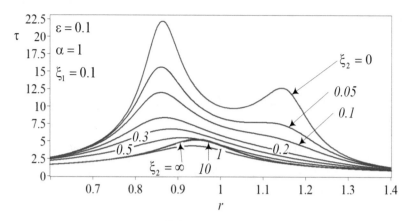

Fig. 8.4 The frequency response $\tau = \frac{X_2}{F/k_1}$ for various ξ_2 and for $\varepsilon = 0.1$, $\xi_1 = 0.4$, and $\alpha = 1$

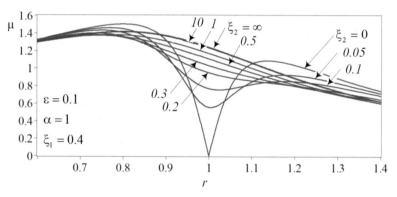

Fig. 8.5 The frequency response $\mu = \frac{X_1}{F/k_1}$ for various ξ_2 and for $\varepsilon = 0.1$, $\xi_1 = 0.4$, and $\alpha = 1$

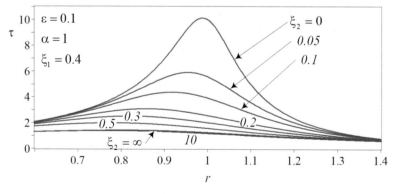

Fig. 8.6 The frequency response $\tau = \frac{X_2}{F/k_1}$ for various ξ_2 and for $\varepsilon = 0.1$, $\xi_1 = 0.4$, and $\alpha = 1$

The relative displacements of m_1 and m_2 in the forced excited system of Fig. 8.2 is

$$z = x_1 - x_2 \tag{8.34}$$

The frequency response of z can be calculated by substituting harmonic solutions for x_1 and x_2.

$$
\begin{aligned}
z &= (A_1 \sin \omega t + B_1 \cos \omega t) - (A_2 \sin \omega t + B_2 \cos \omega t) \\
&= (A_1 - A_2) \sin t\omega + (B_1 - B_2) \cos t\omega \\
&= Z \sin \left(\omega t - \varphi_z \right)
\end{aligned}
\tag{8.35}
$$

The steady-state amplitude of the relative displacement is Z

$$Z = \sqrt{(A_1 - A_2)^2 + (B_1 - B_2)^2} \tag{8.36}$$

$$\left(\frac{Z}{F} \right)^2 = \frac{m_2^2 \omega^4}{Z_1^2 + \omega^2 c_2^2 Z_2^2 + c_1^2 \omega^2 Z_3^2 + \omega^4 c_1 c_2 \left(c_1 c_2 + 2\omega^2 m_2^2 \right)} \tag{8.37}$$

where the parameters Z_1, Z_2, $and\, Z_3$ are the same as in Eqs. (8.5)–(8.7).

$$Z_1 = \left(k_1 - \omega^2 m_1 \right) \left(k_2 - \omega^2 m_2 \right) - \omega^2 m_2 k_2 \tag{8.38}$$

$$Z_2 = k_1 - \omega^2 m_1 - \omega^2 m_2 \tag{8.39}$$

$$Z_3 = \left(k_2 - \omega^2 m_2 \right) \tag{8.40}$$

Using the parameters (8.8)–(8.16), we transform Eq. (8.37) to a dimension form

$$\eta^2 = \left(\frac{Z}{F/k_1} \right)^2 = \frac{r^4}{4\xi_2^2 r^2 \alpha^2 Z_5^2 + Z_4^2 + 8\xi_1 \xi_2 \alpha r^4 Z_5 + 4r^6 \xi_1^2} \tag{8.41}$$

where the parameters Z_4 and Z_5 are the same as given in Eqs. (8.19) and (8.20).

$$Z_4 = \varepsilon \alpha^2 r^2 - \left(r^2 - 1 \right) \left(r^2 - \alpha^2 \right) \tag{8.42}$$

$$Z_5 = r^2 (1 + \varepsilon) - 1 \tag{8.43}$$

Figure 8.7 illustrates an example of the relative displacement frequency response η of the forced excited two DOF system.

Example 286 Transmitted force. The transmitted force to the base is a design measurement for constructing foundation for any mechanical vibrating device. The

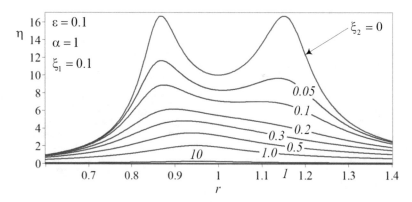

Fig. 8.7 The relative displacement frequency response $\eta = \frac{Z}{F/k_1}$ for various ξ_2 and for $\varepsilon = 0.1$, $\xi_1 = 0.1$, and $\alpha = 1$

frequency response of the transmitted force to the base of the two *DOF* force excited system is calculated and illustrated here.

The transmitted force f_T to the ground for the forced excited system of Fig. 8.2 is the combined forces in spring k_1 and damper c_1.

$$f_T = k_1 x_1 + c_1 \dot{x}_1 \tag{8.44}$$

We determine the frequency response of f_T by substituting for x_1 and x_2.

$$\begin{aligned} z &= k_1 \left(A_1 \sin \omega t + B_1 \cos \omega t\right) + c_1 \omega \left(A_1 \cos \omega t - B_1 \sin \omega t\right) \\ &= \left(A_1 k_1 - \omega B_1 c_1\right) \sin \omega t + \left(B_1 k_1 + \omega A_1 c_1\right) \cos \omega t \\ &= F_T \sin \left(\omega t - \varphi_F\right) \end{aligned} \tag{8.45}$$

The steady-state amplitude of the transmitted force is

$$F_T = \sqrt{\left(A_1 k_1 - \omega B_1 c_1\right)^2 + \left(B_1 k_1 + \omega A_1 c_1\right)^2} \tag{8.46}$$

$$\left(\frac{F_T}{F}\right)^2 = \frac{\left(c_1^2 \omega^2 + k_1^2\right)\left(\left(m_2 \omega^2 - k_2\right)^2 + c_2^2 \omega^2\right)}{Z_1^2 + \omega^2 c_2^2 Z_2^2 + c_1^2 \omega^2 Z_3^2 + \omega^4 c_1 c_2 \left(c_1 c_2 + 2\omega^2 m_2^2\right)} \tag{8.47}$$

where the parameters Z_1, Z_2, and Z_3 are the same as in Eqs. (8.5)–(8.7).

$$Z_1 = \left(k_1 - \omega^2 m_1\right)\left(k_2 - \omega^2 m_2\right) - \omega^2 m_2 k_2 \tag{8.48}$$

$$Z_2 = k_1 - \omega^2 m_1 - \omega^2 m_2 \tag{8.49}$$

$$Z_3 = \left(k_2 - \omega^2 m_2\right) \tag{8.50}$$

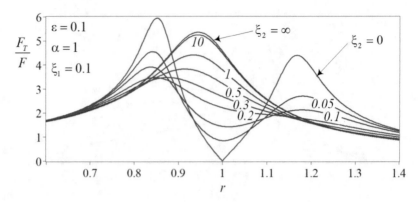

Fig. 8.8 The transmitted force frequency response $\frac{F_T}{F}$ for various ξ_2 and for $\varepsilon = 0.1$, $\xi_1 = 0.1$, and $\alpha = 1$

The parameters (8.8)–(8.16) transform Eq. (8.47) to

$$\left(\frac{F_T}{F}\right)^2 = \frac{(4\xi_1^2 r^2 + 1)\left((r^2 - \alpha^2)^2 + 4\alpha^2\xi_2^2 r^2\right)}{4\xi_2^2 r^2 \alpha^2 Z_5^2 + Z_4^2 + 8\xi_1\xi_2\alpha r^4 Z_5 + 4r^6\xi_1^2} \tag{8.51}$$

where the parameters Z_4 and Z_5 are the same as given in Eqs. (8.19) and (8.20).

$$Z_4 = \varepsilon\alpha^2 r^2 - \left(r^2 - 1\right)\left(r^2 - \alpha^2\right) \tag{8.52}$$

$$Z_5 = r^2(1 + \varepsilon) - 1 \tag{8.53}$$

Figure 8.8 illustrates an example of the transmitted force frequency response of the forced excited two DOF system.

Example 287 Amplitude ratio X_1/X_2. The plot of the frequency response X_1/X_2 is an engineering design graph to understand the excitation frequency ranges at which X_1 is less than or greater than X_2. Here are the mathematical analysis and engineering points of X_1/X_2.

The steady-state amplitude ratio X_1/X_2 of m_1 to m_2 for the system of Fig. 8.2 is

$$\frac{X_1^2}{X_2^2} = \frac{\mu^2}{\tau^2} = \frac{4\alpha^2\xi_2^2 r^2 + \left(r^2 - \alpha^2\right)^2}{4\alpha^2\xi_2^2 r^2 + \alpha^4} \tag{8.54}$$

Figure 8.9 illustrates the amplitude ratio $\mu/\tau = X_1/X_2$ as a function of r for $\alpha = 1$ and various ξ_2. When $\xi_2 = \infty$, there is no relative motion between m_1 and m_2, and we have $\mu/\tau = 1$. The other extreme case of $\xi_2 = 0$ is important to engineers, because when $\xi_2 = 0$, it is possible to stop m_1 from vibrating at a particular frequency ratio. When $\xi_2 = 0$, the amplitude ratio μ/τ simplifies to

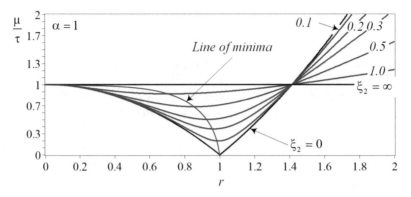

Fig. 8.9 Amplitude ratio $\mu/\tau = X_1/X_2$ of the two DOF forced excited system as a function of r for $\alpha = 1$ and various ξ_2

$$\frac{X_1^2}{X_2^2} = \frac{\mu^2}{\tau^2} = \frac{\left(r^2 - \alpha^2\right)^2}{\alpha^2} \tag{8.55}$$

which provides us with $\mu/\tau = 0$ at $r = \alpha$.

Suppose the primary system is the main system and we attach the secondary system to change the vibration behavior of the primary system. The following points may be useful in designing a vibration absorber to control vibrations of the primary system.

1. If the reduction of μ/τ is important, then the working frequency range should be limited to

$$0 < r < \sqrt{2}\alpha \tag{8.56}$$

or

$$0 < \omega < \sqrt{2}\,\omega_2 \tag{8.57}$$

That is because, besides $r = 0$, there is another node in the graph of μ/τ, at

$$r = \sqrt{2}\,\alpha \tag{8.58}$$

We have $\mu/\tau > 1$ after the node. The nodes are at the intersection of all curves for different values of ξ_2 and can be found by intersecting the extreme cases of $\xi_2 \to 0$ and $\xi_2 \to \infty$.

$$\lim_{\xi_2 \to 0} \frac{\mu}{\tau} = \lim_{\xi_2 \to \infty} \frac{\mu}{\tau} \tag{8.59}$$

$$\frac{r^2 - \alpha^2}{\alpha^2} = \pm 1 \tag{8.60}$$

Therefore, to reduce the amplitude of m_1 and obtain $\mu/\tau < 1$, the excitation frequency must be limited to (8.56).

2. In case the excitation frequency is fixed at a given value, $r = r_0$, we can stop m_1, by designing the parameters of the secondary system thus:

$$\xi_2 = 0 \qquad \alpha = r_0 \tag{8.61}$$

or

$$c_2 = 0 \qquad \omega_2 = \omega_0 \tag{8.62}$$

However, if $\mu/\tau \to 0$, then τ is theoretically very high. Therefore, we need some nonzero ξ_2 to limit the vibration of m_2. Introducing ξ_2 removes the possibility of making $\mu = 0$ by any set of passive parameters.

3. For any given $\xi_2 \neq 0$ and fixed excitation frequency at $r = r_0$, we can minimize μ/τ by designing the parameters of the secondary system.

$$\alpha^2 = \frac{1}{2}\left(1 + \sqrt{8\xi_2^2 + 1}\right) r_0^2 \tag{8.63}$$

That is because the minimum of μ/τ occurs when we have

$$2r^4\xi_2^2 + r^2\alpha^2 - \alpha^4 = 0 \tag{8.64}$$

Solving this equation for α provides us with the relation (8.63), and solving for r determines the frequency at which μ/τ is minimum.

$$r^2 = \frac{2}{1 + \sqrt{8\xi_2^2 + 1}}\alpha^2 \tag{8.65}$$

The minimum value of μ/τ is only a function of the secondary damping ξ_2.

$$\left(\frac{\mu}{\tau}\right)^2_{min} = \frac{(8\xi_2^4 - 4\xi_2^2 - 1)\sqrt{8\xi_2^2 + 1} + 8\xi_2^2 + 1}{8\xi_2^4\sqrt{8\xi_2^2 + 1}} \tag{8.66}$$

The line of minima is shown in Fig. 8.9.

4. For any given value of α, curves of μ/τ for $\xi_2 = 0$ and $\xi_2 = \infty$ determine the boundary of possible and impossible relative frequency responses. The white area of Fig. 8.10 indicates the possible μ/τ, and the shaded area indicates the impossible μ/τ for $\alpha = 1$.

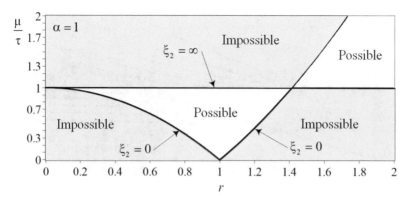

Fig. 8.10 Possible and impossible areas of the amplitude ratio $\mu/\tau = X_1/X_2$ of the two DOF forced excited system, for $\alpha = 1$

Example 288 Natural frequencies. Natural frequencies of any vibrating system are the first and most important characteristic that must be calculated. They indicate the resonance zones that the excitation frequency must avoid. Here is an analytic and general calculation of natural frequencies of the two DOF systems.

Setting damping ratios zero, $\xi_1 = \xi_2 = 0$, the denominators of μ and τ lead to the characteristic equation determining the natural frequencies of the system of Fig. 8.1.

$$\varepsilon\alpha^2 r^2 - \left(r^2 - 1\right)\left(r^2 - \alpha^2\right) = 0 \tag{8.67}$$

The solutions of the equation are

$$r_1^2 = \frac{1}{2}\alpha^2\left(\varepsilon + 1\right) + \frac{1}{2}\left(1 + \sqrt{\alpha^4\left(\varepsilon^2 + 2\varepsilon + 1\right) + 2\alpha^2\left(\varepsilon - 1\right) + 1}\right) \tag{8.68}$$

$$r_2^2 = \frac{1}{2}\alpha^2\left(\varepsilon + 1\right) + \frac{1}{2}\left(1 - \sqrt{\alpha^4\left(\varepsilon^2 + 2\varepsilon + 1\right) + 2\alpha^2\left(\varepsilon - 1\right) + 1}\right) \tag{8.69}$$

The natural frequencies r_1 and r_2 are only functions of the mass ratio ε and stiffness ratio α. So, it is possible to illustrate their changes in a planar graph. Figure 8.11 illustrates the natural frequencies as functions of α for different ε, and Fig. 8.12 illustrates the natural frequencies as functions of ε for different α. The natural frequencies approach $r_1 \to 1$ and $r_2 \to 1$ for any ε when $\alpha \to 0$. Because the natural frequencies are independent of excitations, these results are correct for all two DOF systems of Fig. 8.2.

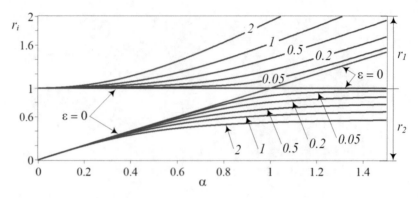

Fig. 8.11 The natural frequencies r_1 and r_2 of the two DOF forced excited system, as function of α for different ε

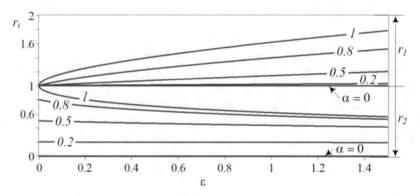

Fig. 8.12 The natural frequencies r_1 and r_2 of the two DOF forced excited system, as function of ε for different α

Example 289 No secondary damping. Extreme cases of every vibrating system are zero damping and infinity damping. These two extreme cases indicate the boundaries of dynamic behavior of the system and after natural frequency determination are the first step in analysis.

When c_2 is zero, there is no damping in the secondary system to kill the vibrations in that part, and the system would be modeled as is shown in Fig. 8.13. Practically, there are many situations in which the damping of the secondary system is very small, and we can ignore it compared to the damping of the primary system. Such a secondary system will not increase the damping of the whole system. However, it introduces more mass and more stiffness. The increased mass can collect more kinetic energy, and more stiffness can collect more potential energy. The interaction of the secondary system will affect the motion of m_1, and its frequency response will be different from Fig. 6.3. The amplitude of the frequency response of m_2 will be controlled by action and reaction of the primary system.

We can recover the equations of motion and frequency responses of this system from Eqs. (8.1)–(8.18) by substituting $c_2 = 0$ or, equivalently, $\xi_2 = 0$.

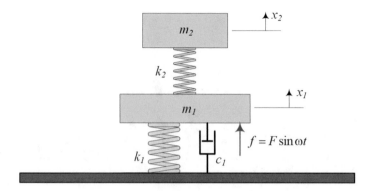

Fig. 8.13 An undamped secondary system (m_2, k_2) is attached to a primary forced excited system (m_1, k_1, c_1)

$$\begin{bmatrix} m_1 & 0 \\ 0 & m_2 \end{bmatrix} \begin{bmatrix} \ddot{x}_1 \\ \ddot{x}_2 \end{bmatrix} + \begin{bmatrix} c_1 & 0 \\ 0 & 0 \end{bmatrix} \begin{bmatrix} \dot{x}_1 \\ \dot{x}_2 \end{bmatrix}$$

$$+ \begin{bmatrix} k_1 + k_2 & -k_2 \\ -k_2 & k_2 \end{bmatrix} = \begin{bmatrix} F \sin \omega t \\ 0 \end{bmatrix} \tag{8.70}$$

$$\left(\frac{X_1}{F} \right)^2 = \frac{(k_2 - \omega^2 m_2)^2}{Z_1^2 + c_1^2 \omega^2 Z_3^2} \tag{8.71}$$

$$\left(\frac{X_2}{F} \right)^2 = \frac{k_2^2}{Z_1^2 + c_1^2 \omega^2 Z_3^2} \tag{8.72}$$

$$\mu^2 = \left(\frac{X_1}{F/k_1} \right)^2 = \frac{(r^2 - \alpha^2)^2}{Z_4^2 + 4r^6 \xi_1^2} \tag{8.73}$$

$$\tau^2 = \left(\frac{X_2}{F/k_2} \right)^2 = \frac{\alpha^4}{Z_4^2 + 4r^6 \xi_1^2} \tag{8.74}$$

$$Z_1 = \left(k_1 - \omega^2 m_1 \right) \left(k_2 - \omega^2 m_2 \right) - \omega^2 m_2 k_2 \tag{8.75}$$

$$Z_3 = \left(k_2 - \omega^2 m_2 \right) \tag{8.76}$$

$$Z_4 = \varepsilon \alpha^2 r^2 - \left(r^2 - 1 \right) \left(r^2 - \alpha^2 \right) \tag{8.77}$$

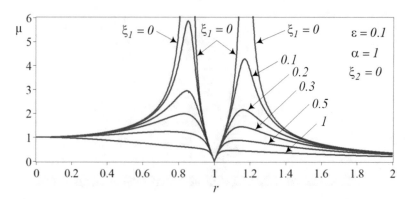

Fig. 8.14 The frequency response $\mu = \frac{X_1}{F/k_1}$ of the primary of the two DOF forced excited system for $\varepsilon = 0.1$, $\alpha = 1$, and $\xi_2 = 0$

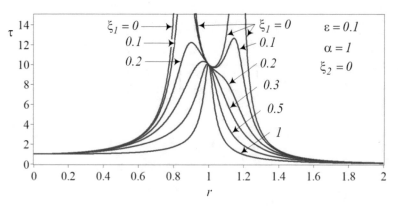

Fig. 8.15 The frequency response $\tau = \frac{X_2}{F/k_1}$ of the secondary of the two DOF forced excited system for $\varepsilon = 0.1$, $\alpha = 1$, and $\xi_2 = 0$

Figure 8.14 illustrates the frequency response μ of the primary system, and Fig. 8.15 illustrates the frequency response τ of the secondary system for $\varepsilon = 0.1$, $\alpha = 1$, and $\xi_2 = 0$. The engineering point of this analysis is having $\mu = 0$ at $r = 1$ for every value of ξ. In other words, attaching the secondary system to the primary, makes the primary to stop stationary at $r = 1$ regardless of the value of damping and the force excitation.

Example 290 Effect of the secondary system on the main system. To better analyze the effect of adding a secondary system to a primary system to make a two DOF system, the undamped system should be studied.

To examine what happens to the frequency response of a forced system after attachment of a secondary system, let us compare the frequency responses of the systems in Figs. 8.16 with Fig. 8.14. The zero damping frequency response and natural frequency of the system of Fig. 8.16a are

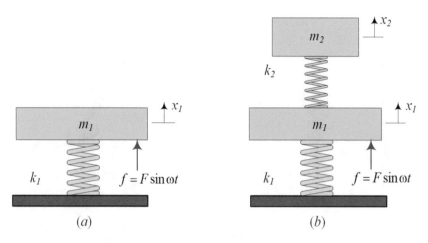

Fig. 8.16 (**a**) A main vibrating system. (**b**) The two DOF forced excited system after attachment of a secondary system to the main

$$\mu_a = \frac{X}{F/k} = \frac{1}{1 - r^2} \qquad r = 1 \tag{8.78}$$

The zero damping frequency response of the main system and natural frequency of Fig. 8.16b are

$$\mu_b = \frac{X_1}{F/k_1} = \frac{r^2 - \alpha^2}{\varepsilon \alpha^2 r^2 - (r^2 - 1)(r^2 - \alpha^2)} \tag{8.79}$$

$$r_1 = \frac{1}{2}\alpha^2 (\varepsilon + 1) + \frac{1}{2}\left(1 + \sqrt{\alpha^4 (\varepsilon^2 + 2\varepsilon + 1) + 2\alpha^2 (\varepsilon - 1) + 1}\right) \tag{8.80}$$

$$r_2 = \frac{1}{2}\alpha^2 (\varepsilon + 1) + \frac{1}{2}\left(1 - \sqrt{\alpha^4 (\varepsilon^2 + 2\varepsilon + 1) + 2\alpha^2 (\varepsilon - 1) + 1}\right) \tag{8.81}$$

When $\alpha = 0$, we have consistency of $r_1 = r = 1$ and $r_2 = 0$.

Figure 8.17 compares the frequency responses of the systems of Fig. 8.16a and b for $\alpha = 1$ and $\varepsilon = 0.1$, when there is no damping in the system. Adding the secondary system shots down the vibrations of the primary system at the most critical frequency $r = 1$. In general, as long as $\xi_2 = 0$, we have $\mu_b = 0$ at $r = \alpha$ for any ξ_1, and this result provides us with the opportunity to shot down the vibrations of the primary system at any desired frequency r_0 by adjusting $\alpha = r_0$. Practically, the primary system is heavier than the secondary system, $\varepsilon < 1$. Furthermore, the frequency response of the main system is low, $\mu_a < 1$, when r is higher than the resonance zone, $r > \sqrt{2}$. Therefore, as an engineering recommendation, we usually add the secondary system to reduce the vibrations of the primary system at a low frequency, $r < 1$.

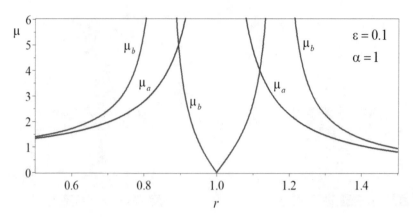

Fig. 8.17 The frequency responses $\mu = \frac{X_1}{F/k_1}$ of the undamped two DOF forced excited systems in 8.16a and b for $\alpha = 1$ and $\xi_2 = 0$

Example 291 Freezing m_1 at a desired frequency. The engineering design application of adding a secondary vibrating system to a forced excited primary is to absorb the vibrations of the primary and save it from oscillation. Here, we show how to adjust a secondary vibration absorber to make a primary system stationary at any particular excitation frequency.

Let us assume that we want to shot down the vibration of the system of Fig. 8.13 at a given frequency, say $r = 0.6$, by adding a secondary system as a vibration absorber. The best design for $r = 0.6$ is to adjust the secondary system at $\alpha = r = 0.6$ for any value of ξ_1. Choosing a mass ratio such as $\varepsilon = m_2/m_1 = 0.02$, we can determine the secondary stiffness.

$$k_2 = k_1 \varepsilon \alpha^2 = 0.02 k_1 \times 0.6^2 = \frac{9}{1250} k_1 \tag{8.82}$$

Figure 8.18 illustrates μ for the design parameters. The vibration of m_1 is zero at the desired frequency $r = 0.6$ for any value of ξ_1. The frequency response of m_2 is shown in Fig. 8.19.

Example 292 No primary damping. No damping in primary is another extreme model for system whose suspension has very low damping compared to the damping of the secondary system. Rubber and elastomer suspension are examples of suspension with low damping.

When c_1 is zero or is very low, we may ignore damping in the primary system and model the system as is shown in Fig. 8.20. A secondary system with damping will increase the damping of the whole system. The interaction of the secondary system will affect the motion of m_1. We can recover the equations of motion and the frequency responses of the system from Eqs. (8.1)–(8.18) by substituting $c_1 = 0$ or, equivalently, $\xi_1 = 0$.

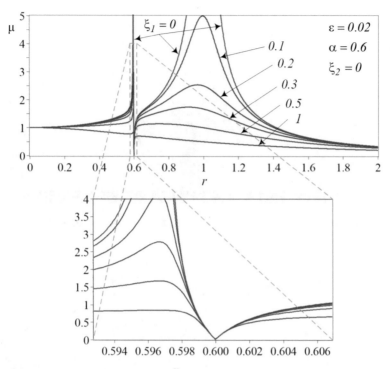

Fig. 8.18 The frequency response $\mu = \frac{X_1}{F/k_1}$ of the primary of the two *DOF* forced excited system for $\varepsilon = 0.02$, $\alpha = 0.6$, and $\xi_2 = 0$

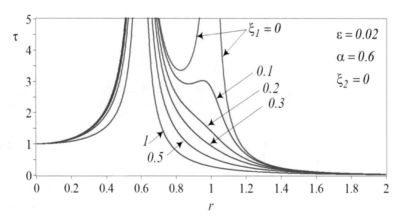

Fig. 8.19 The frequency response $\tau = \frac{X_2}{F/k_1}$ of the secondary of the two *DOF* forced excited system for $\varepsilon = 0.02$, $\alpha = 0.6$, and $\xi_2 = 0$

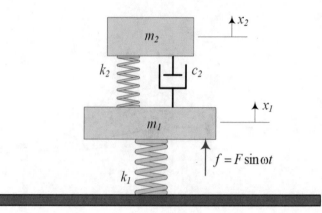

Fig. 8.20 The undamped main system of the two DOF forced excited system

$$\begin{bmatrix} m_1 & 0 \\ 0 & m_2 \end{bmatrix} \begin{bmatrix} \ddot{x}_1 \\ \ddot{x}_2 \end{bmatrix} + \begin{bmatrix} c_2 & -c_2 \\ -c_2 & c_2 \end{bmatrix} \begin{bmatrix} \dot{x}_1 \\ \dot{x}_2 \end{bmatrix}$$

$$+ \begin{bmatrix} k_1 + k_2 & -k_2 \\ -k_2 & k_2 \end{bmatrix} = \begin{bmatrix} F \sin \omega t \\ 0 \end{bmatrix} \tag{8.83}$$

$$\left(\frac{X_1}{F}\right)^2 = \frac{\left(k_2 - \omega^2 m_2\right)^2 + \omega^2 c_2^2}{Z_1^2 + \omega^2 c_2^2 Z_2^2} \tag{8.84}$$

$$\left(\frac{X_2}{F}\right)^2 = \frac{k_2^2 + \omega^2 c_2^2}{Z_1^2 + \omega^2 c_2^2 Z_2^2} \tag{8.85}$$

$$\mu^2 = \left(\frac{X_1}{F/k_1}\right)^2 = \frac{4\alpha^2 \xi_2^2 r^2 + \left(r^2 - \alpha^2\right)^2}{4\xi_2^2 r^2 \alpha^2 Z_5^2 + Z_4^2} \tag{8.86}$$

$$\tau^2 = \left(\frac{X_2}{F/k_1}\right)^2 = \frac{\alpha^2 \left(4\xi_2^2 r^2 + \alpha^2\right)}{4\xi_2^2 r^2 \alpha^2 Z_5^2 + Z_4^2} \tag{8.87}$$

$$Z_1 = \left(k_1 - \omega^2 m_1\right)\left(k_2 - \omega^2 m_2\right) - \omega^2 m_2 k_2 \tag{8.88}$$

$$Z_2 = k_1 - \omega^2 m_1 - \omega^2 m_2 \tag{8.89}$$

$$Z_4 = \varepsilon \alpha^2 r^2 - \left(r^2 - 1\right)\left(r^2 - \alpha^2\right) \tag{8.90}$$

$$Z_5 = r^2 (1 + \varepsilon) - 1 \tag{8.91}$$

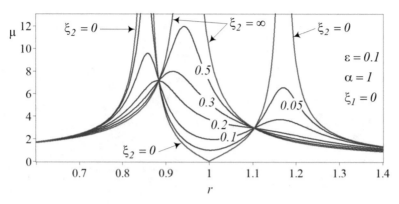

Fig. 8.21 The frequency response $\mu = \frac{X_1}{F/k_1}$ for various ξ_2 and for $\varepsilon = 0.1$ and $\alpha = 1$

When the damping c_1 is also zero, we have an undamped two DOF system with the following frequency responses.

$$\mu = \frac{r^2 - \alpha^2}{\varepsilon \alpha^2 r^2 - \left(r^2 - 1\right)\left(r^2 - \alpha^2\right)} \tag{8.92}$$

$$\tau = \frac{\alpha^2}{\varepsilon \alpha^2 r^2 - \left(r^2 - 1\right)\left(r^2 - \alpha^2\right)} \tag{8.93}$$

When damping is infinite, we have an undamped single DOF system with the following frequency response and natural frequency.

$$\mu = \tau = \frac{1}{r^2\left(1 + \varepsilon\right) - 1} \tag{8.94}$$

$$r_1^2 = r_2^2 = \frac{1}{\varepsilon + 1} \tag{8.95}$$

Having $\xi_2 = \infty$ reduces the system to a single DOF system with mass $m = m_1 + m_2$. Figures 8.21 and 8.22 illustrate μ and τ for $\xi_2 = 0$ and $\xi_2 = \infty$, for a sample mass ratio $\varepsilon = 0.1$ and frequency ratio $\alpha = 1$. By decreasing the damping from $\xi_2 = \infty$, we allow flexibility and relative motion between m_1 and m_2. The two degrees of freedom and relative motion will be more obvious when the damping is lower. Figures 8.21 and 8.22 also illustrate the frequency responses for different ξ_2.

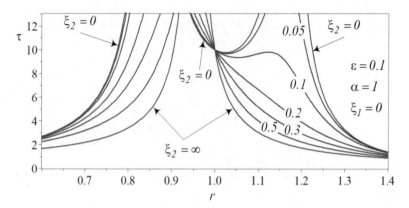

Fig. 8.22 The frequency response $\tau = \frac{X_2}{F/k_1}$ for various ξ_2 and for $\varepsilon = 0.1$ and $\alpha = 1$

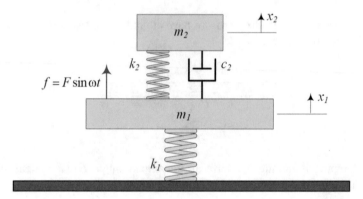

Fig. 8.23 A secondary vibration absorber system (m_2, c_2, k_2) added to a primary vibrating system (m_1, k_1)

8.2 ★ Vibration Absorber

Vibration is an unavoidable phenomena in mechanical systems. In this section, we learn the analysis principles of vibration absorbers. Special attention will be devoted to frequency response analysis, because most of the optimization methods for mechanical systems are based on frequency responses.

Consider a primary system of a mass m_1 supported by a spring k_1, as is shown in Fig. 8.23. There is a harmonic force $f = F \sin \omega t$ applied on m_1. We add a secondary system (m_2, c_2, k_2) to absorb the vibrations of the primary. Such a system is also called a *Frahm absorber*, or *Frahm damper*. Adding a vibration absorber to a primary system generates a new system with *two DOF*. The goal of having a vibration absorber is to design the suspension of the secondary system (c_2, k_2) to reduce the amplitude of vibration of m_1 at a given excitation frequency ω or in a range of frequencies.

To minimize the steady-state vibration amplitude of m_1 when the excitation frequency is variable, we must set k_2 at the optimal value $k_2^\bigstar$

$$k_2^\bigstar = \frac{m_1 m_2}{(m_1 + m_2)^2} k_1 \tag{8.96}$$

and select c_2 within the following range.

$$2m_2\omega_1\xi_1^\bigstar < c_2 < 2m_2\omega_1\xi_2^\bigstar \tag{8.97}$$

The optimal damping ratios $\xi_1^\bigstar$ and $\xi_2^\bigstar$ are

$$\xi_1^\bigstar = \sqrt{\frac{-B - \sqrt{B^2 - 4AC}}{2A}} \tag{8.98}$$

$$\xi_2^\bigstar = \sqrt{\frac{-B + \sqrt{B^2 - 4AC}}{2A}} \tag{8.99}$$

$$A = 16Z_8 - 4r^2 (4Z_4 + 8Z_5) \tag{8.100}$$

$$B = 4Z_9 - 4Z_6 r^2 - Z_7 (4Z_4 + 8Z_5) + 4Z_3 Z_8 \tag{8.101}$$

$$C = Z_3 Z_9 - Z_6 Z_7 \tag{8.102}$$

$$Z_3 = 2\left(r^2 - \alpha^2\right) \tag{8.103}$$

$$Z_4 = \left[r^2 (1 + \varepsilon) - 1\right]^2 \tag{8.104}$$

$$Z_5 = r^2 (1 + \varepsilon)\left[r^2 (1 + \varepsilon) - 1\right] \tag{8.105}$$

$$Z_6 = 2\left[\varepsilon\alpha^2 r^2 - \left(r^2 - \alpha^2\right)\left(r^2 - 1\right)\right]$$
$$\times \left[\varepsilon\alpha^2 - \left(r^2 - \alpha^2\right) - \left(r^2 - 1\right)\right] \tag{8.106}$$

$$Z_7 = \left(r^2 - \alpha^2\right)^2 \tag{8.107}$$

$$Z_8 = r^2\left[r^2 (1 + \varepsilon) - 1\right]^2 \tag{8.108}$$

$$Z_9 = \left[\varepsilon\alpha^2 r^2 - \left(r^2 - 1\right)\left(r^2 - \alpha^2\right)\right]^2 \tag{8.109}$$

Proof The main goal in vibration optimization is to reduce the vibration amplitude of a primary mass, when the system is under a forced vibration. There are two principal methods for decreasing the vibration amplitude of a primary mass: *vibration absorption* and *vibration isolation*. When the suspension of a *primary system* is not easy to optimize, we can add another vibrating system, known as the *vibration absorber* or *secondary system*, to absorb the vibrations of the primary system. The vibration absorber that increases the DOF of the system is an applied method for vibration reduction in frequency domain of systems that have already been designed.

Let us write the equations of motion for the system shown in Fig. 8.23 as

$$m_1\ddot{x}_1 + c_2\left(\dot{x}_1 - \dot{x}_2\right) + k_1 x_1 + k_2\left(x_1 - x_2\right) = F\sin\omega t \tag{8.110}$$

$$m_2\ddot{x}_2 - c_2\left(\dot{x}_1 - \dot{x}_2\right) - k_2\left(x_1 - x_2\right) = 0 \tag{8.111}$$

or in matrix form.

$$\begin{bmatrix} m_1 & 0 \\ 0 & m_2 \end{bmatrix}\begin{bmatrix} \ddot{x}_1 \\ \ddot{x}_2 \end{bmatrix} + \begin{bmatrix} c_2 & -c_2 \\ -c_2 & c_2 \end{bmatrix}\begin{bmatrix} \dot{x}_1 \\ \dot{x}_2 \end{bmatrix}$$
$$+ \begin{bmatrix} k_1 + k_2 & -k_2 \\ -k_2 & k_2 \end{bmatrix}\begin{bmatrix} x_1 \\ x_2 \end{bmatrix} = \begin{bmatrix} F\sin\omega t \\ 0 \end{bmatrix} \tag{8.112}$$

Assuming a steady-state condition, and substituting a set of general harmonic solutions in the equations of motion, we find the following set of equations for A_1, B_1, A_2, and B_2.

$$x_1 = A_1\sin\omega t + B_1\cos\omega t \tag{8.113}$$

$$x_2 = A_2\sin\omega t + B_2\cos\omega t \tag{8.114}$$

$$\begin{bmatrix} a_{11} & -k_2 & -\omega c_2 & \omega c_2 \\ -k_2 & a_{22} & \omega c_2 & -\omega c_2 \\ \omega c_2 & -\omega c_2 & a_{33} & -k_2 \\ -\omega c_2 & \omega c_2 & -k_2 & a_{44} \end{bmatrix}\begin{bmatrix} A_1 \\ A_2 \\ B_1 \\ B_2 \end{bmatrix} = \begin{bmatrix} F \\ 0 \\ 0 \\ 0 \end{bmatrix} \tag{8.115}$$

$$a_{11} = a_{33} = k_2 + k_1 - m_1\omega^2 \tag{8.116}$$

$$a_{22} = a_{44} = k_2 - m_2\omega^2 \tag{8.117}$$

The steady-state amplitudes X_1 and X_2 for vibration of m_1 and m_2 are

$$X_1 = \sqrt{A_1^2 + B_1^2} \tag{8.118}$$

$$X_2 = \sqrt{A_2^2 + B_2^2} \tag{8.119}$$

and they are equal to

$$\left(\frac{X_1}{F}\right)^2 = \frac{\left(k_2 - \omega^2 m_2\right)^2 + \omega^2 c_2^2}{Z_1^2 + \omega^2 c_2^2 Z_2^2} \tag{8.120}$$

$$\left(\frac{X_2}{F}\right)^2 = \frac{k_2^2 + \omega^2 c_2^2}{Z_1^2 + \omega^2 c_2^2 Z_2^2} \tag{8.121}$$

$$Z_1 = \left(k_1 - \omega^2 m_1\right)\left(k_2 - \omega^2 m_2\right) - \omega^2 m_2 k_2 \tag{8.122}$$

$$Z_2 = k_1 - \omega^2 m_1 - \omega^2 m_2 \tag{8.123}$$

Introducing the following parameters

$$\varepsilon = \frac{m_2}{m_1} \tag{8.124}$$

$$\omega_1 = \sqrt{\frac{k_1}{m_1}} \tag{8.125}$$

$$\omega_2 = \sqrt{\frac{k_2}{m_2}} \tag{8.126}$$

$$\alpha = \frac{\omega_2}{\omega_1} \tag{8.127}$$

$$r = \frac{\omega}{\omega_1} \tag{8.128}$$

$$\xi = \frac{c_2}{2m_2\omega_1} \tag{8.129}$$

$$\mu = \frac{X_1}{F/k_1} \tag{8.130}$$

$$\tau = \frac{X_2}{F/k_1} \tag{8.131}$$

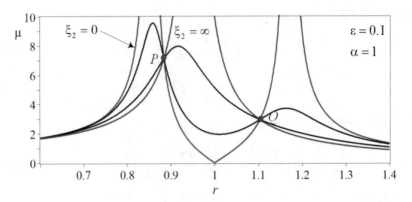

Fig. 8.24 Behavior of frequency response μ for a set of parameters and different damping ratios

we can rearrange the frequency responses (8.120) and (8.121) to the following dimensionless forms.

$$\mu^2 = \left(\frac{X_1}{F/k_1}\right)^2$$

$$= \frac{4\xi_2^2 r^2 + \left(r^2 - \alpha^2\right)^2}{4\xi_2^2 r^2 \left[r^2(1+\varepsilon) - 1\right]^2 + \left[\varepsilon\alpha^2 r^2 - \left(r^2 - 1\right)\left(r^2 - \alpha^2\right)\right]^2} \tag{8.132}$$

$$\tau^2 = \left(\frac{X_2}{F/k_1}\right)^2$$

$$= \frac{4\xi_2^2 r^2 \alpha^2 + \alpha^4}{4\xi_2^2 r^2 \left[r^2(1+\varepsilon) - 1\right]^2 + \left[\varepsilon\alpha^2 r^2 - \left(r^2 - 1\right)\left(r^2 - \alpha^2\right)\right]^2} \tag{8.133}$$

The parameter ε is the *mass ratio* of m_2 and the m_1, ω_1 is the natural frequency of the main system, and ω_2 is the natural frequency of the vibration absorber system. The frequencies ω_1 and ω_2 are not the natural frequencies of the whole system. The parameter α is the natural frequency ratio, r is the excitation frequency ratio, ξ is the damping ratio, and μ and τ are the steady-state amplitudes of the primary and secondary X_1, X_2 divided by the static deflection F/k_1.

Figure 8.24 illustrates the behavior of frequency response μ for

$$\varepsilon = 0.1 \qquad \alpha = 1 \tag{8.134}$$

and damping ratio of

$$\xi = 0 \qquad \xi = 0.2 \qquad \xi = 0.3 \qquad \xi = \infty \tag{8.135}$$

All curves pass through two nodes P and Q, independent of the value of damping ratio ξ. To find the parameters that control the position of the nodes, we find the intersection points of the curves for $\xi = 0$ and $\xi = \infty$. Setting $\xi = 0$ and $\xi = \infty$ yields

$$\mu^2 = \frac{\left(r^2 - \alpha^2\right)^2}{\left(\varepsilon\alpha^2 r^2 - \left(r^2 - 1\right)\left(r^2 - \alpha^2\right)\right)^2} \tag{8.136}$$

$$\mu^2 = \frac{1}{\left(r^2\left(1 + \varepsilon\right) - 1\right)^2} \tag{8.137}$$

When $\xi = 0$, the system is an undamped linear two DOF system with two natural frequencies. The vibration amplitude of the system approaches infinity, $\mu \to \infty$, when the excitation frequency approaches either of the natural frequencies of the system. When $\xi = \infty$, there would be no relative motion between m_1 and m_2, and the system reduces to an undamped linear one DOF system with one natural frequency:

$$\omega_n = \sqrt{\frac{k_1}{m_1 + m_2}} \tag{8.138}$$

or

$$r_n = \frac{1}{\sqrt{1 + \varepsilon}} \tag{8.139}$$

The vibration amplitude of the system approaches infinity, $\mu \to \infty$, when the excitation frequency approaches the natural frequency $\omega \to \omega_{n_i}$ or $r \to 1/\left(1 + \varepsilon\right)$.

Equating Eqs. (8.136) and (8.137) for $\xi = 0$ and $\xi = \infty$, we have

$$\frac{\left(r^2 - \alpha^2\right)^2}{\left(\varepsilon\alpha^2 r^2 - \left(r^2 - 1\right)\left(r^2 - \alpha^2\right)\right)^2} = \frac{1}{\left(r^2\left(1 + \varepsilon\right) - 1\right)^2} \tag{8.140}$$

which simplifies to

$$\varepsilon\alpha^2 r^2 - \left(r^2 - 1\right)\left(r^2 - \alpha^2\right) = \pm\left(r^2 - \alpha^2\right)\left(r^2\left(1 + \varepsilon\right) - 1\right) \tag{8.141}$$

The negative sign is equivalent to

$$r^4\varepsilon = 0 \tag{8.142}$$

which indicates there is a node at $r = 0$. The plus sign produces a quadratic equation for r^2 with two positive solutions r_1 and r_2 corresponding to nodes P and Q:

$$(2 + \varepsilon) r^4 - r^2 \left(2 + 2\alpha^2 (1 + \varepsilon) \right) + 2\alpha^2 = 0 \tag{8.143}$$

$$r_{1,2}^2 = \frac{1}{\varepsilon + 2} \left(\alpha^2 \pm \sqrt{(\varepsilon^2 + 2\varepsilon + 1) \alpha^4 - 2\alpha^2 + 1} + \alpha^2 \varepsilon + 1 \right) \tag{8.144}$$

$$r_1 < r_n < r_2 \tag{8.145}$$

Because the frequency response curves always pass through P and Q, the optimal situation would happen when the nodes P and Q have equal height.

$$\mu (P) = \mu (Q) \tag{8.146}$$

The values of μ^2 at P and Q are independent of ξ. Therefore, we may substitute r_1 and r_2 in Eq. (8.137) for μ corresponding to $\xi = \infty$. However, writing μ from Eq. (8.137), we see that

$$\mu = \frac{1}{r^2 (1 + \varepsilon) - 1} \tag{8.147}$$

produces a positive number for $r < r_n$ and a negative number for $r > r_n$. Therefore, to equate $\mu (P)$ and $\mu (Q)$, we should solve the following for of equations

$$\mu (r_1) = -\mu (r_2) \tag{8.148}$$

or

$$\frac{1}{1 - r_1^2 (1 + \varepsilon)} = \frac{-1}{1 - r_2^2 (1 + \varepsilon)} \tag{8.149}$$

which simplifies to

$$r_1^2 + r_2^2 = \frac{2}{1 + \varepsilon} \tag{8.150}$$

The sum of the roots using Eq. (8.143) is

$$r_1^2 + r_2^2 = \frac{2 + 2\alpha^2 (1 + \varepsilon)}{1 + \varepsilon} \tag{8.151}$$

and, therefore,

$$\frac{2}{1 + \varepsilon} = \frac{2 + 2\alpha^2 (1 + \varepsilon)}{1 + \varepsilon} \tag{8.152}$$

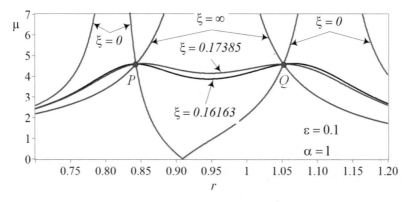

Fig. 8.25 Optimal damping ratio ξ to have maximum of μ at P or Q

which provides us with the optimal value of $\alpha = \alpha^\star$.

$$\alpha^\star = \frac{1}{1+\varepsilon} = \frac{m_1}{m_1 + m_2} \tag{8.153}$$

Equation (8.153) is the required condition to make the height of the nodes P and Q equal and, hence, provides us with the optimal value of α. Having an optimal value for $\alpha^\star$ is equivalent to designing the optimal stiffness k_2 for the secondary suspension, because

$$\alpha = \frac{\omega_2}{\omega_1} = \sqrt{\frac{m_1}{m_2}} \sqrt{\frac{k_2}{k_1}} \tag{8.154}$$

and therefore, Eq. (8.153) provides us with the following condition for optimal stiffness of the vibration absorber, $k_2^\star$.

$$k_2^\star = k_1 \frac{m_1 m_2}{(m_1 + m_2)^2} \tag{8.155}$$

To determine the optimal damping ratio ξ, we force μ to have its maximum at P or Q. Having μ_{Max} at P guarantees that $\mu(r_1)$ is the highest value in a frequency domain around r_1, and having μ_{Max} at Q guarantees that $\mu(r_2)$ is the highest amplitude in a frequency domain around r_2. The position of μ_{Max} is controlled by ξ, so we may determine two optimal values of ξ to make μ_{Max} at $\mu(r_1)$ and at $\mu(r_2)$. An example of this situation is shown in Fig. 8.25.

Using $\alpha^\star$ from (8.153), the nodal frequencies are

$$r_{1,2}^2 = \frac{1}{1+\varepsilon}\left(1 \pm \sqrt{\frac{\varepsilon}{2+\varepsilon}}\right) \tag{8.156}$$

To set the partial derivative $\partial \mu / \partial r$ equal to zero at the nodal frequencies,

$$\left. \frac{\partial \left(\mu^2 \right)}{\partial \left(r^2 \right)} \right|_{r_1^2} = 0 \tag{8.157}$$

$$\left. \frac{\partial \left(\mu^2 \right)}{\partial \left(r^2 \right)} \right|_{r_2^2} = 0 \tag{8.158}$$

let us rewrite μ^2 by the numerator $N(r)$ divided by the denominator $D(r)$ which helps us to find the derivative.

$$\mu^2 = \frac{N(r)}{D(r)} \tag{8.159}$$

$$\frac{\partial \mu^2}{\partial r^2} = \frac{1}{D^2} \left(D \frac{\partial N}{\partial r^2} - N \frac{\partial D}{\partial r^2} \right) = \frac{1}{D} \left(\frac{\partial N}{\partial r^2} - \frac{N}{D} \frac{\partial D}{\partial r^2} \right) \tag{8.160}$$

Differentiating yields

$$\frac{\partial N}{\partial r^2} = 4 \xi^2 + Z_3 \tag{8.161}$$

$$\frac{\partial D}{\partial r^2} = 4 \xi^2 Z_4 + 8 \xi^2 Z_5 + Z_6 \tag{8.162}$$

Equations (8.161) and (8.162), along with (8.156), must be substituted in (8.160) to solve for ξ. After substitution, the equation $\partial \left(\mu^2 \right) / \partial \left(r^2 \right) = 0$ would be

$$\frac{\partial N}{\partial \left(r^2 \right)} - \frac{N}{D} \frac{\partial D}{\partial \left(r^2 \right)} = \left(4 \xi^2 + Z_3 \right) \left(4 \xi^2 Z_8 + Z_9 \right)$$
$$- \left(4 \xi^2 r^2 + Z_7 \right) \left(4 \xi^2 Z_4 + 8 \xi^2 Z_5 + Z_6 \right) = 0 \tag{8.163}$$

because

$$\frac{N}{D} = \frac{4 \xi^2 r^2 + Z_7}{4 \xi^2 Z_8 + Z_9} \tag{8.164}$$

Equation (8.163) is a quadratic equation for ξ^2

$$\left(16 Z_8 - 4 r^2 \left(4 Z_4 + 8 Z_5 \right) \right) \xi^4$$
$$+ \left(4 Z_9 - 4 Z_6 r^2 - Z_7 \left(4 Z_4 + 8 Z_5 \right) + 4 Z_3 Z_8 \right) \xi^2$$
$$+ \left(Z_3 Z_9 - Z_6 Z_7 \right) = A \left(\xi^2 \right)^2 + B \xi^2 + C = 0 \tag{8.165}$$

with the solution

$$\xi^2 = \frac{-B \pm \sqrt{B^2 - 4AC}}{2A} \tag{8.166}$$

where Z_3 to Z_8 are given in Eqs. (8.103)–(8.109) and $A, B, and C$ are from (8.100)–(8.102).

The positive value of ξ from (8.166) for $r = r_1$ and $r = r_2$ provides us with the limiting values for $\xi_1^\star$ and $\xi_2^\star$. Figure 8.25 illustrates the behavior of μ for optimal α and $\xi = 0, \xi_1^\star, \xi_2^\star, \infty$.

If $\xi = 0$, then $\mu = 0$ at $r = 1$, which shows the amplitude of the primary mass reduces to zero if the natural frequency of the primary and secondary systems are equal to the excitation frequency $r = \alpha = 1$. Therefore, the vibration absorber is most effective at $r = \alpha = 1$. ∎

Example 293 ★ Optimal spring and damper for $\varepsilon = 0.1$. Here is a numerical example to calculate the value of optimal stiffness and damping for a practical application.

Consider a vibration absorber with the following mass ratio.

$$\varepsilon = \frac{m_2}{m_1} = 0.1 \tag{8.167}$$

We determine the optimal frequency ratio α from Equation (8.153)

$$\alpha^\star = \frac{1}{1 + \varepsilon} \approx 0.9091 \tag{8.168}$$

and find the nodal frequencies $r_{1,2}^2$ from (8.156).

$$r_{1,2}^2 = \frac{1}{1 + \varepsilon} \left(1 \pm \sqrt{\frac{\varepsilon}{2 + \varepsilon}} \right) = 0.71071, \ 1.1075 \tag{8.169}$$

Now, we set $r = r_1 = \sqrt{0.71071} \approx 0.843$ and evaluate the parameters Z_3 to Z_9 from (8.103)–(8.109)

$$Z_3 = -0.231470544 \qquad Z_4 = 0.0476190476$$
$$Z_5 = -0.1705988426 \qquad Z_6 = 0.0246326501$$
$$Z_7 = 0.01339465321 \qquad Z_8 = 0.03384338136$$
$$Z_9 = 0.0006378406298 \tag{8.170}$$

and then the coefficients A, B, and C from (8.100)–(8.102).

$$A = 3.879887219$$
$$B = -0.08308086729$$
$$C = -0.0004775871233 \qquad (8.171)$$

The first optimal damping ratio $\xi_1^\star$ is then will be calculated.

$$\xi_1^\star = 0.1616320694 \qquad (8.172)$$

Using $r = r_2 = \sqrt{1.1075} \approx 1.05236$, we find

$$Z_3 = 0.562049056 \qquad\qquad Z_4 = 0.04761904752$$
$$Z_5 = 0.2658369375 \qquad\qquad Z_6 = -0.375123324$$
$$Z_7 = 0.07897478534 \qquad\qquad Z_8 = 0.05273670508$$
$$Z_9 = 0.003760704084 \qquad\qquad\qquad\qquad (8.173)$$

$$A = -9.421012739$$
$$B = 0.1167823931$$
$$C = 0.005076228579 \qquad (8.174)$$

The second optimal damping ratio $\xi_2^\star$ is then equal to

$$\xi_2^\star = 0.1738496023 \qquad (8.175)$$

Therefore, the optimal α is $\alpha^\star = 0.9091$, and the optimal ξ is within the following range.

$$0.1616320694 < \xi^\star < 0.1738496023 \qquad (8.176)$$

Example 294 ★ The optimal nodal amplitude. The equal amplitude at nodal frequencies is a function of mass ratio.

Substituting the optimal α from (8.153) in Equation (8.143)

$$r^4 - \frac{2}{2+\varepsilon}r^2 + \frac{2}{(2+\varepsilon)(1+\varepsilon)^2} = 0 \qquad (8.177)$$

provides us with the nodal frequencies $r_{1,2}$.

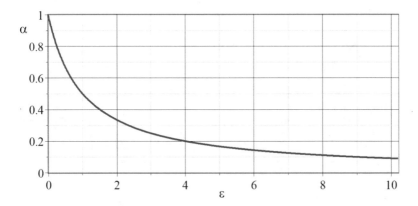

Fig. 8.26 Optimal value of the natural frequency ratio, α, as a function of mass ratio ε

$$r_{1,2}^2 = \frac{1}{1+\varepsilon}\left(1 \pm \sqrt{\frac{\varepsilon}{2+\varepsilon}}\right) \tag{8.178}$$

Applying $r_{1,2}$ in Eq. (8.147) shows that the common nodal amplitude $\mu\left(r_{1,2}\right)$ is only a function of mass ratio ε.

$$\mu = \sqrt{\frac{2+\varepsilon}{\varepsilon}} \tag{8.179}$$

Example 295 ★ Optimal α and mass ratio ε. The optimal α is only a function of mass ratio $\varepsilon = m_2/m_1$. A plot of α versus ε shows how α changes with ε. Practically, we prefer the vibration absorber to be as light as possible, which means ε to be as low as possible. However, lower ε produces higher oscillation amplitude. So, it is a trade-off.

The optimal value of the natural frequency ratio, α, is only a function of mass ratio ε, as determined in Eq. (8.153). Figure 8.26 depicts the behavior of α as a function of ε. The value of optimal α, and, hence, the value of optimal k_2, decreases by increasing $\varepsilon = m_2/m_1$. Therefore, a smaller mass for the vibration absorber needs a softer spring.

Example 296 ★ Nodal frequencies $r_{1,2}$ and mass ratio ε. The frequencies at node for the optimal α are only functions of the mass ratio ε. By selecting ε, the optimal k_2 and the nodal frequencies are determined.

As shown in Eq. (8.156), the nodal frequencies $r_{1,2}$ for optimal α (8.153) are only a function of the mass ratio ε.

$$r_{1,2}^2 = \frac{1}{1+\varepsilon}\left(1 \pm \sqrt{\frac{\varepsilon}{2+\varepsilon}}\right) \tag{8.180}$$

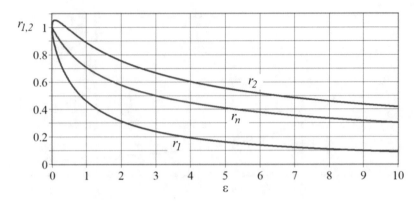

Fig. 8.27 Behavior of nodal frequencies $r_{1,2}$ as a function of mass ratio ε

Figure 8.27 illustrates the behavior of $r_{1,2}$ as a function of ε. If $\varepsilon \rightarrow 0$, the vibration absorber m_2 vanishes, and hence, the system becomes a one DOF primary oscillator. Such a system has only one natural frequency $r_n = 1$. The frequencies $r_{1,2}$ will also approach $r_{1,2} = 1$ by vanishing m_2. However, when $\xi \rightarrow \infty$, the only natural frequency r_n of the resultant one DOF system is as given in Eq. (8.139).

The nodal frequencies $r_{1,2}$ are always on both sides of the one DOF natural frequency r_n, while all of them are decreasing functions of the mass ratio ε.

$$r_1 < r_n < r_2 \tag{8.181}$$

Example 297 ★ Natural frequencies for extreme values of damping. Extreme values of damping are zero and infinity damping. The extreme damping will change the DOF of the system; however, drawing frequency response curves for extreme or any two values of damping determine the nodes of vibrations.

By setting $\xi = 0$ for a given $\varepsilon = 0.1$, we find

$$\mu = \left| \frac{r^2 - 1}{0.1r^2 - \left(r^2 - 1\right)^2} \right| \tag{8.182}$$

and by setting $\xi = \infty$, we find

$$\mu = \left| \frac{1}{1.1r^2 - 1} \right| \tag{8.183}$$

Having $\xi = 0$ is equivalent to no damping. When there is no damping, μ approaches infinity at the real roots of its denominator, r_{n_1} and r_{n_2}, which are the natural frequencies of the system. As an example, the natural frequencies r_{n_1} and r_{n_2}, for $\varepsilon = 0.1$, are

$$0.1r^2 - \left(r^2 - 1\right)^2 = 0 \tag{8.184}$$

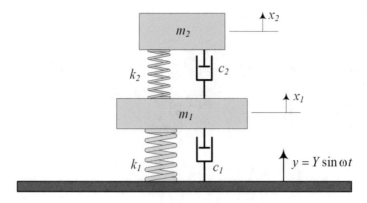

Fig. 8.28 The model of the two DOF base excited vibrating system

$$r_{n_1} = 0.85431 \qquad r_{n_2} = 1.1705 \qquad (8.185)$$

Having $\xi = \infty$ is equivalent to a rigid connection between m_1 and m_2. The system would have only one DOF, and therefore, μ approaches infinity at the only root of the denominator, r_n

$$1.1r^2 - 1 = 0 \qquad (8.186)$$

$$r_n = 0.953 \qquad (8.187)$$

where r_n is always between r_{n_1} and r_{n_2}.

$$r_{n_1} < r_n < r_{n_2} \qquad (8.188)$$

8.3   Base Excitation

Figure 8.28 illustrates a two DOF base excited vibrating system. It is composed by attaching a secondary system (m_2, k_2, c_2) to a primary base excited system (m_1, k_1, c_1). The equations of motion of the system are

$$m_1\ddot{x}_1 + c_1\dot{x}_1 + c_2(\dot{x}_1 - \dot{x}_2) + k_1x_1 + k_2(x_1 - x_2) = k_1y + c_1\dot{y} \qquad (8.189)$$

$$m_2\ddot{x}_2 - c_2(\dot{x}_1 - \dot{x}_2) - k_2(x_1 - x_2) = 0 \qquad (8.190)$$

$$y = Y\sin\omega t \qquad (8.191)$$

and the steady-state amplitudes X_1 and X_2 for vibrations of m_1 and m_2 are

$$\left(\frac{X_1}{k_1 Y}\right)^2 = \frac{\left(k_1^2 + \omega^2 c_1^2\right)\left(\left(k_2^2 + \omega^2 c_2^2\right) + \omega^2 m_2 \left(\omega^2 m_2 - 2k_2\right)\right)}{Z_1^2 + \omega^2 c_2^2 Z_2^2 + c_1^2 \omega^2 Z_3^2 + \omega^4 c_1 c_2 \left(c_1 c_2 + 2\omega^2 m_2^2\right)} \tag{8.192}$$

$$\left(\frac{X_2}{k_1 Y}\right)^2 = \frac{\left(k_1^2 + \omega^2 c_1^2\right)\left(k_2^2 + \omega^2 c_2^2\right)}{Z_1^2 + \omega^2 c_2^2 Z_2^2 + c_1^2 \omega^2 Z_3^2 + \omega^4 c_1 c_2 \left(c_1 c_2 + 2\omega^2 m_2^2\right)} \tag{8.193}$$

$$Z_1 = \left(k_1 - \omega^2 m_1\right)\left(k_2 - \omega^2 m_2\right) - \omega^2 m_2 k_2 \tag{8.194}$$

$$Z_2 = k_1 - \omega^2 m_1 - \omega^2 m_2 \tag{8.195}$$

$$Z_3 = \left(k_2 - \omega^2 m_2\right) \tag{8.196}$$

Introducing the following parameters

$$\varepsilon = \frac{m_2}{m_1} \tag{8.197}$$

$$\xi_1 = \frac{c_1}{2m_1 \omega_1} \tag{8.198}$$

$$\xi_2 = \frac{c_2}{2m_2 \omega_2} \tag{8.199}$$

$$\omega_1 = \sqrt{\frac{k_1}{m_1}} \tag{8.200}$$

$$\omega_2 = \sqrt{\frac{k_2}{m_2}} \tag{8.201}$$

$$\alpha = \frac{\omega_2}{\omega_1} \tag{8.202}$$

$$r = \frac{\omega}{\omega_1} \tag{8.203}$$

$$\mu = \frac{X_1}{Y} \tag{8.204}$$

$$\tau = \frac{X_2}{Y} \tag{8.205}$$

we can transform the frequency responses (8.192) and (8.193) to the following nondimensional forms.

$$\mu^2 = \left(\frac{X_1}{Y}\right)^2 = \frac{\left(4\xi_1^2 r^2 + 1\right)\left(\alpha^2\left(4\xi_2^2 r^2 + \alpha^2\right) + Z_6\right)}{4\xi_2^2 r^2 \alpha^2 Z_5^2 + Z_4^2 + 8\xi_1\xi_2\alpha r^4 Z_5 + 4r^6\xi_1^2} \tag{8.206}$$

$$\tau^2 = \left(\frac{X_2}{Y}\right)^2 = \frac{\alpha^2\left(4\xi_2^2 r^2 + \alpha^2\right)\left(4\xi_1^2 r^2 + 1\right)}{4\xi_2^2 r^2 \alpha^2 Z_5^2 + Z_4^2 + 8\xi_1\xi_2\alpha r^4 Z_5 + 4r^6\xi_1^2} \tag{8.207}$$

$$Z_4 = \varepsilon\alpha^2 r^2 - \left(r^2 - 1\right)\left(r^2 - \alpha^2\right) \tag{8.208}$$

$$Z_5 = r^2(1 + \varepsilon) - 1 \tag{8.209}$$

$$Z_6 = r^2\left(r^2 - 2\alpha^2\right) \tag{8.210}$$

Proof The kinetic and potential energies and the dissipation function of the system are

$$K = \frac{1}{2}m_1\dot{x}_1^2 + \frac{1}{2}m_2\dot{x}_2^2 \tag{8.211}$$

$$P = \frac{1}{2}k_1(x_1 - y)^2 + \frac{1}{2}k_2(x_2 - x_1)^2 \tag{8.212}$$

$$D = \frac{1}{2}c_1(\dot{x}_1 - \dot{y})^2 + \frac{1}{2}c_2(\dot{x}_2 - \dot{x}_1)^2 \tag{8.213}$$

Employing the Lagrange equation (2.547), we find the equations of motion as (8.189) or (8.189) or equivalently in matrix form.

$$\begin{bmatrix} m_1 & 0 \\ 0 & m_2 \end{bmatrix}\begin{bmatrix} \ddot{x}_1 \\ \ddot{x}_2 \end{bmatrix} + \begin{bmatrix} c_1 + c_2 & -c_2 \\ -c_2 & c_2 \end{bmatrix}\begin{bmatrix} \dot{x}_1 \\ \dot{x}_2 \end{bmatrix}$$
$$+ \begin{bmatrix} k_1 + k_2 & -k_2 \\ -k_2 & k_2 \end{bmatrix}\begin{bmatrix} x_1 \\ x_2 \end{bmatrix} = \begin{bmatrix} k_1 y + c_1 \dot{y} \\ 0 \end{bmatrix} \tag{8.214}$$

Substituting a set of harmonic solutions

$$x_1 = A_1 \sin \omega t + B_1 \cos \omega t \tag{8.215}$$

$$x_2 = A_2 \sin \omega t + B_2 \cos \omega t \tag{8.216}$$

provides us with a set of algebraic equations to calculate the coefficients A_1, B_1, A_2, and B_2.

$$
\begin{bmatrix}
a_{11} & -k_2 & -\omega(c_1+c_2) & \omega c_2 \\
-k_2 & a_{22} & \omega c_2 & -\omega c_2 \\
\omega(c_1+c_2) & -\omega c_2 & a_{33} & -k_2 \\
-\omega c_2 & \omega c_2 & -k_2 & a_{44}
\end{bmatrix}
\begin{bmatrix}
A_1 \\ A_2 \\ B_1 \\ B_2
\end{bmatrix}
=
\begin{bmatrix}
k_1 Y \\ c_1 \omega Y \\ 0 \\ 0
\end{bmatrix}
\tag{8.217}
$$

$$
a_{11} = a_{33} = k_2 + k_1 - m_1 \omega^2 \tag{8.218}
$$

$$
a_{22} = a_{44} = k_2 - m_2 \omega^2 \tag{8.219}
$$

The steady-state amplitudes X_1 and X_2 of m_1 and m_2 are

$$
X_1 = \sqrt{A_1^2 + B_1^2} \tag{8.220}
$$

$$
X_2 = \sqrt{A_2^2 + B_2^2} \tag{8.221}
$$

Simplifying for X_1 and X_2, we can write them as (8.192) and (8.193) or as (8.206) and (8.207).

The ratio of the amplitudes of m_1 and m_2 is

$$
\frac{X_1^2}{X_2^2} = 1 + \frac{\omega^2 m_2 \left(\omega^2 m_2 - 2k_2\right)}{\left(k_2^2 + \omega^2 c_2^2\right)} \tag{8.222}
$$

$$
\frac{\mu^2}{\tau^2} = 1 + \frac{r^2 \left(r^2 - 2\alpha^2\right)}{\alpha^2 \left(4\xi_2^2 r^2 + \alpha^2\right)} \tag{8.223}
$$

The primary parameters m_1, k_1, and c_1 do not appear in the amplitude ratio X_1/X_2. Therefore, the values of m_1, k_1, and c_1 will not effect the ratio of X_1/X_2. In the nondimensional form, the ratio of X_1/X_2 is a function of α, ξ_2, and r, and not a function of ε and ξ_1.

The damping ratios of ξ_1 and ξ_2 are, respectively, proportional to c_1 and c_2, and they are defined as the damping ratios of the single DOF subsystems. The parameter $\alpha = \omega_2/\omega_1 = \sqrt{k_2/(\varepsilon k_1)}$ is the stiffness ratio of the secondary to primary system. The frequencies ω_1 and ω_2 are the natural frequencies of the separated primary and secondary subsystems. These are not the natural frequencies of the whole system. The mass ratio ε measures the mass of the system added to the primary one. The frequency responses μ and τ indicate the dynamic amplitude of m_1 and m_2 compared to the amplitude of the harmonic base excitation.

In the base excited system, we assume that the excitation comes from an ideal source such that none of the characteristics or reactions of the system affects the excitation. The source of excitation is assumed to be able to provide infinite energy and power, if needed.   ∎

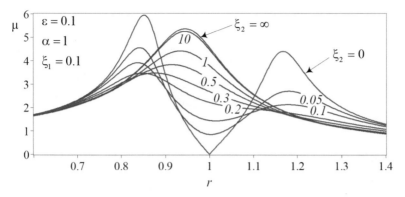

Fig. 8.29 The base excited frequency response $\mu = \frac{X_1}{Y}$ of the two DOF system for various ξ_2 and for $\varepsilon = 0.1, \xi_1 = 0.1$, and $\alpha = 1$

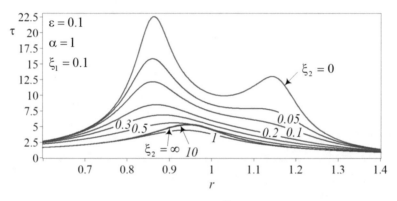

Fig. 8.30 The base excited frequency response $\tau = \frac{X_2}{Y}$ of the two DOF system for various ξ_2 and for $\varepsilon = 0.1, \xi_1 = 0.4$, and $\alpha = 1$

Example 298 Frequency response of the base excited two DOF system. The analytic equations for frequency responses of the system are amplitudes X_i as functions of the excitation frequency ω, as well as being dependent on several parameters. The right way to illustrate the frequency responses is to make the results dimensionless to minimize the number of parameters and generalize the result to be valid for all similar systems. Then we set all parameters to their nominal values except the most important one. Then we plot the frequency response curves for different values of the parameter.

The ratio of X_1 and X_2 is only a function of α, ξ_2, and r. Therefore, it is practical to show the frequency responses (8.206) and (8.207) by choosing ε and ξ_1 and varying α and ξ_2 as functions of r. Showing two examples, Figs. 8.29 and 8.30 depict the frequency responses μ and τ for $\varepsilon = 0.1, \xi_1 = 0.1$, and $\alpha = 1$, and Figs. 8.31 and 8.32 show μ and τ for $\varepsilon = 0.1, \xi_1 = 0.4$, and $\alpha = 1$, using ξ_2 as a parameter.

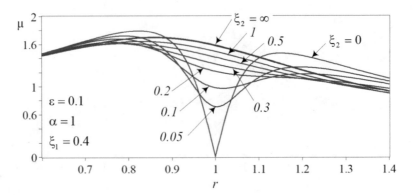

Fig. 8.31 The base excited frequency response $\mu = \frac{X_1}{Y}$ of the two DOF system for various ξ_2 and for $\varepsilon = 0.1$, $\xi_1 = 0.4$, and $\alpha = 1$

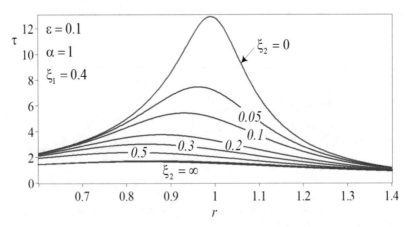

Fig. 8.32 The base excited frequency response $\tau = \frac{X_2}{Y}$ of the two DOF system for various ξ_2 and for $\varepsilon = 0.1$, $\xi_1 = 0.4$, and $\alpha = 1$

Example 299 Relative displacement. Maximum relative displacement in multi-DOF mechanical vibrating system is a design factor. Determination of the relative displacement frequency responses begins from the definition of a new coordinate and substituting the harmonic solution of the other coordinates.

The relative displacement of m_1 and the ground in the two DOF base excited system of Fig. 8.28 is

$$z = x_1 - y \tag{8.224}$$

The frequency response of z can be calculated by substituting for x_1 and y

$$
\begin{aligned}
z &= (A_1 \sin \omega t + B_1 \cos \omega t) - Y \sin \omega t \\
&= (A_1 - Y) \sin \omega t + B_1 \cos \omega t \\
&= Z \sin \left(\omega t - \varphi_{z_1} \right)
\end{aligned} \tag{8.225}
$$

The steady-state amplitude Z of the relative displacement is

$$Z = \sqrt{(A_1 - Y)^2 + B_1^2} \tag{8.226}$$

$$\left(\frac{Z}{Y}\right)^2 = \frac{\omega^4 \left(\omega^2 c_2^2 (m_1 + m_2)^2 + \left(\omega^2 m_2 m_1 - k_2 (m_1 + m_2)\right)^2\right)}{Z_1^2 + \omega^2 c_2^2 Z_2^2 + c_1^2 \omega^2 Z_3^2 + \omega^4 c_1 c_2 \left(c_1 c_2 + 2\omega^2 m_2^2\right)} \tag{8.227}$$

where the parameters Z_1, Z_2, $and\, Z_3$ are the same as given in Eqs. (8.5)–(8.7).

$$Z_1 = \left(k_1 - \omega^2 m_1\right)\left(k_2 - \omega^2 m_2\right) - \omega^2 m_2 k_2 \tag{8.228}$$

$$Z_2 = k_1 - \omega^2 m_1 - \omega^2 m_2 \tag{8.229}$$

$$Z_3 = \left(k_2 - \omega^2 m_2\right) \tag{8.230}$$

Using the parameters (8.197)–(8.205), we transform Eq. (8.227) to a nondimensional form

$$\eta^2 = \left(\frac{Z}{Y}\right)^2 = \frac{r^4 \left((\alpha^4 + 4r^2\alpha^2\xi_2^2)(1 + \varepsilon)^2 - 2r^2\alpha(1 + \varepsilon) + r^4\right)}{4\xi_2^2 r^2\alpha^2 Z_5^2 + Z_4^2 + 8\xi_1\xi_2\alpha r^4 Z_5 + 4r^6\xi_1^2} \tag{8.231}$$

where the parameters Z_4 and Z_5 are the same as given in Eqs. (8.19) and (8.20).

$$Z_4 = \varepsilon\alpha^2 r^2 - \left(r^2 - 1\right)\left(r^2 - \alpha^2\right) \tag{8.232}$$

$$Z_5 = r^2(1 + \varepsilon) - 1 \tag{8.233}$$

Figure 8.33 illustrates an example of the relative displacement frequency response of the base excited two DOF system. Interestingly, the relative displacement can be adjusted to stay zero at different frequency ration than $r = \alpha$ that makes $\mu = 0$.

Example 300 Transmitted force. The transmitted force to the base is always a design factor to construct foundation for a mechanical vibrating device. The frequency response of the transmitted force to the base of the two DOF base excited system is calculated and illustrated here.

The transmitted force f_T to the ground of the base excited system of Fig. 8.28 is the combined forces of the spring k_1 and damper c_1.

$$f_T = k_1(x_1 - y) + c_1(\dot{x}_1 - \dot{y}) \tag{8.234}$$

We determine the frequency response of f_T by substituting for x_1 and y.

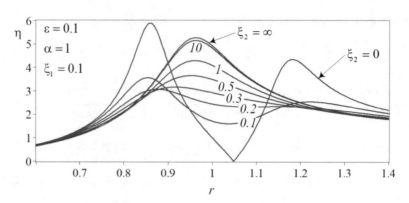

Fig. 8.33 The relative displacement frequency response of the base excited two *DOF* system $\eta = \frac{Z}{Y}$ for various ξ_2 and for $\varepsilon = 0.1$, $\xi_1 = 0.1$, and $\alpha = 1$

$$
\begin{aligned}
f_T &= k_1 \left(A_1 \sin \omega t + B_1 \cos \omega t - Y \sin \omega t\right) \\
&\quad + c_1 \omega \left(A_1 \cos \omega t - B_1 \sin \omega t - Y \cos \omega t\right) \\
&= (k_1 A_1 - c_1 \omega B_1 - k_1 Y) \sin \omega t + (k_1 B_1 + c_1 \omega A_1 - c_1 \omega Y) \cos \omega t \\
&= F_T \sin \left(\omega t - \varphi_F\right)
\end{aligned}
\tag{8.235}
$$

The steady-state amplitude of the transmitted force is

$$
F_T = \sqrt{(k_1 A_1 - c_1 \omega B_1 - Y)^2 + (k_1 B_1 + c_1 \omega A_1 - c_1 \omega Y)^2}
\tag{8.236}
$$

or

$$
\left(\frac{F_T}{Y}\right)^2 = \frac{\omega^6 \left(k_1^2 + c_1^2 \omega^2\right) \left(\omega^2 c_2^2 (m_1 + m_2)^2 + Z_6^2\right)}{Z_1^2 + \omega^2 c_2^2 Z_2^2 + c_1^2 \omega^2 Z_3^2 + \omega^4 c_1 c_2 \left(c_1 c_2 + 2\omega^2 m_2^2\right)}
\tag{8.237}
$$

$$
Z_6 = \left(\omega^2 m_2 m_1 - k_2 (m_1 + m_2)\right)
\tag{8.238}
$$

and the parameters Z_1, Z_2, and Z_3 are the same as in Eqs. (8.5)–(8.7).

$$
Z_1 = \left(k_1 - \omega^2 m_1\right)\left(k_2 - \omega^2 m_2\right) - \omega^2 m_2 k_2
\tag{8.239}
$$

$$
Z_2 = k_1 - \omega^2 m_1 - \omega^2 m_2
\tag{8.240}
$$

$$
Z_3 = \left(k_2 - \omega^2 m_2\right)
\tag{8.241}
$$

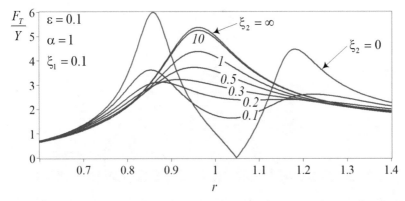

Fig. 8.34 The transmitted force frequency response of the base excited two DOF system $\frac{F_T}{Y}$ for various ξ_2 and for $\varepsilon = 0.1, \xi_1 = 0.1$, and $\alpha = 1$

Using the parameters (8.8)–(8.16), we transform Eq. (8.237) to the following form.

$$\left(\frac{F_T}{Y}\right)^2 = \frac{r^4 \left(4\xi_1^2 r^2 + 1\right)\left(\left(\alpha^4 + 4r^2\alpha^2\xi_2^2\right)(1 + \varepsilon)^2 - Z_7\right)}{4\xi_2^2 r^2\alpha^2 Z_5^2 + Z_4^2 + 8\xi_1\xi_2\alpha r^4 Z_5 + 4r^6\xi_1^2} \tag{8.242}$$

$$Z_4 = \varepsilon\alpha^2 r^2 - \left(r^2 - 1\right)\left(r^2 - \alpha^2\right) \tag{8.243}$$

$$Z_5 = r^2(1 + \varepsilon) - 1 \tag{8.244}$$

$$Z_7 = 2r^2\alpha(1 + \varepsilon) - r^4 \tag{8.245}$$

The parameters Z_4 and Z_5 are the same as in Eqs. (8.19) and (8.20). Figure 8.34 illustrates an example of the transmitted force frequency response of the base excited two DOF system.

Example 301 Amplitude ratio X_1/X_2. The plot of the amplitude frequency response X_1/X_2 depicts an engineering design factor to understand the excitation frequency ranges at which X_1 is less than or greater than X_2.

The amplitude ratio of m_1 to m_2 in system of the base excited system of Fig. 8.28

$$\frac{X_1^2}{X_2^2} = \frac{\mu^2}{\tau^2} = 1 + \frac{r^2\left(r^2 - 2\alpha^2\right)}{\alpha^2\left(4\xi_2^2 r^2 + \alpha^2\right)} \tag{8.246}$$

is the same as Eq. (8.54) for the system of Fig. 8.2.

$$\frac{X_1^2}{X_2^2} = \frac{\mu^2}{\tau^2} = \frac{4\alpha^2\xi_2^2 r^2 + \left(r^2 - \alpha^2\right)^2}{4\alpha^2\xi_2^2 r^2 + \alpha^4} \tag{8.247}$$

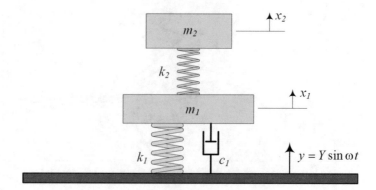

Fig. 8.35 The two DOF base excited vibrating system with no damping in the secondary system

Therefore, all discussions and all results of Example 287 for the forced excitation system are applied for the base excited two DOF system as well.

The natural frequencies of systems are independent of the excitation; therefore, the natural frequencies and discussion of Example 288 are valid for the base excited two DOF system as well. The determinant of the coefficient matrix of Eq. (8.217) is the same as Eq. (8.27).

Example 302 No secondary damping. Examining vibrating systems for the extreme cases of zero damping and infinity damping is a main part of studying the system. These two extreme cases indicate the boundaries of dynamic behavior of the system and are considered a main step in analysis.

Consider the situation when c_2 is too small compared to the primary damper such that we can model the system as is shown in Fig. 8.35. The equations of motion and the frequency responses of the system can be recovered from Eqs. (8.189)–(8.193) by setting $c_2 = 0$

$$
\begin{bmatrix} m_1 & 0 \\ 0 & m_2 \end{bmatrix} \begin{bmatrix} \ddot{x}_1 \\ \ddot{x}_2 \end{bmatrix} + \begin{bmatrix} c_1 & 0 \\ 0 & 0 \end{bmatrix} \begin{bmatrix} \dot{x}_1 \\ \dot{x}_2 \end{bmatrix}
$$

$$
+ \begin{bmatrix} k_1 + k_2 & -k_2 \\ -k_2 & k_2 \end{bmatrix} \begin{bmatrix} x_1 \\ x_2 \end{bmatrix} = \begin{bmatrix} k_1 y + c_1 \dot{y} \\ 0 \end{bmatrix} \tag{8.248}
$$

$$
\left(\frac{X_1}{k_1 Y} \right)^2 = \frac{\left(k_1^2 + \omega^2 c_1^2 \right) \left(k_2^2 + \omega^2 m_2 \left(\omega^2 m_2 - 2k_2 \right) \right)}{Z_1^2 + c_1^2 \omega^2 Z_3^2} \tag{8.249}
$$

$$
\left(\frac{X_2}{k_1 Y} \right)^2 = \frac{\left(k_1^2 + \omega^2 c_1^2 \right) k_2^2}{Z_1^2 + c_1^2 \omega^2 Z_3^2} \tag{8.250}
$$

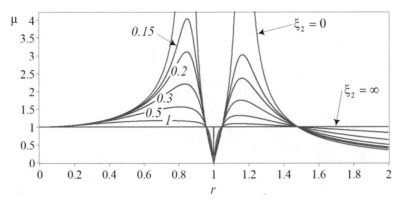

Fig. 8.36 The base excited frequency response $\mu = \frac{X_1}{Y}$ of the primary for $\varepsilon = 0.1$, $\alpha = 1$, and $\xi_2 = 0$

$$Z_1 = \left(k_1 - \omega^2 m_1\right)\left(k_2 - \omega^2 m_2\right) - \omega^2 m_2 k_2 \tag{8.251}$$

$$Z_3 = \left(k_2 - \omega^2 m_2\right) \tag{8.252}$$

or, equivalently, by substituting $\xi_2 = 0$ in Eqs. (8.206) and (8.207).

$$\mu^2 = \left(\frac{X_1}{Y}\right)^2 = \frac{\left(4\xi_1^2 r^2 + 1\right)\left(\alpha^4 + Z_6\right)}{Z_4^2 + 4r^6\xi_1^2} \tag{8.253}$$

$$\tau^2 = \left(\frac{X_2}{Y}\right)^2 = \frac{\alpha^4\left(4\xi_1^2 r^2 + 1\right)}{Z_4^2 + 4r^6\xi_1^2} \tag{8.254}$$

$$Z_4 = \varepsilon\alpha^2 r^2 - \left(r^2 - 1\right)\left(r^2 - \alpha^2\right) \tag{8.255}$$

$$Z_6 = r^2\left(r^2 - 2\alpha^2\right) \tag{8.256}$$

Figure 8.36 illustrates the frequency response μ of the primary system, and Fig. 8.37 illustrates the frequency response τ of the secondary system for the sample numerical values of $\varepsilon = 0.1$, $\alpha = 1$, and $\xi_2 = 0$.

Example 303 Effect of the secondary system on the main system. The effects of adding a secondary system to a primary system to make a two DOF system will be more observable by studying the undamped system.

To examine the effect of a secondary system on a base excited system after attachment the secondary system, let us compare the frequency responses of the systems in Fig. 8.38. The zero damping frequency response and natural frequency of the system of Fig. 8.38a are

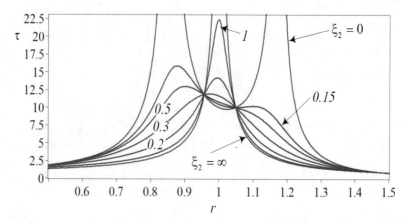

Fig. 8.37 The base excited frequency response $\tau = \frac{X_2}{Y}$ of the secondary for $\varepsilon = 0.1$, $\alpha = 1$, and $\xi_2 = 0$.

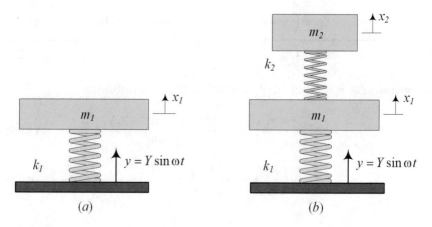

Fig. 8.38 (a) The undamped main base excited vibrating system. (b) The undamped system after attachment of a secondary system to the main

$$\mu_a = \frac{X}{Y} = \frac{1}{1 - r^2} \qquad r = 1 \qquad (8.257)$$

The zero damping frequency response of the main system of Fig. 8.38b is

$$\mu_b = \frac{X_1}{Y} = \frac{r^2 - \alpha^2}{\varepsilon\alpha^2 r^2 - (r^2 - 1)(r^2 - \alpha^2)} \qquad (8.258)$$

These are exactly the same results as we found in Eqs. (8.78) and (8.79) for the two DOF forced excited system of Fig. 8.16. Therefore, the results of Example 290 are valid for the base excited system as well.

Example 304 Freezing m_1 at a desired frequency. The application of adding a secondary vibrating system to a base excited primary is to absorb the vibrations of the primary and save it from oscillation. How to adjust a vibration absorber to make a primary system stationary at any particular excitation frequency is studied here.

Let us examine if it is possible to stop the vibration of m_1 of the system of Fig. 8.35 at a given excitation frequency. Assume the frequency and mass ratio are given.

$$r = 0.6 \qquad \varepsilon = m_2/m_1 = 0.02 \tag{8.259}$$

When $\xi_2 = 0$, we can write the frequency response of m_1 as

$$\mu^2 = \left(\frac{X_1}{Y}\right)^2 = \frac{\left(4\xi_1^2 r^2 + 1\right)\left(\alpha^2 - r^2\right)^2}{Z_4^2 + 4r^6\xi_1^2} \tag{8.260}$$

$$Z_4 = \varepsilon\alpha^2 r^2 - \left(r^2 - 1\right)\left(r^2 - \alpha^2\right) \tag{8.261}$$

which shows that $\mu = 0$ when $r = \alpha$. Therefore, to design a proper secondary system as vibration absorber, we adjust $\alpha = r$. Having $\varepsilon = m_2/m_1 = 0.02$, we can determine the secondary stiffness:

$$k_2 = k_1 \varepsilon\alpha^2 = 0.02 k_1 \times 0.6^2 = \frac{9}{1250} k_1 \tag{8.262}$$

Figure 8.39 illustrates μ for the design parameters. The vibration of m_1 is zero at the desired frequency $r = 0.6$. The frequency response of m_2 is shown in Fig. 8.40.

Example 305 No primary damping. It is an extreme model for system with very low damping in primary compared to the damping of the secondary system. Rubber and elastomer suspension are examples of suspension with low damping.

When the primary damping c_1 is zero or is very low, we can ignore damping in the primary system and model the system as is shown in Fig. 8.41. The interaction of the secondary system will affect the motion of m_1.

We recover the equations of motion and the frequency responses of the system from Eqs. (8.189)–(8.193) by substituting $c_1 = 0$ or, equivalently, $\xi_1 = 0$.

$$\begin{bmatrix} m_1 & 0 \\ 0 & m_2 \end{bmatrix}\begin{bmatrix} \ddot{x}_1 \\ \ddot{x}_2 \end{bmatrix} + \begin{bmatrix} c_2 & -c_2 \\ -c_2 & c_2 \end{bmatrix}\begin{bmatrix} \dot{x}_1 \\ \dot{x}_2 \end{bmatrix} + \begin{bmatrix} k_1 + k_2 & -k_2 \\ -k_2 & k_2 \end{bmatrix}\begin{bmatrix} x_1 \\ x_2 \end{bmatrix} = \begin{bmatrix} k_1 y + c_1 \dot{y} \\ 0 \end{bmatrix} \tag{8.263}$$

$$\left(\frac{X_1}{k_1 Y}\right)^2 = \frac{k_1^2\left(\left(k_2^2 + \omega^2 c_2^2\right) + \omega^2 m_2 \left(\omega^2 m_2 - 2k_2\right)\right)}{Z_1^2 + \omega^2 c_2^2 Z_2^2} \tag{8.264}$$

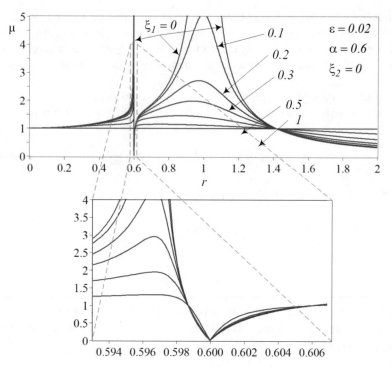

Fig. 8.39 The base excited frequency response $\mu = \frac{X_1}{Y}$ of the primary for $\varepsilon = 0.02$, $\alpha = 0.6$, and $\xi_2 = 0$

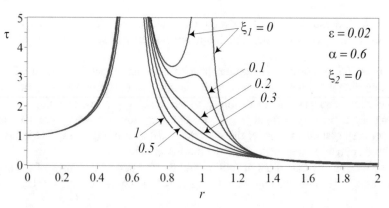

Fig. 8.40 The base excited frequency response $\tau = \frac{X_2}{Y}$ of the secondary for $\varepsilon = 0.02$, $\alpha = 0.6$, and $\xi_2 = 0$

$$\left(\frac{X_2}{k_1 Y}\right)^2 = \frac{k_1^2 \left(k_2^2 + \omega^2 c_2^2\right)}{Z_1^2 + \omega^2 c_2^2 Z_2^2} \tag{8.265}$$

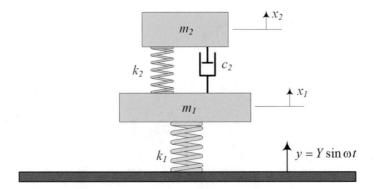

Fig. 8.41 The base excited two DOF system with no primary damping

$$\mu^2 = \left(\frac{X_1}{Y}\right)^2 = \frac{\alpha^2 \left(4\xi_2^2 r^2 + \alpha^2\right) + r^2 \left(r^2 - 2\alpha^2\right)}{4\xi_2^2 r^2 \alpha^2 Z_5^2 + Z_4^2} \tag{8.266}$$

$$\tau^2 = \left(\frac{X_2}{Y}\right)^2 = \frac{\alpha^2 \left(4\xi_2^2 r^2 + \alpha^2\right)}{4\xi_2^2 r^2 \alpha^2 Z_5^2 + Z_4^2} \tag{8.267}$$

$$Z_1 = \left(k_1 - \omega^2 m_1\right)\left(k_2 - \omega^2 m_2\right) - \omega^2 m_2 k_2 \tag{8.268}$$

$$Z_2 = k_1 - \omega^2 m_1 - \omega^2 m_2 \tag{8.269}$$

$$Z_4 = \varepsilon \alpha^2 r^2 - \left(r^2 - 1\right)\left(r^2 - \alpha^2\right) \tag{8.270}$$

$$Z_5 = r^2 (1 + \varepsilon) - 1 \tag{8.271}$$

When the damping c_2 is also zero, we have an undamped two DOF system with frequency responses.

$$\mu^2 = \left(\frac{X_1}{Y}\right)^2 = \frac{\alpha^4 + Z_6}{Z_4^2} \tag{8.272}$$

$$\tau^2 = \left(\frac{X_2}{Y}\right)^2 = \frac{\alpha^4}{Z_4^2} \tag{8.273}$$

When the damping c_2 is infinite, we have an undamped single DOF system with the following frequency response μ and natural frequency r_1.

$$\mu = \tau = \frac{1}{r^2 (1 + \varepsilon) - 1} \tag{8.274}$$

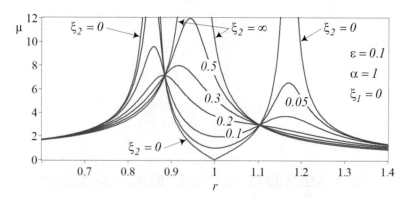

Fig. 8.42 The base excitation frequency response $\mu = \frac{X_1}{Y}$ of the two DOF system for various ξ_2 and for $\varepsilon = 0.1$ and $\alpha = 1$

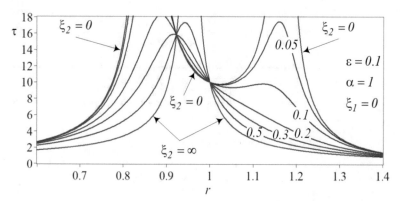

Fig. 8.43 The base excited frequency response $\tau = \frac{X_2}{Y}$ of the two DOF system for various ξ_2 and for $\varepsilon = 0.1$ and $\alpha = 1$

$$r_1^2 = r_2^2 = \frac{1}{\varepsilon + 1} \tag{8.275}$$

Having $\xi_2 = \infty$ reduces the system to a single DOF system with mass $m = m_1 + m_2$. Figures 8.42 and 8.43 illustrate μ and τ for $\xi_2 = 0$ and $\xi_2 = \infty$, the mass ratio $\varepsilon = 0.1$, and the frequency ratio $\alpha = 1$. By decreasing the damping from $\xi_2 = \infty$, we allow flexibility and relative motion between m_1 and m_2. The two degrees of freedom will be more obvious when the damping is lower. Figures 8.42 and 8.43 illustrate the frequency responses for different ξ_2.

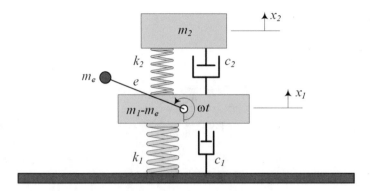

Fig. 8.44 The model of the two DOF eccentric excited vibrating system

8.4 ★ Eccentric Excitation

Figure 8.44 illustrates a two DOF eccentric excited vibrating system. The system is composed by attaching a passive secondary system (m_2, k_2, c_2) to a primary eccentric excited system (m_1, k_1, c_1). The equations of motion of the system are

$$m_1 \ddot{x}_1 + c_1 \dot{x}_1 + c_2 (\dot{x}_1 - \dot{x}_2) + k_1 (x_1 - x_2)$$
$$+ k_2 x_1 = -m_e e \omega^2 \sin \omega t \qquad (8.276)$$
$$m_2 \ddot{x}_2 + c_2 (\dot{x}_2 - \dot{x}_1) + k_2 (x_2 - x_1) = 0 \qquad (8.277)$$

The steady-state amplitudes X_1 and X_2 for vibration of m_1 and m_2 are

$$\left(\frac{X_1}{e}\right)^2 = \frac{m_e^2 \omega^4 \left((k_2 - \omega^2 m_2)^2 + \omega^2 c_2^2\right)}{Z_1^2 + \omega^2 c_2^2 Z_2^2 + c_1^2 \omega^2 Z_3^2 + \omega^4 c_1 c_2 (c_1 c_2 + 2\omega^2 m_2^2)} \qquad (8.278)$$

$$\left(\frac{X_2}{e}\right)^2 = \frac{m_e^2 \omega^4 \left(k_2^2 + \omega^2 c_2^2\right)}{Z_1^2 + \omega^2 c_2^2 Z_2^2 + c_1^2 \omega^2 Z_3^2 + \omega^4 c_1 c_2 (c_1 c_2 + 2\omega^2 m_2^2)} \qquad (8.279)$$

$$Z_1 = \left(k_1 - \omega^2 m_1\right)\left(k_2 - \omega^2 m_2\right) - \omega^2 m_2 k_2 \qquad (8.280)$$

$$Z_2 = k_1 - \omega^2 m_1 - \omega^2 m_2 \qquad (8.281)$$

$$Z_3 = \left(k_2 - \omega^2 m_2\right) \qquad (8.282)$$

Introducing the following parameters

$$\varepsilon = \frac{m_2}{m_1} \tag{8.283}$$

$$\varepsilon_1 = \frac{m_e}{m_1} \tag{8.284}$$

$$\xi_1 = \frac{c_1}{2m_1\omega_1} \tag{8.285}$$

$$\xi_2 = \frac{c_2}{2m_2\omega_2} \tag{8.286}$$

$$\omega_1 = \sqrt{\frac{k_1}{m_1}} \tag{8.287}$$

$$\omega_2 = \sqrt{\frac{k_2}{m_2}} \tag{8.288}$$

$$\alpha = \frac{\omega_2}{\omega_1} \tag{8.289}$$

$$r = \frac{\omega}{\omega_1} \tag{8.290}$$

$$\mu = \frac{X_1}{e} \tag{8.291}$$

$$\tau = \frac{X_2}{e} \tag{8.292}$$

we transform the frequency responses (8.278) and (8.279) to the following dimensionless form.

$$\mu^2 = \left(\frac{X_1}{e\varepsilon_1}\right)^2 = \frac{r^4\left((\alpha^2 - r^2)^2 + 4\xi_2^2\alpha^2 r^2\right)}{4\xi_2^2 r^2\alpha^2 Z_5^2 + Z_4^2 + 8\xi_1\xi_2\alpha r^4 Z_5 + 4r^6\xi_1^2} \tag{8.293}$$

$$\tau^2 = \left(\frac{X_2}{e\varepsilon_1}\right)^2 = \frac{r^4\alpha^2\left(\alpha^2 + 4\xi_2^2 r^2\right)}{4\xi_2^2 r^2\alpha^2 Z_5^2 + Z_4^2 + 8\xi_1\xi_2\alpha r^4 Z_5 + 4r^6\xi_1^2} \tag{8.294}$$

$$Z_4 = \varepsilon\alpha^2 r^2 - \left(r^2 - 1\right)\left(r^2 - \alpha^2\right) \tag{8.295}$$

$$Z_5 = r^2\left(1 + \varepsilon\right) - 1 \tag{8.296}$$

Proof The kinetic and potential energies and the dissipation function of the system are

$$K = \frac{1}{2}m_2\dot{x}_2^2 + \frac{1}{2}(m_1 - m_e)\dot{x}_1^2$$

$$+ \frac{1}{2}m_e(\dot{x}_1 - e\omega\cos\omega t)^2 + \frac{1}{2}m_e(e\omega\sin\omega t)^2 \tag{8.297}$$

$$P = \frac{1}{2}k_1x_1^2 + \frac{1}{2}k_2(x_2 - x_1)^2 \tag{8.298}$$

$$D = \frac{1}{2}c_1\dot{x}_1^2 + \frac{1}{2}c_2(\dot{x}_2 - \dot{x}_1)^2 \tag{8.299}$$

Employing the Lagrange equation (2.547), we find the equations of motion as (8.276), as (8.277), or equivalently as:

$$\begin{bmatrix} m_1 & 0 \\ 0 & m_2 \end{bmatrix}\begin{bmatrix} \ddot{x}_1 \\ \ddot{x}_2 \end{bmatrix} + \begin{bmatrix} c_1 + c_2 & -c_2 \\ -c_2 & c_2 \end{bmatrix}\begin{bmatrix} \dot{x}_1 \\ \dot{x}_2 \end{bmatrix}$$

$$+ \begin{bmatrix} k_1 + k_2 & -k_2 \\ -k_2 & k_2 \end{bmatrix}\begin{bmatrix} x_1 \\ x_2 \end{bmatrix} = \begin{bmatrix} -m_e e\omega^2\sin\omega t \\ 0 \end{bmatrix} \tag{8.300}$$

Substituting a set of harmonic solutions

$$x_1 = A_1\sin\omega t + B_1\cos\omega t \tag{8.301}$$

$$x_2 = A_2\sin\omega t + B_2\cos\omega t \tag{8.302}$$

provides us with a set of algebraic equations to calculate the coefficients A_1, B_1, A_2, and B_2.

$$\begin{bmatrix} a_{11} & -k_2 & -\omega(c_1 + c_2) & \omega c_2 \\ -k_2 & a_{22} & \omega c_2 & -\omega c_2 \\ \omega(c_1 + c_2) & -\omega c_2 & a_{33} & -k_2 \\ -\omega c_2 & \omega c_2 & -k_2 & a_{44} \end{bmatrix}\begin{bmatrix} A_1 \\ A_2 \\ B_1 \\ B_2 \end{bmatrix} = \begin{bmatrix} -m_e e\omega^2 \\ 0 \\ 0 \\ 0 \end{bmatrix} \tag{8.303}$$

$$a_{11} = a_{33} = k_2 + k_1 - m_1\omega^2 \tag{8.304}$$

$$a_{22} = a_{44} = k_2 - m_2\omega^2 \tag{8.305}$$

The steady-state amplitudes X_1 and X_2 are

$$X_1 = \sqrt{A_1^2 + B_1^2} \tag{8.306}$$

$$X_2 = \sqrt{A_2^2 + B_2^2} \tag{8.307}$$

Simplifying X_1 and X_2, we can write them as (8.278) and (8.279) or as dimensionless form of (8.293) and (8.294). The ratio of the amplitudes of m_1 and m_2 is

$$\frac{X_1^2}{X_2^2} = \frac{\left(k_2 - \omega^2 m_2\right)^2 + \omega^2 c_2^2}{k_2^2 + \omega^2 c_2^2} \tag{8.308}$$

$$\frac{\mu^2}{\tau^2} = \frac{\left(\alpha^2 - r^2\right)^2 + 4\xi_2^2 \alpha^2 r^2}{\alpha^4 + 4\xi_2^2 \alpha^2 r^2} \tag{8.309}$$

The primary parameters m_1, k_1, and c_1 do not appear in the amplitude ratio X_1/X_2. Therefore, the values of m_1, k_1, and c_1 will not effect the ratio of X_1/X_2. In the dimensionless form, the ratio of X_1 and X_2 is a function of α, ξ_2, and r, and not a function of ε, ε_1, and ξ_1.

The damping ratios of ξ_1 and ξ_2 are, respectively, proportional to c_1 and c_2, and they are defined as the damping ratios of the single DOF subsystems. The parameter $\alpha = \omega_2/\omega_1 = \sqrt{k_2/(\varepsilon k_1)}$ is the stiffness ratio of the secondary and primary systems. The frequencies ω_1 and ω_2 are the natural frequencies of the separated primary and secondary subsystems. These are not the natural frequencies of the whole system. The mass ratio ε denotes the mass of the added system relative to the primary one. The frequency responses μ and τ are indicating the dynamic amplitude of m_1 and m_2 compared to $e\varepsilon_1$.

In eccentric excited systems, we assume the excitation comes from an ideal source such that none of the characteristics or reactions of the system affects the rotation of m_e. The source can provide as much energy and power as needed.   ■

Example 306 Frequency response of the eccentric two DOF system. The analytic equations of the frequency responses X_i as functions of the excitation frequency ω are calculated in this example. To illustrate the frequency responses, we make the results dimensionless to minimize the number of parameters and generalize the result to be valid for all similar systems. Then we set all parameters to their nominal values except the most important one to plot the frequency response curves for different values of the parameter.

Practically, we show the frequency responses (8.293) and (8.294) by choosing ε and ξ_1 and varying α and ξ_2 as functions of r. Showing two examples, Figs. 8.45 and 8.46 depict the frequency responses μ and τ for $\varepsilon = 0.1$, $\xi_1 = 0.1$, and $\alpha = 1$, and Figs. 8.47 and 8.48 shows μ and τ for $\varepsilon = 0.1$, $\xi_1 = 0.4$, and $\alpha = 1$.

Example 307 Relative displacement. To determine the relative displacement frequency responses, we need to define a new coordinate and substitute the harmonic solution of the relative coordinates. The maximum relative displacement in multi-DOF mechanical vibrating system is always a design factor to be considered.

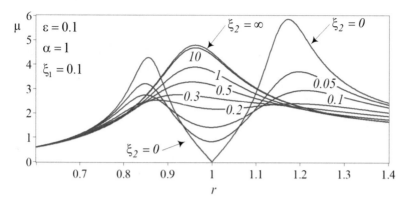

Fig. 8.45 The frequency response $\mu = \frac{X_1}{e\varepsilon_1}$ of the two DOF eccentric excited system for various ξ_2 and for $\varepsilon = 0.1$, $\xi_1 = 0.1$, and $\alpha = 1$

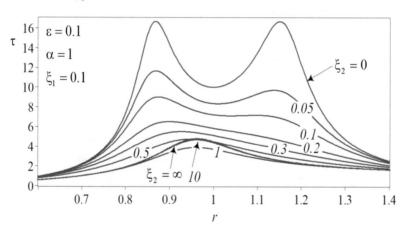

Fig. 8.46 The frequency response $\tau = \frac{X_2}{Y}$ of the two DOF eccentric excited system for various ξ_2 and for $\varepsilon = 0.1$, $\xi_1 = 0.4$, and $\alpha = 1$

The relative displacement of m_1 and m_2 in the eccentric excited system of Fig. 8.44 is

$$z = x_1 - x_2 \tag{8.310}$$

The frequency response of the relative displacement z can be calculated by substituting for x_1 and x_2.

$$
\begin{aligned}
z &= (A_1 \sin \omega t + B_1 \cos \omega t) - (A_2 \sin \omega t + B_2 \cos \omega t) \\
&= (A_1 - A_2) \sin t\omega + (B_1 - B_2) \cos t\omega \\
&= Z \sin (\omega t - \varphi_z)
\end{aligned}
\tag{8.311}
$$

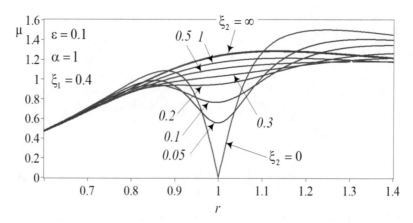

Fig. 8.47 The frequency response $\mu = \frac{X_1}{e\varepsilon_1}$ of the two *DOF* eccentric excited system for various ξ_2 and for $\varepsilon = 0.1, \xi_1 = 0.4$, and $\alpha = 1$

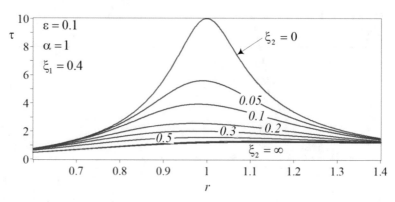

Fig. 8.48 The frequency response $\tau = \frac{X_2}{e\varepsilon_1}$ of the two *DOF* eccentric excited system for various ξ_2 and for $\varepsilon = 0.1, \xi_1 = 0.4$, and $\alpha = 1$

The steady-state amplitude of the relative displacement Z is

$$Z = \sqrt{(A_1 - A_2)^2 + (B_1 - B_2)^2} \tag{8.312}$$

$$\left(\frac{Z}{e}\right)^2 = \frac{m_e^2 m_2^2 \omega^8}{Z_1^2 + \omega^2 c_2^2 Z_2^2 + c_1^2 \omega^2 Z_3^2 + \omega^4 c_1 c_2 \left(c_1 c_2 + 2\omega^2 m_2^2\right)} \tag{8.313}$$

where the parameters Z_1, Z_2, and Z_3 are the same as given in Eqs. (8.5)–(8.7).

$$Z_1 = \left(k_1 - \omega^2 m_1\right)\left(k_2 - \omega^2 m_2\right) - \omega^2 m_2 k_2 \tag{8.314}$$

$$Z_2 = k_1 - \omega^2 m_1 - \omega^2 m_2 \tag{8.315}$$

$$Z_3 = \left(k_2 - \omega^2 m_2\right) \tag{8.316}$$

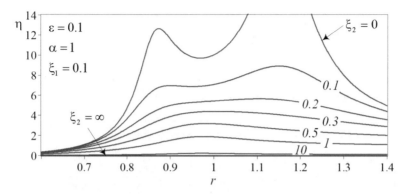

Fig. 8.49 The relative displacement frequency response of the eccentric excited two DOF system $\eta = \frac{Z}{e\varepsilon_1}$ for various ξ_2 and for $\varepsilon = 0.1$, $\xi_1 = 0.1$, and $\alpha = 1$

Using the parameters (8.283)–(8.292), we transform Equation (8.313) to a dimensionless form

$$\eta^2 = \left(\frac{Z}{e\varepsilon_1}\right)^2 = \frac{r^8}{4\xi_2^2 r^2\alpha^2 Z_5^2 + Z_4^2 + 8\xi_1\xi_2\alpha r^4 Z_5 + 4r^6\xi_1^2} \tag{8.317}$$

where the parameters Z_4 and Z_5 are the same as given in Eqs. (8.295) and (8.296).

$$Z_4 = \varepsilon\alpha^2 r^2 - \left(r^2 - 1\right)\left(r^2 - \alpha^2\right) \tag{8.318}$$

$$Z_5 = r^2\left(1 + \varepsilon\right) - 1 \tag{8.319}$$

Figure 8.49 illustrates an example of the relative displacement frequency response of the eccentric excited two DOF system.

Example 308 Transmitted force. To make a proper foundation for any vibrating mechanical device, the transmitted force to the base at any excitation frequency must be known. That means transmitted force frequency response is a design factor and should be calculated and visualized. Here is the analysis for transmitted force to the base of a two DOF eccentric excited system.

The force f_T transmitted to the ground in the eccentric excited system of Fig. 8.44 is the combined forces in spring k_1 and damper c_1.

$$f_T = k_1 x_1 + c_1 \dot{x}_1 \tag{8.320}$$

We determine the frequency response of f_T by substituting for x_1 and x_2.

$$z = k_1 (A_1 \sin \omega t + B_1 \cos \omega t) + c_1 \omega (A_1 \cos \omega t - B_1 \sin \omega t)$$
$$= (A_1 k_1 - \omega B_1 c_1) \sin \omega t + (B_1 k_1 + \omega A_1 c_1) \cos \omega t$$
$$= F_T \sin (\omega t - \varphi_F) \tag{8.321}$$

The steady-state amplitude of the transmitted force is

$$F_T = \sqrt{(A_1 k_1 - \omega B_1 c_1)^2 + (B_1 k_1 + \omega A_1 c_1)^2} \tag{8.322}$$

$$\left(\frac{F_T}{m_e e \omega_1^2} \right)^2 = \frac{\omega^4 \left(c_1^2 \omega^2 + k_1^2 \right) \left((m_2 \omega^2 - k_2)^2 + c_2^2 \omega^2 \right)}{Z_1^2 + \omega^2 c_2^2 Z_2^2 + c_1^2 \omega^2 Z_3^2 + \omega^4 c_1 c_2 \left(c_1 c_2 + 2 \omega^2 m_2^2 \right)} \tag{8.323}$$

and the parameters Z_1, Z_2, and Z_3 are the same as given in Eqs. (8.5)–(8.7).

$$Z_1 = \left(k_1 - \omega^2 m_1 \right) \left(k_2 - \omega^2 m_2 \right) - \omega^2 m_2 k_2 \tag{8.324}$$

$$Z_2 = k_1 - \omega^2 m_1 - \omega^2 m_2 \tag{8.325}$$

$$Z_3 = \left(k_2 - \omega^2 m_2 \right) \tag{8.326}$$

Using the parameters (8.283)–(8.292), we transform Equation (8.323) to

$$\left(\frac{F_T}{m_e e \omega_1^2} \right)^2 = \frac{r^4 (4 \xi_1^2 r^2 + 1) \left((r^2 - \alpha^2)^2 + 4 \alpha^2 \xi_2^2 r^2 \right)}{4 \xi_2^2 r^2 \alpha^2 Z_5^2 + Z_4^2 + 8 \xi_1 \xi_2 \alpha r^4 Z_5 + 4 r^6 \xi_1^2} \tag{8.327}$$

where the parameters Z_4 and Z_5 are the same as in Eqs. (8.295) and (8.296).

$$Z_4 = \varepsilon \alpha^2 r^2 - \left(r^2 - 1 \right) \left(r^2 - \alpha^2 \right) \tag{8.328}$$

$$Z_5 = r^2 (1 + \varepsilon) - 1 \tag{8.329}$$

Figure 8.50 illustrates an example of the transmitted force frequency response of the eccentric excited two DOF system.

Example 309 Amplitude ratio and natural frequencies. Relative amplitude at every excitation frequency is a design factor to understand at which frequency range the primary system is oscillating with higher or lower amplitude with respect to secondary system. The requirement of the application will then show the range of working frequency range.

The amplitude ratio μ/τ of m_1 to m_2 in system of Fig. 8.44 is the same as Eq. (8.54) for the system of Fig. 8.2.

$$\frac{X_1^2}{X_2^2} = \frac{\mu^2}{\tau^2} = \frac{\left(\alpha^2 - r^2 \right)^2 + 4 \xi_2^2 \alpha^2 r^2}{\alpha^4 + 4 \xi_2^2 \alpha^2 r^2} \tag{8.330}$$

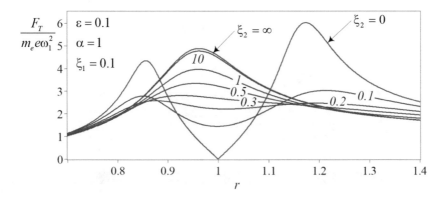

Fig. 8.50 The transmitted force frequency response of the eccentric excited two DOF system $\frac{F_T}{m_e e\omega_1^2}$ for various ξ_2 and for $\varepsilon = 0.1$, $\xi_1 = 0.1$, and $\alpha = 1$

Therefore, the whole discussion and all results of Example 287 for the forced excitation system can be applied for the eccentric excited two DOF system as well.

The natural frequencies of the systems are the roots of the denominator of the frequency responses and are independent of the excitation. Therefore, the natural frequencies and discussion of Example 288 are valid for the eccentric excited two DOF system as well. The determinant of the coefficient matrix of Eq. (8.303) is the same as Eq. (8.27).

Example 310 No secondary damping. Determining the frequency response of vibrating systems for the extreme cases of zero and infinity damping is a main engineering part of examining the system. These two extreme cases indicate the boundaries of dynamic behavior of the system and are considered a main step in analysis.

When c_2 is too small compared to the primary damper, we may ignore the secondary damping and model the system as is shown in Fig. 8.51. The equations of motion and the frequency responses of the system can be recovered from Eqs. (8.276)–(8.294) by substituting $c_2 = 0$

$$\begin{bmatrix} m_1 & 0 \\ 0 & m_2 \end{bmatrix}\begin{bmatrix} \ddot{x}_1 \\ \ddot{x}_2 \end{bmatrix} + \begin{bmatrix} c_1 & 0 \\ 0 & 0 \end{bmatrix}\begin{bmatrix} \dot{x}_1 \\ \dot{x}_2 \end{bmatrix}$$
$$+ \begin{bmatrix} k_1 + k_2 & -k_2 \\ -k_2 & k_2 \end{bmatrix}\begin{bmatrix} x_1 \\ x_2 \end{bmatrix} = \begin{bmatrix} -m_e e\omega^2 \sin \omega t \\ 0 \end{bmatrix} \tag{8.331}$$

$$\left(\frac{X_1}{e}\right)^2 = \frac{m_e^2\omega^4\left(k_2 - \omega^2 m_2\right)^2}{Z_1^2 + c_1^2\omega^2 Z_3^2} \tag{8.332}$$

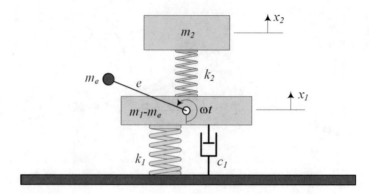

Fig. 8.51 The two DOF eccentric excited vibrating system with no damping in the secondary system

$$\left(\frac{X_2}{e}\right)^2 = \frac{m_e^2\omega^4 k_2^2}{Z_1^2 + c_1^2\omega^2 Z_3^2} \tag{8.333}$$

$$Z_1 = \left(k_1 - \omega^2 m_1\right)\left(k_2 - \omega^2 m_2\right) - \omega^2 m_2 k_2 \tag{8.334}$$

$$Z_3 = \left(k_2 - \omega^2 m_2\right) \tag{8.335}$$

or, equivalently, by substituting $\xi_2 = 0$ in Eqs. (8.293) and (8.294).

$$\mu^2 = \left(\frac{X_1}{e\varepsilon_1}\right)^2 = \frac{r^4\left(\alpha^2 - r^2\right)^2}{Z_4^2 + 4r^6\xi_1^2} \tag{8.336}$$

$$\tau^2 = \left(\frac{X_2}{e\varepsilon_1}\right)^2 = \frac{r^4\alpha^4}{Z_4^2 + 4r^6\xi_1^2} \tag{8.337}$$

$$Z_4 = \varepsilon\alpha^2 r^2 - \left(r^2 - 1\right)\left(r^2 - \alpha^2\right) \tag{8.338}$$

Figure 8.52 illustrates the frequency response μ of the primary systems, and Fig. 8.53 illustrates the frequency response τ of the secondary system for $\varepsilon = 0.1$, $\alpha = 1$, and $\xi_2 = 0$.

Example 311 Effect of the secondary system on the main system. The effects of adding a secondary system to a primary system to make a two DOF system will be observable by studying the undamped system.

To examine the effect of a secondary system on an eccentric excited system after attachment of the secondary system, we compare the frequency responses

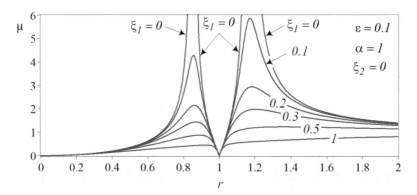

Fig. 8.52 The frequency response $\mu = \frac{X_1}{e\varepsilon_1}$ of the two DOF eccentric excited system for $\varepsilon = 0.1$, $\alpha = 1$, and $\xi_2 = 0$

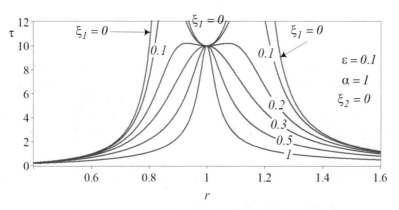

Fig. 8.53 The frequency response $\tau = \frac{X_2}{e\varepsilon_1}$ of the two DOF eccentric excited system for $\varepsilon = 0.1$, $\alpha = 1$, and $\xi_2 = 0$

of the systems of Fig. 8.54. The zero damping frequency response and the natural frequency of the system of Fig. 8.54a are

$$\mu_a = \frac{X}{e\varepsilon} = \frac{1}{1 - r^2} \qquad r = 1 \tag{8.339}$$

The zero damping frequency response of the main system of Fig. 8.54b is

$$\mu_b^2 = \left(\frac{X_1}{e\varepsilon_1}\right)^2 = \frac{r^2 - \alpha^2}{\varepsilon\alpha^2 r^2 - (r^2 - 1)(r^2 - \alpha^2)} \tag{8.340}$$

These are the same results as we found in Eqs. (8.78) and (8.79) for the two DOF forced excited system of Fig. 8.16. Therefore, the results of Example 290 are valid for the eccentric excited system as well.

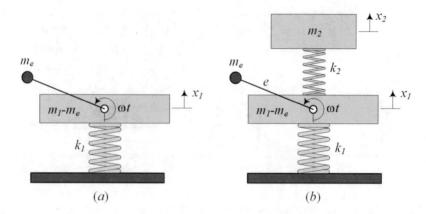

Fig. 8.54 (a) The undamped main eccentric excited vibrating system. (b) The system after attachment of an undamped secondary system to the main

Example 312 Freezing m_1 at a desired frequency. Adding a secondary vibrating system to an eccentric excited primary can absorb the vibrations of the primary and save it from oscillation. How to adjust a vibration absorber to make a primary system stationary at any particular excitation frequency is studied here.

We examine if it is possible to stop the vibration of m_1 of Fig. 8.51 at a given frequency. Assume that the frequency and mass ratio are given and we need to make $X_1 = 0$ at the given r and ε.

$$r = 0.6 \qquad \varepsilon = m_2/m_1 = 0.02 \tag{8.341}$$

When $\xi_2 = 0$, the frequency response of m_1 will be

$$\mu^2 = \left(\frac{X_1}{e\varepsilon_1}\right)^2 = \frac{r^4 \left(\alpha^2 - r^2\right)^2}{Z_4^2 + 4r^6\xi_1^2} \tag{8.342}$$

$$Z_4 = \varepsilon\alpha^2 r^2 - \left(r^2 - 1\right)\left(r^2 - \alpha^2\right) \tag{8.343}$$

It shows that $\mu = 0$ when $r = \alpha$. Therefore, to design a proper secondary system of an eccentric excited two DOF system as vibration absorber, we adjust $\alpha = r$. Having $\varepsilon = m_2/m_1 = 0.02$, we can determine the secondary stiffness.

$$k_2 = k_1\varepsilon\alpha^2 = 0.02k_1 \times 0.6^2 = \frac{9}{1250}k_1 \tag{8.344}$$

Figure 8.55 illustrates μ for the design parameters. The vibration of m_1 is zero at the desired frequency, $r = 0.6$. The frequency response of m_2 is shown in Fig. 8.56.

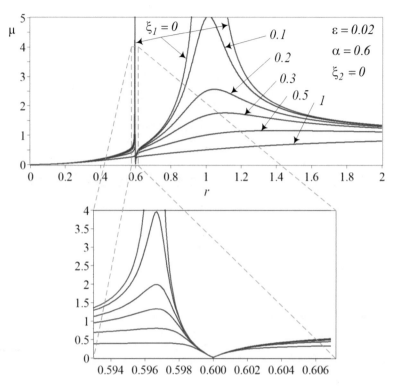

Fig. 8.55 The frequency response $\mu = \frac{X_1}{e\varepsilon_1}$ of the two DOF eccentric excited system for $\varepsilon = 0.02$, $\alpha = 0.6$, and $\xi_2 = 0$

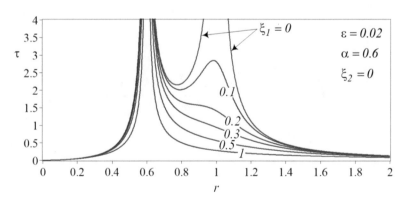

Fig. 8.56 The frequency response $\tau = \frac{X_2}{e\varepsilon_1}$ of the two DOF eccentric excited system for $\varepsilon = 0.02$, $\alpha = 0.6$, and $\xi_2 = 0$

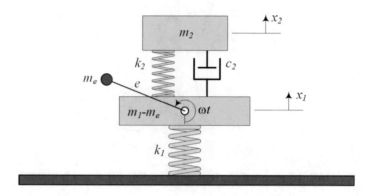

Fig. 8.57 The eccentric excited two DOF system with no primary damping

Example 313 No primary damping. No damping in the primary is an extreme model of systems with very low damping in primary compared to the damping of the secondary system.

If the primary damping c_1 is zero or is very low, we can ignore damping in the primary system and model the system as is shown in Fig. 8.57. The secondary system will change the frequency response of m_1 as a single DOF system. Let us recover the equations of motion and the frequency responses of the system from Eqs. (8.276)–(8.294) by substituting $c_1 = 0$ or, equivalently, $\xi_1 = 0$.

$$
\begin{bmatrix} m_1 & 0 \\ 0 & m_2 \end{bmatrix} \begin{bmatrix} \ddot{x}_1 \\ \ddot{x}_2 \end{bmatrix} + \begin{bmatrix} c_2 & -c_2 \\ -c_2 & c_2 \end{bmatrix} \begin{bmatrix} \dot{x}_1 \\ \dot{x}_2 \end{bmatrix}
$$
$$
+ \begin{bmatrix} k_1 + k_2 & -k_2 \\ -k_2 & k_2 \end{bmatrix} \begin{bmatrix} x_1 \\ x_2 \end{bmatrix} = \begin{bmatrix} -m_e e \omega^2 \sin \omega t \\ 0 \end{bmatrix}
\tag{8.345}
$$

$$
\left(\frac{X_1}{e} \right)^2 = \frac{m_e^2 \omega^4 \left((k_2 - \omega^2 m_2)^2 + \omega^2 c_2^2 \right)}{Z_1^2 + \omega^2 c_2^2 Z_2^2}
\tag{8.346}
$$

$$
\left(\frac{X_2}{e} \right)^2 = \frac{m_e^2 \omega^4 \left(k_2^2 + \omega^2 c_2^2 \right)}{Z_1^2 + \omega^2 c_2^2 Z_2^2}
\tag{8.347}
$$

$$
\mu^2 = \left(\frac{X_1}{e \varepsilon_1} \right)^2 = \frac{r^4 \left((\alpha^2 - r^2)^2 + 4\xi_2^2 \alpha^2 r^2 \right)}{4\xi_2^2 r^2 \alpha^2 Z_5^2 + Z_4^2}
\tag{8.348}
$$

$$
\tau^2 = \left(\frac{X_2}{e \varepsilon_1} \right)^2 = \frac{r^4 \alpha^2 \left(\alpha^2 + 4\xi_2^2 r^2 \right)}{4\xi_2^2 r^2 \alpha^2 Z_5^2 + Z_4^2}
\tag{8.349}
$$

$$Z_1 = \left(k_1 - \omega^2 m_1\right)\left(k_2 - \omega^2 m_2\right) - \omega^2 m_2 k_2 \tag{8.350}$$

$$Z_2 = k_1 - \omega^2 m_1 - \omega^2 m_2 \tag{8.351}$$

$$Z_4 = \varepsilon \alpha^2 r^2 - \left(r^2 - 1\right)\left(r^2 - \alpha^2\right) \tag{8.352}$$

$$Z_5 = r^2\left(1 + \varepsilon\right) - 1 \tag{8.353}$$

When the damping c_2 is also zero, we have an undamped two DOF system with the following frequency responses.

$$\mu^2 = \left(\frac{X_1}{e\varepsilon_1}\right)^2 = \frac{r^4\left(\alpha^2 - r^2\right)^2}{Z_4^2} \tag{8.354}$$

$$\tau^2 = \left(\frac{X_2}{e\varepsilon_1}\right)^2 = \frac{r^4\alpha^4}{Z_4^2} \tag{8.355}$$

When the damping c_2 is infinite, we have an undamped single DOF system with

$$\mu = \tau = \frac{r^2}{r^2\left(1 + \varepsilon\right) - 1} \tag{8.356}$$

and natural frequency of

$$r_1^2 = r_2^2 = \frac{1}{\varepsilon + 1} \tag{8.357}$$

Having $\xi_2 = \infty$ reduces the system to a single DOF system with mass $m = m_1 + m_2$. Figures 8.58 and 8.59 illustrate μ and τ for $\xi_2 = 0$ and $\xi_2 = \infty$, the mass ratio $\varepsilon = 0.1$, and frequency ratio $\alpha = 1$. By decreasing the damping from $\xi_2 = \infty$, the system will be more flexible, and relative motion for m_1 and m_2 appears. Figures 8.58 and 8.59 illustrate the frequency responses for different ξ_2.

8.5 ★ Eccentric Base Excitation

Figure 8.60 illustrates a two DOF eccentric base excited vibrating system. The base has an unbalance mass m_e at a distance e that is rotating with angular velocity ω. The system is composed by attaching a secondary system (m_2, k_2, c_2) to a primary eccentric base excited system. The equations of motion of the system in absolute coordinates are

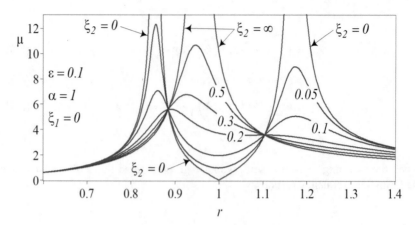

Fig. 8.58 The frequency response $\mu = \frac{X_1}{e\varepsilon_1}$ of the two DOF eccentric excited system, for various ξ_2 and for $\varepsilon = 0.1$ and $\alpha = 1$

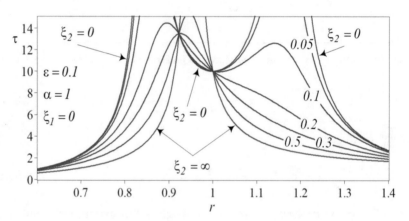

Fig. 8.59 The frequency response $\tau = \frac{X_2}{e\varepsilon_1}$ of the two DOF eccentric excited system, for various ξ_2 and for $\varepsilon = 0.1$ and $\alpha = 1$

$$m_1\ddot{x}_1 + c_1(\dot{x}_1 - \dot{y}) - c_2(\dot{x}_2 - \dot{x}_1)$$

$$+k_1(x_1 - y) - k_2(x_2 - x_1) = 0 \tag{8.358}$$

$$m_2\ddot{x}_2 + c_2(\dot{x}_2 - \dot{x}_1) + k_2(x_2 - x_1) = 0 \tag{8.359}$$

$$m_b\ddot{y} + m_e e\omega^2 \sin\omega t - c_1(\dot{x}_1 - \dot{y}) - k_1(x_1 - y) = 0 \tag{8.360}$$

Using the relative motion of m_2 and m_1 with respect to the base

$$z_1 = x_1 - y \tag{8.361}$$

$$z_2 = x_2 - y \tag{8.362}$$

Fig. 8.60 The model of the two DOF eccentric base excited vibrating system

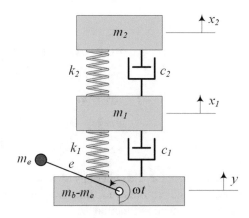

we may develop the equation of motion as

$$m_1\ddot{z}_1 + (c_1(\varepsilon_1 + 1) + c_2)\dot{z}_1 + (k_1(\varepsilon_1 + 1) + k_2)z_1$$
$$-c_2\dot{z}_2 - k_2z_2 = m_e\varepsilon_1 e\omega^2 \sin\omega t \tag{8.363}$$

$$(c_1\varepsilon_2 - c_2)\dot{z}_1 + (k_1\varepsilon_2 - k_2)z_1$$
$$+m_2\ddot{z}_2 + c_2\dot{z}_2 + k_2z_2 = m_e\varepsilon_2 e\omega^2 \sin\omega t \tag{8.364}$$

where

$$\varepsilon_1 = \frac{m_1}{m_b} \qquad \varepsilon_2 = \frac{m_2}{m_b} \tag{8.365}$$

Showing the steady-state amplitudes of z_1 and z_2 by S_1 and S_2, we find

$$\left(\frac{S_1}{m_e e}\right)^2 = \frac{S_{1N}}{S_D}\omega^4 \tag{8.366}$$

$$\left(\frac{S_2}{m_e e}\right)^2 = \frac{S_{2N}}{S_D}\omega^4 \tag{8.367}$$

where

$$S_D = c_2^2\left((a_{11} + a_{21})^2 + (a_{13} + a_{23})^2\right)\omega^2$$
$$+2c_2(k_2 - a_{22})(a_{21}a_{13} - a_{11}a_{23})\omega$$
$$+ (a_{11}a_{22} + k_2a_{21})^2 + (a_{13}a_{22} + k_2a_{23})^2 \tag{8.368}$$

$$S_{1N} = c_2^2(\varepsilon_1 + \varepsilon_2)^2\omega^2 + (\varepsilon_1a_{22} + \varepsilon_2k_2)^2 \tag{8.369}$$

$$S_{2N} = (\varepsilon_1a_{21} - \varepsilon_2a_{11})^2 + (\varepsilon_1a_{23} - \varepsilon_2a_{13})^2 \tag{8.370}$$

and

$$a_{11} = a_{33} = k_1 (\varepsilon_1 + 1) + k_2 - \omega^2 m_1 \qquad (8.371)$$

$$a_{21} = a_{43} = k_1 \varepsilon_2 - k_2 \qquad (8.372)$$

$$a_{22} = a_{44} = k_2 - \omega^2 m_2 \qquad (8.373)$$

$$a_{13} = -a_{31} = -\omega (c_1 (\varepsilon_1 + 1) + c_2) \qquad (8.374)$$

$$a_{23} = -a_{41} = -\omega (c_1 \varepsilon_2 - c_2) \qquad (8.375)$$

Proof We derive the equations of motion of the system by applying the Lagrange method. The required functions are the kinetic energy K, the potential energy P, and the dissipation function D:

$$K = \frac{1}{2} m_1 \dot{x}_1^2 + \frac{1}{2} m_2 \dot{x}_2^2 + \frac{1}{2} (m_b - m_e) \dot{y}^2$$

$$+ \frac{1}{2} m_e (\dot{y} - e\omega \cos \omega t)^2 + \frac{1}{2} m_e (e\omega \sin \omega t)^2 \qquad (8.376)$$

$$P = \frac{1}{2} k_1 (x_1 - y)^2 + \frac{1}{2} k_2 (x_2 - x_1)^2 \qquad (8.377)$$

$$D = \frac{1}{2} c_1 (\dot{x}_1 - \dot{y})^2 + \frac{1}{2} c_2 (\dot{x}_2 - \dot{x}_1)^2 \qquad (8.378)$$

Employing the Lagrange equation (2.500) provides us with the Eqs. (8.358)–(8.360), which we rewrite in matrix form:

$$\begin{bmatrix} m_1 & 0 & 0 \\ 0 & m_2 & 0 \\ 0 & 0 & m_b \end{bmatrix} \begin{bmatrix} \ddot{x}_1 \\ \ddot{x}_2 \\ \ddot{y} \end{bmatrix} + \begin{bmatrix} c_1 + c_2 & -c_2 & -c_1 \\ -c_2 & c_2 & 0 \\ -c_1 & 0 & c_1 \end{bmatrix} \begin{bmatrix} \dot{x}_1 \\ \dot{x}_2 \\ \dot{y} \end{bmatrix}$$

$$+ \begin{bmatrix} k_1 + k_2 & -k_2 & -k_1 \\ -k_2 & k_2 & 0 \\ -k_1 & 0 & k_1 \end{bmatrix} \begin{bmatrix} x_1 \\ x_2 \\ y \end{bmatrix} = \begin{bmatrix} 0 \\ 0 \\ -m_e e\omega^2 \sin \omega t \end{bmatrix} \qquad (8.379)$$

The coefficient matrices of the equations are symmetric because of using the absolute coordinates and the Lagrange method.

Employing the relative coordinates z_1 and z_2:

$$z_1 = x_1 - y \qquad (8.380)$$

$$z_2 = x_2 - y \qquad (8.381)$$

$$x_1 - x_2 = z_1 - z_2 \qquad (8.382)$$

the equations will be

$$m_1\ddot{z}_1 + m_1\ddot{y} + (c_1 + c_2)\,\dot{z}_1 - c_2\dot{z}_2 + (k_1 + k_2)\,z_1 - k_2z_2 = 0 \tag{8.383}$$

$$m_2\ddot{z}_2 + m_2\ddot{y} + c_2\dot{z}_2 - c_2\dot{z}_1 + k_2z_2 + k_2z_1 = 0 \tag{8.384}$$

$$m_b\ddot{y} + m_e e\omega^2 \sin\omega t - c_1\dot{z}_1 - k_1z_1 = 0 \tag{8.385}$$

or

$$\begin{bmatrix} m_1 & 0 & m_1 \\ m_2 & 0 & m_2 \\ 0 & 0 & m_b \end{bmatrix} \begin{bmatrix} \ddot{z}_1 \\ \ddot{z}_2 \\ \ddot{y} \end{bmatrix} + \begin{bmatrix} c_1 + c_2 & -c_2 & 0 \\ -c_2 & c_2 & 0 \\ -c_1 & 0 & 0 \end{bmatrix} \begin{bmatrix} \dot{z}_1 \\ \dot{z}_2 \\ \dot{y} \end{bmatrix}$$

$$+ \begin{bmatrix} k_1 + k_2 & -k_2 & 0 \\ -k_2 & k_2 & 0 \\ -k_1 & 0 & 0 \end{bmatrix} \begin{bmatrix} z_1 \\ z_2 \\ y \end{bmatrix} = \begin{bmatrix} 0 \\ 0 \\ -m_e e\omega^2 \sin\omega t \end{bmatrix} \tag{8.386}$$

Substituting $\ddot{y}$ from Eq. (8.385) in (8.383) and (8.384), we find the two equations of motion in the relative coordinates z_1 and z_2.

$$m_1\ddot{z}_1 + (c_1\,(\varepsilon_1 + 1) + c_2)\,\dot{z}_1 + (k_1\,(\varepsilon_1 + 1) + k_2)\,z_1$$

$$-c_2\dot{z}_2 - k_2z_2 = m_e\varepsilon_1 e\omega^2 \sin\omega t \tag{8.387}$$

$$(c_1\varepsilon_2 - c_2)\,\dot{z}_1 + (k_1\varepsilon_2 - k_2)\,z_1$$

$$+m_2\ddot{z}_2 + c_2\dot{z}_2 + k_2z_2 = m_e\varepsilon_2 e\omega^2 \sin\omega t \tag{8.388}$$

The coefficient matrices of the equations are asymmetric in the relative coordinates.

$$\begin{bmatrix} m_1 & 0 \\ 0 & m_2 \end{bmatrix} \begin{bmatrix} \ddot{z}_1 \\ \ddot{z}_2 \end{bmatrix} + \begin{bmatrix} c_1\,(\varepsilon_1 + 1) + c_2 & -c_2 \\ c_1\varepsilon_2 - c_2 & c_2 \end{bmatrix} \begin{bmatrix} \dot{z}_1 \\ \dot{z}_2 \end{bmatrix}$$

$$+ \begin{bmatrix} k_1\,(\varepsilon_1 + 1) + k_2 & -k_2 \\ k_1\varepsilon_2 - k_2 & k_2\varepsilon_2 \end{bmatrix} \begin{bmatrix} z_1 \\ z_2 \end{bmatrix} = \begin{bmatrix} m_e\varepsilon_1 e\omega^2 \sin\omega t \\ m_e\varepsilon_2 e\omega^2 \sin\omega t \end{bmatrix} \tag{8.389}$$

To find the frequency response, we substitute a set of harmonic solutions

$$z_1 = A_1 \sin\omega t + B_1 \cos\omega t \tag{8.390}$$

$$z_2 = A_2 \sin\omega t + B_2 \cos\omega t \tag{8.391}$$

which provides us with a set of algebraic equations to calculate the coefficients A_1, B_1, A_2, and B_2

$$\begin{bmatrix} a_{11} & -k_2 & a_{13} & \omega c_2 \\ a_{21} & a_{22} & a_{23} & -\omega c_2 \\ -a_{13} & -\omega c_2 & a_{11} & -k_2 \\ -a_{23} & \omega c_2 & a_{21} & a_{22} \end{bmatrix} \begin{bmatrix} A_1 \\ A_2 \\ B_1 \\ B_2 \end{bmatrix} = \begin{bmatrix} m_e \varepsilon_1 e \omega^2 \\ m_e \varepsilon_2 e \omega^2 \\ 0 \\ 0 \end{bmatrix} \tag{8.392}$$

where the elements a_{ij} are given in Eqs. (8.371)–(8.375). Let us denote the steady-state amplitudes of z_1 and z_2 by S_1 and S_2:

$$S_1 = \sqrt{A_1^2 + B_1^2} \tag{8.393}$$

$$S_2 = \sqrt{A_2^2 + B_2^2} \tag{8.394}$$

Simplifying S_1 and S_2, we can write them as (8.366) and (8.367).

The ratio of the amplitudes S_1/S_2 is

$$\frac{S_1^2}{S_2^2} = \frac{c_2^2 (\varepsilon_1 m_1 + m_b)^2 \omega^2 + (\varepsilon_1 m_1 a_{22} + k_2 m_b)^2}{(\varepsilon_1 m_1 a_{21} - a_{11} m_b)^2 + (\varepsilon_1 m_1 a_{23} - a_{13} m_b)^2} \tag{8.395}$$

Introducing the parameters

$$\varepsilon_1 = \frac{m_1}{m_b} \tag{8.396}$$

$$\varepsilon_2 = \frac{m_2}{m_b} \tag{8.397}$$

$$\varepsilon = \frac{m_e}{m_b} \tag{8.398}$$

$$\xi_1 = \frac{c_1}{2 m_1 \omega_1} \tag{8.399}$$

$$\xi_2 = \frac{c_2}{2 m_2 \omega_2} \tag{8.400}$$

$$\omega_1 = \sqrt{\frac{k_1}{m_1}} \tag{8.401}$$

$$\omega_2 = \sqrt{\frac{k_2}{m_2}} \tag{8.402}$$

$$\alpha = \frac{\omega_2}{\omega_1} \tag{8.403}$$

$$r = \frac{\omega}{\omega_1} \tag{8.404}$$

$$\mu = \frac{S_1}{\varepsilon e} \tag{8.405}$$

$$\tau = \frac{S_2}{\varepsilon e} \tag{8.406}$$

we can define the frequency responses $\mu = S_1/(e\varepsilon)$ and $\tau = S_2/(e\varepsilon)$ in dimensionless form.

The frequency responses of the eccentric base excited two DOF system are functions of the 12 parameters and functions: S_1, S_2, m_1, m_2, m_e, m_b, k_1, k_2, c_1, c_2, ω, and e. The responses are functions of *nine* parameters in dimensionless form: μ, τ, ε_1, ε_2, ε, ξ_1, ξ_2, α, and r. Due to the large number of parameters, visualization of the frequency responses of the system, even in special cases, would need too much room. Therefore, we review only a few cases and leave the analysis to the reader, using the frequency response equations (8.366) and (8.367) for situations appearing in practice.

In an eccentric base excited system, we assume that the excitation comes from an ideal source such that none of the characteristics or reactions of the system affects it. The source can provide infinite energy and power if needed. ∎

Example 314 ★ Frequency response of the eccentric base two DOF system. The equations for frequency responses of multi-DOF system are shown by the amplitude X_i as a function of the excitation frequency ω. To visualize the frequency responses, we use the dimensionless form of the equations to minimize the number of parameters and generalize the result. The behavior of the system at any frequency can be understood from the graphs of frequency responses.

We show the frequency responses (8.293) and (8.294) by choosing ε_1, ε_2, and ξ_1 and varying α and ξ_2 as functions of r. As two examples, Figs. 8.61 and 8.62 depict the frequency responses μ and τ for $\varepsilon_1 = 1$, $\varepsilon_2 = 0.1$, $\xi_1 = 0.1$, and $\alpha = 1$, and Figs. 8.63 and 8.64 show μ and τ for $\varepsilon_1 = 1$, $\varepsilon_2 = 0.1$, $\xi_1 = 0.4$, and $\alpha = 1$. These graphs show the frequency responses of the system are very sensitive to ξ_1. The behavior of the frequency responses of the system is significantly different for $\xi_1 = 0.1$ and $\xi_1 = 0.4$.

Example 315 ★ Transmitted force. Transmitted force to the base is a complicated variable for the base eccentric excited system. We may define the force that transfers to the m_b from m_1 as the transmitted force and determine its frequency response. Any other transmitted force between other masses will be calculated similarly.

The transmitted force f_T to the base of the eccentric base excited system of Fig. 8.60 is the combined forces in spring k_1 and damper c_1.

$$f_T = k_1 z_1 + c_1 \dot{z}_1 \qquad z_1 = x_1 - y \tag{8.407}$$

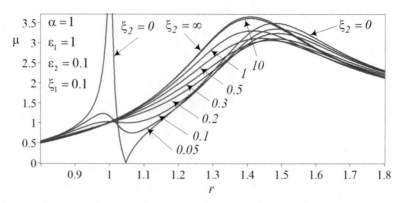

Fig. 8.61 The frequency response $\mu = \frac{S_1}{e\varepsilon}$ of the two DOF eccentric base excited system for $\varepsilon_1 = 1, \varepsilon_2 = 0.1, \alpha = 1$, and $\xi_1 = 0.1$

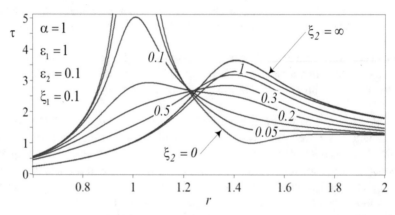

Fig. 8.62 The frequency response $\tau = \frac{S_2}{e\varepsilon}$ of the two DOF eccentric base excited system for $\varepsilon_1 = 1, \varepsilon_2 = 0.1, \alpha = 1$, and $\xi_1 = 0.1$

We determine the frequency response of f_T by substituting a harmonic solution for z_1.

$$z = k_1 (A_1 \sin \omega t + B_1 \cos \omega t) + c_1 \omega (A_1 \cos \omega t - B_1 \sin \omega t)$$
$$= (A_1 k_1 - \omega B_1 c_1) \sin \omega t + (B_1 k_1 + \omega A_1 c_1) \cos \omega t$$
$$= F_T \sin (\omega t - \varphi_F) \tag{8.408}$$

The steady-state amplitude of the transmitted force is F_T.

$$F_T = \sqrt{(A_1 k_1 - \omega B_1 c_1)^2 + (B_1 k_1 + \omega A_1 c_1)^2} \tag{8.409}$$

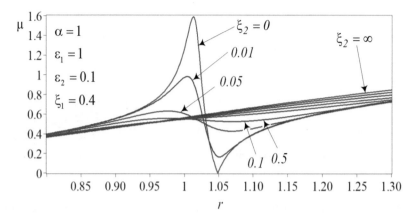

Fig. 8.63 The frequency response $\mu = \frac{S_1}{e\varepsilon}$ of the two DOF eccentric base excited system for $\varepsilon_1 = 1, \varepsilon_2 = 0.1, \alpha = 1$, and $\xi_1 = 0.4$

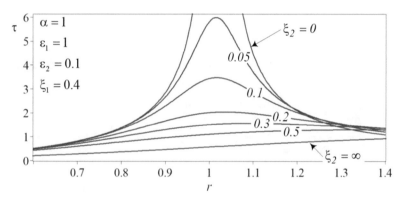

Fig. 8.64 The frequency response $\tau = \frac{S_2}{e\varepsilon}$ of the two DOF eccentric base excited system for $\varepsilon_1 = 1, \varepsilon_2 = 0.1, \alpha = 1$, and $\xi_1 = 0.4$

Figure 8.65 illustrates an example of the transmitted force frequency response $F_T / \left(m_b e \varepsilon \varepsilon_1 \omega_1^2\right)$ of the eccentric excited two DOF system. This graph shows that the transmitted force frequency response has a wide range of resonance zone, for the given data. To use such graphs for design purpose, there should be a book of these graphs for all parameters in their applied range. All such book of graphs come from Eq. (8.409).

Example 316 Relative displacement of m_1 with respect to m_2. Relative displacement of every component of a mechanical system with respect to other components are important variables to be monitored during operation and considered during design process.

By subtraction of z_1 and z_2, we determine the relative motion of m_1 with respect to m_2

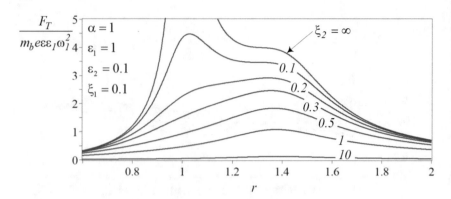

Fig. 8.65 The frequency response of the transmitted force to the ground $F_T / \left(m_b e \varepsilon \varepsilon_1 \omega_1^2\right)$ of the two DOF eccentric base excited system for $\varepsilon_1 = 1$, $\varepsilon_2 = 0.1$, $\alpha = 1$, and $\xi_1 = 0.1$

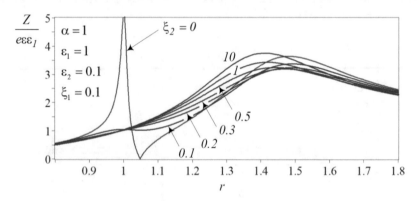

Fig. 8.66 The relative displacement $x_1 - x_2$ frequency response $Z/(e\varepsilon\varepsilon_1)$ of the two DOF eccentric base excited system for $\varepsilon_1 = 1$, $\varepsilon_2 = 0.1$, $\alpha = 1$, and $\xi_1 = 0.1$

$$x_1 - x_2 = z_1 - z_2$$
$$= (A_1 \sin \omega t + B_1 \cos \omega t) - (A_2 \sin \omega t + B_2 \cos \omega t)$$
$$= (A_1 - A_2) \sin \omega t + (B_1 - B_2) \cos \omega t$$
$$= Z \sin (\omega t - \varphi_Z) \tag{8.410}$$

The steady-state amplitude of the relative displacement is Z.

$$Z = \sqrt{(A_1 - A_2)^2 + (B_1 - B_2)^2} \tag{8.411}$$

Figure 8.66 illustrates an example of the relative displacement frequency response $Z/(e\varepsilon\varepsilon_1)$ of the eccentric base excited two DOF system.

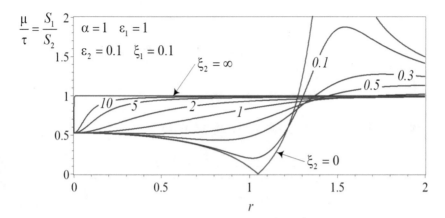

Fig. 8.67 The amplitudes of S_1/S_2 of the eccentric base excited two DOF system for $\alpha = 1$, $\varepsilon_1 = 1$, $\varepsilon_2 = 0.1$, and $\xi_1 = 0.1$

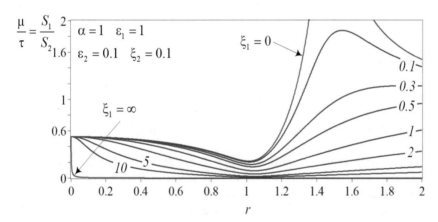

Fig. 8.68 The amplitudes of S_1/S_2 of the eccentric base excited two DOF system for $\alpha = 1$, $\varepsilon_1 = 1$, $\varepsilon_2 = 0.1$, and $\xi_2 = 0.1$

Example 317 ★ **Amplitude ratio.** Relative amplitude for base eccentric excited system has different definitions. We can use the absolute amplitude ratio X_1/X_2 or the relative amplitude ration S_1/S_2 as the amplitude ratio indicator. We use the relative amplitude ratio S_1/S_2 here. The amplitude ratio frequency response is always a design parameter to consider.

The amplitude ratio of S_1/S_2 is shown in Figs. 8.67 and 8.68 for $\alpha = 1$, $\varepsilon_1 = 1$, $\varepsilon_2 = 0.1$, $\xi_1 = 0.1$, and $\xi_1 = 1$ for different ξ_2.

$$\frac{S_1^2}{S_2^2} = \frac{c_2^2 (\varepsilon_1 m_1 + m_b)^2 \, \omega^2 + (\varepsilon_1 m_1 a_{22} + k_2 m_b)^2}{(\varepsilon_1 m_1 a_{21} - a_{11} m_b)^2 + (\varepsilon_1 m_1 a_{23} - a_{13} m_b)^2} \tag{8.412}$$

$$a_{11} = k_1 (\varepsilon_1 + 1) + k_2 - \omega^2 m_1$$

$$a_{21} = k_1 \varepsilon_2 - k_2 \qquad a_{22} = k_2 - \omega^2 m_2$$

$$a_{13} = -\omega (c_1 (\varepsilon_1 + 1) + c_2) \qquad (8.413)$$

$$a_{23} = -\omega (c_1 \varepsilon_2 - c_2)$$

Regardless of the value of ξ_2, the ratio μ/τ begins from

$$\lim_{r \to 0} \frac{\mu}{\tau} = \frac{\varepsilon_1 m_1 k_2 + k_2 m_b}{\varepsilon_1 m_1 (k_1 \varepsilon_2 - k_2) - (k_1 (\varepsilon_1 + 1) + k_2) m_b} \qquad (8.414)$$

$$= \frac{\alpha^2 (\varepsilon_1 + \varepsilon_2)}{\alpha^2 (\varepsilon_1 + \varepsilon_2) + \varepsilon_1} \qquad (8.415)$$

and it ends at one.

$$\lim_{r \to \infty} \frac{\mu}{\tau} = 1 \qquad (8.416)$$

Because of

$$\lim_{r \to \infty} \mu = e \frac{m_e}{m_b} = e\varepsilon \qquad (8.417)$$

the approach of all the frequency ratio curves to $\lim_{r \to \infty} (\mu/\tau) = 1$ means $\lim_{r \to \infty} \mu = \lim_{r \to \infty} \tau = e\varepsilon$.

Furthermore, there is a frequency at which $\mu/\tau = 0$ when $\xi_2 = 0$. At this frequency, $\mu = 0$, and therefore, m_1 stops at its equilibrium. This frequency is called the freezing frequency of m_1 and is found by searching for a frequency r_{m_1} at which $\mu = 0$. The freezing frequency of m_1 is

$$r_{m_1} = \alpha \sqrt{1 + \frac{\varepsilon_2}{\varepsilon_1}} \qquad (8.418)$$

because

$$\lim_{r \to \alpha \sqrt{1 + \frac{\varepsilon_2}{\varepsilon_1}}} \mu = 0 \qquad (8.419)$$

Example 318 ★ No secondary damping. Analysis of vibrating systems for extreme values of zero and infinity damping is a main part of examining the system. These two extreme cases indicate the boundaries of dynamic behavior of the system and are considered a main step in analysis. Here is the behavior of base eccentric two *DOF* system for $.\xi_2 = 0$.

Fig. 8.69 The model of the two *DOF* eccentric base excited vibrating system with no secondary damping

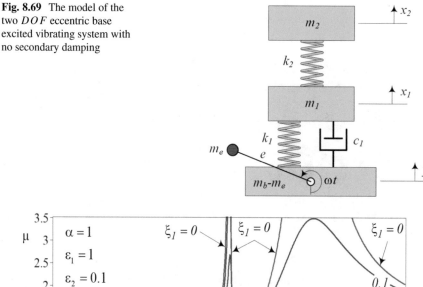

Fig. 8.70 The frequency response $\mu = \frac{S_1}{e\varepsilon}$ of the two *DOF* eccentric base excited system for $\varepsilon_1 = 1$, $\varepsilon_2 = 0.1$, $\alpha = 1$, and $\xi_2 = 0$

In case the damping of the secondary system is too low and we can ignore c_2, the model of the system would be as is shown in Fig. 8.69. The equations of motion and the frequency responses of the system can be recovered from Eqs. (8.358)–(8.395) by substituting $c_2 = 0$. Figures 8.70 and 8.71 illustrate the frequency responses $S_1/(\varepsilon e)$ and $S_2/(\varepsilon e)$ of the relative coordinates $z_1 = x_1 - y$ and $z_2 = x_2 - y$ for $\alpha = 1$, $\varepsilon_1 = 1$, and $\varepsilon_2 = 0.1$.

Example 319 ★ Freezing m_1 with respect to base at a desired frequency. The secondary vibrating system can absorb the vibrations of the primary. The practical adjustment of a vibration absorber to make a given base eccentric vibrating system stationary at any particular excitation frequency is studied here.

Setting $c_2 = 0$ and substituting the parameters (8.371)–(8.375) in the numerator of the frequency response (8.366) yield

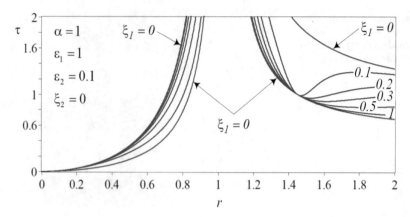

Fig. 8.71 The frequency response $\tau = \frac{S_2}{e\varepsilon}$ of the two DOF eccentric base excited system for $\varepsilon_1 = 1$, $\varepsilon_2 = 0.1$, $\alpha = 1$, and $\xi_2 = 0$

$$S_{1N} = \left(\varepsilon_1\left(k_2 - \omega^2 m_2\right) + \varepsilon_2 k_2\right) \tag{8.420}$$

The numerator will be zero if

$$\omega = \sqrt{\frac{k_2}{m_2}\frac{\varepsilon_1 + \varepsilon_2}{\varepsilon_1}} \tag{8.421}$$

which shows that the oscillation amplitude of m_1 will be zero at the frequency r:

$$r = \alpha\sqrt{1 + \frac{\varepsilon_2}{\varepsilon_1}} \tag{8.422}$$

Let us assume the excitation frequency and the mass ratios are given as $r = 0.6$, $\varepsilon_1 = 1$, and $\varepsilon_2 = 0.1$. To design a proper secondary system of an eccentric base excited two DOF as a vibration absorber and make the amplitude of m_1 zero, we should have

$$\alpha = \frac{r}{\sqrt{1 + \dfrac{\varepsilon_2}{\varepsilon_1}}} \approx 0.572 \tag{8.423}$$

Assuming values of ε_i, say, $\varepsilon_1 = 1$ and $\varepsilon_2 = 0.1$, we can determine the secondary stiffness k_2:

$$k_2 = \frac{\varepsilon_1}{\varepsilon_2}k_1\alpha^2 = 10k_1 \times 0.42426^2 = 1.8k_1 \tag{8.424}$$

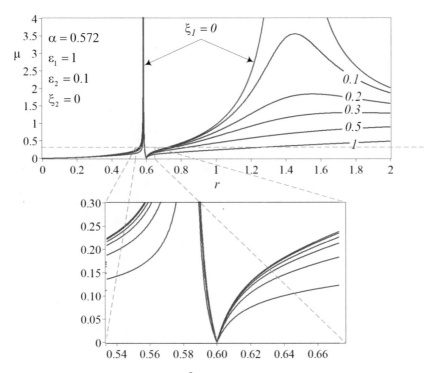

Fig. 8.72 The frequency response $\mu = \frac{S_1}{e\varepsilon}$ of the two *DOF* eccentric base excited system for $\varepsilon_1 = 1, \varepsilon_2 = 0.1, \alpha = 0.572$, and $\xi_2 = 0$

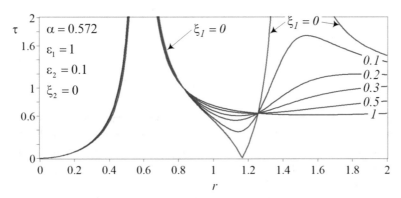

Fig. 8.73 The frequency response $\tau = \frac{S_2}{e\varepsilon}$ of the two *DOF* eccentric base excited system for $\varepsilon_1 = 1, \varepsilon_2 = 0.1, \alpha = 0.572$, and $\xi_2 = 0$

Figure 8.72 illustrates μ for the design parameters. The relative vibration of m_1 is made zero at the given excitation frequency $r = 0.6$. The frequency response τ of m_2 is shown in Fig. 8.73.

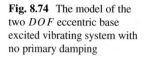

Fig. 8.74 The model of the two *DOF* eccentric base excited vibrating system with no primary damping

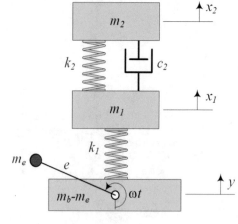

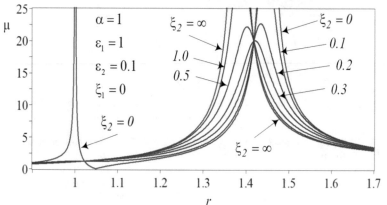

Fig. 8.75 The frequency response $\mu = \frac{S_1}{e\varepsilon}$ of the two *DOF* eccentric base excited system for $\varepsilon_1 = 1$, $\varepsilon_2 = 0.1$, $\alpha = 0.572$, and $\xi_1 = 0$

Example 320 ★ No primary damping. No damping in the primary is an extreme model of systems with very low damping in primary compared to the damping of the secondary system. This model happens very often in real application. Here is to show the effects of elimination of the primary damping on frequency response of the two *DOF* base eccentric excited systems.

Let us assume the primary damping c_1 is zero or is very low to be ignored. The model of such system is shown in Fig. 8.74. The equations of motion and the frequency responses of the system can be recovered from Eqs. (8.358)–(8.395) by substituting $c_1 = 0$. Figures 8.75 and 8.76 illustrate the frequency responses $S_1/(\varepsilon e)$ and $S_2/(\varepsilon e)$ of the relative coordinates $z_1 = x_1 - y$ and $z_2 = x_2 - y$ for a set of nominal values of $\alpha = 1$, $\varepsilon_1 = 1$, $\varepsilon_2 = 0.1$, and different ξ_2.

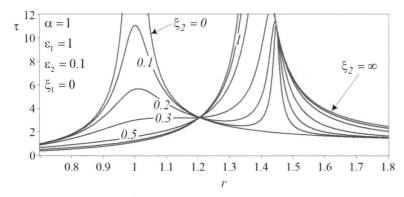

Fig. 8.76 The frequency response $\tau = \frac{S_2}{e\varepsilon}$ of the two *DOF* eccentric base excited system for $\varepsilon_1 = 1, \varepsilon_2 = 0.1, \alpha = 0.572$, and $\xi_1 = 0$

8.6 ★ Comparison for the Frequency Responses of Two *DOF* Systems

There are some common and interesting characteristics among the two *DOF* forced vibrating systems. In this section, we review some practical engineering and design aspects of the four types of forced two *DOF* systems. The response of the system by changing the position of excitation, the existence of nodes and their importance, as well as using the secondary system as vibration absorber will be reviewed here. From an engineering viewpoint, there is a primary excited system, and then a passive unforced secondary system will be attached to the primary system as a vibration absorber. Hence, practically, the excitation is always applied on the primary. However, from an analytic viewpoint, excitation may be applied on both primary and secondary systems.

8.6.1 ★ Position of Excitation

The frequency response of multi-*DOF* systems depends on the position of application of the excitation. Figure 8.77a illustrates the two *DOF* forced vibrating system when the excitation force is applied on the primary system (m_1, k_1, c_1). Figure 8.77b illustrates the same system, when the excitation force is applied on the secondary system (m_2, k_2, c_2). The system of Fig. 8.77a has been studied in Sect. 8.1. Here we analyze the system of Fig. 8.77b.

The equations of motion of the system Fig. 8.77b are

$$m_1\ddot{x}_1 + c_1\dot{x}_1 + c_2(\dot{x}_1 - \dot{x}_2) + k_1x_1 + k_2(x_1 - x_2) = 0 \tag{8.425}$$

$$m_2\ddot{x}_2 - c_2(\dot{x}_1 - \dot{x}_2) - k_2(x_1 - x_2) = F\sin\omega t \tag{8.426}$$

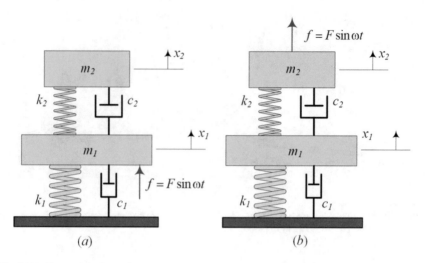

Fig. 8.77 The model of two DOF forced excited systems with different position of excitation

The steady-state amplitudes X_1 and X_2 for vibrations of m_1 and m_2 are

$$\left(\frac{X_1}{F}\right)^2 = \frac{k_2^2 + \omega^2 c_2^2}{Z_1^2 + \omega^2 c_2^2 Z_2^2 + c_1^2 \omega^2 Z_3^2 + \omega^4 c_1 c_2 \left(c_1 c_2 + 2\omega^2 m_2^2\right)} \tag{8.427}$$

$$\left(\frac{X_2}{F}\right)^2 = \frac{\left((k_1 + k_2) - \omega^2 m_1\right)^2 + \omega^2 (c_1 + c_2)^2}{Z_1^2 + \omega^2 c_2^2 Z_2^2 + c_1^2 \omega^2 Z_3^2 + \omega^4 c_1 c_2 \left(c_1 c_2 + 2\omega^2 m_2^2\right)} \tag{8.428}$$

where

$$Z_1 = \left(k_1 - \omega^2 m_1\right)\left(k_2 - \omega^2 m_2\right) - \omega^2 m_2 k_2 \tag{8.429}$$

$$Z_2 = k_1 - \omega^2 m_1 - \omega^2 m_2 \tag{8.430}$$

$$Z_3 = \left(k_2 - \omega^2 m_2\right) \tag{8.431}$$

Introducing the following parameters

$$\varepsilon = \frac{m_2}{m_1} \tag{8.432}$$

$$\xi_1 = \frac{c_1}{2m_1 \omega_1} \tag{8.433}$$

$$\xi_2 = \frac{c_2}{2m_2 \omega_2} \tag{8.434}$$

$$\omega_1 = \sqrt{\frac{k_1}{m_1}} \tag{8.435}$$

$$\omega_2 = \sqrt{\frac{k_2}{m_2}} \tag{8.436}$$

$$\alpha = \frac{\omega_2}{\omega_1} = \sqrt{\frac{k_2}{\varepsilon k_1}} \tag{8.437}$$

$$r = \frac{\omega}{\omega_1} \tag{8.438}$$

$$\mu = \frac{X_1}{F/k_1} \tag{8.439}$$

$$\tau = \frac{X_2}{F/k_1} \tag{8.440}$$

the frequency responses (8.427) and (8.428) transform to the following dimensionless forms.

$$\mu^2 = \left(\frac{X_1}{F/k_1}\right)^2 = \frac{\alpha^2 \left(4\xi_2^2 r^2 + \alpha^2\right)}{4\xi_2^2 r^2 \alpha^2 Z_5^2 + Z_4^2 + 8\xi_1\xi_2\alpha r^4 Z_5 + 4r^6\xi_1^2} \tag{8.441}$$

$$\tau^2 = \left(\frac{X_2}{F/k_1}\right)^2 = \frac{4r^2 \left(\xi_1 + \varepsilon\alpha\xi_2\right)^2 + \left(r^2 - \left(\varepsilon\alpha^2 + 1\right)\right)^2}{4\xi_2^2 r^2 \alpha^2 Z_5^2 + Z_4^2 + 8\xi_1\xi_2\alpha r^4 Z_5 + 4r^6\xi_1^2} \tag{8.442}$$

$$Z_4 = \varepsilon\alpha^2 r^2 - \left(r^2 - 1\right)\left(r^2 - \alpha^2\right) \tag{8.443}$$

$$Z_5 = r^2 \left(1 + \varepsilon\right) - 1 \tag{8.444}$$

Proof To employ the Lagrange method, we need the kinetic and potential energies and the dissipation function of the system which are the same as (8.21)–(8.23):

$$K = \frac{1}{2}m_1\dot{x}_1^2 + \frac{1}{2}m_2\dot{x}_2^2 \tag{8.445}$$

$$P = \frac{1}{2}k_1 x_1^2 + \frac{1}{2}k_2 \left(x_1 - x_2\right)^2 \tag{8.446}$$

$$D = \frac{1}{2}c_1\dot{x}_1^2 + \frac{1}{2}c_2 \left(\dot{x}_1 - \dot{x}_2\right)^2 \tag{8.447}$$

The Lagrange equation (2.547) provides us with the equations of motion (8.425) and (8.426) or

$$
\begin{bmatrix} m_1 & 0 \\ 0 & m_2 \end{bmatrix} \begin{bmatrix} \ddot{x}_1 \\ \ddot{x}_2 \end{bmatrix} + \begin{bmatrix} c_1 + c_2 & -c_2 \\ -c_2 & c_2 \end{bmatrix} \begin{bmatrix} \dot{x}_1 \\ \dot{x}_2 \end{bmatrix}
$$
$$
+ \begin{bmatrix} k_1 + k_2 & -k_2 \\ -k_2 & k_2 \end{bmatrix} = \begin{bmatrix} 0 \\ F \sin \omega t \end{bmatrix} \tag{8.448}
$$

To calculate the frequency response of the system, we substitute a set of harmonic solutions

$$
x_1 = A_1 \sin \omega t + B_1 \cos \omega t \tag{8.449}
$$
$$
x_2 = A_2 \sin \omega t + B_2 \cos \omega t \tag{8.450}
$$

which provides us with a set of algebraic equations to derive the coefficients A_1, B_1, A_2, and B_2.

$$
\begin{bmatrix} a_{11} & -k_2 & -\omega(c_1 + c_2) & \omega c_2 \\ -k_2 & a_{22} & \omega c_2 & -\omega c_2 \\ \omega(c_1 + c_2) & -\omega c_2 & a_{33} & -k_2 \\ -\omega c_2 & \omega c_2 & -k_2 & a_{44} \end{bmatrix} \begin{bmatrix} A_1 \\ A_2 \\ B_1 \\ B_2 \end{bmatrix} = \begin{bmatrix} 0 \\ F \\ 0 \\ 0 \end{bmatrix} \tag{8.451}
$$

$$
a_{11} = a_{33} = k_2 + k_1 - m_1 \omega^2 \tag{8.452}
$$
$$
a_{22} = a_{44} = k_2 - m_2 \omega^2 \tag{8.453}
$$

The steady-state amplitudes X_1 and X_2 of m_1 and m_2 are

$$
X_1 = \sqrt{A_1^2 + B_1^2} \tag{8.454}
$$

$$
X_2 = \sqrt{A_2^2 + B_2^2} \tag{8.455}
$$

which after simplification can be written as (8.427) and (8.428) or equivalently as (8.441) and (8.442).

Practically, the purpose of adding a passive vibrating system to a forced vibrating system is to absorb the vibrations of the forced system and reduce the amplitude of its frequency response. Therefore, a better image of the system of Fig. 8.77b is that the secondary system (m_2, k_2, c_2) was under a forced excitation and we added a primary system (m_1, k_1, c_1) to separate m_2 from the ground by another suspension system. From this viewpoint, the meaning of "primary" and "secondary" do not match with "principal" and "auxiliary." ∎

Example 321 ★ Amplitude ratio. The steady-state amplitude ratio is an engineering indicator to determine the frequency ranges in which m_1 is oscillating with higher amplitude than m_2 or lower.

To examine the effectiveness of the primary system (m_1, k_1, c_1) in magnifying or damping the vibrations of the forced excited secondary (m_2, k_2, c_2), we determine the ratio of the amplitudes of m_1/m_2 of the system of Fig. 8.77b.

$$\frac{X_1^2}{X_1^2} = \frac{k_2^2 + \omega^2 c_2^2}{\left((k_1 + k_2) - \omega^2 m_1\right)^2 + \omega^2 (c_1 + c_2)^2} \tag{8.456}$$

$$\frac{\mu^2}{\tau^2} = \frac{4\alpha^2 \xi_2^2 r^2 + \alpha^4}{4r^2 (\xi_1 + \varepsilon\alpha\xi_2)^2 + \left(r^2 - \left(\varepsilon\alpha^2 + 1\right)\right)^2} \tag{8.457}$$

When the force is applied on m_1 as Fig. 8.77a, the frequency ratio is

$$\frac{X_1^2}{X_2^2} = \frac{\mu^2}{\tau^2} = \frac{4\alpha^2 \xi_2^2 r^2 + \left(r^2 - \alpha^2\right)^2}{4\alpha^2 \xi_2^2 r^2 + \alpha^4} \tag{8.458}$$

which is independent of ξ_1 and ε. However, the frequency ratio μ/τ of Fig. 8.77b depends on all involved parameters: $\xi_1, \xi_2, \varepsilon, \alpha$, and r. The frequency ratio μ/τ of Fig. 8.77b approaches two limits when ξ_2 varies:

$$\lim_{\xi_2 \to 0} \frac{\mu}{\tau} = \frac{\alpha^2}{r^2 - \left(\varepsilon\alpha^2 + 1\right)} \qquad \lim_{\xi_2 \to \infty} \frac{\mu}{\tau} = 1 \tag{8.459}$$

while the frequency ratio μ/τ of Fig. 8.77a approaches the following limits when ξ_2 varies.

$$\lim_{\xi_2 \to 0} \frac{\mu^2}{\tau^2} = \frac{\alpha^2}{\sqrt{4r^2 \xi_1^2 + \left(r^2 - \left(\varepsilon\alpha^2 + 1\right)\right)^2}} \qquad \lim_{\xi_2 \to \infty} \frac{\mu}{\tau} = \frac{1}{\varepsilon} \tag{8.460}$$

The limits for $\xi_2 \to 0$ will be equal when $\xi_1 = 0$, and the limits for $\xi_2 \to \infty$ will be equal when $\varepsilon = 1$.

Example 322 ★ Frequency response of the primary system. Here we examine the effect of the position of application of the exciting force of Fig. 8.77a and b by comparing μ^2 of the two systems.

The frequency responses of the primary systems of Fig. 8.77a and b are

$$\mu_a^2 = \left(\frac{X_1}{F/k_1}\right)^2 = \frac{\alpha^2 \left(4\xi_2^2 r^2 + \alpha^2\right)}{4\xi_2^2 r^2 \alpha^2 Z_5^2 + Z_4^2 + 8\xi_1 \xi_2 \alpha r^4 Z_5 + 4r^6 \xi_1^2} \tag{8.461}$$

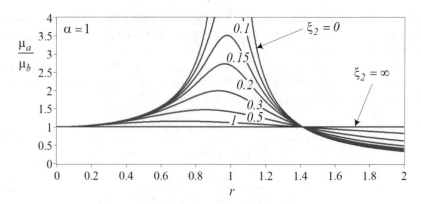

Fig. 8.78 The ratio μ_a/μ_b of the frequency response of m_1 when the force is at m_1 and at m_2 for $\alpha = 1$ and various ξ_2

$$\mu_b^2 = \left(\frac{X_1}{F/k_1}\right)^2 = \frac{4\alpha^2\xi_2^2 r^2 + (r^2 - \alpha^2)^2}{4\xi_2^2 r^2 \alpha^2 Z_5^2 + Z_4^2 + 8\xi_1\xi_2\alpha r^4 Z_5 + 4r^6\xi_1^2} \tag{8.462}$$

The ratio μ_a/μ_b is dependent on the mass ratio ε and the primary damping ξ_1.

$$\frac{\mu_a^2}{\mu_b^2} = \frac{\alpha^2\left(4\xi_2^2 r^2 + \alpha^2\right)}{4\alpha^2\xi_2^2 r^2 + (r^2 - \alpha^2)^2} \tag{8.463}$$

Figure 8.78 depicts μ_a/μ_b for $\alpha = 1$ and various ξ_2. The figure would be similar for any other value of α, which controls the resonance frequency of μ_a/μ_b. The node frequency r_0 of the graph is at the intersection of μ_a/μ_b for $\xi_2 = 0$ and $\xi_2 = \infty$.

$$r_0 = \alpha\sqrt{2} \tag{8.464}$$

At high frequency, $r > r_0$, we have $\mu_a < \mu_b$ and at low frequency, $\mu_a > \mu_b$. Therefore, if the position of excitation could be switched between m_1 and m_2, then we should excite m_2 at low frequency and excite m_1 at high frequency, to get a low amplitude vibration. To have a high amplitude vibration, we excite m_1 at low frequency and excite m_2 at high frequency.

Example 323 ★ Frequency response of the secondary. In case the oscillation of the secondary system is important, their steady-state amplitude must be compared for the position of the excitation force. Here we examine the effect of the position of application of the exciting force of Fig. 8.77a and b by comparing τ^2 of the two systems.

The frequency responses of the secondary systems of Fig. 8.77a and b are

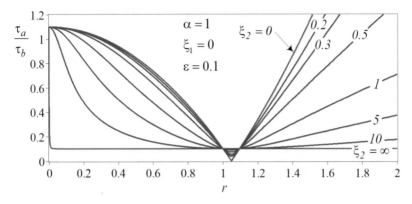

Fig. 8.79 The ratio τ_a/τ_b of the frequency response of m_1 when the force is at m_1 and at m_2 for $\alpha = 1$, $\varepsilon = 0.1$, $\xi_1 = 0$, and various ξ_2

$$\tau_a^2 = \left(\frac{X_2}{F/k_1}\right)^2 = \frac{4r^2 \left(\xi_1 + \varepsilon\alpha\xi_2\right)^2 + \left(r^2 - \left(\varepsilon\alpha^2 + 1\right)\right)^2}{4\xi_2^2 r^2 \alpha^2 Z_5^2 + Z_4^2 + 8\xi_1\xi_2\alpha r^4 Z_5 + 4r^6\xi_1^2} \tag{8.465}$$

$$\tau_b^2 = \left(\frac{X_2}{F/k_1}\right)^2 = \frac{\alpha^2 \left(4\xi_2^2 r^2 + \alpha^2\right)}{4\xi_2^2 r^2 \alpha^2 Z_5^2 + Z_4^2 + 8\xi_1\xi_2\alpha r^4 Z_5 + 4r^6\xi_1^2} \tag{8.466}$$

The ratio τ_a/τ_b of the frequency response of m_2 when the force is at m_1 and when the force is at m_2 is dependent on the mass ratio ε and the primary damping ξ_2, as well as ξ_1, α, and r.

$$\frac{\tau_a^2}{\tau_b^2} = \frac{4r^2 \left(\xi_1 + \varepsilon\alpha\xi_2\right)^2 + \left(r^2 - \left(\varepsilon\alpha^2 + 1\right)\right)^2}{\alpha^2 \left(4\xi_2^2 r^2 + \alpha^2\right)} \tag{8.467}$$

Figure 8.79 depicts a sample of τ_a/τ_b for $\alpha = 1$, $\varepsilon = 0.1$, $\xi_1 = 0$, and various ξ_2. Interestingly, when we change the position of the excitation force from m_1 to m_2, the ratio of steady-state amplitude of X_1 of Fig. 8.77a to the steady-state amplitude X_2 of Fig. 8.77b is always the same:

$$\frac{\mu_a}{\tau_b} = 1 \tag{8.468}$$

The ratio of the amplitude X_2 of Fig. 8.77a to amplitude X_1 of Fig. 8.77b is

$$\frac{\mu_b}{\tau_a} = \frac{4\alpha^2\xi_2^2 r^2 + \left(r^2 - \alpha^2\right)^2}{4r^2 \left(\xi_1 + \varepsilon\alpha\xi_2\right)^2 + \left(r^2 - \left(\varepsilon\alpha^2 + 1\right)\right)^2} \tag{8.469}$$

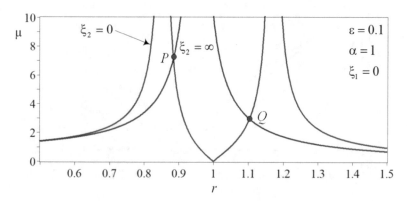

Fig. 8.80 The frequency responses $\mu = \frac{X_1}{F/k_1}$ of the undamped two DOF forced excited systems on primary for $\alpha = 1$, $\varepsilon = 0.1$ and $\xi_1 = 0$ and two cases of $\xi_2 = \infty$ and $\xi_2 = 0$

8.6.2 ★ Nodes

When there is no damping in the primary system of the two DOF systems of Fig. 8.1a, b, and d, there are *four* nodes in the graph of frequency response μ. The first one occurs at $r = 0$, and the fourth one occurs at $r = \infty$. There are also two middle nodes indicated by P and Q as depicted in Fig. 8.80. The middle nodes can be used to control the vibrations of the systems and design an effective vibration isolation. The nodes are the points in (μ, r)-plane where every frequency response curve will pass. The node frequency and node amplitude are independent of the value of the secondary damping, ξ_2. Although the following study can be done for all four systems of Fig. 8.1, here we introduce the method for the system of Fig. 8.77a.

Proof When $\xi_1 = 0$, the frequency response of the primary system of the forced excited two DOF system of Fig. 8.77a is

$$\mu^2 = \left(\frac{X_1}{F/k_1}\right)^2 = \frac{4\alpha^2\xi_2^2 r^2 + \left(r^2 - \alpha^2\right)^2}{4\xi_2^2 r^2 \alpha^2 Z_5^2 + Z_4^2} \tag{8.470}$$

$$Z_4 = \varepsilon\alpha^2 r^2 - \left(r^2 - 1\right)\left(r^2 - \alpha^2\right) \tag{8.471}$$

$$Z_5 = r^2\left(1 + \varepsilon\right) - 1 \tag{8.472}$$

The limits of μ for $\xi_2 \to 0$ and $\xi_2 \to \infty$ are

$$\lim_{\xi_2 \to 0} \mu = \frac{X_1}{F/k_1} = \frac{r^2 - \alpha^2}{\varepsilon\alpha^2 r^2 - \left(r^2 - 1\right)\left(r^2 - \alpha^2\right)} \tag{8.473}$$

$$\lim_{\xi_2 \to \infty} \mu = \frac{X_1}{F/k_1} = \pm\frac{1}{r^2\left(1 + \varepsilon\right) - 1} \tag{8.474}$$

Equating μ at $\xi_2 \to 0$ and at $\xi_2 \to \infty$ provides two equations for the node frequencies r_P and r_Q.

$$\left(r^2 - \alpha^2\right)\left(r^2(1 + \varepsilon) - 1\right) = \varepsilon\alpha^2 r^2 - \left(r^2 - 1\right)\left(r^2 - \alpha^2\right) \tag{8.475}$$

$$-\left(r^2 - \alpha^2\right)\left(r^2(1 + \varepsilon) - 1\right) = \varepsilon\alpha^2 r^2 - \left(r^2 - 1\right)\left(r^2 - \alpha^2\right) \tag{8.476}$$

The first equation provides us with

$$r_P^2 = \frac{1}{\varepsilon + 2}\left(\alpha^2(1 + \varepsilon) + 1 - \sqrt{1 - 2\alpha^2 + \alpha^4(1 + \varepsilon)^2}\right) \tag{8.477}$$

$$r_Q^2 = \frac{1}{\varepsilon + 2}\left(\alpha^2(1 + \varepsilon) + 1 + \sqrt{1 - 2\alpha^2 + \alpha^4(1 + \varepsilon)^2}\right) \tag{8.478}$$

and the second equation gives $r = 0$.

The node frequencies r_P and r_Q are functions of $\varepsilon = m_2/m_1$ and $\alpha = \sqrt{m_1 k_2/(m_2 k_1)}$. Assuming the primary system (m_1, k_1, c_1) is given then, r_P and r_Q are functions of m_2 and k_2.

The natural frequencies of the system are

$$r_1^2 = \frac{1}{2}\left(\alpha^2(1 + \varepsilon) + 1 - \sqrt{1 - 2\alpha^2(1 - \varepsilon) + \alpha^4(1 + \varepsilon)^2}\right) \tag{8.479}$$

$$r_2^2 = \frac{1}{2}\left(\alpha^2(1 + \varepsilon) + 1 + \sqrt{1 - 2\alpha^2(1 - \varepsilon) + \alpha^4(1 + \varepsilon)^2}\right) \tag{8.480}$$

When $\varepsilon = 0$ and, therefore, $\alpha = 1$, the system reduces to a one *DOF* system, and the natural frequencies simplify to

$$r_1 = 0 \qquad r_2 = 1 \tag{8.481}$$

Adding the secondary system makes the system a two *DOF* with two natural frequencies r_1 and r_2 such that

$$0 < r_1 < 1 < r_2 \tag{8.482}$$

Comparing r_1 and r_2 with r_P and r_Q indicates that

$$0 < r_1 < r_P < 1 < r_Q < r_2 \tag{8.483}$$

The node frequencies of the two *DOF* systems in Fig. 8.1 can be analyzed similarly.

∎

Example 324 ★ Node frequencies of two *DOF* base excited system. Some of the extreme cases of the two *DOF* base excited system of Fig. 8.1 are similar. Here is an example of two systems at their extreme cases.

The frequency response of the primary system of the two *DOF* base excited system of Fig. 8.1a for $\xi_1 = 0$ is

$$\mu^2 = \left(\frac{X_1}{Y}\right)^2 = \frac{4\alpha^2\xi_2^2 r^2 + \left(r^2 - \alpha^2\right)^2}{4\xi_2^2 r^2 \alpha^2 Z_5^2 + Z_4^2}$$

$$Z_4 = \varepsilon\alpha^2 r^2 - \left(r^2 - 1\right)\left(r^2 - \alpha^2\right)$$

$$Z_5 = r^2 \left(1 + \varepsilon\right) - 1$$

which is exactly the same as (8.470) for the two *DOF* forced excited system. The node frequency of the base excited system would then be the same as (8.477) and (8.478).

Example 325 ★ Node frequencies of two *DOF* eccentric excited system. Here is another example of equal extreme cases of the two *DOF* base excited systems of Fig. 8.1.

The frequency response of the primary system of the two *DOF* eccentric excited system of Fig. 8.1a for $\xi_1 = 0$ is

$$\mu^2 = \left(\frac{X_1}{e\varepsilon_1}\right)^2 = r^4 \frac{4\alpha^2\xi_2^2 r^2 + \left(r^2 - \alpha^2\right)^2}{4\xi_2^2 r^2 \alpha^2 Z_5^2 + Z_4^2} \tag{8.484}$$

The limits of μ for $\xi_2 \to 0$ and $\xi_2 \to \infty$ are

$$\lim_{\xi_2 \to 0} \mu = \frac{X_1}{e\varepsilon_1} = r^2 \frac{r^2 - \alpha^2}{\varepsilon\alpha^2 r^2 - \left(r^2 - 1\right)\left(r^2 - \alpha^2\right)} \tag{8.485}$$

$$\lim_{\xi_2 \to \infty} \mu = \frac{X_1}{e\varepsilon_1} = \pm \frac{r^2}{r^2 \left(1 + \varepsilon\right) - 1} \tag{8.486}$$

These limits are equal to r^2 multiplied by the limits (8.473) and (8.474). Therefore, the node frequency of the eccentric excited system would be the same as (8.477) and (8.478).

Example 326 ★ The node amplitudes. The amplitude at nodes are the values that the system will reach regardless of the value of damping. Determination of the node amplitude is a step in engineering optimization of frequency response.

We determine the node amplitudes μ_P and μ_Q by substituting the node frequencies (8.477) and (8.478) in (8.474).

$$\mu_P = \frac{\dfrac{1}{\varepsilon + 2}\left(\alpha^2(1+\varepsilon) + 1 - \sqrt{1 - 2\alpha^2 + \alpha^4(1+\varepsilon)^2}\right)}{\dfrac{1+\varepsilon}{\varepsilon + 2}\left(\alpha^2(1+\varepsilon) + 1 - \sqrt{1 - 2\alpha^2 + \alpha^4(1+\varepsilon)^2}\right) - 1} \tag{8.487}$$

$$\mu_Q = \frac{\dfrac{1}{\varepsilon + 2}\left(\alpha^2(1+\varepsilon) + 1 + \sqrt{1 - 2\alpha^2 + \alpha^4(1+\varepsilon)^2}\right)}{\dfrac{1+\varepsilon}{\varepsilon + 2}\left(\alpha^2(1+\varepsilon) + 1 + \sqrt{1 - 2\alpha^2 + \alpha^4(1+\varepsilon)^2}\right) - 1} \tag{8.488}$$

The node amplitudes are functions of $\varepsilon = m_2/m_1$ and $\alpha = \sqrt{m_1 k_2/(m_2 k_1)}$. When the primary system (m_1, k_1, c_1) is given, then μ_P and μ_Q can be controlled by adjusting m_2 and k_2.

Example 327 ★ Natural frequency and mode shapes. Natural frequencies of a linear vibrating system are the first important engineering information that we must calculate to determine the resonance zones of the system. They are determined by calculating eigenvalues of the matrix $A = m^{-1}k$.

Eliminating damping and excitation indicates that the equations of motion of the systems in Fig. 8.1a–d are the same.

$$\begin{bmatrix} m_1 & 0 \\ 0 & m_2 \end{bmatrix}\begin{bmatrix} \ddot{x}_1 \\ \ddot{x}_2 \end{bmatrix} + \begin{bmatrix} k_1 + k_2 & -k_2 \\ -k_2 & k_2 \end{bmatrix} = \begin{bmatrix} 0 \\ 0 \end{bmatrix} \tag{8.489}$$

The natural frequencies and mode shapes of the equations are calculated from the A-matrix.

$$\begin{aligned}[A] &= \begin{bmatrix} m_1 & 0 \\ 0 & m_2 \end{bmatrix}^{-1}\begin{bmatrix} k_1 + k_2 & -k_2 \\ -k_2 & k_2 \end{bmatrix} \\ &= \begin{bmatrix} \dfrac{1}{m_1}(k_1 + k_2) & -\dfrac{k_2}{m_1} \\ -\dfrac{k_2}{m_2} & \dfrac{k_2}{m_2} \end{bmatrix}\end{aligned} \tag{8.490}$$

The natural frequencies are

$$\begin{aligned}\omega_{n_1}^2 = \; &\frac{k_1}{2m_1} + \frac{k_2}{2m_2} + \frac{k_2}{2m_1} \\ &-\frac{1}{2}\sqrt{\left(\frac{k_1}{m_1} - \frac{k_2}{m_2}\right)^2 + \frac{k_2}{m_1}\left(\frac{k_2}{m_1} + 2\frac{k_1}{m_1} + 2\frac{k_2}{m_2}\right)} \end{aligned} \tag{8.491}$$

$$\omega_{n_2}^2 = \frac{k_1}{2m_1} + \frac{k_2}{2m_2} + \frac{k_2}{2m_1}$$

$$+ \frac{1}{2} \sqrt{\left(\frac{k_1}{m_1} - \frac{k_2}{m_2}\right)^2 + \frac{k_2}{m_1}\left(\frac{k_2}{m_1} + 2\frac{k_1}{m_1} + 2\frac{k_2}{m_2}\right)} \qquad (8.492)$$

and their associated eigenvectors are

$$\mathbf{u}_1 = \begin{bmatrix} \dfrac{\frac{1}{2}\left(1 - \frac{k_1 m_2}{k_2 m_1} - \frac{m_2}{m_1}\right)}{+\frac{1}{2}\sqrt{\left(\frac{k_1}{k_2} - \frac{m_2}{m_1}\right)^2 + 2\frac{k_1}{k_2}\frac{m_2^2}{m_1^2} + 2\frac{m_2}{m_1} + 1}} \\ 1 \end{bmatrix} \qquad (8.493)$$

$$\mathbf{u}_2 = \begin{bmatrix} \dfrac{\frac{1}{2}\left(1 - \frac{k_1 m_2}{k_2 m_1} - \frac{m_2}{m_1}\right)}{-\frac{1}{2}\sqrt{\left(\frac{k_1}{k_2} - \frac{m_2}{m_1}\right)^2 + 2\frac{k_1}{k_2}\frac{m_2^2}{m_1^2} + 2\frac{m_2}{m_1} + 1}} \\ 1 \end{bmatrix} \qquad (8.494)$$

Employing the dimensionless parameters (8.8)–(8.16), the natural frequencies and the mode shapes will be

$$r_1^2 = \frac{1}{2}\left(\alpha^2(1+\varepsilon) + 1 - \sqrt{1 - 2\alpha^2(1-\varepsilon) + \alpha^4(1+\varepsilon)^2}\right) \qquad (8.495)$$

$$r_2^2 = \frac{1}{2}\left(\alpha^2(1+\varepsilon) + 1 + \sqrt{1 - 2\alpha^2(1-\varepsilon) + \alpha^4(1+\varepsilon)^2}\right) \qquad (8.496)$$

$$\mathbf{u}_1 = \begin{bmatrix} \dfrac{1}{2}\left(1 - \frac{1}{\alpha} - \varepsilon\right) + \frac{1}{2}\sqrt{\left(\frac{1}{\varepsilon\alpha^2} - \varepsilon\right)^2 + 2\frac{\varepsilon}{\alpha} + 2\varepsilon + 1} \\ 1 \end{bmatrix} \qquad (8.497)$$

$$\mathbf{u}_2 = \begin{bmatrix} \dfrac{1}{2}\left(1 - \frac{1}{\alpha} - \varepsilon\right) - \frac{1}{2}\sqrt{\left(\frac{1}{\varepsilon\alpha^2} - \varepsilon\right)^2 + 2\frac{\varepsilon}{\alpha} + 2\varepsilon + 1} \\ 1 \end{bmatrix} \qquad (8.498)$$

The natural frequencies (8.495) and (8.496) are the same as (8.68) and (8.69). When $\varepsilon = 0$ and, therefore, $\alpha = 1$, the system reduces to the one DOF primary system with mass $m = m_1 + m_2$, and the natural frequencies simplify to

$$r_1 = 0 \qquad r_2 = 1 \qquad (8.499)$$

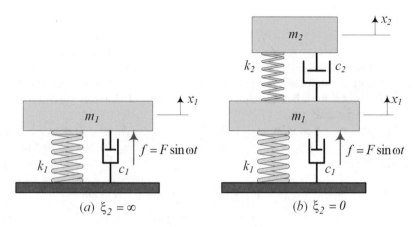

Fig. 8.81 The two DOF system. (**a**) when $\xi_2 = \infty$. (**b**) when $\xi_2 = 0$

Adding the secondary system makes the system a two DOF one with two natural frequencies r_1 and r_2 such that $0 < r_1 < 1 < r_2$. Figures 8.11 and 8.12 illustrate how the natural frequencies change when α and ε vary.

Figure 8.81b illustrates the two DOF systems when $\xi_2 = 0$, and Fig. 8.81a illustrates the systems when $\xi_2 = \infty$. The frequency responses X_1 of the systems are compared in Fig. 8.80 to remind that the two natural frequencies of two DOF systems fall on two sides of the natural frequency of the one DOF system of Fig. 8.81a.

8.7 Summary

There are many practical engineering situations where we model a vibrating system by attaching a secondary vibrating system m_2, k_2, c_2 to a primary system m_1, k_1, c_1. The secondary system makes the whole system to have two degrees of freedom (DOF) and affects the primary system, making it to deviate from its basic frequency responses of one DOF systems.

The harmonic excitation of the primary system can only be one of the following types:

1. base excitation;
2. eccentric excitation;
3. eccentric base excitation;
4. forced excitation.

Base Excitation

The *base excitation* is the most practical model for vertical vibration of mechanical systems. The *eccentric excitation* is a model for equipment attached to a rotary motor on a suspension. The *eccentric base excitation* is a model for vibration of any double equipment mounted on an engine. While the *forced excitation* has almost no practical application, it is the simplest two DOF forced vibrating system to serve as a good educational base.

The equations of motion of a two DOF forced vibrating system such that a secondary system (m_2, k_2, c_2) is attached to a primary forced excited system (m_1, k_1, c_1) are

$$m_1\ddot{x}_1 + c_1\dot{x}_1 + c_2(\dot{x}_1 - \dot{x}_2) + k_1 x_1 + k_2(x_1 - x_2) = F\sin\omega t \qquad (8.500)$$

$$m_2\ddot{x}_2 - c_2(\dot{x}_1 - \dot{x}_2) - k_2(x_1 - x_2) = 0 \qquad (8.501)$$

The steady-state amplitudes X_1 and X_2 for vibration of m_1 and m_2 are

$$\left(\frac{X_1}{F}\right)^2 = \frac{\left(k_2 - \omega^2 m_2\right)^2 + \omega^2 c_2^2}{Z_1^2 + \omega^2 c_2^2 Z_2^2 + c_1^2 \omega^2 Z_3^2 + \omega^4 c_1 c_2\left(c_1 c_2 + 2\omega^2 m_2^2\right)} \qquad (8.502)$$

$$\left(\frac{X_2}{F}\right)^2 = \frac{k_2^2 + \omega^2 c_2^2}{Z_1^2 + \omega^2 c_2^2 Z_2^2 + c_1^2 \omega^2 Z_3^2 + \omega^4 c_1 c_2\left(c_1 c_2 + 2\omega^2 m_2^2\right)} \qquad (8.503)$$

where

$$Z_1 = \left(k_1 - \omega^2 m_1\right)\left(k_2 - \omega^2 m_2\right) - \omega^2 m_2 k_2 \qquad (8.504)$$

$$Z_2 = k_1 - \omega^2 m_1 - \omega^2 m_2 \qquad (8.505)$$

$$Z_3 = \left(k_2 - \omega^2 m_2\right) \qquad (8.506)$$

Introducing the following parameters,

$$\varepsilon = \frac{m_2}{m_1} \qquad \xi_1 = \frac{c_1}{2m_1\omega_1} \qquad \xi_2 = \frac{c_2}{2m_2\omega_2} \qquad (8.507)$$

$$\omega_1 = \sqrt{\frac{k_1}{m_1}} \qquad \omega_2 = \sqrt{\frac{k_2}{m_2}} \qquad \alpha = \frac{\omega_2}{\omega_1} = \sqrt{\frac{k_2}{\varepsilon k_1}} \qquad (8.508)$$

$$r = \frac{\omega}{\omega_1} \qquad \mu = \frac{X_1}{F/k_1} \qquad \tau = \frac{X_2}{F/k_1} \qquad (8.509)$$

we transform the frequency responses to the following dimensionless forms.

$$\mu^2 = \left(\frac{X_1}{F/k_1}\right)^2 = \frac{4\alpha^2\xi_2^2 r^2 + \left(r^2 - \alpha^2\right)^2}{4\xi_2^2 r^2 \alpha^2 Z_5^2 + Z_4^2 + 8\xi_1\xi_2\alpha r^4 Z_5 + 4r^6\xi_1^2} \tag{8.510}$$

$$\tau^2 = \left(\frac{X_2}{F/k_1}\right)^2 = \frac{\alpha^2\left(4\xi_2^2 r^2 + \alpha^2\right)}{4\xi_2^2 r^2 \alpha^2 Z_5^2 + Z_4^2 + 8\xi_1\xi_2\alpha r^4 Z_5 + 4r^6\xi_1^2} \tag{8.511}$$

$$Z_4 = \varepsilon\alpha^2 r^2 - \left(r^2 - 1\right)\left(r^2 - \alpha^2\right) \tag{8.512}$$

$$Z_5 = r^2\left(1 + \varepsilon\right) - 1 \tag{8.513}$$

Eccentric Excitation

A two DOF base excited vibrating system is made by attaching a secondary system (m_2, k_2, c_2) to a primary base excited system (m_1, k_1, c_1). The equations of motion of the system are

$$m_1\ddot{x}_1 + c_1\dot{x}_1 + c_2\left(\dot{x}_1 - \dot{x}_2\right) + k_1 x_1 + k_2\left(x_1 - x_2\right) = k_1 y + c_1\dot{y} \tag{8.514}$$

$$m_2\ddot{x}_2 - c_2\left(\dot{x}_1 - \dot{x}_2\right) - k_2\left(x_1 - x_2\right) = 0 \tag{8.515}$$

$$y = Y\sin\omega t \tag{8.516}$$

and the steady-state amplitudes X_1 and X_2 for vibrations of m_1 and m_2 are

$$\left(\frac{X_1}{k_1 Y}\right)^2 = \frac{\left(k_1^2 + \omega^2 c_1^2\right)\left(\left(k_2^2 + \omega^2 c_2^2\right) + \omega^2 m_2\left(\omega^2 m_2 - 2k_2\right)\right)}{Z_1^2 + \omega^2 c_2^2 Z_2^2 + c_1^2\omega^2 Z_3^2 + \omega^4 c_1 c_2\left(c_1 c_2 + 2\omega^2 m_2^2\right)} \tag{8.517}$$

$$\left(\frac{X_2}{k_1 Y}\right)^2 = \frac{\left(k_1^2 + \omega^2 c_1^2\right)\left(k_2^2 + \omega^2 c_2^2\right)}{Z_1^2 + \omega^2 c_2^2 Z_2^2 + c_1^2\omega^2 Z_3^2 + \omega^4 c_1 c_2\left(c_1 c_2 + 2\omega^2 m_2^2\right)} \tag{8.518}$$

$$Z_1 = \left(k_1 - \omega^2 m_1\right)\left(k_2 - \omega^2 m_2\right) - \omega^2 m_2 k_2 \tag{8.519}$$

$$Z_2 = k_1 - \omega^2 m_1 - \omega^2 m_2 \tag{8.520}$$

$$Z_3 = \left(k_2 - \omega^2 m_2\right) \tag{8.521}$$

Introducing the following parameters

$$\varepsilon = \frac{m_2}{m_1} \qquad \xi_1 = \frac{c_1}{2m_1\omega_1} \qquad \xi_2 = \frac{c_2}{2m_2\omega_2} \tag{8.522}$$

$$\omega_1 = \sqrt{\frac{k_1}{m_1}} \qquad \omega_2 = \sqrt{\frac{k_2}{m_2}} \qquad \alpha = \frac{\omega_2}{\omega_1} \tag{8.523}$$

$$r = \frac{\omega}{\omega_1} \qquad \mu = \frac{X_1}{Y} \qquad \tau = \frac{X_2}{Y} \tag{8.524}$$

we can transform the frequency responses to the following nondimensional forms.

$$\mu^2 = \left(\frac{X_1}{Y}\right)^2 = \frac{\left(4\xi_1^2 r^2 + 1\right)\left(\alpha^2\left(4\xi_2^2 r^2 + \alpha^2\right) + Z_6\right)}{4\xi_2^2 r^2 \alpha^2 Z_5^2 + Z_4^2 + 8\xi_1\xi_2\alpha r^4 Z_5 + 4r^6\xi_1^2} \tag{8.525}$$

$$\tau^2 = \left(\frac{X_2}{Y}\right)^2 = \frac{\alpha^2\left(4\xi_2^2 r^2 + \alpha^2\right)\left(4\xi_1^2 r^2 + 1\right)}{4\xi_2^2 r^2 \alpha^2 Z_5^2 + Z_4^2 + 8\xi_1\xi_2\alpha r^4 Z_5 + 4r^6\xi_1^2} \tag{8.526}$$

$$Z_4 = \varepsilon\alpha^2 r^2 - \left(r^2 - 1\right)\left(r^2 - \alpha^2\right) \tag{8.527}$$

$$Z_5 = r^2\left(1 + \varepsilon\right) - 1 \tag{8.528}$$

$$Z_6 = r^2\left(r^2 - 2\alpha^2\right) \tag{8.529}$$

Eccentric Excitation

A two DOF eccentric excited vibrating system is composed by attaching a passive secondary system (m_2, k_2, c_2) to a primary eccentric excited system (m_1, k_1, c_1). The equations of motion of the system are

$$m_1\ddot{x}_1 + c_1\dot{x}_1 + c_2\left(\dot{x}_1 - \dot{x}_2\right) + k_1\left(x_1 - x_2\right)$$
$$+k_2 x_1 = -m_e e\omega^2 \sin\omega t \tag{8.530}$$
$$m_2\ddot{x}_2 + c_2\left(\dot{x}_2 - \dot{x}_1\right) + k_2\left(x_2 - x_1\right) = 0 \tag{8.531}$$

The steady-state amplitudes X_1 and X_2 for vibration of m_1 and m_2 are

$$\left(\frac{X_1}{e}\right)^2 = \frac{m_e^2\omega^4\left(\left(k_2 - \omega^2 m_2\right)^2 + \omega^2 c_2^2\right)}{Z_1^2 + \omega^2 c_2^2 Z_2^2 + c_1^2\omega^2 Z_3^2 + \omega^4 c_1 c_2\left(c_1 c_2 + 2\omega^2 m_2^2\right)} \tag{8.532}$$

$$\left(\frac{X_2}{e}\right)^2 = \frac{m_e^2 \omega^4 \left(k_2^2 + \omega^2 c_2^2\right)}{Z_1^2 + \omega^2 c_2^2 Z_2^2 + c_1^2 \omega^2 Z_3^2 + \omega^4 c_1 c_2 \left(c_1 c_2 + 2\omega^2 m_2^2\right)} \tag{8.533}$$

$$Z_1 = \left(k_1 - \omega^2 m_1\right)\left(k_2 - \omega^2 m_2\right) - \omega^2 m_2 k_2 \tag{8.534}$$

$$Z_2 = k_1 - \omega^2 m_1 - \omega^2 m_2 \tag{8.535}$$

$$Z_3 = \left(k_2 - \omega^2 m_2\right) \tag{8.536}$$

Introducing the following parameters

$$\varepsilon = \frac{m_2}{m_1} \qquad \varepsilon_1 = \frac{m_e}{m_1} \tag{8.537}$$

$$\xi_1 = \frac{c_1}{2m_1 \omega_1} \qquad \xi_2 = \frac{c_2}{2m_2 \omega_2} \tag{8.538}$$

$$\omega_1 = \sqrt{\frac{k_1}{m_1}} \qquad \omega_2 = \sqrt{\frac{k_2}{m_2}} \qquad \alpha = \frac{\omega_2}{\omega_1} \tag{8.539}$$

$$r = \frac{\omega}{\omega_1} \qquad \mu = \frac{X_1}{e} \qquad \tau = \frac{X_2}{e} \tag{8.540}$$

we transform the frequency responses to the following dimensionless form.

$$\mu^2 = \left(\frac{X_1}{e\varepsilon_1}\right)^2 = \frac{r^4 \left(\left(\alpha^2 - r^2\right)^2 + 4\xi_2^2 \alpha^2 r^2\right)}{4\xi_2^2 r^2 \alpha^2 Z_5^2 + Z_4^2 + 8\xi_1 \xi_2 \alpha r^4 Z_5 + 4r^6 \xi_1^2} \tag{8.541}$$

$$\tau^2 = \left(\frac{X_2}{e\varepsilon_1}\right)^2 = \frac{r^4 \alpha^2 \left(\alpha^2 + 4\xi_2^2 r^2\right)}{4\xi_2^2 r^2 \alpha^2 Z_5^2 + Z_4^2 + 8\xi_1 \xi_2 \alpha r^4 Z_5 + 4r^6 \xi_1^2} \tag{8.542}$$

$$Z_4 = \varepsilon \alpha^2 r^2 - \left(r^2 - 1\right)\left(r^2 - \alpha^2\right) \tag{8.543}$$

$$Z_5 = r^2 (1 + \varepsilon) - 1 \tag{8.544}$$

Eccentric Base Excitation

A two DOF eccentric base excited vibrating system has a base with an unbalance mass m_e at a distance e that is rotating with angular velocity ω. The system is composed by attaching a secondary system (m_2, k_2, c_2) to a primary eccentric base excited system. The equations of motion of the system in absolute coordinates are

$$m_1\ddot{x}_1 + c_1 (\dot{x}_1 - \dot{y}) - c_2 (\dot{x}_2 - \dot{x}_1)$$

$$+k_1 (x_1 - y) - k_2 (x_2 - x_1) = 0 \tag{8.545}$$

$$m_2\ddot{x}_2 + c_2 (\dot{x}_2 - \dot{x}_1) + k_2 (x_2 - x_1) = 0 \tag{8.546}$$

$$m_b\ddot{y} + m_e e\omega^2 \sin \omega t - c_1 (\dot{x}_1 - \dot{y}) - k_1 (x_1 - y) = 0 \tag{8.547}$$

Using the relative motion of m_2 and m_1 with respect to the base

$$z_1 = x_1 - y \qquad z_2 = x_2 - y \tag{8.548}$$

we may develop the equation of motion as

$$m_1\ddot{z}_1 + (c_1 (\varepsilon_1 + 1) + c_2) \dot{z}_1 + (k_1 (\varepsilon_1 + 1) + k_2) z_1$$

$$-c_2\dot{z}_2 - k_2 z_2 = m_e\varepsilon_1 e\omega^2 \sin \omega t \tag{8.549}$$

$$(c_1\varepsilon_2 - c_2) \dot{z}_1 + (k_1\varepsilon_2 - k_2) z_1$$

$$+m_2\ddot{z}_2 + c_2\dot{z}_2 + k_2 z_2 = m_e\varepsilon_2 e\omega^2 \sin \omega t \tag{8.550}$$

where

$$\varepsilon_1 = \frac{m_1}{m_b} \qquad \varepsilon_2 = \frac{m_2}{m_b} \tag{8.551}$$

Showing the steady-state amplitudes of z_1 and z_2 by S_1 and S_2, we find

$$\left(\frac{S_1}{m_e e}\right)^2 = \frac{S_{1N}}{S_D}\omega^4 \qquad \left(\frac{S_2}{m_e e}\right)^2 = \frac{S_{2N}}{S_D}\omega^4 \tag{8.552}$$

where

$$S_D = c_2^2 \left((a_{11} + a_{21})^2 + (a_{13} + a_{23})^2\right) \omega^2$$

$$+2c_2 (k_2 - a_{22}) (a_{21}a_{13} - a_{11}a_{23}) \omega$$

$$+ (a_{11}a_{22} + k_2a_{21})^2 + (a_{13}a_{22} + k_2a_{23})^2 \tag{8.553}$$

$$S_{1N} = c_2^2 (\varepsilon_1 + \varepsilon_2)^2 \omega^2 + (\varepsilon_1 a_{22} + \varepsilon_2 k_2)^2 \tag{8.554}$$

$$S_{2N} = (\varepsilon_1 a_{21} - \varepsilon_2 a_{11})^2 + (\varepsilon_1 a_{23} - \varepsilon_2 a_{13})^2 \tag{8.555}$$

and

$$a_{11} = a_{33} = k_1 (\varepsilon_1 + 1) + k_2 - \omega^2 m_1 \tag{8.556}$$

$$a_{21} = a_{43} = k_1\varepsilon_2 - k_2 \tag{8.557}$$

$$a_{22} = a_{44} = k_2 - \omega^2 m_2 \tag{8.558}$$

$$a_{13} = -a_{31} = -\omega \left(c_1 \left(\varepsilon_1 + 1 \right) + c_2 \right) \tag{8.559}$$

$$a_{23} = -a_{41} = -\omega \left(c_1 \varepsilon_2 - c_2 \right) \tag{8.560}$$

Vibration Absorber

Consider a primary system of a mass m_1 supported by a spring k_1. There is a harmonic force $f = F \sin \omega t$ applied on m_1. We add a secondary system (m_2, c_2, k_2) to absorb the vibrations of the primary. Adding a vibration absorber to a primary system generates a new system with two DOF. The goal of having a vibration absorber is to design the suspension of the secondary system (c_2, k_2) to reduce the amplitude of vibration of m_1 at a given excitation frequency ω or in a range of frequencies.

$$m_1 \ddot{x}_1 + c_2 \left(\dot{x}_1 - \dot{x}_2 \right) + k_1 x_1 + k_2 \left(x_1 - x_2 \right) = F \sin \omega t \tag{8.561}$$

$$m_2 \ddot{x}_2 - c_2 \left(\dot{x}_1 - \dot{x}_2 \right) - k_2 \left(x_1 - x_2 \right) = 0 \tag{8.562}$$

To minimize the steady-state vibration amplitude of m_1 when the excitation frequency is variable, we must set k_2 at the optimal value $k_2^\star$

$$k_2^\star = \frac{m_1 m_2}{(m_1 + m_2)^2} k_1 \tag{8.563}$$

and select c_2 within the following range.

$$2 m_2 \omega_1 \xi_1^\star < c_2 < 2 m_2 \omega_1 \xi_2^\star \tag{8.564}$$

The optimal damping ratios $\xi_1^\star$ and $\xi_2^\star$ are

$$\xi_1^\star = \sqrt{\frac{-B - \sqrt{B^2 - 4AC}}{2A}} \tag{8.565}$$

$$\xi_2^\star = \sqrt{\frac{-B + \sqrt{B^2 - 4AC}}{2A}} \tag{8.566}$$

$$A = 16 Z_8 - 4 r^2 \left(4 Z_4 + 8 Z_5 \right) \tag{8.567}$$

$$B = 4 Z_9 - 4 Z_6 r^2 - Z_7 \left(4 Z_4 + 8 Z_5 \right) + 4 Z_3 Z_8 \tag{8.568}$$

$$C = Z_3 Z_9 - Z_6 Z_7 \tag{8.569}$$

$$Z_3 = 2\left(r^2 - \alpha^2\right) \tag{8.570}$$

$$Z_4 = \left[r^2\left(1 + \varepsilon\right) - 1\right]^2 \tag{8.571}$$

$$Z_5 = r^2\left(1 + \varepsilon\right)\left[r^2\left(1 + \varepsilon\right) - 1\right] \tag{8.572}$$

$$Z_6 = 2\left[\varepsilon\alpha^2 r^2 - \left(r^2 - \alpha^2\right)\left(r^2 - 1\right)\right]$$
$$\times \left[\varepsilon\alpha^2 - \left(r^2 - \alpha^2\right) - \left(r^2 - 1\right)\right] \tag{8.573}$$

$$Z_7 = \left(r^2 - \alpha^2\right)^2 \tag{8.574}$$

$$Z_8 = r^2\left[r^2\left(1 + \varepsilon\right) - 1\right]^2 \tag{8.575}$$

$$Z_9 = \left[\varepsilon\alpha^2 r^2 - \left(r^2 - 1\right)\left(r^2 - \alpha^2\right)\right]^2 \tag{8.576}$$

8.8   Key Symbols

a, b, c, d	Constant parameters
A, B	Coefficients of frequency response
$[A]$	$=[m]^{-1}[k]$, characteristic matrix, part of $[T]$
$[B]$	$=[A]^{-1} = [k]^{-1}[m]$, inverse characteristic matrix
c	Damping
$[c]$	Damping matrix
C	Constant coefficient, constant of integration, amplitude
D	Dissipation function
e	Exponential function
E	Mechanical energy
f	Force, function
f_T	Transmitted force
F	Force amplitude, force magnitude, constant force
F_T	Transmitted force amplitude
$\mathbf{F}$	Forces
g	Functions of displacement, gravitational acceleration
$\mathbf{g}$	Gravitational acceleration vector
i	Counting index, unit of imaginary numbers
I	Mass moment

$\mathbf{I}$, $[\mathbf{I}]$	Identity matrix
k	Stiffness
k_{ij}	Stiffness matrix elements
k_s	Sprung stiffness
k_u	Unsprung stiffness
$[k]$	Stiffness matrix
$[k']$	Diagonal stiffness matrix
l	Length
$\mathcal{L}$	Lagrangian
m	Mass
K	Kinetic energy
$[m]$	Mass matrix
$[m']$	Diagonal mass matrix
p	Principal coordinate
$\mathbf{p}$	Principal coordinates
P	Potential energy
$\mathbf{P}$	Principal force
q	Generalized coordinate, temporal function of x
$\mathbf{q}$	Generalized coordinates
$\mathbf{Q}$	Generalized force
r	Frequency ratio, radial coordinate
S	Amplitude
t	Time
u	Components of $\mathbf{u}$
$\mathbf{u}$	Eigenvector
$[U]$	Modal matrix
$v \equiv \dot{x}$	Velocity
$v_0 = \dot{x}_0$	Initial velocity
W	Work
x	Displacement coordinate, displacement variable
$\mathbf{x}$	Displacement coordinates
x_0	Initial displacement
X	Displacement amplitude
$\mathbf{X}$	Displacement amplitudes
y	Displacement excitation, displacement, relative displacement
$\mathbf{y}$	Displacements
Y	Amplitude of displacement excitation
z	Variable, relative displacement
Z	Amplitude of z
Z	Short name of expressions

Greek

α	Stiffness parameter, stiffness ratio
ε	Mass ratio, coupling mass element
λ	Eigenvalue
η	Frequency response of relative motion

θ Angular coordinate, angular variable
τ Frequency response of secondary coordinate
φ Phase angle
μ Frequency response of primary coordinate
ξ Damping ratio
ω Excitation frequency, angular frequency
ω_0 Critical excitation frequency
ω_i, ω_n Natural frequency

Symbol
$\mathcal{L}$ Lagrangian
DOF Degree of freedom

Exercises

1. Possible node.
   It seems that there is a node in Fig. 8.15.

   (a) Magnify the area around the node.
   (b) Plot the frequency response τ for $\xi_1 = 0$ and $\xi_1 = \infty$. In case the plot for $\xi_1 = \infty$ is not available, plot τ for a large enough ξ_1.
   (c) Determine the height of the node, which means the frequency response at the node, τ_0.
   (d) ★ Determine what parameter that controls the value of the frequency response at the node and plot τ_0 versus the parameter.
   (e) ★ Determine if it is possible to minimize τ_0 and find the value of the parameter that minimizes τ_0, if any.

2. Nodes height control.
   There are two nodes in Fig. 8.17. Call the nodes P for the lower frequency and Q for the higher frequency.

   (a) Determine the frequencies at which the nodes occur.
   (b) Determine the node frequency responses.
   (c) ★ Determine what parameter controls the value of the frequency response at the nodes and plot μ_P and μ_Q versus the parameter.
   (d) ★ Determine if it is possible to equalize μ_P and μ_Q and find the value of the parameter that does that, if it exists.
   (e) ★ Determine and plot the horizontal distance of μ_P and μ_Q.

3. Slope at $r = 0.6$.
   Determine the slope of μ at $r = 0.6$ for both directions, as shown in Fig. 8.18. The value frequency derivative, $dX/dr of$, of a frequency response determines the high and low sensitivity zones of change of amplitude to change of excitation frequency.

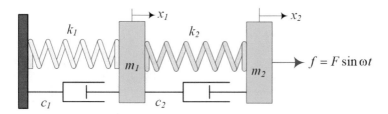

Fig. 8.82 A two DOF vibrating system with an applied harmonic force on the secondary system

4. Zero slope at the second resonance.
   In Fig. 8.19, determine the value of ξ_1 at which the frequency response curve at the second resonance zone is flat.
5. Nodes height control.
   There are two nodes in Fig. 8.21. Name the nodes P for the lower frequency and Q for the higher frequency. Determine α such that $\mu_P = 2\mu_Q$.
6. Slope at $r = 1$.
   Consider the frequency response of Fig. 8.21 and determine the slope of μ at $r = 1$ for the curve of $\xi_2 = 0$.
7. ★ Playing with max and min.
   Consider the system with the frequency response of Fig. 8.21.

   (a) Determine the nodes' frequencies.
   (b) Determine ξ_2 such that the first maximum of μ occurs at the first node.
   (c) Determine ξ_2 such that the first maximum of μ occurs at the second node.
   (d) Determine the minimum ξ_2 such that μ does not show a minimum.
   (e) Determine ξ_2 such that μ shows only one maximum.
   (f) Prove that it is not true to say, when μ has only one maximum, that the system has one DOF.

8. Forced excitation on the secondary system.
   A two DOF vibrating system is shown in Fig. 8.82, in which a harmonic force is applied to the secondary system.

   (a) Determine the equations of motion.
   (b) Determine the frequency responses of the system.
   (c) Assume $c_1 = 0$ and plot the frequency responses for zero and infinite damping c_2.
   (d) Assume $c_2 = 0$ and plot the frequency responses for zero and infinite damping c_1.

9. ★ Zero amplitude of forced excitation.
   In the forced excited two DOF systems, we can stop m_1 at $r = r_0$ by adjusting $\alpha = r_0$. Use the free body diagram of Fig. 8.83 to show that the net force f_1 on m_1 is

$$f_1 = F \sin \omega t + k_2 (x_2 - x_1) - k_1 x_1 \qquad (8.577)$$

Fig. 8.83 Free body diagram
of m_1 in system of Fig. 8.16b

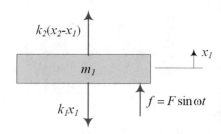

(a) Assume no damping and substitute x_1 and x_2

$$x_1 = A_1 \sin \omega t + B_1 \cos \omega t = X_1 \sin (\omega t - \varphi_1) \qquad (8.578)$$
$$x_2 = A_2 \sin \omega t + B_2 \cos \omega t = X_2 \sin (\omega t - \varphi_2) \qquad (8.579)$$

$$X_1 = \frac{F}{k_1} \frac{r^2 - \alpha^2}{\varepsilon \alpha^2 r^2 - (r^2 - 1)(r^2 - \alpha^2)} \qquad (8.580)$$

$$X_2 = \frac{F}{k_1} \frac{\alpha^2}{\varepsilon \alpha^2 r^2 - (r^2 - 1)(r^2 - \alpha^2)} \qquad (8.581)$$

and prove that $f_1 = 0$.
(b) Prove that $f_1 = 0$ regardless of the value of ξ_2, as long as $\xi_1 = 0$.

10. Node of secondary frequency response.

(a) Determine the value of τ at the node of Fig. 8.22.
(b) Determine ξ_2 such that the frequency response curve at the second resonance is just flat.

11. Transmitted force frequency response.

(a) Determine ξ_2 such that the frequency response curve of Fig. 8.8 at the zone around $r = 1$ is flat.
(b) ★ Is it possible to make the two maxima of F_T/F equal? Prove your answer.

12. Node of base excited system.
Figure 8.36 depicts the frequency response of the main system of a two DOF base excited system.

(a) Determine the node frequency. Is it $r = \sqrt{2}$?
(b) ★ Is there any ξ_1 to make the two maxima of μ equal?

13. ★ Playing with max and min of τ.
Consider the system with the frequency response of Fig. 8.37.

(a) Replot the figure for $0.6 < r < 1.4$.
(b) Determine the nodes' frequencies.
(c) Determine ξ_1 such that the first maximum of τ occurs at the first node.
(d) Determine ξ_1 such that the first maximum of τ occurs at the second node.
(e) Determine the minimum ξ_1 such that τ does not show a minimum.
(f) Determine ξ_1 such that τ shows only one maximum.

14. Nodes height control.
There are two nodes in Fig. 8.42. Call the nodes P for the lower frequency and Q for the higher frequency.

(a) Determine the frequencies at which the nodes occur.
(b) Determine the node frequency responses.
(c) ★ Determine what parameter controls the value of the frequency response at the nodes and plot μ_P and μ_Q versus the parameter.
(d) ★ Determine if it is possible to equalize μ_P and μ_Q and find the value of the parameter that does that, if it exists.
(e) ★ Determine and plot the horizontal distance of μ_P and μ_Q.

15. Slope at $r = 1$.
Consider the frequency response of Fig. 8.29 and determine the slopes of μ at $r = 1$ for the curve of $\xi_2 = 0$.

16. Eccentric excitation at secondary system.
Figure 8.84 illustrates a two DOF system where an eccentric excitation at the secondary system makes it vibrate.

(a) Determine the frequency responses of the system
(b) Plot the frequency responses for $c_1 = 0$ and $c_2 = 0$.
(c) ★ Keep $c_1 = 0$ and see if there is any node in any of the frequency responses.
(d) ★ Keep $c_2 = 0$ and see if there is any node in any of the frequency responses.

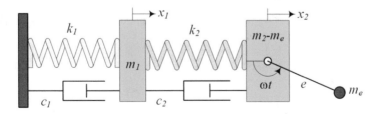

Fig. 8.84 A two DOF system with an eccentric excitation at the secondary system

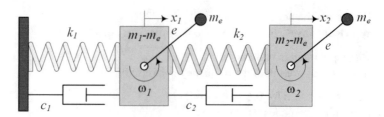

Fig. 8.85 A two DOF system with two eccentric excitations at the primary and secondary systems

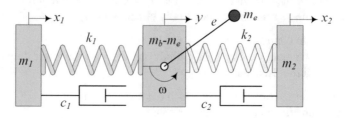

Fig. 8.86 An eccentric excitations with two attached secondary systems

17. ★ Two eccentric excitations.

Figure 8.85 illustrates a two DOF system with two eccentric excitations at the primary and secondary systems. Determine the frequency responses of the system.

18. Possible node.

It seems Fig. 8.53 has a node at $r = 1$.

(a) Magnify the area around the point.
(b) Plot the frequency response τ for $\xi_1 = 0$ and $\xi_1 = \infty$. In case the plot for $\xi_1 = \infty$ is not available, plot τ for a large enough ξ_1.
(c) Determine the height of the node, τ_0.
(d) ★ Determine what parameter controls the value of frequency response at the node and plot τ_0 versus the parameter.
(e) ★ Determine if it is possible to minimize τ_0 and find the value of the parameter that minimizes τ_0, if any.

19. Nodes height control.

(a) There are two nodes in Fig. 8.58. Call the nodes P for the lower frequency node and Q for the higher frequency one.
(b) Determine α such that $\mu_P = 2\mu_Q$.
(c) Determine α such that $2\mu_P = \mu_Q$.
(d) Determine α such that $\mu_P = \mu_Q$.

20. ★ Eccentric excitation with two secondary.

Determine the frequency responses of the system in Fig. 8.86.

Appendix A
Frequency Response Curves

There are four types of one DOF harmonically excited systems as shown in Fig. 6.1:

1. base excitation,
2. eccentric excitation,
3. eccentric base excitation,
4. forced excitation.

The frequency responses of the four systems can be summarized, labeled, and shown as follows (Figs. A.1, A.2, A.3, A.4, A.5, A.6, A.7, A.8):

$$S_0 = \frac{X_F}{F/k} \tag{A.1}$$

$$= \frac{1}{\sqrt{\left(1 - r^2\right)^2 + (2\xi r)^2}} \tag{A.2}$$

$$S_1 = \frac{\dot{X}_F}{F/\sqrt{km}} \tag{A.3}$$

$$= \frac{r}{\sqrt{\left(1 - r^2\right)^2 + (2\xi r)^2}} \tag{A.4}$$

$$S_2 = \frac{\ddot{X}_F}{F/m} = \frac{Z_B}{Y} = \frac{X_E}{e\varepsilon_E} = \frac{Z_R}{e\varepsilon_R} \tag{A.5}$$

$$= \frac{r^2}{\sqrt{\left(1 - r^2\right)^2 + (2\xi r)^2}} \tag{A.6}$$

© The Author(s), under exclusive license to Springer Nature Switzerland AG 2022
R. N. Jazar, *Advanced Vibrations*, https://doi.org/10.1007/978-3-031-16356-2

$$S_3 = \frac{\dot{Z}_B}{\omega_n Y} = \frac{\dot{X}_E}{e\varepsilon_E \omega_n} = \frac{\dot{Z}_R}{e\varepsilon_R \omega_n} \tag{A.7}$$

$$= \frac{r^3}{\sqrt{(1-r^2)^2 + (2\xi r)^2}} \tag{A.8}$$

$$S_4 = \frac{\ddot{Z}_B}{\omega_n^2 Y} = \frac{\ddot{X}_E}{e\varepsilon_E \omega_n^2} = \frac{\ddot{Z}_R}{e\varepsilon_R \omega_n^2} \tag{A.9}$$

$$= \frac{r^4}{\sqrt{(1-r^2)^2 + (2\xi r)^2}} \tag{A.10}$$

$$G_0 = \frac{F_{T_F}}{F} = \frac{X_B}{Y} \tag{A.11}$$

$$= \frac{\sqrt{1 + (2\xi r)^2}}{\sqrt{(1-r^2)^2 + (2\xi r)^2}} \tag{A.12}$$

$$G_1 = \frac{\dot{X}_B}{\omega_n Y} \tag{A.13}$$

$$= \frac{r\sqrt{1 + (2\xi r)^2}}{\sqrt{(1-r^2)^2 + (2\xi r)^2}} \tag{A.14}$$

$$G_2 = \frac{\ddot{X}_B}{\omega_n^2 Y} = \frac{F_{T_B}}{kY} = \frac{F_{T_E}}{e\omega_n^2 m_e} = \frac{F_{T_R}}{e\omega_n^2 m_e}\left(1 + \frac{m_b}{m}\right) \tag{A.15}$$

$$= \frac{r^2\sqrt{1 + (2\xi r)^2}}{\sqrt{(1-r^2)^2 + (2\xi r)^2}} \tag{A.16}$$

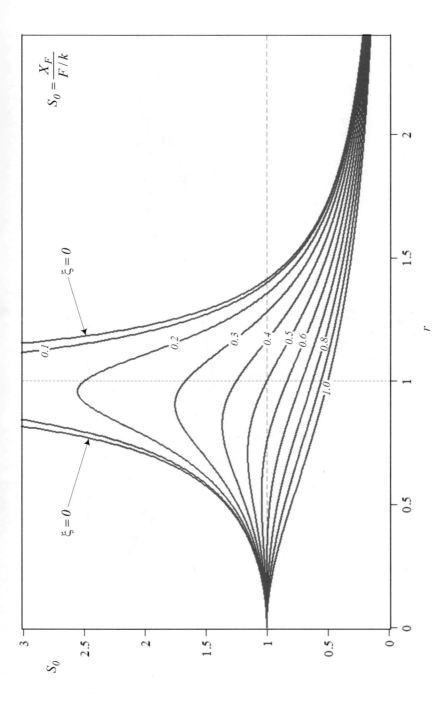

Fig. A.1 Frequency response for S_0

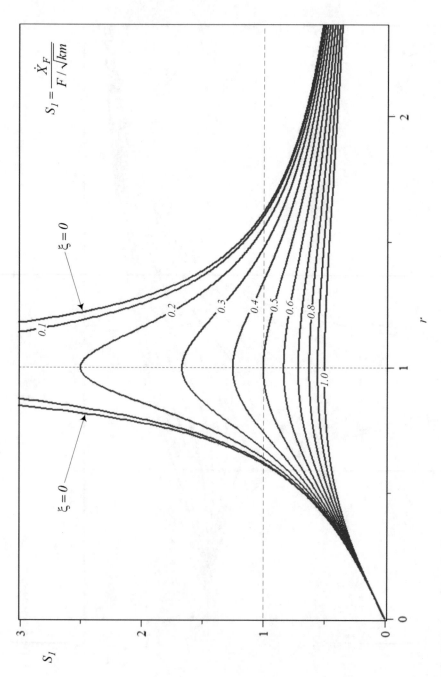

Fig. A.2 Frequency response for S_1

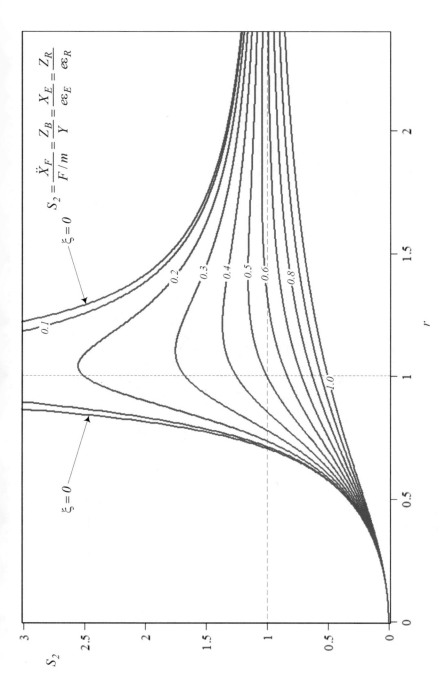

Fig. A.3 Frequency response for S_2

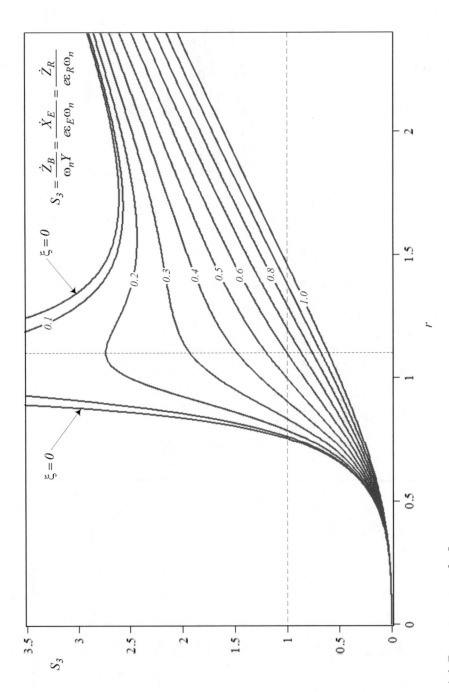

Fig. A.4 Frequency response for S_3

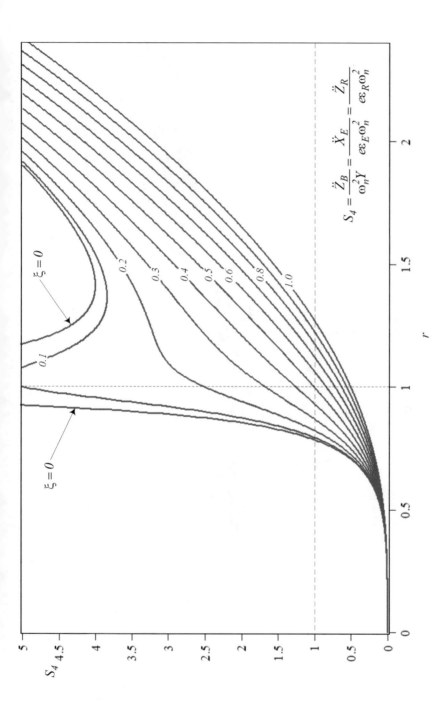

Fig. A.5 Frequency response for S_4

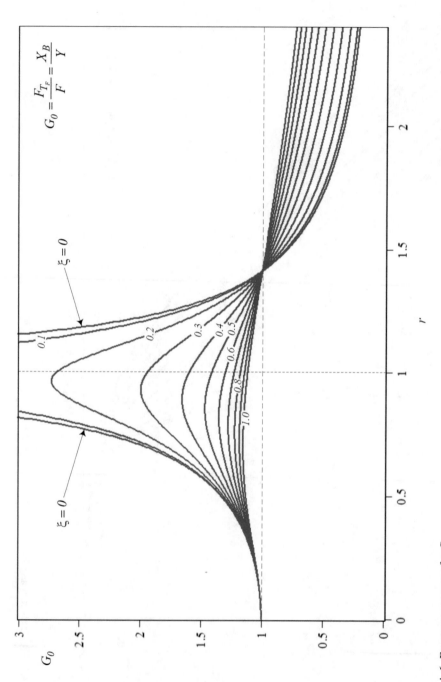

Fig. A.6 Frequency response for G_0

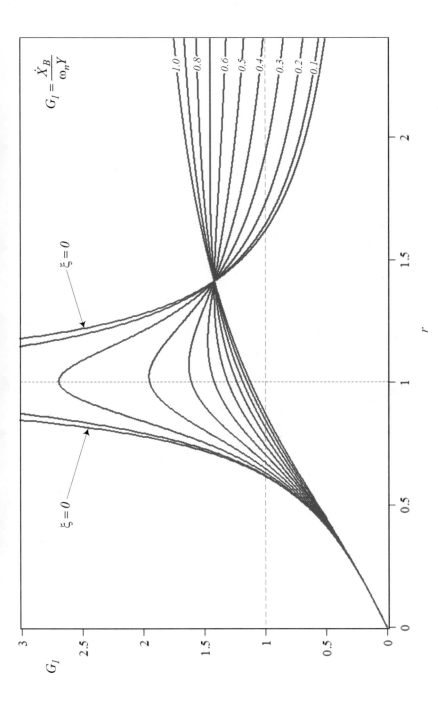

Fig. A.7 Frequency response for G_1

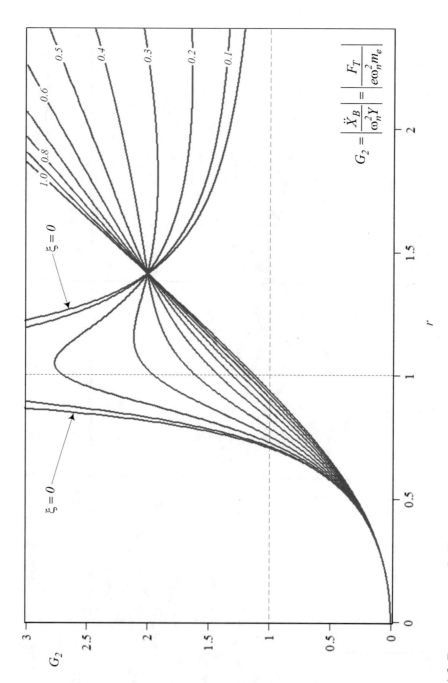

Fig. A.8 Frequency response for G_2

Appendix B
Matrix Calculus

An $m \times n$ matrix $\mathbf{A}$ consists of mn numbers arranged in m rows and n columns. The element in row i and column j of the matrix $\mathbf{A}$ is denoted by a_{ij}.

$$\mathbf{A} = \begin{bmatrix} a_{11} & a_{12} & \cdots & a_{1n} \\ a_{21} & a_{22} & \cdots & a_{2n} \\ \cdots & \cdots & \cdots & \cdots \\ a_{m1} & a_{m2} & \cdots & a_{mn} \end{bmatrix} = \begin{bmatrix} a_{ij} \end{bmatrix} \tag{B.1}$$

An $m \times 1$ matrix is a column vector $\mathbf{v}$ of dimension m, and a $1 \times n$ matrix is a row vector $\mathbf{w}$ of dimension n. An $m \times n$ matrix is called a square matrix if $m = n$.

$$\mathbf{v} = \begin{bmatrix} v_{11} \\ v_{21} \\ \cdots \\ v_{m1} \end{bmatrix} \tag{B.2}$$

$$\mathbf{w} = \begin{bmatrix} w_{m1} & w_{m2} & \cdots & w_{mn} \end{bmatrix} \tag{B.3}$$

Two matrices $\mathbf{A} = \begin{bmatrix} a_{ij} \end{bmatrix}$ and $\mathbf{B} = \begin{bmatrix} b_{ij} \end{bmatrix}$ are equal only if $a_{ij} = b_{ij}$.

Diagonal Matrix
A diagonal matrix is a square matrix $\mathbf{\Lambda} = \begin{bmatrix} \lambda_{ij} \end{bmatrix}$ such that $\lambda_{ij} = 0$ for $i \neq j$. The main diagonal of a square matrix are elements from top left to bottom right.

$$\mathbf{\Lambda} = \begin{bmatrix} \lambda_{11} & 0 & \cdots & 0 \\ 0 & \lambda_{22} & \cdots & 0 \\ \cdots & \cdots & \cdots & \cdots \\ 0 & 0 & \cdots & \lambda_{nn} \end{bmatrix} \tag{B.4}$$

© The Author(s), under exclusive license to Springer Nature Switzerland AG 2022
R. N. Jazar, *Advanced Vibrations*, https://doi.org/10.1007/978-3-031-16356-2

A diagonal matrix may also be abbreviated as:

$$\mathbf{\Lambda} = \{\lambda_{11}, \lambda_{22}, \lambda_{22}, \cdots, \lambda_{nn}\} \tag{B.5}$$

A scalar matrix $\mathbf{S} = \begin{bmatrix} s_{ij} \end{bmatrix}$ is a diagonal matrix whose diagonal elements are equal $s_{ii} = s$.

$$\mathbf{S} = \begin{bmatrix} s & 0 & \cdots & 0 \\ 0 & s & \cdots & 0 \\ \cdots & \cdots & \cdots & \cdots \\ 0 & 0 & \cdots & s \end{bmatrix} \tag{B.6}$$

If all elements of the diagonal of $\mathbf{S}$ are unity, the matrix is the identity matrix, or unit matrix, and shown by $\mathbf{I}$.

$$\mathbf{I} = \begin{bmatrix} 1 & 0 & \cdots & 0 \\ 0 & 1 & \cdots & 0 \\ \cdots & \cdots & \cdots & \cdots \\ 0 & 0 & \cdots & 1 \end{bmatrix} \tag{B.7}$$

A null matrix is a matrix of any shape in which every entry is zero.

$$\mathbf{0} = \begin{bmatrix} 0 & 0 & \cdots & 0 \\ 0 & 0 & \cdots & 0 \\ \cdots & \cdots & \cdots & \cdots \\ 0 & 0 & \cdots & 0 \end{bmatrix} \tag{B.8}$$

Trace

The trace of a square matrix $\mathbf{A} = \begin{bmatrix} a_{ij} \end{bmatrix}$ is the sum of its diagonal elements.

$$\mathrm{tr}\mathbf{A} = a_{11} + a_{22} + \cdots + a_{nn} \tag{B.9}$$

$$\mathrm{tr}\,[\mathbf{AB}] = \mathrm{tr}\,[\mathbf{BA}] \tag{B.10}$$

Matrix Arithmetic

$$k\mathbf{A} = k\begin{bmatrix} a_{ij} \end{bmatrix} = \begin{bmatrix} ka_{ij} \end{bmatrix} \tag{B.11}$$

$$\mathbf{A} + \mathbf{B} = \mathbf{B} + \mathbf{A} = \mathbf{C} \tag{B.12}$$

$$c_{ij} = a_{ij} + b_{ij} \tag{B.13}$$

$$\mathbf{AB} = \mathbf{C} \qquad c_{ij} = \sum_m a_{im}b_{mj} \tag{B.14}$$

$$\mathbf{AB} \neq \mathbf{BA} \tag{B.15}$$

$$[\mathbf{AB}]\,\mathbf{C} = \mathbf{A}\,[\mathbf{BC}] = \mathbf{D} \tag{B.16}$$

$$d_{ij} = \sum_{m,n} (a_{im}b_{mn})\,c_{jn} = \sum_{m,n} a_{im}\left(b_{mn}c_{jn}\right) \tag{B.17}$$

$$[\mathbf{A} + \mathbf{B}]\,\mathbf{C} = \mathbf{AC} + \mathbf{BC} = \mathbf{D} \tag{B.18}$$

$$d_{ij} = \sum_m (a_{im} + b_{im})\,c_{nj} = \sum_m \left(a_{im}c_{nj} + b_{im}c_{nj}\right) \tag{B.19}$$

Determinants

The determinant of a 2×2 matrix $\mathbf{A}$, with elements a_{ij} comprising real or complex numbers, or functions,

$$\mathbf{A} = \begin{bmatrix} a_{11} & a_{12} \\ a_{21} & a_{22} \end{bmatrix} = [a_{ij}] \tag{B.20}$$

is denoted by $|\mathbf{A}|$ or by $\det \mathbf{A}$ and is defined by

$$\det \mathbf{A} = |\mathbf{A}| = a_{11}a_{22} - a_{12}a_{21} \tag{B.21}$$

The determinant of a 3×3 matrix $\mathbf{A}$,

$$\mathbf{A} = \begin{bmatrix} a_{11} & a_{12} & a_{13} \\ a_{21} & a_{22} & a_{23} \\ a_{31} & a_{32} & a_{33} \end{bmatrix} = [a_{ij}] \tag{B.22}$$

with elements a_{ij} comprising real or complex numbers, or functions, is denoted by $|\mathbf{A}|$ or by $\det \mathbf{A}$ and is defined by

$$\det \mathbf{A} = a_{11}a_{22}a_{33} + a_{12}a_{23}a_{31} + a_{21}a_{32}a_{13}$$

$$-a_{13}a_{22}a_{31} - a_{23}a_{32}a_{11} - a_{12}a_{21}a_{33} \tag{B.23}$$

as determinant of the matrix of order $(n-1)$ derived by deletion of its ith row and jth column

$$A = \begin{bmatrix} a_{11} & a_{12} & \cdots & a_{1n} \\ a_{21} & a_{22} & \cdots & a_{2n} \\ \cdots & \cdots & \cdots & \cdots \\ a_{n1} & a_{n2} & \cdots & a_{nn} \end{bmatrix} = [a_{ij}] \qquad (B.24)$$

we need first, to define the minor M_{ij} associated with the element a_{ij} as determinant the order $(n-1)$ derived from by deletion of its ith row and jth column; and second, to define the cofactor C_{ij} associated with the element a_{ij} as

$$C_{ij} = (-1)^{i+j} M_{ij} \qquad (B.25)$$

The determinant of A may be based on the expansion of elements of a row or of a column of A. Expansion of det A by elements of the ith row is

$$\det A = \sum_{j=1}^{n} a_{ij} C_{ij} \qquad i = 1, 2, 3, \cdots, n \qquad (B.26)$$

and expansion of det A by elements of the jth column is

$$\det A = \sum_{i=1}^{n} a_{ij} C_{ij} \qquad j = 1, 2, 3, \cdots, n \qquad (B.27)$$

The expansion method to determine the determinant is called the Laplace expansion method.

As an example, let us calculate det A, where

$$A = \begin{bmatrix} -2 & 1 & 3 \\ 1 & 2 & -1 \\ 4 & -1 & -2 \end{bmatrix} \qquad (B.28)$$

Expanding $|A|$ by elements of its second row gives

$$|A| = C_{21} + 2C_{22} - C_{23} \qquad (B.29)$$

$$C_{21} = (-1)^{2+1} \begin{vmatrix} 1 & 3 \\ -1 & -2 \end{vmatrix} = -1 \qquad (B.30)$$

$$C_{22} = (-1)^{2+2} \begin{vmatrix} -2 & 3 \\ 4 & -2 \end{vmatrix} = -8 \qquad (B.31)$$

$$C_{22} = (-1)^{2+3} \begin{vmatrix} -2 & 1 \\ 4 & -1 \end{vmatrix} = 2 \tag{B.32}$$

Therefore,

$$|\mathbf{A}| = -1 + 2(-8) - 2 = -19 \tag{B.33}$$

Alternatively, expanding $|\mathbf{A}|$ by elements of its third column gives

$$|\mathbf{A}| = 3C_{13} - C_{23} - 2C_{33} \tag{B.34}$$

$$C_{21} = (-1)^{1+3} \begin{vmatrix} 1 & 2 \\ 4 & -1 \end{vmatrix} = -9 \tag{B.35}$$

$$C_{22} = (-1)^{2+3} \begin{vmatrix} -2 & 1 \\ 4 & -1 \end{vmatrix} = 2 \tag{B.36}$$

$$C_{22} = (-1)^{3+3} \begin{vmatrix} -2 & 1 \\ 1 & 2 \end{vmatrix} = -5 \tag{B.37}$$

Therefore,

$$|\mathbf{A}| = 3(-9) - 2 - 2(-5) = -19 \tag{B.38}$$

If $\mathbf{A} = [a_{ij}]$ and $\mathbf{B} = [b_{ij}]$ are two $n \times n$ matrices, then the following results are true for determinants of $\mathbf{A}$ and $\mathbf{B}$.

1. If any two adjacent rows (or columns) of $|\mathbf{A}|$ are interchanged, the sign of the resulting determinant is changed.
2. If any two rows (or columns) of $|\mathbf{A}|$ are identical, then $|\mathbf{A}| = 0$.
3. The value of a determinant is not changed if any multiple of a row (or column) is added to any other row (or column) of $|\mathbf{A}|$.
4. $|c\mathbf{A}| = c^n |\mathbf{A}|, c \in \mathbb{N}$.
5. $|\mathbf{A}^T| = |\mathbf{A}|$
6. $|\mathbf{A}| |\mathbf{B}| = |\mathbf{AB}|$
7. $|\mathbf{A}^{-1}| = 1/|\mathbf{A}|$

Transpose, Symmetric, and Skew Matrices
The matrix $\mathbf{A}^T = [a_{ji}]$ obtained from $\mathbf{A} = [a_{ij}]$ by changing rows to columns is the transpose of $\mathbf{A}$.

$$\mathbf{A} = [a_{ij}] \qquad \mathbf{A}^T = [a_{ji}] \tag{B.39}$$

$$\left[\mathbf{A}^T\right]^T = \mathbf{A} \tag{B.40}$$

$$[\mathbf{A} + \mathbf{B} + \cdots + \mathbf{C}]^T = \mathbf{A}^T + \mathbf{B}^T + \cdots + \mathbf{C}^T \tag{B.41}$$

$$[\mathbf{AB}\cdots\mathbf{C}]^T = \mathbf{C}^T \cdots \mathbf{B}^T \, \mathbf{A}^T \tag{B.42}$$

$$[\mathbf{AB}]^T = \left[\sum_n a_{in}b_{nj}\right]^T = \left[\sum_n b_{ni}a_{jn}\right]^T = \mathbf{B}^T \, \mathbf{A}^T \tag{B.43}$$

A matrix $\mathbf{A}$ such that $\mathbf{A}^T = \mathbf{A}$ is a symmetric matrix. A matrix $\mathbf{A}$ such that $\mathbf{A}^T = -\mathbf{A}$ is a skew, skew-symmetric, or alternating matrix.

Inverse

The inverse of a square matrix $\mathbf{A}$ is indicated by $\mathbf{A}^{-1}$,

$$\mathbf{A}^{-1} = \frac{\mathrm{adj}\mathbf{A}}{|\mathbf{A}|} \qquad \mathbf{A}^{-1}\,\mathbf{A} = \mathbf{A}\,\mathbf{A}^{-1} = \mathbf{I} \tag{B.44}$$

where $\mathrm{adj}\mathbf{A}$ is the adjoint of the matrix $\mathbf{A}$ and C_{ij} are the cofactors associated with element a_{ij} of $\mathbf{A}$.

$$\mathrm{adj}\mathbf{A} = \left[C_{ij}\right]^T \tag{B.45}$$

$$[\mathbf{AB}\cdots\mathbf{C}]^{-1} = \mathbf{C}^{-1}\cdots\mathbf{B}^{-1}\,\mathbf{A}^{-1} \tag{B.46}$$

$$\left[\mathbf{A}^{-1}\right]^{-1} = \mathbf{A} \tag{B.47}$$

$$\left[\mathbf{A}^{-1}\right]^T = \left[\mathbf{A}^T\right]^{-1} \tag{B.48}$$

$$[k\mathbf{A}]^{-1} = k^{-1}\mathbf{A}^{-1} \tag{B.49}$$

Non-square matrices of order $m \times n$ and $m \neq n$ do not have inverse. However, if $\mathbf{A}$ is $m \times n$ and the rank of $\mathbf{A}$ is equal to n, then $\mathbf{A}$ has a left inverse. An $n \times m$ matrix $\mathbf{B}$ such that $\mathbf{BA} = \mathbf{I}$. If $\mathbf{A}$ has rank m, then it has a right inverse. An $n \times m$ matrix $\mathbf{B}$ such that $\mathbf{AB} = \mathbf{I}$.

Sherman-Morrison-Woodbury formula.

$$\left[A + v^T w \right]^{-1} = A^{-1} - A^{-1} v \left[I + v^T A^{-1} w \right]^{-1} v^T A^{-1} \tag{B.50}$$

Differentiation and Integration
If $A(t) = \left[a_{ij}(t) \right]$ and $B(t) = \left[b_{ij}(t) \right]$ are two $n \times n$ matrices, then

$$\frac{d}{dt} A(t) = \left[\frac{d}{dt} a_{ij}(t) \right] \tag{B.51}$$

$$\frac{d}{dt} \left[aA(t) \pm bB(t) \right] = a\frac{d}{dt} A(t) \pm b\frac{d}{dt} B(t) \tag{B.52}$$

$$\frac{d}{dt} \left[A(t) B(t) \right] = \left[\frac{d}{dt} A(t) \right] B(t) + A(t) \left[\frac{d}{dt} B(t) \right] \tag{B.53}$$

$$\frac{d}{dt} \left[A(t) B(t) \right]^T = \left[\frac{d}{dt} B(t) \right]^T A^T(t) + B^T(t) \left[\frac{d}{dt} A(t) \right]^T \tag{B.54}$$

$$\int_{t_0}^{t} A(s)\, ds = \left[\int_{t_0}^{t} a_{ij}(s)\, ds \right] \tag{B.55}$$

$$\int_{t_0}^{t} aA(s) \pm bB(s)\, ds = a \int_{t_0}^{t} A(s)\, ds \pm b \int_{t_0}^{t} B(s)\, ds \tag{B.56}$$

If $|A| \neq 0$, then

$$\frac{d}{dt} A^{-1}(t) = -A^{-1}(t) \left[\frac{d}{dt} A(t) \right] A^{-1}(t) \tag{B.57}$$

$$\frac{d}{dt} A^T(t) = -A^T(t) \left[\frac{d}{dt} A(t) \right] A^T(t) \tag{B.58}$$

If $A(x) = \left[a_{ji}(x) \right]$, then

$$\frac{d |A|}{dx} = \sum_{i,j=1}^{n} \frac{d a_{ij}}{dx} C_{ij} \tag{B.59}$$

where C_{ij} are cofactors associated with element a_{ij} of A.

The Matrix Exponential

If $\mathbf{A}$ is a square matrix, and z is any complex variable, then the matrix exponential $\exp(\mathbf{A}z)$ is defined as

$$e^{\mathbf{A}z} = \mathbf{I} + \mathbf{A}z + \frac{\mathbf{A}^2 z^2}{2!} + \cdots + \frac{\mathbf{A}^n z^n}{n!} = \sum_{k=0}^{\infty} \frac{\mathbf{A}^k z^k}{k!} \tag{B.60}$$

$$e^0 = \mathbf{I} \qquad e^{\mathbf{I}z} = \mathbf{I}\, e^z \qquad e^{\mathbf{A}(z_1 + z_2)} = e^{\mathbf{A}z_1} e^{\mathbf{A}z_2} \tag{B.61}$$

$$e^{-\mathbf{A}z} = \left[e^{\mathbf{A}z} \right]^{-1} \qquad e^{\mathbf{A}z} e^{\mathbf{B}z} = e^{(\mathbf{A}+\mathbf{B})z} \tag{B.62}$$

$$\frac{d^k}{dz^k} e^{\mathbf{A}z} = \mathbf{A}^k\, e^{\mathbf{A}z} = e^{\mathbf{A}z}\, \mathbf{A}^k \tag{B.63}$$

Appendix C
Ordinary Differential Equations

The mathematical formulation of problems in engineering and science usually leads to differential equations.

Definition 328 A **differential equation** is an equation containing one or more derivative of the function under consideration. The equation

$$\frac{dx}{dt} = f(x, t) \tag{C.1}$$

is a differential equation of the function x, as a function of t.

Definition 329 The **order** of a differential equation is the order of the highest derivative therein. The general differential equation

$$F\left(t, x(t), \frac{dx(t)}{dt}, \frac{d^2x(t)}{dt^2}, \ldots, \frac{d^n x(t)}{dt^n}\right) = 0 \tag{C.2}$$

is an nth-order differential equation on the unknown $x(t)$.

Definition 330 A **first-order ordinary differential equation** is an equation that involves at most the first derivative of an unknown function, x. If x is a function of t, then the first-order ordinary differential equation is shown as

$$\frac{dx}{dt} = f(x, t) \tag{C.3}$$

where $f(x, t)$ is a given function of the variables x and t which is defined in some interval $a < t < b$.

© The Author(s), under exclusive license to Springer Nature Switzerland AG 2022
R. N. Jazar, *Advanced Vibrations*, https://doi.org/10.1007/978-3-031-16356-2

Definition 331 An **explicit** solution of the first-order ordinary differential equation

$$\frac{dx}{dt} = f(x, t) \tag{C.4}$$

in some interval $a < t < b$ is any function $x = x(t)$ that identically satisfied the differential equation (C.4), i.e., its substitution into the equation reduces the differential equation to an identity everywhere within that interval.

Definition 332 Conditions specified at a single point are called **initial conditions** if the solution of differential equation is subjected to be satisfied. The differential equation together with those initial conditions is an **initial-value problem**. Conditions at two points are called **boundary conditions**, and the differential equation together with the boundary conditions is a **boundary-value problem**.

Definition 333 A first-order ordinary differential equation of the form

$$\frac{dx}{dt} = f(x) \tag{C.5}$$

where $f(x)$ is a function of x alone, and not t, is called **autonomous**. Therefore, a first-order ordinary differential equation of the form

$$\frac{dx}{dt} = f(x, t) \tag{C.6}$$

where $f(x, t)$ is a function of x and t is called **nonautonomous**.

Definition 334 If the function $x(t) = cte$ is a solution of the differential equation

$$\frac{dx}{dt} = f(x, t) \tag{C.7}$$

then the solution is called an **equilibrium solution** of (C.7).

Definition 335 The equilibrium solution $x(t) = cte$ of the differential equation

$$\frac{dx}{dt} = f(x, t) \tag{C.8}$$

is called **stable** if all solutions of the differential equation that start near this equilibrium solution remain near this equilibrium solution as $t \to \infty$. An equilibrium solution is called **unstable** if all solutions of the differential equation that start near this equilibrium solution move from this equilibrium solution as $t \to \infty$.

Theorem 336 *If $f(x, t)$ and $\partial f / \partial x$ are defined and continuous in a finite rectangular region of $x - t$ plane including the point (x_0, t_0), then the differential equation*

$$\frac{dx}{dt} = f(x, t) \tag{C.9}$$

has a unique solution $x(t)$, passing through the point

$$x(t_0) = x_0 \tag{C.10}$$

This solution is valid for all x and t which the solution remains inside the rectangle.

Definition 337 A particular of the differential equation,

$$\frac{dx}{dt} = f(x, t) \tag{C.11}$$

which we cannot find from the family of regular solutions by selecting a finite value for the arbitrary constant c of the family of solution, is called a **singular solution**.

Definition 338 The set of differential equations

$$\frac{dx}{dt} = P(x, y, t) \qquad \frac{dy}{dt} = Q(x, y, t) \tag{C.12}$$

is called a **system of first-order differential equations**, where $P(x, y, t)$ and $Q(x, y, t)$ are the given functions of x, y, t. When P and Q do not contain t explicitly, the system (C.12) is called **autonomous**; otherwise, it is called **nonautonomous**.

Definition 339 The system of autonomous differential equations

$$\frac{dx}{dt} = P(x, y, t) \qquad \frac{dy}{dt} = Q(x, y, t) \tag{C.13}$$

has an **equilibrium point** at (x_0, y_0), if

$$P(x_0, y_0) = 0 \qquad Q(x_0, y_0) = 0 \tag{C.14}$$

Appendix D
Trigonometric Formula

Numeric Values and Functions

$$\pi \simeq 3.14159265359 \cdots \simeq \frac{355}{113} \simeq \frac{103993}{33102} \tag{D.1}$$

$$1 \text{ rad} \simeq 57.296 \deg \simeq 57° \; 17' \; 44.8'' \tag{D.2}$$

Trigonometric Functions

θ is an acute angle of a right-angled triangle; the hypotenuse h is the long side that connects the two acute angles. The side b adjacent to θ is the side of the triangle that connects θ to the right angle. The third side a is opposite to θ.

$$h = hypotenuse \qquad a = opposite \qquad b = adjacent \tag{D.3}$$

$$\sin \theta = \frac{a}{h} \qquad \cos \theta = \frac{b}{h} \qquad \tan \theta = \frac{a}{b} = \frac{\sin \theta}{\cos \theta} \tag{D.4}$$

$$\csc \theta = \frac{1}{\sin \theta} = \frac{h}{a} \qquad \sec \theta = \frac{1}{\cos \theta} = \frac{h}{b} \qquad \cot \theta = \frac{b}{a} \tag{D.5}$$

Definitions in Terms of Exponentials

$$\cos z = \frac{e^{iz} + e^{-iz}}{2} = \cosh(iz) \tag{D.6}$$

$$\sin z = \frac{e^{iz} - e^{-iz}}{2i} = -i \sinh(iz) \tag{D.7}$$

$$\tan z = \frac{e^{iz} - e^{-iz}}{i\left(e^{iz} + e^{-iz}\right)} = \frac{1}{i}\tanh z \qquad \text{(D.8)}$$

$$e^{iz} = \cos z + i\sin z \qquad \text{(D.9)}$$

$$e^{-iz} = \cos z - i\sin z \qquad \text{(D.10)}$$

$$(\cos\theta + i\sin\theta)^n = \cos n\theta + i\sin n\theta \qquad \text{(D.11)}$$

Angle Sum and Difference

$$\sin(\alpha \pm \beta) = \sin\alpha\cos\beta \pm \cos\alpha\sin\beta \qquad \text{(D.12)}$$

$$\cos(\alpha \pm \beta) = \cos\alpha\cos\beta \mp \sin\alpha\sin\beta \qquad \text{(D.13)}$$

$$\sinh(\alpha \pm \beta) = \sinh\alpha\cosh\beta \pm \cosh\alpha\sinh\beta \qquad \text{(D.14)}$$

$$\cosh(\alpha \pm \beta) = \cosh\alpha\cosh\beta \mp \sinh\alpha\sinh\beta \qquad \text{(D.15)}$$

$$\tan(\alpha \pm \beta) = \frac{\tan\alpha \pm \tan\beta}{1 \mp \tan\alpha\tan\beta} \qquad \text{(D.16)}$$

$$\cot(\alpha \pm \beta) = \frac{\cot\alpha\cot\beta \mp 1}{\cot\beta \pm \cot\alpha} \qquad \text{(D.17)}$$

$$\tanh(\alpha \pm \beta) = \frac{\tanh\alpha \pm \tanh\beta}{1 \mp \tanh\alpha\tanh\beta} \qquad \text{(D.18)}$$

Periodicity

$$\sin(\alpha + 2n\pi) = \sin\alpha \qquad \text{(D.19)}$$

$$\cos(\alpha + 2n\pi) = \cos\alpha \qquad \text{(D.20)}$$

$$\tan(\alpha + 2n\pi) = \tan\alpha \qquad \text{(D.21)}$$

Symmetry

$$\sin(-\alpha) = -\sin\alpha \tag{D.22}$$

$$\cos(-\alpha) = \cos\alpha \tag{D.23}$$

$$\tan(-\alpha) = -\tan\alpha \tag{D.24}$$

Displacement

$$\sin(\frac{\pi}{2} - \alpha) = \cos\alpha \tag{D.25}$$

$$\cos(\frac{\pi}{2} - \alpha) = \sin\alpha \tag{D.26}$$

$$\tan(\frac{\pi}{2} - \alpha) = \cot\alpha \tag{D.27}$$

$$\sec(\frac{\pi}{2} - \alpha) = \csc\alpha \tag{D.28}$$

$$\csc(\frac{\pi}{2} - \alpha) = \sec\alpha \tag{D.29}$$

$$\cot(\frac{\pi}{2} - \alpha) = \tan\alpha \tag{D.30}$$

Multiple Angles

$$\sin(2\alpha) = 2\sin\alpha\cos\alpha = \frac{2\tan\alpha}{1 + \tan^2\alpha} \tag{D.31}$$

$$\cos(2\alpha) = 2\cos^2\alpha - 1 = 1 - 2\sin^2\alpha = \cos^2\alpha - \sin^2\alpha \tag{D.32}$$

$$\tan(2\alpha) = \frac{2\tan\alpha}{1 - \tan^2\alpha} \tag{D.33}$$

$$\cot(2\alpha) = \frac{\cot^2\alpha - 1}{2\cot\alpha} \tag{D.34}$$

$$\sin(3\alpha) = -4\sin^3\alpha + 3\sin\alpha \tag{D.35}$$

$$\cos(3\alpha) = 4\cos^3\alpha - 3\cos\alpha \tag{D.36}$$

$$\tan(3\alpha) = \frac{-\tan^3\alpha + 3\tan\alpha}{-3\tan^2\alpha + 1} \tag{D.37}$$

$$\sin(4\alpha) = -8\sin^3\alpha\cos\alpha + 4\sin\alpha\cos\alpha \tag{D.38}$$

$$\cos(4\alpha) = 8\cos^4\alpha - 8\cos^2\alpha + 1 \tag{D.39}$$

$$\tan(4\alpha) = \frac{-4\tan^3\alpha + 4\tan\alpha}{\tan^4\alpha - 6\tan^2\alpha + 1} \tag{D.40}$$

$$\sin(5\alpha) = 16\sin^5\alpha - 20\sin^3\alpha + 5\sin\alpha \tag{D.41}$$

$$\cos(5\alpha) = 16\cos^5\alpha - 20\cos^3\alpha + 5\cos\alpha \tag{D.42}$$

$$\sin(n\alpha) = 2\sin((n-1)\alpha)\cos\alpha - \sin((n-2)\alpha) \tag{D.43}$$

$$\cos(n\alpha) = 2\cos((n-1)\alpha)\cos\alpha - \cos((n-2)\alpha) \tag{D.44}$$

$$\tan(n\alpha) = \frac{\tan((n-1)\alpha) + \tan\alpha}{1 - \tan((n-1)\alpha)\tan\alpha} \tag{D.45}$$

Half-Angle

$$\cos\frac{\alpha}{2} = \pm\sqrt{\frac{1+\cos\alpha}{2}} \tag{D.46}$$

$$\sin\frac{\alpha}{2} = \pm\sqrt{\frac{1-\cos\alpha}{2}} \tag{D.47}$$

$$\tan \frac{\alpha}{2} = \frac{1 - \cos \alpha}{\sin \alpha} = \frac{\sin \alpha}{1 + \cos \alpha} = \pm \sqrt{\frac{1 - \cos \alpha}{1 + \cos \alpha}} \tag{D.48}$$

$$\sin \alpha = \frac{2 \tan \frac{\alpha}{2}}{1 + \tan^2 \frac{\alpha}{2}} \tag{D.49}$$

$$\cos \alpha = \frac{1 - \tan^2 \frac{\alpha}{2}}{1 + \tan^2 \frac{\alpha}{2}} \tag{D.50}$$

Powers of Functions

$$\cos^2 \alpha + \sin^2 \alpha = 1 \tag{D.51}$$

$$\cosh^2 \alpha - \sinh^2 \alpha = 1 \tag{D.52}$$

$$\sin^2 \alpha = \frac{1}{2} (1 - \cos(2\alpha))$$

$$\sin \alpha \cos \alpha = \frac{1}{2} \sin(2\alpha) \tag{D.53}$$

$$\cos^2 \alpha = \frac{1}{2} (1 + \cos(2\alpha)) \tag{D.54}$$

$$\sin^3 \alpha = \frac{1}{4} (3 \sin(\alpha) - \sin(3\alpha)) \tag{D.55}$$

$$\sin^2 \alpha \cos \alpha = \frac{1}{4} (\cos \alpha - 3 \cos(3\alpha)) \tag{D.56}$$

$$\sin \alpha \cos^2 \alpha = \frac{1}{4} (\sin \alpha + \sin(3\alpha)) \tag{D.57}$$

$$\cos^3 \alpha = \frac{1}{4} (\cos(3\alpha) + 3 \cos \alpha) \tag{D.58}$$

$$\sin^4 \alpha = \frac{1}{8} (3 - 4 \cos(2\alpha) + \cos(4\alpha)) \tag{D.59}$$

$$\sin^3 \alpha \cos \alpha = \frac{1}{8} \left(2 \sin(2\alpha) - \sin(4\alpha) \right) \tag{D.60}$$

$$\sin^2 \alpha \cos^2 \alpha = \frac{1}{8} \left(1 - \cos(4\alpha) \right) \tag{D.61}$$

$$\sin \alpha \cos^3 \alpha = \frac{1}{8} \left(2 \sin(2\alpha) + \sin(4\alpha) \right) \tag{D.62}$$

$$\cos^4 \alpha = \frac{1}{8} \left(3 + 4 \cos(2\alpha) + \cos(4\alpha) \right) \tag{D.63}$$

$$\sin^5 \alpha = \frac{1}{16} \left(10 \sin \alpha - 5 \sin(3\alpha) + \sin(5\alpha) \right) \tag{D.64}$$

$$\sin^4 \alpha \cos \alpha = \frac{1}{16} \left(2 \cos \alpha - 3 \cos(3\alpha) + \cos(5\alpha) \right) \tag{D.65}$$

$$\sin^3 \alpha \cos^2 \alpha = \frac{1}{16} \left(2 \sin \alpha + \sin(3\alpha) - \sin(5\alpha) \right) \tag{D.66}$$

$$\sin^2 \alpha \cos^3 \alpha = \frac{1}{16} \left(2 \cos \alpha - 3 \cos(3\alpha) - 5 \cos(5\alpha) \right) \tag{D.67}$$

$$\sin \alpha \cos^4 \alpha = \frac{1}{16} \left(2 \sin \alpha + 3 \sin(3\alpha) + \sin(5\alpha) \right) \tag{D.68}$$

$$\cos^5 \alpha = \frac{1}{16} \left(10 \cos \alpha + 5 \cos(3\alpha) + \cos(5\alpha) \right) \tag{D.69}$$

$$\tan^2 \alpha = \frac{1 - \cos(2\alpha)}{1 + \cos(2\alpha)} \tag{D.70}$$

Products of sin and cos

$$\cos \alpha \cos \beta = \frac{1}{2} \cos(\alpha - \beta) + \frac{1}{2} \cos(\alpha + \beta) \tag{D.71}$$

$$\cos n\theta \cos \theta = \frac{1}{2} \cos(n + 1)\theta + \frac{1}{2} \cos(n - 1)\theta \tag{D.72}$$

$$\cos m\theta \cos n\theta = \frac{1}{2}\cos(m+n)\theta + \frac{1}{2}\cos(m-n)\theta \qquad \text{(D.73)}$$

$$\sin \alpha \sin \beta = \frac{1}{2}\cos(\alpha-\beta) - \frac{1}{2}\cos(\alpha+\beta) \qquad \text{(D.74)}$$

$$\sin \alpha \cos \beta = \frac{1}{2}\sin(\alpha-\beta) + \frac{1}{2}\sin(\alpha+\beta) \qquad \text{(D.75)}$$

$$\cos \alpha \sin \beta = \frac{1}{2}\sin(\alpha+\beta) - \frac{1}{2}\sin(\alpha-\beta) \qquad \text{(D.76)}$$

$$\sin(\alpha+\beta)\sin(\alpha-\beta) = \cos^2 \beta - \cos^2 \alpha = \sin^2 \alpha - \sin^2 \beta \qquad \text{(D.77)}$$

$$\cos(\alpha+\beta)\cos(\alpha-\beta) = \cos^2 \beta + \sin^2 \alpha \qquad \text{(D.78)}$$

Sum of Functions

$$\sin \alpha \pm \sin \beta = 2\sin \frac{\alpha \pm \beta}{2} \cos \frac{\alpha \pm \beta}{2} \qquad \text{(D.79)}$$

$$\cos \alpha + \cos \beta = 2\cos \frac{\alpha + \beta}{2} \cos \frac{\alpha - \beta}{2} \qquad \text{(D.80)}$$

$$\cos \alpha - \cos \beta = -2\sin \frac{\alpha + \beta}{2} \sin \frac{\alpha - \beta}{2} \qquad \text{(D.81)}$$

$$\sinh \alpha \pm \sinh \beta = 2\sinh \frac{\alpha \pm \beta}{2} \cosh \frac{\alpha \pm \beta}{2} \qquad \text{(D.82)}$$

$$\cosh \alpha + \cosh \beta = 2\cosh \frac{\alpha + \beta}{2} \cosh \frac{\alpha - \beta}{2} \qquad \text{(D.83)}$$

$$\cosh \alpha - \cosh \beta = -2\sinh \frac{\alpha + \beta}{2} \sinh \frac{\alpha - \beta}{2} \qquad \text{(D.84)}$$

$$\tan \alpha \pm \tan \beta = \frac{\sin(\alpha \pm \beta)}{\cos \alpha \cos \beta} \qquad \text{(D.85)}$$

$$\cot \alpha \pm \cot \beta = \frac{\sin(\beta \pm \alpha)}{\sin \alpha \sin \beta} \qquad \text{(D.86)}$$

$$\frac{\sin\alpha + \sin\beta}{\sin\alpha - \sin\beta} = \frac{\tan\frac{\alpha+\beta}{2}}{\tan\frac{\alpha-+\beta}{2}} \tag{D.87}$$

$$\frac{\sin\alpha + \sin\beta}{\cos\alpha - \cos\beta} = \cot\frac{-\alpha + \beta}{2} \tag{D.88}$$

$$\frac{\sin\alpha + \sin\beta}{\cos\alpha + \cos\beta} = \tan\frac{\alpha + \beta}{2} \tag{D.89}$$

$$\frac{\sin\alpha - \sin\beta}{\cos\alpha + \cos\beta} = \tan\frac{\alpha - \beta}{2} \tag{D.90}$$

Trigonometric Relations

$$\sin^2\alpha - \sin^2\beta = \sin(\alpha + \beta)\sin(\alpha - \beta) \tag{D.91}$$
$$= \cos^2\beta - \cos^2\alpha \tag{D.92}$$

$$\cos^2\alpha - \cos^2\beta = -\sin(\alpha + \beta)\sin(\alpha - \beta) \tag{D.93}$$

$$\cos^2\alpha - \sin^2\beta = \cos(\alpha + \beta)\cos(\alpha - \beta) \tag{D.94}$$
$$= \cos^2\beta - \sin^2\alpha \tag{D.95}$$

Inverse of Trigonometric Functions

$$\sin x = y \qquad \cos x = y \qquad \tan x = y \tag{D.96}$$

$$\arcsin y = (-1)^k x + k\pi \tag{D.97}$$
$$\arccos y = \pm x + 2k\pi \tag{D.98}$$
$$\arcsin y = x + k\pi \tag{D.99}$$
$$k = 0, 1, 2, 3, \cdots$$

$$-1 \leq y \leq 1 \quad -\frac{\pi}{2} \leq \arcsin y \leq \frac{\pi}{2}$$
$$-1 \leq y \leq 1 \quad 0 \leq \arccos y \leq \pi$$
$$-\infty \leq y \leq \infty \quad -\frac{\pi}{2} \leq \arctan y \leq \frac{\pi}{2}$$
$$-\infty \leq y \leq \infty \quad 0 \leq \text{arccot}\, y \leq \pi$$

(D.100)

$$1 \leq y \quad 0 \leq \text{arccsc}\, y \leq \frac{\pi}{2}$$
$$y \leq -1 \quad -\frac{\pi}{2} \leq \text{arccsc}\, y \leq 0$$
$$1 \leq y \quad 0 \leq \text{arcsec}\, y \leq \frac{\pi}{2}$$
$$y \leq -1 \quad \frac{\pi}{2} \leq \text{arcsec}\, y \leq \pi$$

(D.101)

Derivatives of Trigonometric Functions

$$\frac{d}{d\theta} \sin \theta = \cos \theta \tag{D.102}$$

$$\frac{d}{d\theta} \cos \theta = -\sin \theta \tag{D.103}$$

$$\frac{d}{d\theta} \tan \theta = 1 + \tan^2 \theta = \sec^2 \theta \tag{D.104}$$

$$\frac{d}{d\theta} \csc \theta = -\csc \theta \cot \theta \tag{D.105}$$

$$\frac{d}{d\theta} \sec \theta = \sec \theta \tan \theta \tag{D.106}$$

$$\frac{d}{d\theta} \cot \theta = -\left(1 + \cot^2 \theta\right) = -\csc^2 \theta \tag{D.107}$$

Integrals of Trigonometric Functions

$$\int \sin \theta \, d\theta = -\cos \theta + C \tag{D.108}$$

$$\int \cos \theta \, d\theta = \sin \theta + C \tag{D.109}$$

$$\int \tan \theta \, d\theta = -\ln |\cos \theta| + C \tag{D.110}$$

$$\int \csc \theta \, d\theta = \frac{1}{2} \ln \left(-\frac{\cos \theta - 1}{\cos \theta + 1} \right) + C$$

$$= -\ln |\csc \theta + \cot \theta| + C \tag{D.111}$$

$$\int \sec \theta \, d\theta = \frac{1}{2} \ln \left(\frac{\sin \theta + 1}{\sin \theta - 1} \right) + C$$

$$= \ln |\sec \theta + \tan \theta| + C \tag{D.112}$$

$$\int \cot \theta \, d\theta = \frac{1}{2} \ln (2 - 2\cos 2\theta) + C = \ln |\sin \theta| + C \tag{D.113}$$

$$\int \cos^2 x \, dx = \frac{1}{2} x - \frac{1}{4} \sin 2x \tag{D.114}$$

$$\int \sin^2 x \, dx = \frac{1}{2} x + \frac{1}{4} \sin 2x \tag{D.115}$$

$$\int \cos mx \cos nx \, dx = \frac{\sin (m - n) x}{2 (m - n)} + \frac{\sin (m + n) x}{2 (m + n)} \quad m^2 \neq n^2 \tag{D.116}$$

$$\int \sin mx \sin nx \, dx = \frac{\sin (m - n) x}{2 (m - n)} - \frac{\sin (m + n) x}{2 (m + n)} \quad m^2 \neq n^2 \tag{D.117}$$

$$\int \sin mx \cos nx \, dx = -\frac{\cos (m - n) x}{2 (m - n)} - \frac{\cos (m + n) x}{2 (m + n)} \quad m^2 \neq n^2 \tag{D.118}$$

$$\int_{-\pi}^{\pi} \cos mx \cos nx \, dx = \pi \delta_{mn} \quad m^2 \neq n^2 \quad m, n \in \mathbb{N} \tag{D.119}$$

$$\int_{-\pi}^{\pi} \sin mx \sin nx \, dx = \pi \delta_{mn} \quad m^2 \neq n^2 \quad m, n \in \mathbb{N} \tag{D.120}$$

$$\int_{-\pi}^{\pi} \sin mx \cos nx \, dx = 0 \quad m^2 \neq n^2 \quad m, n \in \mathbb{N} \tag{D.121}$$

$$\int x \cos x \, dx = \cos x + x \sin x \tag{D.122}$$

$$\int x \sin x \, dx = \sin x - x \cos x \tag{D.123}$$

$$\int x^2 \cos x \, dx = 2x \cos x + \left(x^2 - 2 \right) \sin x \tag{D.124}$$

$$\int x^2 \sin x \, dx = 2x \sin x - \left(x^2 - 2 \right) \cos x \tag{D.125}$$

$$\int \cos mx \cos nx \, dx = \frac{\sin (m - n) x}{2 (m - n)} + \frac{\sin (m + n) x}{2 (m + n)} \quad m^2 \neq n^2 \qquad \text{(D.126)}$$

$$\int \sin mx \sin nx \, dx = \frac{\sin (m - n) x}{2 (m - n)} - \frac{\sin (m + n) x}{2 (m + n)} \quad m^2 \neq n^2 \qquad \text{(D.127)}$$

$$\int \sin mx \cos nx \, dx = -\frac{\cos (m - n) x}{2 (m - n)} - \frac{\cos (m + n) x}{2 (m + n)} \quad m^2 \neq n^2 \qquad \text{(D.128)}$$

$$\int_{-\pi}^{\pi} \cos mx \cos nx \, dx = \pi \delta_{mn} \qquad m^2 \neq n^2 \qquad m, n \in \mathbb{N} \qquad \text{(D.129)}$$

$$\int_{-\pi}^{\pi} \sin mx \sin nx \, dx = \pi \delta_{mn} \qquad m^2 \neq n^2 \qquad m, n \in \mathbb{N} \qquad \text{(D.130)}$$

$$\int_{-\pi}^{\pi} \sin mx \cos nx \, dx = 0 \qquad m^2 \neq n^2 \qquad m, n \in \mathbb{N} \qquad \text{(D.131)}$$

$$\int x \cos x \, dx = \cos x + x \sin x \qquad \text{(D.132)}$$

$$\int x \sin x \, dx = \sin x - x \cos x \qquad \text{(D.133)}$$

$$\int x^2 \cos x \, dx = 2x \cos x + \left(x^2 - 2 \right) \sin x \qquad \text{(D.134)}$$

$$\int x^2 \sin x \, dx = 2x \sin x - \left(x^2 - 2 \right) \cos x \qquad \text{(D.135)}$$

Appendix E
Algebraic Formula

$$(a \pm b)^2 = a^2 \pm 2ab + b^2 \tag{E.1}$$

$$(a \pm b)^3 = a^3 \pm 3a^2b + 3ab^2 \pm b^2 \tag{E.2}$$

$$(a + b + c)^2 = a^2 + b^2 + c^2 + 2\,(ab + bc + ca) \tag{E.3}$$

$$(a + b + c)^3 = a^3 + b^3 + c^3 + 3a^2\,(b + c)$$
$$+3b^2\,(c + a) + 3c^2\,(a + c) + 6abc \tag{E.4}$$

$$a^2 - b^2 = (a - b)\,(a + b) \tag{E.5}$$

$$a^2 + b^2 = (a - ib)\,(a + ib) \tag{E.6}$$

$$a^3 - b^3 = (a - b)\left(a^2 + ab + b^2\right) \tag{E.7}$$

$$a^2 + b^2 = (a + b)\left(a^2 - ab + b^2\right) \tag{E.8}$$

Appendix F
Unit Conversions

General Conversion Formulas

$$N^a \, m^b \, s^c \approx 4.448^a \times 0.3048^b \times lb^a \, ft^b \, s^c$$
$$\approx 4.448^a \times 0.0254^b \times lb^a \, in^b \, s^c$$
$$lb^a \, ft^b \, s^c \approx 0.2248^a \times 3.2808^b \times N^a \, m^b \, s^c$$
$$lb^a \, in^b \, s^c \approx 0.2248^a \times 39.37^b \times N^a \, m^b \, s^c$$

Conversion Factors

Acceleration

$$1 \, ft/s^2 \approx 0.3048 \, m/s^2 \quad 1 \, m/s^2 \approx 3.2808 \, ft/s^2$$

Angle

$$1 \, deg \approx 0.01745 \, rad \quad 1 \, rad \approx 57.307 \, deg$$

Area

$$1 \, in^2 \approx 6.4516 \, cm^2 \qquad 1 \, cm^2 \approx 0.155 \, in^2$$
$$1 \, ft^2 \approx 0.09290304 \, m^2 \quad 1 \, m^2 \approx 10.764 \, ft^2$$
$$1 \, acre \approx 4046.86 \, m^2 \qquad 1 \, m^2 \approx 2.471 \times 10^{-4} \, acre$$
$$1 \, acre \approx 0.4047 \, hectare \quad 1 \, hectare \approx 2.471 \, acre$$

© The Author(s), under exclusive license to Springer Nature Switzerland AG 2022
R. N. Jazar, *Advanced Vibrations*, https://doi.org/10.1007/978-3-031-16356-2

Damping

$$1\,\mathrm{N\,s/m} \approx 6.85218 \times 10^{-2}\,\mathrm{lb\,s/ft} \quad 1\,\mathrm{lb\,s/ft} \approx 14.594\,\mathrm{N\,s/m}$$
$$1\,\mathrm{N\,s/m} \approx 5.71015 \times 10^{-3}\,\mathrm{lb\,s/in} \quad 1\,\mathrm{lb\,s/in} \approx 175.13\,\mathrm{N\,s/m}$$

Energy and Heat

$$1\,\mathrm{Btu} \approx 1055.056\,\mathrm{J} \quad 1\,\mathrm{J} \approx 9.4782 \times 10^{-4}\,\mathrm{Btu}$$
$$1\,\mathrm{cal} \approx 4.1868\,\mathrm{J} \quad 1\,\mathrm{J} \approx 0.23885\,\mathrm{cal}$$
$$1\,\mathrm{kW\,h} \approx 3600\,\mathrm{kJ} \quad 1\,\mathrm{MJ} \approx 0.27778\,\mathrm{kW\,h}$$
$$1\,\mathrm{ft\,lbf} \approx 1.355818\,\mathrm{J} \quad 1\,\mathrm{J} \approx 0.737562\,\mathrm{ft\,lbf}$$

Force

$$1\,\mathrm{lb} \approx 4.448222\,\mathrm{N} \quad 1\,\mathrm{N} \approx 0.22481\,\mathrm{lb}$$

Fuel Consumption

$$1\,\mathrm{l/100\,km} \approx 235.214583\,\mathrm{mi/gal} \quad 1\,\mathrm{mi/gal} \approx 235.214583\,\mathrm{l/100\,km}$$
$$1\,\mathrm{l/100\,km} = 100\,\mathrm{km/l} \quad 1\,\mathrm{km/l} = 100\,\mathrm{l/100\,km}$$
$$1\,\mathrm{mi/gal} \approx 0.425144\,\mathrm{km/l} \quad 1\,\mathrm{km/l} \approx 2.352146\,\mathrm{mi/gal}$$

Length

$$1\,\mathrm{in} \approx 25.4\,\mathrm{mm} \quad 1\,\mathrm{cm} \approx 0.3937\,\mathrm{in}$$
$$1\,\mathrm{ft} \approx 30.48\,\mathrm{cm} \quad 1\,\mathrm{m} \approx 3.28084\,\mathrm{ft}$$
$$1\,\mathrm{mi} \approx 1.609347\,\mathrm{km} \quad 1\,\mathrm{km} \approx 0.62137\,\mathrm{mi}$$

Mass

$$1\,\mathrm{lb} \approx 0.45359\,\mathrm{kg} \quad 1\,\mathrm{kg} \approx 2.204623\,\mathrm{lb}$$
$$1\,\mathrm{slug} \approx 14.5939\,\mathrm{kg} \quad 1\,\mathrm{kg} \approx 0.068522\,\mathrm{slug}$$
$$1\,\mathrm{slug} \approx 32.174\,\mathrm{lb} \quad 1\,\mathrm{lb} \approx 0.03.1081\,\mathrm{slug}$$

Moment and Torque

$$1\,\mathrm{lb\,ft} \approx 1.35582\,\mathrm{N\,m} \quad 1\,\mathrm{N\,m} \approx 0.73746\,\mathrm{lb\,ft}$$
$$1\,\mathrm{lb\,in} \approx 8.85075\,\mathrm{N\,m} \quad 1\,\mathrm{N\,m} \approx 0.11298\,\mathrm{lb\,in}$$

Mass Moment

$$1\,\mathrm{lb\,ft^2} \approx 0.04214\,\mathrm{kg\,m^2} \quad 1\,\mathrm{kg\,m^2} \approx 23.73\,\mathrm{lb\,ft^2}$$

Power

$$1\,\text{Btu}/\text{h} \approx 0.2930711\,\text{W} \qquad 1\,\text{W} \approx 3.4121\,\text{Btu}/\text{h}$$
$$1\,\text{hp} \approx 745.6999\,\text{W} \qquad\quad 1\,\text{kW} \approx 1.341\,\text{hp}$$
$$1\,\text{hp} \approx 550\,\text{lb}\,\text{ft}/\text{s} \qquad\quad 1\,\text{lb}\,\text{ft}/\text{s} \approx 1.8182 \times 10^{-3}\,\text{hp}$$
$$1\,\text{lb}\,\text{ft}/\text{h} \approx 3.76616 \times 10^{-4}\,\text{W} \quad 1\,\text{W} \approx 2655.2\,\text{lb}\,\text{ft}/\text{h}$$
$$1\,\text{lb}\,\text{ft}/\text{min} \approx 2.2597 \times 10^{-2}\,\text{W} \quad 1\,\text{W} \approx 44.254\,\text{lb}\,\text{ft}/\text{min}$$

Pressure and Stress

$$1\,\text{lb}/\text{in}^2 \approx 6894.757\,\text{Pa} \quad 1\,\text{MPa} \approx 145.04\,\text{lb}/\text{in}^2$$
$$1\,\text{lb}/\text{ft}^2 \approx 47.88\,\text{Pa} \qquad 1\,\text{Pa} \approx 2.0886 \times 10^{-2}\,\text{lb}/\text{ft}^2$$
$$1\,\text{Pa} \approx 0.00001\,\text{atm} \qquad 1\,\text{atm} \approx 101325\,\text{Pa}$$

Stiffness

$$1\,\text{N}/\text{m} \approx 6.85218 \times 10^{-2}\,\text{lb}/\text{ft} \quad 1\,\text{lb}/\text{ft} \approx 14.594\,\text{N}/\text{m}$$
$$1\,\text{N}/\text{m} \approx 5.71015 \times 10^{-3}\,\text{lb}/\text{in} \quad 1\,\text{lb}/\text{in} \approx 175.13\,\text{N}/\text{m}$$

Temperature

$$^\circ\text{C} = (^\circ\text{F} - 32)/1.8$$
$$^\circ\text{F} = 1.8\,^\circ\text{C} + 32$$

Velocity

$$1\,\text{mi}/\text{h} \approx 1.60934\,\text{km}/\text{h} \qquad 1\,\text{km}/\text{h} \approx 0.62137\,\text{mi}/\text{h}$$
$$1\,\text{mi}/\text{h} \approx 0.44704\,\text{m}/\text{s} \qquad 1\,\text{m}/\text{s} \approx 2.2369\,\text{mi}/\text{h}$$
$$1\,\text{ft}/\text{s} \approx 0.3048\,\text{m}/\text{s} \qquad 1\,\text{m}/\text{s} \approx 3.2808\,\text{ft}/\text{s}$$
$$1\,\text{ft}/\text{min} \approx 5.08 \times 10^{-3}\,\text{m}/\text{s} \quad 1\,\text{m}/\text{s} \approx 196.85\,\text{ft}/\text{min}$$

Volume

$$1\,\text{in}^3 \approx 16.39\,\text{cm}^3 \qquad 1\,\text{cm}^3 \approx 0.0061013\,\text{in}^3$$
$$1\,\text{ft}^3 \approx 0.02831685\,\text{m}^3 \quad 1\,\text{m}^3 \approx 35.315\,\text{ft}^3$$
$$1\,\text{gal} \approx 3.785\,\text{l} \qquad\qquad 1\,\text{l} \approx 0.2642\,\text{gal}$$
$$1\,\text{gal} \approx 3785.41\,\text{cm}^3 \qquad 1\,\text{l} \approx 1000\,\text{cm}^3$$

Bibliography

Chapter 1: Vibration Kinematics

Agyris, J., & Mlejnek, H. P. (1991). *Computational mechanics*. New York: Elsevier.

Axler, S., Bourdon, P., & Ramey, W. (2001). *Harmonic function theory*. New York: Springer.

Balachandran, B., & Magrab, E. B., (2003). *Vibrations*. Pacific Grove, CA: Brooks/Cole.

Ballou G. M. (2008). *Handbook for sound engineers* (4th edn.). Oxford, UK: Focal Press, Elsevier.

Benson, D. J. (2007). *Music: A mathematical offering*. London, UK: Cambridge University Press.

Benaroya, H. (2004). *Mechanical vibration: Analysis, uncertainities, and control*. New York: Marcel Dekker.

Buzdugan, G., Mihailescu, E., & Rades, M. (1986). *Vibration measurement*. Dordrecht: Springer.

Coddington, E. A. (1961). *Ordinary differential equations*. New Jersey: Prentice Hall.

Carslaw, H. S. (1921). *An interoduction to the theory of Fourier's series and integrals*. London UK: Macmillan.

Chihara, T. S. (1978). *An introduction to orthogonal polynomials*. New York, New York: Gordon and Breach, Science Publishers, Inc.

Dass, H. K., & Verma E. R. (2014). *Higher mathematical physics*. New Delhi, India: S CHAND & Company.

Del Pedro, M., & Pahud, P. (1991). *Vibration mechanics*. The Netherland: Kluwer Academic Publishers.

Den Hartog, J. P. (1934). *Mechanical vibrations*. New York: McGraw-Hill.

Dimarogonas, A. (1996). *Vibration for engineers*. New Jersey: Prentice Hall.

Esmailzadeh, E., & Jazar, R. N. (1997). Periodic solution of a Mathieu-Duffing type equation. *International Journal of Nonlinear Mechanics, 32*(5), 905–912.

Esmailzadeh, E., Mehri, B., & Jazar, R. N. (1996). Periodic solution of a second order, autonomous, nonlinear system. *Journal of Nonlinear Dynamics, 10*(4), 307–316.

Fauvel, J., Flood, R., & Wilson, R. (2003). *Music and mathematics: From Pythagoras to fractals*. New York, NY: Oxford University Press Inc.

Fourier, J. B. J. (1878). *The analytical theory of heat*, Translated by: Freeman A., digitally printed version 2009. New york, NY: Cambridge University Press.

Grigorieva, E. (2015). *Methods of solving nonstandard problems*. New York, NY: Springer.

Gunther, G. (2012). *The physics of music and color*. New York: Springer.

Harris, C. M., & Piersol, A. G. (2002). *Harris' shock and vibration handbook*. New York, NY: McGraw-Hill.

Holmes, M. H. (2009). *Introduction to the foundations of applied mathematics*. New York, NY: Springer.

© The Author(s), under exclusive license to Springer Nature Switzerland AG 2022
R. N. Jazar, *Advanced Vibrations*, https://doi.org/10.1007/978-3-031-16356-2

Inman, D. (2007). *Engineering vibrations*. New York, NY: Prentice Hall.

Jazar, R. N. (1997). *Analysis of nonlinear parametric vibrating systems*. Ph. D. Thesis, Mechanical Engineering Department, Tehran: Sharif University of Technology.

Jazar, R. N. (2004). Stability chart of parametric vibrating systems using energy-rate method. *International Journal of Non-Linear Mechanics, 39*(8), 1319–1331.

Jazar, R. N. (2011). *Advanced dynamics: Rigid body, multibody, and aero-space applications*. New York, NY: Wiley.

Jazar, R. N. (2013). *Advanced vibrations: A modern approach*. New York, NY: Springer.

Jazar, R. N. (2017). *Vehicle dynamics: Theory and application* (3rd edn.). New York, NY: Springer.

Jazar, R. N. (2020). *Approximation methods in science and engineering*. New York, NY: Springer.

Jazar, R. N. (2021). *Perturbation methods in science and engineering*. New York: Springer.

Jazar, R. N., Kazemi, M., & Borhani, S., (1992). *Mechanical Vibrations* (in Persian). Tehran: Ettehad Publications.

Jazar, R. N., Mahinfalah M., Mahmoudian N., & Aagaah M. R. (2008). Energy-rate method and stability chart of parametric vibrating systems, *Journal of the Brazilian Society of Mechanical Sciences and Engineering, 30*(3), 182–188.

Jazar, R. N., Mahinfalah M., Mahmoudian N., & Aagaah M. R., (2009). Effects of nonlinearities on the steady state dynamic behavior of electric actuated microcantilever-based resonators. *Journal of Vibration and Control, 15*(9), 1283–1306.

Jazar, R. N., Mahinfalah M., Mahmoudian N., Aagaah M. R., & Shiari B., (2006). Behavior of Mathieu equation in stable regions. *International Journal for Mechanics and Solids, 1*(1).

Jedrzejewski, F. (2006). *Mathematical theory of music*. Delatour, France: ircam Centre Pompidou.

Meirovitch, L. (1967). *Analytical methods in vibrations*. New York: Macmillan.

Meirovitch, L. (1997). *Principles and techniques of vibrations*. New Jersey: Prentice Hall.

Meirovitch, L. (2002). *Fundamentals of vibrations*. New York: McGraw-Hill.

Mickens, R. E. (2019). *Generalized trigonometric and hyperbolic functions*. Boca Raton, FL: CRC Press.

Mickens, R. E. (2004). *Mathematical methods for the natural and engineering sciences*. Singapore: World Scientific Publishing.

Minofsky, V. P. (1975). *Problems in higher mathematics*. Moscow: Mir Publishers.

Ogata, K. (2004). *System dynamics*. New Jersey: Prentice Hall.

Osinski, Z. (1998). *Damping of vibrations*. Rotterdam, Netherlands: A. A. Balkema Publishers.

Palm, W. J. (2006). *Mechanical vibration*. New York: John Wiley.

Prestini, E. (2004). *The evolution of applied harmonic analysis: Models of the real world*. New York, NY: Springer.

Rao, S. S. (2018). *Mechanical vibrations*. Harlow, UK: Pearson Prentice Hall.

Rayleigh, J. W. S. (1945). *The theory of sound*. New York: Dover Publication.

Roseau, M. (1987). *Vibrations in mechanical systems*. Berlin: Springer-Verlag.

Shabana, A. A. (1997). *Vibration of discrete and continuous systems*. New York: Springer-Verlag.

Sansone, G. (1991). *Orthogonal functions*. Translated From the Italian by Diamond A. H. New York: Dover Publications, Inc.

Shima, H., & Nakayama, T. (2010). *Higher mathematics for physics and engineering*. Berlin, Heidelberg: Springer-Verlag.

Simmons, G. F. (2017). *Differential equations with applications and historical notes*. Boca Raton, FL: CRC Press.

Snowdon, J. C. (1968). *Vibration and shock in damped mechanical systems*. New York: John Wiley.

Titchmarsh, E. C. (1948). *Introduction to the theory of Fourier integrals*. London UK: Oxford University Press.

Thomson, W. T., & Dahleh, M. D. (1997). *Theory of vibration with applications*. New Jersey: Prentice Hall.

Tongue, B. H. (2001). *Principles of vibration*. New York: Oxford University Press.

Tse, F. S., Morse, I. E., & Hinkle, R. T. (1978). *Mechanical vibrations theory and applications*. Boston, MA: Allyn and Bacon Inc.

Chapter 2: Vibration Dynamics

Angeles, J. (2011). *Dynamic response of linear mechanical systems: Modeling, analysis and simulation*. New York: Springer.

Dimarogonas, A. (1996). *Vibration for engineers*. New Jersey: Prentice Hall.

Esmailzadeh, E. (1978). Design synthesis of a vehicle suspension system using multi-parameter optimization. *Vehicle System Dynamics, 7*, 83–96.

Esmailzadeh, E., & Jazar, R. N. (1997). Periodic solution of a mathieu-duffing type equation. *International Journal of Nonlinear Mechanics, 32*(5), 905–912.

Esmailzadeh, E., & Jazar, R. N. (1998). Periodic behavior of a cantilever with end mass subjected to harmonic base excitation. *International Journal of Nonlinear Mechanics, 33*(4), 567–577.

Jazar, R. N. (1997). *Analysis of nonlinear parametric vibrating systems*. Ph. D. Thesis, Mechanical Engineering Department, Tehran: Sharif University of Technology.

Jazar, R. N. (2011). *Advanced dynamics: Rigid body, multibody, and aero-space applications*. New York, NY: Wiley.

Jazar, R. N. (2012a). Nonlinear modeling of squeeze-film phenomena. In: L. Dai, R. Jazar (Eds.), *Nonlinear approaches in engineering applications*. New York, NY: Springer.

Jazar, R. N. (2012b). Nonlinear mathematical modeling of microbeam. In: L. Dai, R. Jazar (Eds.), *Nonlinear approaches in engineering applications*. New York, NY: Springer.

Jazar, R. N. (2012c). Derivative and coordinate frames. *Journal of Nonlinear Engineering, 1*(1), 25–34.

Jazar, R. N. (2013). *Advanced vibrations: A modern approach*. New York, NY: Springer.

Jazar, R. N. (2017). *Vehicle dynamics: Theory and application* (3rd edn.). New York, NY: Springer.

Jazar, R. N. (2019). *Advanced vehicle dynamics*. New York: Springer.

Jazar, R. N. (2020). *Approximation methods in science and engineering*. New York, NY: Springer.

Jazar, R. N. (2021). *Perturbation methods in science and engineering*. New York: Springer.

Jazar, R. N. (2022). *Applied robotics: Kinematics, dynamics, and control* (3rd edn.). New York, NY: Springer.

Jazar, R. N., Kazemi, M., & Borhani, S. (1992). *Mechanical vibrations* (in Persian). Tehran: Ettehad Publications.

Jazar, R. N., Mahinfalah, M., Mahmoudian, N., Aagaah, M. R., & Shiari, B. (2006). Behavior of Mathieu equation in stable regions. *International Journal for Mechanics and Solids, 1*(1).

Jazar, R. N., Mahinfalah, M., Mahmoudian, N., & Aagaah, M. R. (2008a). Energy-rate method and stability chart of parametric vibrating systems. *Journal of the Brazilian Society of Mechanical Sciences and Engineering, 30*(3), 182–188.

Jazar, R. N., Mahinfalah, M., Mahmoudian, N., & Aagaah, M. R. (2009). Effects of nonlinearities on the steady state dynamic behavior of electric actuated microcantilever-based resonators. *Journal of Vibration and Control, 15*(9), 1283–1306.

Korenev, B. G., & Reznikov, L. M. (1993). *Dynamic vibration absorbers: Theory and technical applications*. West Sussex, England: Wiley.

Meirovitch, L. (1967). *Analytical methods in vibrations*. New York: Macmillan.

Meirovitch, L., (1997). *Principles and techniques of vibrations*. New Jersey: Prentice Hall.

Meirovitch, L. (2002). *Fundamentals of vibrations*. New York: McGraw-Hill.

Ogata, K. (2004). *System dynamics*. New Jersey: Prentice Hall.

Osinski, Z. (1998). *Damping of vibrations*. Rotterdam, Netherlands: A. A. Balkema Publishers.

Palm, W. J. (2006). *Mechanical vibration*. New York: John Wiley.

Roseau, M. (1987). *Vibrations in mechanical systems*. Berlin: Springer-Verlag.

Snowdon, J. C. (1968). *Vibration and shock in damped mechanical systems*. John Wiley, New York.

Tse, F. S., Morse, I. E., & Hinkle, R. T. (1978). *Mechanical vibrations theory and applications*. Boston, MA: Allyn and Bacon Inc.

Chapter 3: One Degree of Freedom

Angeles, J. (2011). *Dynamic response of linear mechanical systems: Modeling, analysis and simulation*. New York: Springer.

Alkhatib, R., Jazar, R. N., & Golnaraghi, M. F. (2004), Optimal design of passive linear mounts with genetic algorithm method. *Journal of Sound and Vibration, 275*(3–5), 665–691.

Celia, C. W., Nice, A. T. F., & Elliott K. F. (1982). *Advanced mathematics*. London, UK: Macmillan Education LTD.

Cheng, F. Y. (2000). *Matrix analysis of structural dynamics: applications and earthquake engineering*. Madison Avenue, New York: Marcel Dekker, Inc.

Christopherson, J., & Jazar, R. N. (2006), Dynamic behavior comparison of passive hydraulic engine mounts, part 1: Mathematical analysis. *Journal of Sound and Vibration, 290*, 1040–1070.

Christopherson, J., & Jazar, R. N. (2006), Dynamic behavior comparison of passive hydraulic engine mounts, part 2: Finite element analysis. *Journal of Sound and Vibration, 290*, 1071–1090.

Coddington, E. A. (1961). *Ordinary differential equations*. New Jersey: Prentice Hall.

Del Pedro, M., & Pahud, P. (1991). *Vibration mechanics*. The Netherland: Kluwer Academic Publishers.

Den Hartog, J. P. (1934). *Mechanical vibrations*. New York: McGraw-Hill.

Deshpande, S., Mehta, S., & Jazar, R. N. (2006), Optimization of secondary suspension of piecewise linear vibration isolation systems. *International Journal of Mechanical Sciences, 48*(4), 341–377.

Dimarogonas, A. (1996). *Vibration for engineers*. New Jersey: Prentice Hall.

Esmailzadeh, E. (1978). Design synthesis of a vehicle suspension system using multi-parameter optimization. *Vehicle System Dynamics, 7*, 83–96.

Esmailzadeh, E., & Jazar, R. N. (1997). Periodic solution of a Mathieu-Duffing type equation. *International Journal of Nonlinear Mechanics, 32*(5), 905–912.

Esmailzadeh, E., & Jazar, R. N. (1998). Periodic behavior of a Cantilever with end mass subjected to harmonic base excitation. *International Journal of Nonlinear Mechanics, 33*(4), 567–577.

Esmailzadeh, E., Mehri, B., & Jazar, R. N. (1996). Periodic solution of a second order, autonomous, nonlinear system. *Journal of Nonlinear Dynamics, 10*(4), 307–316.

Géradin, M., & Rixen D. J. (2015). *Mechanical vibrations: Theory and application to structural dynamics*. Chichester, West Sussex, UK: John Wiley & Sons.

Graham, K. S. (2000). *Fundamentals of mechanical vibrations* (2nd edn.). New York, NT: McGraw-Hill.

Harris, C. M., & Piersol, A. G. (2002). *Harris' shock and vibration handbook*. New York: McGraw-Hill.

Hirsch, M. W. & Smale, S. (1974). *Differential equations, dynamic systems, and linear algebra*. New York: Academic Press.

Inman, D. (2007). *Engineering vibrations*. New York: Prentice Hall.

Jazar, R. N. (1997). *Analysis of nonlinear parametric vibrating systems*. Ph. D. Thesis, Mechanical Engineering Department, Tehran: Sharif University of Technology.

Jazar, R. N., & Golnaraghi, M. F. (2002a). Engine mounts for automotive applications: A survey. *The Shock and Vibration Digest, 34*(5), 363–379.

Jazar, R. N., & Golnaraghi M. F. (2002b). Nonlinear modeling, experimental verification, and theoretical analysis of a hydraulic engine mount. *Journal of Vibration and Control, 8*(1), 87–116.

Jazar, R. N., Kazemi, M., & Borhani, S. (1992). *Mechanical vibrations* (in Persian). Tehran: Ettehad Publications.

Jazar, R. N., Narimani, A., Golnaraghi, M. F., & Swanson, D. A. (2003). Practical frequency and time optimal design of passive linear vibration isolation mounts. *Journal of Vehicle System Dynamics, 39*(6), 437–466.

Jazar, R. N., Alkhatib, R., & Golnaraghi, M. F. (2006a). Root mean square optimization criterion for vibration behavior of linear quarter car using analytical methods. *Journal of Vehicle System Dynamics, 44*(6), 477–512.

Jazar, R. N., Houim, R., Narimani, A., & Golnaraghi, M. F. (2006b). Frequency response and jump avoidance in a nonlinear passive engine mount. *Journal of Vibration and Control, 12*(11), 1205–1237.

Jazar, R. N., Mahinfalah M., Mahmoudian N., & Aagaah M. R. (2008). Energy-rate method and stability chart of parametric vibrating systems. *Journal of the Brazilian Society of Mechanical Sciences and Engineering, 30*(3), 182–188.

Meirovitch, L. (1967). *Analytical methods in vibrations*. New York: Macmillan.

Meirovitch, L. (1997). *Principles and techniques of vibrations*. New Jersey: Prentice Hall.

Meirovitch, L. (2002). *Fundamentals of vibrations*. New York: McGraw-Hill.

Narimani, A., Golnaraghi, M. F., & Jazar, R. N. (2004a). Frequency response of a piecewise linear system. *Journal of Vibration and Control, 10*(12), 1775–1894.

Narimani, A., Jazar, R. N., & Golnaraghi, M. F. (2004b). Sensitivity analysis of frequency response of a piecewise linear system in frequency island. *Journal of Vibration and Control, 10*(2), 175–198.

Ogata, K. (2004). *System dynamics*. New Jersey: Prentice Hall.

Osinski, Z. (1998). *Damping of vibrations*. Rotterdam, Netherlands: A. A. Balkema Publishers.

Palm, W. J. (2006). *Mechanical vibration*. New York: John Wiley.

Rao, S. S. (2018). *Mechanical vibrations*. Harlow, UK: Pearson Prentice Hall.

Roseau, M. (1987). *Vibrations in mechanical systems*. Berlin: Springer-Verlag.

Snowdon, J. C. (1968). *Vibration and shock in damped mechanical systems*. New York: John Wiley.

Tse, F. S., Morse, I. E., & Hinkle, R. T. (1978). *Mechanical vibrations theory and applications*. Boston, MA: Allyn and Bacon Inc.

Chapter 4: Multi Degrees of Freedom

Angeles, J. (2011). *Dynamic response of linear mechanical systems: Modeling, analysis and simulation*. New York: Springer.

Coddington, E. A. (1961). *Ordinary differential equations*. New Jersey: Prentice Hall.

Dimarogonas, A. (1996). *Vibration for engineers*. New Jersey: Prentice Hall.

Géradin, M., & Rixen, D. J. (2015). *Mechanical vibrations: Theory and application to structural dynamics*. Chichester, West Sussex, United Kingdom: John Wiley & Sons.

Graham, K. S. (2000). *Fundamentals of mechanical vibrations* (2nd edn.). New York, NT: McGraw-Hill.

Harris, C. M., & Piersol, A. G. (2002). *Harris' shock and vibration handbook*. New York: McGraw-Hill.

Hirsch, M. W., & Smale, S. (1974). *Differential equations, dynamic systems, and linear algebra*. New York: Academic Press.

Jazar, R. N., Kazemi, M., & Borhani, S. (1992). *Mechanical vibrations* (in Persian). Tehran: Ettehad Publications.

Meirovitch, L. (1967). *Analytical methods in vibrations*. New York: Macmillan.

Meirovitch, L. (1997). *Principles and techniques of vibrations*. New Jersey: Prentice Hall.

Meirovitch, L. (2002). *Fundamentals of vibrations*. New York: McGraw-Hill.

Muller, P. C., & Schierlen, W. O. (1977). *Forced linear vibrations*. New York: Springer-Verlag.

Ogata, K. (2004). *System dynamics*. New Jersey: Prentice Hall.

Osinski, Z. (1998). *Damping of vibrations*. Rotterdam, Netherlands: A. A. Balkema Publishers.

Palm, W. J. (2006). *Mechanical vibration*. New York: John Wiley.

Roseau, M. (1987). *Vibrations in mechanical systems*. Berlin: Springer-Verlag.

Tse, F. S., Morse, I. E., & Hinkle, R. T. (1978). *Mechanical vibrations theory and applications*. Boston, MA: Allyn and Bacon Inc.

Warburton, G. B. (1976). *The dynamical behaviour of structures*. Oxford, UK: Pergamon Press.

Chapter 5: First-Order Systems

Angeles, J. (2011). *Dynamic response of linear mechanical systems: Modeling, analysis and simulation*. New York: Springer.

Coddington, E. A. (1961). *Ordinary differential equations*. New Jersey: Prentice Hall.

Del Pedro, M., & Pahud, P. (1991). *Vibration mechanics*. The Netherland: Kluwer Academic Publishers.

Dettman, J. W. (1986). *Linear algebra and differential equations*. New York: Dover Publications.

Dimarogonas, A. (1996). *Vibration for engineers*. New Jersey: Prentice Hall.

Hirsch, M. W., & Smale, S. (1974). *Differential equations, dynamic systems, and linear algebra*. New York: Academic Press.

Jazar, R. N. (2006). Mathematical modeling and simulation of thermoelastic effects in flexural microcantilever resonators dynamics. *Journal of Vibration and Control, 12*(2), 139–163.

Jazar, R. N., & Golnaraghi, M. F. (2002). Engine mounts for automotive applications: A survey. *The Shock and Vibration Digest, 34*(5), 363–379.

Jazar, R. N., Kazemi, M., & Borhani, S. (1992). *Mechanical vibrations* (in Persian). Tehran: Ettehad Publications.

Meirovitch, L. (1967). *Analytical methods in vibrations*. New York: Macmillan.

Meirovitch, L. (1997). *Principles and techniques of vibrations*. New Jersey: Prentice Hall.

Meirovitch, L. (2002). *Fundamentals of vbrations*. New York: McGraw-Hill.

Ogata, K. (2004). *System dynamics*. New Jersey: Prentice Hall.

Palm, W. J. (2006). *Mechanical vibration*. New York: John Wiley.

Roseau, M. (1987). *Vibrations in mechanical systems*. Berlin: Springer-Verlag.

Snowdon, J. C. (1968). *Vibration and shock in damped mechanical systems*. New York: John Wiley.

Chapter 6: One Degree of Freedom Systems

Alkhatib, R., Jazar, R. N., & Golnaraghi, M. F. (2004). Optimal design of passive linear mounts with genetic algorithm method. *Journal of Sound and Vibration, 275*(3–5), 665–691.

Angeles, J. (2011). *Dynamic response of linear mechanical systems: Modeling, analysis and simulation*. New York: Springer.

Celia, C. W., Nice, A. T. F., & Elliott, K. F. (1982). *Advanced mathematics*. London, UK: Macmillan Education LTD.

Cheng, F. Y. (2000). *Matrix analysis of structural dynamics: applications and earthquake engineering*. New York: Marcel Dekker, Inc., Madison Avenue.

Del Pedro, M., & Pahud, P. (1991). *Vibration mechanics*. The Netherland: Kluwer Academic Publishers.

Den Hartog, J. P. (1934). *Mechanical vibrations*. New York: McGraw-Hill.

Dimarogonas, A. (1996). *Vibration for engineers*. New Jersey: Prentice Hall.

Esmailzadeh, E. (1978). Design synthesis of a vehicle suspension system using multi-parameter optimization. *Vehicle system dynamics, 7*, 83–96.

Esmailzadeh, E., & Jazar, R. N. (1998). Periodic behavior of a Cantilever with end mass subjected to harmonic base excitation. *International Journal of Nonlinear Mechanics, 33*(4), 567–577.

Esmailzadeh, E., Mehri, B., & Jazar, R. N. (1996). Periodic solution of a second order, autonomous, nonlinear system. *Journal of Nonlinear Dynamics, 10*(4), 307–316.

Golnaraghi, M. F., & Jazar, R. N. (2001). Development and analysis of a simplified nonlinear model of a hydraulic engine mount. *Journal of Vibration and Control, 7*(4), 495–526.

Harris, C. M., & Piersol, A. G. (2002). *Harris' shock and vibration handbook.* New York: McGraw-Hill.

Inman, D. (2007). *Engineering vibrations.* New York: Prentice Hall.

Jazar, R. N. (1997). *Analysis of nonlinear parametric vibrating systems.* Ph. D. Thesis, Mechanical Engineering Department, Tehran: Sharif University of Technology.

Jazar, R. N., Mahinfalah, M., & Deshpande, S. (2007). Design of a piecewise linear vibration isolator for jump avoidance. *IMechE Part K: Journal of Multi-Body Dynamics, 221*(K3), 441–450.

Korenev, B. G., & Reznikov, L. M. (1993). *Dynamic vibration absorbers: Theory and technical applications.* West Sussex, England: Wiley.

Meirovitch, L. (1967). *Analytical methods in vibrations.* New York: Macmillan.

Meirovitch, L. (1997). *Principles and techniques of vibrations.* New Jersey: Prentice Hall.

Meirovitch, L. (2002). *Fundamentals of vibrations.* New York: McGraw-Hill.

Narimani, A., Golnaraghi, M. F., & Jazar, R. N. (2004). Frequency response of a piecewise linear system. *Journal of Vibration and Control, 10*(12), 1775–1894.

Narimani, A., Jazar, R. N., & Golnaraghi, M. F. (2004). Sensitivity analysis of frequency response of a piecewise linear system in frequency island. *Journal of Vibration and Control, 10*(2), 175–198.

Ogata, K. (2004). *System dynamics.* New Jersey: Prentice Hall.

Osinski, Z. (1998) *Damping of vibrations.* Rotterdam, Netherlands: A. A. Balkema Publishers.

Palm, W. J. (2006). *Mechanical vibration.* New York: John Wiley.

Rao, S. S. (2018). *Mechanical vibrations.* Harlow, UK: Pearson Prentice Hall.

Roseau, M. (1987). *Vibrations in mechanical systems.* Berlin: Springer-Verlag.

Stahl, P., & Jazar, R. N. (2005). Frequency response analysis of piecewise nonlinear vibration isolator. In *International design engineering technical conferences and computers and information in engineering conference, Orlando, Florida, November.*

Snowdon, J. C. (1968). *Vibration and shock in damped mechanical systems.* New York: John Wiley.

Tse, F. S., Morse, I. E., & Hinkle, R. T. (1978). *Mechanical vibrations theory and applications.* Boston, MA: Allyn and Bacon Inc.

Chapter 7: Multi Degrees of Freedom Systems

Alkhatib, R., Jazar, R. N., & Golnaraghi, M. F. (2004), Optimal design of passive linear mounts with genetic algorithm method. *Journal of Sound and Vibration, 275*(3–5), 665–691.

Celia, C. W., Nice, A. T. F., & Elliott, K. F. (1982). *Advanced mathematics.* London, UK: Macmillan Education LTD.

Cheng, F. Y. (2000). *Matrix analysis of structural dynamics: applications and earthquake engineering.* Madison Avenue, New York: Marcel Dekker, Inc.

Esmailzadeh, E. (1978). Design synthesis of a vehicle suspension system using multi-parameter optimization. *Vehicle system dynamics, 7,* 83–96.

Harris, C. M., & Piersol, A. G. (2002). *Harris' shock and vibration handbook.* New York: McGraw-Hill.

Inman, D. (2007). *Engineering vibrations.* New York: Prentice Hall.

Jazar, R. N., Kazemi, M., & Borhani, S. (1992). *Mechanical vibrations* (in Persian). Tehran: Ettehad Publications.

Meirovitch, L. (1967). *Analytical methods in vibrations.* New York: Macmillan.

Meirovitch, L. (1997). *Principles and techniques of vibrations*. New Jersey: Prentice Hall.
Meirovitch, L. (2002). *Fundamentals of vibrations*. New York: McGraw-Hill.
Rao, S. S. (2018). *Mechanical vibrations*. Harlow, UK: Pearson Prentice Hall.
Ogata, K. (2004). *System dynamics*. New Jersey: Prentice Hall.
Roseau, M. (1987). *Vibrations in mechanical systems*. Berlin: Springer-Verlag.
Snowdon, J. C. (1968). *Vibration and shock in damped mechanical systems*. New York: John Wiley.
Tse, F. S., Morse, I. E., & Hinkle, R. T. (1978). *Mechanical vibrations theory and applications*. Boston, MA: Allyn and Bacon Inc.

Chapter 7: Multi Degrees of Freedom Systems

Alkhatib, R., Jazar, R. N., & Golnaraghi, M. F. (2004), Optimal design of passive linear mounts with genetic algorithm method. *Journal of Sound and Vibration, 275*(3–5), 665–691.
Celia, C. W., Nice, A. T. F., & Elliott, K. F. (1982). *Advanced mathematics*. London, UK: Macmillan Education LTD.
Cheng, F. Y. (2000). *Matrix analysis of structural dynamics: applications and earthquake engineering*. Madison Avenue, New York: Marcel Dekker, Inc.
Den Hartog, J. P. (1934). *Mechanical vibrations*. New York: McGraw-Hill.
Esmailzadeh, E. (1978). Design synthesis of a vehicle suspension system using multi-parameter optimization. *Vehicle System Dynamics, 7*, 83–96.
Harris, C. M., & Piersol, A. G. (2002). *Harris' shock and vibration handbook*. New York: McGraw-Hill.
Inman, D. (2007). *Engineering vibrations*. New York: Prentice Hall.
Jazar, R. N., Kazemi, M., & Borhani, S. (1992). *Mechanical vibrations* (in Persian). Tehran: Ettehad Publications.
Meirovitch, L. (1967). *Analytical methods in vibrations*. New York: Macmillan.
Meirovitch, L. (1997). *Principles and techniques of vibrations*. New Jersey: Prentice Hall.
Meirovitch, L. (2002). *Fundamentals of vibrations*. New York: McGraw-Hill.
Palm, W. J. (2006). *Mechanical vibration*. New York: John Wiley.
Roseau, M. (1987). *Vibrations in mechanical systems*. Berlin: Springer-Verlag.
Snowdon, J. C. (1968). *Vibration and shock in damped mechanical systems*. New York: John Wiley.
Tse, F. S., Morse, I. E., & Hinkle, R. T. (1978). *Mechanical vibrations theory and applications*. Boston, MA: Allyn and Bacon Inc.

Index

© The Author(s), under exclusive license to Springer Nature Switzerland AG 2022
R. N. Jazar, *Advanced Vibrations*, https://doi.org/10.1007/978-3-031-16356-2

Printed in the United States
by Baker & Taylor Publisher Services